DNA Insertion Elements, Plasmids, and Episomes

DNA Insertions Meeting, Cold Spring Harbor Laboratory, 1976.

DNA
Insertion Elements, Plasmids, and Episomes

Edited by

A.I. Bukhari
J.A. Shapiro
S.L. Adhya

Cold Spring Harbor Laboratory 1977

DNA
Insertion
Elements,
Plasmids,
and Episomes

International Standard Book Number 0-87969-118-2
Library of Congress Catalog Card Number 76-57904
Printed in United States of America

Cover: An example of stem-loop structures seen when DNA from *E. coli* is denatured and reannealed. Inverted duplications, separated by unique sequences, snapped back to form these structures. In the cover picture, the stem has a length equal to that of IS*1*, 0.8 kb, and the loop consists of 13.7 kb. *(Photograph by L.T. Chow and T.R. Broker, Cold Spring Harbor Laboratory)*

Contents

Preface xiii

Introduction: New Pathways in the Evolution of Chromosome Structure 3
J.A. Shapiro, S.L. Adhya and A.I. Bukhari

The Nomenclature Problem 13

Nomenclature of Transposable Elements in Prokaryotes 15
A. Campbell, D. Berg, D. Botstein, E. Lederberg, R. Novick, P. Starlinger and W. Szybalski

SECTION I IS Elements

Mutations Caused by the Integration of IS*1* and IS*2* into the *gal* Operon 25
P. Starlinger

Specific Sites for Integration of IS Elements within the Transferase Gene of the *gal* Operon of *E. coli* K12 31
D. Pfeifer, P. Habermann and D. Kubai-Maroni

The *gal3* Mutation of *E. coli* 37
A. Ahmed

Repeated DNA Sequences in Plasmids, Phages, and Bacterial Chromosomes 49
H. Ohtsubo and E. Ohtsubo

IS*1* and IS*2* in *E. coli*: Implications for the Evolution of the Chromosome and Some Plasmids 65
H. Saedler

The Organization of Putative Insertion Sequences on the *E. coli* Chromosome 73
L.T. Chow

Chromosomal Rearrangements in the *gal* Region of *E. coli* K12 after Integration of IS*1* 81
H.-J. Reif and H. Saedler

Polarity of Insertion Mutations Is Caused by Rho-mediated Termination of Transcription 93
A. Das, D. Court, M. Gottesman and S. Adhya

Isolation of Mutations in Insertion Sequences That Relieve IS-induced Polarity 99
P.K. Tomich and D.I. Friedman

The Isolation of IS*1* and IS*2* DNA 109
F. Schmidt, J. Besemer and P. Starlinger

Physical Mapping of IS*1* by Restriction Endonucleases 115
N.D.F. Grindley

Preliminary Observations

A. On Mutations Affecting IS*1*-mediated Deletion Formation in *E. coli* 125
P. Nevers, H.-J. Reif and H. Saedler

B. On the Presence of a Promoter in IS*2* 129
B. Rak

C. On the Polarity of Insertion Mutations 133
J. Besemer

SECTION II Transposons and Plasmids

Formation of Conjugative Drug Resistance (R) Plasmids 139
S. Mitsuhashi, H. Hashimoto, S. Iyobe and M. Inoué

Recombination between *Pseudomonas aeruginosa* Plasmids of Incompatibility Groups P-1 and P-2 147
G.A. Jacoby and A.E. Jacob

Transposition of a Plasmid DNA Sequence Which Mediates Ampicillin Resistance: General Description and Epidemiologic Considerations 151
F. Heffron, C. Rubens and S. Falkow

Deletions Affecting Transposition of an Antibiotic Resistance Gene 161
F. Heffron, P. Bedinger, J. Champoux and S. Falkow

Promotion of Insertions and Deletions by Translocating Segments of DNA Carrying Antibiotic Resistance Genes 169
J. Brevet, D.J. Kopecko, P. Nisen and S.N. Cohen

Transposition of Tn*1* to a Broad-host-range Drug Resistance Plasmid 179
J.P. Hernalsteens, R. Villarroel-Mandiola, M. Van Montagu and J. Schell

Translocation and Illegitimate Recombination by the Tetracycline Resistance Element Tn*10* 185
D. Botstein and N. Kleckner

Insertion and Excision of the Transposable Kanamycin Resistance Determinant Tn*5* 205
D.E. Berg

Transposition and Deletion of Tn*9*: A Transposable Element Carrying the Gene for Chloramphenicol Resistance 213
J.L. Rosner and M.M. Gottesman

Physical Structure and Deletion Effects of the Chloramphenicol Resistance Element Tn*9* in Phage Lambda 219
L.A. MacHattie and J.B. Jackowski

Structure and Location of Antibiotic Resistance Determinants in Bacteriophages P1Cm and P7 (ϕAmp) 229
T. Yun and D. Vapnek

Amplification of the Tetracycline Resistance Determinant on Plasmid pAMα1 in *Streptococcus faecalis* 235
D.B. Clewell and Y. Yagi

SECTION III Bacteriophage Mu: A Transposable Element

The Mechanism of Bacteriophage Mu Integration 249
A.I. Bukhari, E. Ljungquist, F. de Bruijn and H. Khatoon

Characterization of Covalently Closed Circular DNA Molecules Isolated after Bacteriophage Mu Induction 263
B.T. Waggoner, M.L. Pato and A.L. Taylor

Mu-mediated Illegitimate Recombination as an Integral Part of the Mu Life Cycle 275
A. Toussaint, M. Faelen and A.I. Bukhari

On the *kil* Gene of Bacteriophage Mu 287
P. van de Putte, G. Westmaas, M. Giphart and C. Wijffelman

Bacteriophage Mu Genome: Structural Studies on Mu DNA and Mu Mutants Carrying Insertions 295
L.T. Chow and A.I. Bukhari

Electron Microscope Studies of Nondefective Bacteriophage Mu Mutants Containing Deletions or Substitutions 307
L.T. Chow, R. Kahmann and D. Kamp

Structure and Packaging of Mu DNA 315
E. Bade, H. Delius and B. Allet

DNA Partial Denaturation Mapping Studies of Packaging of Bacteriophage Mu DNA 319
M.M. Howe, M. Schnös and R.B. Inman

Asymmetric Hybridization of Mu Strands with Short Fragments Synthesized during Mu DNA Replication 329
C. Wijffelman and P. van de Putte

Mapping of Restriction Sites in Mu DNA 335
R. Kahmann, D. Kamp and D. Zipser

SECTION IV Nonhomologous Recombination and the λ Paradigm

Flexibility in Attachment-site Recognition by λ Integrase 343
L. Enquist and R. Weisberg

Isolation of *Escherichia coli* Mutants Unable to Support Lambda Integrative Recombination 349
H.I. Miller and D.I. Friedman

A Mutant of *Escherichia coli* Deficient in a Host Function Required for Phage Lambda Integration and Excision 357
J.G.K. Williams, D.L. Wulff and H.A. Nash

Integrative Recombination of Bacteriophage λ – The Biochemical Approach to DNA Insertions 363
H.A. Nash, K. Mizuuchi, R.A. Weisberg, Y. Kikuchi and M. Gellert

The Integrase Promoter of Bacteriophage Lambda 375
A. Campbell, L. Heffernan, S.-L. Hu and W. Szybalski

Position Effects of Insertion Sequences IS*2* near the Genes for Prophage λ Insertion and Excision 381
J. Zissler, E. Mosharrafa, W. Pilacinski, M. Fiandt and W. Szybalski

The Phage λ Integration Protein (Int) Is Subject to Control by the *c*II and *c*III Gene Products 389
D. Court, S. Adhya, H. Nash and L. Enquist

Temperate Coliphage P2 as an Insertion Element 395
R. Calendar, E.W. Six and F. Kahn

Recombination Models for the Inverted DNA Sequences of the Gamma-Delta Segment of *E. coli* and the G Segments of Phages Mu and P1 403
T.R. Broker

An Electron Microscope Study of Actively Recombining Plasmid DNA Molecules 409
H. Potter and D. Dressler

SECTION V Eukaryotic Systems

An Introductory Note on the Controlling Elements in Maize 425

The Position Hypothesis for Controlling Elements in Maize 429
P.A. Peterson

The Case for DNA Insertion Mutations in *Drosophila* 437
M.M. Green

"Flip-Flop" Control and Transposition of Mating-type Genes in Fission Yeast 447
R. Egel

The Cassette Model of Mating-type Interconversion 457
J.B. Hicks, J.N. Strathern and I. Herskowitz

The Origin and Complexity of Inverted Repeat DNA Sequences in *Drosophila* 463
R.F. Baker and C.A. Thomas, Jr.

Nucleotide Sequence Arrangements in the Genome of Herpes Simplex Virus and Their Relation to Insertion Elements 471
W.C. Summers and J. Skare

The Structure of the Adeno-associated Virus Genome 477
B.J. Carter, L.M. de la Maza and F.T. Jay

SECTION VI Genetic Rearrangements; Techniques and Applications

Mapping Ribosome Protein Genes in *E. coli* by Means of Insertion Mutations 487
S.R. Jaskunas and M. Nomura

Chromosomal Rearrangements Resulting from Recombination between Ribosomal RNA Genes 497
C.W. Hill, R.H. Grafstrom and B.S. Hillman

Potential of RP4::Mu Plasmids for In Vivo Genetic Engineering of Gram-negative Bacteria 507
J. Dénarié, C. Rosenberg, B. Bergeron, C. Boucher, M. Michel and M. Barate de Bertalmio

In Vivo Genetic Engineering: The Mu-mediated Transposition of Chromosomal DNA Segments onto Transmissible Plasmids 521

M. Faelen, A. Toussaint, M. Van Montagu, S. Van den Elsacker, G. Engler and J. Schell

Construction and Use of Gene Fusions Directed by Bacteriophage Mu Insertions 531
M.J. Casadaban, T.J. Silhavy, M.L. Berman, H.A. Shuman, A.V. Sarthy and J.R. Beckwith

In Vivo Genetic Engineering: Exchange of Genes between a Lambda Transducing Phage and ColE1 Factor 537
K. Shimada, Y. Fukumaki and Y. Takagi

Translocation of Ampicillin Resistance from R Factor onto ColE1 Factor Carrying Genes for Synthesis of Guanine 543
S. Maeda, K. Shimada and Y. Takagi

Selected Translocation of DNA Segments Containing Antibiotic Resistance Genes 549
P.J. Kretschmer and S.N. Cohen

Detection of Transposable Antibiotic Resistance Determinants with Phage Lambda 555
D.E. Berg

Properties of the Plasmid RK2 as a Cloning Vehicle 559
R.J. Meyer, D. Figurski and D.R. Helinski

Insertion of Mu DNA Fragments into Phage λ In Vitro 567
D.D. Moore, J.W. Schumm, M.M. Howe and F.R. Blattner

The *E. coli* Gamma-Delta Recombination Sequence Is Flanked by Inverted Duplications 575
T.R. Broker, L.T. Chow and L. Soll

SECTION VII Appendices

APPENDIX A IS Elements

1. IS Elements in *Escherichia coli*, Plasmids, and Bacteriophages 583
   Compiled by W. Szybalski
2. Sequence of the Ends of IS*1* Element 591
   Contributed by H. Ohtsubo and E. Ohtsubo
3. DNA Sequences at the Ends of IS*1* 595
   Contributed by N.D.F. Grindley
4. Nucleotide Sequences at Two Sites for IS*2* DNA Insertion 597
   Contributed by R.E. Musso and M. Rosenberg

APPENDIX B Bacterial Plasmids

1. Tables

a. Introduction 601
By J.A. Shapiro

b. Plasmids Studied in *Escherichia coli* and Other Enteric Bacteria 607
Compiled by A.E. Jacob, J.A. Shapiro, L. Yamamoto, D.I. Smith, S.N. Cohen and D. Berg

c. Plasmids Studied in *Pseudomonas aeruginosa* and Other Pseudomonads 639
Compiled by G.A. Jacoby and J.A. Shapiro

d. Plasmids of *Staphylococcus aureus* 657
Compiled by R.P. Novick, S. Cohen, L. Yamamoto and J.A. Shapiro

e. Plasmids of Other Gram-positive Bacteria 663
Compiled by A.E. Jacob, J.A. Shapiro and L. Yamamoto

f. Plasmids Constructed In Vitro and In Vivo 665
Compiled by S.N. Cohen

2. Maps

a. F, the *E. coli* Sex Factor 671
Contributed by J.A. Shapiro

b. Special Sequences in the Structure of Cointegrate Drug Resistance Plasmids Related to F 672
Contributed by S.N. Cohen

c. *Eco*RI, *Hin*dIII, and *Bam*HI Cleavage Map of R538-1 674
Contributed by D. Vapnek

d. Tn*7* Insertion Map of RP4 675
Contributed by P.T. Barth and N.J. Grinter

e. Physical Map of RP4 678
Contributed by A. DePicker, M. Van Montagu and J. Schell

f. Restriction Enzyme Map of RK2 680
Contributed by R. Meyer, D. Figurski and D.R. Helinski

g. Restriction Map of the ColE1 Derivative pCR1 681
Contributed by K. Armstrong and D.R. Helinski

h. Restriction Map of ColE1 and pNT1 Plasmids 682
Contributed by H. Ohmori and J.-I. Tomizawa

i. The Circular Restriction Map of pBR313 684
Contributed by F. Bolivar, R.L. Rodriguez, M.C. Betlach and H.W. Boyer

j. The Circular Restriction Map of pBR322 686
Contributed by F. Bolivar, R.L. Rodriguez, P.J. Greene, M.C. Betlach, H.L. Heyneker, H.W. Boyer, J.H. Crosa and S. Falkow

3. Bibliography 689
Compiled by J.A. Shapiro, A.E. Jacob, R.P. Novick and L. Yamamoto

APPENDIX C Temperate Bacteriophages

1. The Genomes of Temperate Viruses of Bacteria 705
Contributed by E. Ljungquist and A.I. Bukhari
2a. Genetic, Physical, and Restriction Map of Bacteriophage λ 713
Contributed by S. Gottesman and S. Adhya
b. Restriction Enzyme Cleavage Maps of Bacteriophage λ with a Focus on the Attachment-site Region 719
Contributed by D. Kamp and R. Kahmann
3. Genetic and Physical Structure of Bacteriophage P1 DNA 721
Contributed by M.B. Yarmolinsky
4. Genetic and Physical Map of Bacteriophage P2 733
Contributed by D.K. Chattoraj
5. Genetic, Physical, and Restriction Map of Bacteriophage P22 737
Contributed by M. Susskind and D. Botstein
6. Genetic, Physical, and Restriction Map of Bacteriophage ϕ80 741
Contributed by P. Youderian
7. Genetic and Physical Map of Bacteriophage Mu 745
Contributed by B. Allet, F. Blattner, M. Howe, M. Magazin, D. Moore, K. O'Day, D. Schultz and J. Schumm
8. Bacteriophage Mu: Methods for Cultivation and Use 749
Contributed by A.I. Bukhari and E. Ljungquist

APPENDIX D Restriction Endonucleases

1. Restriction and Modification Enzymes and Their Recognition Sequences 757
Compiled by R.J. Roberts

Index 769

Preface

The Cold Spring Harbor Laboratory was the scene of an exciting meeting on DNA Insertions from May 18 to May 21, 1976. The meeting brought together biologists whose interests ranged from *Escherichia coli* to maize and from bacteriophage λ to herpesviruses. It is not possible in a scientific monograph to capture entirely the spirit of the meeting—the excitement, the inspired discussions. This book is an experiment in publishing the proceedings of a meeting. The book presents not only papers that were discussed at the meeting but also other articles and information that we hope readers will find useful. We have made an effort to scan a broad spectrum of studies and to focus on many facets of the insertion phenomena. The introduction outlines the topics covered and explains the organization of the book.

The DNA Insertions meeting and this monograph would not have been possible without the encouragement and support of J.D. Watson, Director of the Cold Spring Harbor Laboratory. The meeting was supported by funds from the National Science Foundation (PCM 76-06478), the National Cancer Institute (CA-13106), and the Cold Spring Harbor Laboratory.

We are indebted to all of our colleagues who participated in the meeting: to D. Botstein, A. Campbell, H. Lewis, M. Malamy, M. Meselson, R. Novick, J. Sambrook, A.L. Taylor, and R. Weisberg for chairing the sessions, and to S. Brenner for ending the meeting with his spirited summary. We are also grateful to B. McClintock, W. Szybalski, S. Cohen, and other colleagues for their help and advice. We are especially indebted to the authors who contributed their papers and who patiently listened to the various editorial demands.

We wish to express our appreciation to Gladys Kist of the meetings office for the meeting arrangements, to Bob Yaffee for artwork, and to Roberta Salant, Annette Zaninovic, and Judith Atkin of the publications office for help with the preparation of the book. Special thanks are due to Nancy Ford, Director of Publications, who guided the development of this book.

A.I. Bukhari
J.A. Shapiro
S.L. Adhya

DNA Insertion Elements, Plasmids, and Episomes

Introduction: New Pathways in the Evolution of Chromosome Structure

J. A. Shapiro
Department of Microbiology
University of Chicago
Chicago, Illinois 60637

S. L. Adhya
Laboratory of Molecular Biology
National Cancer Institute
National Institutes of Health
Bethesda, Maryland 20014

A. I. Bukhari
Cold Spring Harbor Laboratory
Cold Spring Harbor, New York 11724

This book deals with a new class of genetic elements: DNA insertions. The existence of DNA insertion elements has become clear from a decade of research on the expression and structure of bacterial genomes. Recent developments indicate that DNA insertion elements serve a special evolutionary function. They mediate the integration of one segment of genetic information into another. This process is independent of previously recognized mechanisms for the interaction of DNA molecules. There are suggestions that DNA insertion elements exist not only in prokaryotic cells but also in eukaryotic cells. Thus we face the discovery of new pathways for the reassortment of genetic information and the evolution of chromosome structure in both higher and lower organisms.

Because the study of DNA insertion elements is so young, hypotheses and problems far outnumber accepted concepts. The papers in this book constitute a first attempt to define these problems and state working models to resolve them. They represent the state of knowledge about insertion elements as it stood at the time of the DNA Insertions Meeting at Cold Spring Harbor Laboratory in May, 1976. The purpose of this introduction is to explain the developments which prompted that meeting and present our view of how several different lines of research converged to define an exciting new field in genetics.

About a decade ago, the first examples of DNA insertion elements in bacterial systems were recognized in laboratories whose primary interest was the regulation of gene expression, not chromosome structure. Analysis of various operons in *Escherichia coli* revealed an unexpected class of pleiotropic mutations. They were often located in structural genes for pathway enzymes, were neither base substitution, frameshift, nor deletion mutations, and exerted strong polar effects on the expression of other cistrons distal to the operon promoter. Eventually it was shown that these mutations resulted from the insertion of fairly large segments (700 to 1400 base pairs) of DNA into the structural or regulatory genes of these operons. Such insertions were observed in a number of genetic systems, including the *E. coli gal, lac,* and ribosomal protein operons and bacteriophage λ, P1, and P2. Further analysis of independent insertions by electron microscope heteroduplex methods revealed that many of them appear to be identical. Thus a limited number of specific DNA sequences can insert themselves into different sites in the bacterial genome to shut off gene activity. To date, repeated examples of four such insertion sequences (IS) have been documented; these have been labeled IS*1*, IS*2*, IS*3*, and IS*4*. They contain, respectively, about 800, 1300, 1200, and 1400 base pairs. A complete tabulation of these and other unclassified *E. coli* insertion mutations can be found in Appendix A1.

The discovery that the same sequences[1] can each appear repeatedly at one or more sites in the bacterial genome leads to an important conclusion: insertion mutations are not the result of random noise in the usually accurate mechanisms for replication and segregation of the genetic material. Indeed, further studies turned up many interesting characteristics of insertion sequences. The bacterial chromosome contains multiple copies of at least some IS elements. IS*1* and IS*4* exert polar effects when linearly inserted in either of the two possible ways (orientations I and II), whereas IS*2* and IS*3* are known to be polar only in one direction (orientation I). Moreover, IS*2* seems to contain a site which acts as a promoter for gene expression when present in the nonpolar orientation II. Insertion can occur at many sites but is not random, and there are "hot spots" for IS-induced mutations. The presence of some IS elements can lead to a high frequency of spontaneous deletions, which begin at the site of insertion. The one aspect of IS physiology that has received the most attention so far is the effect of insertion mutations on gene expression. We know that at least some of them are polar because they contain sites for the action of the transcription termination factor rho. Although no clear picture is yet available for all the phenotypic consequences of insertion mutations, this short list indicates that IS elements have a complex structure. Presumably, this structure is the result of natural selection for some function(s) useful to the bacterial cell. One such function could involve the control of gene expression. The papers in Section I describe the history, physiology, and structure of IS mutations.

It would be unlikely that these complex genetic elements evolved simply as mutator elements. Some clues about the normal function of IS elements have emerged from a different area of research in bacterial genetics—the study of

[1] It should be remembered that sequence identity is established by heteroduplex formation. This does not exclude small differences in nucleotide sequences.

plasmid structure. Part of the impetus for plasmid research is to understand the replication of the genetic material. Plasmids are suitable objects for this research because they are independently replicating genetic units (replicons) and are smaller and easier to handle than many other replicons (such as bacterial or eukaryotic chromosomes). Because of a large background of genetic data, the *E. coli* sex factor, F, and related drug resistance plasmids became the objects of detailed electron microscopic examination. This examination produced a very interesting picture of plasmid organization. At various sites on the plasmids, specific DNA sequences are repeated. These sequences are not randomly located. On F and its derivatives, they mark the sites where the plasmid integrates into the bacterial chromosome to form Hfr strains. On F-related drug resistance plasmids, they mark the boundaries between regions determining antibiotic resistance and regions containing genes for conjugal transfer and replication. Thus these repeated sequences serve to define either (1) the boundaries between different blocks of genetic information integrated into a single replicon or (2) the sites where two replicons fuse. These boundary sequences were similar in size to the polar insertions, and heteroduplex experiments between plasmids and phages with IS elements gave a spectacular result—at least three of the plasmid boundary sequences ($\lambda\mu$, $\epsilon\zeta$, and $\alpha\beta$) are the same as IS*1*, IS*2*, and IS*3*. Some of the plasmid maps in Appendix B2 illustrate these structural features, which are also discussed in detail in Sections I and II.

The identification of IS elements with plasmid boundary sequences made it possible to assign a special evolutionary function to them: i.e., to provide sites for the integration of DNA from different sources into a single replicon. One way to visualize this process is to think of IS elements as joints or connectors for the modular construction of chromosomes. Different genetic "modules" (segments of DNA encoding replication functions, conjugal transfer, antibiotic resistance, special metabolic capabilities, and so forth) could reassort into optimal replicon structures through the action of DNA insertion elements. Such a function would obviously offer selective advantages to organisms like bacteria which are subject to rapid changes in environmental pressures. In addition, the existence of boundary sequences for specific genetic determinants could facilitate their replication independent of other segments of DNA present in the same chromosome. (This role has in fact been proposed for IS elements in the selective amplification of resistance genes in response to antibiotic challenge.)

If insertion elements act as sites for joining DNA segments, the question arises as to how they do this. One way would be to provide genetic homology between interacting segments. From the structural studies of F-chromosome interactions, this appeared to be a plausible mechanism. For example, integration of F, containing an IS*3* sequence, results in the presence of two IS*3* sequences on either side of the integrated sex factor. This indicates that integration occurred by reciprocal recombination into an IS*3* element on the chromosome (Fig. 1). However, an old observation on Hfr formation suggested that this explanation

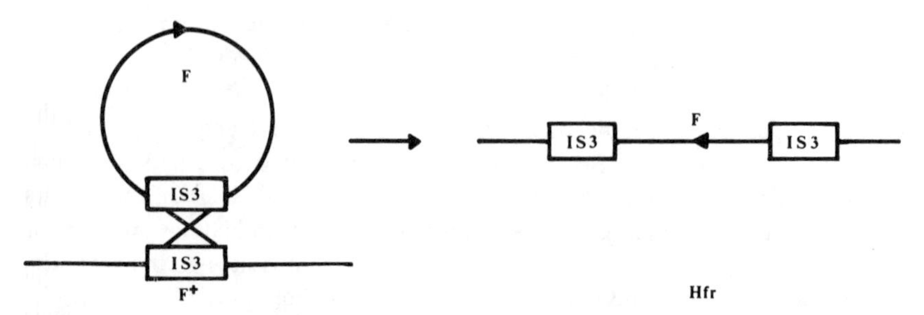

Figure 1
Hfr formation by reciprocal recombination between IS elements in F and the chromosome.

might not be correct, as F integrates into the host chromosome of *rec*$^-$ mutants lacking normal homologous recombination functions.

The discovery of transposable antibiotic resistance has made it possible to study further the behavior of DNA insertions. Analysis of the transfer of resistance determinants between bacterial strains suggested that some of these genes can hop (or transpose) from one replicon to another, for example, from a drug resistance plasmid to the chromosome, to another plasmid, or to the genome of a temperate bacteriophage. This process is widespread. It occurs in both gram-negative and gram-positive bacteria and is known to involve determinants for resistance to at least seven different antibiotics. Sometimes the drug resistance determinant inserts into a known genetic locus and causes a detectable polar mutation. Insertions occur at many sites but are not random, and the presence of an inserted drug resistance gene often leads to an increased frequency of spontaneous deletions. These transposable antibiotic resistance determinants have been called "transposons," and a proposal for a rational nomenclature system of these and other insertion elements follows this Introduction.

Clearly, drug resistance transposons are functionally equivalent in many ways to the IS elements. Physical examination of replicons carrying transposons has shown that this equivalence is more than a coincidence. In the cases that have been examined so far, transposition events have the following structural characteristics: (1) insertion of a transposon does not cause observable deletion of material from the recipient site, and (2) the inserted segment consists of a resistance gene (or genes) bounded by repeated terminal sequences. Sometimes these terminal repeats are parallel, but generally they are in opposing orientations. In this second case, heteroduplexes between the replicon and its homolog with the inserted transposon have a characteristic "lollipop" structure (Fig. 2). The known terminal repeats vary in length from about 80 to over 1700 base pairs, and some of them are the same as IS elements already identified. For example, the parallel repetitious sequences bounding Tn*9* (encoding chlor-

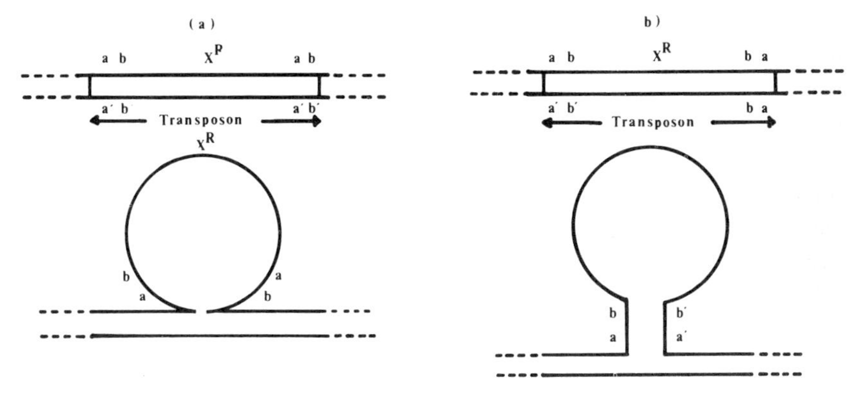

Figure 2
Structure of antibiotic resistance transposons. The top pair of figures shows the arrangement of terminal repetitious sequences around the resistance determinant (X^R) in parallel (*a*) or opposed (*b*) orientation. The bottom pair of figures shows the structure of heteroduplex molecules formed by hybridizing homologous DNA strands with and without the inserted transposon.

amphenicol resistance) hybridize with IS*1*, and the inverted repetitious sequences bounding Tn*10* (encoding tetracycline resistance) hybridize with IS*3*. The terminal repetitious sequences do not change as a transposon moves from one site to another. Hence it is logical to conclude that the IS elements play a role in the transposition process, and we can assign to them a specific function of promoting the movement of drug resistance transposons. Evidently, this function confers a selective advantage to cultures subjected to varying antibiotic challenges.

Confirmation of the hypothesis that IS elements serve to join different DNA segments is not the only advance resulting from the discovery of drug resistance transposons. These elements also present significant experimental advantages because the movement of each transposon is easily traced by scoring the resistance phenotype associated with it. Thus quantitative genetic and physiologic experimentation on insertion are possible. One of the most important observations is that transposition occurs at comparable frequencies in cells with and without normal homologous recombination pathways. This means that DNA insertion is a process which involves specific enzymatic functions as well as specific DNA sequences. How these two components function and what genes code for the specific enzymes are the subjects of active research.

The papers in Section II deal with plasmid structure and transposable antibiotic resistance. Some of the results presented there confirm the critical role of the terminal sequences in the transposition process and indicate that transposons code for diffusible products involved in their insertion. It would be surprising if transposable loci were limited to antibiotic resistance determinants, and preliminary data on the evolution of plasmids in bacteria other than *E. coli* and its close relatives indicate that transposable elements may exist for other markers as well. Because of the important role that plasmid research has played (and will continue to play) in the development of our ideas on DNA insertion elements, we have included in the appendices (B1) an extensive table on natural and hybrid plasmids from a wide variety of bacteria.

In brief, the first two sections of this book show how studies of genetic regulation, replicon structure, and antibiotic resistance have combined in an elegant way to reveal a new pathway for the evolution of prokaryotic genomes. Systems for studying details of how this pathway functions and evidence that it may also exist in eukaryotic organisms are discussed in Sections III, IV, and V.

As the papers in Section III illustrate, the temperate bacteriophage Mu provides a special system for studying DNA insertions. Insertion is a way of life for Mu. For other temperate phages, such as λ or P2, integration into the host genome is a dispensable function. But for Mu, it is essential to the viral life cycle. All mutants of this phage which do not integrate also do not replicate, and a striking feature of Mu is that its linear genome always exists in association with host DNA on both sides—even in mature virus particles. Like the IS elements, Mu causes polar mutations by inserting its DNA into host genes. Mu integration appears to be more promiscuous than that of IS elements or transposons, and, at least in *E. coli,* the phage can insert itself irrespective of the host sequences it encounters. There are other important similarities between the integration of Mu and that of drug resistance transposons. In both cases, for example, imprecise excision of the inserted DNA from a particular chromosome site is at least ten times more common than precise excision. Moreover, the existence of host DNA at the ends of Mu DNA in phage particles means that every integration of Mu is, in effect, a transposition event from one "chromosomal" location to another. There is also genetic evidence that the Mu genome inserts at several sites in the bacterial genome as it replicates following induction. Perhaps the most compelling demonstration that Mu is really a highly evolved insertion element is the finding that Mu can serve to integrate pieces of the *E. coli* chromosome into plasmids from unrelated organisms. By doing this, Mu carries out the "joining" function which we assert is central to the evolutionary role of bacterial DNA insertion elements. Because Mu is a virus which can be easily propagated and purified, it will be a valuable model for analyzing the behavior of DNA insertion elemens. Perhaps, once the molecular mechanisms of Mu integration and replication are known, it will be easier to understand how IS elements and transposons function.

Currently, there is one system of integrative recombination that we understand in some detail—that of bacteriophage λ. The papers on the integration of λ and other prophages make up Section IV of this book. The conceptual breakthrough in our understanding of the λ system came from a proposal by Allan Campbell in 1962. The essential features of the Campbell model are (1) circularization of the infecting λ DNA by joining of single-stranded cohesive ends and (2) insertion of the viral circle into the bacterial chromosome by reciprocal recombination at specific loci (called *att*) in both host and phage DNA. Experimental evidence has confirmed the accuracy of this model for λ integration and has added some important details to it. First, recombination between the viral and host *att* loci does not depend on the bacterial homologous recombination system but involves phage proteins: the *int* gene product is required for integration, and both the *int* and *xis* gene products are needed for excision from the chromosome following induction (Fig. 3). Second, the phage *att* and bacterial *att* are not homologous, so that the *int*-promoted integration reaction and the *int* plus *xis*-promoted excision reaction have different substrates. Third, λ integration ordinarily shows a high degree of specificity for one site, but when this site is removed by deletion, integration occurs at a large number of other specific sites on the bacterial chromosome. (Some phages, such as P2, ordinarily have more than one preferred integration site.) There are some intriguing parallels between the behavior of λ and that of the IS elements: (1) one of the preferred secondary λ integration sites appears to be the same as one of the "hot spots" for IS insertions in the *gal* operon; (2) both the formation of spontaneous deletions from an inserted IS element and λ*int* function (which participates in the formation of *att* deletions in the phage chromosome) are temperature-sensitive; and (3) removal of prophage λ genes from secondary attachment sites is frequently imprecise. Undoubtedly, understanding the molecular mechanisms of λ integration and excision will help develop ideas on the movement of DNA insertion elements. We are probably on the threshold of understanding the λ system since both integrative and excisive λ recombination can now be studied in vitro.

It should be noted, however, that there are two facets of the λ paradigm. One is the role of specific sequences and enzymes. The other is the role of the circular form of DNA in integration. Use of specific sequences and enzymes is typical of many systems (other phages, IS elements, transposons) and shows us that nonhomologous recombination (where enzymes recognize and join sequences that are not joined by homologous recombination enzymes) is far

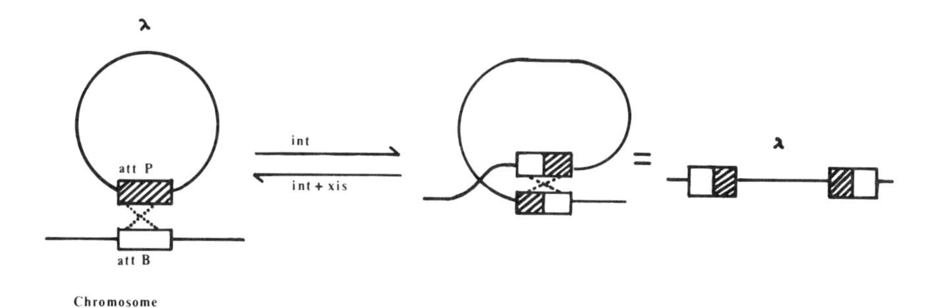

Figure 3
Integration and excision of λ DNA. Note that none of the four boxed *att* regions is homologous with any other.

from being an "illegitimate" cellular function. It may be that the formation of DNA rings is true only of a certain subset of temperate bacteriophages and DNA insertion elements. Genetic and physical evidence for the circularization of certain temperate phage genomes has been sought without success. In the case of Mu, this search has been quite extensive. There is also no evidence for the existence of autonomous circular forms of IS elements and drug resistance transposons. Although this situation may simply reflect the newness of the subject and the difficulty of performing the right experiments, it is important to be cautious when applying a beautiful model, such as λ integration, to related but distinct phenomena. A detailed analysis of how other temperate bacteriophages and the bacterial chromosomes interact will help to sort out the general and specific features of all these systems. For this reason, data on a number of temperate phages from various bacterial species are presented in Appendix C.

So far, we have discussed only prokaryotic organisms. If they have special pathways for the interaction of DNA segments, then similar phenomena almost certainly exist in eukaryotic organisms as well. Section V contains a small part of a growing body of genetic and physical data which support this assumption. Indeed, the first examples of DNA insertion elements were probably detected before DNA was identified as the genetic material. These are the so-called "controlling elements" in corn (maize). They surfaced through extended cytogenetic analysis of unstable phenotypes. The instabilities were traced to a number of genetic elements which had the following interesting properties: (1) they could move or transpose from one chromosomal site to another (sometimes on the same chromosome, sometimes on a different one); (2) the presence of a controlling element at a given site could influence the activity of nearby genes (generally by blocking expression of the dominant phenotype); and (3) the presence of a controlling element affected interaction of different chromosomes by promoting deletions, translocations, or other rearrangements. In sum, controlling elements are transposable genetic elements that affect gene activity and chromosome structure. Although parallels have been drawn between controlling elements and regulatory genes, it is apparent that their true prokaryotic counterparts are IS elements, transposons, and phage Mu. (It may be true, of course, that these elements do play a role in the regulation of gene activity, and transposable elements carrying "start" and "stop" signals suggest many especially interesting models for the control of cellular differentiation.) Similar kinds of cytogenetic studies in *Drosophila* mutants with unstable phenotypes have also revealed the existence of analogous elements. A third eukaroytic system where DNA insertion elements can explain genetic results is the regulation of yeast sexuality. Models involving the insertion or transposition of DNA segments have been proposed for mating-type interconversions.

In other eukaryotic systems, physical and biochemical studies have provided some interesting parallels with transposable elements and temperate bacteriophages. Some of these are the following: (1) electron microscope and biochemical studies of DNA from eukaryotic cells and viruses reveal many inverted repetitious sequences and lollipop structures; (2) the DNA of herpes simplex virus contains at least two sets of inverted repetitious sequences which bound invertible segments of the viral genome in a manner similar to the G-segment structure of bacteriophage Mu; (3) the genomes of tumor viruses integrate (in part or intact) into host chromosomes during transformation; (4) the DNA transcripts of certain RNA tumor viruses appear to integrate during viral replication; (5) specific DNA sequences in eukaryotic cells (such as ribosomal RNA genes) undergo selective replication at certain developmental stages; and (6) certain developmental phenomena (such as the generation of lymphocyte clones for specific immunoglobulins) may involve the joining of specific nonhomologous DNA segments. Although the similarity of these phenomena to the insertion phenomena of bacteria remains speculative, there can be little doubt that the prokaryotic systems will be the sources of useful models for solving eukaryotic mysteries.

Like other scientific advances, the discovery of DNA insertion elements provides new technological tools. These tools are useful for manipulating and analyzing the bacterial genome. Section VI contains papers illustrating this technology. It also contains papers which describe briefly experimental methods for studying and identifying DNA insertion elements. We hope that these papers, along with the reference material in the appendices, will help accomplish the main objective of the first meeting on DNA insertions—to stimulate further work on what may well be the most important topic in genetics today.

The Nomenclature Problem

Genetic nomenclature is an intricate problem which geneticists are constantly seeking to resolve. Yet, as a genetic system expands, the problem becomes even more complex, and a nonspecialist soon finds it exceedingly difficult to reconcile the conflicting conventions. The schemes for prokaryotic gene symbols and allele numbers now have to accommodate the insertion mutations. Notations of insertion mutations should be generally applicable, as the same insertions can be found in plasmids, in bacteriophage genomes, and in bacterial chromosomes. The rules followed in this book for insertion mutations are those recommended and drafted by a committee which originated at the meeting on DNA insertions. These rules are explained by Campbell et al. in the following pages.

We have tried to be consistent in the use of nomenclature rules. Some problems were encountered, however, because of differences in the nomenclature generally followed by bacterial geneticists and that favored by bacteriophage geneticists. The rules for bacterial genetics, adopted by the *Journal of Bacteriology* and many other journals, stem from a proposal by Demerec et al. (*Genetics,* vol. 54, pp. 61-76 [1966]). The conventions of phage geneticists are exemplified in *The Bacteriophage Lambda* (ed. A. D. Hershey, Cold Spring Harbor Laboratory [1971]). In the Demerec system, the gene symbol and the allele number are closed up and italicized. For example, *lacZ88* means a specific mutation, numbered 88, in the *Z* gene of the *lac* operon of *E. coli.* Phage geneticists frequently do not italicize the allele numbers and sometimes use letters to identify alleles. For example, *int*-*c*60 identifies a mutation, called *c*60, which affects the *int* gene of λ; an amber mutation, number 7, in the *S* gene is written as *S*7 or *Sam*7.[1]

Some insertion mutations in bacterial genes historically have been referred to in this manner. Thus *galT*-S101 refers to a mutation, number S101, in the *T* gene of the *gal* operon. To avoid confusion between gene symbols and alleles, we have tried to conform, wherever possible, to the Demerec rule that mutation numbers should not include letters. For example, *galT*-S101 will be referred to as *galT101,* and *lacZ*MS348 as *lacZ348.* Wherever changes have been made, they are explained in footnotes. No attempt has been made to change the classic phage notations used in the individual contributions to this book. To resolve ambiguities that may arise, the reader should consult the proposal by Campbell

[1] Note that in the proposal by Campbell et al., all mutation letters and numbers are italicized regardless of the convention.

et al. and references therein. It should be noted that transposons are generally referred to by numbers as recommended by Campbell et al. In some contexts, however, their previous names have been retained. For example, a chloramphenicol resistance determinant is called Tn*9*, but sometimes the terms Tn*9*, Tn*C*, and *cam* are used interchangeably.

Nomenclature of Transposable Elements in Prokaryotes

A. Campbell
Department of Biological Sciences
Stanford University
Stanford, California 94305

D. Berg
Department of Biochemistry
University of Wisconsin
Madison, Wisconsin 53706

D. Botstein
Department of Biology
Massachusetts Institute of Technology
Cambridge, Massachusetts 02139

E. Lederberg
Department of Medical Microbiology
Stanford University School of Medicine
Stanford, California 94305

R. Novick
Public Health Research Institute
of the City of New York
New York, New York 10016

P. Starlinger
Institut für Genetik
Universität zu Köln
D-5000 Köln, W. Germany

W. Szybalski
McArdle Laboratory for Cancer Research
University of Wisconsin
Madison, Wisconsin 53706

At the Cold Spring Harbor meeting on DNA insertions, a committee was elected to draft a set of rules for nomenclature. These rules are modified from a proposal by D. Berg and W. Szybalski. The final form is based on discussions held at the meeting and on previously accepted conventions for the Mu prophage (Howe and Bade 1975), for bacterial plasmids (Novick et al. 1976), and for bacterial genetics (Demerec et al. 1966). In the following report, the rules are presented in **boldface type**. Adherence to these rules in all publications about inserting elements is highly desirable.

Interspersed with the rules are some comments, which include discussion and explanation of the rules, and also suggestions for possible usage in some cases not covered by the rules. The rules are intended to be minimal and applicable to current or clearly imminent situations. We have not tried to anticipate problems that may arise far in the future. We have also avoided prescribing rules for problems that arise currently but rarely. Many of these are best treated on an individual basis relevant to specific needs.

TRANSPOSABLE ELEMENTS

Definition

Transposable elements[1] are DNA segments which can insert into several sites in a genome.

Comments

The definition applies to elements ranging from phage Mu, which seems to insert at random, to phage λ, which usually, but not always, inserts at a single site. Some (such as λ) have an identified free phase, whereas others (such as the IS or Tn elements) are presently known only in the inserted state.

CLASSES OF ELEMENTS

IS Elements

IS (simple insertion sequences) elements contain no known genes unrelated to insertion function and are generally shorter than 2 kb.

Symbols

IS*1*, **IS*2***, **IS*3***, **IS*4***, **IS*5***, etc. (Fiandt et al. 1972; Hirsch et al. 1972; Malamy et al. 1972).

Comments

At present, the IS elements listed above are not known to contain any genes, either related or unrelated to insertion function. Our intention in specifying "genes unrelated to insertion function" is that even if insertion genes should be discovered in some IS elements, they still would be classified as simple insertion sequences. The term "gene" as used here means a DNA segment coding for a functional product and does not include, for example, the "stop signals" responsible for polar effects of IS elements.

Operationally, the failure to detect other genes in IS elements is closely tied to the absence of a recognizable phenotype conferred by their presence (other than phenotypes caused by interruption of chromosomal operons by insertion). This in turn relates to the fact that IS elements have an endogenous origin. They are always present in the cell, not only at the site under examination, but at other, unidentified sites as well. Several copies of each of the common

[1] The term "transposable element" is derived from the work of McClintock (1952) on mobile genetic elements in maize. In prokaryotes, an acceptable synonym is "translocation sequence" (Novick et al. 1976).

IS elements are present in the *E. coli* genome. The phenotype of *E. coli* lacking, for example, IS*1* is unknown.

Because several copies of each element are present in the genome, it is currently impossible to trace the lineage of a given IS element. When IS*2* appears at a new site, we cannot say which of the several preexisting IS*2*'s is ancestral to it. It must therefore clearly be borne in mind that a name such as IS*2* is a generic term referring to all insertions that appear identical by hydridization or heteroduplexing but which could have individual differences in base sequence.

There is presently no advantage to giving each IS*2* in a different location a specific as well as a generic name. Such a designation would provide no additional information not already given by the mutation number (see below) that identifies the insertion event. This situation may change in the future if IS elements are introduced into species from which they are naturally absent. IS*2*, for example, is said to be absent from *Salmonella typhimurium.* If a single copy of IS*2* is introduced into *Salmonella typhimurium,* then all subsequent transpositions within that species must be derived from that specific IS*2*.

If and when such experiments are performed, all IS*2*'s derived from an identified IS*2* insertion should be assigned a specific designation (e.g., IS*2*.1) to distinguish them from IS*2* insertions derived from another, potentially different source.

Tn Elements

Tn (more complex transposable elements, often containing IS elements) elements behave formally like IS elements but contain additional genes unrelated to insertion function; they are generally larger than 2 kb.

Symbols

Tn*1*, Tn*2*, Tn*3*, etc. A different number is assigned to each independent isolate from nature *even if* it is apparently identical to some previous isolate. If useful in a particular context, gene functions carried by a Tn element may be indicated in parentheses following the name; e.g. Tn*6* (Km). ("Tn*6* is a transposable element that confers kanamycin resistance.")

Comments

Unlike IS elements, the lineage of Tn elements generally can be followed unambiguously. Hence Tn*1*, Tn*2*, etc., are specific rather than generic terms. Every isolate labeled Tn*2* has a pedigree in the laboratory that traces back to the original Tn*2* isolate.

Designations for some Tn elements described in the published literature are given in Table 1. In Table 1, "plasmid origin" indicates the natural plasmid

Table 1
Tn Elements

Transposable element[a]	Plasmid origin	Resistance markers[b]	Reference
Tn*1*	RP4	Ap	Hedges and Jacob (1974)
Tn*2*	RSF1030	Ap	Heffron et al. (1975b)
Tn*3*	R1	Ap	Kopecko and Cohen (1975)
Tn*4*	R1	Ap Sm Su	Kopecko and Cohen (1975)
Tn*5*	JR67	Km	Berg et al. (1975)
Tn*6*	JR72	Km	Berg et al. (1975)
Tn*7*	R483	Tp Sm	Barth et al. (1975)
Tn*9*	pSM14	Cm	Gottesman and Rosner (1975)
Tn*10*	R100	Tc	Foster et al. (1975); Kleckner et al. (1975)

[a]The symbol Tn*8* has not yet been assigned.
[b]Ap = ampicillin, Sm = streptomycin, Su = sulfonamide; Km = kanamycin; Tp = trimethoprim; Tc = tetracycline.

from which the Tn element was derived. For example, Gottesman and Rosner (1975) studied transpositions of the chloramphenicol resistance determinant Tn*9* from P1Cm to λ. Phage P1Cm originated from growth of phage P1 in a bacterium whose chloramphenicol resistance was derived from plasmid R14 (Kondo and Mitsuhashi 1964). The R14 plasmid is now called pSM14 (Novick 1974). The Tn*3* element of Kopecko and Cohen (1975) came from pSC50, a derivative of R1. Tn*3* has also been transferred directly out of R1 (Heffron et al. 1975a).

The designations assigned in Table 1 should be used in all publications on the elements listed there.

Episomes

Episomes are complex, self-replicating elements, often containing IS and Tn elements.

ORGANISMS AND GENOMES WITH INSERTED ELEMENTS

Insertion in a Particular Organism

When an inserting element has been introduced into a previously described bacterial strain, or has been transposed within a strain, the new strain should be given an isolation number. The genotype of a strain that has acquired an inserting element can be denoted by the genotype of the parent strain, followed by the name of the element in parentheses:

R126 = W3350(λ) = F⁻ *galK2 galT1*(λ).

Plasmids bearing IS or Tn elements should be designated according to the rules developed for plasmid nomenclature (Novick et al. 1976). In particular, each plasmid line derived from an independent insertion event should be assigned a new plasmid number.

Insertion at a Particular Site

If location is known, specify gene or region in which element is inserted, followed by a number designating the particular insertion mutation, then by a double colon, and finally by name of inserted element:

galT236::IS*1* (IS*1* within gene *galT*)
hisG1341::Tn*10* (Tn*10* within gene *hisG*)
galPO-E-490::IS*2* (IS*2* between *galPO* and *galE*)
λ*P-Qb1*::IS*2* (IS*2* between genes *P* and *Q*)
F8 42kb-7::Tn*2* (Tn*2* at 42 kb on F8)

Designation of new insertion mutations should be by number. Previously published symbols may be used for mutations already named in the literature, e.g., MS348, *r*32, *bi*2, *crg*, etc. Old symbols should be changed only when it is generally agreed that they violate the rules "excessively" (Demerec et al. 1966). As in other areas of prokaryotic genetics, each published allele number must be unique. Individual laboratories engaged in parallel mutant isolations should endeavor by private communication to avoid assignment of the same number to different mutations prior to publication.

Comments

The rules are intentionally nonspecific as to designations for insertions of known elements at unknown sites or unknown elements at known sites.

For many purposes, strains carrying known elements at unknown sites are most conveniently treated as described above for insertions in a particular organism, postponing a designation of the specific insertion event until information on location is available. Some research groups may find it useful to give each insertion of a specific element a unique identification number, regardless of map location. (For example, "*hisG1341* is the same as Tn*10* insertion number 547. Tn*10* insertion number 548 is at an unknown location and confers no nutritional requirements.") Where gene location is unknown, such identification numbers may be used (together with strain designations) in publications. It is preferable not to express such numbers in a form closely resembling official genetic symbols.

Insertion of an uncharacterized element at a known site should be indicated simply by a gene designation and mutant number. The insertional nature of the mutation should not be part of its name unless or until the insertion has been identified; e.g., *biop131* (insertion of unknown DNA into the biotin promoter).

Where the orientation of the insertion is known, it may be desired to include that orientation in the strain description. Orientation may be specified either (a) with respect to direction along the genome or (b) with respect to the polarity of the operon.

Following (with a slight modification) the conventions adopted for phage Mu, genome orientation is indicated by a plus (+) or minus (-) sign following the symbol of the element.

For example, *proA*::Mu•+. (A plus orientation means that if one moves around the *E. coli* map in a clockwise motion, the immunity end of the Mu prophage will be reached first and the S end last [Howe and Bade 1975].) For IS and Tn elements, the plus orientation is arbitrary and the minus orientation is opposite to it.

Orientation with respect to the operon may be indicated with roman numerals I and II. Orientation I is assigned arbitrarily, usually when a new IS element is first characterized in an operon with known polarity:

lacZ348::IS*1*•I
xis int-c60::IS2•II.

For IS*1*, both orientations are polar. For IS*2*, orientation II is constitutive or antipolar (Fiandt et al. 1972; Saedler et al. 1974).

Mutations in Inserted Elements

Mutations are listed after the symbol for the element:

hisG1327::Tn*10 tet43*
y r32::IS2 *rip2.*

Complex Genome Designations

A strain having two insertions and other mutations would be written

λ *Wam43 b2 xis int-c60*::IS2 *cI857 P-Qb1*::IS2 *Sam7.*

Comments

If polarity is indicated, the above strain would be

λ *Wam43 b2 xis int-c60*::IS2•+ *cI857 P-Qb1*::IS2•+ *Sam7* (genome polarity)

or

λ *Wam43 b2 xis int-c60*::IS2•II *cI857 P-Qb1*::IS2•I *Sam7* (operon polarity).

When both the inserted element and the genome into which it is inserted carry mutations, the inserted element plus its mutations may be set off by parentheses:

λ *Wam43 y r32*::(IS2 *rip2*)•I *Sam7.*

CENTRAL REGISTRY

To avoid duplication of numbers, all new IS and Tn elements should be checked with a central registry before numerals are assigned to them in publications.

A registry for Tn elements will be maintained by Esther Lederberg, Department of Medical Microbiology, Stanford University Medical School, Stanford, California 94305, as part of the Plasmid Reference Center.

Acknowledgments

The authors are grateful to Stanley Cohen and to many other individuals for suggestions and information.

REFERENCES

Barth, P. T., N. Datta, R. W. Hedges and N. J. Grintner. 1976. Transposition of a deoxyribonucleic acid sequence encoding trimethoprim and streptomycin resistances from R483 to other replicons. *J. Bact.* **125**:800.

Berg, D. E., J. Davies, B. Allet and J. Rochaix. 1975. Transposition of R factor genes to bacteriophage λ. *Proc. Nat. Acad. Sci.* **72**:3628.

Demerec, M., E. A. Adelberg, A. J. Clark and P. E. Hartman. 1966. A proposal for a uniform nomenclature in bacterial genetics. *Genetics* **54**:61.

Fiandt, M., W. Szybalski and M. H. Malamy. 1972. Polar mutations in *lac, gal,* and phage λ consist of a few IS DNA sequences inserted with either orientation. *Mol. Gen. Genet.* **119**:223.

Foster, T. J., T. G. B. Howe and K. M. V. Richmond. 1975. Translocation of the tetracycline resistance determinant from R100-1 to the *Escherichia coli* K-12 chromosome. *J. Bact.* **124**:1153.

Gottesman, M. and J. L. Rosner. 1975. Acquisition of a determinant for chloramphenicol resistance by coliphage λ. *Proc. Nat. Acad. Sci.* **72**:5041.

Hedges, R. W. and A. E. Jacob. 1974. Transposition of ampicillin resistance from RP4 to other replicons. *Mol. Gen. Genet.* **132**:31.

Heffron, F., C. Rubens and S. Falkow. 1975a. Translocation of a plasmid DNA sequence which mediates ampicillin resistance: Molecular nature and specificity of insertion. *Proc. Nat. Acad. Sci.* **72**:3623.

Heffron, F., R. Sublett, R. W. Hedges, A. Jacob and S. Falkow. 1975b. Origin of the TEM beta lactamase gene found on plasmids. *J. Bact.* **122**:250.

Hirsch, J.-J., P. Starlinger and P. Brachet. 1972. Two kinds of insertions in bacterial genes. *Mol. Gen. Genet.* **119**:191.

Howe, M. and E. Bade. 1975. Molecular biology of bacteriophage Mu. *Science* **190**:624.

Kleckner, N., R. Chan, B. Tye and D. Botstein. 1975. Mutagenesis by insertion of a drug-resistance element carrying an inverted repetition. *J. Mol. Biol.* **97**:561.

Kondo, E. and S. Mitsuhashi. 1964. Drug resistance of enteric bacteria. IV. Active transducing bacteriophage P1*CM* produced by the combination of R factor with bacteriophage P1. *J. Bact.* **88**:1260.

Kopecko, D. J. and S. N. Cohen. 1975. Site-specific *recA*-independent recombination between bacterial plasmids: Involvement of palindromes at the recombinational loci. *Proc. Nat. Acad. Sci.* **72**:1373.

Malamy, M. H., M. Fiandt and W. Szybalski. 1972. Electron microscopy of polar insertions in the *lac* operon of *Escherichia coli. Mol. Gen. Genet.* **119**:207.

McClintock, B. 1952. Chromosome organization and gene expression. *Cold Spring Harbor Symp. Quant. Biol.* **16**:13.

Novick, R. P. 1974. Bacterial plasmids. In *Handbook of microbiology* (ed. A. I. Laskin and H. A. Lechevalier), vol. IV, p. 537. C. R. C. Press, Cleveland, Ohio.

Novick, R. P., R. C. Clowes, S. N. Cohen, R. Curtiss III, N. Datta and S. Falkow. 1976. Uniform nomenclature for bacterial plasmids: A proposal. *Bact. Rev.* **40**:168.

Saedler, H., H. J. Reif, S. Hu and N. Davidson. 1974. IS*2*, a genetic element for turn-off and turn-on of gene activity in *E. coli. Mol. Gen. Genet.* **132**:265.

SECTION I
IS Elements

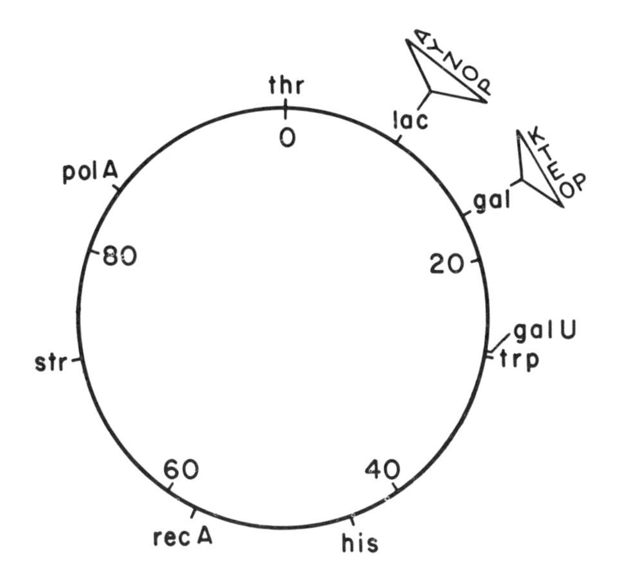

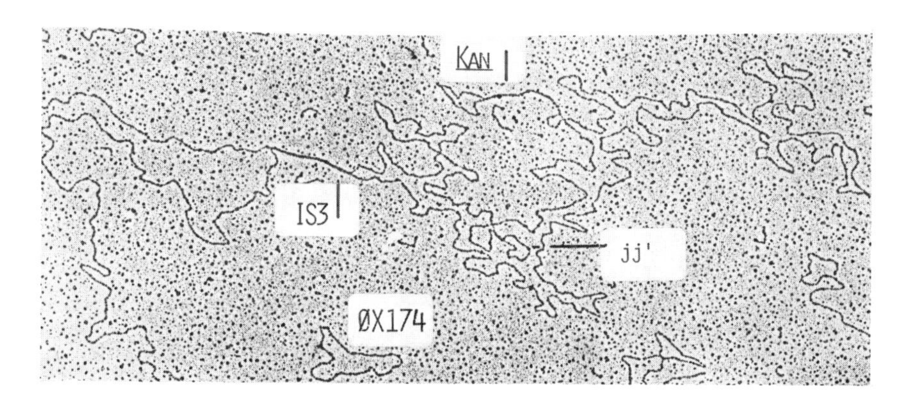

The discovery of the IS*1* and IS*2* elements stemmed from studies on polar mutations in the *gal* operon of *E. coli*. Several papers in this section discuss properties of insertions in the *gal* operon. The position of the *gal* operon on the linkage group of *E. coli* is shown in the section-title figure (top). The linkage group is calibrated from 0 to 100 minutes (Bachmann et al., *Bact. Revs.* 40:116 [1976]). The *gal* operon consists of the operator-promoter region (*OP*) and three structural genes *E, T,* and *K*. The *galU* gene encoding uridine diphosphoglucose pyrophosphorylase is not a part of the *gal* operon and is located at minute 27 on the genetic map. The three *gal* operon genes encode enzymes involved in the following reactions (Adhya and Shapiro, *Genetics* 62:231 [1969]):

$$\text{galactose (Gal)} + \text{ATP} \xrightarrow{\text{galactose kinase } (galK)} \text{Gal-1-P} + \text{ADP};$$

$$\text{Gal-1-P} + \text{uridine diphospho(UDP)glucose} \underset{\text{transferase } (galT)}{\overset{\text{galactose-1-P-uridyl}}{\rightleftharpoons}} \text{UDPGal} + \text{glucose-1-P};$$

$$\text{UDPGal} \underset{\text{epimerase } (galE)}{\overset{\text{uridine diphosphogalactose-4-}}{\rightleftharpoons}} \text{UDP glucose.}$$

The sum of these three reactions is

$$\text{Gal} + \text{ATP} \longrightarrow \text{glucose-1-P} + \text{ADP}.$$

The linkage map also shows the position of the *lac* operon, which consists of a promoter (*P*), an operator (*O*), a gene for the enzyme β-galactosidase (*Z*), a gene for *lac* permease (*Y*), and a gene for thiogalactoside transacetylase (*A*). The *lac* operon has been very useful for analyzing the properties of insertion elements such as IS*3* (Malamy et al., *Mol. Gen. Genet.* 119:207 [1972]), Tn*5* (Berg, this volume), and bacteriophage Mu (Bukhari et al., this volume).

The bottom half of the figure is from a study by L. T. Chow and T. R. Broker (Cold Spring Harbor Laboratory) on the physical location of the *lac* operon on the chromosome of *E. coli.* The electron micrograph illustrates nested stem-loop structures in denatured and reannealed DNA from *E. coli* strain DB01162-36 (isolated by D. E. Berg), which carries the kanamycin transposon in the *lac* promoter. The DNA includes the *lac* region of the chromosome, as deduced from the characteristic stem-loop structure of the kanamycin resistance marker (Berg et al., *Proc. Nat. Acad. Sci.* 72:3628 [1975]). The sequences comprising the inverted repetition (1.3 kb) flanking the largest single-stranded loop (70.9 kb, exclusive of the *kan* insertion) have been identified previously on F$'$ 13 (*lac*$^+$) to be IS*3* (Hu et al., *J. Bact.* 122:749 [1975]). The smallest stem-loop structure, jj$'$, was also identified on F$'$ 13 and allows the molecule illustrated to be oriented with respect to the *E. coli* map. In other heteroduplexes in the same DNA preparation (not shown), a sequence of 1.3 kb, believed to be an IS*2*, has been located 3.7 kb from jj$'$ on the side opposite to that of the *kan* insertion (Chow and Broker, unpubl.). This confirms the prediction of Hu et al. (see above) that an IS*2* sequence at that *E. coli* map position is the site at which F factor integrates to form Hfr 13. The order of sequences and the distances between adjacent sequences are IS*3*—46.9 kb (*lacAYZ-OP*:)*kan*—8.2 kb (*lacI*)—jj$'$—3.7 kb—IS*2*—8.4 kb—IS*3*. The *kan*, jj$'$, IS*2*, and IS*3* measure, respectively, 5.6, 2.4, 1.3, and 1.3 kb. Thus the *lac* operon has been mapped precisely with respect to outside elements.

Mutations Caused by the Integration of IS*1* and IS2 into the *gal* Operon

P. Starlinger
Institut für Genetik
Universität zu Köln
D-5000 Köln 41, W. Germany

Insertion of an IS element into a gene abolishes the function of the gene and thus causes a mutation. If the gene is a part of an operon, the insertion can result in severe polar effects on the expression of genes located distal to the mutation with respect to the promoter (*P*)-operator (*O*) region. The first clear demonstration of the insertion of IS elements came from studies of the spontaneously isolated strong polar mutations of the *gal* operon of *Escherichia coli.* These mutations were originally described by Lederberg (1960), who found that some of her Gal⁻ mutants showed reduced activity for three different enzymes involved in galactose utilization. Similar mutations were subsequently isolated by Shapiro (1967), Saedler and Starlinger (1967a,b), Jordan et al. (1967), and Adhya and Shapiro (1969). Isolation of these mutations was facilitated by the availability of a method for selecting galactokinase-negative mutants. The enzyme galactokinase is coded in *galK,* the most distal gene of the *gal* operon. The method for selecting galactokinase-deficient cells is based on the observation that cells which lack the products of any of the *galT, galE,* or *galU* genes but which synthesize even small amounts of galactokinase grow poorly in the presence of galactose. This is because of the intracellular accumulation of galactose-1-phosphate (see Lederberg 1960). The method for isolating the polar mutations of the *gal* operon was to plate *galU⁻* cells in the presence of galactose (Saedler and Starlinger 1967a; Adhya and Shapiro 1969). Since the *galU* gene is not a part of the *gal* operon, the effects of the new mutations on the functioning of the *gal* operon could be analyzed without interference from the *galU* mutations.

Deletion mapping of the mutations causing the galactokinase-negative phenotype showed that only some of the mutations had occurred in the *galK* gene (Pfeifer and Oellermann 1967; Shapiro and Adhya 1969). Other mutations mapped in genes *galT, galE,* and the control region of the *gal* operon. These mutants had become galactose-resistant owing to a severe polar effect of the mutations. The polarity resulted in residual synthesis of the galactokinase at levels as low as 0.1% to 0.5% of the fully induced wild-type level of the enzyme. This residual synthesis is much lower than that observed with polar amber mutations in *galE* or *galT* (Jordan and Saedler 1967; Adhya and Shapiro 1969).

Characterization of the Insertion Mutations

The mutants were found to revert to Gal$^+$ spontaneously in the range of 10^{-6} to 10^{-8} per cell plated from an overnight culture grown from a small inoculum. This indicated that the mutations were not deletions. The reversion rate, however, was not enhanced by mutagens. This suggested that the mutations were also not point mutations. The revertible polar mutations could then be chromosomal aberrations, including inversions, duplications, or transpositions of DNA from other regions of the chromosome. To distinguish between these possibilities, it was necessary to measure the amount of DNA within the galactose operon. Inversions should not alter this amount as compared to the wild type. Duplications and transpositions should lead to "extra" DNA in the *gal* operon. In the case of duplications of the *gal* operon, however, this DNA should not introduce new sequences within the operon, whereas DNA transposed from another region of the chromosome would introduce a new sequence into the *gal* operon.

Measurements of the amount of *gal* operon DNA would be possible only if this operon were part of a DNA molecule of defined length. Such molecules are available as the chromosomes of λd*gal* or λp*gal* transducing phages. The mutations to be tested were thus recombined with λd*gal* phages by the process of homogenate formation, and the resultant mutant phages were tested for an increase in density due to extra DNA (Jordan et al. 1968; Shapiro 1969). All of the mutations tested caused an increase in the density of λ*gal* transducing phages to a variable extent. It was also shown that reversion to wild type of the *gal* operon carried by the phages was accompanied by a reduction in density to that of the parental λd*gal* phages.

To determine whether the extra DNA was introduced by a duplication or as the result of an insertion of outside DNA, Michaelis et al. (1969) transcribed the DNA extracted from mutant λd*gal* phages in vitro with RNA polymerase, hybridized this RNA repeatedly to DNA of λd*gal* phages lacking the mutation but otherwise isogenic, and tested for the presence of a fraction that could specifically hybridize to the DNA of mutant phage. Such RNA was found. This indicated that the extra DNA was not normally a part of the *gal* operon and thus had not arisen by duplication. Experiments in which RNA transcribed from the DNA of λd*gal* phages containing one of the mutations was cross-hybridized to the DNA carrying another of these mutations suggested that the inserted sequences were specific and were related to each other.

The introduction of the heteroduplex technique (Davis and Davidson 1968; Westmoreland et al. 1969) allowed electron microscopic visualization of the inserted sequences. This technique also allowed the detection of homologies between different inserted DNA segments (Fiandt et al. 1972; Hirsch et al. 1972a,b; Malamy et al. 1972). The authors cited adopted the designation IS for inserted sequences. In addition to the mutants characterized by Jordan et al. (1968) and Shapiro (1969), they examined two other groups of mutations caused by insertions: mutations isolated in bacteriophage λ (Brachet et al. 1970) and polar mutations isolated in the *lac* operon of *E. coli* (Malamy 1967, 1970). Most of the mutations were found to be caused by the insertion of two distinct sequences, with lengths of approximately 800 and 1350 base pairs. These sequences were called IS*1* and IS*2*, respectively. Two other classes, IS*3* and IS*4*, were represented by one member each.

IS*1* and IS*2* can be found in either orientation with respect to the rest of the chromosome. Hybridization of identical strands of the DNA of two mutant phages carrying the same IS element, but in opposite orientations with respect to the chromosome, leads to the formation of heteroduplexes between the IS-element DNAs. These heteroduplexes extend into single-stranded tails. This allows visualization of the homology between members of the same class (Westmoreland et al. 1969). This technique was used by Shapiro et al. (1969) for the isolation of pure *lac* operon DNA.

All members of the same class were found to yield linear heteroduplex molecules of the same length found for insertion loops in heteroduplex molecules prepared from phage chromosomes that differed only in the presence of the IS element on one of the molecules. This finding indicated complete homology of the members of one IS class at the level of resolution of the electron microscope and also the absence of circular permutation of the IS element in the different mutants tested.

Specificity of IS Integration

The integration sites of the IS elements within the *gal* operon exhibit a degree of specificity intermediate between the very pronounced specificity of integration of bacteriophage λ (Gottesman and Weisberg 1971) and the absence of any specificity for the integration of bacteriophage Mu (Bukhari and Zipser 1972). It resembles the intermediate specificity for the integration of λ in the chromosomes of bacteria that carry a deletion of the normal attachment site for λ (Shimada et al. 1972).

IS*1* (in both orientations) and IS*2* (in one orientation) have been detected in the control region of the *gal* operon of *E. coli* (Fiandt et al. 1972; Hirsch et al. 1972a,b; Saedler et al. 1972). The sites of integration in these mutants were found to be identical by genetic mapping (Saedler et al. 1972) and heteroduplex mapping (Hirsch et al. 1972a). However, the mutants fell into two classes with respect to the polarity of mutations, that is, the degree to which the expression of the *galE* gene is inhibited. One class consisted of insertions in the *galOP* region (*galOP128*::IS*1*, *galOP306*::IS*1*, *galOP308*::IS*2*), which caused a coordinate reduction in the synthesis of all of the three enzymes of the *gal* operon. The other class (*galOP141*::IS*1*, *galOP490*::IS*2*) caused an almost complete loss of *galE* expression (Saedler et al. 1972; de Crombrugghe et al. 1973). The difference in *galE* expression between *galOP128*::IS*1* and *galOP141*::IS*2* mutants had been inferred by Adhya and Shapiro (1969) from their studies on the *gal* operon. The reason for this difference in the effects of insertions, all of which are located in the *galOP* region and cannot be otherwise resolved, is not clear. It is possible that there are two, independent, but very close, insertion sites for IS elements in the *gal* control region. Alternatively, it could be that a small region is recognized for IS integration, and that the exact insertion point can be anywhere in that region. Perhaps this question can only be answered by sequencing the junctions of the IS elements and the host genes.

CONCLUDING REMARKS

Studies on many of the spontaneously isolated polar mutations of the *gal* operon of *E. coli* have established that these mutations are caused by the insertion of specific DNA sequences. These sequences have been termed the IS elements. The elements appear to insert themselves at specific sites within the genes. The presence of an IS element in an operon usually causes a reduction in the expression of genes located distally to the insertion with respect to the promoter region. This polarity is apparently the result of a block in mRNA synthesis. The distal genes can be shown not to be transcribed in vivo (Starlinger et al. 1973). The mechanism of this termination of transcription is only beginning to be understood and is discussed in detail in other papers in this section.

Note Added in Proof

H. Chadwell has hybridized isolated IS*1* DNA and IS*2* DNA (see Schmidt et al., this volume) to single strands of λd*gal* DNA carrying the respective insertions in the leader sequence of the *gal* operon and elongated the IS DNA at the 3′ terminus with DNA polymerase I using ribosubstitution conditions. The resulting oligonucleotides allow the conclusion that *galOP141*::IS*1*, *galOP142*::IS*1* (newly isolated by J. Besemer, pers. comm.), and *galOP308*::IS*2* are inserted in the leader sequence of the *gal* operon in positions varying by a few nucleotides. This raises the possibility that the integration mechanism resembles restriction enzymes of type I, which have a recognition site but not a defined cleavage site.

Acknowledgment

Work done in the author's laboratory was supported by the Deutsche Forschungsgemeinschaft through SFB 74.

REFERENCES

Adhya, S. L. and J. A. Shapiro. 1969. The galactose operon of *E. coli* K-12. I. Structural and pleiotropic mutations of the operon. *Genetics* **62**:231.

Brachet, P., H. Eisen and A. Rambach. 1970. Mutations of coliphage λ affecting the expression of replicative functions O and P. *Mol. Gen. Genet.* **108**:266.

Bukhari, A. I. and D. Zipser. 1972. Random insertion of Mu-1 DNA within a single gene. *Nature New Biol.* **236**:240.

Davis, R. W. and N. Davidson. 1968. Electron-microscopic visualization of deletion mutations. *Proc. Nat. Acad. Sci.* **60**:243.

de Crombrugghe, S., S. Adhya, M. Gottesman and I. Pastan. 1973. Effect of rho on transcription of bacterial operons. *Nature New Biol.* **241**:260.

Fiandt, M., W. Szybalski and M. H. Malamy. 1972. Polar mutations in *lac, gal* and phage λ consist of a few DNA sequences inserted with either orientation. *Mol. Gen. Genet.* **119**:223.

Gottesman, M. E. and R. A. Weisberg. 1971. Prophage insertion and excision. In *The bacteriophage lambda* (ed. A. D. Hershey), p. 113. Cold Spring Harbor Laboratory, Cold Spring Harbor, New York.

Hirsch, H. J., H. Saedler and P. Starlinger. 1972a. Insertion mutations in the control region of the galactose operon of *E. coli.* II. Physical characterization of the mutations. *Mol. Gen. Genet.* **115**:266.

Hirsch, H. J., P. Starlinger and P. Brachet. 1972b. Two kinds of insertions in bacterial genes. *Mol. Gen. Genet.* **119**:191.

Jordan, E. and H. Saedler. 1967. Polarity of amber mutations and suppressed amber mutations in the galactose operon of *E. coli. Mol. Gen. Genet.* **100**:283.

Jordan, E., H. Saedler and P. Starlinger. 1967. Strong polar mutations in the transferase gene of the galactose operon of *E. coli. Mol. Gen. Genet.* **100**:296.

——. 1968. O° and strong polar mutations in the *gal* operon are insertions. *Mol. Gen. Genet.* **102**:353.

Lederberg, E. M. 1960. Genetic and functional aspects of galactose metabolism in *Escherichia coli* K-12. In *Microbial genetics. 10th Symposium of the Society for General Microbiology,* p. 115. University Press, Cambridge.

Malamy, M. H. 1967. Frameshift mutations in the lactose operon of *E. coli. Cold Spring Harbor Symp. Quant. Biol.* **31**:189.

——. 1970. Some properties of insertion mutations in the *lac* operon. In *The lactose operon* (ed. J. R. Beckwith and D. Zipser), p. 359. Cold Spring Harbor Laboratory, Cold Spring Harbor, New York.

Malamy, M. H., M. Fiandt and W. Szybalski. 1972. Electron microscopy of polar insertions in the *lac* operon of *Escherichia coli. Mol. Gen. Genet.* **119**:207.

Michaelis, G., H. Saedler, P. Venkov and P. Starlinger. 1969. Two insertions in the galactose operon having different sizes but homologous DNA sequences. *Mol. Gen. Genet.* **104**:371.

Pfeifer, D. and R. Oellermann. 1967. Mapping of *gal* mutants by transducing λd*gal* phages carrying deletions of the gal^- region. *Mol. Gen. Genet.* **99**:248.

Saedler, H. and P. Starlinger. 1967a. O° mutations in the galactose operon in *E. coli.* I. Genetic characterization. *Mol. Gen. Genet.* **100**:178.

——. 1967b. O° mutations in the galactose operon in *E. coli.* II. Physiological characterization. *Mol. Gen. Genet.* **100**:190.

Saedler, H., J. Besemer, B. Kemper, B. Rosenwirth and P. Starlinger. 1972. Insertion mutations in the control region of the *gal* operon of *E. coli.* I. Biological characterization of the mutations. *Mol. Gen. Genet.* **115**:258.

Shapiro, J. A. 1967. The structure of the galactose operon in *E. coli* K-12. Ph.D. thesis, Cambridge University, Cambridge, England.

——. 1969. Mutations caused by the insertion of genetic material into the galactose operon of *Escherichia coli*. *J. Mol. Biol.* **40**:93.

Shapiro, J. A. and S. L. Adhya. 1969. The galactose operon of *E. coli* K-12. II. A deletion analysis of operon structure and polarity. *Genetics* **62**:249.

Shapiro, J. A., L. MacHattie, L. Eron, G. Ihler, K. Ippen and J. Beckwith. 1969. Isolation of pure *lac* operon DNA. *Nature* **224**:768.

Shimada, K., R. A. Weisberg and M. Gottesman. 1972. Prophage λ at unusual chromosomal locations. I. Location of the secondary attachment sites and the properties of lysogens. *J. Mol. Biol.* **63**:483.

Starlinger, P., H. Saedler, B. Rak, E. Tillmann, P. Venkov and L. Waltschewa. 1973. mRNA distal to polar nonsense and insertion mutations in the *gal* operon of *E. coli*. *Mol. Gen. Genet.* **122**:779.

Westmoreland, B. C., W. Szybalski and H. Ris. 1969. Mapping of deletions and substitutions in heteroduplex DNA molecules of bacteriophage lambda by electron microscopy. *Science* **163**:1343.

Specific Sites for Integration of IS Elements within the Transferase Gene of the *gal* Operon of *E. coli* K12

D. Pfeifer and P. Habermann
Institut für Genetik
Universität zu Köln
D-5000 Köln 41, W. Germany

D. Kubai-Maroni
Department of Zoology
Duke University
Durham, North Carolina 27706

Mutations caused by the integration of IS elements at different sites in the galactose operon have been described (Jordan et al. 1967; Saedler and Starlinger 1967a,b; Adhya and Shapiro 1969; Shapiro and Adhya 1969; Saedler et al. 1972). These investigations did not suggest any strong site specificity. There were indications for "hot spots" of integration at two sites only. Shapiro and Adhya (1969) mapped eight strong polar mutations in the same deletion group of the *galT* gene. One of these mutants, S101, was later designated tentatively as IS*4* by Fiandt et al. (1972). Shimada et al. (1973) showed that these polar mutants did not recombine with each other or with mutants caused by the integration of bacteriophage λ into *galT*. Saedler et al. (1972) described a set of mutants caused by the integration of IS elements into the control region of the *gal* operon. These mutations did not recombine with each other. Electron microscopy of heteroduplex molecules showed that these mutants contain IS*1* in both orientations and IS*2* in one orientation (Hirsch et al. 1972a,b).

Isolation and Mapping of Insertions in *galT*

The discovery that IS elements consist of several well-defined sequences, and that at least two of these can be integrated into the same site in the *galOP* region, led us to a reinvestigation of the question of site specificity of IS elements. For this purpose, we isolated 150 galactose-resistant mutants in a *galU* background of an *E. coli* K12 strain lysogenic for phage λ. LFT (low-frequency transducing) lysates were prepared and used for a preliminary mapping of the mutants. Forty of them mapped in *galT* and were used for further studies. HFT (high-frequency transducing) lysates of λ carrying these mutations were then prepared and used to transduce an *E. coli* strain carrying a deletion of the *gal* operon. The mutations in this background were mapped with a set of λd*gal* phages carrying deletions of known lengths, as described by Pfeifer and Oellermann (1967) and Pfeifer et al. (1974).

Figure 1 shows that all but one (*galT15*) of the 40 mutations map in the same region defined by deletion group 9. In crosses of the mutants with each other, no recombinants were observed. Furthermore, they did not recombine with the IS mutant S101. We believe, therefore, that our deletion group 9 corresponds to the deletion group that defines the position of the cluster of eight polar mutants in *galT* described by Shapiro and Adhya (1969).

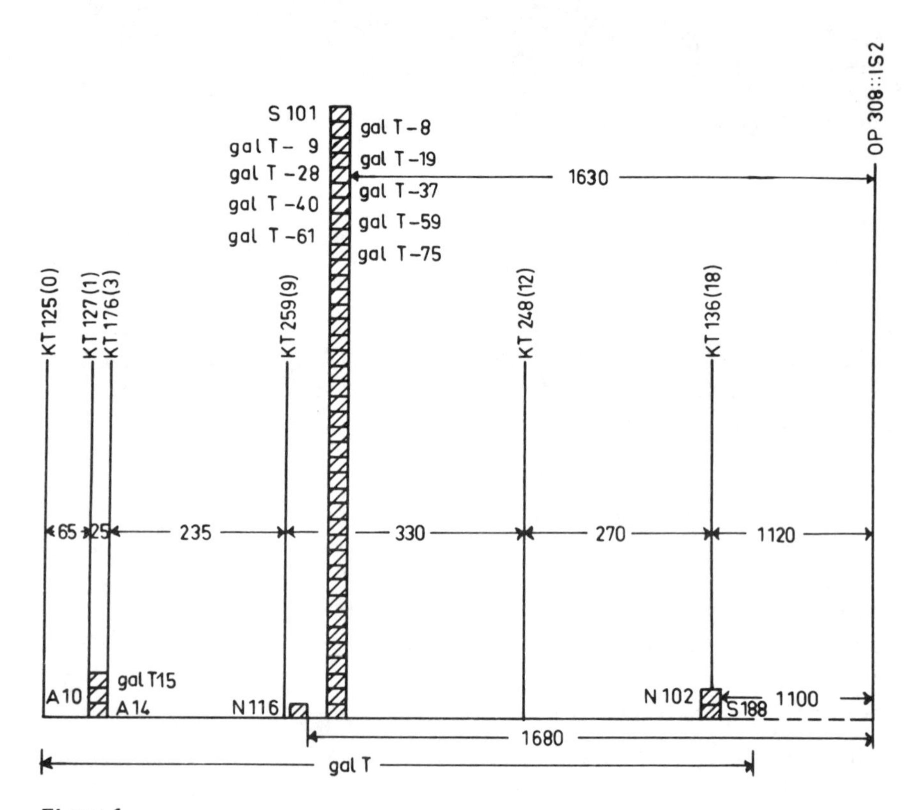

Figure 1
A physical map of IS mutations within the transferase gene of *E. coli* K12. The vertical lines with the designations KT125-KT136 mark deletion end points in *galT* of λd*gal* del phages. The distances between successive deletion end points and between the insertion mutation *galOP308*::IS*2* and the insertion mutations N102 and N116 are expressed in number of nucleotides, according to the data of Pfeifer et al. (1974). The numbers in parentheses to the right of the designations for each deletion end point refer to the number of the respective deletion group as defined by deletion mapping (Pfeifer and Rosypal 1969). Each cross-hatched box represents one IS mutation. The designations of several IS mutations are shown to the left of the boxes. (See text for further explanation.)

Electron Microscope Heteroduplex Studies

So far, ten of the 40 mutations, of which nine belong to the class mapping in deletion group 9, were used for heteroduplex mapping. Heteroduplex molecules were prepared from λd*gal* phages carrying the unknown insertions and λd*galOP308*::IS*2*. All of the nine insertions belonging to the cluster in deletion group 9 were about the size of IS*2*. The one mutation mapping in deletion group 1 has the length of IS*1*.

Heteroduplex molecules were also prepared by annealing the denatured DNA of λd*gal* carrying one of the mutations, *galT8*, with the DNA of the other mutant phages of the sample investigated. When *galT8* was heteroduplexed with *galT37* or *galT59*, an insertion loop was seen at the junction between phage and bacterial DNA, but no irregularity was found in the region of gene *galT*. This suggested that the two mutants were caused by the same insertion in the same orientation at the same site. Heteroduplex molecules prepared from the DNA of mutant *galT8* and any of the mutants *galT9*, *galT19*, *galT28*, *galT40*, *galT61*, or *galT75* showed a substitution loop in *galT*. Both branches of the substitution loop were of equal length, and this corresponded to the length of the insertion loop in the heteroduplex molecules described above (Fig. 2). This indicated that the two mutants used for the preparation of the heteroduplex molecules were caused by the integration of the same IS element, but in opposite orientation. This conclusion was supported by the finding of duplex molecules corresponding in length to the insertions. Single-stranded tails resulted when identical rather than complementary strands of the DNA of mutants *galT8* and *galT75* were used (Fig. 3).

A regular duplex structure without any loops was seen in heteroduplexes between *galT75* and *galT101*.[1] This links our series of mutations to the cluster isolated by Shapiro and Adhya (1969). Mutant S101, belonging to this cluster, had previously been shown to be different from IS*3* and probably different from IS*2* (Fiandt et al. 1972). The latter IS element, however, was only tested in one orientation. Extensive tests using λd*gal308*::IS*2* and λ*y-r*32::IS*2* revealed no homology between the cluster of IS mutants and IS*2*. This indicates that the cluster of mutants described here is caused by a new IS element, designated IS*4*. It is not known whether all of these mutants are integrated into one single site or whether they are integrated into a region and thus are all very near to each other but in different sites. No differences in position were detected by either genetic recombination or heteroduplex mapping, but the resolution of both of these methods did not exclude different sites separated by a few nucleotides.

The finding of IS*4* in both orientations in deletion group 9 of gene *galT* indicates that IS*4* is polar in both orientations. It therefore resembles IS*1* and is unlike IS*2*. To our knowledge, IS*4* has been isolated only at this one site. Furthermore, no IS elements other than IS*4* have been detected at this site. In several galactose-resistant strains (including the one described here), we have

[1] *galT101* is S101 mutation (Shapiro and Adhya 1969).

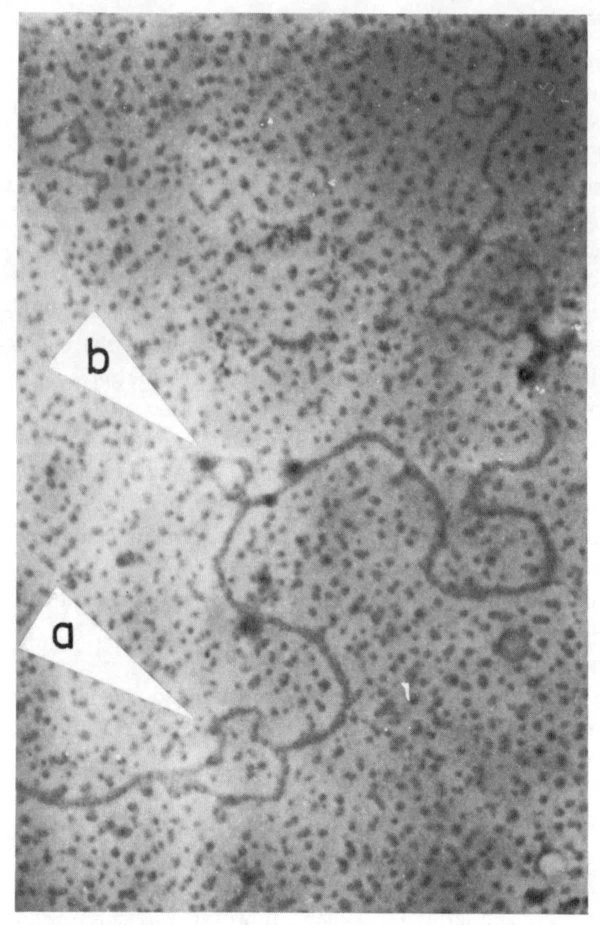

Figure 2
Heteroduplex molecules prepared from r strands of λd*galT8* and l strands of *galT75*. The insertion/deletion loop (*a*) and the substitution loop (*b*) are interpreted in the text.

found an abundance of IS*4*. These results are different from earlier studies on the isolation of IS elements, described by Jordan et al. (1967) and Saedler and Starlinger (1967a), in which predominantly IS*1* and IS*2* were found. The reason for this difference is not clearly understood.

Site Specificity of IS*1*

In addition to mapping IS*4* insertions, we have mapped several mutants caused by the integration of IS*1*. The *galT102*::IS*1* (Jordan et al. 1967) and

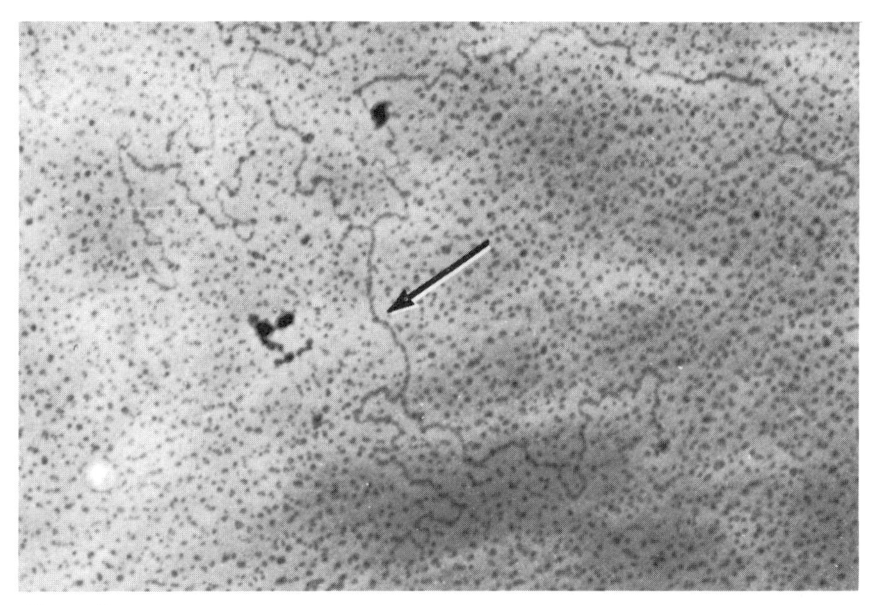

Figure 3
Heteroduplex molecule prepared from r strands of λd*galT8* and λd*gal75.* This IS duplex is indicated by an arrow.

galT188::IS*1*[2] (Adhya and Shapiro 1969) mutations are both located in the same place in deletion group 18 and do not recombine with each other. A second spot for the integration of IS*1* is probably in deletion group 1, which contains the newly isolated *galT15* and two strong polar mutants, A10 and A14 (Jordan et al. 1967). These three insertions are probably IS*1*, as judged by their strong polarity and strong tendency to revert, which distinguishes them from IS*2* and IS*4*, which rarely revert. In addition, *galT15* has been shown to have the length of IS*1* in heteroduplex molecules. These three mutants also fail to recombine with each other.

The IS elements as a whole appear to insert themselves at many different sites, but the site specificity of individual IS elements might be quite pronounced.

Acknowledgment

This work was supported by the Deutsche Forschungsgemeinschaft through SFB 74.

[2]*galT102* is N102 mutation (Jordan et al. 1967); *galT188* is S188 mutation (Adhya and Shapiro 1969).

REFERENCES

Adhya, S. L. and J. A. Shapiro. 1969. The galactose operon of *E. coli* K-12. I. Structural and pleiotropic mutations of the operon. *Genetics* **62**:231.

Fiandt, M., W. Szybalski and M. H. Malamy. 1972. Polar mutations in *lac, gal* and phage λ consist of a few DNA sequences inserted with either orientation. *Mol. Gen. Genet.* **119**:223.

Hirsch, H. J., H. Saedler and P. Starlinger. 1972a. Insertion mutations in the control region of the galactose operon of *E. coli.* II. Physical characterization of the mutations. *Mol. Gen. Genet.* **115**:266.

Hirsch, H. J., P. Starlinger and P. Brachet. 1972b. Two kinds of insertions in bacterial genes. *Mol. Gen. Genet.* **119**:191.

Jordan, E., H. Saedler and P. Starlinger. 1967. Strong polar mutations in the transferase gene of the galactose operon in *E. coli. Mol. Gen. Genet.* **100**:296.

Pfeifer, D. and R. Oellermann. 1967. Mapping of the *gal* mutants by transducing λd*gal* phages carrying deletions of the *gal* region. *Mol. Gen. Genet.* **99**:248.

Pfeifer, D. and S. Rosypal. 1969. Deletion mapping of galactose mutants with transducing λ phages. *Mol. Gen. Genet.* **104**:116.

Pfeifer, D., H. J. Hirsch, D. Bergmann and M. Hamlaoui. 1974. Non-random distribution of endpoints of deletions in the *gal* region of *E. coli. Mol. Gen. Genet.* **132**:203.

Saedler, H. and P. Starlinger. 1967a. O^o mutations in the galactose operon in *E. coli.* I. Genetic characterization. *Mol. Gen. Genet.* **100**:178.

——. 1967b. O^o mutations in the galactose operon in *E. coli.* II. Physiological characterization. *Mol. Gen. Genet.* **100**:190.

Saedler, H., J. Besemer, B. Kemper, B. Rosenwirth and P. Starlinger. 1972. Insertion mutations in the control region of the *gal* operon of *E. coli.* I. Biological characterization of the mutations. *Mol. Gen. Genet.* **115**:258.

Shapiro, J. A. and S. L. Adhya. 1969. The galactose operon of *E. coli* K-12. II. A deletion analysis of operon structure and polarity. *Genetics* **62**:249.

Shimada, K., R. A. Weisberg and M. E. Gottesman. 1973. Prophage λ at unusual chromosomal locations. II. Mutations induced by bacteriophage λ in *Escherichia coli* K-12. *J. Mol. Biol.* **80**:297.

The *gal3* Mutation of *E. coli*

A. Ahmed
Department of Genetics
University of Alberta
Edmonton, Alberta, Canada T6G 2E9

Perhaps no other mutation belonging to the Lederberg collection of gal^- mutants of *E. coli* K12 has attracted greater interest than the *gal3* mutation (Morse et al. 1956). The main reason for this interest has been the unique properties exhibited by this mutation.

The pleiotropic nature of *gal3* was apparent quite early in the studies of Lederberg (1960). The simultaneous reduction of the levels of the three enzymes (kinase, transferase, and epimerase) coded by the *gal* operon, the inability to complement other *gal* mutants, and the difficulties encountered in the construction of stable transducing lines with bacteriophage lambda suggested that the *gal3* mutation exerted multiple effects. Yet, by all genetic criteria, the mutational alteration appeared not to be a deletion.

Properties of the *gal3* Mutation

The essential features of *gal3* (Morse et al. 1956; Hill and Echols 1966; Morse 1967; Adhya and Shapiro 1969; Shapiro and Adhya 1969) can be summarized as follows: (1) *gal3* was originally isolated as a mutation of spontaneous origin; (2) it reverts spontaneously to gal^+ at a rate of 3.7×10^{-8} per bacterium per division, which is about 20 times higher than the rate for other *gal* mutants of the same set; (3) the reversion rate is not significantly enhanced by treatment with chemical mutagens such as 2-aminopurine, *N*-methyl-*N'*-nitro-*N*-nitrosoguanidine or substituted acridines such as ICR-191A; (4) it maps as a point mutant in the operator-promoter (*OP*) region of the *gal* operon; (5) it has an extreme polar effect on the synthesis of the three enzymes specified by the *gal* operon; (6) the polarity is not relieved by the action of amber and ochre suppressors, although polarity suppressors such as *suA* were not tested; and (7) it shows drastically reduced levels of *gal*-specific mRNA even after induction of the *gal* operon. These observations immediately suggest that *gal3* may belong to an entirely different class of polar mutations.

The most intriguing feature of the *gal3* mutation, however, is the production of several different kinds of spontaneous gal^+ revertants. These reversions fall into at least three distinct categories on the basis of their stability, mode of enzyme synthesis, and efficiency of transduction by bacteriophage λ. The first class, which is also the most frequent, consists of stable revertants which are inducible for the *gal* enzymes and are normally transduced by λ at a frequency comparable to the wild-type strains. These revertants are indistinguishable from the wild-type *gal*$^+$ in all respects. The second class consists of stable revertants which synthesize the *gal* enzymes in a constitutive manner, usually at a high level comparable to the fully induced wild type. These reversions are transduced at a considerably lower frequency (1-2% of the wild type) by λ. The third class, which is also the least frequent, consists of unstable revertants which constantly segregate the original *gal3* mutation at a high frequency (6%). These reversions show typically low levels of *gal* enzymes synthesized in a constitutive manner and poor transducibility by λ. A very rough estimate of the relative frequencies of occurrence of the three classes of revertants would be 65:35:1, respectively, bearing in mind that these frequencies vary depending upon a number of genotypic and environmental factors.

Nature and Identification of the *gal3* Mutation

With respect to its reversion behavior and polarity, the *gal3* mutation resembles the class of insertion mutations known to arise by the linear integration of foreign DNA segments (Jordan et al. 1968; Shapiro 1969). By transferring the *gal3* mutation onto a λ*gal* genome and examining the λ*gal*$^+$/λ*gal3* DNA heteroduplexes (Fig. 1), we were able to show that the *gal3* mutation was caused by the addition of a segment of 1100-1200 base pairs in the *OP* region of the *gal* operon (Ahmed and Scraba 1975).

It is now clearly established that the insertion mutations arise by the integration of but a few specific DNA sequences, designated as IS*1* through IS*5* (Hirsch et al. 1972; Fiandt et al. 1972). The sizes and properties of these insertion sequences (IS) are summarized in Appendix A1. The *gal3* insertion has been identified as IS*2* by electron microscopy of heteroduplexes with known insertion sequences (M. Fiandt, A. Ahmed and W. Szybalski, in prep.). The insertion is located in orientation I (Hirsch et al. 1972; Fiandt et al. 1972), in which it is known to be polar in all cases examined to date.

Reversion of the *gal3* Mutation

Because of its location in the *OP* region, the IS*2* sequence confers on *gal3* the unique properties observed during its reversion. A scheme to explain the various classes of *gal*$^+$ revertants from *gal3* is shown in Figure 2 (Ahmed 1975; Ahmed and Johansen 1975). This scheme is based on the assumption that the insertion causes a disruption of the *gal* operator sequence without causing substantial damage to the promoter. Thus the insertion is believed to interfere with the binding of the repressor but not of the RNA polymerase. On induction, synthesis of *gal* mRNA is initiated at the promoter, but it is quickly terminated on encountering the site for rho action located on IS*2*, which appears to be the basis for the extreme polarity of the *gal3* mutation.

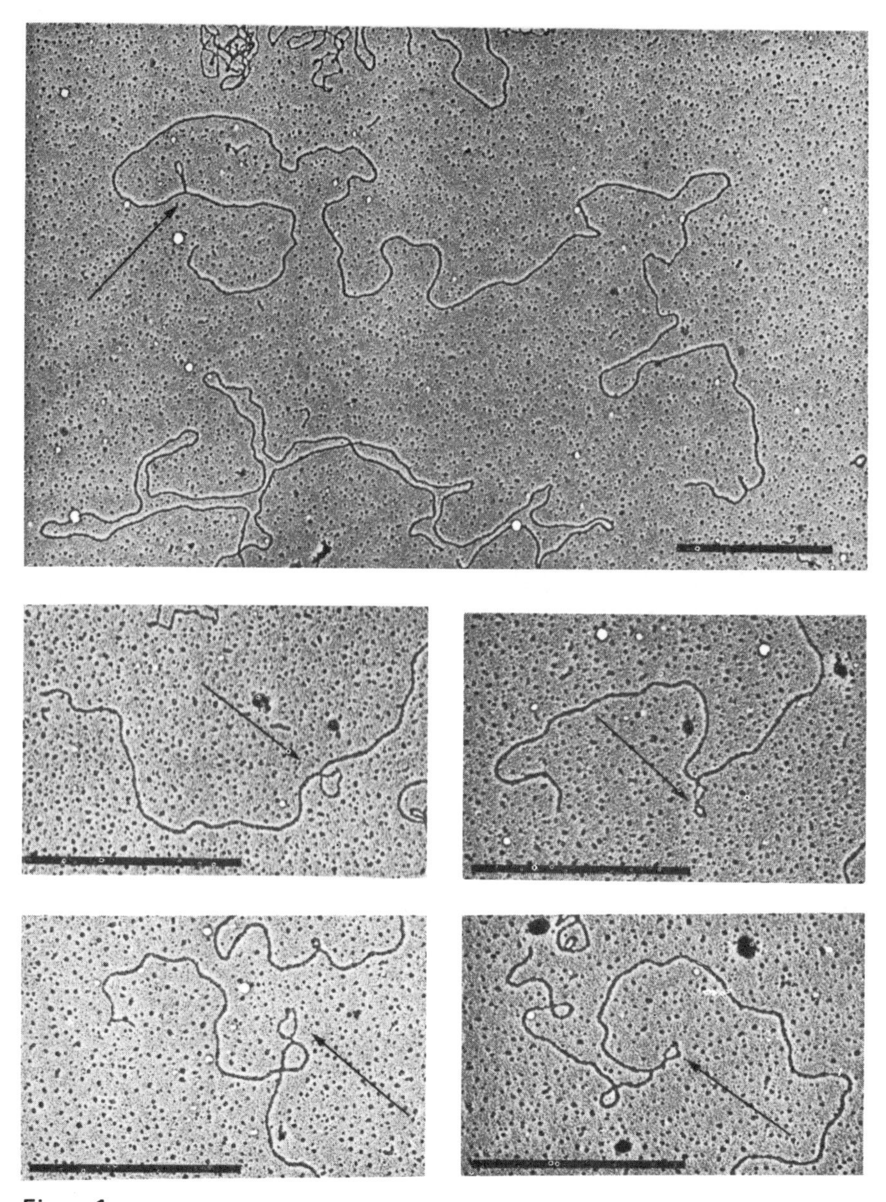

Figure 1
Electron micrographs of $\lambda gal^+/\lambda gal3$ heteroduplexes. The single-stranded loop corresponds to the *gal3* insertion (IS*2*). Each bar represents 1.0 μm. (Reprinted, with permission, from Ahmed and Scraba 1975.)

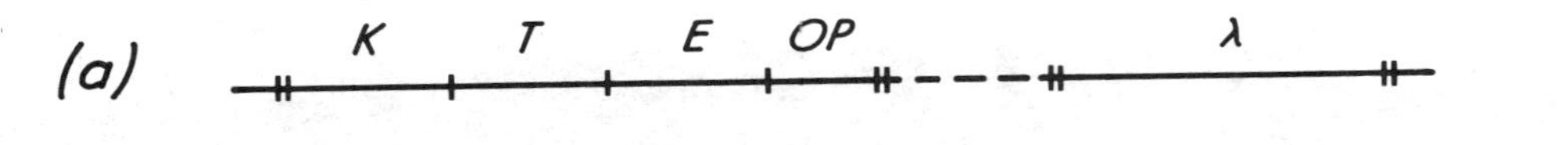

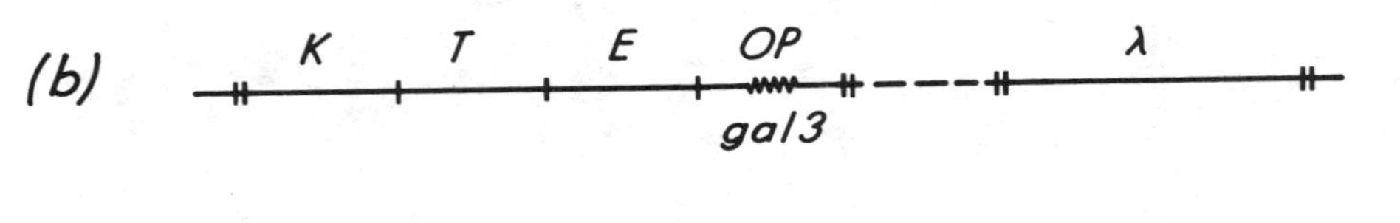

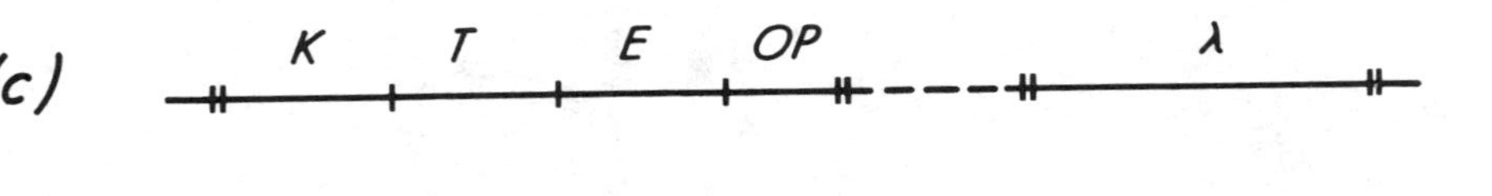

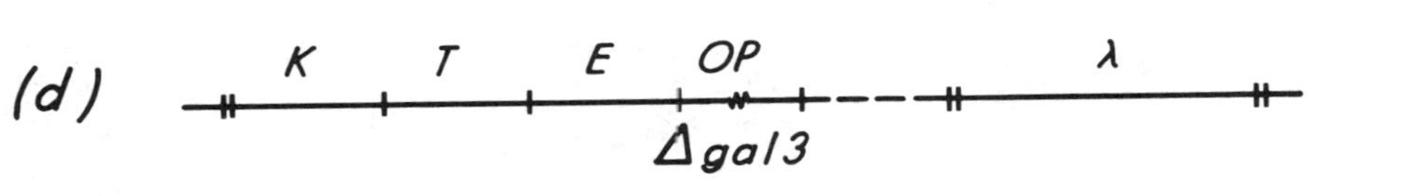

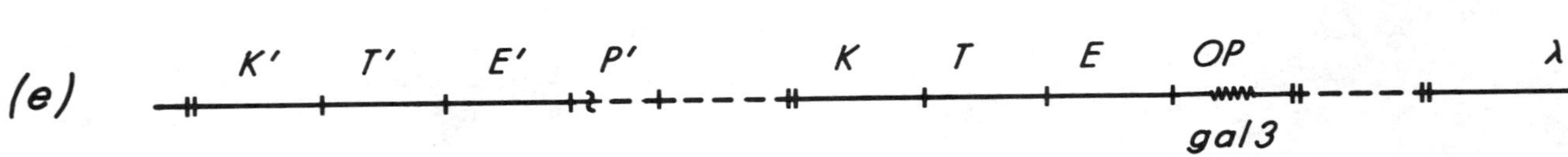

Figure 2
Proposed scheme for the origin of different kinds of gal^+ revertants from *gal3*: (*a*) wild-type gal^+; (*b*) *gal3* mutation caused by IS*2* insertion into the *galOP* region; (*c*) stable, inducible reversion caused by precise excision of the insertion; (*d*) stable, constitutive reversion caused by partial deletion or imprecise excision of the insertion; and (*e*) unstable, constitutive reversion caused by tandem duplication of the *gal* operon. The second *gal* copy ($K'T'E'$) is shown connected to a new promoter (P'). *K*, *T*, and *E* refer to the structural genes for kinase, transferase, and epimerase enzymes, and *OP* denotes operator-promoter of the *gal* operon.

The stable, inducible reversions of *gal3* arise by precise excision of the insertion sequence (Fig. 2c). This leads to the restoration of the gal^+ base sequence so that these revertants are indistinguishable from the wild type in their stability, inducibility, and efficiency of transduction by bacteriophage λ. When such reversions are selected from a λ*gal3* phage, the specific increase in buoyant density (which is normally associated with *gal3*) is found to be eliminated. Moreover, electron microscopy of heteroduplexes between these revertants and wild-type λgal^+ indicates that the insertion is excised with remarkable precision (Ahmed 1975). The process of excision of the IS*2* insertion is independent of the *recA* function. This conclusion is based on the observation that stable, inducible reversions arise with approximately equal frequency in both $recA^+$ and $recA^-$ strains.

In contrast, imprecise excision of the insertion from the *OP* region results in the production of stable, constitutive reversions (Fig. 2d). In one such case, which has been studied by electron microscopy (Ahmed and Johansen 1975), the reversion was found to have been caused by the removal of $\frac{3}{4}$ of the IS*2* sequence. It appears that the deletion removed the site(s) for transcription termination so that transcription initiating at the *gal* promoter reads through the IS*2* fragment. Since the operator sequence remains disrupted due to the presence of the IS fragment, enzyme synthesis is constitutive and approaches the fully induced wild-type level.

In principle, removal of a portion of the IS could occur either by small deletions within the insertion or by defective excision of the insertion. In view of the frequent occurrence of stable, constitutive reversions, it would seem that the excision system is capable of making errors at a high frequency. Like normal excision, this process is also independent of the *recA* function.

Unstable Reversions

The most interesting revertants produced by *gal3* are the unstable, constitutive revertants which are transduced at a significantly reduced frequency by λ. It has been proposed that one class of these reversions arises by tandem duplications of the *gal* operon (Ahmed 1975). The mechanism envisioned is that, during replication or repair, the entire *gal* operon harboring *gal3* is first copied up to the promoter of a nearby operon (*gal3 ETK . . . P′*, as shown in Fig. 2e); a second round of copying then ensues from a site beyond the insertion (for instance, from the *gal3.E* junction) to create a tandem duplication (*gal3 ETK . . . P′E′T′K′*). Such an event would result in connecting the second copy of the *gal* operon (*E′T′K′*) to a different promoter (*P′*). Consequently, the expression of these genes would be subject to different signals of regulation, which could account for the low levels of constitutive enzyme synthesis. Due to intramolecular recombination between the duplicate segments, the revertants would produce *gal3* segregants at a high frequency. Since the functional copy of the *gal* operon (i.e., *. . . P′E′T′K′*) would be physically separated from λ*att* by a nonfunctional copy (i.e., *gal3 ETK . . .*), the frequency of *gal*$^+$ transduction by λ would be considerably reduced.

Although the duplication hypothesis provides satisfactory explanations for various properties of the unstable reversions, direct physical proof for duplications is still lacking due to the inability to establish λ high-frequency transducing (HFT) lines bearing these reversions. However, two lines of indirect genetic evidence support this model. First, although these unstable reversions are not packaged by λ, they can be transduced by phage P1 which can package more DNA (Ahmed 1975). Second, the unstable reversions are stabilized in a *recA* background. Since tandem duplications are known to be unstable due to the occurrence of inter- and intramolecular recombination, it follows that they are stabilized in the absence of the *recA* function (Folk and Berg 1971; Bellet et al. 1971). It was found that in the presence of a *recA* mutation, the instability of an unstable, constitutive reversion was eliminated, although the constitutive expression was not affected. As a further check, it was found that introduction of an F′ *recA*$^+$ episome (F108; Low 1972) into this strain caused a return of the instability. This striking dependence of reversion instability on the *recA* function supports the view that one class of these reversions may arise by tandem duplications.

The constitutive revertants could also arise by the inversion of IS*2*, as proposed by Saedler et al. (1974). They have postulated that IS*2* carries a promoter which in one orientation (orientation II) can cause constitutive expression of the *gal* operon. Frequent inversions of IS*2* or alternating cycles of excision and reinsertion of IS*2* in opposite orientations could provide a plausible mechanism for the appearance of unstable, constitutive revertants.

Unstable reversions of *gal3* arising by inversions of IS*2* would be expected to display properties somewhat different from those arising by tandem duplications. For instance, both the origin and instability of inversion reversions would be independent of the *recA* function. On the other hand, duplication reversions may not require *recA* for their appearance, but their instability would be directly dependent upon it. Starting with a *gal3 recA* strain, we have isolated an unstable, constitutive reversion (Rev.*331*) whose instability is entirely independent of the *recA* function. A similar class of *gal3* reversions has been reported previously (Morse and Pollock 1969). These revertants might represent the class expected to arise by inversions of IS*2*. Although physical studies will provide the definitive answer, it is clear that the unstable, constitutive reversions of *gal3* arise by more than one kind of genetic event.

Inhibition of λ*gal* Production in Constitutive Revertants

A novel property of the constitutive reversions of *gal3* is the severe inhibition of the production of transducing particles. A lysate of a wild-type *gal*$^+$ (λ) strain, prepared by induction, contains a high titer of λ particles ($\sim 5 \times 10^9$/ml) and a few λ*gal* transducing particles formed by defective excision (Campbell 1969). The λ*gal*/λ ratio of such low-frequency transducing (LFT) lysates is typically around 5×10^{-7}. A similar ratio is found in stable, inducible revertants of *gal3*. Surprisingly, lysates prepared by the induction of λ from constitutive revertants of *gal3* are characteristically deficient in normal transducing particles. The *gal*^{c}*200* reversion, like fifty other constitutive revertants tested, exhibits a 50-fold reduction in the relative proportion of λ*gal* particles (Table 1). Other characteristics of this phenomenon are that (a) the inhibition is exercised selectively at a step in the formation of λ*gal* particles since the titer of λ particles remains normal; (b) both the stable and unstable constitutive reversions display this effect; and (c) the inhibition can be alleviated by a deletion [for instance, Δ*31*(*chlD-pgl*)] which extends up to the right terminus of the IS*2* fragment (Table 1).

As shown in Table 1, the inhibition of transducing particle formation by constitutive reversions is even more pronounced in HFT lysates. The λ*gal*/λ ratio in an HFT lysate of strain 46.1, which produces λ*gal*$^{+}$ and λ on induction, is 2.1×10^{-2}. Similar ratios are exhibited by various inducible revertants, although some reduction is normally observed. However, the presence of constitutive reversions on the (λ*gal*) genome causes a drastic reduction in the ratio to about 2.5×10^{-5}, which represents a 700-fold decrease in the production of λ*gal* particles. Again this inhibition is expressed specifically on the formation of λ*gal*, not λ, particles.

Direct evidence for the inhibition of λ*gal* formation was obtained by CsCl density gradient centrifugation of ^{3}H-labeled HFT lysates (Fig. 3). In these experiments, special precautions were taken to minimize possible losses of λ*gal*

Table 1

Influence of *gal3* Reversions on the Production of λ*gal* Particles

Strain	Genotype[a]	Description	Lysate	Ratio[b] λ*gal*/λ
X407 (λ)	*gal*$^{+}$ (λ)	inducible wild type	LFT	4.8×10^{-7}
Stable 3	*gal*$^{+}$ (λ)	inducible reversion	LFT	3.5×10^{-7}
EJ200	*gal*^{c}*200* (λ)	constitutive reversion	LFT	9.7×10^{-9}
Δ31	*gal*^{c}*200* Δ*31*(*chlD-pgl*) (λ)	constitutive reversion (with *chl-pgl* deletion)	LFT	4.7×10^{-7}
46.1	(λ*gal*$^{+}$) (λ)	inducible wild type	HFT	2.1×10^{-2}
109	(λ*gal3*^{i}) (λ)	inducible reversion	HFT	7.6×10^{-3}
A/14-2	(λ*gal3*^{c}) (λ)	constitutive reversion	HFT	2.5×10^{-5}

[a] All HFT-producing revertants were selected from a *gal3*(λ*gal3*) (λ) strain derived from 46.1. Several independent inducible and constitutive revertants were tested. λ*gal3*^{i} and λ*gal3*^{c} represent inducible and constitutive reversions, respectively, on λ*gal*. Strain 46.1 carries the parent λ*gal*$^{+}$.

[b] Titers of λ in all lysates were approximately $4\text{–}5 \times 10^{9}$ pfu/ml.

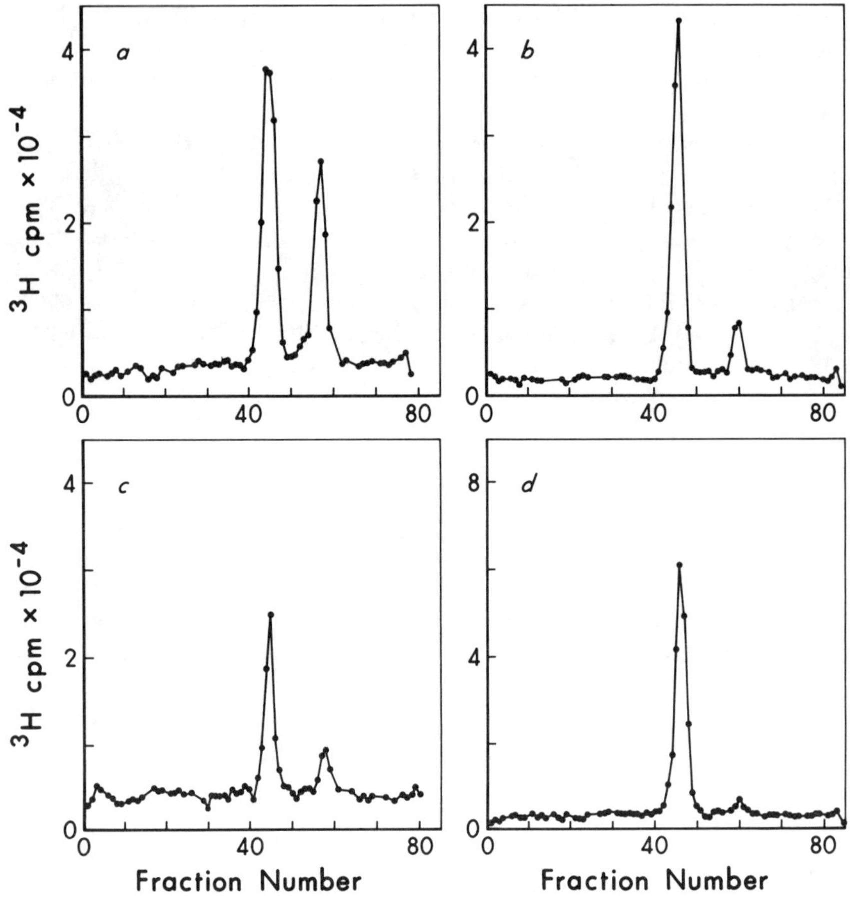

Figure 3
CsCl density gradient centrifugation of HFT lysates from strains harboring λ*gal3* or various constitutive reversions (λ*gal3*^{c}) of *gal3*. The phages were labeled with [^{3}H] thymidine after induction. (*a*) λ, λ*gal3*; (*b*) λ, λ*gal3*^{c} A/14-1; (*c*) λ, λ*gal3*^{c} A/11-2; and (*d*) λ, λ*gal3*^{c} A/14-8. In each profile, the denser band (*left*) corresponds to λ, and the lighter band (*right*) corresponds to λ*gal*.

particles during various steps. The banding profile of the ^{3}H-lysate of a strain which produces λ and λ*gal3* phages on induction is shown in Figure 3a. Similar profiles were obtained for strains producing λ and λ*gal*$^+$ or λ*gal* bearing inducible reversions. In sharp contrast, the banding profiles of three constitutive reversions (Fig. 3b-d) show a marked reduction in the content of λ*gal* particles. Thus the λ*gal3* particles are formed efficiently, but the formation of λ*gal3*^{c} (containing constitutive reversions) is inhibited.

The molecular basis of this remarkable phenomenon is not yet clear. The inhibition of λ*gal* formation both in LFT and HFT lysates suggests that the block occurs after defective excision, possibly at a step during replication or packaging of λ*gal* DNA. Since the block is specific for constitutive revertants (i.e., there is an apparent requirement for IS*2* *and* constitutive transcription), we are led to believe that transcription along IS*2* may expose a specific base sequence which is recognized and cut by a sequence-specific endonuclease. The nicking may either lead to selective degradation of the λ*gal* DNA pool by various exonucleases or it may interfere with the polarized packaging of the polymeric DNA. According to this hypothesis, the deletion Δ*31*(*chlD-pgl*) enables packaging of the constitutive reversion (*gal*^{c}*200*) by specifically eliminating the site on IS*2* that is recognized by the endonuclease (Table 1).

Termination of Deletions at *gal3*

In order to circumvent difficulties encountered in the construction of permanent λ*gal* lines bearing constitutive reversions, we decided to delete the *chlD-pgl* region which lies between *gal* and λ*att*. Our reasoning was that removal of this extra DNA might facilitate packaging of these reversions. During these studies, we made the unexpected observation that extended deletions tend to terminate preferentially at the right terminus of the *gal3* insertion (Ahmed and Johansen 1975).

A number of spontaneous *chlD-pgl* deletions were selected as chlorate-resistant mutants from a *gal3* (λ) parent. Fifteen of these deletions were found to have retained the *gal3* mutation, but their exact end points at the left (according to the representation in Fig. 4) were not known. The right extremities of these deletions were clearly variable since some terminated before (λ), whereas others removed (λ) and other adjoining genes. When *gal*$^{+}$ revertants were selected from these *gal3* Δ(*chlD-pgl*) deletions, it was surprising to find that the revertants were always constitutive. No inducible revertants were ever found. This was unusual because *gal3* normally reverts to produce both inducible and constitutive revertants.

A systematic study revealed that the frequency of spontaneous *chlD-pgl* deletions is increased 10- to 15-fold in the presence of the *gal3* insertion. Reif and Saedler (1975) observed a 30- to 2000-fold increase in deletion frequency in the presence of IS*1* but did not detect any effect of IS*2*. An even more profound influence is exercised by *gal3* on the end points of *chlD-pgl* deletions, but this effect is expressed only in the presence of prophage λ. Deletions arising from *gal*$^{+}$, *gal*$^{+}$ (λ), and *gal3* strains seem to have random end points at both the left and right. Thus deletions obtained from *gal3* either delete the *gal* operon or retain *gal3* together with its property of reverting to produce both inducible and constitutive revertants. In contrast, deletions originating from a *gal3* (λ) strain produce constitutive revertants only, indicating that during the process of deletion formation in the presence of λ, the *OP* region of the *gal* operon is also

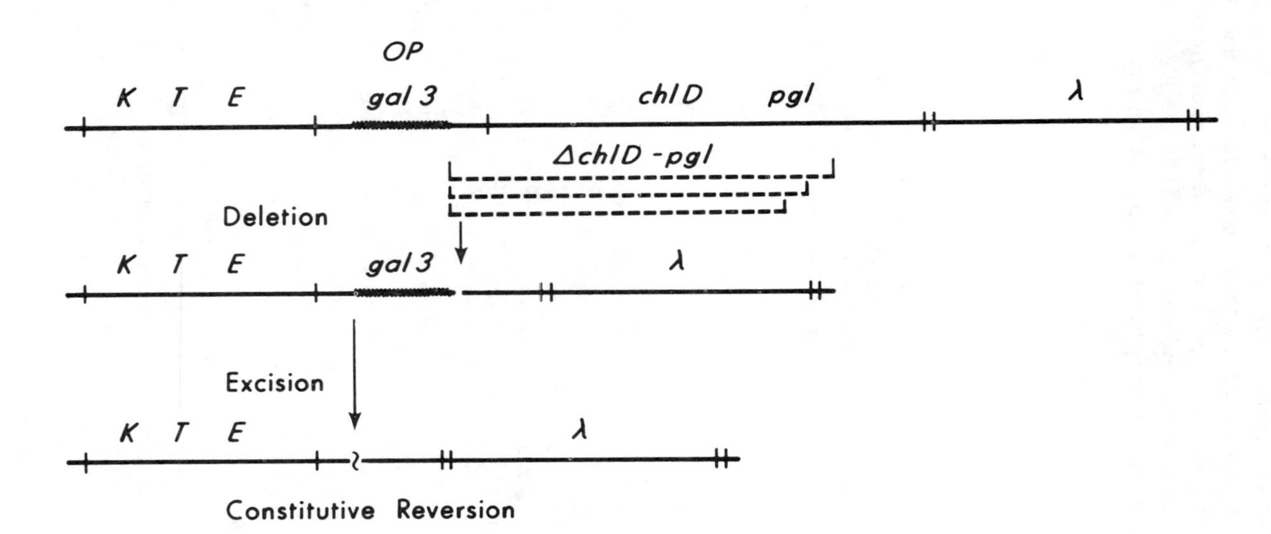

Figure 4
Scheme proposed to explain the exclusive appearance of stable, constitutive reversions from *gal3* strains carrying *chlD-pgl* deletions. The deletions have a fixed end point at the left (i.e., at the right terminus of IS*2*) but variable end points at the right. Ordinary excisions of the insertion would yield constitutive revertants by fusing the *gal* structural genes to new promoters on the right. These deletions are believed to arise by the action of a specific endonuclease at the right end of IS*2*, which is followed by exonucleolytic degradation proceeding towards the right.

impaired. It is clear that the λ function is required specifically at the time of deletion formation because constitutive revertants arise equally well from deletions whether they have retained or deleted the prophage. We proposed a simple scheme (Fig. 4) to explain the origin of exclusively constitutive revertants from these deletions (Ahmed and Johansen 1975). According to this scheme, the presence of IS*2* and prophage λ confers a unique specificity on spontaneous deletions to terminate preferentially at the right extremity of the IS*2* sequence. In this process, a part of the *galOP* region is also removed. Thus the *chlD-pgl* deletions have a fixed end point at their left, but their right end point is variable. It is easy to see that simple excisions of the insertion from such deletions would produce only constitutive revertants by connecting the *gal* operon to new promoters. Apparently, this remarkable specificity resides in the base sequence at the right terminus of the insertion. A similar specificity may exist at the left terminus as well.

It is possible that the ends of insertion sequences constitute preferred sites for the action of a sequence-specific endonuclease. An initial endonucleolytic nick made at one end of the IS, followed by exonucleolytic degradation to various extents outward, would produce deletions of the kind actually found. A similar explanation has been proposed for the *int*-promoted deletions of λ (Davis and Parkinson 1971). These deletions extend to various lengths on either side from a central crossover point within the *att* site of λ. However, despite the striking similarity, the two kinds of deletion are probably promoted by different enzymatic functions since preliminary studies suggest that the end-point specificity of *chlD-pgl* deletions is maintained in λ*int* lysogens of *gal3.*

GENERAL COMMENTS

The *gal3* mutation, caused by the insertion of IS*2*, exemplifies the properties that are beginning to be associated specifically with the insertion sequences. Thus the IS*2* element can act as a vehicle for turning off gene expression in a reversible manner, can act as a hot spot for recombination and for linking unrelated genomes, and can act as a preferred site for deletions. It would not be too surprising to find that the diverse genetic phenomena exhibited by IS*2* (such as its integration and excision, the specificity of deletion end points, and the inhibition of transducing particle formation) reflect different functional aspects of a single enzyme system.

Acknowledgments

I thank Bev Clark and Ron Lee for their assistance, and Eric Johansen for reading the manuscript. This work was supported by a grant from the National Research Council of Canada.

REFERENCES

Adhya, S. L. and J. A. Shapiro. 1969. The galactose operon of *E. coli* K-12. I. Structural and pleiotropic mutations of the operon. *Genetics* **62**:231.

Ahmed, A. 1975. Mechanisms of reversion of the *gal3* mutation of *Escherichia coli. Mol. Gen. Genet.* **136**:243.

Ahmed, A. and E. Johansen. 1975. Reversion of the *gal3* mutation of *Escherichia coli*: Partial deletion of the insertion sequence. *Mol. Gen. Genet.* **142**:263.

Ahmed, A. and D. Scraba. 1975. The nature of the *gal3* mutation of *Escherichia coli. Mol. Gen. Genet.* **136**:233.

Bellett, A., H. G. Busse and R. L. Baldwin. 1971. Tandem genetic duplications in a derivative of phage lambda. In *The bacteriophage lambda* (ed. A. D. Hershey), p. 501. Cold Spring Harbor Laboratory, Cold Spring Harbor, New York.

Campbell, A. M. 1969. *Episomes.* Harper and Row, New York.

Davis, R. W. and J. Parkinson. 1971. Deletion mutants of bacteriophage lambda. III. Physical structure of *attφ. J. Mol. Biol.* **56**:403.

Fiandt, M., W. Szybalski and M. H. Malamy. 1972. Polar mutations in *lac, gal* and phage λ consist of a few IS-DNA sequences inserted with either orientation. *Mol. Gen. Genet.* **119**:223.

Folk, W. R. and P. Berg. 1971. Duplication of the structural gene for glycyl-transfer RNA synthetase in *Escherichia coli. J. Mol. Biol.* **58**:595.

Hill, C. W. and H. Echols. 1966. Properties of a mutant blocked in inducibility of messenger RNA for the galactose operon of *Escherichia coli. J. Mol. Biol.* **19**:38.

Hirsch, H. J., P. Starlinger and P. Brachet. 1972. Two kinds of insertions in bacterial genes. *Mol. Gen. Genet.* **119**:191.

Jordan, E., H. Saedler and P. Starlinger. 1968. O^0 and strong polar mutations in the *gal* operon are insertions. *Mol. Gen. Genet.* **102**:353.

Lederberg, E. M. 1960. Genetic and functional aspects of galactose metabolism in *Escherichia coli* K-12. In *Microbial genetics. 10th Symposium of the Society for General Microbiology*, p. 115. University Press, Cambridge.

Low, K. B. 1972. *Escherichia coli* K-12 F-prime factors, old and new. *Bact. Rev.* **36**:587.

Morse, M. L. 1967. Reversion instability of an extreme polar mutant of the galactose operon. *Genetics* **56**:331.

Morse, M. L. and B. F. Pollock. 1969. Reversion instability in the galactose operon of *Escherichia coli. J. Bact.* **99**:567.

Morse, M. L., E. M. Lederberg and J. Lederberg. 1956. Transductional heterogenotes in *Escherichia coli. Genetics* **41**:758.

Reif, H.-J. and H. Saedler. 1975. *IS1* is involved in deletion formation in the *gal* region of *E. coli* K-12. *Mol. Gen. Genet.* **137**:17.

Saedler, H., H.-J. Reif, S. Hu and N. Davidson. 1974. *IS2*, a genetic element for turn-off and turn-on of gene activity in *E. coli. Mol. Gen. Genet.* **132**:265.

Shapiro, J. A. 1969. Mutations caused by the insertion of genetic material into the galactose operon of *Escherichia coli. J. Mol. Biol.* **40**:93.

Shapiro, J. A. and S. L. Adhya. 1969. The galactose operon of *E. coli* K-12. II. A deletion analysis of operon structure and polarity. *Genetics* **62**:249.

Repeated DNA Sequences in Plasmids, Phages, and Bacterial Chromosomes

H. Ohtsubo and E. Ohtsubo
Department of Microbiology
Health Sciences Center
State University of New York at Stony Brook
Stony Brook, New York 11794

The electron microscope heteroduplex method is a powerful tool for studying the sequence organization of DNA molecules and, as such, has been used to characterize several kinds of repeated DNA sequences present in bacterial plasmids and certain bacteriophages. The heteroduplex analysis of plasmids containing DNA segments of the *E. coli* chromosome has also revealed that the *E. coli* chromosome contains repeated sequences, some of which are the same sequences found in plasmids. These repeated sequences appear either in the same orientation (direct repeats) or in an inverted orientation (inverted repeats) in a genome.

The first part of this article presents a review of the properties and functions of several kinds of particular sequences that have been found as repeated sequences in bacterial plasmids, phages, and chromosomes. The second part describes an approach, based on direct isolation of inverted repeat sequences, for detecting repeated sequences in plasmids. The method can also be used for isolating inverted repeat sequences in bacterial chromosomes. The isolation method is generally useful not only for characterizing repeated sequences which are already known, but also for detecting new repeated sequences. Obviously, isolated DNA fragments of repeated sequences can also be subjected to extensive biochemical analysis.

Repeated Sequences in the *E. coli* System

Table 1 summarizes the particular sequences that appear as repeated sequences in plasmids (F, R100, R6, R1, and their derivatives), bacteriophages (P1 and Mu), and the *E. coli* chromosome. Other inverted repeat sequences (not listed in Table 1) are found in the ColV factor (Sharp et al. 1973) and in translocation elements, Tn*2* (Ap) (Heffron et al. 1975) and Tn*5* (Km) (Berg et al. 1975).

In electron microscope heteroduplex studies, the sequences given in Table 1 have been observed either as "inversion loops" (Sharp et al. 1972) with duplex stems in a molecule formed between inverted repeats on a single strand, or as "out-of-register structures" (Ohtsubo et al. 1974a) formed in out-of-register order between direct repeat sequences on two single strands derived from a genome.

Table 1
Repeated Sequences and Their Involvements in Genetic Fusion Phenomena

Molecular length (kb)	Sequence notation	Sequence involvement
1.29	$\alpha\beta$(IS*3*)	insertions[a]; formations of Hfr's and F′[b]; deletions[c]; translocations[d]
5.70	$\gamma\delta$	formation of Hfr's[e,f], F′[b], and ϕ80 transducing phages[g]; segregations of a part of F′ genomes[e,f]
1.22	$\epsilon\zeta$(IS*2*)	insertions[c,e]; formations of Hfr's[b] and ϕ80 transducing phages[g]; deletions[h]
0.96	$\eta\theta$	transposition (?) of KM(NM)[i]
0.55	$\iota\kappa$	segregations of the Tc region[j]
0.71	$\lambda\mu$(IS*1*)	insertions[a,k]; deletions and segregations[a,k,e,l]; transpositions[m]
0.34		inverted repeats found in F152-1[i]
1.90	*r*16(*rrn*)	F′ formation (?)[n]
3.20	*r*23(*rrn*)	F′ formation (?)[n]
0.20	j(j′)	inverted repeats found in the *E. coli lac* region[i]
0.62		inversions of a segment of P1[o]
~0.02		inversions of the G segment in Mu[p]

The locations of particular sequences listed in this table are schematically represented in Figs. 1-4. [a]Hu et al. 1975c; [b]Hu et al. 1975a; [c]Hu et al. 1975b; [d]Kleckner et al. 1975; [e]Ohtsubo et al. 1974a; [f]Palchaudhuri et al. 1976; [g]Ohtsubo et al. 1974b; [h]Saedler et al. 1974; [i]Sharp et al. 1973; [j]E. Ohtsubo, unpubl.; [k]Ptashne and Cohen 1975; [l]S. Mickel, E. Ohtsubo and W. Bauer, in prep.; [m]MacHattie and Jackowski, this volume; [n]Deonier et al. 1974; [o]Lee et al. 1974; [p]Hsu and Davidson 1974.

For convenience, we have assigned Greek-letter notations to some of the sequences. The notation $\alpha\beta$, for example, denotes this sequence in an order on the map of a genome. The notation $\beta'\alpha'$ denotes the complementary sequence in the inverted order in the genome (Ohtsubo et al. 1974a; Davidson et al. 1974).

Schematic representations of F and R100 plasmids and F′ plasmids carrying *lac* and *ilv-argE* regions of the *E. coli* chromosome are shown in Figures 1–4. All of the particular sequences listed in Table 1, except for the inverted repeats in phages P1 and Mu and in an F derivative (F152-1), are present as a single copy or as repeated sequences in either inverted or direct order. These figures also show that some sequences appear at the same time both in plasmids and in the *E. coli* chromosome.

Recently it has been shown that the sequences $\alpha\beta$, $\epsilon\zeta$, and $\lambda\mu$ are identical to the insertion (IS) sequences IS*3*, IS*2*, and IS*1*, respectively, which cause strongly polar mutations when inserted into the *gal* and *lac* operons of the *E. coli* chromosome and into phage λ genes (Hirsch et al. 1972; Malamy et al. 1972;

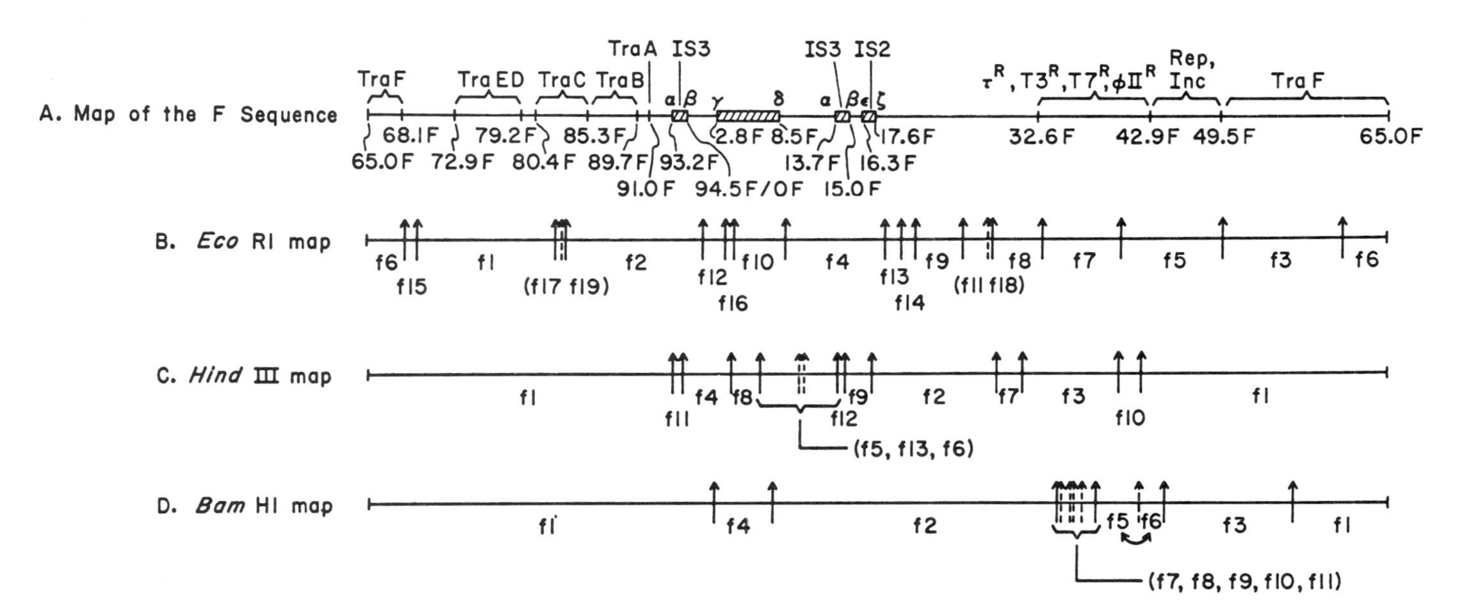

Figure 1

The genetic and physical map of F. Although F is actually a circular duplex, the structure is displayed in a linear representation by cutting the circle at a point on a single strand. All numbers followed by the letter "F" are distances in kb from a selected origin. Special sequences in F are indicated by Greek letters (see sequence notations in Table 1). The positions of several genes of F are also shown. Cleavage sites on F are shown at the bottom of the figure. F produces 19 fragments, f1–f19, by *Eco*RI, 13 fragments, f1–f13, by *Hin*dIII, and 11 fragments, f1–f11, by *Bam*HI digestion (H. Ohtsubo, J. Childs and E. Ohtsubo, unpubl.).

Hu et al. 1975a,b). The other sequences listed in Table 1 have not yet been identified as insertion sequences.

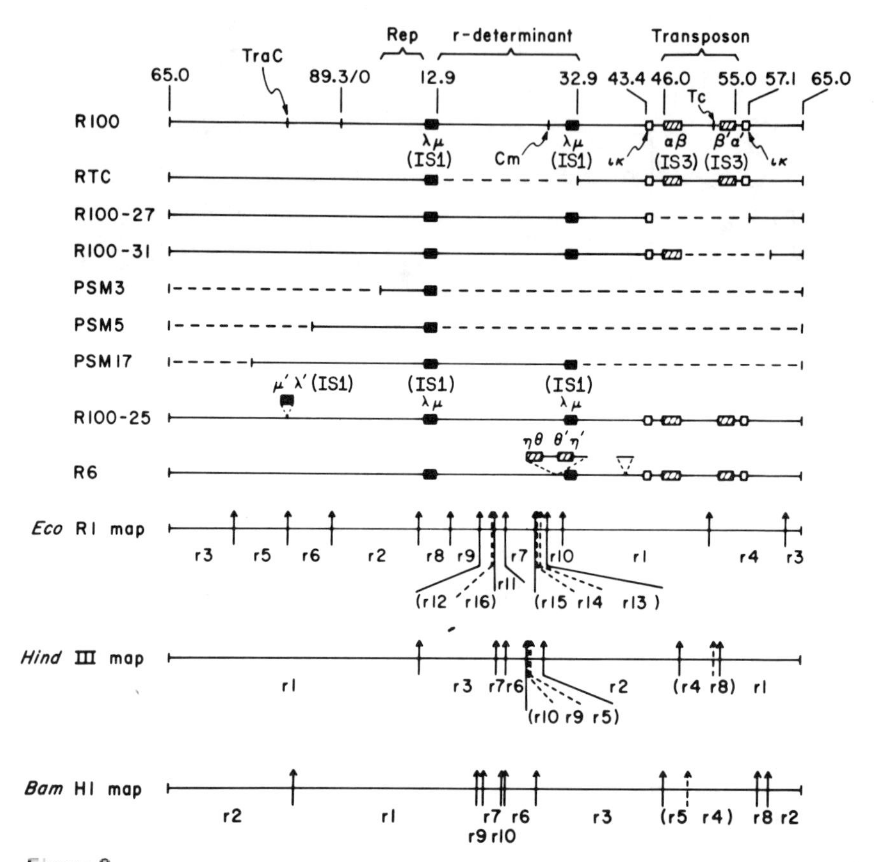

Figure 2

Physical structures of R100 and its derivatives. The molecules are actually circular duplexes, but the structures are displayed in a linear representation by cutting the circle on a single strand at a point that lacks interesting features. All the numbers are distances in kb from a selected origin. Deleted regions in plasmids derived from R100 are shown by broken lines. Insertions in R100-25 and R6 are also shown. Special repeated sequences are indicated by the Greek letters (see sequence notations in Table 1). The two copies of the sequences in an inverted orientation in a molecule form duplexes after renaturation (e.g., between $\alpha\beta$ and $\alpha'\beta'$). Approximate map positions of the genes responsible for genetic transfer (TraC), resistance to chloramphenicol (Cm), and resistance to tetracycline (Tc) are also shown on the R100 map. Restriction enzyme cleavage sites on R100 are shown at the bottom of the figure; the same restriction sites are on R6, homologous to R100. R100 produces 16 fragments, r1–r16, by *Eco*RI digestion; R100 produces 10 fragments, r1–r10, by *Hin*dIII and by *Bam*HI digestion (E. Ohtsubo, S. Shaw and H. Ohtsubo, unpubl.).

Electron microscope heteroduplex studies have also revealed that some of the sequences listed in Table 1 are effectively involved in the genetic fusion that occurs by recombination between two different genomes or two different DNA sequences in a genome. Table 1 summarizes phenomena, such as hybrid molecule formations, deletions or segregations of a part of plasmid genomes, transpositions of a DNA segment, and inversions, in which particular sequences are involved. In molecules resulting from genetic fusion, the particular sequences are usually seen at the junction points where genetic recombination has occurred.

A typical example is the heteroduplex results of studies on the R factor R100 (same as *NR1* and R222) and its derivatives. As shown in Figure 2, R100 carries three sets of repeated sequences. The mutant derivatives of R100 carry deletions at the junctions of which the particular sequences $\alpha\beta$(IS*3*), $\lambda\mu$(IS*1*), and $\iota\kappa$ are seen. Two mutants, RTC and R100-27, are thought to be derived through recombination between repeated sequences of $\lambda\mu$(IS*1*) and $\iota\kappa$, respectively (E. Ohtsubo, unpubl.). Other mutants are thought to be derived by recombination occurring at the end points of $\alpha\beta$(IS*3*) (note the structure of R100-31 in Fig. 2) (Hu et al. 1975b) and $\lambda\mu$(IS*1*) (e.g., PSM3 in Fig. 2) (S. Mickel, E. Ohtsubo and W. Bauer, in prep.). Translocation of the tetracyline resistance (Tc) region in the R factor has been shown clearly to be due to the presence of $\alpha\beta$(IS*3*) sequences flanking the Tc region (see Fig. 2) (Tye et al. 1974; Kleckner et al. 1975).

These heteroduplex results suggest that the genetic recombination in some of these cases occurs between particular DNA sequences repeated in a genome or between particular DNA sequences present in two different genomes. The heteroduplex results suggest also that there are many cases of recombinatorial events in which the end points of the particular sequences are "hot spots" that can recombine with any one of several sites on a genome.

Indications that the end points of the particular sequences are effective hot spots for recombination are of interest. There may be special characteristics of the nucleotide sequences at the ends of the particular sequences.

As reviewed briefly above, genetic fusion phenomena are not simply the result of "illegitimate" recombination events (Franklin 1971) but are due to specific events related to the involvement of particular DNA sequences. These sequences are thought to have played important roles in evolution.

Isolation of Inverted Repeat Sequences in Plasmids and Bacterial Chromosomes

Isolation of Inverted Repeats, including $\alpha\beta$(IS3), $\epsilon\zeta$(IS2), and $\lambda\mu$(IS1), in Plasmids

We here offer an alternate approach to the electron microscope heteroduplex method for detecting the repeated sequences that occur as inverted repeats in the *E. coli* plasmids. The method involves the direct isolation of DNA fragments

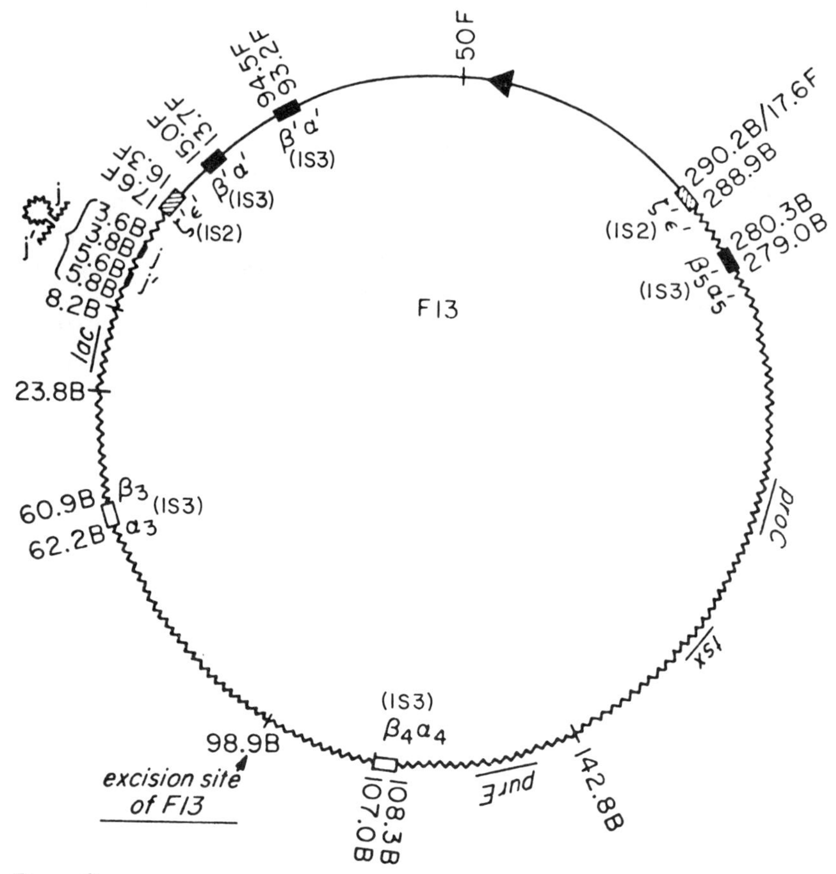

Figure 3
Structure of F13 carrying the *lac* region of the *E. coli* chromosome. Solid line is F DNA; sawtoothed line is bacterial chromosome DNA. Numbers denoted by the letter F are distances in kb from the selected origin on F (see Fig. 1); numbers denoted by the letter B are distances in kb from the selected origin for the chromosomal DNA in F13. Special repeated sequences are indicated by Greek letters (see sequence notations in Table 1). Different copies of IS*3* are indicated by numbers ($\alpha_3\beta_3$, $\alpha_4\beta_4$, etc). The positions of several genes on the chromosomal part are shown in the schematic representations. (Modified from Hu et al. 1975a.)

of inverted repeats and is based on the observation that the inverted repeat DNA sequences rapidly form duplexes after denaturation and renaturation of a genome containing these sequences (Sharp et al. 1972; Sharp et al. 1973; Schmidt et al. 1975). The procedures for the isolation (Ohtsubo and Ohtsubo 1976) involve (1) denaturation of intact plasmid DNA, (2) a rapid, 30-second renaturation of inverted repeat sequences in the genome, (3) digestion of the

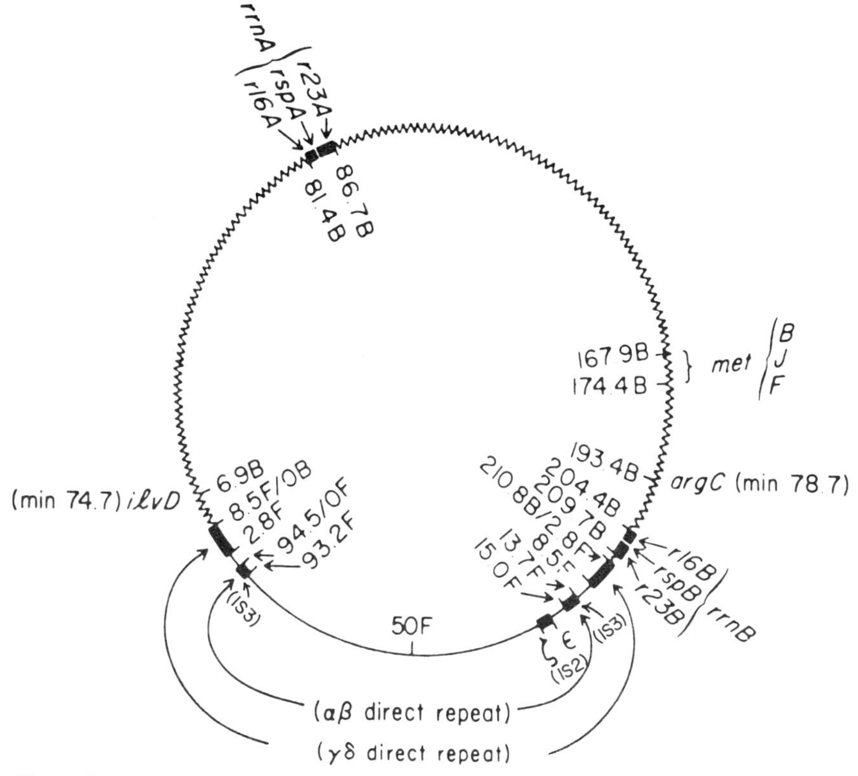

Figure 4
Structure of an F′ (F14), as determined by Ohtsubo et al. (1974a). Solid line is F DNA; sawtoothed line is bacterial chromosome DNA containing the *ilv-arg* region. Numbers denoted by the letter F are distances in kb from the selected origin on F (see Fig. 1); numbers denoted by the letter B are distances in kb from the selected origin for the chromosomal DNA in F14. Special sequences in F are indicated by Greek letters (see sequence notations in Table 1). There are two loci for ribosomal ribonucleic acid genes on F14, designated *rrnA* and *rrnB*.

single-stranded portion by S1 nuclease to recover duplex DNA, and (4) detection and purification of the duplexes using 1.4% agarose gel electrophoresis.

Figure 5 shows the results of gel electrophoresis after S1 nuclease treatment of various R plasmids carrying the inverted repeats of αβ(IS*3*), λμ(IS*2*), and ηθ (the physical structures of these plasmids are shown in Fig. 2) and of an F derivative, F8(N33), carrying inverted repeats of εζ(IS*2*).

When R100, RTC, and F8(N33) carrying one kind of inverted repeat were examined, a single DNA band was observed in agarose gels (Fig. 5, A-a,c, B). When the plasmids R100-25 and R6, which carried two different sets of inverted repeats, were examined, two bands appeared in the gels (Fig. 5A-d,e). However, R100-27, which carried no inverted repeats, did not produce any characteristic

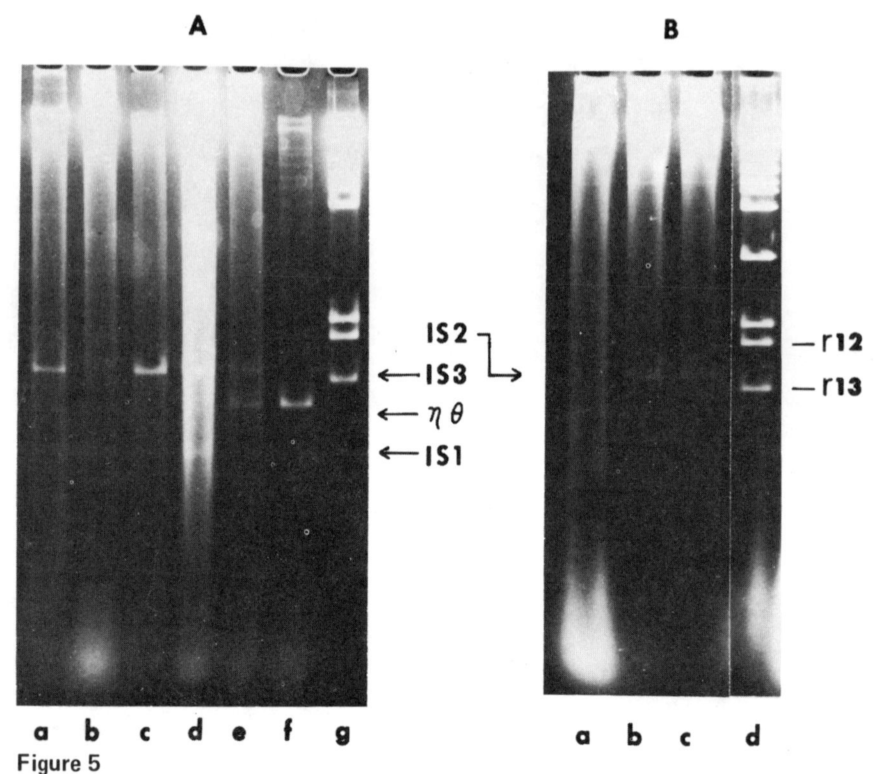

Figure 5

Electrophoresis of DNA in 1.4% agarose gels. (*A*) Generation of characteristic DNA bands of inverted repeat sequences present in various R plasmids: (*a*) R100, (*b*) R100-27, (*c*) RTC, (*d*) R100-25 (see also Fig. 6B), (*e*) R6, and (*f*) DNA fragments of R6 cleaved by *Eco*R1. These DNAs were denatured, renatured, and treated with S1 nuclease before gel electrophoresis (Ohtsubo and Ohtsubo 1976). (*B*) Generation of inverted repeat sequences of IS*2* in F8(N33) under the same conditions, except that denatured F8(N33) DNA was incubated for renaturation for (*a*) 0 sec, (*b*) 30 sec, and (*c*) 3 min. Column *g* in *A* and column *d* in *B* are the *Eco*RI digests of R6 and R100, respectively, which were used as length standards. The lengths of the fragments r12 and r13 are approximately 1.5 and 1.1 kb, respectively.

band (Fig. 5A-b). These results indicate that the inverted repeat sequences present in the plasmids can be isolated by our procedure.

As schematically shown in Figure 2, the R plasmids used here are known to have direct repeats of the sequences $\lambda\mu$(IS*2*) (700 bases) and $\iota\kappa$ (about 550 bases) in their genomes. It is apparent from the results above that these direct repeats can not be isolated by the method used here.

Our isolation procedure requires that the two copies of an inverted repeat be linked together on a single strand. This is demonstrated by the following

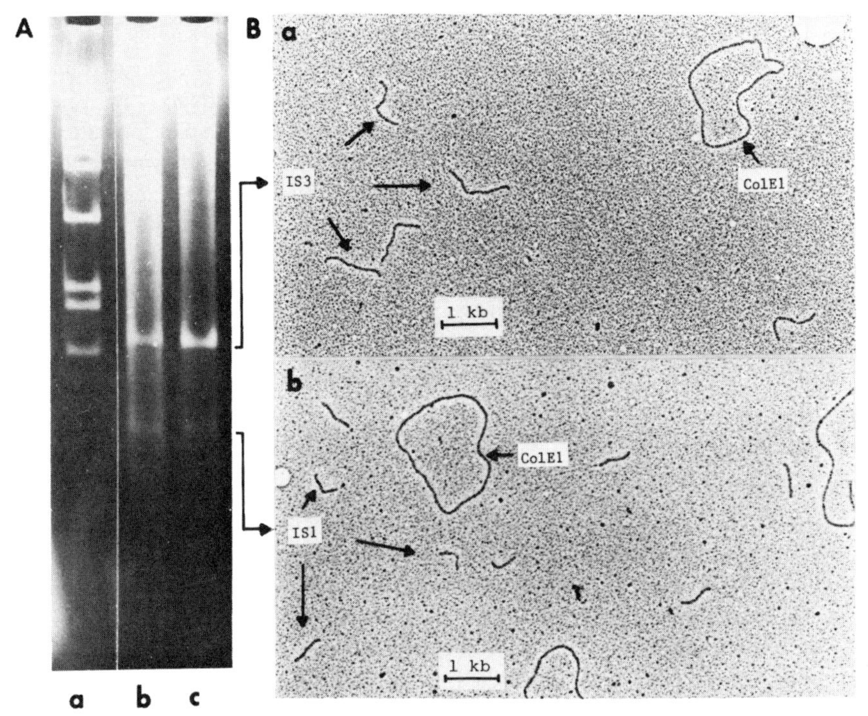

Figure 6
Electrophoresis (*A*) in 1.4% agarose gels of (*b, c*) R100-25 DNA which had been denatured, renatured, and treated with S1 and of (*a*) *Eco*RI-digested DNA fragments of R100 which were used as length standards. (*B*) Electron micrographs of isolated inverted repeat sequences of (*a*) IS*3* and (*b*) IS*1* derived from R100-25. The DNA samples were eluted from gels by gel electrophoresis (Ohtsubo and Ohtsubo 1976). The DNA samples for electron microscopy were prepared using an aqueous spreading technique (Davis et al. 1971). In the micrographs, the ColE1 double-stranded molecules (approx. 6.34 kb long) shown were used as length standards. The bar represents a length of 1 kb for double-stranded DNA.

experiment. An R plasmid, R6, was digested by *Eco*RI endonuclease, which cleaves the linkage between two repeats of $\alpha\beta$(IS*3*) but leaves the linkage of the two copies of the 1.0-kb long sequence ($\eta\theta$) intact (see *Eco*RI cleavage map, Fig. 2). The *Eco*RI-digested R6 DNA was then denatured, renatured, and treated with S1 nuclease. As expected, only one band (1.0 kb long) was seen after gel electrophoresis (Fig. 5A-f).

Electron micrographs of purified $\alpha\beta$(IS*3*) and $\lambda\mu$(IS*1*), which were eluted from gels by extensive gel electrophoresis, are shown in Figure 6. The preparation contained duplex DNAs which were quite homogeneous in size.

In the experiment described here, mutant plasmids with known physical structures are very useful, not only for the isolation of several kinds of repeated sequences, but also for the characterization of some of these sequences as either IS*1*, IS*2*, or IS*3*. However, we have also been characterizing the isolated sequences by using restriction endonucleases. Preliminary studies indicate that *Hin*dIII cleaves IS*2* and IS*3* (in F, but not in R) but not IS*1*, whereas other restriction endonucleases, such as *Eco*RI and *Bam*HI, do not cleave any of these sequences (see cleavage maps, Figs. 1 and 2). Other restriction enzymes, such as *Hae*III, *Hpa*II, *Hha*, *Alu*, and *Hin*f, cleave the IS*1* sequence at least at one site (H. Ohtsubo and E. Ohtsubo, unpubl.). This approach can be applied directly to characterize the duplex DNA of inverted repeats isolated by the method described here.

Isolation of Inverted Repeat Sequences in Bacterial Chromosomes

The method described in the previous section was used to detect inverted repeat sequences in two *E. coli* K12 derivatives (W3310 F^- and JE2252 HfrC) and a *Shigella* strain (*S. dysenteriae*) closely related to the *E. coli* strains.

The bacterial DNA was isolated and used without any extensive shearing. The average length of the DNAs in the preparation was estimated to be about 70 kb. Note that the larger the bacterial DNA fragment, the higher the probability of inverted repeats being included in the fragment. In the procedure for the isolation of inverted repeats, we used hydroxylapatite as an additional step to concentrate duplex DNA fragments produced after S1 nuclease treatment, since it is assumed that not many inverted repeat sequences are present in bacterial chromosomes.

The results of gel electrophoresis and schematic representations of the gel bands are shown in Figure 7. For comparison, estimated band positions of the particular sequences listed in Table 1 are also shown in Figure 7. The following information was obtained from the gel electrophoresis results: (1) Both *E. coli* K12 strains (F^- and Hfr) produced gel bands of similar size, although some of the gel bands differed in intensity. (2) The *Shigella* strain produced a family of gel bands, some of which are similar in size to those produced from the *E. coli* K12 strains, although they too differ in intensity. Note that the *Shigella* strain produced an extremely dense DNA band, the length of which is similar to that of $\lambda\mu$(IS*1*) (see Fig. 7). (3) *E. coli* K12 and *Shigella* strains produced duplex DNA fragments having sequences similar to some of those listed in Table 1.

These duplex DNA fragments must be mostly inverted repeat sequences present in bacterial chromosomes. The *E. coli* strains used did not carry any plasmids, and the *Shigella* strain harbored two plasmids, the molecular lengths of which are about 2 and 30 kb. Plasmids of *Shigella* were isolated but they were found to have no inverted repeat sequences when examined in the electron microscope.

The intensity of gel bands may reflect the number of copies of inverted repeat sequences present in the chromosome. Direct observation of inverted repeats in

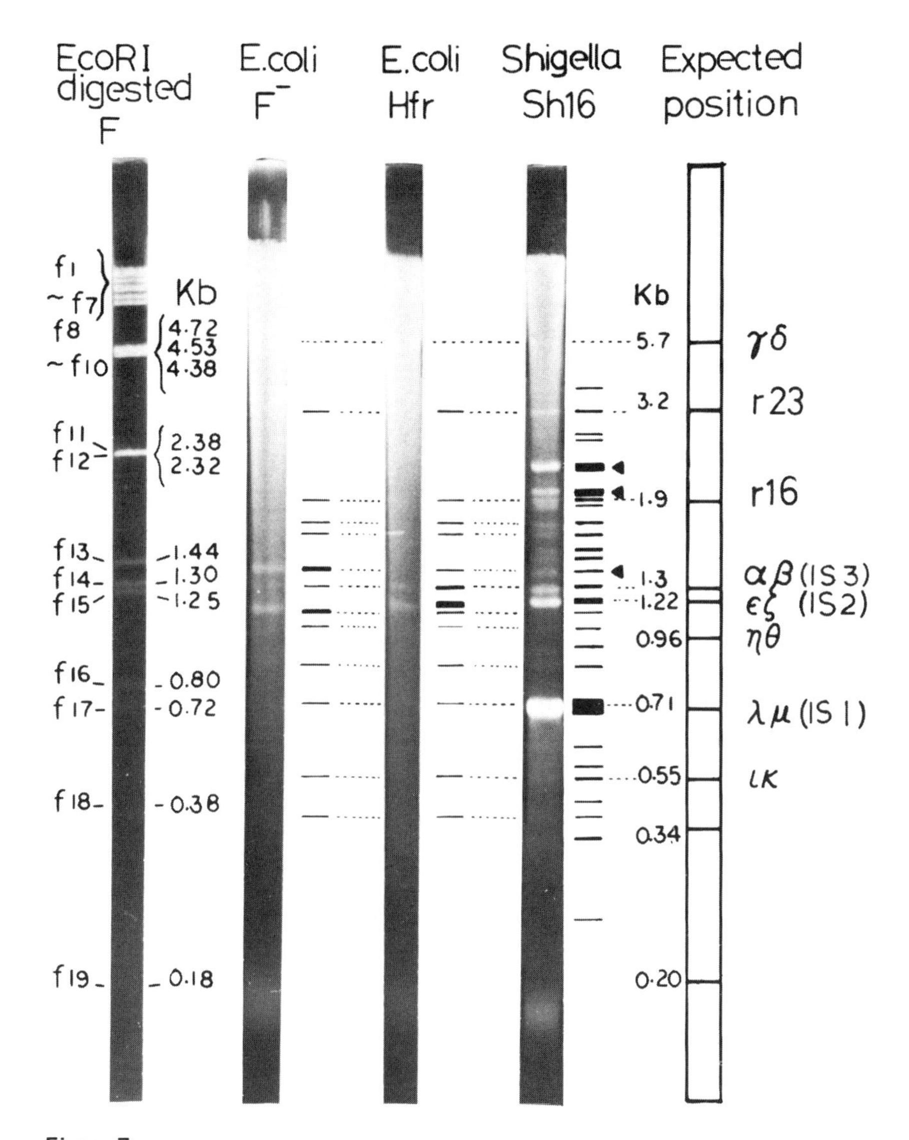

Figure 7

Electrophoresis in 1.4% agarose gels showing generation of characteristic duplex DNA bands from *E. coli* (F$^-$ and Hfr) and *Shigella dysenteriae* chromosomal DNA after denaturation, renaturation, and S1 nuclease treatment. Schematic representations of these gel bands are shown to the right of each gel. Expected band positions of special sequences listed in Table 1 are also schematically shown in the figure. The three DNA bands marked with closed triangles in the gel for the *Shigella dysenteriae* strain are, in order from the top, the open circular, the linear, and the covalently closed circular duplex configurations of a plasmid 2 kb in length.

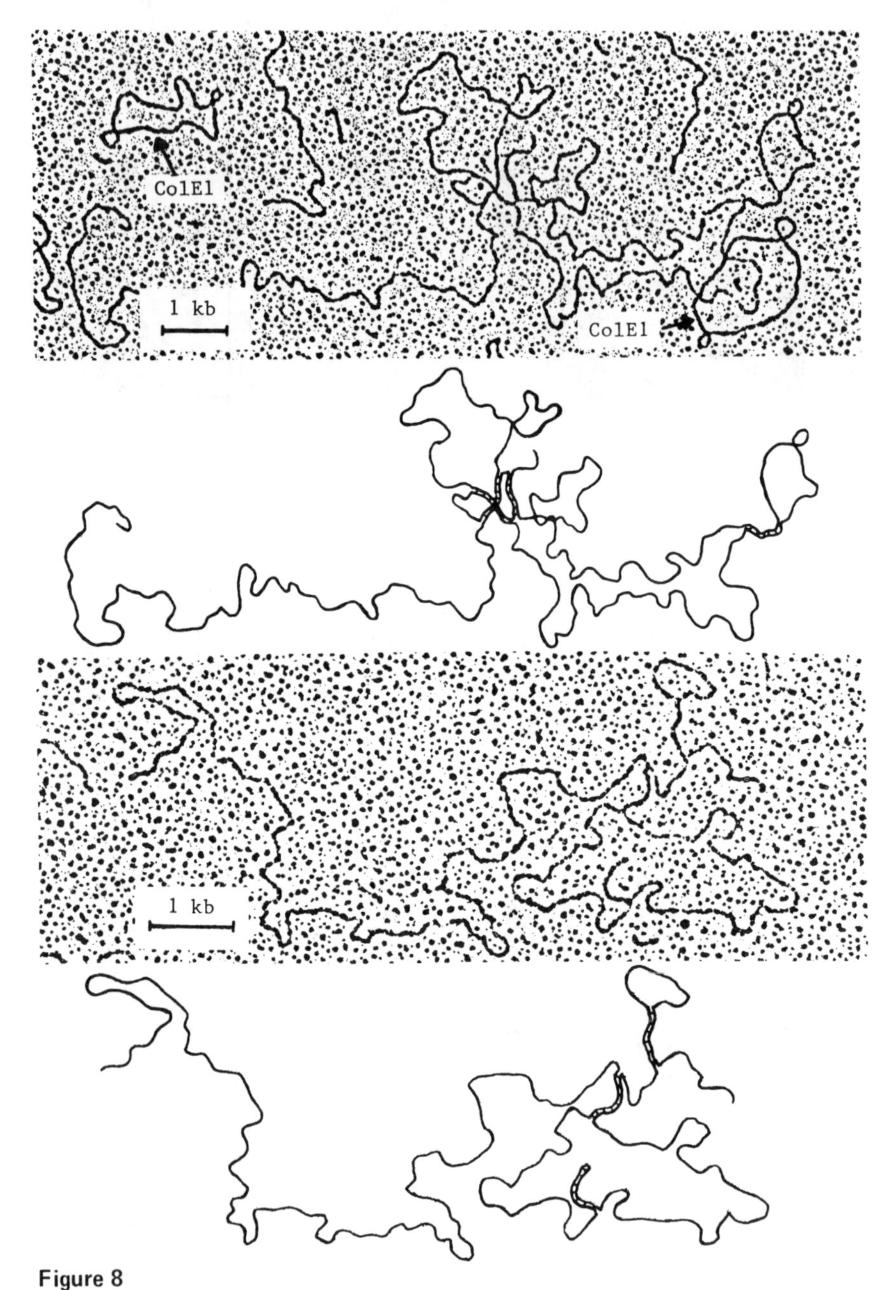

Figure 8

Electron micrographs of *S. dysenteriae* DNA showing many inversion loops with duplex stems of about 700 bases in length. These molecules are easily seen after denaturation and renaturation of the DNA, using the formamide spreading technique of Sharp et al. (1972). Tracings for interpretation of each DNA molecule are also shown. The bar represents a length of 1 kb for double-stranded DNA.

E. coli DNA molecules with electron microscope heteroduplex methods confirmed that there were few copies of inverted repeats and these were spaced by rather long DNA segments. Direct observation of inverted repeats in *Shigella* DNA molecules viewed in the electron microscope confirmed that there were many copies of inverted repeats corresponding in length to the $\lambda\mu$(IS*1*) sequence seen in Figure 7. These repeats were spaced by short DNA segments and sometimes appeared as tandem repeats. Electron micrographs of some of these molecules are shown in Figure 8.

The intensity of gel bands may also depend on the distance between two copies of the inverted repeat sequences. The differences in intensity between F^- and Hfr strains of *E. coli* for some of the gel bands perhaps suggest that integration of F may increase the distance between inverted repeats and may introduce a particular sequence of F which may form a duplex with the same sequence inverted in the *E. coli* chromosome.

As described above, the *Shigella* strain produced an extremely intense DNA band corresponding in length to the $\lambda\mu$(IS*1*) sequence. Preliminary experiments showed that *Salmonella typhimurium* and *Proteus mirabilis* produced gel-band patterns similar to that for the *E. coli* F^- strain but did not produce an unusually large number of fragments as seen in the *Shigella* strain. At present, we do not know whether many of the repeats found in the *Shigella* chromosomal DNA are the same as the $\lambda\mu$(IS*1*) sequence that is involved in genetic recombination (reviewed above). However, it is interesting to consider that the presence of highly repeated sequences in the *Shigella* strain may be associated with the observation that, in general, *Shigella* strains generate drug resistance plasmids with a higher frequency than *E. coli* strains (S. Mitsuhashi, pers. comm.).

Acknowledgment

This research has been supported by U.S. Public Health Service Grant #GM 22007-02 from the National Institute of General Medical Sciences.

REFERENCES

Berg, D. E., J. Davis, B. Allet and J. D. Rochaix. 1975. Transposition of R factor genes to bacteriophage λ. *Proc. Nat. Acad. Sci.* 72:3628.

Davidson, N., R. C. Deonier, S. Hu and E. Ohtsubo. 1974. Electron microscope heteroduplex studies of sequence relations among plasmids of *Escherichia coli*. X. Deoxyribonucleic acid sequence organization of F and of F-primes and the sequences involved in Hfr formation. In *Microbiology, 1974* (ed. D. Schlessinger), p. 56. American Society for Microbiology, Washington, D.C.

Davis, R. W., M. Simon and N. Davidson. 1971. Electron microscope heteroduplex methods for mapping regions of base sequence homology in nucleic acids. In *Methods in enzymology*, (ed. L. Grossman and K. Moldave), vol. XXI, p. 413. Academic Press, New York.

Deonier, R. C., E. Ohtsubo, H.-J. Lee and N. Davidson. 1974. Electron microscope heteroduplex studies of sequence relations among plasmids of *Escherichia coli.* VII. Mapping of ribosomal RNA genes in plasmid F14. *J. Mol. Biol.* **89**:819.

Franklin, N. C. 1971. Illegitimate recombination. In *The bacteriophage lambda* (ed. A. D. Hershey), p. 175. Cold Spring Harbor Laboratory, Cold Spring Harbor, New York.

Heffron, F., C. Rubens and S. Falkow. 1975. The translocation of a plasmid DNA sequence which mediates ampicillin resistance. Molecular nature and specificity of insertion. *Proc. Nat. Acad. Sci.* **72**:3623.

Hirsch, H. J., P. Starlinger and P. Brachet. 1972. Two kinds of insertions in bacterial genes. *Mol. Gen. Genet.* **119**:191.

Hsu, M.-T. and N. Davidson. 1974. Electron microscope heteroduplex study of the heterogeneity of Mu phage and prophage DNA. *Virology* **58**:229.

Hu, S., E. Ohtsubo and N. Davidson. 1975a. Electron microscope heteroduplex studies of sequence relations among plasmids of *E. coli.* XI. The structure of F13 and related F-primes. *J. Bact.* **122**:749.

Hu, S., E. Ohtsubo, N. Davidson and H. Saedler. 1975b. Electron microscope heteroduplex studies of sequence relations among bacterial plasmids. XII. Identification and mapping of the insertion sequences IS1 and IS2 in F and R plasmids. *J. Bact.* **122**:764.

Hu, S., K. Ptashne, S. N. Cohen and N. Davidson. 1975c. $\alpha\beta$ Sequence of F in *IS3*. *J. Bact.* **123**:687.

Kleckner, N., R. K. Chan, B.-K. Tye and D. Botstein. 1975. Mutagenesis by insertion of a drug-resistance element carrying an inverted repetition. *J. Mol. Biol.* **97**:561.

Lee, H.-J., E. Ohtsubo, R. Deonier and N. Davidson. 1974. Electron microscope heteroduplex studies of sequence relations among bacterial plasmids. V. ilv^+ deletion mutants of F14. *J. Mol. Biol.* **89**:585.

Malamy, M. H., M. Fiandt and W. Szybalski. 1972. Electron microscopy of polar insertions in the *lac* operon of *Escherichia coli. Mol. Gen. Genet.* **119**:207.

Ohtsubo, E., R. C. Deonier, H.-J. Lee and N. Davidson. 1974a. Electron microscope heteroduplex studies of sequence relations among plasmids of *Escherichia coli.* IV. The F sequences in F14. *J. Mol. Biol.* **89**:565.

Ohtsubo, E., H.-J. Lee, R. C. Deonier and N. Davidson. 1974b. Electron microscope heteroduplex studies of sequence relations among plasmids of *Escherichia coli.* VI. Mapping of F14 sequences homologous to ϕ80d*met*BJF and ϕ80d*arg*ECBH phages. *J. Mol. Biol.* **89**:599.

Ohtsubo, H. and E. Ohtsubo. 1976. Isolation of inverted repeat sequences, including IS1, IS2, and IS3, in *Escherichia coli* plasmids. *Proc. Nat. Acad. Sci.* (in press).

Palchaudhuri, S., E. Ohtsubo and W. K. Maas. 1976. Fusion of two F-prime factors in *Escherichia coli* studied by electron microscope heteroduplex analysis. *Mol. Gen. Genet.* (in press).

Ptashne, K. and S. N. Cohen. 1975. Occurrence of insertion sequence regions on plasmid deoxyribonucleic acid as direct and inverted nucleotide sequence duplications. *J. Bact.* **122**:776.

Saedler, H., H.-J. Reif, S. Hu and N. Davidson. 1974. IS2, a genetic element for turn-off and turn-on of gene activity in *E. coli. Mol. Gen. Genet.* **132**:265.

Schmidt, C. W., J. E. Manning and N. Davidson. 1975. Interspersion of repetitive and nonrepetitive DNA sequences in the *Drosophila melanogaster* genome. *Cell* **46**:141.

Sharp, P. A., S. N. Cohen and N. Davidson. 1973. Electron microscope heteroduplex studies of sequence relations among plasmids of *Escherichia coli.* II. Structure of drug resistance (R) factors and F factors. *J. Mol. Biol.* **75**:235.

Sharp, P. A., M.-T. Hsu, E. Ohtsubo and N. Davidson. 1972. Electron microscope heteroduplex studies of sequence relations among plasmids of *E. coli.* I. Structure of F-prime factors. *J. Mol. Biol.* **71**:471.

Tye, B.-K., R. K. Chan and D. Botstein. 1974. Packaging of an oversized transducing genome by *Salmonella* phage P22. *J. Mol. Biol.* **85**:485.

IS*1* and IS2 in *E. coli:* Implications for the Evolution of the Chromosome and Some Plasmids

H. Saedler
Institut für Biologie III
Universität Freiburg
D-7800 Freiburg, i. Br., W. Germany

The presence of insertion elements in the chromosome of *E. coli* originally became apparent by the transposition of these sequences (IS elements) from their natural positions into functional genes, resulting in a recognizable mutant phenotype. The elements have been numbered according to the order in which they were detected. IS*1* is about 800 nucleotide pairs long, and IS*2*, IS*3*, and IS*4* are each approximately 1400 base pairs long (see previous articles in this book and, for review, Starlinger and Saedler 1976).

In this report I will concentrate on the topics listed below. These are mainly concerned with IS*1* and IS*2* since these are the elements with which I am most familiar.

1. IS elements as natural components of the *E. coli* chromosome (Saedler and Heiss 1973).
2. The presence of IS elements in certain R factors (Hu et al. 1975b).
3. IS*2* as a constituent of F^+ DNA (Saedler and Heiss 1973; Hu et al. 1975b).
4. Chromosomal rearrangements mediated by IS*1* (Reif and Saedler 1975).
5. IS*2* as a genetic element for turn-off and turn-on of gene activity (Saedler et al. 1974).

IS*1* and IS*2* in the Chromosome of *E. coli* K12

Once specialized transducing phages carrying IS elements were available, it became possible to probe other DNA molecules for the presence of IS elements by DNA-DNA hybridization. Using a filter-binding assay to detect hybridization of chromosomal [^{3}H]DNA to filter-bound IS DNA, about eight copies of IS*1* and about five copies of IS*2* were found to be constituents of the *E. coli* chromosome (Saedler and Heiss 1973).

The same technique was used to analyze various other DNAs. The chromosome of *Salmonella typhimurium,* for example, was shown to contain IS*1* but not IS*2*. The F^+ and R1*drd*19 plasmid DNAs were also shown to carry IS elements as natural components. This information was used for a detailed analysis of the positions of the IS elements on these plasmids.

IS*1* in R(fi^{+}) Plasmids

The R factors of the fi^{+} class are composed of two units, each capable of replicating autonomously if dissociated from one another. The resistance transfer factor (RTF unit) codes all the functions necessary for cell-to-cell contact, thus allowing the transfer of the plasmid. The drug resistance determinant (r determinant) carries most of the antibiotic resistance genes. Rownd and Mickel (1971) showed that R factors can dissociate into the RTF unit and the r determinant in *Proteus mirabilis.* The structure of these complex plasmids is shown in Figure 1. This is a diagrammatic representation of all the features observed on R1*drd*19, R6, and R100-1 by the heteroduplex technique (Hu et al. 1975b). In all three plasmids, an IS*1* element separates the RTF unit from the r determinant at each junction. Both IS*1* elements are oriented in the same direction (Hu et al. 1975b; Ptashne and Cohen 1975). This finding suggests a model for the formation and dissociation of R factors as well as for the amplification of the antibiotic resistance genes. Dissociation may occur by recombination between IS*1* sequences on the plasmid generating two units, each containing an IS*1*. Fusion results from the reverse reaction. Amplification of the antibiotic resistance genes on a plasmid might result from recombination between the homologous IS*1* elements of different r-determinant molecules, leading to plasmid molecules with repeated r-determinant units.

In addition to IS*1*, other IS elements have also been observed on R factors. For example, in R6 and R100-1, IS*2* is found at a position within the transfer

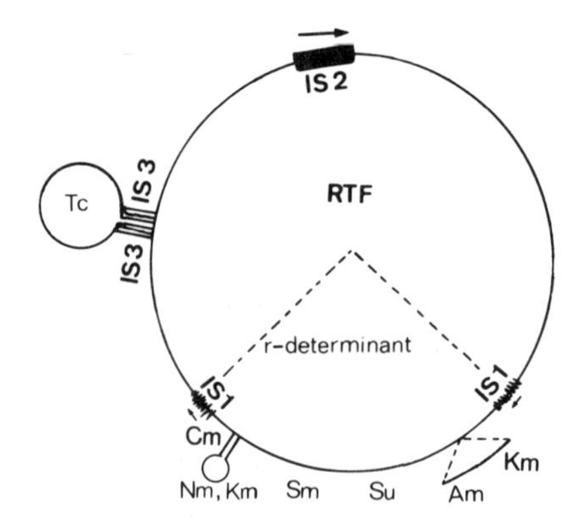

Figure 1
The composite structure shown is based on heteroduplex studies by Sharp et al. (1973), Hu et al. (1975b), and Ptashne and Cohen (1975). It includes features from R1*drd*19, R6, and R100-1 plasmids.

genes where it does not cause a transfer-defective mutation but rather contributes to the transfer-positive character of the plasmids (Hu et al. 1975b). The role of IS*2* in the expression of neighboring genes will be discussed later in this report. Also, two IS*3* elements have been seen to bracket the tetracycline resistance gene in inverted orientation (Ptashne and Cohen 1975). The occurrence of IS elements on the R plasmids and their ability to transpose the antibiotic resistance genes to various other DNA molecules suggest a possible role for these elements in the formation and evolution of R plasmids (for review, see Cohen and Kopecko 1976).

IS*2* in F Factor

IS*2* has been shown by DNA-DNA hybridization to occur in F^+ plasmid DNA (Saedler and Heiss 1973). Detailed heteroduplex mapping by electron microscopy showed that IS*2* is a natural component of the F^+ plasmid located at coordinates 16.3 to 17.6 (Hu et al. 1975b). The role of IS*2* in the formation of Hfr13 and of F′ plasmids has been discussed by Davidson et al. (1975) and Hu et al. (1975a).

IS*1*-induced Rearrangements of Chromosomal Sequences

The formation of new chromosomal sequences can result from translocation, duplication, inversion, or deletion of genetic material. All these processes may be involved in the evolution of plasmids as well as of chromosomes. In addition to perhaps being involved in the formation and evolution of R factors (as pointed out above), the IS elements may also be responsible for chromosomal rearrangements. Nonadjacent chromosomal regions can be brought together by deletion of the intermittant genetic material, resulting in a new chromosomal order. This reaction has been studied extensively in IS*1*-induced deletion formation (Reif and Saedler 1975). The IS*1* element serves as a generator for deletions. In this process, apparently, the termini of the integrated IS*1* element are most important (see Reif and Saedler, this volume). It is not yet clear, however, which enzymes are involved in this rather unusual type of recombination. Apparently, the normal recombination pathway of *E. coli* is not used. Recently, however, a mutant has been isolated which is deficient in IS*1*-induced deletion formation (Nevers et al., this volume). Such mutants may be helpful in the analysis of such illegitimate recombinational events.

IS*2* as a Genetic Element for Turn-off and Turn-on of Gene Activity

The *gal* operon in the F8 *gal* plasmid is 2.6 kb pairs away from the IS*2* that occurs naturally in F DNA. In this plasmid, the *gal* enzymes are under their normal repressor control, i.e., they are inducible. However, if an IS*2* element is integrated into the control region of the *gal* operon (*galOP308*::IS*2*), the expression of the *gal* enzymes is reduced about 100-fold. The small residual amount of activity still observed is no longer under *gal* repressor control; i.e., it is constitutive. The orientation of IS*2* which prevents expression of the *gal* genes has been termed orientation I. It should be noted that in polar mutations (which are due to the integration of IS*2*), the element is always found integrated in the same orientation with respect to the direction of transcription of the operon into which the element has integrated (see Starlinger and Saedler 1976). It so happens that the IS*2* element naturally occurring in the F8 *gal* plasmid is in opposite orientation with respect to the IS*2* element integrated into the control region of the *gal* operon.

Different classes of Gal$^+$ revertants are obtained from the Gal$^-$ mutants containing the plasmid F8 *galOP308*::IS*2*. One class includes inducible revertants resulting from accurate excision of the IS*2* element, which restores the wild-type *gal* operon. In addition, at least two categories of Gal$^+$ constitutive revertants are found. These classes of high- and intermediate-level constitutive revertants have been characterized (Saedler et al. 1974), and inferences resulting from these studies are discussed below.

*IS*2 *Carries a Strong Promoter*

Heteroduplex analysis of the high-level constitutive revertants showed that the *gal* structural genes are fused to the IS*2* that occurs naturally on F. This IS*2* is oriented opposite to the polar IS*2* inserted in the *galOP* region of the parental plasmid and is thus in orientation II with respect to the direction of transcription of the *gal* operon. The *gal* genes on the fused plasmid are now expressed two to three times more efficiently than they are with the *gal*-specific promoter. The structures of all the plasmids mentioned thus far, as well as some other derivatives, and the corresponding expression of galactokinase are shown in Figure 2.

These data indicate that the IS*2* element naturally occurring on F carries a strong promoter. There are two lines of evidence supporting the inference that an IS*2* element carries a strong promoter. Gal$^-$ segregants from the high-level Gal$^+$ constitutive revertants can be isolated with a frequency of about 1%. In these Gal$^-$ segregants, IS*2* has been excised accurately from the plasmids so that the *gal* structural genes are fused to F DNA sequences adjacent to the parental IS*2* element. Since the expression of the *gal* genes is reduced about 200-fold, these F DNA sequences apparently do not contain a promoter from which transcription of the *gal* genes could be initiated. Therefore, this Gal$^-$ segregant plasmid supports the assumption that the expression of the *gal* genes in the fused plasmid is initiated within the IS*2* element in orientation II (Fig. 2). The segregant plasmid is devoid of the *gal* control region as well as of any IS*2* sequence. If another IS*2* is integrated in orientation II to the right of *galE* in such a plasmid, the cells will be converted to a constitutive Gal$^+$ phenotype,

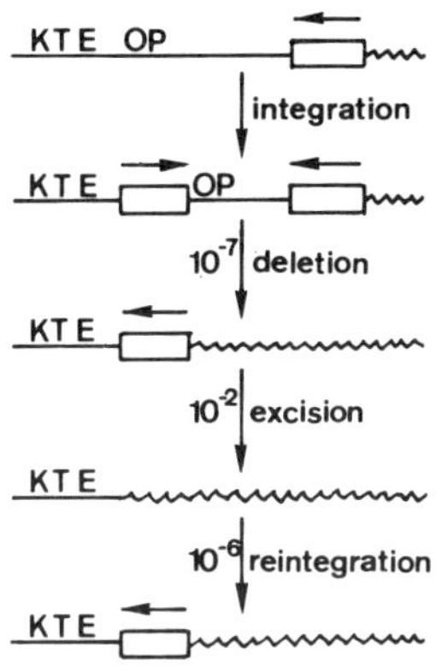

relevant structure of Fgal plasmids	Gal phenotye	orientation of IS 2	% kinase activity – inducer	+ inducer
KTE OP	+	0	10	100
↓ integration				
KTE OP	–	I	1	2
10^{-7} ↓ deletion				
KTE	+	II	221	325
10^{-2} ↓ excision				
KTE	–	0	1	1
10^{-6} ↓ reintegration				
KTE	+	II	200	300

Figure 2
The wavy line represents F DNA, whereas the solid line is chromosomal genetic material. The physical structure of the plasmids and the enzyme data are from Saedler et al. (1974).

assuming that every IS*2* element carries a strong promoter. Such constitutive Gal^+ revertants do appear with a frequency of approximately 10^{-6} after prolonged incubation of the cells on minimal galactose plates. These revertants are all of the high-level constitutive class. Two plasmids isolated from these revertants and analyzed with the heteroduplex technique were shown to have acquired IS*2* in orientation II to the right of the *galE* gene. One such revertant is included in Figure 2. However, these plasmids also carry long deletions of F material adjacent to the newly integrated IS*2* element. This observation and the finding that revertants appear only after prolonged incubation might reflect a two-step process: deletion of some F material to create an integration site for IS*2* in orientation II, followed by integration of the element into the newly formed site. Whatever the precise mechanism of this reversion event might be, the finding that IS*2* is integrated in orientation II indicates that at least some copies of IS*2* carry a strong promoter. IS*2*, therefore, seems to be a transposable element for turn-off and turn-on of gene activity depending on its orientation of integration. In orientation I it severely reduces the expression of adjacent genes, whereas in orientation II it allows expression at a high level.

IS2 Can Undergo Internal Changes

In addition to being able to integrate in two possible orientations and thereby cause either turn-off or turn-on of adjacent gene activity, IS*2* can undergo internal changes which affect gene expression. Heteroduplex analysis has shown that the structures of intermediate constitutive revertants obtained from F8 *galOP308*::IS*2* do not differ detectably from the structures of their Gal⁻ segregants or that of the parental Gal⁻ plasmid.

The structures of these plasmids and the expression of galactokinase are shown in Figure 3. The event leading to intermediate constitutivity apparently occurs within the IS*2* element itself since the constitutivity is transferred together with the plasmid. It is possible that this is a recombinational event within IS*2* rather than a mutational event. At present, it is not clear why the reverse reaction (reversion of the Gal⁻ plasmid to a Gal⁺ phenotype) is three to four orders of magnitude lower.

In the intermediate constitutive revertants, IS*2* remains fixed in orientation I but apparently undergoes an internal change which modifies the expression of the adjacent genes (Fig. 3). Partial or complete inversion of the promoter carried by the IS*2* element might be responsible for the difference in gene expression of these plasmids.

*IS*2 *Element in the* E. coli *Chromosome Appears to Undergo Inversion*

Intermediate- and high-level constitutive revertants are also obtained from *galOP308*::IS*2* located on the *E. coli* chromosome (Saedler et al. 1972; Saedler et al. 1974). Genetic experiments exclude gene duplication, loss of the IS*2* element, and fusion to another promoter. Therefore it was suggested that the high-level constitutive revertants may have arisen by in situ inversion of the total IS*2* element. Work is currently in progress to provide positive physical evidence for this hypothesis.

*Chromosomal Rearrangements Can Be Mediated by IS*2

Previously, Reif and Saedler (1975) reported that IS*2* integrated in orientation I does not cause an increase in deletion formation above the spontaneous level. The frequency of deletion in the *gal* region in the absence of any IS element is

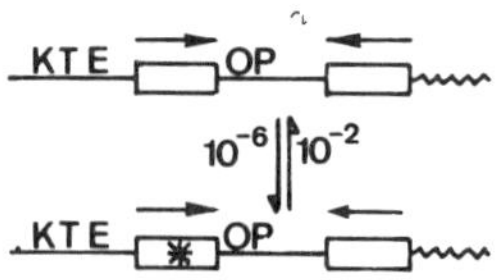

relevant structure of Fgal plasmids	Gal phenotye	% kinase activity −inducer	% kinase activity +inducer
KTE OP	−	1	2
(10^{-6} ↓ ↑ 10^{-2})			
KTE * OP	+	20	33

Figure 3
The star within one IS*2* element indicates an alteration within that element. (Data from Saedler et al. 1974.)

about 10^{-7}. However, in high-level constitutive revertants of a *gal*::IS*2* mutation (see above), chromosomal deletions adjacent to the element have been observed with a frequency of about 10^{-5} (H.-J. Reif and H. Saedler, unpubl.).

Unlike the IS*1*-induced deletion formation, this system does not seem to be dependent on the growth temperature of the cells. Therefore, the mechanism involved in IS*2* deletion formation appears to be different from that involved in IS*1* deletion formation. However, since the deletions formed in the IS*2* system have not yet been characterized further, it is not clear whether in this system the termini of the integrated element play a role similar to that in the IS*1* system.

CONCLUDING REMARKS

We have seen that IS elements not only carry signals influencing gene expression, but they apparently can also cause rearrangements of gene sequences; for example, IS*2* could be used in evolution as a prefabricated element to allow efficient expression of newly evolved or previously silent genes if the element becomes translocated to these genes. IS*1*, on the other hand, might achieve a similar effect on gene expression (although it carries no promoter) by fusing a silent gene to another promoter in its vicinity or by transposing the adjacent gene to a more distant promoter.

Whatever the role of IS elements in the evolution of chromosomes and plasmids may be, it is noteworthy that formally analogous elements are also observed in eukaryotic organisms such as *Zea mays, Drosophila melanogaster,* and others. Whether these eukaryotic elements have any relationship to the known IS elements of prokaryotic origin is open to speculation.

Acknowledgments

This work was supported by the Deutsche Forschungsgemeinschaft. I thank Dr. P. Nevers for carefully reading the manuscript.

REFERENCES

Cohen, S. N. and D. J. Kopecko. 1976. Structural evolution of bacterial plasmids: Role of translocating genetic elements and DNA sequence insertions. *Fed. Proc.* **35**:2031.

Davidson, N., R. C. Deonier, S. Hu and E. Ohtsubo. 1975. The DNA sequence organization of F and F-primes and the sequences involved in Hfr formation. *Microbiology* **1**:56.

Hu, S., E. Ohtsubo and N. Davidson. 1975a. Electron microscope heteroduplex studies of sequence relations among plasmids of *E. coli.* XI. The structure of F13 and related F-primes. *J. Bact.* **122**:749.

Hu, S., E. Ohtsubo, N. Davidson and H. Saedler. 1975b. Electron microscope heteroduplex studies of sequence relations among bacterial plasmids. XII. Identification and mapping of the insertion sequences IS1 and IS2 in F and R plasmids. *J. Bact.* **122**:764.

Ptashne, K. and S. N. Cohen. 1975. Occurrence of insertion sequence (IS) regions on plasmid deoxyribonucleic acid as direct and inverted nucleotide sequence duplications. *J. Bact.* **122**:776.

Reif, H.-J. and H. Saedler. 1975. IS1 is involved in deletion formation in the *gal* region of *E. coli* K-12. *Mol. Gen. Genet.* **137**:17.

Rownd, R. and S. Mickel. 1971. Dissociation of the RTF and r-determinants of the R-factor NR 1 in *Proteus mirabilis. Nature New Biol.* **234**:40.

Saedler, H. and B. Heiss. 1973. Multiple copies of the insertion DNA sequences IS1 and IS2 in the chromosome of *E. coli* K-12. *Mol. Gen. Genet.* **122**:267.

Saedler, H., H.-J. Reif, S. Hu and N. Davidson. 1974. IS2, a genetic element for turn-off and turn-on of gene activity in *E. coli. Mol. Gen. Genet.* **132**:265.

Saedler, H., J. Besemer, B. Kemper, B. Rosenwirth and P. Starlinger. 1972. Insertion mutations in the control region of the *gal* operon in *E. coli. Mol. Gen. Genet.* **115**:258.

Sharp, P. A., S. N. Cohen and N. Davidson. 1973. Electron microscope heteroduplex studies of sequence relations among plasmids of *E. coli.* II. Structure of drug resistance (R) factors and F-factors. *J. Mol. Biol.* **75**:235.

Starlinger, P. and H. Saedler. 1976. IS-elements in microorganisms. In *Current topics in microbiology and immunology,* vol. 75, p. 111. Springer Verlag, Berlin.

The Organization of Putative Insertion Sequences on the *E. coli* Chromosome

L. T. Chow
Cold Spring Harbor Laboratory
Cold Spring Harbor, New York 11724

Insertion sequences (IS) are a small class of short DNA segments of discrete lengths and base arrangements found in *E. coli* and in F, F′, and R episomes (Saedler and Heiss 1973; Hu et al. 1975; Davidson et al. 1975). They can be transposed from one site to another and are presumably involved in the integration and excision of the episomes and in the translocation of drug resistance determinants. The ends of insertion sequences can also be "hot spots" for intramolecular recombination in F- and R-related plasmids (Davidson et al. 1975; Hu et al. 1975; Ohtsubo and Ohtsubo, this volume). All these recombinational events are independent of the *E. coli recA* recombination system. Although the number of copies per *E. coli* chromosome has been estimated at eight for IS*1* and five for IS*2* (Saedler and Heiss 1973), their distribution is not known.

To get a better insight into the organization of the IS elements on the *E. coli* chromosome, I denatured *E. coli* DNA (strain CSH50 lysogenic for bacteriophages λ and Mu) and reannealed it to the extent that 10% to 20% of the DNA was renatured. Using electron microscopy, I have found short DNA duplexes with a single-stranded fork at each end. The lengths of these duplexes fall into discrete size classes. Some correspond to those of IS*1* and of IS*2*, IS*3*, or IS*4*. Most of the observed sequences forming the duplex are arranged pairwise in inverted order with respect to each other on the same strand of DNA so that they form a stem-loop structure (Fig. 1A,B,C). Duplex regions of the same lengths established between independent strands were seen much less frequently. Pairs of short duplexes separated by noncomplementary "spacer" sequences (Fig. 1D) were rarely seen and then only in structures formed from independent strands. No multiple stem-loop structures were found. The low frequency of observation of structures formed between two different strands could be accounted for by the bimolecular kinetics of the hetereoduplex formation, as compared to the rapid intrastrand snap-back of the inverted duplications. The *E. coli* strain studied has no plasmid or episome by pedigree, as was confirmed by EM analysis of the extracted native DNA. Therefore, the short duplexes in self-renatured samples must reflect the sequence arrangements on the *E. coli* chromosome. The lengths and standard deviations of the duplex regions, their separation, and the number of observations of each are presented

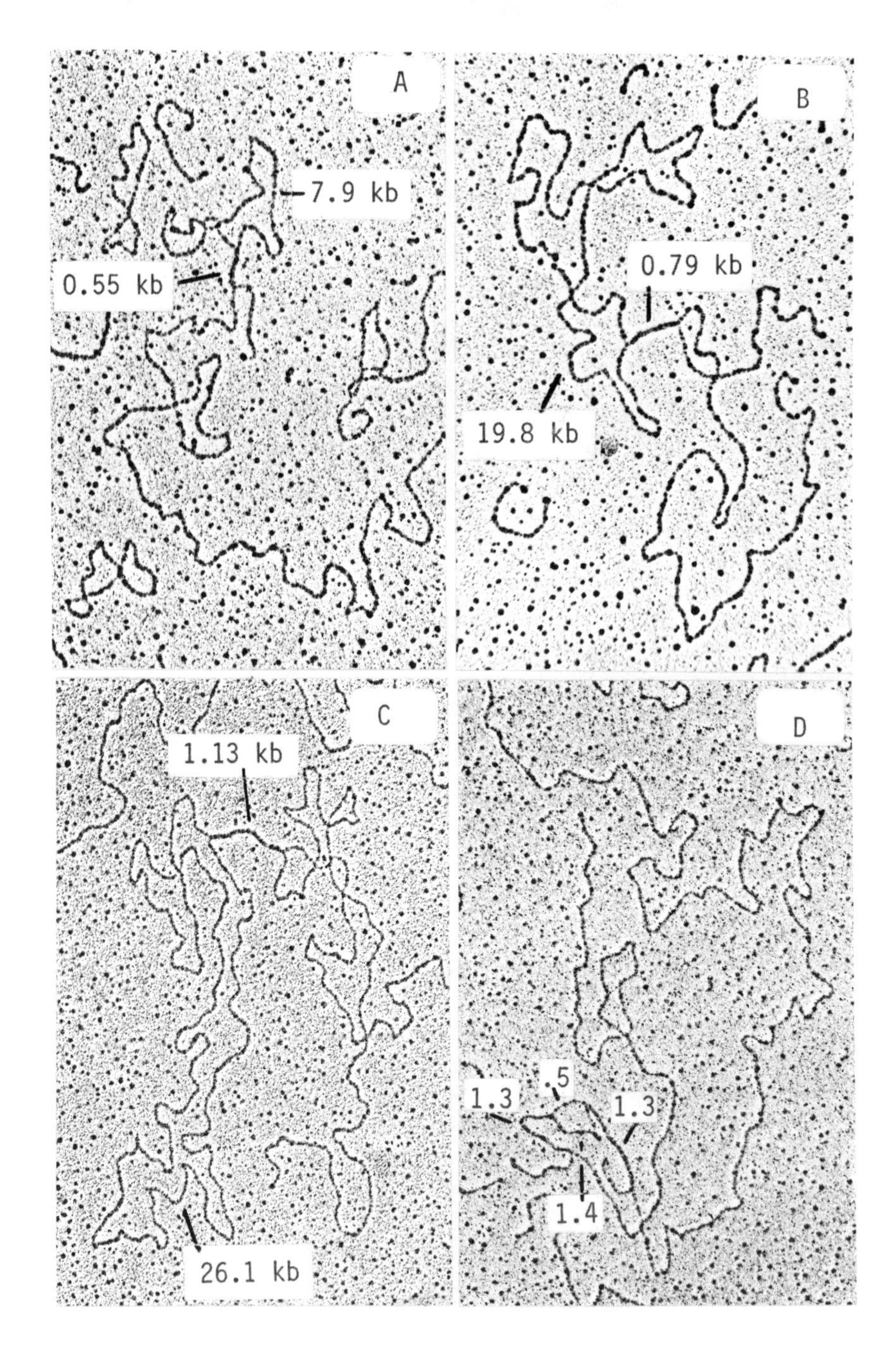

Figure 1

Heteroduplexes formed by association of putative insertion sequences in *E. coli* strain CSH50. *A, B,* and *C* are intramolecular stem-loop structures formed by the "snap-back" of pairs of IS sequences arranged in opposite orientations. *D* is an intermolecular heteroduplex showing two pairs of IS sequences, each separated by a small "spacer." All numbers are in kilobases.

in Table 1. Two striking features revealed by the data in Table 1 are discussed below.

Discrete Lengths of Duplexes

The lengths of the duplexes correspond to those of IS*1*, of IS*2*, IS*3*, or IS*4*, of ribosomal DNA, and of the $\gamma\delta$ sequence. Although the $\gamma\delta$ sequence has not yet been classified as an insertion sequence, it has some of the properties of an insertion sequence: for instance, it is located in the integrative region of the F sex factor; it probably promoted the integration of F to generate Hfr AB313; and its ends do serve as hot spots for the specific segregation and general deletion of F′ episomes (Ohtsubo et al. 1974; Davidson et al. 1975). The pair of putative $\gamma\delta$ sequences observed is in direct order on the same strand, separated by 77.3 kb. This was deduced from a structure that included a duplex of 5.7 kb with a fork of two single strands at each end. The ends of one of the single strands from each fork hybridized to form a closed loop 77.3 kb in length

Table 1
Lengths, Distributions, and Number of Observations of Putative Insertion Sequences in the *E. coli* CSH50 Chromosome

Tentative assignment	Duplex (kb)	Spacer (kb)	
		inverted[a]	direct[a]
?	0.5 ± 0.08 (10)	7.5 ± 0.4 (4)	
		25.2 (1)	
		28.4 (1)	
IS*1*	0.75 ± 0.04 (19)	21.7 ± 0.9 (17)	
		13.7 (1)	
?	1.0 ± 0.08 (16)	12.8 (1)	
		15.9 (1)	
		21.8 ± 0.7 (10)	
		27.1 ± 1.3 (3)	
IS*2*, *3*, or *4*	1.3 ± 0.08 (19)	5.30 (1)	{ 0.5 (2)[b]
		26.3 ± 0.6 (3)	{ 1.3 ± 0.09 (2)
		28.9 ± 0.9 (7)	
		159.0 (1)	
?	2.8 (1)		
?	3.4 ± 0.14 (6)		
?	4.6 ± 0.26 (4)		
Ribosomal genes	5.2 ± 0.08 (7)		[0.2 ± 0.05 (2)][c]
$\gamma\delta$	5.6 ± 0.13 (6)		77.3 (1)

[a]The numbers in parentheses indicate the number of observations.
[b]The heteroduplex with the pair of spacers indicated has been seen twice (see Fig. 1D).
[c]Internal spacer between the 16S and 23S ribosomal RNA genes.

(see Fig. 4 of Chow and Davidson [1973] for schematic representations of similar "out-of-register" circular structures).

Other duplexes resulting from the snap-back of inverted repetitions have lengths of 0.5 and 1.0 kb. Both have been found independently in *E. coli* by Ohtsubo and Ohtsubo (this volume) using a different technique. Since neither has been identified as an insertion sequence, their functions are not clear. Due to the possibility of fortuitous lateral aggregations, duplex regions shorter than 0.4 kb have been excluded from the data analysis.

The observed bimolecular duplexes of 2.8 kb, 3.4 kb, and 4.6 kb might have resulted from gene duplication or gene translocation. They could also have formed from cryptic lambda or Mu prophages in the bacterial DNA and the lambda or Mu phage DNAs included during the denaturation and reannealing.

The Nonrandom Length Distribution of Single-stranded DNA between Inverted Duplications

Most of the single-stranded *E. coli* DNA were between 50 and 150 kb in length, with some longer than 200 kb. However, most of the "spacers" (snap-back loops) were either 22 kb or 27.5 ± 1.5 kb. In addition, there was a nonrandom pattern of stem-loop combinations. For example, segments of 22 kb were always observed between 0.75-kb and 1.0-kb inverted repetitions, whereas 27.5 ± 1.5-kb segments were most frequently associated with 1.0-kb and 1.3-kb duplexes.

The single observation of some stem-loop size combinations and the very frequent observations of others imply that the frequent examples do not generally represent redundant observations of the same sequence from the *E. coli* chromosome. Rather, they apparently indicate that different segments of the genome have evolved with similar unit organization of 22-kb or 27.5 ± 1.5-kb segments flanked by inverted duplications of insertion sequences and other elements.

Estimation of Amount of DNA Involved in Organization

From the data in Table 1, it is possible to approximate the number of copies of these putative insertion sequences and the fractions of the *E. coli* chromosome involved in this organization. Liberal, conservative, or moderate values are obtained depending upon which of the three following assumptions is made: (1) All the stem-loop structures observed in the EM search summarized in Table 1 represent distinct and independent segments of the *E. coli* chromosome. (2) Each stem-loop size combination reveals only one occurrence, irrespective of the number of times it is observed. (3) A few (1-3) observations of a particular size combination represent redundant detection of the same segment, whereas a larger number of observations of a combination indicates similar arrangements of distinct segments. The number of these more frequent structures was divided by three to approximate an average of three sightings of each different stem-loop sequence. The values obtained using each of the three methods of calculation are presented in Table 2.

The amount of unique sequence involved as estimated by Method 1 is too high because such stem-loop structures were found rather infrequently. Method 2 probably gives too low a value because there is no obvious reason for the large bias in the number of observations of the various size combinations. Method 3 gives a moderate and probably more realistic estimate of the amount of the *E. coli* chromosome involved in the sequence organization that generates the stem-loop structure upon denaturation and renaturation. However, these estimates include only insertion sequences located relatively close to each other in opposite orientations. Therefore, the number of copies of insertion sequences present on the chromosome may be systematically underestimated by such an electron microscope study.

At present, I do not know the significance of the apparent organization of roughly 18% of the *E. coli* chromosome into units of 22-kb and 27.5 ± 1.5-kb DNA segments flanked by inverted repetitious sequences. Such DNA segments may have the potential to translocate, like Tn*10* (Kleckner et al. 1975) and Tn*5* (Berg et al. 1975) elements, or to invert, like Mu G (Hsu and Davidson, 1974) and P1 G segments (Lee et al. 1974; Chow and Bukhari, this volume). Translocation or inversion could have a regulatory function for gene expression. Alternatively, such organization could be required for folding the chromosome into a compact nucleoid or it might reflect the evolution of the bacterial chromosome from DNA segments pooled from donor ancestors by translocation.

Table 2
Estimates of the Number of Copies of Putative Insertion Sequences and the Amount of *E. coli* DNA Bounded by Inverted Sequences

		Estimate of number of copies[a]		
Tentative assignment	Duplex (kb)	method 1	method 2	method 3
?	0.5	12	6	6
IS*1*	0.75	36	4	14
?	1.0	30	8	12
IS*2*, *3*, or *4*	1.3	24	8	10
Fraction of *E. coli* DNA involved[b]		.30	.10	.16

[a]The estimates of the number of copies of insertion sequences are made as described in the text. Only those present as inverted duplications and revealed by EM as single-stranded, stem-loop structures in renatured samples were included. Stems shorter than 0.4 kb were excluded.

[b]The fraction of *E. coli* DNA bounded by inverted sequences was calculated by summing the lengths of the stem-loop combinations estimated by each method and dividing by 4.39×10^3 kb, the length of the bacterial chromosome.

SUMMARY

Total *E. coli* DNA (strain CSH50 F^-, R^-, lysogenized with λp*lac5* and Mu) was denatured, reannealed, and observed by electron microscopy. The single-strand lengths ranged from about 50 to 150 kb. In some molecules, a short duplex region with a single-stranded fork at each end was observed. The duplex lengths were 0.75 kb, 1.3 kb, 5.22 kb, and 5.62 kb, corresponding to those of IS*1*, of IS*2*, IS*3*, or IS*4*, of ribosomal RNA genes, and of the γδ sequence, respectively. Duplexes of 1.0 kb and 0.5 kb were also found. Most of the duplexes were observed as intramolecular stem-loop structures and were therefore interpreted to be sequence duplications in inverted order on the same DNA strand. The most frequent separations of the putative inverted insertion sequences were around 22 kb and 27.5 ± 1.5 kb. Short duplexes formed by the hybridization of insertion sequences on separate strands were found at much lower frequencies. Pairs of duplex segments arranged in direct order, with "spacers" of single-stranded DNA between the pairs, were rarely observed.

Acknowledgments

I am grateful to Dr. Regine Kahmann for supplying the DNA of *E. coli* CSH50. I thank Dr. Thomas Broker for critical reading of the manuscript. I am also grateful to Ms. Marie Moschitta for typing the manuscript. This work has been supported by a Cancer Center Grant (CA13106) to Cold Spring Harbor Laboratory.

REFERENCES

Berg, D. E., J. Davies, B. Allet and J. D. Rochaix. 1975. Transposition of R factor genes to bacteriophage lambda. *Proc. Nat. Acad. Sci.* **72**:3628.

Chow, L. T. and N. Davidson. 1973. Electron microscope mapping of the distribution of ribosomal genes of the *Bacillus subtilis* chromosome. *J. Mol. Biol.* **75**:265.

Davidson, N., R. C. Deonier, S. Hu and E. Ohtsubo. 1975. Electron microscope heteroduplex studies of sequence relations among plasmids of *Escherichia coli.* X. Deoxyribonucleic acid sequence and organization of F and F-primes and the sequences involved in Hfr formation. In *Microbiology 1974* (ed. D. Schlessinger), p. 56. American Society for Microbiology, Washington, D.C.

Hsu, M.-T. and N. Davidson. 1974. Electron microscope heteroduplex study of the heterogeneity of Mu phage and prophage DNA. *Virology* **58**:229.

Hu, S., E. Ohtsubo, N. Davidson and H. Saedler. 1975. Electron microscope heteroduplex studies of sequence relations among bacterial plasmids. Identification and mapping of the insertion sequences IS1 and IS2 in F and R plasmids. *J. Bact.* **122**:764.

Kleckner, N., R. K. Chan, B.-K. Tye and D. Botstein. 1975. Mutagenesis by insertion of a drug-resistance element carrying an inverted repetition. *J. Mol. Biol.* **97**:561.

Lee, H.-J., E. Ohtsubo, R. C. Deonier and N. Davidson. 1974. Electron microscope heteroduplex studies of sequence relations among plasmids of *Escherichia coli.* V. *ilv*$^+$ deletion mutants of F14. *J. Mol. Biol.* **89**:585.

Ohtsubo, E., R. C. Deonier, H.-J. Lee and N. Davidson. 1974. Electron microscope heteroduplex studies of sequence relations among plasmids of *E. coli.* IV. The F sequences in F14. *J. Mol. Biol.* **89**:565.

Saedler, H. and B. Heiss. 1973. Multiple copies of the insertion DNA sequences IS1 and IS2 in the chromosome of *E. coli* K-12. *Mol. Gen. Genet.* **122**:267.

Chromosomal Rearrangements in the *gal* Region of *E. coli* K12 after Integration of IS*1*

H.-J. Reif* and H. Saedler†
Institut für Genetik
Universität zu Köln
D-5000 Köln 41, W. Germany

Formation of *gal* Deletions in *gal*:IS*1* Mutants

The DNA elements IS*1* and IS*2* are known to integrate into the *gal* operon of *Escherichia coli* (*E. coli*) K12 and lead to strong polar *gal* mutations (Jordan et al. 1967, 1968; Shapiro 1969; Hirsch et al. 1972a,b; Starlinger and Saedler 1972, 1976).

This reaction is reversible, and a functional *gal* operon is restored by exact excision of the insertion (Jordan et al. 1967). Furthermore, it has been shown that integration of IS*1* into the *gal* operon confers genetic instability in the region of the insertion such that *gal* deletions occur at a high frequency. Whereas in a wild-type *gal* operon deletions are formed with a frequency of approximately 10^{-7} per cell, some *gal*::IS*1* mutations give rise to deletions of adjacent DNA with frequencies up to 10^{-3} per cell (Reif and Saedler 1975).

IS*1*-induced deletion formation has been analyzed in some detail, and a model is proposed (Fig. 1) which attributes reversion and stimulated deletion formation to site-specific illegitimate recombinational events. The model is based on Campbell's (1962) model for the formation of specialized transducing phages.

Some essential features of IS*1*-induced deletion formation are as follows:

1. The exact excision of IS*1*, as well as the enhanced formation of deletions, is independent of the bacterial *recA* system and is thus defined as an illegitimate recombination event.
2. Exact excision by recombination between both ends of IS*1* is much less frequent than deletion formation. Whereas reversion of different *gal*::IS*1* mutations takes place at frequencies in the range of 10^{-6} to 10^{-7} per cell, the deletion formation is much more frequent and is usually in the range of 3×10^{-4} to 10^{-6} per cell.
3. Deletion formation is strongly dependent on growth temperature. When cells are grown at 42°C, the frequency of deletion formation relative to that of cells grown at 32°C is reduced. The extent of the reduction varies with

Present addresses: *Department of Biological Sciences, Stanford University, Stanford California 94305; †Institut für Biologie III, Universität Freiburg, Schaenzlestr. 9-11, 7800 Freiburg i. Br., W. Germany.

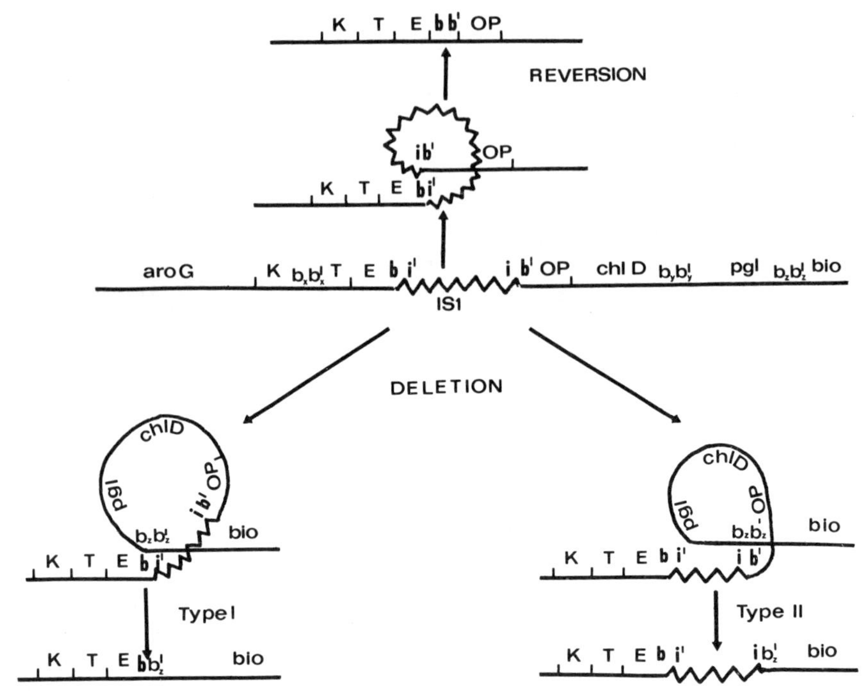

Figure 1
Model for reversion and enhanced deletion formation in *gal*::IS*1* mutations. The letters i and i′ are the termini of IS*1*, and $b_x b_x'$ are sequences in the *gal* region that can serve as substrate for deletion formation.

different strains and different deletions and is in the range of 10- to 1000-fold. The reversion of *gal*::IS*1* mutations by exact excision of the DNA element is independent of growth temperature.

4. There are several classes of deletions formed. One end point of most deletions is at the position where IS*1* was inserted in the *gal* operon. The other end point occurs at different sites, but these do not appear to be distributed randomly. This was shown by mapping of deletions terminating in the *gal* operon, a region well defined by deletion groups. There also appear to be preferential points for deletion termination outside the *gal* operon, although there are very few markers available for mapping these regions.
5. Several *gal*::IS*1* mutants, which differ with respect to the location of IS*1* within the *gal* operon as well as in the orientation of the insertions, were analyzed for their deletion patterns. It was observed that each *gal*::IS*1* mutant yielded a unique deletion pattern with respect to deletion frequencies and temperature dependence of formation of different deletion classes. This observation led us to postulate that the new DNA sequences created by integration of the IS element determine the deletion pattern.

To explain the above results, our model (Fig. 1) proposes that two types of deletions can occur depending on which of the two termini of the integrated IS*1* interacts with a given (potential) deletion end point adjacent to IS*1*. Interaction of the second deletion end point with the proximal terminus of IS*1* should result in a deletion that retains IS*1* in the *gal* region (type-II deletion), whereas interaction of the same deletion end point with the terminus of IS*1* distal to it should lead to a deletion that includes the insertion element (type-I deletion).

Retention of IS*1* in Deletions

In order to determine if both types of deletions described in the previous section occur, two types of experiments were performed: (1) recombination analysis and (2) studies of DNA heteroduplex molecules in the electron microscope.

Recombination Analysis

The recombination analysis is based on the observation that IS*1*-induced deletions can be found extending either to the left or to the right of the position of IS*1* in the *gal* operon (Reif and Saedler 1975). If the deletions are of type II (Fig. 1), then there should be common (IS*1*) sequences at the ends of both classes of deletions. In such common IS*1* sequences bordering the deletions, recombination to restore the parental *gal*::IS*1* operon should be possible in a cross between two such deletions. Such a cross is illustrated in Figure 2. Recom-

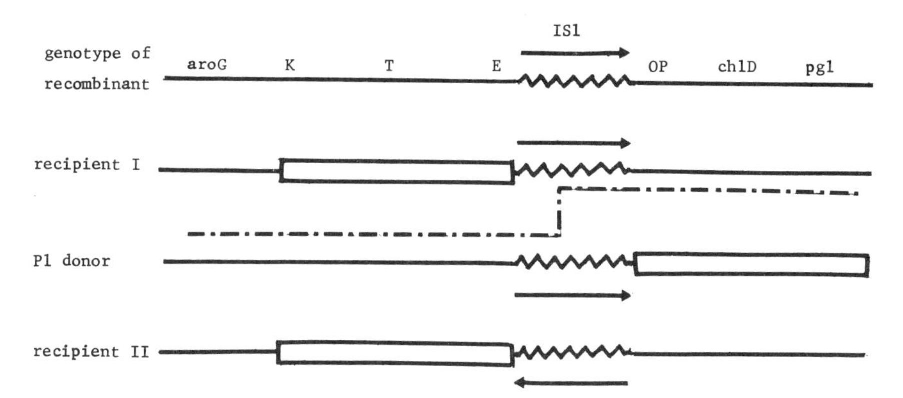

Figure 2
Recombination test. The symbol ʌʌʌ represents IS*1* DNA. The *galOP141*::IS*1* and *galOP306*::IS*1* mutations are located at the same position but differ in the orientation of the integrated IS*1*. The orientation of IS*1* is indicated by an arrow. Recombination to *gal*$^+$ can occur only when the IS elements have the same orientation. Since the recipient strains carry an *suA* allele, the polar effect of IS*1* integrated in the *galOP* region is partially relieved and the expression of the *gal* structural genes is possible, thus allowing growth of the recombinants on minimal galactose medium.

Table 1
Recombination Analysis

P1 donor (*gal-chlD* type deletions) from strains		Recipient *gal* deletions from strains 141			306					
		b1	c1	f1	h5	a10	c1	i18	d6	e8
141	2-9	0	0	0	0	0	0	0	0	0
	3-5	16	33	16	0	0	0	0	0	0
	1-3	29	38	19	0	0	0	0	0	0
	3-3	21	59	12	0	0	0	0	0	0
306	1-2	0	0	0	8	11	4	21	6	11
	2-5	0	0	0	15	17	6	26	8	13
	3-2	0	0	0	8	43	23	10	28	13
	2-1	0	0	0	19	24	16	18	15	15
	3-12	0	0	0	0	4	3	4	n.t.[a]	n.t.

Strains 141, b1, . . . etc., and strains 306, h5, . . . etc., are independent isolates of IS*1*-induced *gal* deletions derived from strains *galOP141*::IS*1* and *galOP306*::IS*1*, respectively. These deletion strains, which lack all or part of the *gal* structural genes, were transduced with P1*kc* phage grown on independent isolates of IS*1*-induced *galOP-chlD* or *galOP-pgl* deletions (first column, 141, 2-9, . . . etc., and 306, 1-2, . . . etc.) derived from strains *galOP141* and *galOP306*::IS*1* as indicated (cf. Fig. 2).

The values listed are averages of at least two independent experiments. They represent recombinants per 10^9 P1 phages. Multiplicity of infection in the transductions was 1. The transduction mixture was plated on minimal galactose agar on which the selected genotype, *galOP*::IS*1*, can grow because an *suA* allele in the recipient relieves the insertion polarity. *Gal*$^+$ colonies were isolated and tested for *chlD* and *pgl* markers as well as for their phenotype on MacConkey *gal* indicator plates. All *gal*$^+$ strains tested showed the Gal$^+$ phenotype characteristics of *galOP141, suA* and *galOP306, suA*. Strains 141 and 306 differ only in the orientation of the IS*1* element in the *gal* operon (Hirsch et al. 1972a,b).

[a]n.t. = not tested.

bination to the parental genotype cannot occur if the two crossed deletions are derived from two parents that differ with regard to the orientation of IS*1* in the *gal* operon.

The results of such crosses are given in Table 1. No recombination occurred in the control experiments. However, when deletions derived from the same parent, but which extended either to the right or to the left of the position of IS*1*, were crossed with each other, recombination to the parental genotype was detected in all but one case. This indicates that in most IS*1*-induced deletions the insertion is retained at its position in the *gal* region.

Analysis of DNA Heteroduplex Molecules

The results of the genetic experiments were confirmed by DNA heteroduplex analysis. If IS*1* is present in IS*1*-induced deletions, this extra DNA should appear in an electron micrograph.

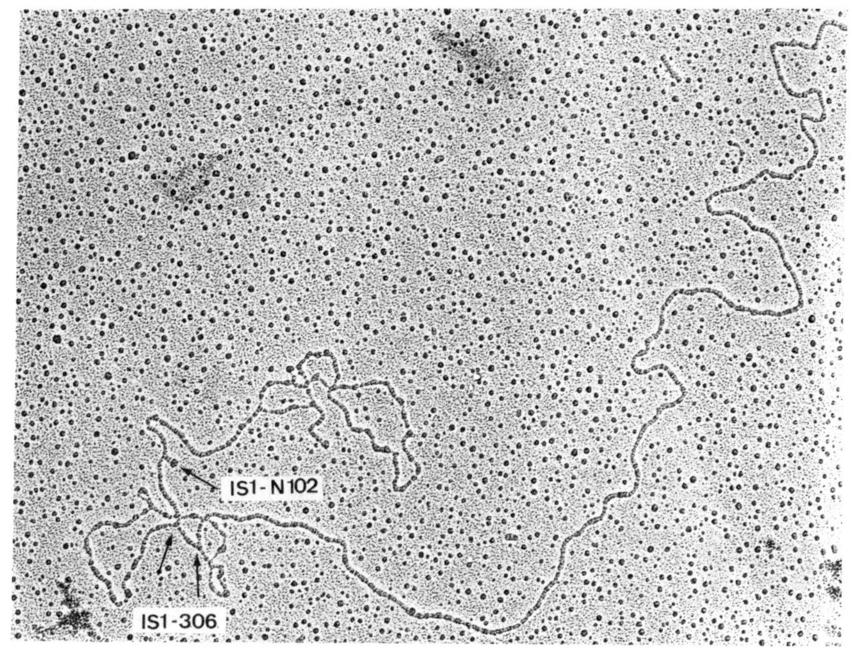

Figure 3
Electron micrograph of a heteroduplex molecule λd*galT102*::IS*1*/λdΔ(*galOP306* ::IS*1-pgl*)*nadA*. (λd*galT102* is N102 mutation.)

We hybridized DNA from a specialized transducing λd*gal-nadA* phage carrying an IS*1*-induced deletion with the DNA of a λd*gal* phage. The DNA heteroduplex molecules should show a single-stranded DNA loop corresponding to the IS*1*-induced deletion. Presence of IS*1* at the border of the deletion would be expected to result in a single-stranded large loop opposed by another single-stranded region of the size of IS*1*, whereas absence of IS*1* from the deletion should result in only a single-stranded deletion loop. Figure 3 shows one of the DNA heteroduplexes actually found. It is apparent that a large part, if not all, of the insertion is retained in the *gal* operon after deletion formation.

Continued Genetic Instability in IS*1*-induced Deletions

Since the above experiments clearly show the presence of the insertion element in the IS*1*-induced deletions, we asked whether the deletions show the same genetic instability as the *gal*::IS*1* parent. To test this, the frequencies of formation of further secondary deletions in the primary deletion strains were measured by selection for loss of outside markers such as *tolAB* and *chlD*. In nearly all primary deletion strains tested, further secondary deletions occurred with high frequencies on either side of the retained insertion (H.-J. Reif and H. Saedler, unpubl.).

In only a few cases were secondary deletions found less frequently than in the parental *gal*::IS*1* strain. In these few deletion strains, however, the formation of secondary *gal-tolAB* deletions is more frequent than in the gal^+ wild type, where this class of deletion was not detected.

Similar results were found with primary deletions from all *gal*::IS*1* mutants investigated. Furthermore, it could be shown that even the secondary deletion strains are not stable, since, in those strains, further deletions in the *gal* region are formed with elevated frequencies typical of IS*1*-induced deletions.

Thus we conclude that deletions, if they remove anything at all from the IS*1* sequence in *gal*, do not remove those parts of IS*1* responsible for deletion production. We do not know whether type-I deletions occur infrequently or whether they were just not found with the selection screening procedure employed. Therefore we are not certain whether the model presented for IS*1*-dependent deletion formation (which demands the occurrence of both types of deletions) requires modification. The number of copies of IS*1* in different *E. coli* strains is quite constant, even though the insertion elements apparently are excised and transposed (Saedler and Heiss 1973). Thus there may exist an efficient integration system within the bacterial cell which counteracts loss of IS*1*. In a type-I deletion process, a deletion as well as a DNA fragment containing IS*1* is formed. It is conceivable that since this fragment contains IS*1*, it frequently reintegrates into the chromosome. Such an event would not have been detected with the methods employed for isolation of deletions.

Deletions Fusing the *gal* Genes to a New Promoter

Deletions extending to the right of the integrated IS*1* sequence were isolated from *galOP*::IS*1* mutants. These deletions remove the *galOP* region together with the adjacent *chlD* genes and sometimes additional, more distant genes. These strains are Gal^-, but the *galE* gene is expressed to a low extent since phage U3 (Watson and Paigen 1971; Reif and Saedler 1975) can grow on most of these strains

It was observed that some of the strains in which deletions removed the *galOP* region can revert with low frequency (10^{-7} revertants per cell) to a Gal^+ phenotype. Those reversions must be caused by the fusion of the *gal* genes to a new promoter in the vicinity of the *gal* operon. This type of reversion is not found in all IS*1*-induced *galOP* deletions. Long deletions terminating beyond the *uvrB* gene and some *galOP-pgl* deletions also did not revert, although they did display the low *galE* expression observed in the other *galOP* deletions. By contrast, all *galOP-chlD* deletions tested showed reversion. The revertants were not tested quantitatively for enzyme content or inducibility. However, the *gal* enzymes do not seem to be fully expressed, since the strains grow only slowly on minimal galactose plates and display only a pink color on MacConkey galactose indicator plates compared to the dark red color of gal^+ wild-type colonies.

When analyzed for further deletion formation, the revertants had regained genetic stability. No *tolAB*⁻, *gal*⁻ colonies were detected, even though more than 10^9 cells were plated under selective conditions. Thus this class of revertants is as stable as the *gal*⁺ wild type and may have lost the IS*1* element from the *gal* region.

There are several events that could conceivably account for the formation of the revertants:

1. An exact excision of IS*1* allows expression of the *gal* genes that were already fused to a new promoter by the primary deletion.
2. A secondary type-II deletion excises IS*1* and adjacent DNA, fusing the *gal* genes to a new promoter.
3. Integration of IS*2* in orientation II can lead to expression of the *gal* genes fused to the promoter on IS*2* (Saedler et al. 1974).

Since a *gal* operon with either IS*1* and/or IS*2* in orientation II would be genetically unstable (H. Saedler and H.-J. Reif, unpubl.), we believe that possibilities 1 and 2 are more likely. It is hoped that DNA heteroduplex studies presently underway will reveal the nature of the reversions.

Formation of Deletions in *galOP*::IS*1* *galT*::IS*1* Double Mutants

In many transposons, insertion sequences are found in two copies bordering the transposed region (see also other articles in this book). We therefore analyzed the deletion formation in strains with two IS*1* elements in the *gal* operon. We constructed strains that contained a *galOP*::IS*1* insertion and a second copy of IS*1* integrated in the *galT* gene. Deletions were isolated after selection of strains that had lost the *galE* gene bordered by the insertions. The deletions were analyzed for outside markers.

Two types of *galT*::IS*1*, *galOP*::IS*1* double mutants were constructed: the *galE* gene and part of the *galT* gene were flanked by two IS*1* elements either as direct or as inverted repeats. The frequencies of deletion formation were measured in *recA* derivatives (results are shown in Table 2). In the three strains investigated, deletions in the *gal* region occurred with frequencies of roughly 0.1%. This is the highest frequency for IS*1*-induced deletion formation yet observed. The high frequency is independent of the relative orientation of the two IS*1* elements in *gal*. However, differences between the two types of strains were observed with respect to the effect of growth temperature on deletion formation. In strains with IS*1* in *galT* and *galOP* as direct repeats, deletion formation did not show a growth-temperature dependence, whereas in strains with IS*1* integrated in *gal* as inverted repeats, growth of the cells at 42°C reduced the formation of deletions 18-fold. This reduction is not as pronounced as that found in *gal*::IS*1* mutations (Reif and Saedler 1975), but obviously the temperature effect is still present.

Table 2
Frequencies of Deletions in *galOP*::IS*1*, *galT*::IS*1*, *recA* Double Mutants

Strain	IS*1* orientation	*Gal* deletions per cell × 10^3		
		32°C	42°C	32°C/42°C
141, 116 *recA*	→ →	0.8	0.3	2.5
306, 102 *recA*	← ←	3.5	1.6	2.1
141, 102 *recA*	← →	0.9	0.05	18

The data shown are averages of three independent experiments. Column 2 gives the relative orientation of the two IS*1* elements in the *gal* operon.

The frequencies of deletion formation of roughly 0.1% in the double mutants after growth at 32°C are higher than the value expected from addition of the deletion frequencies of the corresponding single mutants. This indicates that the two copies of IS*1* may interact, resulting in even more pronounced genetic instability than observed in a single *gal*::IS*1* mutation. Further evidence for this interpretation comes from preliminary mapping of deletions derived from *galOP*::IS*1*, *galT*::IS*1* strains. These deletions were tested for loss of outside markers such as *aroG, nadA, chlD,* and *pgl.* Contrary to the results with *gal*::IS*1* single mutations, which mostly yield deletions covering markers outside the *gal* operon (Reif and Saedler 1975), the majority of the deletions from the double mutants terminate within or near the *gal* operon and do not cover adjacent markers (Table 3).

This result and the reduced or absent temperature dependence indicate that different substrates or enzymes for deletion formation may be involved. It might be speculated that the temperature independence of deletion formation in strains with the insertions as direct repeats in the *gal* operon is based on the interaction of the termini of both IS*1* elements themselves. Exact excision (i.e., reaction between the ends of one IS*1* element) is independent of growth temperature. It is conceivable that a similar reaction takes place in *galT*::IS*1*, *galOP*::IS*1* double mutants, with IS*1* in direct repeat using one end from the *galT*::IS*1* and the other from the *galOP*::IS*1* instead of both ends of one insertion. Such an exact excision reaction would in this case lead to a deletion of the *galE* gene located between the two insertions. A similar reaction cannot take place between inversely repeated IS*1*. This could be the reason for the residual influence of growth temperature in this strain.

We did not determine whether the deleted material transposes to another site on the chromosome. Moreover, as mentioned earlier, the selection and screening method employed would not have detected cells in which transposition of deleted material had occurred.

Table 3

Ratio of Frequencies of Deletions Terminating Inside and Outside the *gal* Operon

		Ratio Δint./Δext.	
Strain	Orientation of IS*1* copies	measured 32°C (42°C)	calculated 32°C (42°C)
141, 116	→ →	4.0 (4.9)	0.43 (0.12)
306, 102	← ←	2.4 (99)	0.23 (0.28)
141, 102	← →	0.8 (1.5)	0.12 (0.1)

Ratio of frequencies of deletions terminating within or outside the *gal* operon from *galOP*::IS*1*, *galT*::IS*1* double mutants. The relative orientation of the integrated IS*1* elements is given in column 2. The calculated values are based on previously published results (Reif and Saedler 1975) and are derived by addition of deletion frequencies of the corresponding *gal*::IS*1* single mutants.

Genetic Instability of Spontaneously Formed Deletions

Since IS*1* occurs in several copies at different sites in the *E. coli* chromosome (Saedler and Heiss 1973), it is conceivable that some spontaneous deletions may be caused by IS*1*, either by integration of the DNA element and subsequent deletion, or by an integrative deletion, as described by Fan (1969) for integration of F′ factors into the chromosome.

If this type of event occurred, some of the spontaneous deletions would be expected to have properties of IS*1*-induced deletions and to show temperature-sensitive genetic instability.

Spontaneous deletions arising with a frequency of 3×10^{-7} per cell in a galactose-sensitive *galT* mutant have been described previously (Reif and Saedler 1975). Eleven of these deletions which have termini within the *gal* operon were analyzed for genetic instability by measuring deletion formation. Four of them showed a high level of formation of both *tolAB-nadA* and *chlD-pgl* deletions at 32°C, typical of *gal*::IS*1* strains. The enhanced deletion formation was temperature-dependent and decreased 40–250-fold after growth of the cells at 42°C. The majority of the secondary deletions showed continued instability. Highly unstable deletions were not found among ten spontaneous *gal-chlD* deletions derived from a gal^+ wild-type strain nor in several *gal-uvrB* deletions generated by induction of a λ prophage (Pfeifer et al. 1974).

Further experiments are required to verify the hypothesis that IS*1*-induced deletion formation accounts for the formation of some spontaneous deletions. Since they were found in only one of several strains analyzed, the possibility remains that the isolation of genetically unstable deletions was due to the genetic constitution of the particular strain used.

CONCLUSION

Following integration of IS*1* into the *gal* operon of *E. coli* K12, deletions in the neighborhood of IS*1* occur with high frequency. Nearly all deletions terminate at the position of IS*1* but do not remove it. This results in continued genetic instability in that chromosomal region and causes further deletion formation to occur. Only infrequently is IS*1* removed from the *gal* operon and genetic stability regained.

The chromosomal rearrangements generated by successive deletion formation can, on occasion, result in a fusion of the *gal* genes to a new promoter.

The highest frequencies yet observed for IS*1*-induced deletions occur when two copies of IS*1* are integrated into the *gal* operon and loss of the genes bordered by the two copies of IS*1* is selected.

A small fraction of spontaneous deletions show characteristics of IS*1*-induced deletions. This may indicate that insertion sequences are also involved in the formation of some of the spontaneous deletions.

Acknowledgments

We are grateful to Drs. K. Brown and C. Yanofsky for reading and improving the manuscript. This work was supported by the Deutsche Forschungsgemeinschaft through SFB 74.

REFERENCES

Campbell, A. 1962. Episomes. *Adv. Genet.* **11**:101.

Fan, D. P. 1969. Deletions in limited homology recombination in *E. coli. Genetics* **61**:351.

Hirsch, H. J., H. Saedler and P. Starlinger. 1972a. Insertion mutations in the control region of the *gal* operon in *E. coli.* II. Physical characterization of the mutations. *Mol. Gen. Genet.* **115**:266.

Hirsch, H. J., P. Starlinger and P. Brachet. 1972b. Two kinds of insertions in bacterial genes. *Mol. Gen. Genet.* **119**:191.

Jordan, E., H. Saedler and P. Starlinger. 1967. Strong polar mutations in the transferase gene of the *gal* operon in *E. coli. Mol. Gen. Genet.* **100**:296.

———. 1968. O° and strong polar mutations in the *gal* operon are insertions. *Mol. Gen. Genet.* **102**:353.

Pfeifer, D., J. J. Hirsch, D. Bergmann and M. Hamlaoui. 1974. Nonrandom distribution of endpoints of deletions of the *gal* region. *Mol. Gen. Genet.* **132**: 203.

Reif, H.-J. and H. Saedler. 1975. IS1 is involved in deletion formation in the *gal* region of *E. coli* K12. *Mol. Gen. Genet.* **137**:17.

Saedler, H. and B. Heiss. 1973. Multiple copies of the insertion DNA sequences IS1 and IS2 in the chromosome of *E. coli* K12. *Mol. Gen. Genet.* **122**:267.

Saedler, H., H.-J. Reif, S. Hu and N. Davidson. 1974. IS2, a genetic element for turn-off and turn-on of gene activity in *E. coli. Mol. Gen. Genet.* **132**:265.

Shapiro, J. A. 1969. Mutations caused by the insertion of genetic material into the *gal* operon of *E. coli. J. Mol. Biol.* **40**:93.

Starlinger, P. and H. Saedler. 1972. Insertion mutations in microorganisms. *Biochimie* **54**:177.

——. 1976. IS-elements in microorganisms. In *Current topics in microbiology and immunology,* vol. 75, p. 111, Springer-Verlag, Berlin.

Watson, G. and K. Paigen. 1971. Isolation and characterization of an *E. coli* bacteriophage requiring cell wall galactose. *J. Virol.* **8**:669.

Polarity of Insertion Mutations Is Caused by Rho-mediated Termination of Transcription

A. Das, D. Court, M. Gottesman and S. Adhya
Laboratory of Molecular Biology
National Cancer Institute
Bethesda, Maryland 20014

Polarity created by the presence of an IS element in an operon is manifested by a severe decrease in the expression of gene(s) located distal to the insertion mutation. IS elements were first described in the *gal* operon of *E. coli* (Jordan et al. 1968; Adhya and Shapiro 1969; Shapiro 1969). The *gal* operon contains three structural genes, *E, T,* and *K* (see Fig. 1); insertions in *E* or *T* are polar on *K*.

In vivo, the *gal* operon is transcribed as a unit, with termination of transcription occurring beyond *K* (Hill and Echols 1966; Guha et al. 1968). In vitro, termination of transcription within the operon can be produced by the addition of high concentrations of the transcription termination factor rho. Furthermore, the presence of IS*2* (but not IS*1*) insertions causes the transcription termination at the insertion, even at low levels of rho.

Although various other models have been proposed to explain polarity (see Zipser 1969; Morse and Guertin 1972), we will present evidence from our work on *gal* that IS polarity, like nonsense polarity, is caused principally by premature termination of transcription at rho-sensitive sites within operons.

Transcription Termination and Polarity

We have proposed a general model which describes how transcription termination and polarity are related (Adhya et al. 1974; S. Adhya, M. Gottesman, A. Das, B. de Crombrugghe and D. Court, in prep.). Briefly, the model states that after ribosomes dissociate from the nascent mRNA beyond a polypeptide chain termination codon (UAG, UAA, and UGA), transcription continues to the next rho site on the DNA. There rho acts to terminate transcription. If ribosomes reinitiate protein synthesis beyond the nonsense codon, and before the rho site, rho cannot work, and transcription elongates beyond the site. Polarity results when a nonsense mutation triggers ribosomal dissociation in a region devoid of translation restart codons. In this case, cistrons beyond the next rho site are not transcribed.

Two lines of evidence support the idea that polarity is the result of rho-mediated transcription termination: (1) A phage function which overrides known rho-sensitive termination signals also releases IS or nonsense polarity. (2) Mutations in the bacterial gene that codes for rho abolish polarity.

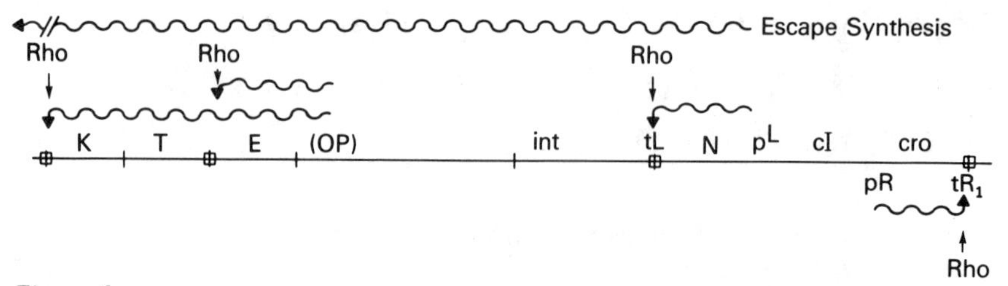

Figure 1
Transcription map of *gal* operon and prophage lambda. *E, T,* and *K* represent genes of the *gal* operon. Arrows show directions of transcription initiated at three different promoters: P_{gal}, λp_L and λp_R. Rho protein provokes transcription termination at the sites represented by ⊞. The "escape mRNA" is extended at least as far as the *gal* operon.

λN *Gene Product Overrides Transcription Termination Barriers*

The expression of the coliphage λ genome is almost entirely dependent on its *N* gene product. Rho provokes termination of transcription at specific sites on λ DNA template in vitro (Roberts 1969; see Fig. 1). In vivo, in the absence of *N* function, transcription also stops at these sites (Heinemann and Spiegelman 1971; M. Rosenberg and D. Court, in prep.). In the presence of *N* function, prophage transcription overrides these rho-sensitive termination barriers and continues into the neighboring bacterial genes. This phenomenon, known as escape synthesis, leads to constitutive expression of the *gal* enzymes whose synthesis is otherwise controlled by the *gal* operator-promoter (Adhya et al. 1974, 1975). IS elements in *gal*, which are polar when transcribed from the *gal* promoter, are nonpolar when transcribed from the λ promoter in the presence of *N*. Assuming that *N* functions uniquely as an antiterminator of transcription, it follows that insertional polarity is caused by transcription termination.

Absence of Termination in rho *Mutants*

One of the ways to identify elements that cause insertional polarity is to look for bacterial mutations that suppress this polarity. We have isolated suppressor mutations which relieve polarity caused by an IS*2* insertion in the *gal* operon (see Fig. 2). Some, if not all, of these suppressors arise by a mutation in *rho*: (1) These mutations, which render the cell temperature-sensitive (ts) for growth, map between the *ilv* and *cya* loci of the chromosome. (2) Rho protein, purified from one of these *ts* mutants, is defective in transcription termination on a λ DNA template.

Results presented in Table 1 show that the rho mutation suppresses the polar effects not only of an IS*2* element but also of all other types of polar mutations tested in the *gal* operon. Other polarity suppressor mutations have been described which also affect rho (Richardson et al. 1975; Korn and Yanofsky

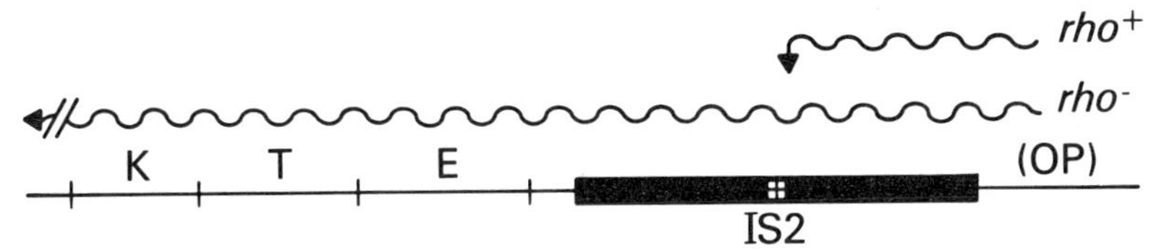

Figure 2
Schematic representation of the principle of isolation of *rho* mutant. It has been shown that an IS*2* element carries a strong rho-sensitive transcription termination site (de Crombrugghe et al. 1973). A strain (gal_3) carrying an IS*2* between the operator-promoter loci and the *E* gene is Gal^- because the transcription terminates within the IS*2* at the site indicated by ⊞. A mutation in *rho* (the gene for rho protein) may make the strain Gal^+ by allowing the transcription to continue beyond ⊞ into *E, T,* and *K* genes. Following this principle, *rho* mutants have been isolated as nitrosoguanidine-induced Gal^+ revertants of gal_3 (Das et al. 1976).

1976; Ratner 1976). These mutations are not lethal and suppress IS*1* but not IS*2* polarity (Reyes et al. 1976). Note that IS*1* (700 base pairs) and IS*2* (1400 base pairs) have different base sequences (Fiandt et al. 1972).

IS2 Polarity

The IS*2* element carries within itself a strong rho-sensitive site (de Crombrugghe et al. 1973). Within the framework of our model, we assume that the rho site in IS*2* is preceded by one or more nonsense codons which stop translation and allow rho action.

Table 1
Release of Polarity by *rho* Mutation

Relevant genotype	Type of polar mutation	Level of galactokinase
gal^+ rho^+		10.0
gal^+ rho^-		11.5
$galE_{am57}$ rho^+	amber	0.5
$galE_{am57}$ rho^-	amber	11.4
galOP306 rho^+	IS*1*-I	0.1
galOP306 rho^-	IS*1*-I	16.0
galOP128 rho^+	IS*1*-II	0.1
galOP128 rho^-	IS*1*-II	14.0
gal_3 rho^+	IS*2*-I	0.1
gal_3 rho^-	IS*2*-I	12.0

The *ts15* mutation (Das et al. 1976) was transferred into strain SA500 (F^- his^- str^r sup^o) carrying the various *gal* mutations shown. Induction of galactokinase, the product of *galK* gene, and galactokinase assay were as described previously (Adhya et al. 1974). Roman numerals I and II signify the two orientations of IS elements.

*IS*1 *Polarity*

The cause of the polar effect of IS*1*, which shows polarity in both orientations, is less obvious. The IS*1* insertion does not contain a detectable rho-sensitive termination site in vitro (de Crombrugghe et al. 1973). Although the possibility remains that IS*1* DNA carries a transcription termination site which needs another factor besides rho, we feel that IS*1* exerts its polar effect by introducing nonsense codons. After polypeptide chain termination at these codons, transcription is terminated at the next rho site in the operon, creating polarity on cistrons distal to that site. We believe that IS*1* elements might be more polar than a nonsense mutation because they introduce more than one nonsense triplet and do not contain a translational reinitiation signal. The former would explain why a tRNA suppressor does not detectably release IS*1* polarity (S. Adhya, unpubl. obs.). Since the IS*1* element exerts its characteristic strong polarity only on cistrons distal to the next rho termination site, the pattern of polarity of these elements in a given operon should reflect the distribution of rho sites in that operon. This may explain why the IS*1* elements are nonpolar in a ribosomal protein operon (Jaskunas and Nomura, this volume).

REFERENCES

Adhya, S. and J. Shapiro. 1969. The galactose operon of *E. coli* K12. I. Structural and pleiotropic mutations. *Genetics* **62**:231.

Adhya, S., M. Gottesman and B. de Crombrugghe. 1974. Release of polarity in *Escherichia coli* by gene *N* of phage λ: Termination and antitermination of transcription. *Proc. Nat. Acad. Sci.* **71**:2534.

Adhya, S., D. Court, B. de Crombrugghe and M. Gottesman. 1975. Changes in transcription of host genes after bacteriophage lambda induction. In *Proceedings of the 10th FEBS Meeting,* FEBS, p. 69.

Das, A., D. Court and S. Adhya. 1976. Isolation and characterization of conditional lethal mutants of *Escherichia coli* defective in transcription termination factor rho. *Proc. Nat. Acad. Sci.* **73**:1959.

de Crombrugghe, B., S. Adhya, M. Gottesman and I. Pastan. 1973. Effect of rho on transcription of bacterial operons. *Nature New Biol.* **241**:260.

Fiandt, M., W. Szybalski and M. H. Malamy. 1972. Polar mutations in *lac, gal* and phage λ consist of a few IS-DNA sequences inserted with either orientation. *Mol. Gen. Genet.* **119**:223.

Guha, A., M. Tabuczynski and W. Szybalski. 1968. Orientation of transcription of the galactose operon as determined by hybridization of *gal* mRNA with the separated DNA strands of coliphage λdg. *J. Mol. Biol.* **35**:207.

Heinemann, S.F. and W. G. Spiegelman. 1971. Role of the gene *N* product in phage lambda. *Cold Spring Harbor Symp. Quant. Biol.* **35**:315.

Hill, C. W. and H. Echols. 1966. Properties of a mutant blocked in inducibility of messenger RNA for the galactose operon of *Escherichia coli. J. Mol. Biol.* **19**:38.

Jordan, E., H. Saedler and P. Starlinger. 1968. O° and strong polar mutations in the *gal* operon are insertions. *Mol. Gen. Genet.* **102**:353.

Korn, L. J. and C. Yanofsky. 1976. Polarity suppressors increase expression of the wild-type tryptophan operon of *E. coli. J. Mol. Biol.* **103**:395.

Morse, D. E. and M. Guertin. 1972. Amber suA mutations which relieve polarity. *J. Mol. Biol.* **63**:605.

Ratner, D. 1976. Evidence that mutations in the *suA* polarity suppressing gene directly affect termination factor rho. *Nature* **259**:151.

Reyes, O., M. Gottesman and S. Adhya. 1976. Suppression of polarity of insertion mutations in the *gal* operon and *N* mutation in bacteriophage lambda. *J. Bact.* **126**:1108.

Richardson, J. P., C. Grimley and C. Lowery. 1975. Transcription termination factor rho activity is altered in *Escherichia coli* with suA mutations. *Proc. Nat. Acad. Sci.* **72**:1725.

Roberts, J. W. 1969. Termination factor for RNA synthesis. *Nature* **224**:1168.

Shapiro, J. A. 1969. Mutations caused by the insertion of genetic material into the galactose operon of *Escherichia coli. J. Mol. Biol.* **40**:93.

Zipser, D. 1969. Polar mutations and operon function. *Nature* **221**:21.

Isolation of Mutations in Insertion Sequences That Relieve IS-induced Polarity

P. K. Tomich and D. I. Friedman
Department of Microbiology
University of Michigan
Ann Arbor, Michigan 48109

Insertion sequences (IS) are known to cause polarity in a variety of bacterial operons (Jordan et al. 1968; Shapiro 1969; Malamy 1970). The polarity induced by one insertion (IS*2* in orientation I) appears to be caused by a Rho-sensitive termination site (de Crombrugghe et al. 1973). A variant of phage λ (λ*r*32) which carries IS*2* in the *y* region (Fig. 1) has been isolated by Brachet et al. (1970).

We exploited the following characteristics of λ*r*32 in the isolation of mutations that relieve IS-induced polarity: (1) The IS sequence is located in the *y* region, resulting in λ*r*32 making a clear plaque (Brachet et al. 1970). (2) The added DNA results in a phage particle that is more sensitive to inactivation by divalent metal chelating agents (Parkinson and Huskey 1971). (3) The *r*32-IS*2* element exerts observable polarity on the expression of λ functions only when the phage *N* function is not expressed (Brachet et al. 1970). The *N* function appears to act by permitting transcription which initiates at λ-specific promoters to transcend termination barriers; i.e., *N* acts as an antiterminator (Roberts 1969). There have been a series of studies consistent with a model of *N* action (Friedman et al. 1973) which postulates that *N* is specifically recognized at the early λ promoters p_R and p_L where it modifies RNA polymerase. The modified polymerase can then transcend termination signals (Franklin 1974; Adhya et al. 1974; Friedman et al. 1976b). In the absence of *N* activity, λ*r* variants, unlike r^+, cannot express genes lying distal to the IS insertion, particularly replication genes *O* and *P* (see Fig. 1).

The basis for our selection of mutations *r*elieving *I*S *p*olarity (*rip*) is the observation that λ*r* variants do not grow in host mutants that reduce *N* expression (*nus* mutants) under conditions where λr^+ phage do grow. Two such mutants

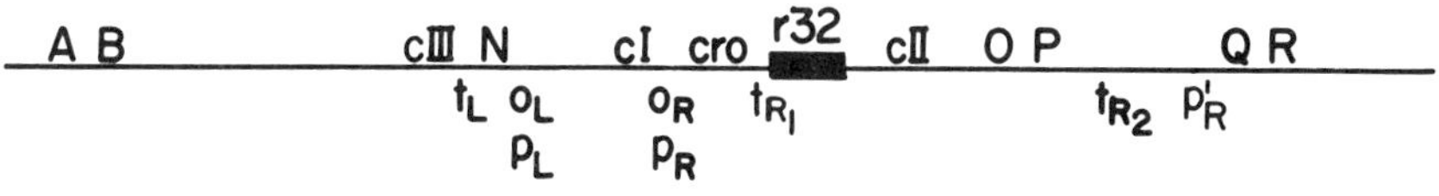

A B cIII N cI cro r32 cII O P Q R
t_L o_L o_R t_{R_1} t_{R_2} p'_R
p_L p_R

Figure 1
Genetic map of λ showing the location of the *r*32-IS*2* element (t represents transcription termination signals).

were used in this study: K-95, which carries the *nusA*1 mutation (mapping at minute 68[1]) (Friedman and Baron 1974), and K-450, which carries the *nusB*5 mutation (mapping at minute 11[1]) (Friedman et al. 1976a) (see Table 1). These hosts exhibit similar phenotypes; λ plates at low temperatures (e.g., 32°C) but not at high temperatures (e.g., 42°C).

Presumably, the additional termination site in the insertion sequence, coupled with reduced *N* activity, prevents efficient expression of the genes located distal to the insertion. Therefore, variants of λ*r* that have lost the IS-associated polarity should plate on *nus* bacteria at 32°C.

Selection Procedure

λ*c*I857*r*32 was mutagenized by making a plate lysate using the mutator host strain K-481 (*mutD*). The *c*I857 mutation results in the synthesis of a temperature-sensitive repressor; a phage carrying this mutation plates turbid at low temperatures and clear at high temperatures (Sussman and Jacob 1962). The lysate grown on the *mutD* strain was plated on a mixed lawn of *nusA*1 and *nusB*5 at 32°C. Mutants that formed plaques on this mixed lawn and maintained

Table 1
Bacterial and Phage Strain

Bacteria[a]	Relevant genotype	Source
K-95	*nusA*1	this laboratory
K-450	*nusB*5	this laboratory
K-481	*mutD*	L. Enquist
K-125	Su°, $(\lambda^{++})_n$	this laboratory
SA500	*gal*$^+$	S. Adhya
OR1008	*galOP308*	S. Adhya
FZ14	*dnaB*ts; *recA*	I. Herskowitz
K-377	a λ lysogen derived from FZ14	this laboratory
K-203	C600 Su^+_2, $(\lambda^{++})_n$	this laboratory
Phage	**Genotype**	**Source**
λ*r*32	λ*int*6*c*I857*r*32	this laboratory
	(derived from *Nam*7, 53λ*c*I857*r*32)	W. Szybalski
λ*r*32*rip*2	λ*int*6*c*I857*r*32*rip*2	this laboratory
$\lambda imm^{21}P^-$	$\lambda imm^{21}Pam3$	M. Yarmolinsky
$\lambda imm^{434}Q^-R^-$	$\lambda bio11imm^{434}Qam117Ram60$	this laboratory
$\lambda N^-v_1v_3$	$\lambda Nam7,53v_1v_3$	M. Ptashne
λ*c*I*c*17		M. Yarmolinsky

[a] All bacterial strains were derivatives of *E. coli* K12.

[1] This mapping is based on the 100-min *E. coli* linkage map (Bachmann et al. 1976).

a clear morphology at 32°C were screened for pyrophosphate sensitivity. These characteristics were taken as presumptive evidence that the mutant still carried at least part of the *r*32 insertion.

A priori we might expect to find several classes of λ*r*32 variants in which the IS-induced polarity has been overcome. First are those in which the IS element has been lost. As discussed above, this class is eliminated by our screening procedure. Second are variants that carry mutations in λ genes. Results from previous studies in this laboratory (unpubl.) indicate that these mutations are most likely to be found in the *N* gene, and they can be selected against by using a mixed lawn of the *nusA*1 and *nusB*5 mutants. Third are variants that carry constitutive promoters in the insertion. Since previous studies have shown that the presence of a constitutive promoter in the *y* region interferes with λ growth in the *nus* hosts (Friedman et al. 1976b), the selection process should select against this class. A fourth class of λ*r*32 mutants that should be effectively selected by this process are those carrying mutations in the IS which relieve insertion polarity. This report will focus on one such mutant selected in this manner, namely, λ*r*32*rip*2.

Evidence That the *rip* Mutation Does Not Result in a New Promoter Activity

If the Rip effect is due to the generation of a constitutive promoter in IS*2*, then the genes directly distal to that sequence (e.g., replication genes *O* and *P*) should be constitutively expressed. That this is not the case is demonstrated by the following complementation experiment. A Su° λ lysogen (K-125) was coinfected with either λ*r*32*rip*2 or λ*r*32 and with a hybrid phage carrying the immunity of phage 21, λimm^{21}*Pam*3. Under these conditions, none of the phages can replicate alone; the λ phages cannot because they are repressed, and the λimm^{21} phage cannot because it lacks *P* gene function. However, if the *rip*2 mutation results in the formation of activation of a promoter in the inserted DNA, then, even in the presence of repression, the λ*r*32 should express *P* function, permitting replication of the λimm^{21} phage in the complementation experiment. The validity of this test is shown by the observation that a λ variant, λ*c*I*c*17[2], which carries a constitutive promoter in the *y* gene, can supply *P* function to the λ$imm^{21}P^-$ phage (Table 2). However, the λ*r*32*rip*2 phage cannot supply *P* function to the imm^{21} phage (Table 2), demonstrating that there cannot be a constitutive promoter in the inserted DNA.

Formally, the *rip* mutation might generate another type of promoter that would not be selected against by our procedure, one permitting *N* utilization. In the presence of *N*, transcription initiating from such a promoter should be

[2] A multiple λ lysogen, K-125, was used to overcome the virulent effect of λ*c*I*c*17 (Shimada et al. 1972).

Table 2
Complementation for *P* Gene Product

Multiple lambda lysogen superinfected by	$\lambda imm^{21}Pam3$ per cell
$\lambda cIc17 + \lambda imm^{21}Pam3$	15 ± 3 (3)
$\lambda cI857r32 + \lambda imm^{21}Pam3$	3 ± 1 (3)
$\lambda cI857r32rip2 + \lambda imm^{21}Pam3$	2 ± 2 (3)

The multiple λ lysogen, K-125, was grown at 37°C in tryptone broth containing 0.2% maltose, 0.01 M $MgSO_4$ and 0.10% B_1 (TB). Logarithmically growing cells (~10^8/ml) were infected with the appropriate phage at a multiplicity of 5. In each case, phage and adsorption was >85%. Infected cells were diluted into TB and incubated at 37°C, aliquots were removed at 0 and 120 min postinfection, treated with chloroform, and titered on a Su^+_2 λ lysogen, K-203. Each infection was repeated three times.

able to transcend termination signals and therefore would not be expected to interfere with the growth of λ in *nus* mutants (Friedman et al. 1976b).

We tested for the presence of such a promoter, using a modification of an assay system developed by Couturier et al. (1973; Friedman and Ponce-Campos 1975). This assay utilizes the level of λ endolysin synthesis as a measure of *N* activity. Although the gene (*R*) which codes for endolysin is under the control of the *Q* regulation function, *Q* synthesis is under *N* control (Herskowitz and Signer 1971; Couturier et al. 1973). Therefore, the amount of synthesis of endolysin indirectly reflects the level of *N* expressed. Accordingly, a λ lysogen is coinfected with the λ variant to be tested and with $\lambda imm^{434}Q^-R^-$. The system works as follows: The λ immunity represses transcription from the early p_R promoter (see Fig. 1) and insures that any observed endolysin synthesis results from transcription initiating from an alternative promoter. Since the 434 phage is Q^-R^-, it can supply *N* in *trans* without supplying *Q* or directing the synthesis of endolysin. Moreover, the fact that $\lambda N^- v_1 v_3$, a phage in which the p_R promoter cannot be repressed, is able to synthesize endolysin when *N* is supplied in *trans* (Table 3) demonstrates that the system is functional. As shown in Table 3, neither λ*r*32 nor λ*r*32*rip*2 can direct endolysin synthesis in this assay system. Therefore, we conclude that neither of these phages carries an *N*-utilizing promoter.

Mapping of the *rip*2 Mutation

To demonstrate conclusively that *rip*2 actually maps in the IS*2* element, we crossed the *rip*2 mutation into an IS*2* element located in a bacterial operon. The bacterial strain OR1008 carries an IS*2* element in the galactose operator-promoter region (*galOP308*). The *rip*2 mutation was crossed into the *gal* IS*2* element in a two-step process devised by M. Gottesman, S. Adhya and O. Reyes (pers. comm.) (see Fig. 2).

First, the bacteria were lysogenized with the λ*int*6cI857*r*32*rip*2 mutant. Since

Table 3
Expression of Endolysin under Conditions of Replication Inhibition and Repression

K-377 at 39°C superinfected by	Endolysin units
$\lambda cI857r32$	<1
$\lambda cI857r32rip2$	<1
$\lambda imm^{434}Q^-R^-$	<1
$\lambda N^-\nu_1\nu_3$	<1
$\lambda cI857r32 + \lambda imm^{434}Q^-R^-$	<1
$\lambda cI857r32rip2 + \lambda imm^{434}Q^-R^-$	<1
$\lambda N^-\nu_1\nu_3 + \lambda imm^{434}Q^-R^-$	27.6

The details of these experiments are presented in the text. In each case, phage adsorption was >90%. Extracts were made from aliquots removed 90 min postinfection. Endolysin activity was measured as described by Friedman and Ponce-Campos (1975).

the phage is *int*⁻, it will not lysogenize in the usual manner, but it should lysogenize by a Rec-promoted recombination between the $\lambda r32$ and any IS*2* element on the bacterial chromosome. Lysogens carrying the $\lambda cI857r32$ derivatives in the *galOP* region can be detected by replica-plating onto MacConkey galactose plates and pulse-heating at 42°C. If the phage has integrated into the Gal operator-promoter region, galactose genes will be expressed by transcription initiating from the λ promoter, and the indicator plate will turn red. This occurs because the λ prophage carries a temperature-sensitive repressor that denatures at 42°C (Sussman and Jacob 1962). Lysogens of either

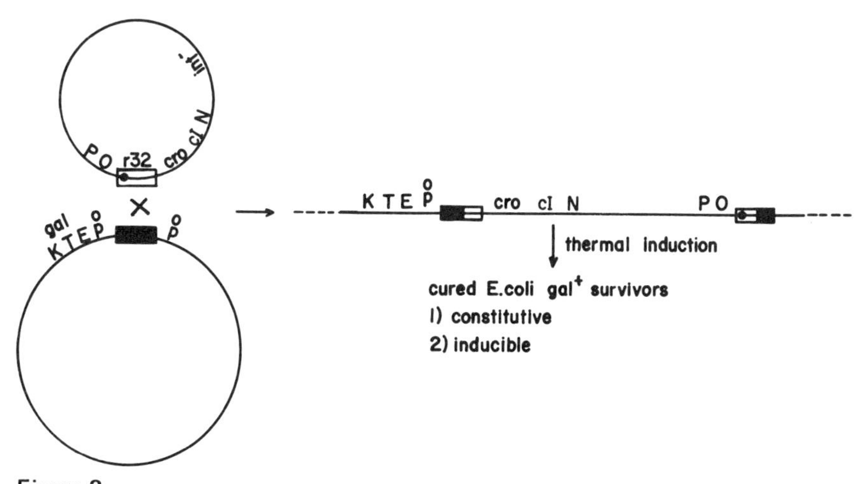

Figure 2
Mechanism of recombining IS sequence in λ with IS sequence in galactose operon. See text for discussion of method.

λ*int*6*c*I857*r*32 or λ*int*6*c*I857*r*32*rip*2 were selected using the EMBO plate method (Gottesman and Yarmolinsky 1968), and they occurred at a frequency of $\sim 10^{-6}$. Lysogenization at the normal *att* site on the *E. coli* chromosome is much rarer, as shown by the observation that, under identical conditions, no lysogens were found when λ*int*6*c*I857 ($< 2 \times 10^{-8}$) was used.

Second, bacteria cured of the λ*r*32*rip*2 phage should have lost the phage and regenerated the IS element by recombination. Depending upon where the recombination occurs and provided that *rip* is located in the IS, some of the regenerated bacterial IS elements should contain the *rip* mutation (Fig. 2). Cured bacteria are easily obtained by looking for heat-resistant survivors. Since the prophage carries a heat-sensitive repressor, all bacteria carrying the prophage will be killed at higher temperatures. If the original IS is generated, then the bacterium will be Gal^-. If an IS carrying *rip*2 is generated, then the bacterium would be expected to be Gal^+.

Based on these arguments, our analysis of cured lysogens demonstrates that the *rip*2 mutation can be crossed from one IS*2* to another. The frequency of Gal^+ cured cells is $\sim 10^4$-fold higher for cured λ*int*6*r*32*rip*2 lysogens than for λ*int*6*r*32 lysogens (Table 4). This implies that the Gal^+ survivors are a direct result of the *rip* mutation.

If the *rip*2 mutation merely acts to relieve the polarity induced by the IS element, then synthesis of galactose enzymes by the cured Gal^+ derivatives should be under the control of the *gal* operator-promoter; i.e., they should be induced by the gratuitous inducer α-D-fucose. Evidence of this is presented in Table 5. First, the data show that the two control strains function properly: SA500, which contains no insertion sequences, exhibits a 15- to 20-fold induction, and a derivative strain, OR1008, which carries the *galOP308* insertion, shows no induction.

Second, fucose-inducible Gal^+ bacteria cured of the λ prophage (Isolate 1) can be isolated from the λ*r*32*rip*2 lysogen. The relatively low inducibility observed is not unexpected. Although *rip*2 relieves polarity, the remaining insertion sequence in the galactose operator-promoter might affect inducibility

Table 4
Galactose-utilizing Survivors

Strain	Gal^+/Gal^- surviving at 40°C
OR1008	$<10^{-6}$
OR1008(λ*c*I857*r*32)	10^{-5}–10^{-4}
OR1008(λ*c*I857*r*32*rip*2)	20–80%

Bacteria were grown at 32°C to a density of 5×10^7 cells/ml in glucose-M-9 minimal medium supplemented with casamino acids. Each culture was shifted to 41°C and grown with aeration for 90 min. Aliquots were plated on MacConkey galactose plates at 42°C.

Table 5
Inducibility of Galactose Operon

Culture	Galactokinase activity	
	no fucose	+5 mM fucose
SA500	1.9 ± 0.4	26 ± 7
OR1008	0.1 ± 0.1	0.5 ± 0.3
Isolate 1	1.3 ± 1.0	8.4 ± 4.1
Isolate 2	4.5 ± 1.0	4.5 ± 0.1

Bacteria were grown overnight in M-9 medium containing 0.2% glycerol and 0.2% casamino acids. Cultures were diluted and grown in duplicate for 2 hr. One duplicate was grown in the presence of α-D-fucose. Cells were harvested and galactokinase assayed according to the method of Wilson and Hogness (1966).

of the operon. That there are still insertion sequences in the *galOP* region of Isolate 1 is shown by the following experiment. Isolate 1 can be lysogenized with λ*int*6*r*32, yielding Gal⁻ lysogens. This means that the phage is integrating in the *gal* operon, thus showing that there must be homology between the λ*r*32 phage and the galactose operon, an homology which can only be due to the IS*2* element common to both the phage and the galactose operon.

Finally, another class of Gal⁺ cured bacteria was obtained; those represented by Isolate 2 and which are constitutive for galactokinase synthesis. This class of bacteria is also isolated only when the λ*r*32 prophage carries the *rip*2 mutation. One plausible explanation for this is that this class results from deletions that remove bacterial and prophage genes fusing the *gal* operon with the IS element to another promoter. Accordingly, the presence of the *rip*2 mutation relieves the IS polar effect. Such a deletion effect has been observed at high levels for an IS*1* element (Reif and Saedler 1975) and occurs at low levels for IS*2* (Ahmed, this volume).

SUMMARY

We have devised a selection procedure for isolating mutations in insertion sequences that relieve IS-induced polarity. The procedure involves the use of λ variants (λ*r*) carrying IS in an early operon. This report focuses on the characterization of one mutation, *rip*2. Genetic studies have located the mutation in the IS element, and functional studies have shown that the polarity relief does not result from the generation of a promoter in the insertion sequence. One likely explanation for this is that the *rip*2 mutation eliminates the transcription termination signal in the IS*2* element.

Acknowledgments

We would like to thank Oscar Reyes for supplying bacterial strains and Justine Posner for excellent help in preparing the manuscript. We also thank Ranka Ponce-Campos and Paula Holton for technical assistance. These studies were supported by grants from the National Institute of Allergy and Infectious Diseases (1R0 1 A111459-01 and 1 F32 AI05144-01) and from the National Science Foundation (GB-41719).

REFERENCES

Adhya, S., M. Gottesman and B. de Crombrugghe. 1974. Release of polarity in *Escherichia coli* by gene *N* of phage λ: Termination and antitermination of transcription. *Proc. Nat. Acad. Sci.* **71**:2534.

Bachmann, B., K. B. Lew and A. L. Taylor. 1976. Recalibrated linkage map of *Escherichia coli* K12. *Bact. Rev.* **40**:116.

Brachet, P., H. Eisen and A. Rambach. 1970. Mutations of coliphage λ affecting the expression of replicative functions O and P. *Mol. Gen. Genet.* **108**:266.

Couturier, M., C. Dambly and R. Thomas. 1973. Control of development in temperate bacteriophages. *Mol. Gen. Genet.* **120**:231.

de Crombrugghe, B., S. Adhya, M. Gottesman and I. Pastan. 1973. Effect of rho on transcription of bacterial operons. *Nature New Biol.* **241**:260.

Franklin, N. C. 1974. Altered reading of genetic signals fused to the operon of bacteriophage λ: Genetic evidence for modification of polymerase by the protein product of the *N* gene. *J. Mol. Biol.* **89**:33.

Friedman, D. I. and L. S. Baron. 1974. Genetic characterization of a bacterial locus involved in the activity of the *N* function of phage λ. *Virology* **58**:141.

Freidman, D. I. and R. Ponce-Campos. 1975. Differential effect of phage regulator functions on transcription from various promoters: Evidence that the P22 gene 24 and the λ gene *N* products distinguish three classes of promoters. *J. Mol. Biol.* **98**:537.

Friedman, D. I., M. F. Baumann and L. S. Baron. 1976a. Cooperative effects of bacterial mutations affecting λ*N* gene expression. I. Isolation and characterization of *nus*B mutant. *Virology* **73**:119.

Friedman, D. I., G. S. Wilgus and R. J. Mural. 1973. Gene *N* regulator function of phage λ*imm*21: Evidence that a site of *N* action differs from a site of *N* recognition. *J. Mol. Biol.* **81**:505.

Friedman, D. I., C. A. Jolly, R. J. Mural, R. Ponce-Campos and M. F. Baumann. 1976b. Growth of λ variants with added or altered promoters in *N*-limiting bacterial mutants: Evidence that an *N* recognition site lies in the P_R promoter. *Virology* **71**:61.

Gottesman, M. and M. Yarmolinsky. 1968. Integration-negative mutants of bacteriophage lambda. *J. Mol. Biol.* **31**:487.

Herskowitz, I. and E. R. Signer. 1971. Control of transcription from the *r* strand of bacteriophage lambda. *Cold Spring Harbor Symp. Quant. Biol.* **35**:355.

Jordan, E., H. Saedler and P. Starlinger. 1968. Strong polar mutations in the transferase gene of the galactose operon in *E. coli*. *Mol. Gen. Genet.* **102**:353.

Malamy, M. H. 1970. Some properties of insertion mutations in the *lac* operon of *E. coli.* In *The lactose operon* (ed. J. R. Beckwith and D. Zipser), p. 359. Cold Spring Harbor Laboratory, Cold Spring Harbor, New York.

Parkinson, J. S. and R. Huskey. 1971. Deletion mutants of bacteriophage lambda. I. Isolation and initial characterization. *J. Mol. Biol.* **56**:369.

Reif, H.-J. and H. Saedler. 1975. IS*1* is involved in deletion formation in the *gal* region of *E. coli* K12. *Mol. Gen. Genet.* **137**:17.

Roberts, J. W. 1969. Termination factor for RNA synthesis. *Nature* **224**:1168.

Shapiro, J. A. 1969. Mutations caused by the insertion of genetic material into the galactose operon of *Escherichia coli. J. Mol. Biol.* **40**:93.

Shimada, K., R. A. Weisberg and M. E. Gottesman. 1972. Prophage lambda at unusual chromosomal location. I. Location of the secondary attachment sites and the properties of the lysogens. *J. Mol. Biol.* **63**:483.

Sussman, R. and F. Jacob. 1962. Sur un systéme de répression thermosensible chez le bactériophage λ d' *Escherichia coli. Compt. Rend. Acad. Sci.* **254**: 1517.

Wilson, D. and D. Hogness.1966. Galactokinase and uridine diphospho-galactose 4-epimerase from *Escherichia coli.* In *Methods in enzymology* (ed. E. Neufeld and V. Ginsburg), vol. 8, p. 229. Academic Press, New York.

The Isolation of IS*1* and IS2 DNA

F. Schmidt, J. Besemer and P. Starlinger
Institut für Genetik
Universität zu Köln
D-5000 Köln-41, W. Germany

A variety of questions concerning the structure and function of IS DNA will be solved only by the technique of DNA sequence analysis. The specificity of integration and the structure of the integration site are among these problems. Since pure IS DNA would be a useful tool in such a study, we have undertaken to purify IS*1* and IS*2* DNA (Schmidt et al. 1976). To achieve this, λ and λd*gal* phages carrying IS*1* or IS*2* DNA were used. Since λ mutants carrying the IS elements in either orientation are available, we were able to employ the technique introduced by Shapiro et al. (1969) for the isolation of *lac* operon DNA.

PURIFICATION PROCEDURE

Heteroduplex molecules were prepared from identical strands of λ or λd*gal* DNA carrying the same IS element but in opposite orientation relative to λ DNA. The heteroduplex molecules were digested with endonuclease S1, as shown in Figure 1. Since single-strand preparations without IS elements subjected to the same digestion procedure are not digested to 100% TCA solubility, it must be concluded that some of the undigested material is not of IS origin. Therefore, the IS DNA had to be purified further.

This was done either by sucrose gradient centrifugation or by polyacrylamide gel electrophoresis. Peaks were pooled and recentrifuged, or bands were cut out, eluted, and rerun to show the purity of the material (see Figs. 2 and 3). As a control, the purification procedure, including annealing and S1 digestion, was repeated with DNA single strands that did not carry an IS element. When this DNA was subjected to sucrose gradient centrifugation or gel electrophoresis, no distinct peak or band could be detected.

Samples of tritiated IS*1* and IS*2* DNA were analyzed on sucrose gradients together with ^{14}C-labeled polyoma DNA (Fig. 2). The molecular weights of IS*1* and IS*2* were calculated by Studier's formula (Studier 1965). Lengths of 730 and 1220 base pairs were found for IS*1* and IS*2* DNA, respectively.

Samples of IS*1* and IS*2* DNA were run on analytical gels together with *Hin*dII cleavage products of λdv1. From these experiments (Fig. 3), the lengths of IS*1* and IS*2* were determined to be 850 and 1300 base pairs, respectively.

IS*1* and IS*2* DNA were also examined in the electron microscope. Electron

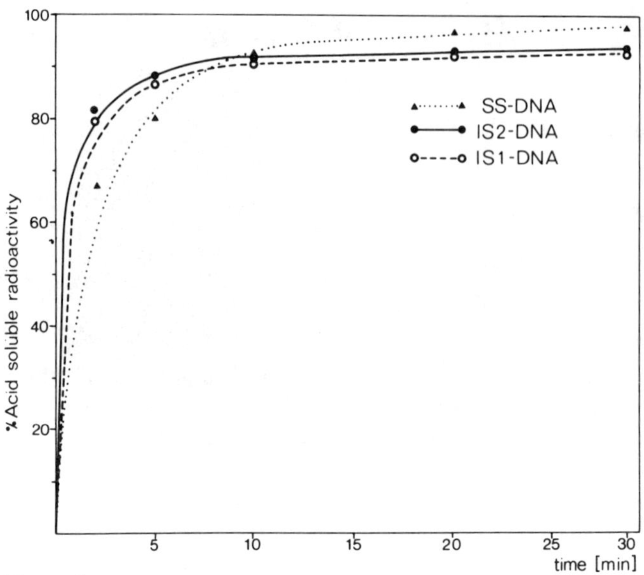

Figure 1
Digestion of single-stranded and heteroduplex DNA with endonuclease S1. The heteroduplex DNA is formed by annealing l strands of λd*galOP141*::IS*1* and λcII*r*14::IS*1* to yield IS*1* DNA and by annealing r strands of λd*galOP308*::IS*2* and λ*y-r*32::IS*2* to yield IS*2* DNA. The single strands were prepared according to the method of Shapiro et al. (1969). The heteroduplexes were formed as described by Hirsch et al. (1972). Digestion with endonuclease S1 was carried out in 0.4 N NaCl at 37°C and pH 5.0. Four units of enzyme were used, and the incubation time was 30 min.

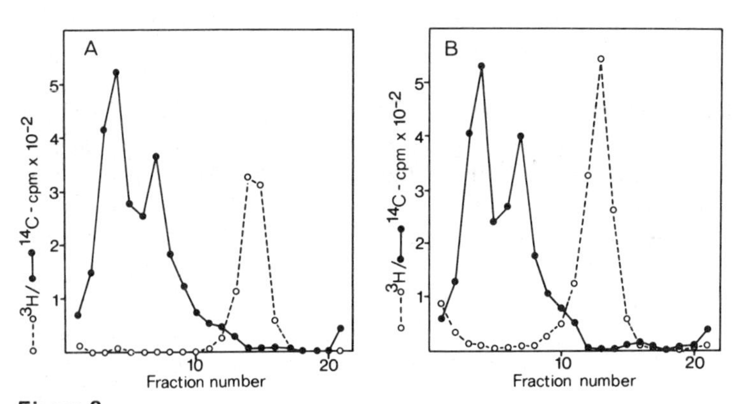

Figure 2
Sedimentation of IS*1* (*A*) and IS*2* (*B*) DNA in sucrose gradients. Either ^{3}H-labeled IS*1* or IS*2* DNA was run together with ^{14}C-labeled polyoma DNA. A 3.2-ml aliquot of a linear 5-20% sucrose gradient in 0.1 M NaCl, 0.1 M Tris HCl, pH 8, and 0.01 M EDTA was centrifuged for from 3 to 7 hr at 65,000 rpm at 4°C in an SW 65 K rotor in the Spinco ultracentrifuge.

Figure 3
Electrophoresis of IS*1* and IS*2* DNA on polyacrylamide gels. Samples of IS*1* and IS*2* DNA purified by preparative gel electrophoresis were electrophoresed together with *Hin*dII digestion products of λdvl (Streeck and Hobom 1975). (*a*) *Hin*dII cleavage products of λdvl; (*b*) same as *a* plus IS*1* DNA + IS*2* DNA; (*c*) IS*2* DNA; (*d*) IS*1* DNA. Conditions of electrophoresis and determination of molecular weight were as described by Streeck and Hobom (1975), with the exception of the buffer used for electrophoresis, which was as described by Peacock and Dingman (1968).

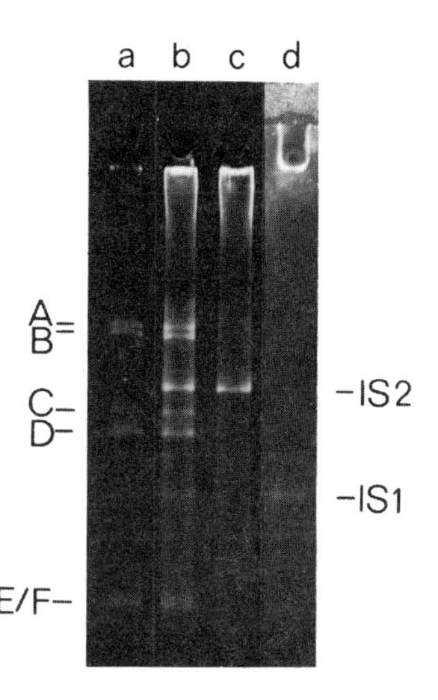

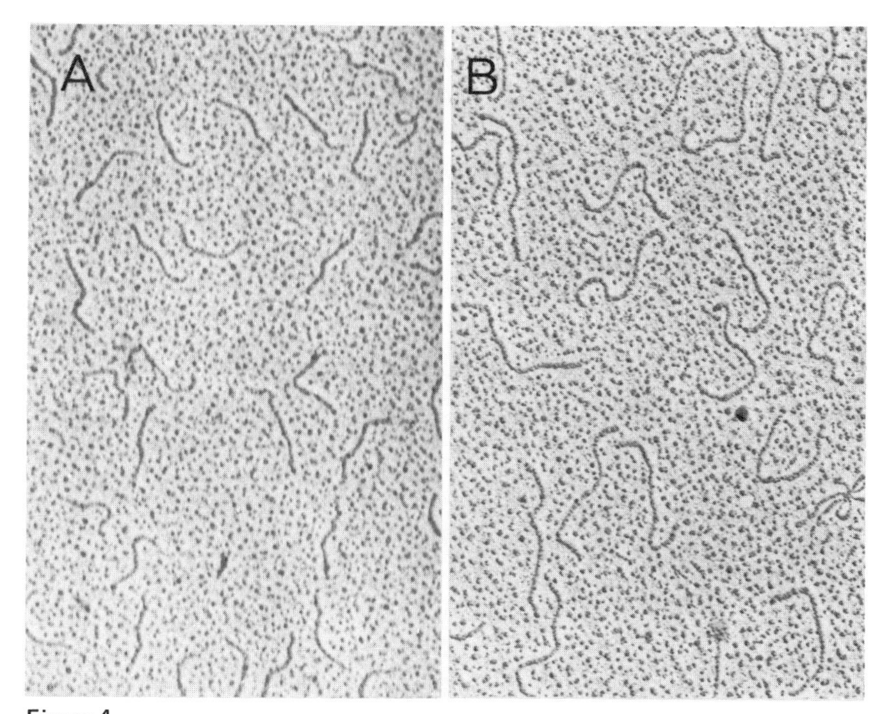

Figure 4
Electron micrographs of IS*1* (*A*) and IS*2* (*B*) DNA. Magnification, 17,800X. The IS DNA was purified by sucrose gradient centrifugation.

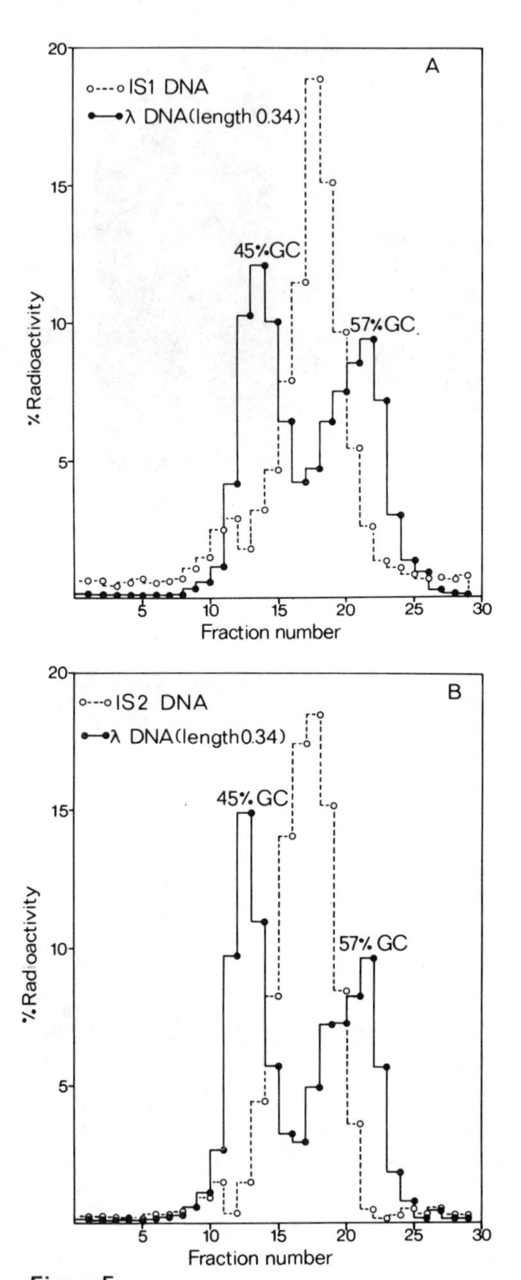

Figure 5

Equilibrium centrifugation of Hg complexes of IS*1* (*A*) and IS*2* (*B*) DNA in $(Cs)_2SO_4$. ^{3}H-labeled IS DNA was centrifuged together with sheared fragments of ^{32}P-labeled λ DNA. The λ DNA was sheared to an average fragment size of 0.34, as described by Skalka et al. (1968).

micrographs of IS*1* and IS*2* are shown in Figure 4. Length calibrations were obtained by measuring open circular SV40 DNA molecules. From these measurements, the length of isolated IS*1* DNA was found to be 820 ± 65 base pairs and that of isolated IS*2* DNA, 1350 ± 70 base pairs.

CHARACTERIZATION OF IS*1* AND IS*2* DNA

The base compositions of IS*1* and IS*2* were estimated from the densities of Hg complexes of these DNAs, labeled with [^{3}H] thymidine, together with fragments of ^{32}P-labeled λ DNA with an average length of 0.34 of the unbroken λ molecule. These complexes were sedimented to equilibrium in cesium sulfate gradients according to the method of Skalka et al. (1968). The results are shown in Figure 5. From the densities of both IS DNAs in an intermediate position between the two λ peaks, base compositions of approximately 50% GC were estimated.

Experiments with whole λ and λd*gal* DNA containing IS*1* and IS*2* in different positions had shown that IS*1* was not cleaved by the restriction nucleases *Eco*RI, *Hin*dII, or *Hin*dIII (J. Besemer and R. Streeck, data not shown).

Isolated IS*2* DNA is not cleaved by *Eco*RI but is cleaved once each by *Hin*dII and *Hin*dIII. *Hin*dII cleaves DNA subterminally, at approximately 11% from one of its ends. *Hin*dIII cleaves IS*2* once, approximately in the middle of the molecule.

The 5′ termini of IS*1* and IS*2* DNA were phosphorylated and labeled with ^{32}P by T4-induced polynucleotide kinase. Results are shown in Table 1. It is seen that G is the predominant nucleotide found.

Table 1
5′-Terminal Nucleotides of IS*1* and IS*2* DNA

	T7 – S1	T7 + S1	IS*1*(a)	IS*1*(b)	IS*2*(c)	IS*2*(d)
C	7.2	7.4	7.6	9.4	9.9	11.3
A	42.8	42	11.6	22.4	9.8	18.3
G	8.7	8.1	54.1	47.8	51.3	45.2
T	35.7	38.3	22.7	15.5	33.7	19.3
Total	94.4	95.8	96.0	95.1	93.7	94.2

IS*1*, IS*2*, and T7 DNA were each phosphorylated at the 5′ terminus, digested to mononucleotides, and separated by paper electrophoresis essentially as described by Maniatis et al. (1975). The paper was cut into 1-cm strips and counted. When the radioactivity of the strips was plotted against strip number, four distinct peaks were seen in the position where marker nucleotides were found. The radioactivity of the peaks was added and is given as percent of total radioactivity. The difference between the sum of the four peaks and total radioactivity is due to material retained at the origin and traces of radioactivity between the peaks.

CONCLUSIONS

From the results described, we see that a material homogeneous in size can be isolated from appropriate heteroduplex molecules after digestion with nuclease S1. The lengths of the molecules obtained are not different from those described for heteroduplex molecules prior to digestion. We can conclude, therefore, that the IS elements used for the pairwise production of the heteroduplex molecules are sufficiently similar to each other to prevent cleavage by endonuclease S1. Although it is still uncertain whether a single base mismatch allows the cleavage with S1, these experiments exclude the possibility of longer unpaired regions.

The length homogeneity at the nucleotide level cannot be assessed from the length measurements. The experiments with polynucleotide kinase show more than 50% of the label in G. The remaining label is distributed among the other three nucleotides. This indicates that, in our preparations, both IS*1* and IS*2* DNA have a G residue at the 5′ terminus. Because of the high radioactivity in the other three nucleotides, this will have to be confirmed by sequence analysis.

This material is apparently of sufficient purity to serve as a starting material for sequence studies.

Acknowledgments

We thank Dr. R. Streeck, München, for the digestion of DNA samples with *Hin*d enzymes and for the gifts of enzymes and λdvl cleavage products; Dr. E. Winnacker for the gift of labeled polyoma DNA; U. Reif for the gift of SV40 DNA; and Drs. B. Kemper and H. Chadwell for advice on nucleic acid end group analysis. This work was supported by the Deutsche Forschungsgemeinschaft through SFB 74.

REFERENCES

Hirsch, H. J., P. Starlinger, and P. Brachet. 1972. Two kinds of insertions in bacterial genes. *Mol. Gen. Genet.* **119**:191.

Maniatis, T., A. Jeffrey, and D. G. Kleid. 1975. Nucleotide sequence of the rightward operator of phage λ. *Proc. Nat. Acad. Sci.* **75**:1184.

Peacock, A. C. and C. W. Dingman. 1968. Molecular weight estimation and separation of ribonucleic acid by electrophoresis in agarose acrylamide composite gels. *Biochemistry* **7**:668.

Schmidt, F., J. Besemer and P. Starlinger. 1976. The isolation of IS*1* and IS*2* DNA. *Mol. Gen. Genet.* **145**:145.

Shapiro, J., L. MacHattie, L. Eron, K. Ippen, G. Ihler and J. Beckwith. 1969. Isolation of pure *lac* operon DNA. *Nature* **224**:768.

Skalka, A., E. Burgi and A. D. Hershey. 1968. Segmental distribution of nucleotides in the DNA of bacteriophage λ. *J. Mol. Biol.* **34**:1.

Streeck, R. E. and G. Hobom. 1975. Mapping of cleavage sites for restriction endonucleases in λdv plasmids. *Eur. J. Biochem.* **57**:595.

Studier, F. W. 1965. Sedimentation studies of the size and shape of DNA. *J. Mol. Biol.* **11**:373.

Physical Mapping of IS*1* by Restriction Endonucleases

N. D. F. Grindley
Department of Molecular Biophysics and Biochemistry
Yale University
New Haven, Connecticut 06510

As a preliminary to nucleotide sequence studies of the IS*1* insertion, I have constructed a restriction endonuclease cleavage map of λd*gal* carrying either of two independent IS*1* insertions in the *galT* gene. The λd*gal* DNA, with and without the IS*1* insertions, was prepared from the λ, λd*gal* double lysogens MS4-0, MS4-2, and MS4-3 (Fiandt et al. 1972). MS4-2 and MS4-3 carry the IS*1* insertions *galT104* and *galT188*,[1] respectively (Shapiro 1969); the insertions are in opposite polarity to each other.

The banding patterns of DNA fragments produced by cleavage of the three λd*gal* phage DNAs with a variety of restriction endonucleases[2] were compared on 1.4% agarose gels (Sugden et al. 1975). Analysis of these patterns indicates that the IS*1* element is not cut by *Eco*RI, *Hpa*I, *Hin*dIII, *Hin*cII, *Bam*I, *Bgl*I, *Bgl*II, *Xba*I, or *Xho*I; in each case, one DNA fragment of λd*gal* is replaced in λd*gal104* or λd*gal188* digests by a new band which is about 800 nucleotide pairs (n.p.) longer. Similar analyses show that the restriction endonucleases *Pst*I and *Bal*I each cut IS*1* once; a single band of λd*gal* DNA is replaced by two new bands in the IS*1*-containing DNAs.

Figure 1 shows the DNA banding patterns of the three phage DNAs following digestion with *Hin*II. Band D (about 2900 n.p.) of λd*gal* is replaced by a new band of about 3700 n.p. in digests of both λd*gal104* and λd*gal188*; the three patterns are otherwise identical. The *Hin*II D fragment therefore contains that region of the *galT* gene into which both insertions have been made; the two new bands of 3700 n.p. contain the same chromosomal region plus the IS*1* element of about 800 n.p. (see Fig. 2a).

[1] *galT104* and *galT188* are S104 and S188 mutations, respectively (Shapiro 1969).
[2] Restriction endonucleases are named as proposed by Smith and Nathans (1973). *Bam*HI is from *Bacillus amyloliquefaciens* H, *Bgl*I and *Bgl*II from *Bacillus globiggi*, *Bal*I from *Brevibacterium albidum*, *Eco*RI from *Escherichia coli* RY13, *Hae*III from *Haemophilus aegyptius*, *Hin*cII from *Haemophilus influenzae Rc*, *Hin*dIII from *Haemophilus influenzae Rd*, *Hpa*I and *Hpa*II from *Haemophilus parainfluenzae*, *Pst*I from *Providencia stuartii* 164, *Xba*I from *Xanthomonas badrii*, and *Xho*I from *Xanthomonas holicola*.

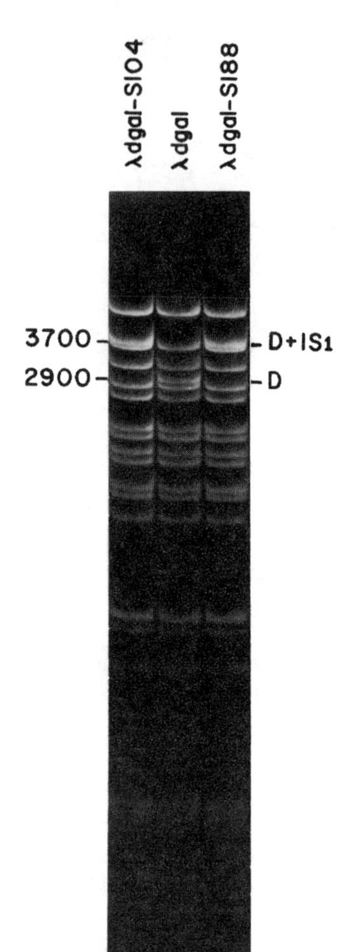

Figure 1
Electrophoresis of *Hin*II digests of DNA from λd*gal* and the two IS*1* mutants λd*gal104* and λd*gal188* on a 1.4% agarose gel. The gel contained ethidium bromide and was photographed under UV illumination. The *Hin*II D and *Hin*II D + IS*1* bands are labeled.

I have isolated the *Hin*II bands D and D + IS*1* from preparative 1.4% agarose or 3.5% polyacrylamide gels, labeled them with ^{32}P by nick translation with DNA polymerase I, and subjected them to further digestion with other restriction endonucleases. The reasoning used for analysis of the patterns obtained is as follows:

1. Bands common to λd*gal*, λd*gal104*, and λd*gal188* are from cuts within *gal* DNA and contain only *gal* sequences.
2. Bands common to λd*gal104* and λd*gal188* but absent from λd*gal* contain only IS*1* sequences.
3. Bands common to λd*gal* and one of the two insertion sequence mutants should contain the site of IS*1* insertion of the other mutant, assuming, of

course, that both insertions are not into the same restriction fragment; if they are into the same fragment, then λd*gal* will give a single, unique fragment containing both the insertion sites.

4. Bands unique to λd*gal104* or λd*gal188* should contain the boundaries between IS*1* and *gal* DNA; there should be two such bands from each insertion mutant, assuming there is at least one cut within IS*1*.

The actual sequence of successive restriction endonuclease digestions is shown schematically in Figure 2.

*Hae*II Cleavage Map of the *Hin*II Fragments Containing IS*1* Insertions

The banding patterns of DNA fragments obtained from *Hae*III digestion of the *Hin*II D and D + IS*1* fragments are shown in Figure 3. Using the reasoning outlined above, we can see that there are four bands (common to all three digests) containing *gal* sequences alone, namely, A_0, C_0, D_0, and E_0, and three bands (common to λd*gal104* and λd*gal188*) containing IS*1* sequences alone, namely, E_i, F_i, and G_i. The one remaining band from λd*gal*, B_0, is unique and thus contains both sites of IS*1* insertion. Each of the insertion mutants gives rise to two additional bands, B_2 and C_2 from λd*gal104* and B_3 and D_3 from λd*gal188*; these are unique bands and contain the junctions between *gal* and IS*1* DNA sequences.

A tentative map of the *Hae*III fragments is shown in Figure 2b. This map is based on analysis of *Hae*III partial digestion products and on *Hae*III digestion of the fragments obtained following cleavage of each *Hin*II D + IS*1* fragment with *Pst*I or *Bal*I, each of which makes a single cut in IS*1* and does not cut the *Hin*II D fragment of λd*gal*. (*Bal*I cuts a subset of the *Hae*III cleavage sites [R. J. Roberts, pers. comm.].)

The relative positions of the *Pst*I and *Bal*I cleavage sites on the maps are consistent with the known opposed polarities of the IS*1* elements in these two λd*gal* derivatives. The size of IS*1*, as calculated from the *Hae*III fragments, is $B_2 + E_i + F_i + G_i + C_2 - B_0 = B_3 + E_i + F_i + G_i + D_3 - B_0 = 820$ n.p.

*Hpa*II Cleavage Map of the *Hae*III Fragments Containing Sites of IS*1* Insertion

To locate further the IS*1*-*gal* junctions and to obtain smaller DNA fragments suitable for sequencing, I have mapped the *Hae*III fragments B_0 from λd*gal*, B_2 and C_2 from λd*gal104*, and B_3 and D_3 from λd*gal188* by cleavage with *Hpa*II. *Hpa*II digestion of these and also their parental *Hin*II fragments gives the banding patterns shown in Figure 4. The *Hpa*II cleavage map, shown in Figure 2c, is based on the following analysis.

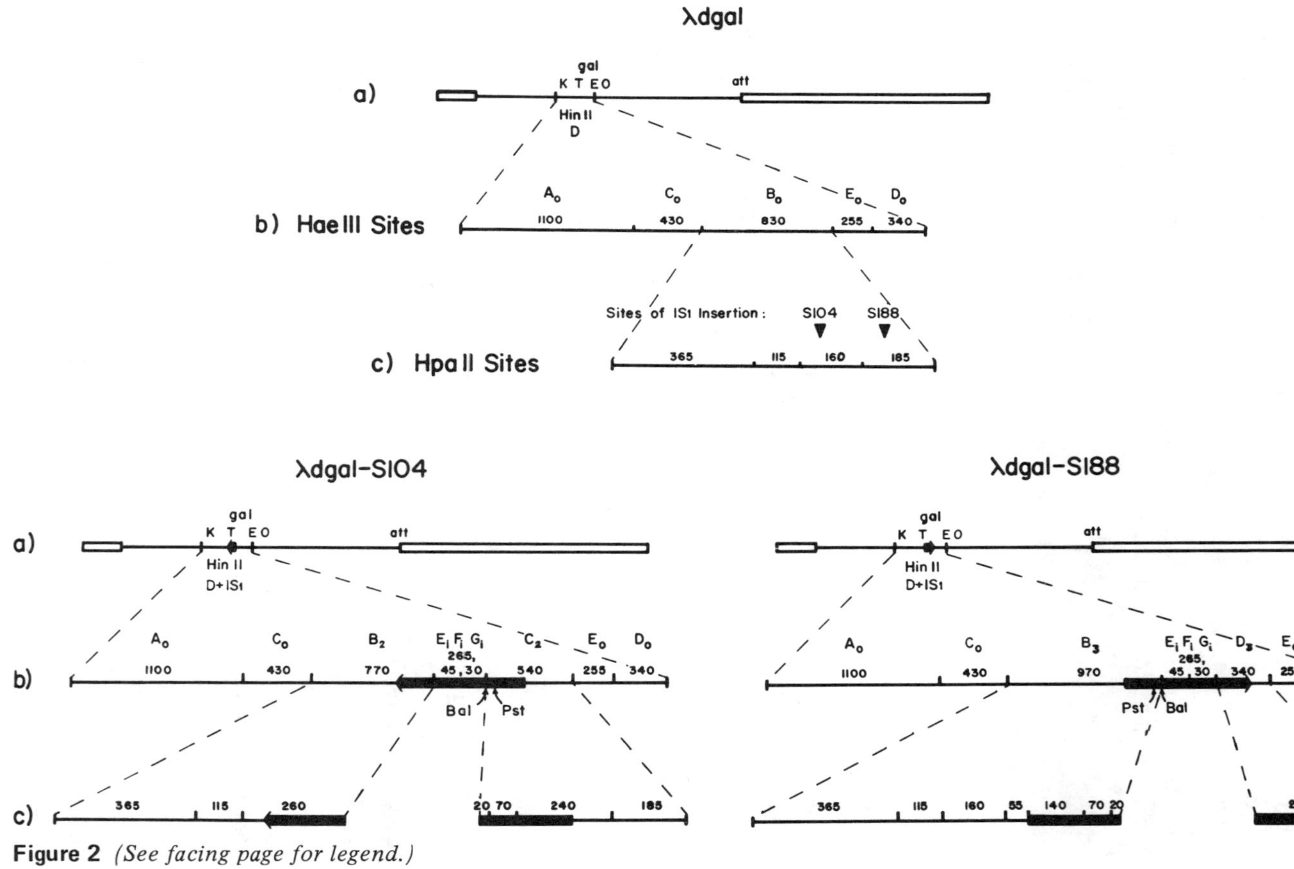

Figure 2 *(See facing page for legend.)*

The Hae*III Fragments B_2 and C_2 from* λ*d*gal104

*Hpa*II digestion produces three fragments from B_2 and four from C_2. Inspection of the banding patterns indicates that two of the B_2 bands (365 and 115 n.p.) are common to the digest of B_0 of λd*gal* and contain *gal* sequences alone, whereas the third band (260 n.p.) is unique and thus contains a *gal*-IS*1* boundary. Similarly, C_2 gives rise to one band containing only *gal* sequences (185 n.p.), two bands containing only IS*1* sequences (70 and 20 n.p., in common with the digest of the *Hae*III B_3 fragment of λd*gal188*), and one unique band (240 n.p.) containing the other *gal*-IS*1* junction of S104. The only band obtained from B_0 that is not found in the digests of B_2 or C_2 is the 160-n.p. band; this must contain the site of IS*1* insertion in *galT104*.

Two further observations allow the construction of the map in Figure 2c. First, the 20-n.p. band from C_2 is missing from the *Hpa*II digest of λd*gal104* *Hin*II D + IS*1*; therefore, it is a *Hpa*II-*Hae*III fragment and must come from the end of the *Hae*III fragment C_2. Second, the observation of a partial digestion product of about 280 n.p. from *Hpa*II cleavage of B_0 (and B_3) suggests that the 160-n.p. band of B_0 is adjacent to the 115-n.p. band rather than to the 365-n.p. band.

The Hae*III Fragments B_3 and D_3 of* λ*d*gal188

*Hpa*II gives seven bands from B_3 and two from D_3 (which is obtained by subtracting the D_0 banding pattern from that of the doublet D_3 + D_0). This total of nine bands is two more than the seven bands derived from B_2 and C_2 (which should contain exactly the same sequences as B_3 and D_3 but in a slightly different arrangement). This suggests that two new *Hpa*II recognition sites have been created in λd*gal188* or that two sites have been destroyed in λd*gal104*. Of the seven bands from B_3, three (365, 160, and 115 n.p.) contain *gal* sequences alone, whereas two (70 and 20 n.p.) consist only of IS*1* sequences The remaining two bands (140 and 55 n.p.) are unique. Examination of the two bands obtained from the digest of D_3 (210 and 135 n.p.) shows that neither of

Figure 2
Restriction endonuclease cleavage maps of IS*1* insertions in the *galT* gene of *E. coli.* (*a*) Mature phage DNA of λd*gal* and IS*1* insertion mutants S104 and S188, showing approximate location of the *Hin*II D and D + IS*1* fragments; λ sequences are shown as the double lines, *E. coli* chromosomal sequences as the single line. (*b*) Expanded view of the *Hin*II D and D + IS*1* fragments showing the sites cut by *Hae*III. Also shown are IS*1* (*heavy black arrow*) and the sites within IS*1* cleaved by *Pst*I and *Bal*I. (*c*) Expanded view of the *Hae*III fragments which contain the sites of IS*1* insertion, showing the sites cut by *Hpa*II. The positions of the two IS*1* insertions into the *galT* sequences of B_0 are indicated. Fragment sizes are in nucleotide pairs.

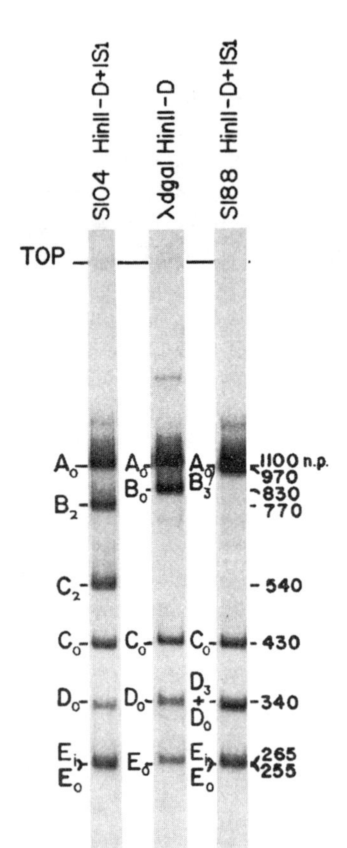

Figure 3

Autoradiogram showing *Hae*III digests of the *Hin*II D fragment of λd*gal* and of the *Hin*II D + IS*1* fragments of λd*gal104* and λd*gal188* after electrophoresis on a 5% polyacrylamide gel. The sizes of the fragments in nucleotide pairs were determined by comparison of their relative mobilities to those of *Hin*II fragments of φX174 (Lee and Sinsheimer 1974) on 3.5% and 5% polyacrylamide gels using the Tris-borate-EDTA buffer of Peacock and Dingman (1969) and on 7% gels using the Tris-borate-$MgCl_2$ buffer of Maniatis and Ptashne (1973). The fragments A_0 and B_3 are readily separated on 3.5% gels, whereas E_0 and E_i may be resolved on 7% gels.

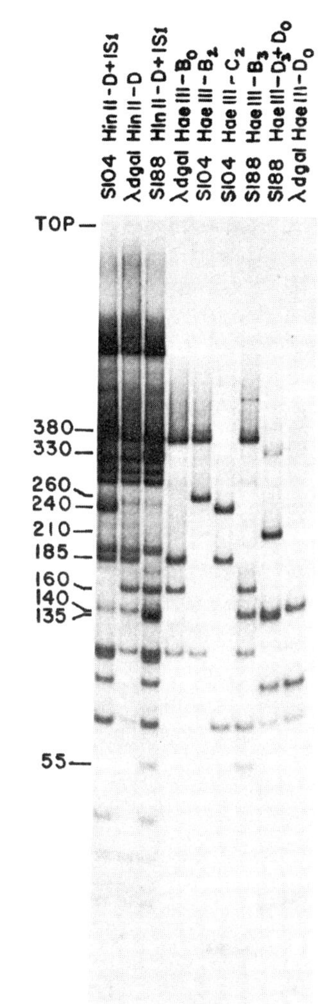

Figure 4
Autoradiogram showing *Hpa*II digests of the *Hin*II D and D + IS*1* fragments of λd*gal* and λd*gal104* and λd*gal188* and of the *Hae*III digestion products B_0, B_2, C_2, B_3, D_3 + D_0, and D_0 after electrophoresis on a 7% polyacrylamide gel.

these is found in the digests of B_0, B_2, or C_2; therefore, these are also unique fragments. Thus four unique fragments are generated by *Hpa*II digestion of the λd*gal188* *Hae*III fragments B_3 and D_3 instead of the expected two. This observation is supported by the finding that *Hpa*II digestion of the *Hin*II D + IS*1* fragment of λd*gal188* also gives four unique bands (330, 140, 140, and 55 n.p., see Fig. 4).

The simplest explanation of these results is that in λd*gal188,* insertion of IS*1* into *galT* has generated two new *Hpa*II recognition sites. Each of the two bands normally expected to carry the *gal*-IS*1* junctions would then be cut by *Hpa*II into two further bands, one containing *gal* sequences, the other IS*1* sequences. The *Hpa*II fragment of λd*gal* that contains the site of the S188 insertion is the

185-n.p. band of B_0. All the *gal* sequences of this 185-n.p. fragment should therefore be found in one of the unique *Hpa*II fragments from B_3 (140 and 55 n.p.) plus one of the two fragments from D_3 (210 and 135 n.p.). The only arrangement consistent with this argument is that the 55-n.p. fragment of B_3 and the 135-n.p. fragment of C_3 contain these *gal* sequences. If this argument is correct, then the 210-n.p. band of D_3 and the 140-n.p. band of B_3 contain only IS*1* sequences and are a measure of the distances of the ends of IS*1* from the nearest *Hpa*II site within IS*1*. We can thus place the IS*1*-*gal* junctions in S104, as shown to scale in Figure 2.

The restriction maps of the two IS*1* insertions into *galT* allow certain conclusions to be drawn. The S188 mutation has been located to the operator proximal side of S104 by genetic mapping (Shapiro 1969). From this we can conclude that the transcription of the *Hin*II D fragment from the *gal* operator-promoter proceeds from the D_0 end towards A_0. Recently, Schmidt et al. (1976) have presented data suggesting that guanosyl residues occur at both 5′ termini of isolated IS*1* elements. The only sequence arrangement consistent with this finding and with the creation of two *new Hpa*II sites (5′-NC↓CGGN-3′; Garfin and Goodman 1974) by the S188 insertion, without loss of a host *Hpa*II site, is that the IS*1* inserted into the host sequence 5′-CCGCGG-3′ to give

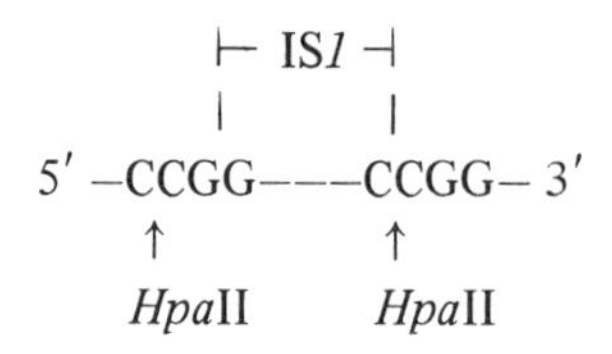

Hopefully, nucleotide sequence studies will clarify whether the palindromic nature, or the specific base sequence of this region, or neither plays a role in the IS*1* insertion event.

Acknowledgments

I thank Michael Malamy for the λ lysogens and Richard Roberts for allowing me to use his collection of restriction endonucleases. I also thank David Goldfarb for technical assistance, David Goldberg for many helpful discussions, and Joan Steitz and June Grindley for critical reading of the manuscript. This work was supported by National Institutes of Health Grant AI10243 to Joan Steitz. The author was a postdoctoral fellow of the Science Research Council of Britain.

REFERENCES

Fiandt, M., W. Szybalski and M. H. Malamy. 1972. Polar mutations in *lac, gal* and phage λ consist of a few IS-DNA sequences inserted with either orientation. *Mol. Gen. Genet.* **119**:223.

Garfin, D. E. and H. M. Goodman. 1974. Nucleotide sequences at the cleavage sites of two restriction endonucleases from *Haemophilus parainfluenzae. Biochem. Biophys. Res. Comm.* **59**:108.

Lee, A. S. and R. L. Sinsheimer. 1974. A cleavage map of bacteriophage ϕX174 genome. *Proc. Nat. Acad. Sci.* **71**:2882.

Maniatis, T. and M. Ptashne. 1973. Multiple repressor binding at the operators in bacteriophage λ. *Proc. Nat. Acad. Sci.* **70**:1531.

Peacock, A. C. and C. W. Dingman. 1968. Molecular weight estimation and separation of ribonucleic acid by electrophoresis in agarose-acrylamide composite gels. *Biochemistry* **7**:668.

Schmidt, F., J. Besemer and P. Starlinger. 1976. The isolation of IS*1* and IS*2* DNA. *Mol. Gen. Genet.* **145**:145.

Shapiro, J. A. 1969. Mutations caused by the insertion of genetic material into the galactose operon of *Escherichia coli. J. Mol. Biol.* **40**:93.

Smith, H. O. and O. Nathans. 1973. A suggested nomenclature for bacterial host modification and restriction systems and their enzymes. *J. Mol. Biol.* **81**:419.

Sugden, B., B. DeTroy, R. J. Roberts and J. Sambrook. 1975. Agarose slab-gel electrophoresis equipment. *Anal. Biochem.* **68**:36.

Preliminary Observations

A. On Mutations Affecting IS*1*-mediated Deletion Formation in *E. coli*

P. Nevers, H.-J. Reif* and H. Saedler
Institut für Biologie III
Universität Freiburg
7800 Freiburg i. Br., W. Germany

Deletion formation, inversion, and duplication are little-understood processes which have been classified as illegitimate recombination (Campbell 1962) since they do not require extensive DNA homology and are independent of the normal Rec pathways. The chromosomal rearrangements brought about by illegitimate recombination may play an important role in evolution. The deletion of genetic material, for example, brings sequences which were formerly separated into close proximity, creating a new order of genetic information at the point of fusion. We wish to report the isolation of a mutant of *E. coli* that is defective in one process of illegitimate recombination, the process of deletion formation.

Deletions are normally formed in *E. coli* with very low frequencies, about 10^{-6} to 10^{-9} per cell (e.g., Franklin 1971), which seriously handicaps studies of the functions involved. However, the presence of an integrated IS*1* element enhances deletion of sequences adjacent to the insertion more than 1000-fold (Reif and Saedler 1975), thus providing a suitable basis for analysis. IS*1*-mediated deletion formation has been shown to be independent of *recA,* but it does vary with the growth temperature of the bacteria; 10 to 100 times more deletions are formed at 32°C than at 42°C (Reif and Saedler 1975).

For the isolation of mutants defective in deletion formation, we selected a strain in which the element is integrated in the *galT* gene of the *gal* operon, *galT116*::IS*1*.[1] As shown previously (Reif and Saedler 1975), this strain is capable of deleting material to the right of the insertion, including part of *galT, galE, galOP,* and *chlD,* with a frequency of 10^{-4}. Single colonies of an NG-mutagenized strain were plated on NA-full medium plates (Miller 1972) and incubated at 32°C to permit maximal deletion formation. Approximately 2000 clones were then screened for the loss of both the *galE* and *chlD* functions by replica-plating the clones onto plates containing 0.2% $NaClO_3$ and spread with 10^9 U3 bacteriophage, followed by anaerobic incubation at 37°C. Since loss of *galE* confers resistance to U3 (Watson and Paigen 1971) and loss of

*Present address: Department of Biological Sciences, Stanford University, Stanford, California 94305.

[1] *galT116* is N116 mutation (Reif and Saedler 1975).

chlD permits anaerobic growth on $NaClO_3$ (Adhya et al. 1968), those clones in which deletion of both genes has occurred with a high frequency will produce a considerable number of resistant papillae under the conditions described above. About 500 clones that showed little or no growth were regarded as possible mutants, patched on NA plates at 32°C, and retested in the same manner. Of the original 2000 clones, one called del1 was found to be defective in IS*1*-mediated deletion formation. Repetition of the entire procedure with a strain of different genetic background yielded a second mutant, del2.

After confirming that the two mutants, del1 and del2, were not deficient in growth, the frequencies of deletion formation in the wild-type parent strains and in the two mutants were compared quantitatively. For this purpose, the ability of the strains to produce deletions extending from the right end of the IS*1* element and including the genes *galE* and *pgl* was measured. Single colonies of the strains to be tested were grown overnight at 32°C in supplemented M9 minimal medium (Miller 1972) containing casamino acids, glycerol, nicotinic acid, and biotin to permit deletion formation. Appropriate dilutions of the bacteria were then plated on NA seeded with at least 10^8 U3 bacteriophage and incubated at 37°C. Surviving U3^r colonies, which include some *galE* deletions, were replica-plated onto minimal plates containing maltose, nicotinic acid, biotin, and 0.01% casamino acids and tested with KI solution, as described by Adhya et al. (1968), in order to identify those that were *pgl*$^-$ as well. Colonies that appeared to be *galE*$^-$ and *pgl*$^-$ were tested further on agar containing nitrate (Adhya et al. 1968) for the loss of the *chlD* function. Colonies in which all three functions were missing were considered to be deletions.

As shown in Table 1, Δ(*galT-pgl*) deletions are formed with a frequency of $1.1\text{-}1.6 \times 10^{-4}$ in wild-type strains. In the mutant del1, this frequency is consistently reduced by a factor of 5 to 6, and in the mutant del2 by a factor of 25 to 30.

As already mentioned, the frequency of IS*1*-mediated deletion formation is dependent on the growth temperature of the bacteria. The two mutants, del1 and del2, were examined to see whether the residual activity in the mutants also showed this effect. The frequency with which Δ(*galT-pgl*) deletions are formed at two different growth temperatures, 32°C and 42°C, was measured as described above. From the data in Table 1 it can be seen that both mutants also exhibit a significant reduction in deletion formation at the higher temperature.

Reif and Saedler (1975) showed that an integrated IS*1* element can become accurately excised, resulting in *gal*$^+$ revertants. In wild-type strains, this occurs with a frequency of about $6\text{-}10 \times 10^{-7}$. In both mutants, the value for reversion to *gal*$^+$ is only slightly less, about 3×10^{-7}. Thus this process does not seem to be seriously affected in the mutants. The difference in the effect of the mutation on excision as opposed to deletion formation indicates that these processes may be governed by two different enzyme systems. This is depicted in Figure 1.

The mutant del2 can be complemented by F143, as shown in Table 1. This indicates that the del2 mutation is located in the vicinity of *recA*. However, a

Table 1
Properties of the Mutants

Strain	Frequency of Δ(*galT-pgl*)/cell ($\times 10^7$) 32°C	42°C	Frequency of excision of IS*1* (reversion to *gal*$^+$) ($\times 10^7$) 32°C	UV sensitivity	Recombination proficiency
galT116::IS*1*, del$^+$	1600	12	13	nt	nt
galT116::IS*1*, del1	290	25	3.5	nt	nt
galT116::IS*1*, del$^+$	1110	70	6.4	r	+
galT116::IS*1*, del2	42	3	3.1	r	+
galT116::IS*1*, del2/F143del$^+$	759	44	8.8	nt	nt

The mutants del1 and del2 are compared with their respective wild-type parental strains with regard to deletion formation, excision, UV sensitivity, and recombination proficiency. nt = not tested, r = resistant.

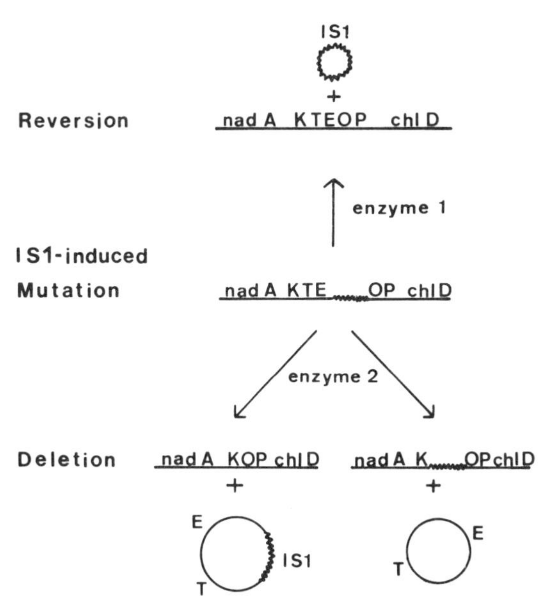

Figure 1
Schematic representation of the two different enzymatic pathways responsible for excision of the IS*1* element on the one hand and IS*1*-mediated deletion formation on the other. Exact excision of the element results in restoration of the *gal*$^+$ genotype. During deletion formation, the IS*1* element theoretically can either be removed with the deleted material or remain in the chromosome. The wavy line represents the IS*1* element.

survival curve of the mutant established after increasing doses of UV irradiation did not differ from that of its wild-type parent strain, demonstrating that del2 is not UV-sensitive. Furthermore, the mutant was able to recombine with an F′ factor carrying a point mutation in *galE*, producing *gal*$^+$ recombinants with the same frequency as its wild-type parent. Therefore, del2 is probably not identical to *recA*. Further work is required to locate the mutation precisely.

The isolation of mutants of this type represents an important step in identifying the functions involved in deletion formation.

Acknowledgment

This work was supported by the Deutsche Forschungsgemeinschaft.

REFERENCES

Adhya, S., P. Cleary and A. Campbell. 1968. A deletion analysis of prophage λ and adjacent genetic regions. *Proc. Nat. Acad. Sci.* **61**:956.

Campbell, A. 1962. Episomes. *Adv. Genet.* **11**:101.

Franklin, N. C. 1971. Illegitimate recombination. In *The bacteriophage lambda* (ed. A. D. Hershey), p. 175. Cold Spring Harbor Laboratory, Cold Spring Harbor, New York.

Miller, J. H. 1972. *Experiments in molecular genetics.* Cold Spring Harbor Laboratory, Cold Spring Harbor, New York.

Reif, H.-J. and H. Saedler. 1975. IS1 is involved in deletion formation in the gal region of *E. coli* K12. *Mol. Gen. Genet.* **137**:17.

Watson, G. and K. Paigen. 1971. Isolation and characterization of an *E. coli* bacteriophage requiring cell wall galactose. *J. Virol.* **8**:669.

B. On the Presence of a Promoter in IS*2*

B. Rak
Institut für Genetik
Universität zu Köln
D-5000 Köln, W. Germany

IS*2* causes polar mutations when inserted in orientation I. Saedler et al. (1974) have proposed that IS*2* carries a promoter in orientation II. This proposal is based on studies of a constitutive Gal$^+$ revertant of mutant *galOP308*::IS*2*·I (IS*2* in orientation I). This revertant (F108), carrying the *gal* operon on the F8 *gal* episome, was shown to have the *gal* genes fused to a nearby IS*2* element in orientation II. If IS*2* carries a promoter, then the *gal* message initiated in IS*2* might be expected to be covalently linked to IS*2*-specific RNA. To test this possibility, I have examined F108 and some constitutive revertants of the *galOP308*::IS*2* mutant (the physical structure of the *gal* operon in this mutant has not been examined directly). These revertants produce the *gal* enzymes as well as the *gal* messenger RNA at 2-3 times the wild-type level.

The results I have obtained indicate that, in constitutive revertants of *galOP308*::IS*2*, some IS*2*-coded RNA is covalently linked to the *gal* mRNA. These results stem from experiments involving two-step hybridization of mRNA to IS*2* and *gal* DNA. The procedure of these experiments (described in Rak 1976) was as follows: The r strands of the λ transducing phages λd*gal308*::IS*2* and λ*r*32::IS*2* were separated from the complementary strands and purified. The r strand of λd*gal308*::IS*2* (IS*2* in orientation I in *gal*) carries the strand of IS*2* which we call the *b* strand; the *b* strand is linked to the *gal* strand that is not transcribed (is not the "sense" strand). The r strand of λ*r*32 contains the complementary strand, the *a* strand of IS*2*. Lambda *r*32 contains no *gal* DNA. RNA, pulse-labeled with [^{3}H]uridine, was isolated from F108 (Saedler et al. 1974) and R20, a constitutive revertant of the *galOP308*::IS*2* mutant (Saedler et al. 1972). This pulse-labeled RNA was first hybridized to the r strand of the λd*gal308*::IS*2* and λ*r*32 DNAs. Out of an input of about 10^7 cpm, about 0.5-1% of the counts hybridized the r strands. This RNA was eluted and then hybridized to λd*gal* DNA (the l strand containing the "sense" strand of *gal*). Only a very small amount of RNA from the λd*gal308* r strand hybridized to the λd*gal* l strand (~ 1-2% of the RNA hybridizing at the first step). However, about 15-20% of the RNA that hybridized to the λ*r*32 r strand also hybridized to the λd*gal* l strand, whereas this value became $< 1\%$ when the RNA originated from a *gal* wild-type cell.

I interpret these results to mean that, in the strains examined, the *gal* message is covalently bound to IS*2* RNA, and that this RNA is a transcript of the *a* strand of IS*2*. Since the IS*2* RNA is complementary to the *a* strand, it binds only to λ*r*32 r strand (containing the IS*2* *a* strand) and not to the λd*gal308* r strand (containing the IS*2* *b* strand). The *gal* message does not bind to the λd*gal308* r strand because this strand does not contain the "sense" strand of *gal*.

If this interpretation is correct, it would support the idea that IS*2* carries a promoter. However, the interpretation is complicated–IS*2* is supposed to carry a promoter in orientation II, in which case the *b* strand of IS*2* should be linked to the *gal* sense strand and the IS*2* RNA should be complementary to the *b* strand. (The original *galOP308* mutation contains IS*2* in orientation I and the strand linked to the *gal* sense strand is the *a* strand; in Gal^+ revertants, the orientation is apparently reversed, and thus the *b* strand should be linked to the sense strand.) Assuming that the IS*2* message is covalently linked to the *gal* message, we have to postulate that, in the Gal^+ revertants, some portion of IS*2* in orientation I is left in between the *gal* genes and IS*2* in orientation II. At least in F108, this part of IS*2* cannot be large since Saedler et al. (1974) did not detect such an arrangement electron microscopically.

I have estimated the *gal* mRNA cotransferred with IS*2* RNA, in the experiments described above, to be about 1000 nucleotides long (Rak 1976). The RNase T1 fingerprints of IS*2*-specific RNA, obtained from RNA-DNA hybrids by digesting the *gal* mRNA tail, indicate that the IS*2* RNA in IS*2*-*gal* mRNA is not more than 20 nucleotides long. Preliminary sequence analysis shows that the 5′ end of this RNA consists of a uridine 5′-phosphate (Rak 1976). In all experiments, the 5′-terminal sequence is found to be pUCUG. Uridine has not been found to occur at a high frequency at the beginning of bacterial messages (Chamberlin 1974). It can be speculated, therefore, that the terminus of the IS*2*-*gal* message in my experiments results from a cleavage event, and that initiation of transcription may actually occur in IS*2* in orientation II.

A direct confirmation of the exact location of the promoter on IS*2* must await further experimentation.

Acknowledgments

I am grateful to Drs. P. Starlinger, H. Saedler and H. Chadwell for discussions. This work was supported by the Deutsche Forschungsgemeinschaft through SFB 74.

REFERENCES

Chamberlin, M. J. 1974. The selectivity of transcription. *Annu. Rev. Biochem.* **43**:721.

Rak, B. 1976. *Gal*-mRNA initiated within IS2. I. Hybridization studies. *Mol. Gen. Genet.* **149**:135.

Saedler, H., H.-J. Reif, S. Hu and N. Davidson. 1974. IS2, a genetic element for turn-off and turn-on of gene activity in *E. coli. Mol. Gen. Genet.* **132**:265.
Saedler, H., J. Besemer, B. Kemper, B. Rosenwirth and P. Starlinger. 1972. Insertion mutations in the control region of the *gal* region of *E. coli.* I. Biological characterization of the mutations. *Mol. Gen. Genet.* **115**:258.

C. On the Polarity of Insertion Mutations

J. Besemer
Institut für Genetik
Universität zu Köln
D-5000 Köln, W. Germany

It has been proposed that the polarity of polar mutations results from the action of the transcription termination factor rho at sites distal to the nonsense codon (Richardson et al. 1975). A similar model has been proposed by Adhya et al. (1976) for insertion mutations.

It is known that IS*1* causes polar mutations in both orientations, whereas IS*2* apparently is polar only in orientation I (Saedler et al. 1974). Two interesting points should be kept in mind while considering the polar effects of these insertion elements. First, in the *gal* operon of *Escherichia coli,* the insertions appear to exert stronger polar effects than nonsense codons. Second, residual expression of the downstream *gal* genes is constitutive.

Suppressor mutations that relieve the polar effects of IS*1* and IS*2* in the *lac* operon and the *gal* operon have been described (Malamy 1970; Malamy et al. 1972; Das et al., this volume). These suppressor mutations affect the *suA* gene, which presumably encodes the transcription termination factor rho. I have examined the effect of similar suppressor mutations on the polarity of *galOP*::IS*1* and *galOP*::IS*2* mutations. The suppressor mutations were isolated by selecting for Gal^+ revertants after nitrosoguanidine mutagenesis of *galOP*::IS*1* and *galOP*::IS*2* mutants. Four such Gal^+ revertants were shown to carry suppressor mutations that were cotransduced with the *ilv* operon at a frequency of about 80% by phage P1. Since the *suA* mutations are known to be very close to *ilv,* it is likely that the four mutations are in the *suA* gene.

The suppressor mutations reduced polarity of IS*1* insertions more than that of an IS*2* insertion tested. For example, the level of galactokinase (coded in the *K* gene, the most distal gene of the *gal* operon) in *galOP306*::IS*1*·I and *galOP128*::IS*1*·II mutants increased from about 0.5% in SuA^+ background to 20% and 15%, respectively, of the wild-type level in the presence of *suA18* mutations. In the *galOP308*::IS*2*·I mutant, it increased from about 1% to about 3.5%. The synthesis of galactokinase in the presence of *suA18* mutations, however, remains at least partially constitutive. Thus, in a wild-type *gal* operon, the galactokinase activity increases about 30-fold in the presence of 10^{-3} M D-fucose either in *suA*$^+$ or *suA18* strains, but, in the *galOP* insertion mutants, the effect of inducer does not exceed threefold.

A somewhat less striking effect of suppressor mutations on IS*2* polarity as compared to IS*1* polarity can be ascribed to the presence of a strong rho-sensitive termination signal on IS*2*. The presence of such a signal on IS*2* but not on IS*1* has been demonstrated by de Crombrugghe et al. (1973) in an in vitro system. The apparent absence of a termination signal on IS*1* is intriguing. If there is no such signal, then the polarity of IS*1* in either orientation must be ascribed to rho-sensitive sites distal to the site of IS*1* insertion. However, in view of the observation that IS*1* insertions are generally more polar than nonsense mutations in the *gal* operon, the possibility of a rho-independent termination signal in IS*1* in both orientations should be considered.

The presence of termination signals on both IS*1* and IS*2* (orientation I) would provide a plausible explanation for the constitutivity of the enzyme synthesis discussed above. This explanation is based on the idea that RNA polymerase molecules are not always released at a stop signal. They pass on for continued transcription at a frequency which is much lower than the frequency at which the polymerase molecules are started at the *gal* promoter. As a result, polymerases will pile up at the stop signal. In the case of *galOP*::IS mutations, the piling up of polymerases proximal to the stop signal would cause a reduction in the binding of the repressor to the operator, as the operator may be covered by polymerases. Complete constitutivity would be expected when the stop is at least as effective as the repressor action at the *gal* operator; moderate inducibility (constitutivity) would be expected when the stop signal is weakened; for instance, as a consequence of a defect in the rho factor. This is in fact observed when residual galactokinase synthesis in *galT*::IS*1* insertion mutants in suA^+ and suA^- cells is compared. Synthesis of galactokinase is inducible only in cells carrying *suA18* mutations. Induction factors of 5-10 are obtained. Furthermore, piling up of polymerases at a transcription stop signal may have some influence on the expression of genes located proximal to the insertion mutation. In the case of insertions in *galT* (the middle gene of the operon), the transcription of epimerase gene (the first gene of the *gal* operon) would be reduced. This effect would be expected to decrease with increasing distance of the insertion from the border of gene *E.* Such an effect for strong polar mutations in the *galT* gene has been observed by Jordan et al. (1967) and Starlinger et al. (1973). However, the properties of these mutations as simple insertions have been verified only in a few cases. A detailed description of experimental results leading to the piling up model is being published elsewhere (Besemer and Herpers 1977).

REFERENCES

Adhya, S. L., M. Gottesman, B. de Crombrugghe and D. Court. 1976. Control of transcription termination. In *RNA polymerase* (ed. R. Losick and M. Chamberlin), p. 719. Cold Spring Harbor Laboratory, Cold Spring Harbor, New York.

Besemer, J. and M. Herpers. 1977. Suppression of polarity of insertion mutations within the *gal* operon of *E. coli. Mol. Gen. Genet.* **151**:295.

de Crombrugghe, B., S. L. Adhya, M. Gottesman and I. Pastan. 1973. Effect of rho on transcription of bacterial operons. *Nature New Biol.* **241**:260.

Jordan, E., H. Saedler and P. Starlinger. 1967. Strong polar mutations in the transferase gene of the galactose operon in *E. coli. Mol. Gen. Genet.* **100**:296.

Malamy, M. H. 1970. Some properties of insertion mutations in the *lac* operon. In *The lactose operon* (ed. J. R. Beckwith and D. Zipser), p. 359. Cold Spring Harbor Laboratory, Cold Spring Harbor, New York.

Malamy, M. H., M. Fiandt and W. Szybalski. 1972. Electron microscopy of polar insertions in the *lac* operon of *Escherichia coli. Mol. Gen. Genet.* **119**:207.

Richardson, J. P., C. Grimley and C. Lowery. 1975. Transcription termination factor rho activity is altered in *Escherichia coli* with *suA* gene mutations. *Proc. Nat. Acad. Sci.* **72**:1725.

Saedler, H., H.-J. Reif, S. Hu and N. Davidson. 1974. IS2, a genetic element for turn-off and turn-on of gene activity in *E. coli. Mol. Gen. Genet.* **132**:265.

Starlinger, P., H. Saedler, B. Rak, E. Tillman, P. Venkov and L. Waltschewa. 1973. mRNA distal to polar nonsense and insertion mutations in the *gal* operon of *E. coli. Mol. Gen. Genet.* **122**:279.

SECTION II
Transposons and Plasmids

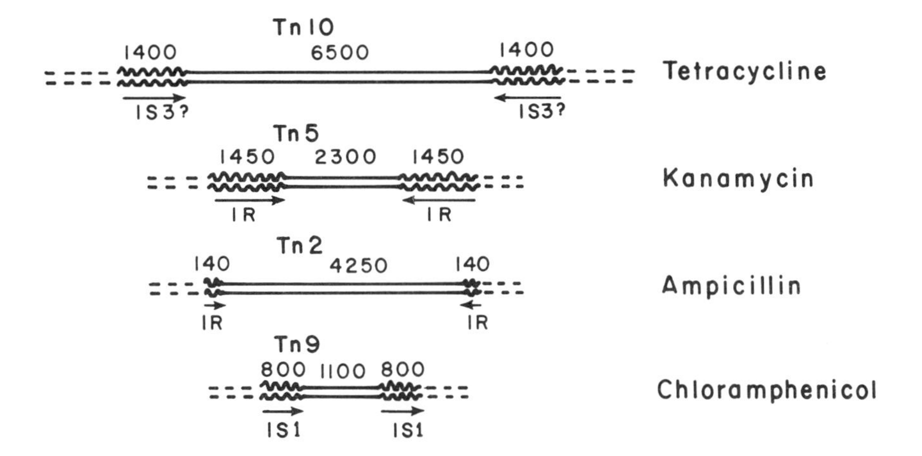

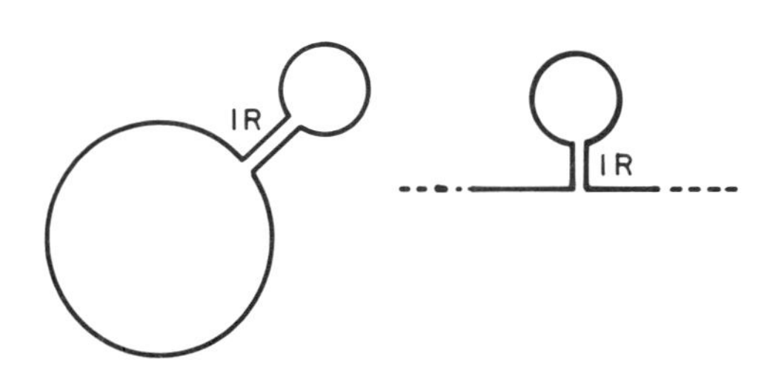

Many plasmid genes determining drug resistance have been found to be located in discrete DNA units which can "dissociate" from the plasmids and insert themselves into another DNA molecule. They can undergo further translocation from the new DNA host to different DNA replicons. These movable DNA units have variously been called drug resistance determinants, translocatable elements, transposable elements, translocons, and transposons. The term transposon is used here mainly to distinguish the complex units from the simple transposable elements such as IS*1* and IS*3*. The clear-cut cases of transposons studied so far have all involved drug resistance markers.

The section-title figure shows the representative transposons. The drug to which resistance is encoded in the transposon is indicated to the right of the drawing. A fundamental characteristic of transposons appears to be the presence of repetitious sequences at the ends. With the exception of Tn*9*, the transposons so far examined have clearly detectable repetitious sequences in inverted order. The jagged lines in the figure show the repetitious sequences at the ends of a transposon. The arrows indicate the relative orientations of the repetitious sequences. The solid lines indicate the DNA carrying the drug resistance genes, flanked by the repetitious sequences. The broken line is intended to indicate DNA into which the transposon is inserted (the transposons are not known to have autonomous existence).

The sizes of the DNA segments are given in base pairs. The size estimates are approximate and, with the exception of Tn*10*, are based on information presented in papers in this volume. The Tn*10* size is based on a recently refined estimate of D. Ross, N. Kleckner and D. Botstein (unpubl.). This size refinement stems from an upgrading by about 10% of the length of ϕX174 DNA, a molecule used as a length standard in many studies. According to Sanger et al. (*Nature* 256:687 [1977]), the size of ϕX174 is about 5375 nucleotides.

It can be seen in the figure that the 1400-base-pair inverted sequences in Tn*10* are labeled as IS*3*. [Recently, the endonuclease cleavage data of E. Ohtsubo and H. Ohtsubo have raised the possibility that 1400-base-pair inverted repeats in Tn*10* may not be IS*3*.] The unidentified inverted repeats are labeled IR. In Tn*9*, IS*1* is present in direct order, IS*1* has small inverted repeats at the ends (Appendices A2 and A3). Single-stranded DNA containing a transposon with inverted repeats gives rise to the typical "lollipop" structures shown at the bottom of the figure. The circle shows a single-stranded DNA of an hypothetical plasmid carrying a transposon.

Other transposons characterized by the end of 1976, but not shown in the figure, are Tn*1* and Tn*3* (both determining ampicillin resistance), Tn*4* (determining resistance to ampicillin, sulfonamide and streptomycin), and Tn*6* (determining kanamycin resistance). Tn*1* and Tn*3* are about the same size as Tn*2*, and all three may have sequences in common. Tn*4* is about 20,500 base pairs in length, with 140 base pairs as inverted repeats (Tn*3* is a part of Tn*4*). Tn*6* is approximately 4100 base pairs in length.

Formation of Conjugative Drug Resistance (R) Plasmids

S. Mitsuhashi, H. Hashimoto, S. Iyobe and M. Ionué
Department of Microbiology
School of Medicine, Gunma University
Maebashi, Japan

The drug resistance transfer factors, or the R plasmids, were discovered in Japan as a result of epidemiologic and genetic studies of the rapid emergence of multiple drug resistance in *Shigella* strains. These studies indicated that multiple resistance could be transferred from one strain to another by mixed cultivation. Spontaneous elimination of multiple resistance could occur during storage (Mitsuhashi 1971). R factors are of particular importance because they are widely distributed in bacteria and because the inheritance of multiple resistance by pathogenic strains can severely impede the medical treatment of bacterial infections.

It is generally accepted that R plasmids consist of two major genetic elements. One element, called transfer factor (T), is responsible for the autonomy of the plasmid as a replicon as well as its conjugal transferability (Mitsuhashi et al. 1969). The other element contains information for drug resistance. Some plasmids carry drug resistance markers but cannot be conjugally transferred from one bacterial strain to another. These nonconjugative plasmids are called the r factors, to distinguish them from the conjugative R plasmids. The drug resistance markers can be transferred from both r and R plasmids to other genetic elements. This mechanism of transfer is apparently independent of the normal means of conjugal transfer of the R plasmids and is referred to as transposition or translocation. We have studied the apparent transposition of various drug resistance determinants from both the R and r factors to bacterial chromosomes, bacteriophages, and other plasmids.

Translocation of the Drug Resistance Genes from R or r Plasmids

Studies on transduction of an R plasmid with bacteriophage epsilon (ϵ) in *Salmonella* strains revealed that the gene for tetracycline (Tc) resistance could be consistently separated from the markers for resistance to chloramphenicol (Cm), streptomycin (Sm), and sulfonamide (Su) (Harada et al. 1963). The Tc resistance gene, *tet,* in the resulting transductants was transferred to the

Table 1
Formation of Recombinants between Resistance Determinants and Bacteriophages

Origin	Bacteriophage	Resistance determinant	Recombinant	Reference
E. coli Rms14$^+$	P1	*cml*	P1*cml*	Kondo and Mitsuhashi (1964)
E. coli (P1*cml*)	P1*cml*	*cml*	P1d*cml*	Kondo and Mitsuhashi (1966)
E. coli Rms10$^+$	P1	*cml.str.sul*	P1d*cml.str.sul*	Egawa and Mitsuhashi (1964)
S. newington Rms10$^+$	ϵ	*tet*	ϵd*tet*	Kameda et al. (1965)
S. newington Rms10$^+$	ϵ	*cml.str.sul*	ϵd*cml.str.sul*	Kameda et al. (1965)
S. aureus MS353(Tc^r)(S1)[a]	S1	*tet*	Slp*tet*	Inoué et al. (1975); Inoué and Mitsuhashi (1975); Mitsuhashi et al. (1976)
S. aureus MS353 (Cm^r)(S1)[b]	S1	*cml*	Slp*cml*	Inoué et al. (1975); Mitsuhashi et al. (1976)

[a] *S. aureus* MS353(Tc^r) was obtained by transduction from MS3878 r-ms7(Tc)$^+$ with bacteriophage S1. MS353(Tc^r)(S1) carries a nontransferable plasmid encoding Tc resistance and is lysogenized with S1.

[b] *S. aureus* MS353(Cm^r) was obtained by transduction from MS3878 r-ms6(Cm)$^+$ with bacteriophage S1. MS353(Cm^r)(S1) carries a nontransferable plasmid encoding Cm resistance and is lysogenized with S1.

Escherichia coli chromosome by conjugation and was observed to be integrated between *pro* and *lac* (Harada et al. 1967b). Similarly, the genes encoding resistance to Cm, Sm, and Su could be shown to be associated with the chromosomes of *Salmonella* and *E. coli* (Egawa and Mitsuhashi 1964; Kameda et al. 1965). The genes for resistance markers are referred to here as *cml*[1] (for Cm^r), *str* (for Sm^r), and *sul* (for Su^r).

The *cml* marker could be located at several sites on the *E. coli* chromosome; near *metB* (Iyobe et al. 1969), *lac, ara, gal, mtl,* or *pro* (Iyobe et al. 1970). We were also able to transfer the *cml* marker to phage P1 by transducing Cm resistance from an R factor into *E. coli* and then inducing Cm-resistant P1 lysogens with UV light. Among the phage in the resultant lysates were plaque-forming specialized P1 transducing particles carrying *cml* (Kondo and Mitsuhashi 1964). Furthermore, we obtained P1-defective particles that could simultaneously transduce resistances to Cm, Sm, and Su at a high frequency with a helper P1 phage (Egawa and Mitsuhashi 1964). Defective transducing particles were also obtained from the *Salmonella* phage epsilon (ϵ) (Kameda et al. 1965).

Similar experiments were done with *Staphylococcus aureus* strains. *S. aureus* clinical isolates are multiply resistant owing to the presence of various types of nonconjugative resistance (r) plasmids. These plasmids generally encode resistance to a single drug, but several plasmids coexist in a cell (Mitsuhashi et al. 1973, 1976; Inoué et al. 1975). The Tc^r determinant in *S. aureus* MS3878 carrying the nonconjugative plasmid r-ms7(Tc) was transduced out to an *S. aureus* strain with bacteriophage S1. Upon infection of the Tc-resistant transductants with bacteriophage S1, we obtained bacteriophage S1p*tet*. Similarly, we obtained S1p*cml* phage as a consequence of transduction of the r-ms6(Cm) plasmid with S1. Table 1 summarizes the results of the recombination of drug resistance determinants with bacteriophages P1, ϵ, and S1 in *E. coli, S. newington,* and *S. aureus,* respectively.

To test the acquisition of drug resistance determinants by various episomes and plasmids, the *tet* determinant from an R plasmid was integrated into the *E. coli* chromosome (Harada et al. 1967b), and the *E. coli* strain was infected with F, F-*lac,* R(Cm), and T plasmids. We obtained the following recombinants between the *tet* determinant and the plasmids: F-*tet*, F-*lac.tet* (Harada et al. 1964, 1967a), R(Cm)-*tet* (Kameda et al. 1969), and T-*tet* (Mitsuhashi et al. 1969). Similarly, from Cm-resistant *E. coli* W3630 in which the *cml* determinant on an R plasmid was integrated in the chromosome (Iyobe et al. 1969), we obtained the recombinants F-*cml* and R(Tc)-*cml* (Iyobe et al. 1969, 1970). From *E. coli* W3110 carrying a P1 *cml* prophage, we obtained the recombinant F-*cml,* which possesses both F functions and an ability to confer Cm resistance (Kondo and Mitsuhashi 1964). Table 2 lists the recombinants between resistance determinants and plasmids.

[1] In other papers, the symbol *cam* is used instead of *cml.*

Table 2
Formation of Recombinants between Resistance Determinants and Plasmids

Origin	Infected with plasmid	Recombinant	Reference
E. coli W1177 *tet*	F	F-*tet*	Harada et al. (1964)
E. coli W1177 *tet*	F-*lac*	F-*lac.tet*	Harada et al. (1967)
E. coli W1177 *tet*	R(Cm)	R(Cm)-*tet*	Harada et al. (1967)
E. coli W1177 *tet*	T	T-*tet*	Kameda et al. (1969, 1970); Mitsuhashi et al. (1969)
E. coli W3110 P1*cml*	F	F-*cml*	Kondo and Mitsuhashi (1966)
E. coli W3630 *cml*	F	F-*cml*	Kondo and Mitsuhashi (1966)
E. coli W3630 *cml*	R(Tc)	R(Tc)-*cml*	Iyobe et al. (1969)

Cointegration of r(Km) with T-*tet* Plasmid

The r-ms3(Km) plasmid is a nonconjugative plasmid encoding kanamycin (Km) resistance (gene symbol *kan*) and is thermosensitive for its own maintenance in host bacteria. When T-*tet* was introduced into *E. coli* W3630 r(Km)$^+$, three types of transconjugants were obtained after selection on a plate containing both Tc and Km: those (termed T-*tet.kan*1) which conjugally transferred Tc.Km resistance together at a frequency of 10^{-1}; those (termed T-*tet.kan*2) which conjugally transferred Tc.Km resistance together at a frequency of 10^{-3}, a frequency similar to that of the parent T-*tet* plasmid; and those in which the Tc resistance was transferred at a high frequency but Km resistance was transferred at a low frequency, both resistance determinants existing separately. These results are summarized in Table 3.

The DNA molecules of the T-*tet*, r-*kan*, and T-*tet.kan*1 plasmids were purified by ethidium bromide-cesium chloride density-gradient centrifugation. The lengths of the plasmids were measured by electron microscopy. The contour lengths of r-*kan* and T-*tet* plasmid DNAs were 11.7 ± 0.3 and 26.4 ± 0.7 μm, respectively. The contour length of a recombinant T-*tet.kan*1 plasmid DNA was 38.6 ± 0.9 μm, indicating that the T-*tet.kan*1 plasmid was formed by the cointegration of both the r-*kan* and T-*tet* plasmids (Table 4).

DISCUSSION

Conjugative (R) and nonconjugative (r) plasmids are widely distributed among both gram-positive and gram-negative bacteria. Because these plasmids carry genes determining drug resistance, they are of particular importance in clinical medicine and livestock hygiene. The drug resistance determinants can be trans-

Table 3

Isolation of Recombinants between r-ms3(Km) and T-*tet* Plasmid

			Properties of the transconjugants		
				transfer frequency[b] of resistance to	
Plasmid in donor	Selective drug	Transfer frequency[a]	resistance pattern	Tc	Km
T-*tet* + r(Km)	Tc + Km	1.1×10^{-7}	Tc.Km	10^{-1}	10^{-1} (4/29)[d]
T-*tet* + r(Km)	Tc + Km	1.1×10^{-7}	Tc.Km	10^{-3}	10^{-3} (7/29)
T-*tet* + r(Km)	Tc + Km	1.1×10^{-7}	Tc.Km	10^{-3}	$< 10^{-6}$ (18/29)
T-*tet*	Tc	1.2×10^{-3}	Tc	10^{-3}	(30/30)
r(Km)	Km	$< 10^{-8}$[c]	Tc	10^{-3}	(30/30)

[a] Donor, W3630; recipient, ML1410 NA^r; time of mixed culture, 1 hr for T-*tet*.
[b] Recipient, W3630; time of mixed culture, 1 hr.
[c] Time of mixed culture, 18 hr.
[d] Denominator, number of transconjugants tested; numerator, number of transconjugants obtained.

Table 4
Summary of Physical Properties of Plasmid DNA Molecules from W3630 Carrying Each Plasmid

Plasmid	Amount of satellite peak relative to major peak (%)[a]	Mean contour length ± standard deviation (μm)[b]	Molecular weight (× 10^6 daltons)[c]	Number of copies per chromosome[d]
r(Km)	1.8	11.7 ± 0.3 (27)	24.2	0.8
T-*tet*	2.7	26.4 + 0.7 (24)	54.6	1.2
T-*tet.kan*1	3.2	38.6 ± 0.9 (22)	79.9	1.0

[a]Determined by dye-buoyant density centrifugation of *E. coli* W3630 carrying each plasmid.
[b]Number of plasmid DNA molecules measured is given in parentheses.
[c]Molecular weight was calculated by accepting that 1 μm of DNA is equivalent to 2.07 × 10^6 daltons (Lang 1970).
[d]Isolated as covalently closed circular molecules. Based on the data that the molecular weight of the *E. coli* chromosome is 2.5 × 10^9 daltons.

located, both from R and r plasmids, to the bacterial chromosome. These drug resistance determinants can be easily recombined into the genomes of bacteriophages and into the F factor, the F-*lac* episome, and the T plasmid DNA. Thus active bacteriophages possessing drug resistance genes, such as P1*cml* (Kondo and Mitsuhashi 1964), S1p*tet* (Inoué and Mitsuhashi 1975), S1p*cml* (Mitsuhashi et al. 1976), P1*tet*, and P1*tet.cml* (Mise and Arber 1975), have been obtained. Defective transducing phage strains carrying different drug resistance determinants have also been isolated.

The tendency of drug resistance determinants to dissociate from the plasmids and recombine with different DNA molecules can now be related to the presence of specific insertion sequences. These sequences, called IS elements, apparently mediate the integration of plasmids into the host chromosome and the recombination within or among plasmid DNAs. As discussed elsewhere in this volume, many of the drug resistance determinants are transposed as well-defined units. The movement of these well-defined units from one site to another can, in some cases, be clearly ascribed to the presence of IS elements.

Earlier we proposed the idea that nonconjugative plasmids may have evolved from bacteriophages by recombination events that resulted in the acquisition of drug resistance genes but caused a loss of phage virulence (Mitsuhashi et al. 1973). The formation of plasmids encoding multiple resistance may be the result of recombination between different drug resistance plasmids or of the acquisition of a further resistance gene(s) by an r plasmid encoding single drug resistance.

SUMMARY

The drug resistance plasmids can be conjugative (R) or nonconjugative (r). An R plasmid consists of two major genetic segments, a transfer (T) factor and a drug resistance determinant. The transfer factor in turn contains a genetic element responsible for replication and a genetic element responsible for conjugal transferability. The r plasmids generally carry a single drug resistance marker and lack a mechanism for conjugal transfer. The drug resistance determinants on R (or r) plasmids may be translocated to a bacterial chromosome or to the genome of a bacteriophage. The translocated resistance determinants thus obtained may be further picked up by F, F-*lac,* and T plasmids.

The drug resistance determinants apparently are translocated as well-defined DNA units.

REFERENCES

Egawa, R. and S. Mitsuhashi. 1964. Newly isolated Hft lysate of an R factor. *Japan. J. Bact.* **19**:237.

Harada, K., M. Kameda, M. Suzuki and S. Mitsuhashi. 1963. Drug resistance of enteric bacteria. II. Transduction of transmissible drug-resistance (R) factor with phage epsilon. *J. Bact.* **86**:1332.

——. 1964. Drug resistance of enteric bacteria. III. Acquisition of transferability of nontransferable R(Tc) factor with F factor and formation of F.R(Tc). *J. Bact.* **88**:1257.

——. 1967a. Drug resistance of enteric bacteria. X. Combination of defective R(Tc) factor with other episomes. *Japan. J. Microbiol.* **11**:143.

Harada, K., M. Kameda, M. Suzuki, S. Shigehara and S. Mitsuhashi. 1967b. Drug resistance of enteric bacteria. VIII. Chromosomal location of nontransferable R factor in *Escherichia coli. J. Bact.* **93**:1246.

Inoué, M. and S. Mitsuhashi. 1975. A bacteriophage S1 deviative that transduces tetracycline resistance to *Staphylococcus aureus. Virology* **68**:544.

Inoué, M., T. Okubo, H. Oshima and S. Mitsuhashi. 1975. Staphylococcal plasmids carrying tetracycline and chloramphenicol resistance. In *Microbial drug resistance* (ed. S. Mitsuhashi and H. Hashimoto), p. 153. University of Tokyo Press, Tokyo, and University Park Press, Baltimore.

Iyobe, S., H. Hashimoto and S. Mitsuhashi. 1969. Drug resistance of enteric bacteria. XVII. Integration of chloramphenicol resistance gene of an R factor on *Escherichia coli* chromosome. *Japan. J. Microbiol.* **13**:225.

——. 1970. Integration of chloramphenicol resistance genes of an R factor into various sites of an *Escherichia coli* chromosome. *Japan. J. Microbiol.* **14**:463.

Kameda, M., K. Harada, M. Suzuki and S. Mitsuhashi. 1965. Drug resistance of enteric bacteria. V. High frequency of transduction of R factors with bacteriophage epsilon. *J. Bact.* **90**:1174.

——. 1969. Formation of transferable drug resistance factor by recombination between resistance determinants and transfer factors. *Japan. J. Microbiol.* **13**: 225.

Kameda, M., M. Suzuki, T. Nakajima, K. Harada and S. Mitsuhashi. 1970. Genetic properties of a T factor and T-r recombinants. *Japan. J. Microbiol.* **14**:339.

Kondo, E. and S. Mitsuhashi. 1964. Active transducing bacteriophage P1 *CM* by the combination of R factors with bacteriophage P1. *J. Bact.* **88**:1266.

——. 1966. Drug resistance of enteric bacteria. VI. Introduction of bacteriophage P1 *CM* into *Salmonella typhi* and formation of P1*dCM* and F-*CM* elements. *J. Bact.* **91**:1974.

Lang, D. 1970. Molecular weights of coliphage and coliphage DNA. III. Contour length and molecular weight of DNA from bacteriophages T4, T5 and T7, and from bovine papilloma virus. *J. Mol. Biol.* **54**:557.

Mise, K. and W. Arber. 1975. Bacteriophage P1 carrying drug resistance gene(s) of the R factor NR1. In *Microbial drug resistance* (ed. S. Mitsuhashi and H. Hashimoto), p. 165. University of Tokyo Press, Tokyo, and University Park Press, Baltimore.

Mitsuhashi, S. 1971. Epidemiology of bacterial drug resistance. In *Transferable drug resistance factor R* (ed. S. Mitsuhashi), p. 1. University of Tokyo Press, Tokyo, and University Park Press, Baltimore.

Mitsuhashi, S., M. Kameda, K. Harada and M. Suzuki. 1969. Formation of recombinants between non-transmissible drug-resistance determinants and transfer factors. *J. Bact.* **97**:1520.

Mitsuhashi, S., M. Inoué, H. Kawbe, H. Oshima and T. Okubo. 1973. Genetic and biochemical studies of drug resistance in staphylococci. In *Contributions to microbiology and immunology: Staphylococci and staphylococcal infections* (ed. J. Jeljaszewicz), vol. 1, p. 144. Karger, Basel.

Mitsuhashi, S., M. Inoué, H. Oshima, T. Okubo and T. Saito. 1976. Epidemiologic and genetic studies of drug resistance in staphylococci. In *3rd International Symposium on Staphylococci,* Warszawa, p. 255. Gustave Fisher Verlag, Stuttgart.

Recombination between *Pseudomonas aeruginosa* Plasmids of Incompatibility Groups P-1 and P-2

G. A. Jacoby
Massachusetts General Hospital
Boston, Massachusetts 02114

A. E. Jacob
Bacteriology Department
Royal Postgraduate Medical School
London W12 OHS, England

Plasmids found in *Pseudomonas aeruginosa* can be divided broadly into two categories: those that are transmissible to *Escherichia coli* and to other enterobacteria and those that are transmissible only between *Pseudomonas* strains (Bryan et al. 1973). They can be classified further into at least eight incompatibility groups (Shahrabadi et al. 1975; Jacoby 1977) (see Appendix B1c). Plasmids of group P-1 are transmissible to *E. coli* (where they belong to incompatibility group P), have molecular weights of 30-52 million daltons and confer susceptibility to phages PRR1, PRD1, and PR4 (Datta et al. 1971; Grinsted et al. 1971; Olsen and Shipley 1973; Olsen et al. 1974; Stanisich 1974; Hedges and Jacob 1975; Stanisich et al. 1976). Plasmids of group P-2 are not transmissible to *E. coli,* have molecular weights of 23-25 million daltons, and do not allow propagation of PRR1, PRD1, or PR4 (Bryan et al. 1973; Jacoby 1974a; Shahrabadi et al. 1975; Jacoby 1977). Despite these differences, P-1 and P-2 plasmids have similar contents of guanine plus cytosine and share extensive DNA homology (Shahrabadi et al. 1975). Conditions have been found that allow for the selection of P-1-P-2 recombinant plasmids. Recombinant formation is Rec-independent and may have a counterpart in the evolution of plasmids in nature.

The first technique for recombinant plasmid formation utilizes the transmission barrier imposed by the limited host range of P-2 plasmids. A *P. aeruginosa* donor strain carrying both a P-1 plasmid (RP1, RP4, or R751) (Datta et al. 1971; Grinsted et al. 1972; Jobanputra and Datta 1974) and a P-2 plasmid (pMG1, pMG2, pMG5, RPL11) (Jacoby 1974a,b; Korfhagen and Loper 1975) was mated with *E. coli* K12, and selection was imposed for resistance markers on the P-2 plasmid. Transconjugants were obtained, at frequencies of from 10^{-7} to 10^{-9}, in which P-2 markers were expressed and serially transmissible in *E. coli* together with P-1 markers. These plasmids had P-1 incompatibility properties, conferred susceptibility to phages PRR1, PRD1, and PR4 active on P-1- but not on P-2-carrying strains, and behaved on sucrose gradient centrifugation as a unimolecular species with a molecular weight higher than that of the P-1 parent (Jacoby et al. 1976).

For example, in a cross from a *P. aeruginosa* donor carrying RP4 (Cb, Km, Tc)[1] and pMG2 (Gm, Sm, Su, Hg) to *E. coli,* RP4-pMG2 recombinants were obtained that had an additional DNA segment of 12 million daltons and that expressed Gm, Sm, Su, and Hg as well as Cb, Km, and Tc resistances. P-1 plasmids with identical resistance markers on a DNA segment of similar size have recently been isolated from natural sources (Stanisich et al. 1976).

The resistance genes of P-2 plasmids are thus functional in *E. coli* when translocated to a P-1 plasmid, and the failure of P-2 plasmid transmission to enterobacteria suggests that their replication is host-specific. This specificity was also implied by the behavior of an RP4-RPL11 recombinant which was stably inherited in *E. coli* but was quite unstable in *P. aeruginosa.* As separate plasmids, RP4 and RPL11 coexist stably. A plausible explanation is that the RP4-RPL11 hybrid carries both P-1 and P-2 replication systems which cannot function normally as part of a single structure in a host where P-2 replication genes are expressed.

Although recombinant plasmids formed with RP1 or RP4 expressed all the resistance markers of the P-1 parent, translocation of P-2 DNA into R751 usually led to loss of Tp resistance. If a *P. aeruginosa* donor carrying R751 (Tp) and pMG2 (Gm, Sm, Su, Hg) was mated with *E. coli* and if selection was imposed simultaneously for Tp and Sm resistance, recombinants were obtained at a frequency of 10^{-2}- to 10^{-3}-fold lower than if selection were for Sm or Gm resistance alone (Jacoby et al. 1976). Since general homology between P-1 and P-2 DNA would not be expected to lead to insertions at a particular site, interaction between insertion sequences may be involved. In support of this possibility is the observation that recombinant plasmid formation is independent of a functional Rec gene in both donor and recipient (Table 1).

The second technique for recombinant plasmid formation relies on fertility inhibition. Many P-2 plasmids inhibit the fertility of RP4 and other P-1-group plasmids, as indicated by the loss of susceptibility to phages PRR1, PRD1, and PR4 and by decreased RP4 transfer from a (P-2) (RP4) host (Jacoby 1977). On transfer to *P. aeruginosa,* the recombinant RP4-pMG2 was compatible with pMG5, a P-2 plasmid that inhibits RP4 fertility. In a *P. aeruginosa* × *P. aeruginosa* cross, a strain carrying pMG5 (Su, Tm) and RP4-pMG2 (Cb, Km, Tc, Gm, Sm, Su) transferred Tm resistance (pMG5) alone at a frequency of 10^{-2} but transferred Cb, Gm, or Sm resistance at an inhibited frequency of 10^{-4} to 10^{-5}. Some Cb, Gm, or Sm transconjugants lost resistance markers of the original RP4-pMG2 hybrid and acquired Tm resistance. The new combinations of resistances were serially cotransmissible regardless of the marker used

[1] Abbreviations: Cb, carbenicillin; Cm, chloramphenicol; Gm, gentamicin; Km, kanamycin; Sm, streptomycin; Su, sulfonamide; Tc, tetracycline; Tm, tobramycin; Tp, trimethoprim; Hg, mercuric ion; Tn*A*, transposon *A*.

Table 1
Influence of Rec Phenotype on Recombinant Plasmid Formation

P. aerguinosa donor[a]	*E. coli* recipient	Transconjugant frequency[b]	
		Cb^r	Sm^r
Rec^+PA0303 (RP1) (pMG1)	Rec^+KL16	3×10^{-3}	9×10^{-9}
Rec^+PA0303 (RP1) (pMG1)	Rec^-KL16-99	4×10^{-3}	2×10^{-6}
Rec^-JC9010 (RP1) (pMG1)	Rec^+KL16	4×10^{-3}	1×10^{-8}
Rec^-JC9010 (RP1) (pMG1)	Rec^-KL16-99	4×10^{-3}	1×10^{-6}

[a] *P. aeruginosa* strains PA0303 (FP^- *arg18 rec*$^+$) or JC9010 (FP^- *arg18 rec*$^-$) (Jacoby 1974a) containing RP1 and pMG1 were prepared, mixed at a concentration of 2×10^8 cells/ml with an equal concentration of *E. coli* KL16 (Hfr *thi1 relA1 recA*$^+$) or KL16-99 (Hfr *thi1 relA1 recA1 deoB13*) (Bachmann 1972; Bachmann et al. 1976), and incubated for 18 hr at 37°C. Transconjugants were selected on media containing 100 μg/ml carbenicillin or 25 μg/ml streptomycin and thiamine but lacking arginine.

[b] The frequency of transconjugants is expressed relative to the number of recipients at the end of mating.

for selection, again suggesting that recombination between P-1 and P-2 plasmids had been obtained. In this case, the recombinants had the host-range and incompatibility properties of P-2 plasmids. Evidently, fertility inhibition provided a transmission barrier that facilitated DNA translocation from the inhibited (RP4-pMG2) to the inhibiting (pMG5) plasmid. In *E. coli*, Tn*A* (Tn*1*) transposition has been detected under similar conditions (Hedges and Jacob 1974).

Fertility inhibition may have evolved as a mechanism to facilitate conditions for genetic exchange between plasmids.

SUMMARY

Transposition of DNA between plasmids belonging to different incompatibility groups in *Pseudomonas aeruginosa* can be observed when host-range limitation or fertility inhibition restricts plasmid transmissibility. Recombination between plasmids belonging to incompatibility groups P-1 and P-2 is Rec-independent and can be site-specific, suggesting that insertion sequences are involved.

Acknowledgments

G. A. Jacoby was supported by a grant from the National Science Foundation, and A. E. Jacob by a grant from the Medical Research Council of the United Kingdom.

REFERENCES

Bachmann, B. J. 1972. Pedigrees of some mutant strains of *Escherichia coli* K12. *Bact. Rev.* **36**:525.

Bachmann, B. J., K. B. Low and A. L. Taylor. 1976. Recalibrated linkage map of *Escherichia coli* K-12. *Bact. Rev.* **40**:116.

Bryan, L. E., S. D. Semaka, H. M. van den Elizen, J. E. Kinnear and R. L. S. Whitehouse. 1973. Characteristics of R931 and other *Pseudomonas aeruginosa* R factors. *Antimicrob. Agents Chemother.* **3**:625.

Datta, N., R. W. Hedges, E. J. Shaw, R. B. Sykes and M. H. Richmond. 1971. Properties of an R factor from *Pseudomonas aeruginosa. J. Bact.* **108**:1244.

Grinsted, J., J. R. Saunders, L. C. Ingram, R. B. Sykes and M. H. Richmond. 1972. Properties of an R factor which originated in *Pseudomonas aeruginosa* 1822. *J. Bact.* **110**:529.

Hedges, R. W. and A. E. Jacob. 1974. Transposition of ampicillin resistance from RP4 to other replicons. *Mol. Gen. Genet.* **132**:31.

——. 1975. Mobilization of plasmid-borne drug resistance determinants for transfer from *Pseudomonas aeruginosa* to *Escherichia coli. Mol. Gen. Genet.* **140**:69.

Jacoby, G. A. 1974a. Properties of R plasmids determining gentamicin resistance by acetylation in *Pseudomonas aeruginosa. Antimicrob. Agents Chemother.* **6**: 239.

——. 1974b. Properties of an R plasmid in *Pseudomonas aeruginosa* producing amikacin (BB-K8), butirosin, kanamycin, tobramycin and sisomicin resistance. *Antimicrob. Agents Chemother.* **6**:807.

——. 1977. Classification of plasmids in *Pseudomonas aeruginosa.* In *Microbiology, 1977* (ed. D. Schlessinger). American Society for Microbiology, Washington, D.C. (In press.)

Jacoby, G. A., A. E. Jacob and R. W. Hedges. 1976. Recombination between plasmids of incompatibility groups P-1 and P-2. *J. Bact.* **127**:1278.

Jobanputra, R. S. and N. Datta. 1974. Trimethoprim R factors in enterobacteria from clinical specimens. *J. Med. Microbiol.* **7**:169.

Korfhagen, T. R. and J. C. Loper. 1975. RPL11, an R factor of *Pseudomonas aeruginosa* determining carbenicillin and gentamicin resistance. *Antimicrob. Agents Chemother.* **7**:69.

Olsen, R. H. and P. Shipley. 1973. Host range and properties of the *Pseudomonas aeruginosa* R factor R1822. *J. Bact.* **113**:772.

Olsen, R. H., J. Siak and R. H. Gray. 1974. Characteristics of PRD1, a plasmid-dependent broad host range DNA bacteriophage. *J. Virol.* **14**:689.

Shahrabadi, M. S., L. E. Bryan and H. M. van den Elizen. 1975. Further properties of P-2 R-factors of *Pseudomonas aeruginosa* and their relationship to other plasmid groups. *Can. J. Microbiol.* **21**:592.

Stanisich, V. A. 1974. The properties and host range of male-specific bacteriophages of *Pseudomonas aeruginosa. J. Gen. Microbiol.* **84**:332.

Stanisich, V. A., P. M. Bennett and J. M. Ortiz. 1976. A molecular analysis of transductional marker rescue involving P-group plasmids in *Pseudomonas aeruginosa. Mol. Gen. Genet.* **143**:333.

Transposition of a Plasmid DNA Sequence Which Mediates Ampicillin Resistance: General Description and Epidemiologic Considerations

F. Heffron, C. Rubens and S. Falkow
Department of Microbiology
University of Washington
Seattle, Washington 98195

Plasmid-mediated resistance to penicillins and cephalosporins is almost universally associated with the elaboration of penicillinase (β-lactamase). Two general classes of the enzymes are recognized: TEM and O (Hedges et al. 1974; Dale and Smith 1974). The TEM enzyme is by far the most common R-plasmid-mediated β-lactamase and is found on a wide variety of naturally occurring plasmids. Two forms of TEM β-lactamase, TEM-1 and TEM-2, have been distinguished by isoelectric focusing (Matthew and Hedges 1976); out of 77 naturally occurring plasmids that mediate ampicillin resistance (Ap), 70 produced TEM-1, and 7 produced TEM-2. Surprisingly, regardless of the bacterial species or geographic source of the R plasmid, the TEM proteins have been found to be essentially identical. R plasmids are a heterogeneous class of genetic elements of quite diverse origins (Falkow 1975). Yet, regardless of size, or overall guanine + cytosine content, or incompatibility group, all plasmids encoding for a TEM β-lactamase possess a common sequence of DNA of about 3×10^6 daltons (Heffron et al. 1975b).

There is now strong evidence that the structural gene encoding for both the TEM-1 and TEM-2 β-lactamase resides in a 3×10^6-dalton sequence of DNA, Tn*A*,[1] that is capable of transposition from replicon to replicon independently of the normal *rec* functions of the host cell (Hedges and Jacob 1974; Heffron et al. 1975a,b; Kopecko and Cohen 1975; Bennett and Richmond 1976). This finding has been of considerable importance in the epidemiology of certain infectious diseases. Paralleling the discovery that the Ap gene resides in a transposable sequence, other workers have found that the genes for tetracycline resistance (Kleckner et al. 1975), several forms of kanamycin resistance (Berg et al. 1975), chloramphenicol resistance (Gottesman and Rosner 1975), and trimethoprim-streptomycin resistance (Barth et al. 1976) are transposable as well. The following will present our general findings with Tn*A* and some of the epidemiologic implications of transposable antibiotic resistance.

[1] Tn*A* is now termed Tn*1*, Tn*2*, or Tn*3* depending on the plasmid of origin. However, these sequences are very similar and appear identical in DNA heteroduplexes in the electron microscope. In this paper, the generic term Tn*A* is still used for the most part. However, the Tn*A* from RP4 (R64-1) is Tn*1*, that from RSF1030 is Tn*2*, and that from R1*drd*19 is Tn*3*.

Demonstration of the Molecular Nature and Specificity of Tn*A* Insertion

The wide distribution of ampicillin resistance (Ap^r) on many different plasmids and the rapid appearance of R plasmids specifying Ap^r shortly after bacterial populations are exposed to the penicillins at subtherapeutic and therapeutic levels may be taken as evidence of the efficiency of Tn*A* transposition in nature (Heffron et al. 1975b; Rubens et al. 1976). Consequently, we devised a relatively simple laboratory model which permitted us to monitor directly the transposition of Tn*A* from one plasmid to another.

A bacterial strain was constructed carrying a large Ap^r self-transmissible plasmid (R1*drd*19 or R64-1) and a small (5.5×10^6-dalton) nonconjugative R plasmid, RSF1010, which specified resistance to sulfonamide (Su) and streptomycin (Sm). The total plasmid complement can be isolated easily from this strain by dye-buoyant density centrifugation, and the large and small plasmids can be separated from each other by sedimentation through a sucrose gradient. Each of the fractions from this sucrose gradient is then added to $CaCl_2$-treated *E. coli* cells. Under appropriate conditions, such treated cells take up plasmid DNA, which may then replicate. RSF1010 plasmids carrying Tn*A* (recombinant plasmids) were isolated as an intermediate peak of Ap^r transformants between the DNA of the large conjugative plasmid and the DNA of the smaller parental RSF1010 plasmid.

The frequency of transposition of Tn*1* or Tn*3* from the large conjugative plasmid to RSF1010 was approximately the same whether recombinant plasmids were isolated from rec^+ or rec^- hosts, indicating that Tn*A* transposition is independent of the normal *recA* function of *E. coli.* Table 1 shows the results obtained for 200 recombinant RSF1010 plasmids isolated from isogenic rec^+ and rec^- bacterial hosts. Whereas RSF1010 normally specifies SuSm resistance, the recombinant RSF1010 derivatives carrying Tn*A* may be sensitive to one or both of these antibiotics. In all, four phenotypic classes were observed: Group I, $Su^rSm^rAp^r$; Group II, $Su^rSm^sAp^r$, Group III, $Su^sSm^sAp^r$; and Group IV, $Su^sSm^rAp^r$. No significant difference was found between the distributions of phenotypes obtained from rec^+ and rec^- host cells.

The DNA from each of 38 RSF1010 plasmids carrying Tn*A* was isolated and the molecular mass of each measured. All of the plasmids were $8.7 \pm 0.4 \times 10^6$

Table 1
Phenotypes of Recombinants Isolated from Transforming *E. coli* C600 with Plasmid DNA Isolated from rec^+ or rec^- Host

DNA from	Total colonies	Group I $Su^rSm^rAp^r$	Group II $Su^rSm^sAp^r$	Group III $Su^sSm^sAp^r$	Group IV $Su^sSm^rAp^r$	R1*drd*[a]
rec^-	200	173	15	3	6	3
rec^+	200	166	25	4	5	0

[a]Plasmid in clone characterized to be R1*drd*19 and not an RSF1010 recombinant.

daltons. Like the parental RSF1010 plasmid, the recombinant RSF1010 plasmids possessed only a single site susceptible to cleavage by the restriction endonuclease *Eco*RI. Hence, it would appear that the acquisition of Tn*A* by RSF1010 is always associated with an increase in molecular weight of about 3×10^6 daltons. Moreover, the finding of only a single *Eco*RI site in both the parental and recombinant RSF1010 plasmids permitted us to examine directly the molecular consequences of the addition of Tn*A* to RSF1010.

Recombinant plasmid DNA and RSF1010 DNA were mixed, cleaved with *Eco*RI, denatured, renatured, and viewed in the electron microscope. Examination of such heteroduplexes (Fig. 1a) showed that the recombinant plasmids were formed by a single insertion of about 3×10^6 daltons of DNA into RSF1010. This insertion was shown to be identical to the Tn*A* sequence by heteroduplexing the recombinant plasmids with an unrelated plasmid, RSF1030, which had been employed initially to define the Tn*A* sequence (Heffron et al. 1975b). Examination of single-stranded molecules of RSF1010 recombinants revealed yet another facet of the Tn*A* insertion sequence: these single-stranded molecules show intrastrand reannealing of a 140-base-pair (bp) sequence flanking Tn*A*. Both of these short sequences are carried along during the transposition process and do not result from the presence of one such sequence in the recipient RSF1010 and another introduced on insertion of Tn*A*. Hence, Tn*A* is a 3.2×10^6-dalton segment of inserted DNA bounded by inverted complementary sequences of 140 bp.

The examination of a number of different RSF1010/RSF1010::Tn*A* heteroduplexes, after cleavage of plasmids with the restriction endonuclease *Eco*RI, clearly showed that Tn*A* could be inserted into several sites on the RSF1010 genome. In order to determine the site of Tn*A* insertion precisely, it was necessary to have an internal molecular marker to distinguish the left and right *Eco*RI ends. For this purpose, R684, a plasmid closely related to but slightly larger than RSF1010 was used (Barth and Grinter 1974). RSF1010/R684 heteroduplexes are homologous except for a small, single-stranded insertion loop (Fig. 1b). Figure 1c shows a heteroduplex between R684 and a recombinant RSF1010 plasmid. Two single-stranded loops were observed, one corresponding to Tn*A* and the other corresponding to the marker insertion from R684. The *Eco*RI site closest to the R684 marker insertion was arbitrarily designated the right-hand end of the molecule. In this way, the site of Tn*A* insertion for 38 (21 derived from a *rec*$^+$ host, 17 from a *rec*$^-$ host) independent RSF1010 recombinant plasmids could be determined unequivocally to within 100 base pairs by heteroduplex analysis. Figure 2 shows the distribution of these insertion sites on the RSF1010 genome. Statistical analysis of this distribution discloses two significant points: first, the distribution of Tn*A* insertion sites is not different in recombinant plasmids isolated from *rec*$^+$ or *rec*$^-$ hosts, and second, the distribution of Tn*A* sites is nonrandom. This latter finding means that even though 19 distinct, nonoverlapping Tn*A* insertion sites can be recognized (determined statistically by computing the 99% confidence interval for the site of

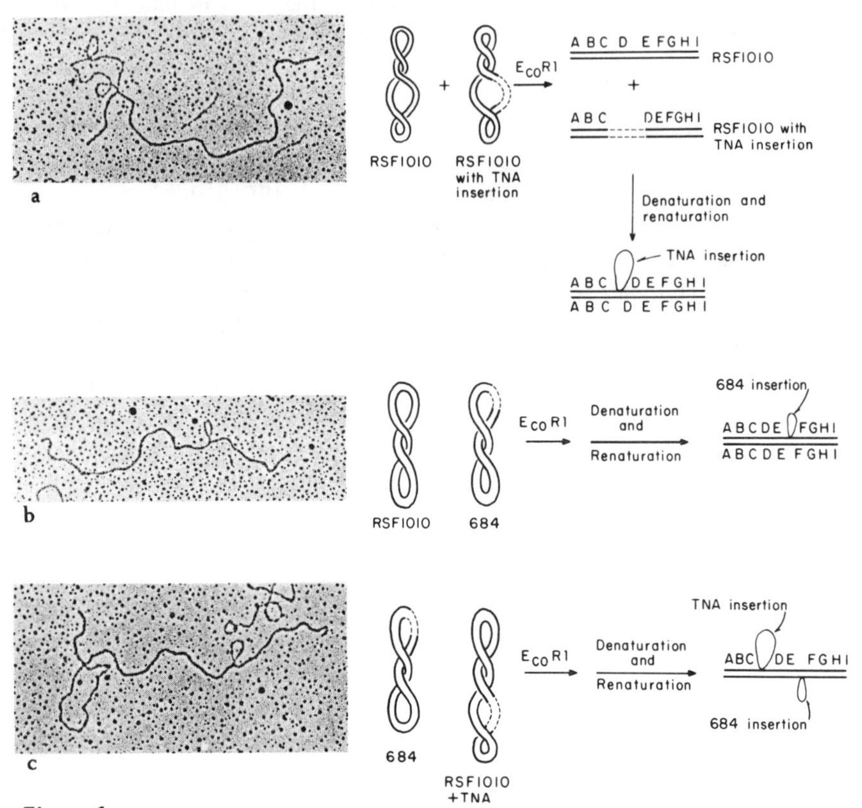

Figure 1

Demonstration of the insertion of Tn*A* into the RSF1010 plasmid. (*a*) *Eco*RI-cleaved RSF1010 DNA and the DNA of a recombinant RSF1010::Tn*A* plasmid are shown heteroduplexed together. The insertion of Tn*A* is seen as a single loop of single-stranded DNA. (*b*) Heteroduplex of RSF1010/R684 DNA, showing the single small insertion loop of R684 which serves to distinguish the left and right ends of the *Eco*RI-cleaved linear molecule. The *Eco*RI-generated end closest to the R684 insertion marker is arbitrarily designated as the right-hand end of the molecule. (*c*) The same RSF1010::Tn*A* recombinant plasmid as in *a* is shown here heteroduplexed with R684. Two noninteracting insertion loops corresponding to Tn*A* (*left*) and R684 (*right*) are seen.

insertion and identifying those insertions whose confidence intervals do not overlap), Tn*A* insertion appears to show some degree of specificity. Obviously, the underlying specificity of insertion is not high and presumably reflects either insertions at a relatively common short sequence of nucleotides or, alternatively, insertions clustered as noted for mutational "hot spots."

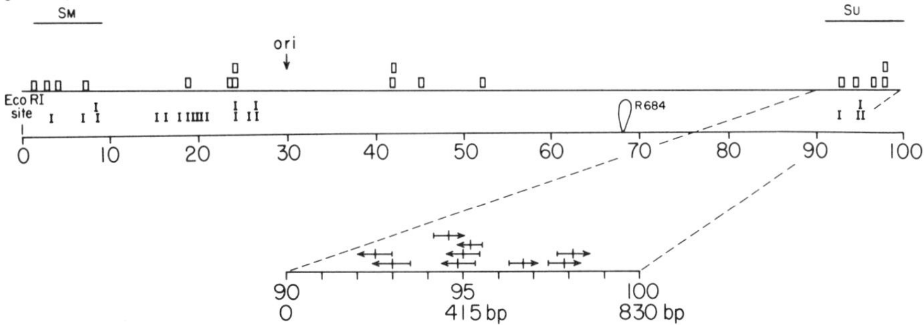

Figure 2

Map of the site of Tn*A* insertions into the plasmid RSF1010. The left and right *Eco*RI ends are distinguished by the insertion found in R684. Insertions obtained in a *rec*⁻ host are represented by □; those obtained in a *rec*⁺ host are represented by I. The site of Tn*A* insertion has been determined by measuring 25–50 separate heteroduplex molecules and is statistically accurate (99% confidence interval) to within ± 100 base pairs. The expanded section of the right-hand 10% of the map shows the orientation and confidence interval for nine insertions that affect sulfonamide resistance (Su). The orientation of the insertion is represented by the direction of the arrow. For insertions in orientation M (for mutagenic), the arrow points toward the nearer *Eco*RI site, whereas for insertions in orientation P (polar), the arrow points toward the center of the molecule. The length of the arrow in each case represents the 99% confidence interval about the mean. The origin of replication (ori) for RSF1010 has been determined in pulse-label experiments (J. de Graaff, F. Heffron, J. H. Crosa and S. Falkow, in prep.). Abbreviations: kb, kilobase; bp, base pair; Sm, streptomycin; Su, sulfonamide.

Tn*A* Inserts in Two Orientations and Is Mutagenic

Whether derived from a rec^+ or rec^- host, there was an excellent correlation between the site of Tn*A* insertion and the phenotype of the recombinant plasmid. Insertions of Tn*A* into the left-hand end of the RSF1010, from 1% to 9% of the molecular length, were all of the Group II phenotype ($Su^rSm^sAp^r$; Table 1). Presumably, this represents instances in which Tn*A* has been inserted in the gene specifying streptomycin phosphotransferase, thereby causing its inactivation. Insertions of Tn*A* located between 92% and 99% from the left-hand end of the molecule belonged to either the Group III phenotype ($Su^sSm^sAp^r$; Table 1) or the Group IV phenotype ($Su^sSm^rAp^r$; Table 1). Clearly, insertions in this region of the molecule always lead to a loss in sulfonamide resistance, thus suggesting that Tn*A* insertion had occurred within the structural gene for sulfonamide resistance. However, in some cases, Tn*A* insertion appeared to have only a mutagenic effect on Su (Group IV), whereas in other cases (e.g., Group III) it not only inactivated Su but also had a polar effect on Sm as well. These differences were all the more intriguing since, in several cases, a recombinant plasmid exhibiting the Group III phenotype and a recombinant plasmid exhibiting the Group IV phenotype possessed Tn*A* insertions that mapped at the same site. Further studies showed that the difference between Group III and Group IV recombinant plasmids was in the orientation of insertion of Tn*A* within RSF1010 (Rubens et al. 1976). This conclusion was based on several lines of evidence. For example, as illustrated in Figure 3, heteroduplexes between Group III and Group IV plasmids invariably show two, noninteracting, single-stranded loops corresponding in position and size to Tn*A*. Additional heteroduplex analyses, as well as analyses of endonuclease digests of recombinant plasmids of the Group III and Group IV phenotypes, have revealed that Tn*A* inserts without circular permutation via a specific region, presumably the 140-bp inverted repeat, in either of two orientations. In one orientation, which we designate M, Tn*A* insertion leads to a loss of Su^r but permits expression of Sm^r; in the opposite orientation, which we designate P, Tn*A* insertion within the Su gene has a polar effect on Sm as well. This latter finding suggests that the

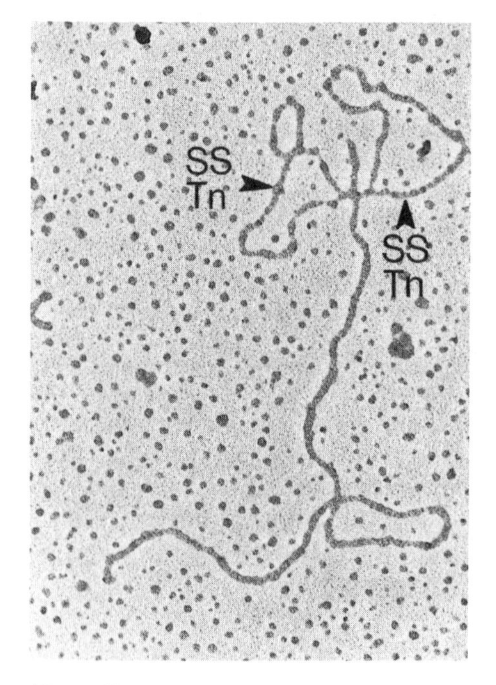

Figure 3
A DNA heteroduplex showing two Tn*A* insertions, in opposite orientations, which have occurred at sites within 20 bp of each other (at a position corresponding to 95% in the map shown in Fig. 2).

SuSm genes comprise a single transcriptional unit in which the Sm structural gene(s) is distal. As might be expected, all recombinant plasmids of the Group III phenotype were found to have Tn*A* inserted in orientation P, and all Group IV plasmids had Tn*A* inserted in orientation M. The orientation of nine distinct Tn*A* insertions into the Su region is summarized in the expanded sequence shown in Figure 2. The two orientations occur with approximately equal frequency in this region of RSF1010. The finding that Tn*A* may insert into recipient DNA in two orientations which have different phenotypic effects is analogous to the observation that the insertion of IS*2* can prevent or promote gene expression within an operon depending upon its orientation of integration (Saedler et al. 1974).

Tn*A* insertions outside the Su and Sm regions of RSF1010 all exhibit the Group I phenotype ($Su^rSm^rAp^r$). Nevertheless, two members of this phenotypic class which have insertions at 42 ± 1% from the left-hand end of RSF1010 possess a significantly higher level of resistance to these three drugs. Further examination of these plasmids has shown that they have an increased copy number (roughly double) per cell over that of the parental RSF1010 plasmid or other Tn*A*-containing derivatives. Perhaps Tn*A* insertion at this specific site affects a plasmid gene which controls the rate of initiation of plasmid replication or some other aspect of replication control. This idea is supported by the recent finding (J. de Graaff, F. Heffron, J. H. Crosa and S. Falkow, in prep.) that the origin of RSF1010 replication is located only some 0.7×10^6 daltons from the 42% site of Tn*A* insertion (see Fig. 2).

As noted earlier, two forms of the TEM β-lactamase, TEM-1 and TEM-2, have been distinguished by isoelectric focusing. Transposable sequences mediating both TEM-1 and TEM-2 have been examined. We have not been able to measure any significant difference in size between the Tn*A* transposable sequence (Tn*3*) derived from R1*drd*19 which encodes for a TEM-1 β-lactamase and the Tn*A* transposable sequence (Tn*1*) derived from RP4 which mediates a TEM-2 β-lactamase. Nevertheless, DNA-DNA duplex studies indicate that Tn*3* and Tn*1* may differ by as much as 13% in their polynucleotide sequences. Moreover, Tn*3* and Tn*1* yield slightly different fragment patterns when cleaved with the *Hin*cII restriction endonuclease (Rubens et al. 1976). The differences between the Tn*3* and Tn*1* transposons are not large, and, in general, one is struck more by their high degree of homology than by their differences. As noted by Matthew and Hedges (1976), the origin of the transposon probably predates the divergence of the TEM enzyme; furthermore, we suppose it predates the other divergent regions as well. Given the wide dissemination of Tn*A* in different bacterial species and on diverse plasmid species (Heffron et al. 1975b; Matthew and Hedges 1976), one might expect to find a considerable degree of heterogeneity among different Tn*A* transposons. However, this does not appear to be the case. Thus many sequences other than those coding for the TEM β-lactamase have tended to be conserved and may be necessary for transposition to occur. At any rate, we have not noted any difference in the specificity of insertion or orientation preference of Tn*3* and Tn*1*.

Epidemiologic Considerations

The ability of drug resistance genes to be transposed between DNA molecules independent of bacterial *rec* functions surely helps to explain the rapid evolution of R plasmids that possess varied permutations of antibiotic resistance determinants. It also helps to explain why apparently identical resistance genes are common to a wide variety of R plasmids of otherwise disparate evolutionary origin (Heffron et al. 1975b; Falkow 1975). There is no firm evidence showing that transposition occurs at the same frequency in nature as that observed in the laboratory. However, given the enormous selective pressure of the therapeutic use of antibiotics in human and veterinary medicine and of their virtually uncontrolled subtherapeutic use in animal feeds, there is a high probability of genetic events involving antibiotic resistance genes occurring (Falkow 1975). The evolutionary "success" of R plasmids in modern society probably reflects the efficiency of transposition in nature. Plasmids to which antibiotic resistance genes have been transposed in the laboratory have their precise counterparts in nature. We have found that naturally occurring $Su^rSm^rAp^r$ plasmids isolated from *Salmonella typhimurium, Proteus mirabilis,* and *Escherichia coli* are precisely identical to several of the RSF1010::Tn*A* recombinant plasmids constructed in our laboratory (F. Heffron, C. Rubens and S. Falkow, in prep.).

Since plasmids are so widespread among bacterial species (Falkow 1975), we must consider their potential for continuously acquiring and disseminating drug resistance determinants. We know that Ent plasmids and plasmids that contribute to pathogenicity (So et al. 1975) may acquire transposable antibiotic resistance genes from an R plasmid (M. So, F. Heffron and S. Falkow, in prep.). The transposition events may be widespread in the enteric species. Transposition also provides a plausible mechanism by which previously uniformly antibiotic-sensitive species well outside the enteric sphere may be converted to antibiotic resistance by the addition of a transposable DNA sequence to the preexisting indigenous plasmid pool. Such a sequence of events has been postulated to explain the sudden emergence of ampicillin and tetracycline resistance in certain strains of *Haemophilus influenzae* type b, a causative agent of meningitis and other acute infections in children (Elwell et al. 1975). It has been firmly established that the ampicillin resistance genes in these *Haemophilus* strains reside in a transposon which is indistinguishable from Tn*A* found in enteric species (de Graaff et al. 1976). To what extent transposable antibiotic resistance genes may cross species barriers is not known. The enteric R plasmids are already considered to be promiscuous genetic elements capable of inhabiting over 36 bacterial genera (Falkow 1975), but perhaps their transposable antibiotic resistance genes have an even more widespread potential for dissemination in microorganisms.

SUMMARY

Ampicillin resistance resides in a 3×10^6-dalton DNA segment, Tn*A*, that is bounded by inverted complementary sequences of about 140 bp. Like IS sequences, insertions of Tn*A* are mutagenic and occur, without circular permutation, in two different orientations. Insertions in one orientation (P) are strongly polar on distal gene expression, whereas insertions in the other orientation (M) are less polar and may even have a promoter effect.

Insertion of Tn*A* into the plasmid RSF1010 is precise in that an identical sequence is inserted each time and there is no apparent loss of recipient DNA. We have recently noted, however, that deletions accompanying Tn*A* insertion may occur in RSF1010 at a low rate ($\leqslant 3\%$ of the time), and that Tn*A* insertion into the F plasmid is accompanied by deletion formation at a relatively high rate ($\geqslant 20\%$).

Furthermore, Tn*A* insertion occurs independently of the *E. coli recA* function. Tn*A* is able to insert into a minimum of 19 distinct sites on the genome of the plasmid RSF1010. Insertion in nonrandom, suggesting that Tn*A* may recognize a specific, but fairly common, nucleotide sequence at or near the site of insertion into the recipient genome.

The TEM β-lactamase, determined by a gene that forms part of Tn*A*, is found in two forms, TEM-1 and TEM-2. Both of these enzymes are associated with transposable DNA sequences, called Tn*3* and Tn*1*, respectively, which are indistinguishable in size, specificity of insertion, and orientation preference. Tn*3* and Tn*1* show only minor differences in polynucleotide sequence ($\leqslant 13\%$). Furthermore, recent studies have shown that Tn*A* transposons from a variety of sources also have few differences in polynucleotide sequence. Given the wide dissemination of Tn*A* (particularly Tn*3*) in different bacterial species and on diverse plasmids, the conservation of Tn*A* sequences is more striking than their differences. The apparent conservation of Tn*A* sequences is supported by the finding, reported elsewhere in this volume (Heffron et al.), that deletion mutants affecting Tn*A* transposition map across 2×10^6 daltons of this 3.2×10^6-dalton sequence.

Epidemiologic studies show that transposition of antibiotic resistance in nature has played a key role in the evolution of R plasmids. Indeed, several plasmids constructed in the laboratory by Tn*A* transposition have been shown to have precise counterparts in nature.

Acknowledgments

This work was supported by a grant from the National Science Foundation (PCM 75-14174) and by contract DADA17-72-C-2149 from the U. S. Army Research and Development Command.

REFERENCES

Barth, P. T. and N. J. Grinter. 1974. Comparison of the deoxyribonucleic acid molecular weights and homologies of plasmids conferring linked resistance to streptomycin and sulfonamides. *J. Bact.* **120**:618.

Barth, P. T., N. Datta, R. W. Hedges and N. J. Grinter. 1976. Transposition of a deoxyribonucleic acid sequence encoding trimethoprim and streptomycin resistances from R483 to other replicons. *J. Bact.* **125**:800.

Bennett, P. M. and M. H. Richmond. 1976. The transposition of a discrete piece of DNA carrying an *amp* gene between replicons in *Escherichia coli. J. Bact.* **126**:1.

Berg, D. E., J. Davies, B. Allet, and J. Rochaix. 1975. Transposition of R factor genes to bacteriophage. *Proc. Nat. Acad. Sci.* **72**:3628.

Dale, J. W. and J. T. Smith. 1974. R factor-mediated beta-lactamases that hydrolyze oxacillin: Evidence for two distinct groups. *J. Bact.* **119**:351.

de Graaff, J., L. Elwell and S. Falkow. 1976. Molecular nature of two beta-lactamase specifying plasmid isolates from *Haemophilus influenzae* type b. *J. Bact.* **126**:439.

Elwell, L. P., J. de Graaff, D. Seiber and S. Falkow. 1975. Plasmid-linked ampicillin resistance in *Haemophilus influenzae* type b. *Infect. Immunol.* **12**:404.

Falkow, S. 1975. *Infectious multiple drug resistance.* Pion Limited, London.

Gottesman, M. M. and J. L. Rosner. 1975. Acquisition of a determinant for chloramphenicol resistance by coliphage lambda. *Proc. Nat. Acad. Sci.* **72**: 5041.

Hedges, R. W., N. Datta, P. Kontomichalou and J. T. Smith. 1974. Molecular specificities of R factor-determined beta-lactamases: Correlation with plasmid compatibility. *J. Bact.* **117**:56.

Hedges, R. W. and A. Jacob. 1974. Transposition of ampicillin resistance from RP4 to other replicons. *Mol. Gen. Genet.* **132**:31.

Heffron, F., C. Rubens and S. Falkow. 1975a. Translocation of a plasmid DNA sequence which mediates ampicillin resistance: Molecular nature and specificity of insertion. *Proc. Nat. Acad. Sci.* **72**:3623.

Heffron, F., R. Sublett, R. W. Hedges, A. Jacob and S. Falkow. 1975b. Origin of the TEM beta-lactamase gene found on plasmids. *J. Bact.* **122**:250.

Kleckner, N., R. K. Chan, B.-K. Tye and D. Botstein. 1975. Mutagenesis by insertion of a drug-resistance element carrying an inverted repetition. *J. Mol. Biol.* **97**:561.

Kopecko, D. J. and S. N. Cohen. 1975. Site-specific *recA* independent recombination between bacterial plasmids: Involvement of palindromes at the recombinational loci. *Proc. Nat. Acad. Sci.* **72**:1373.

Matthew, M. and R. W. Hedges. 1976. Analytical isoelectric focusing of R factor-determined beta-lactamases: Correlation with plasmid compatibility. *J. Bact.* **125**:713.

Rubens, C., F. Heffron and S. Falkow. 1976. Transposition of a plasmid DNA sequence which mediates ampicillin resistance: Independence from host *rec* functions and orientation of insertion. *J. Bact.* (in press).

Saedler, H., H.-J. Reif, S. Hu and N. Davidson. 1974. IS2. A genetic element for turn-off and turn-on of gene activity in *E. coli. Mol. Gen. Genet.* **132**:265.

So, M., J. F. Crandall, J. H. Crosa and S. Falkow. 1975. Extrachromosomal determinants which contribute to bacterial pathogenesis. In *Microbiology, 1974* (ed. D. Schlessinger), p. 16. American Society for Microbiology, Washington, D. C.

Deletions Affecting Transposition of an Antibiotic Resistance Gene

F. Heffron, P. Bedinger, J. Champoux and S. Falkow
Department of Microbiology, School of Medicine
University of Washington
Seattle, Washington 98195

A number of plasmid-mediated antibiotic resistance genes have been found to reside in discrete DNA sequences capable of *recA*-independent transposition (Hedges and Jacob 1974; Berg et al. 1975; Gottesman and Rosner 1975; Hedges et al. 1975; Heffron et al. 1975a,b; Kleckner et al. 1975; Cohen and Kopecko 1976). One such transposable sequence (Tn*A*) carrying a gene for ampicillin resistance (Hedges and Jacob 1974; Heffron et al. 1975a,b; Bennett and Richmond 1976; Cohen and Kopecko 1976) is 4.8 kilobases (kb) in size and is flanked by short inverted repeat sequences of about 140 base pairs (bp). Tn*A*[1] resembles the transposable IS sequences of *Escherichia coli* in its genetic effects (Jordan et al. 1968; Shapiro 1969; Hirsch et al. 1969; Malamy et al. 1972). It seems likely that the transposable antibiotic resistance determinants are evolutionarily related to IS sequences. However, the origin of such elements and their mechanism of transposition remain unclear. In this paper, we describe the isolation and characterization of deletions of Tn*A* in an attempt to define better the structural and biochemical basis of transposition.

The plasmid pMB8 (received from M. Betlach and H. W. Boyer) is a 1.8×10^6-dalton derivative of ColE1 which specifies immunity to colicin E1 (Col^{imm}) but not to colicin biosynthesis. pMB8, like ColE1, has a strict dependence upon DNA polymerase I (polI) for its maintenance and replication within *E. coli* host cells (Kingsbury and Helinski 1970). We have transposed Tn*A* (Tn*3*) from the conjugative plasmid R1*drd*19 to pMB8, using methods described previously (Heffron et al. 1975b). The resulting plasmid, pMB8::Tn*3*, is 5.0×10^6 daltons in size. Since Tn*3* constitutes more than 60% of the total DNA sequences of this plasmid, it provided an excellent model for the isolation of deletions within Tn*3*.

A method for obtaining random deletions in circular DNA in vitro by enzymatic digestion has been described recently (Carbon et al. 1975). Although employing the same general reasoning, we developed a different sequence of enzymatic reactions. Covalently closed circular (CCC) pMB8::Tn*3* DNA was treated with pancreatic DNase I in the presence of Mg^{++} to introduce, on average, 0.3 single-strand breaks per molecule. After separation of the nicked

[1] Tn*A* is now termed Tn*1*, Tn*2*, or Tn*3* depending on the plasmid of origin. However, these sequences are virtually identical, and, in this paper, the generic term Tn*A* is still at times used. (See note to preceding article by Heffron et al., this volume.)

(OC) molecules from the unnicked CCC molecules by dye-isopycnic centrifugation, the OC DNA was treated with *E. coli* exonuclease III to introduce a small (on average $\leqslant$ 600 bp) single-stranded gap. The gapped DNA was in turn treated with S1 endonuclease to remove the short, single-stranded sequence and leave linear, permuted plasmid DNA. The linear DNA was purified by sedimentation through a neutral sucrose gradient and used to transform $CaCl_2$-treated *E. coli* C600 containing F::Tn*K*(3).[2] The transformants were selected for ampicillin resistance.

As noted earlier, pMB8 has a strict dependence upon the polI function of *E. coli.* We exploited this property in our search for mutants that were transposition-defective. Individual Ap^r transformant clones of C600 [F::Tn*K*(3); pMB8::Tn*3*] were inoculated into brain-heart infusion broth, and donor cells (5×10^6/ml) in the exponential phase of growth were subsequently mixed with recipient cells (6×10^7/ml) of *E. coli* W3110 (F^- $polI^-$ nal^r) for 6 to 7 hours. The mating mixture was plated (0.02 ml) on MacConkey agar containing 250 μg/ml of carbenicillin and 30 μg/ml of nalidixic acid. Since pMB8::Tn*3* cannot be stably maintained in $polI^-$ cell lines, Ap^rnal^r transconjugant clones appearing on this selective medium represent those *E. coli* W3110 cells that have received F::Tn*K*(3) carrying a transposed Tn*3* segment. More complete details of this technique and its application for "marking" cryptic plasmids will be published elsewhere (M. So, F. Heffron and S. Falkow, in prep.). Transposition of Tn*3* from pMB8::Tn*3* normally occurs at a frequency of about 7×10^{-4} per F^+ transconjugant. Examination of 600 clones transformed with enzymatically treated pMB::Tn*3* revealed clones that failed to yield Ap^rnal^r transconjugant clones when mixed with the $polI^-$ recipient cells. Thus Tn*3* transposition could not be detected in these clones, all 25 of which were subsequently found to contain deleted pMB8::Tn*3* derivatives.

Using the four restriction endonucleases, *Bam*HI, *Eco*RI, *Hin*cII, and *Hae*II, we have determined a map of pMB::Tn*3* (Fig. 1, top line) by sequentially digesting plasmid DNA with pairs of enzymes and examining the resultant linear fragments by agarose gel electrophoresis. As anticipated, treatment of DNA from the pMB8::Tn*3* deletion mutants with these enzymes showed agarose gel electrophoresis fragment patterns which differed from the parental plasmid and permitted precise mapping of many of the deletions. The mapping of the deletion mutants was further facilitated by electron microscope heteroduplex analysis. Both pMB8 and pMB8::Tn*3* contain a single *Eco*RI restriction site. Different pairs of deletion mutants were cleaved to linear fragments with *Eco*RI, denatured and renatured, and then viewed in the electron microscope. As illustrated in Figure 2, such heteroduplexes appear as a duplex with two, well-defined, single-stranded deletion loops.

The results obtained from mapping several of the deletion mutants are shown

[2] F::Tn*K*(3) is a derivative of F which contains a 3.2-kb kanamycin insertion into *Eco*RI fragment f14 of F. Tn*K*(3) is our isolate of a transposon carrying kanamycin resistance and has not yet been assigned a specific number.

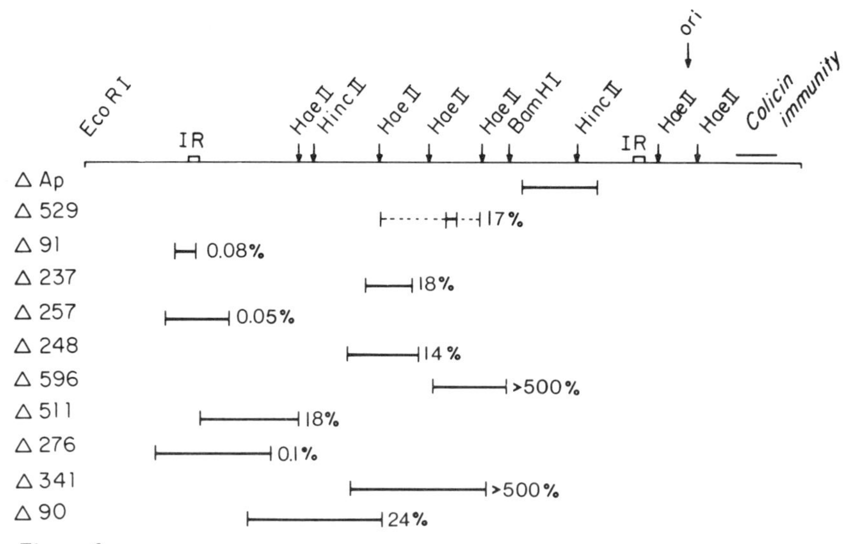

Figure 1
Deletions affecting transposition of Tn*A* and the degree to which they may be complemented. The upper line drawing represents *Eco*RI-cleaved pMB8::Tn*3*. This plasmid is 7.4 kb in length and specifies colicin immunity and ampicillin resistance. The two bars labeled IR represent the short, inverted repeat sequences that flank Tn*3* and accompany its insertion; ori designates the origin of replication. Sites cleaved by restriction enzymes are represented by arrows adjacent to the abbreviation for the enzyme. Deletions are represented by solid bars for the DNA sequence that has been deleted. The deletion Δ529 is only 60 bp and has been determined to lie in the region shown dotted. All the deletions are transposition-negative, with the possible exception of the one designated ΔAp, which is ampicillin-sensitive and was isolated in another compatible plasmid RSF1010::Tn*1* but not in pMB8::Tn*3*. Complementation values are given at the end of the solid bar representing the deletion and were obtained as described in the text.

in the lower part of Figure 1. All of these deletions lead to a loss in the ability of Tn*A* to transpose (with the exception of ΔAp, isolated in RSF1010::Tn*1*, as indicated in the legend to Fig. 1). Some of the deletions affect little more than one of the inverted repeat sequences, clearly establishing, for the first time, the requirement of the inverted terminal repeat for transposition.

All of the deletion mutants isolated thus far have encompassed only a single inverted repeat sequence on the "left"-hand end of the Tn*A* map. The absence of deletions in Tn*A* affecting the "right"-hand inverted repeat probably reflects the fact that the genetic selection employed requires preservation of the structural gene for ampicillin resistance and the origin of replication in any transformant. Thus, although it is clear that a deletion affecting the left-hand

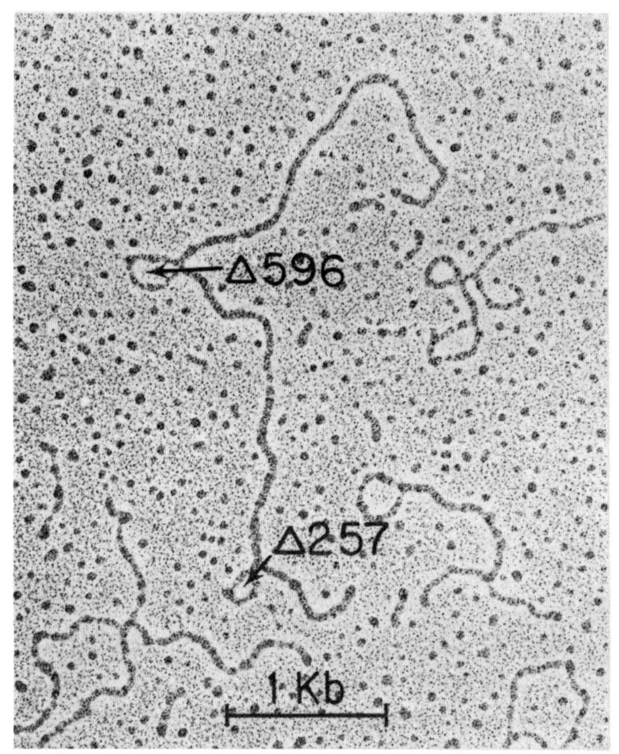

Figure 2
Visualization of transposition-negative deletions. Two *Eco*RI-cleaved transposition mutants are shown heteroduplexed against each other. The deletion nearer the *Eco*RI end is Δ257, which deletes one inverted repeat sequence. The deletion near the center of the molecule is Δ596.

inverted repeat abolishes translocation, at present we do not know whether deletion of the right-hand repeat sequence would have a similar effect.

Perhaps the most interesting finding was that deletions encompassing only a small central region of Tn*3* were transposition-defective, thus suggesting that an essential transposition function was encoded in this region. To pursue this possibility further, complementation studies were attempted. *E coli recA*$^-$ strains were constructed containing three coexisting plasmids: F::Tn*K*(3), one of the pMB8::Tn*3* deletion mutants, and RSF1010::Tn*1* ΔAp. The latter plasmid encodes streptomycin resistance and contains a deletion encompassing the structural gene for the TEM β-lactamase. The premise was that functions deleted in the pMB8 transposition-negative mutants might be complemented, in *trans,* by a protein produced by RSF1010::Tn*1* ΔAp. We supposed this would permit transposition of the Tn*3* segment of pMB8 to F::Tn*K*(3). The F::Tn*K*(3)::Tn*3* would in turn be detected as Ap^r transconjugants by transfer to W3110 F$^-$ polI nal^r, as described earlier. The degree of complementation for any deletion mutant was determined by comparison with a transposition-positive control strain containing F::Tn*K*(3), RSF1010::Tn*1* ΔAp, and the parental pMB8::Tn*3* plasmid. Negative controls included strains containing F::Tn*K*(3) and the transposition-negative deletion but not the RSF1010 derivative. The results of these complementation studies are shown in Figure 1. The deletions Δ91, Δ257, and Δ276, which included one of the inverted segments, were not complemented at a significantly greater frequency than were the negative controls. The failure of the inverted segment to be complemented again underlines the structural requirement for this region in transposition. This finding is also consistent with our working hypothesis that the inverted segment serves as a specific recognition site which is essential for insertion and/or excision of Tn*3* during transposition. The deletions Δ90, Δ237, Δ248, Δ511, and Δ529 are located in the central region of Tn*3* and were complemented at a frequency of between 14% and 24% of the transposition-positive control. These results suggest that essential functions, presumably one or more proteins, are encoded in the central region of Tn*3*. Deletions Δ341 and Δ596 gave anomalous results. Although these deletions span the central region of Tn*3* and normally have a transposition-negative phenotype, they appear to transpose at a much higher frequency than normal when complemented. Moreover, although transconjugant clones arising from complementation of centrally located deletions were colicin-sensitive as expected, the Ap^r transconjugants from complementation studies with Δ341 and Δ596 were usually colicin-immune (59/60). In addition, these latter transconjugants could in turn transfer F Km^{r}Ap^r and colicin immunity to an F$^-$ recipient at a very high frequency. It appears likely, therefore, that these transconjugants contain a substantial fraction (or all) of the pMB8 genome in addition to Tn*3*, a phenomenon currently being investigated more thoroughly.

The simplest model to explain the origin of transposable sequences such as Tn*A* is to assume that IS sequences have at some time in the past become inserted at either side of an antibiotic resistance gene. We can imagine that the IS sequences subsequently transposed and carried along the intervening DNA. Several investigators (Heffron et al. 1975a; Kleckner et al. 1975; Saunders 1975; Cohen and Kopecko 1976) have emphasized that there might be a fundamental relationship between inverted repeat sequences and transposable segments of DNA. The results of this study directly confirm this view. However, our results also make clear that, at least with respect to Tn*3*, the DNA flanked by the inverted repeat DNA sequences is not merely "excess baggage" that happens to include a selectively advantageous antibiotic resistance gene. Rather, the central region of Tn*3* encodes for essential transposition functions. Tn*A* thus begins to emerge as a highly evolved genetic element that may control and even mediate its own *recA*-independent recombination and transposition.

SUMMARY

We have isolated deletions in the Tn*A* (Tn*3*) transposon that determines resistance to ampicillin. The deletions that remove the left-end inverted repeat sequences render Tn*3* nontransposable. This finding clearly establishes the role of inverted repeat sequences in the transposition of Tn*A*. The deletions that span the center of Tn*3* also cause a defect in transposition. This defect can be overcome by complementation with another Tn*A* (Tn*1*). This result demonstrates that Tn*A* controls its own transposition.

Acknowledgments

This work was supported by a grant from the National Science Foundation (PCM 75-14174) and by contract DADA17-72-C-2149 from the U. S. Army Research and Development Command.

REFERENCES

Bennett, P. M. and M. H. Richmond. 1976. The translocation of a discrete piece of DNA carrying an Amp gene between replicons in *Escherichia coli*. *J. Bact.* **126**:1.

Berg, D. E., J. Davies, B. Allet and J. Rochaix. 1975. Transposition of R-factor genes to bacteriophage λ. *Proc. Nat. Acad. Sci.* **72**:3628.

Carbon, J., T. E. Shenk and P. Berg. 1975. Biochemical procedure for production of small deletions in simian virus 40 DNA. *Proc. Nat. Acad. Sci.* **72**:1392.

Cohen, S. N. and D. J. Kopecko. 1976. Structural evolution of bacterial plasmids: Role of translocating genetic elements and DNA sequence insertion. *Fed. Proc.* **35**:2031.

Gottesman, M. M. and J. L. Rosner. 1975. Acquisition of a determinant for chloramphenicol resistance by coliphage lambda. *Proc. Nat. Acad. Sci.* **72**:5041.

Hedges, R. W. and A. E. Jacob. 1974. Transposition of ampicillin resistance from RP4 to other replicons. *Mol. Gen. Genet.* **132**:31.

Hedges, R. W., A. E. Jacob, N. Datta and J. N. Coetzee. 1975. Properties of plasmids produced by recombination between R factors of groups J and FII. *Mol. Gen. Genet.* **140**:289.

Heffron, F., C. Rubens and S. Falkow. 1975a. Translocation of a plasmid DNA sequence which mediates ampicillin resistance: Molecular nature and specificity of insertion. *Proc. Nat. Acad. Sci.* **72**:3623.

Heffron, F., R. Sublett, R. Hedges, A. Jacob and S. Falkow. 1975b. Origin of the TEM beta-lactamase gene found on plasmids. *J. Bact.* **122**:250.

Hirsch, H. J., P. Starlinger and P. Bracket. 1972. Two kinds of insertions in bacterial genes. *Mol. Gen. Genet.* **119**:191.

Kingsbury, D. T. and D. R. Helinski. 1970. DNA polymerase as a requirement for the maintenance of the bacterial plasmid colicinogenic factor E1. *Biochem. Biophys. Res. Comm.* **41**:1538.

Kleckner, N., R. K. Chan, B.-K. Tye and D. Botstein. 1975. Mutagenesis by insertion of a drug-resistance element carrying an inverted repetition. *J. Mol. Biol.* **97**:561.

Jordan, E., H. Saedler and P. Starlinger. 1968. O° and strong polar mutations in the *gal* operon are insertions. *Mol. Gen. Genet.* **102**:353.
Malamy, M. H., M. Fiandt and W. Szybalski. 1972. Electron microscopy of polar insertions in the *lac* operon of *Escherichia coli. Mol. Gen. Genet.* **119**:207.
Saunders, J. R. 1975. Transposable resistance genes. *Nature* **258**:3844.
Shapiro, J. A. 1969. Mutations caused by insertion of genetic material into the galactose operon of *Escherichia coli. J. Mol. Biol.* **40**:93.

Promotion of Insertions and Deletions by Translocating Segments of DNA Carrying Antibiotic Resistance Genes

J. Brevet, D. J. Kopecko, P. Nisen and S. N. Cohen
Department of Medicine
Stanford University School of Medicine
Stanford, California 94305

The role that illegitimate recombination plays in the structural evolution of plasmid, phage, and bacterial genomes has become increasingly evident (Cohen and Kopecko 1976; Starlinger and Saedler 1976). Such recombination involves the joining of segments at DNA sites that lack extensive genetic homology, whereas "ordinary" or "general" genetic recombination involves homologous DNA sequences. In *Escherichia coli,* illegitimate recombination is independent of the *recA* gene product and occurs with a variety of structurally defined genetic elements that are capable of insertion into and excision from multiple sites. Studies showing the *recA*$^-$ independent translocation of genes that express resistance to ampicillin (Ap) (Kopecko and Cohen 1975; Heffron et al. 1975), tetracycline (Tc) (Kleckner et al. 1975; Foster et al. 1975), kanamycin (Km) (Berg et al. 1975), chloramphenicol (Cm) (Gottesman and Rosner 1975), and trimethoprim-streptomycin (Tp-Sm) (Barth et al. 1976) have been reported recently.

In an earlier paper (Kopecko and Cohen 1975), we were concerned primarily with the *recA* independence of the recombinational event and with the presence of inverted repeats at the boundaries of the ampicillin (Ap) resistance translocating DNA segments. During those studies, however, we observed on the pSC50 plasmid another translocating DNA segment (m.w. 13.6 $\times$ 10^6 daltons) that included genes for resistance to streptomycin (Sm) and sulfonamide (Su) as well as a gene for resistance to Ap. The present report describes electron microscope heteroduplex studies of DNA sequence relationships between the two translocating genetic elements and the recombinant and parental plasmids. The plasmids used in these studies are shown in Table 1. Our findings suggest that evolutionary divergence of several groups of related plasmids has occurred segmentally by a series of illegitimate recombinational events involving the translocation, insertion, and/or deletion of structurally defined regions of DNA.

Table 1
Plasmids Used

Plasmid name	Description	Plasmid phenotype	Reference
pSC101		Tc	Cohen et al. (1973)
pSC105	101::*Eco*RI Km fragment	Tc Km-Nm	Cohen et al. (1973)
pSC190	(type II) 101::Tn*3*	Tc Ap	Kopecko and Cohen (1975); D. J. Kopecko, J. Brevet and S. N. Cohen (in prep.)
pSC179	(type I) 101::Tn*4*	Tc Sm Su Ap	Kopecko and Cohen (1975); D. J. Kopecko, J. Brevet and S. N. Cohen (in prep.)
pSC120	(type I) 101::Tn*4*	Tc Sm Su Ap	Kopecko and Cohen (1975); D. J. Kopecko, J. Brevet and S. N. Cohen (in prep.)
pSC50		Cm Sm Su Ap	Kopecko and Cohen (1975); D. J. Kopecko, J. Brevet and S. N. Cohen (in prep.)
R6		Tc Cm Km-Nm Sm Su	Sharp et al. (1973)
pSC206		Cm Km-Nm Sm Su Ap	D. J. Kopecko, J. Brevet and S. N. Cohen (in prep.)
RTF1		segregant of R1 lacking drug resistance determinants	Cohen and Miller (1970)
R1-19		Cm Sm Su Ap Km	Cohen and Miller (1970)

Structure of the Tn*3* and Tn*4* Translocating Segments and Mapping of Their Insertion Sites on pSC101

In cells containing pSC50 and pSC101 plasmids, two types of recombinant plasmids were found which resulted from the translocation of segments of pSC50 onto pSC101. Type-II plasmids consist of the entire pSC101 genome plus a 3.2-megadalton translocating DNA segment, the Tn*3* (Tn*A*) transposon that determines resistance to ampicillin (Kopecko and Cohen 1975; D. J. Kopecko, J. Brevet and S. N. Cohen, in prep.). This translocating unit contains a 140-nucleotide-long inverted repeat sequence at each of its ends (labeled IRb in Fig. 1A) (Kopecko and Cohen 1975; Heffron et al. 1975). Type-I recombinant plasmids consist of the entire pSC101 genome plus a 13.6-megadalton

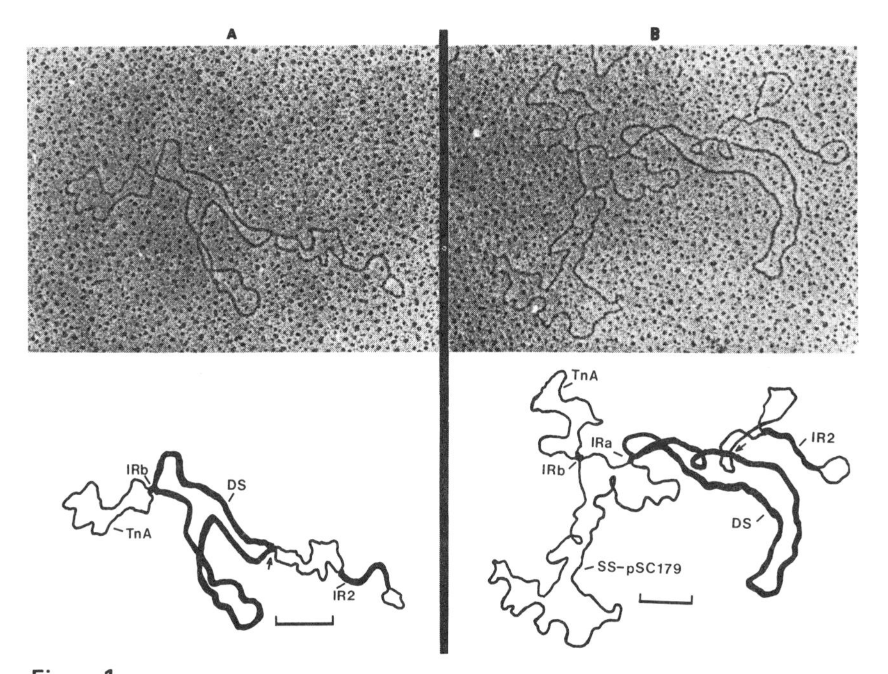

Figure 1
Mapping of Tn*3* and Tn*4* insertion sites on pSC101.

(*A*) pSC105/pSC190 heteroduplex. The DS segment indicates the pSC101 sequences common to both strands. The distance between the stem (IRb) of the Tn*3* segment and the single-strand segment (SS-105) derived from pSC105 plasmid (see arrow) represents the distance from the insertion site of Tn*3* to the *Eco*RI site of pSC101.

(*B*) pSC105/pSC179 heteroduplex. The distance from IRa to the SS-pSC105 single-strand deletion loop is the distance from the insertion site of Tn*4* to the *Eco*RI site of pSC101. The inverted repeat sequences (IRb) of the Tn*3* segment have annealed in this molecule. This photomicrograph also shows the inverted repeat sequence (IRa) that forms the stem of Tn*4*.

DNA segment derived from pSC50. This segment, which has been termed Tn*4* (Tn*S*), determines resistance to streptomycin, sulfonamide, and ampicillin (Fig. 1B). Like Tn*3*, Tn*4* also contains a 140-nucleotide-long inverted repeat sequence at each of its ends (labeled IRa in Fig. 1B), but the homology of these two palindromes (IRa and IRb) cannot be readily determined by heteroduplex techniques.

Figure 1B shows that in addition to the short, linear duplex segment of DNA (labeled IRa), a second duplex region (labeled IRb) can be observed *within* the single-strand segment Tn*4*. This second duplex region demarcates a loop the size of the Tn*3* translocating segment of DNA found in type-II recombinant DNA. The identity of the Tn*3*-sized loop found in type-I recombinant molecules and that found on the type-II recombinant plasmids was determined from analysis of heteroduplexes between the two molecules (data not shown).

Analyses of different combinations of heteroduplexes, together with the previously reported agarose gel electrophoresis patterns for *Eco*RI-digested pSC120 and pSC179 (two different type-I recombinant plasmids) (Kopecko and Cohen 1975), enabled construction of the map shown in Figure 2. The diagram shows the sites of insertion of Tn*4* into pSC101 to form the pSC120 and pSC179 recombinant plasmids and indicates the location of Tn*3* within Tn*4*. By heteroduplex analysis between these two different recombinant plasmids, the two Tn*4* segments were determined to be identical.

The measured distance between the *Eco*RI site of pSC101 and the insertion sites of Tn*3* and Tn*4* on the various recombinant plasmids are given in Figure 3,

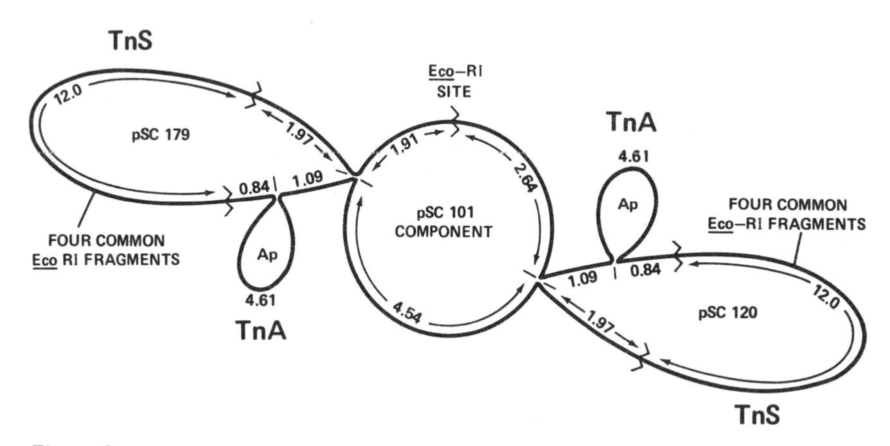

Figure 2

Composite map of *Eco*RI restriction-endonuclease-generated fragments of pSC120 and pSC179 plasmids. The insertion sites of Tn*4* on the pSC101 plasmids are shown for both pSC120 and pSC179 in this schematic diagram. In addition, the position of Tn*3* within the Tn*4* translocating segment is indicated.

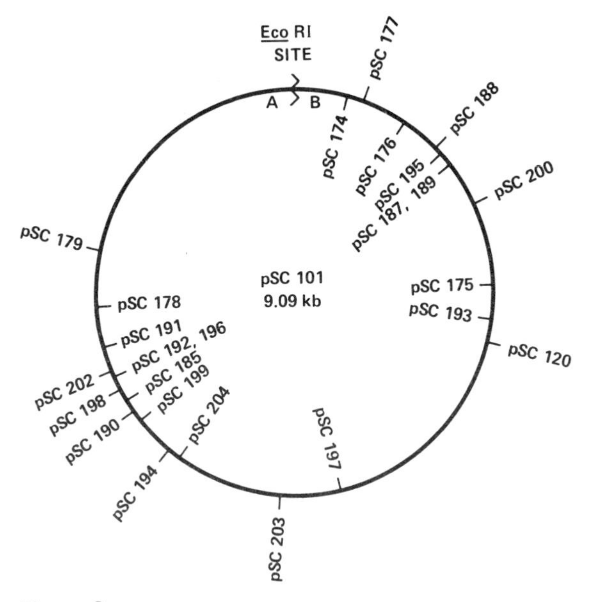

Figure 3
Insertion sites of Tn*3* and Tn*4* on pSC101. A circular genome of pSC101 plasmid indicates the *Eco*RI site and the recipient sites for the translocating units. The insertion sites were mapped from heteroduplexes between pSC105 and individual recombinant plasmids of both types independently isolated in either $recA^+$ or $recA^-$ cells as described previously (Kopecko and Cohen 1975).

a map of the translocational sites on pSC101. The large number of recipient sites on the pSC101 plasmid for the translocating elements suggests that insertion of Tn*3* and Tn*4* is either random, or, alternatively, that a frequently recurring specific DNA nucleotide sequence is required for integration of these elements. Evidence in support of the latter alternative will be discussed elsewhere (D. J. Kopecko, J. Brevet and S. N. Cohen, in prep.).

DNA Sequence Relationship of the Tn*3* and Tn*4* Translocating DNA Segment on the pSC50 Parent Plasmid

To map the location of Tn*3* and Tn*4* translocating DNA segments on the pSC50 parent plasmid, the plasmid pSC206 was obtained by recombination between plasmids pSC50 and R6. pSC206 contains a Km-Nm resistance segment of R6 that has been recombined with pSC50 at a site equivalent to its location on R6. This site (pSC50 kb coordinate 35.6) is in a region of DNA sequence homology between the R6 and pSC50 plasmids (Sharp et al. 1973). The Km-Nm hairpin loop structure present on the inserted DNA segment serves as a convenient marker in heteroduplex analysis for locating the Tn*3* and Tn*4* translocating elements on the pSC50 parent plasmid because the remainder of the pSC206 plasmid is entirely homologous with pSC50.

The location of the DNA segment added to pSC50 to yield pSC206 was first mapped in relation to the IS*1* region of the plasmid by analysis of a heteroduplex (Fig. 4A) between pSC206 and its RTF derivative (Sharp et al. 1973; Hu et al. 1975; Ptashne and Cohen 1975). The distance between the inserted segment on pSC206 and Tn*3* was then measured on a single-strand molecule pSC206 (Fig. 4B), using the IR2 inverted repeat and the short stem of Tn*3* as markers. This distance was confirmed by heteroduplex analysis of pSC206 and a type-II recombinant plasmid (data not shown). Assignment of Tn*4* to specific kb coordinates of pSC50 was accomplished by heteroduplex analysis of the pSC206 plasmid and a type-I plasmid (data not shown). From the data obtained by these measurements, it was possible to draw the map of the R determinant of

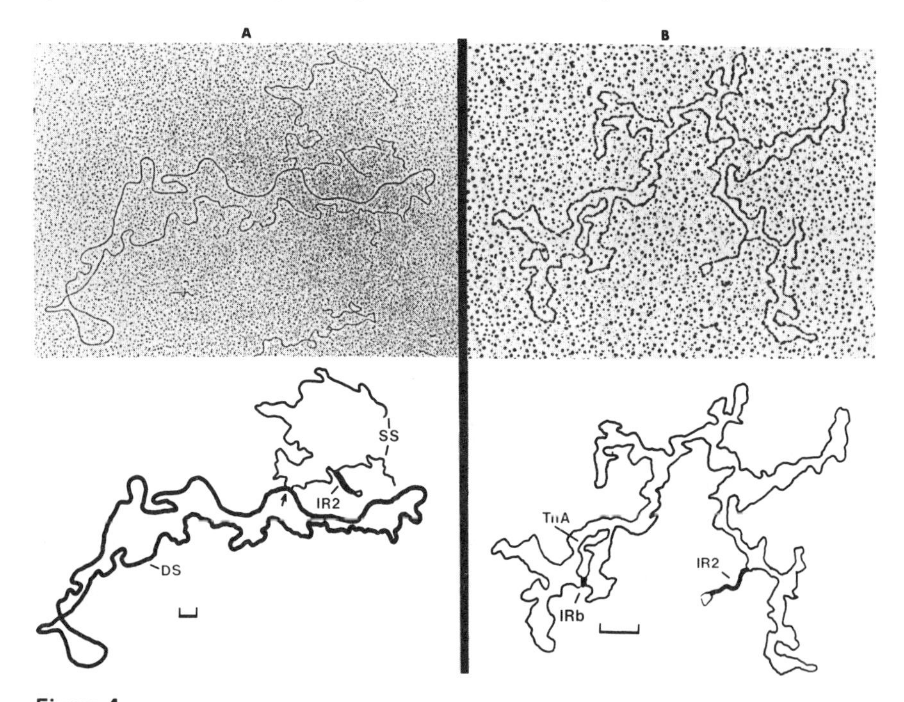

Figure 4
Heteroduplex mapping of Tn*3* and Tn*4* segments on pSC50.

(*A*) pSC206/RTF-1 heteroduplex. The double-stranded region of homology indicates the RTF segment of pSC206. The arrow points to the base of the deletion loop that is the R-determinant segment of pSC206. The distance between the stem of the Km-Nm inverted repeat (IR2) and the arrow indicates the distance to the end of the RTF segment.

(*B*) The single-strand, self-annealed pSC206 molecule has enabled the measurement of the distance between the stem of the Tn*3* segment (IRb) and the inverted repeat (IR2) of the Km-Nm resistance loop.

pSC50 and to map the Tn*3* and Tn*4* translocating DNA segment onto pSC50 (Fig. 5). These heteroduplexes provide additional evidence that Tn*3* is located within the Tn*4* segment.

These data confirm several points of special interest: Within experimental limits, the right end of Tn*4* is located at the terminus of one of the duplicated copies of IS*1*, the IS region that has been shown to be the site of dissociation of the RTF and R-determinant components of several cointegrated plasmids (Hu et al. 1975; Ptashne and Cohen 1975). The left end of Tn*4* is located at the site of insertion of the Km-Nm resistance segment of R6 onto pSC50 to form pSC206.

Earlier heteroduplex studies by Sharp et al. (1973) mapped the Km-Nm gene of R1-19 in the general vicinity of the Ap resistance determinant of that plasmid. In our current studies, more precise localization of the R1-19 segment carrying Km resistance was possible by using the pSC206 plasmid in the formation of heteroduplex (data not shown).

This analysis showed that a DNA segment 9.5 kb in length was deleted from R1-19 in the formation of Km-Nm-sensitive pSC50, and that the site of the deletion was at 14.7 kb on the pSC50 plasmid. The location of the deletion at the right terminus of Tn*4* and the size of the deletion indicate that it consists of the *entire* segment of DNA contained between the right terminus of Tn*4* and the left terminus of IS*1*-b. This interpretation is consistent with data presented above suggesting that the ends of Tn*4* and IS*1* are in juxtaposition in pSC50.

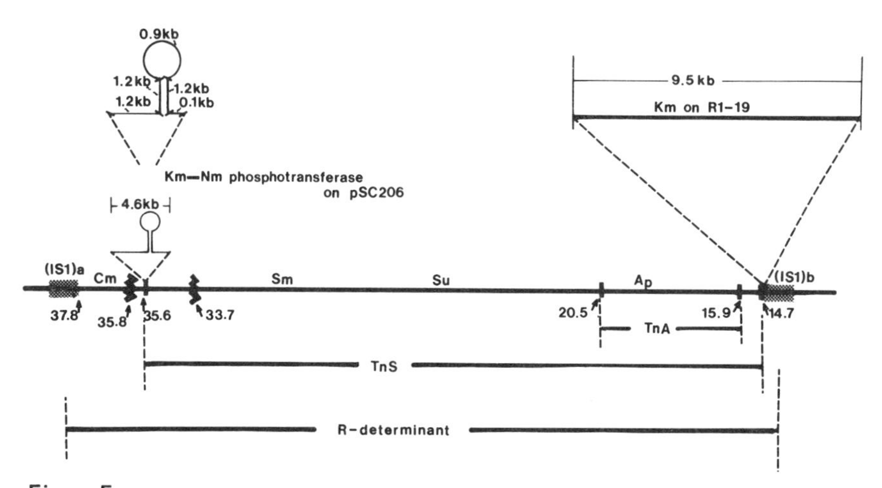

Figure 5

Composite map of the R-determinant region of the pSC50 plasmid showing the Km gene segment that has recombined with pSC50 to form pSC206 and the Km gene segment that was deleted from R1-19 in the formation of pSC50. The left terminus of IS*1*-b was given a kb coordinate of 14.7 (Hu et al. 1975), and the other kb coordinates were assigned using this value.

The above interpretations depend on contour length measurements obtained during DNA heteroduplex analysis. Despite the consistency of the measured distances observed in examinations of various heteroduplex combinations, the measurements are subject to some experimental error. However, our interpretation was confirmed by an experiment in which Tn*4* was translocated onto pSC101 from R1-19 to determine whether the segment of R1-19 carrying Km resistance was located within Tn*4* or outside of this translocating unit. In this experiment, carried out in *recA*$^-$ hosts, recombinant plasmids carrying both Sm and Ap resistance were selected and then tested for expression of resistance to Km. All 50 of the Tc^r, Ap^r, and Sm^r recombinants tested failed to express resistance to Km, suggesting that the Km resistance segment of R1-19 is not contained within the large translocating segment (i.e., Tn*4*) derived from that plasmid. Moreover, examination of the DNA of a recombinant plasmid by electron microscopy and agarose gel electrophoresis indicated that the segment of DNA translocated to pSC101 from R1-19 was identical in size to, and entirely homologous with, the Tn*4* unit derived from pSC50.

Since the Tn*4* translocating segment carrying the Ap, Sm, and Su drug resistance determinants of pSC50 extends over nearly the entire R-determinant region of this plasmid (with the exception of the first 2.2 kb of that region), the Cm resistance genes of the plasmid (which are not contained within Tn*4*) must be located between the coordinates 35.6-37.8 in the region adjacent to IS*1*-a.

Tn*3* Translocating DNA Segment Leads to the Deletion of DNA Sequences

In studies of translocation of Tn*3* to the pSC101-related plasmid pSC105, we observed that the insertion of Tn*3* segments in such a plasmid could lead to the deletion of a variable length of DNA sequences in both directions. These deletions, which will be described in detail elsewhere (P. Nisen, D. J. Kopecko and S. N. Cohen, in prep.), occurred after the insertion of Tn*3* into pSC105 and were not associated with excision of Tn*3*. They occurred precisely at the terminus of Tn*3* and were seen in both directions.

CONCLUSION

The results presented here indicate that the R-determinant components of R6/R6-5 plasmid and of R1/pSC50 plasmid consist of DNA segments that appear to have been joined as a consequence of a series of illegitimate recombinational events. Certain antibiotic resistance genes are located on translocating elements, and such segments can insert within other translocating units. The termini of translocating antibiotic resistance segments appear to represent recombinational "hot spots" at which deletion or insertion of other segments of DNA occur. For example, deletion of the Km-Nm resistance DNA segment between the right end of Tn*4* and the IS*1* region has presumably occurred during the evolutionary divergence of the R1/pSC50 group of plasmids. A second locus of structural divergence between these groups of related plasmids exists at the left end of Tn*4*: a DNA segment containing the Km-Nm phosphotransferase gene exists in R6 and R6-5, but it is not present in R1/pSC50 or in another related plasmid, R100. It is not known whether the R100 plasmids, which lack both Km-Nm resistance genes, represent a common ancestor of the R6/R6-5 and R1-19/pSC50 group, whether they are an intermediate form, or whether they are the most recent product of plasmid evolution.

Although *recA*-independent translocation of plasmid DNA regions does not necessarily involve demonstrable inverted repeats, it would appear from our data that such structures have a prominent role in translocation-recombination or deletion of certain DNA segments that carry antibiotic resistance genes. However, the mechanisms by which the two limbs of the inverted repeat or palindrome can promote translocation, deletion, or insertion of DNA remain unknown.

Acknowledgments

These investigations were supported by grant AI 08619 from the National Institutes of Health, grant BMS 75-14176 from the National Science Foundation, and grant VC139A from the American Cancer Society to S. N. Cohen.

REFERENCES

Barth, P. T., N. Datta, R. W. Hedges and N. J. Grinter. 1976. Transposition of a deoxyribonucleic acid sequence encoding trimethoprim and streptomycin resistances from R483 to other replicons. *J. Bact.* **125**:800.

Berg, D. E., J. Davies, B. Allet and J. Rochaix. 1975. Transposition of R factor genes to bacteriophage λ. *Proc. Nat. Acad. Sci.* **72**:3628.

Cohen, S. N. and D. J. Kopecko. 1976. Structural evolution of bacterial plasmids: Role of translocating genetic elements and DNA sequence insertions. *Fed. Proc.* **35**:2031.

Cohen, S. N. and C. A. Miller. 1970. Non-chromosomal antibiotic resistance in bacteria. II. Molecular nature of R-factors isolated from *Proteus mirabilis* and *Escherichia coli*. *J. Mol. Biol.* **50**:671.

Cohen, S. N., A. C. Y. Chang, H. W. Boyer and R. B. Helling. 1973. Construction of biologically functional bacterial plasmids *in vitro*. *Proc. Nat. Acad. Sci.* **70**:3240.

Foster, T. J., T. G. B. Howe and K. M. V. Richmond. 1975. Translocation of the tetracycline resistance determinant from R100-1 to the *Escherichia coli* K-12 chromosome. *J. Bact.* **124**:1153.

Gottesman, M. M. and J. L. Rosner. 1975. Acquisition of a determinant for chloramphenicol resistance by coliphage lambda. *Proc. Nat. Acad. Sci.* **72**:5041.

Heffron, F., C. Rubens and S. Falkow. 1975. Translocation of a plasmid DNA sequence which mediates ampicillin resistance: Molecular nature and specificity of insertion. *Proc. Nat. Acad. Sci.* **72**:3623.

Hu, S., E. Ohtsubo, N. Davidson and H. Saedler. 1975. Electron microscope heteroduplex studies of sequence relations among bacterial plasmids. Identification and mapping of the insertion sequences IS1 and IS2 in F and R plasmids. *J. Bact.* **122**:764.

Kleckner, N., R. K. Chan, B.-K. Tye and D. Botstein. 1975. Mutagenesis by insertion of a drug-resistance element carrying an inverted repetition. *J. Mol. Biol.* **97**:561.

Kopecko, D. J. and S. N. Cohen. 1975. Site-specific *recA*-independent recombination between bacterial plasmids: Involvement of palindromes at the recombinational loci. *Proc. Nat. Acad. Sci.* **72**:1373.

Ptashne, K. and S. N. Cohen. 1975. Occurrence of insertion sequence (IS) regions on plasmid deoxyribonucleic acid as direct and inverted nucleotide sequence duplications. *J. Bact.* **122**:776.

Sharp, P. A., S. N. Cohen and N. Davidson. 1973. Electron microscope heteroduplex studies of sequence relations among plasmids of *Escherichia coli.* II. Structure of drug resistance (R) factors and F factors. *J. Mol. Biol.* **75**:235.

Starlinger, P. and H. Saedler. 1976. IS-elements in microorganisms. In *Current topics in microbiology and immunology,* vol. 75, p. 111. Springer Verlag, Berlin.

Transposition of Tn*1* to a Broad-host-range Drug Resistance Plasmid

J. P. Hernalsteens
Laboratorium Genetische Virologie
Vrije Universiteit Brussel
B-1640 St. Genesius Rode, Belgium

R. Villarroel-Mandiola, M. Van Montagu and J. Schell
Laboratorium Histologie en Genetika
Rijksuniversiteit Gent
B-9000 Gent, Belgium

The β-lactamase genes (*bla*$^+$) of many drug resistance plasmids are located on transposons which can move between different replicons (Hedges and Jacob 1974). The integration of these transposons can cause polar mutations (Heffron et al. 1975). We report a rapid and easy method for mutagenizing bacterial plasmids by one of these elements, Tn*1* from the RP4 plasmid (Datta et al. 1971). The method is based on the observation that Tn*1* can be transferred from a chromosomal site in a donor strain to a recipient by means of a transmissible plasmid. Tn*1* is first translocated onto the plasmid, which can then be recovered in recipient cells by mating.

The markers on bacterial strains and plasmids are listed in Table 1. Strain J53 (RP4) carries a copy of Tn*1* inserted in its chromosome (Hedges and Jacob (1974). Sodium dodecyl sulfate (SDS) curing of RP4 (Ingram et al. 1972) yields kanamycin- and tetracycline-sensitive clones that are all carbenicillin-resistant (100/100 tested), and the carbenicillin resistance marker (Tn*1*) is cotransducible by P1 with the *met* marker of J53 (data not shown).

The cured strain is called J53::Tn*1*. This strain could not donate the *bla*$^+$ marker to strain W3110 after overnight conjugation on a Millipore filter. In contrast, J53::Tn*1* (R702) proved an efficient donor of Tn*1*. Upon mating with W3110, 1.6% of all R702 recipients had received Tn*1*, and 4000/4000 ampicillin-resistant exconjugants selected had also received R702 plasmid markers. We tested 400 *bla*$^+$ recipients for subsequent transfer of Tn*1* by the rapid plate method of Davies (1975). Transfer was detected in 397 cases and was always accompanied by transfer of R702 markers. Hence, Tn*1* must be mobilized from J53::Tn*1* by transposition to R702.

Some of the R702::Tn*1* plasmids isolated in these experiments had mutations in various resistance markers. Table 2 summarizes the results of crosses with 15 independent clones of the J53::Tn*1* (R702) donor. We found mutations in all

Table 1
Bacteria and Plasmids Used

Bacteria (*E. coli* K12)		Plasmids (Inc P)	
no.	relevant markers	no.	resistance pattern
J53	*pro⁻met⁻*F⁻	RP4	Ap Km Tc
W3110	prototrophic F⁻	R702	Km Tc Sm Su Hg
C600*rif*^{R}	*thr⁻leu⁻thi⁻rif*^{R} F⁻		

Both plasmids were kindly provided by Dr. N. Datta. The following antibiotic concentrations were used to score resistance markers: kanamycin, 25 μg/ml; tetracycline, 10 μg/ml; streptomycin, 20 μg/ml; carbenicillin, 200 μg/ml; sulfathiazole, 100 μg/ml; and $HgCl_2$, 4×10^{-5} M. All antibiotics were used in LB medium, except sulfathiazole, which was used in minimal A medium (Miller 1972).

of the resistance markers as well as in the transfer genes. To prove that insertion of Tn*1* had caused these mutations, we performed pairwise crosses between three independently obtained kanamycin-sensitive mutants. Table 3 shows that all combinations produced kanamycin-resistant recombinants and that only about 0.1% of them retained Tn*1*. Thus the kanamycin-sensitive and ampicillin-resistant phenotypes are due to the same genetic event. No revertants of any Tn*1*-induced mutations were found for insertion in the kanamycin (4 isolates),

Table 2
Induction of R702 Mutations by Tn*1*

Phenotype	Frequency	No. of independent isolates
Km^s	4.3×10^{-3}	11
Tc^s	2.7×10^{-3}	7
Hg^s	3.1×10^{-3}	8
Sm^s	6.3×10^{-4}	1
Su^s	2.6×10^{-3}	10
Sm^sSu^s	4.4×10^{-3}	7
Tra^-	7.5×10^{-3}	3

After crossing J53::Tn*1* (R702) with W3110 on Millipore filters, the Tn*1*-containing recipients were selected on appropriately supplemented minimal medium containing 200 μg/ml carbenicillin. These transconjugants were directly replica-plated onto plates of the same selective medium containing each of the antibiotics for which R702 confers resistance to determine their resistance patterns. The transfer test was done with purified clones using C600*rif*^{R} as recipient and selecting on LB medium containing 100 μg/ml rifampicin and 25 μg/ml kanamycin by the method of Davies (1975).

Table 3
Recombination between Kanamycin-sensitive Mutants

Strain	Frequency of kanamycin-resistant cells	Frequency of ampicillin resistance in recombinants
1 X 3	1.8×10^{-6}	1.0×10^{-3}
2 X 3	1.1×10^{-7}	$<4 \times 10^{-3}$
1 X 2	1.0×10^{-7}	1.6×10^{-3}

Strains 1, 2, and 3 are independent W3110 (R702 *aphA*::Tn*1*) isolates (Km^s). For two of the three strains, 10^9 cells were crossed by overnight conjugation on a Millipore membrane filter on LB agar. The frequency of resistant cells is expressed per total cells after conjugation. Kanamycin-resistant revertants were never detected in control experiments with only one strain. The recombinants were tested in situ for ampicillin resistance by the method of Novick and Richmond (1965).

tetracycline (7 isolates), mercury (8 isolates), or streptomycin (8 isolates) determinants. (The limit of detection was 1 revertant in 10^{10} cells.) This suggests that either integration or excision of Tn*1* is not precise. The high stability of Tn*1*-induced mutation differs from that of mutations caused by insertion of other transposons, such as Tn*10* (Kleckner et al. 1975), where revertants can be found.

Two out of ten sulfonamide-sensitive insertion mutants of R702 did revert at frequencies of about 10^{-4} and 10^{-5}. Two kinds of revertants appeared: (1) ampicillin-sensitive with wild-type levels of sulfonamide resistance (4/20 and 3/20, respectively, for the two mutants), and (2) ampicillin-resistant with generally lower than wild-type sulfonamide resistance levels. SDS curing showed that the Tn*1* *bla*$^+$ marker is still on the plasmid in the second class of revertants. We conclude that the two reverting mutants resulted from polar Tn*1* insertions outside the structural gene for the sulfonamide resistance product. In this case, precise excision would not be required to relieve polarity. Support for this hypothesis comes from the observation that seven of eight streptomycin-sensitive mutants are also leaky for sulfonamide resistance (Table 2). Heffron et al. (1975) have reported insertion of a *bla*$^+$ transposon into an SmSu operon on the small, nonconjugative plasmid RSF1010.

The method of using transposon-induced mutations can probably be applied to many bacterial species. An example is the oncogenic phytopathogen *Agrobacterium tumefaciens.* Table 4 shows that a variety of broad-host-range plasmids can be introduced into this species. Using techniques similar to those described above, we have isolated antibiotic-sensitive and transfer-deficient R702::Tn*1* derivatives in *A. tumefaciens.* We are now investigating the insertion of Tn*1* into the oncogenic *Agrobacterium* plasmids.

Table 4
Transfer of Drug Resistance Plasmids from *E. coli* K12 to *Agrobacterium tumefaciens* Strains

Plasmid	Incompatibility group	Resistances in *E. coli*	*A. tumefaciens* recipient	Transfer frequency per recipient	Resistances in *Agrobacterium*	Stability of recipient
RP4	P	Cb Km Tc	B6S3	5.0×10^{-1}	Km Tc	stable
			C58	9.6×10^{-1}	Cb Km Tc	stable
R702	P	Hg Km Sm Su Tc	B6S3	6.0×10^{-1}	Hg Km Sm Su Tc	stable
			C58	4.3×10^{-1}	Hg Km Sm Su Tc	stable
S-a	W	Cm Km Sm Su	B6S3	9.8×10^{-3}	Cm Km Sm Su	stable
			C58	3.8×10^{-2}	Cm Km Sm Su	stable
R55	C	Cb Cm Gm Km Su	B6S3	$< 10^{-9}$	–	–
			C58	2.6×10^{-8}	Cm Gm Km Su	stable
R128	N	Ap Tc Su	B6S3	$< 10^{-9}$	–	–
			C58	9.6×10^{-5}	Su	unstable
R15	N	Sm Su	B6S3	$< 10^{-9}$	–	–
			C58	$< 10^{-9}$	–	–
R1*drd*19	FII	Cb Cm Km Sm Su	B6S3	$< 10^{-9}$	–	–
			C58	$< 10^{-9}$	–	–
R64*drd*11	Iα	Sm Tc	B6S3	$< 10^{-9}$	–	–
			C58	$< 10^{-9}$	–	–

Conjugation was performed overnight on a Millipore filter using about 10^9 cells of both donor and recipient. Plasmid-containing transconjugants were selected using kanamycin (25 μg/ml), tetracycline (2.5 μg/ml), or sulfathiazole (100 μg/ml). The presence of other resistance markers was determined with Difco Bacto-sensitivity disks.

Acknowledgments

We thank Mrs. F. Deboeck for excellent technical assistance. This investigation was supported by a grant from the "Kankerfonds van de ASLK."

REFERENCES

Datta, N., R. W. Hedges, E. J. Shaw, R. B. Sykes and M. H. Richmond. 1971. Properties of an R factor from *Pseudomonas aeruginosa. J. Bact.* **108**:1244.

Davies, J. 1975. Genetic methods for the study of antibiotic resistance plasmids. In *Methods in enzymology* (ed. J. H. Hash), vol. 43, p. 41. Academic Press, New York.

Hedges, R. W. and A. E. Jacob. 1974. Transposition of ampicillin resistance from RP4 to other replicons. *Mol. Gen. Genet.* **132**:31.

Heffron, F., C. Rubens and S. Falkow. 1975. Translocation of a plasmid DNA sequence which mediates ampicillin resistance: Molecular nature and specificity of insertion. *Proc. Nat. Acad. Sci.* **72**:3623.

Ingram, L., R. B. Sykes, J. Grinsted, J. R. Saunders and M. H. Richmond. 1972. A transmissible resistance element from a strain of *Pseudomonas aeruginosa* containing no detectable extrachromosomal DNA. *J. Gen. Microbiol.* **72**:269.

Kleckner, N., R. K. Chan, B.-K. Tye and D. Botstein. 1975. Mutagenesis by insertion of a drug-resistance element carrying an inverted repetition. *J. Mol. Biol.* **97**:561.

Miller, J. H. 1972. *Experiments in molecular genetics,* pp. 432-433. Cold Spring Harbor Laboratory, Cold Spring Harbor, New York.

Novick, R. P. and M. H. Richmond. 1965. Nature and interactions of the genetic elements governing penicillinase synthesis in *Staphylococcus aureus. J. Bact.* **90**:467.

Translocation and Illegitimate Recombination by the Tetracycline Resistance Element Tn*10*

D. Botstein and N. Kleckner
Department of Biology
Massachusetts Institute of Technology
Cambridge, Massachusetts 02139

Work in a number of laboratories has revealed the presence in bacteria of genetic elements that are capable of insertion, as intact units, into a large number of sites on the bacterial chromosome. Most of the elements are also implicated in other types of illegitimate recombination events. Despite the fact that they vary tremendously in size and genetic complexity (from the 800-base-pair IS*1* to bacteriophage lambda with 47,000 base pairs and 50 genes), these elements do have a number of important features in common, in addition to the ability to insert at many sites (Starlinger and Saedler 1972, 1976; Shimada et al. 1973; Hedges and Jacob 1974; Kleckner et al. 1975; Berg et al. 1975; Heffron et al. 1975; Gottesman and Rosner 1975; Kopecko and Cohen 1975; Couturier 1976; see also other papers in this volume).

For the past several years, we have been studying a genetic element which is derived from a drug resistance plasmid and which is capable of translocation to many sites on bacterial, phage, and plasmid chromosomes. This element, now called Tn*10*, carries on it a gene(s) which confers tetracycline (Tc) resistance to bacteria. This property has made a number of genetic experiments feasible. The results of these experiments (along with the results of EM studies) are summarized in this paper.

Tn*10* has been implicated in several different kinds of illegitimate recombination events: insertion into many different sites, excision (both precise and imprecise) from these sites, and formation of deletions and other chromosomal aberrations. Although we do not yet understand the mechanisms involved in any of the illegitimate recombination events associated with Tn*10*, the detailed genetic properties we have thus far been able to discover should limit the kinds of models that can be entertained to explain these phenomena.

Evidence That Tn*10* Translocates Repeatedly as a Discrete Entity

We have shown that Tn*10* can translocate as a discrete entity into many different positions on several different kinds of prokaryotic genomes. This process can happen repeatedly, and we have followed, by genetic and/or physical means, the translocation of Tn*10* from a drug resistance plasmid (R222) to the genome of the *Salmonella* phage P22 (where at least 20 different sites of insertion exist); from P22 into the *Salmonella* chromosome itself (at least 100 different sites); from *Salmonella* to the genome of coliphage lambda (at least 5 sites); and finally from lambda to the chromosome of *Escherichia coli* (at least 20 sites). At three points in this process (R222, P22, and lambda), the physical form of the insertion could be easily monitored by electron microscopy of the DNA, and in all cases we could observe that translocation of Tn*10* involved the translocation of the same segment of DNA from one genome to the next.

Originally, Tn*10* was found on a drug resistance plasmid. Sharp et al. (1973) found on plasmid R100 a stretch of DNA (9.2 kb in length) which they were able to associate with the gene(s) specifying tetracycline resistance. This stretch of DNA had an unusual structure: it consisted of a 6.4-kb segment bounded by a 1.4-kb inverted repetition. Exactly the same structure was found by Tye et al. (1974) on the DNA of two P22 specialized transducing variants which transduce tetracycline resistance at high frequency. These P22 phages had been isolated and characterized by Watanabe et al. (1972) and Chan et al. (1972) from lysates of P22 grown on *Salmonella* strains harboring the drug resistance plasmid R222 (probably identical to R100). Experiments by Chan and Botstein (1976) showed that the genes acquired by one of these specialized transducing phages are inserted at a site not adjacent to the prophage attachment site, making it exceedingly unlikely that the insertions arose by inexact excision of an integrated prophage (the normal way in which transducing variants of P22 and lambda arise).

Direct demonstration that the insertion of Tn*10* occurs at more than one site in a genome was first accomplished by means of a heteroduplex experiment involving the DNA of two independent P22*tet*^{R} transducing phages (Tc10/Tc106) (Fig. 1) (Kleckner et al. 1975). Here the Tn*10* insertion is clearly visible, and the emanation of the stems from different sites on the P22 genome is evident. The partial pairing of the loops shows that the two insertions are inserted in the same orientation with respect to the P22 genome.

Translocation of the Tn*10* element to the *Salmonella* chromosome can be readily detected by infecting a suitable *Salmonella* host with a P22*tet*^{R} phage under conditions where the ability of the phage to perpetuate itself in the host has been impaired by various combinations of phage and host mutations. Selection for tetracycline-resistant cells yields "transductants" which contain no phage genes but which have Tn*10* inserted into the bacterial chromosome. In some cases, the insertion is within known genes, causing recognizable mutations. Approximately 1% of the *tet*^{R} transductants in such experiments acquired new nutritional requirements, and simple genetic tests confirmed that these requirements were the result of insertion within genes required for the biosynthesis of the appropriate nutrients (Kleckner et al. 1975).

Transfer of such an insertion mutation from *Salmonella* to *E. coli* was accomplished by conjugation experiments, and the resultant *E. coli* insertion auxotroph was used as a host for a lambda mutant that carries a large deletion of nonessential genes and is thus capable of acquiring a large insertion. The result of such an experiment was a transducing variant of lambda that had acquired Tn*10*. As shown in Figure 2, the DNA of this phage again has an insertion identical in form to that of Tn*10* found in the R factor and in P22. (D. Barker and N. Kleckner, unpubl.). Thus the idea that insertion mutations in *Salmonella* are the result of insertion of Tn*10* was directly confirmed. In turn, the lambda *tet*^{R} phage can be used to make insertion mutations in *E. coli.*

It is clear from this series of experiments that Tn*10* is a discrete entity capable of repeated translocation as an intact unit from genome to genome without change of form and without loss of ability to translocate further.

Tn*10* and IS*3*

Ptashne and Cohen (1975) have shown that the inverted repetitions of Tn*10* are homologous to the IS*3* insertion sequence. IS*3* was originally detected as a polar insertion in the *lac* operon (Malamy 1967, 1970; Malamy et al. 1972), and sequences homologous to this IS*3* insertion have been found on F and on several F′ and R factors (Ptashne and Cohen 1975; Hu et al. 1975). On this basis, IS*3* is believed to be a genetic element capable of independent insertion as a discrete unit, although in no case has a particular IS*3* sequence been shown directly to move from one position to another. Tn*10* thus appears to be a composite element in which two intact IS*3* elements cooperate to effect the translocation of genetic material between them which encodes the tetracycline resistance determinant. MacHattie and Jackowski (this volume) show that sequences homologous to IS*1* flank the chloramphenicol determinant in Tn*9*; here the IS*1* is directly repeated, not inverted, but it again seems likely that the IS*1* sequences are the operative elements in the translocation of Tn*9*.

Specificity of Tn*10* Insertion Events

From the distribution of the auxotrophic mutations that can be recovered by insertion of Tn*10* into the *Salmonella* and *E. coli* chromosomes, it is clear that Tn*10* inserts at a large number of different sites. Several hundred Tn*10*-generated auxotrophs have been characterized in these organisms (Kleckner et al. 1975 and unpubl.; J. Roth et al., unpubl.), and more than 30 distinguishable requirements are represented. This result does not exclude the possibility of highly preferred locations at positions where insertion does not result in a detectable mutation, but it does indicate a large number of possible insertion sites.

We have examined the specificity of Tn*10* insertion at higher resolution by mapping genetically 85 independent insertions of Tn*10* into the histidine operon of *Salmonella,* 55 of which were isolated in a *rec*$^+$ strain and 30 in a *recA*$^-$ strain. The distribution of the 55 insertions found in the *rec*$^+$ strain is shown in Figure 3, which summarizes data from complementation tests, two-factor

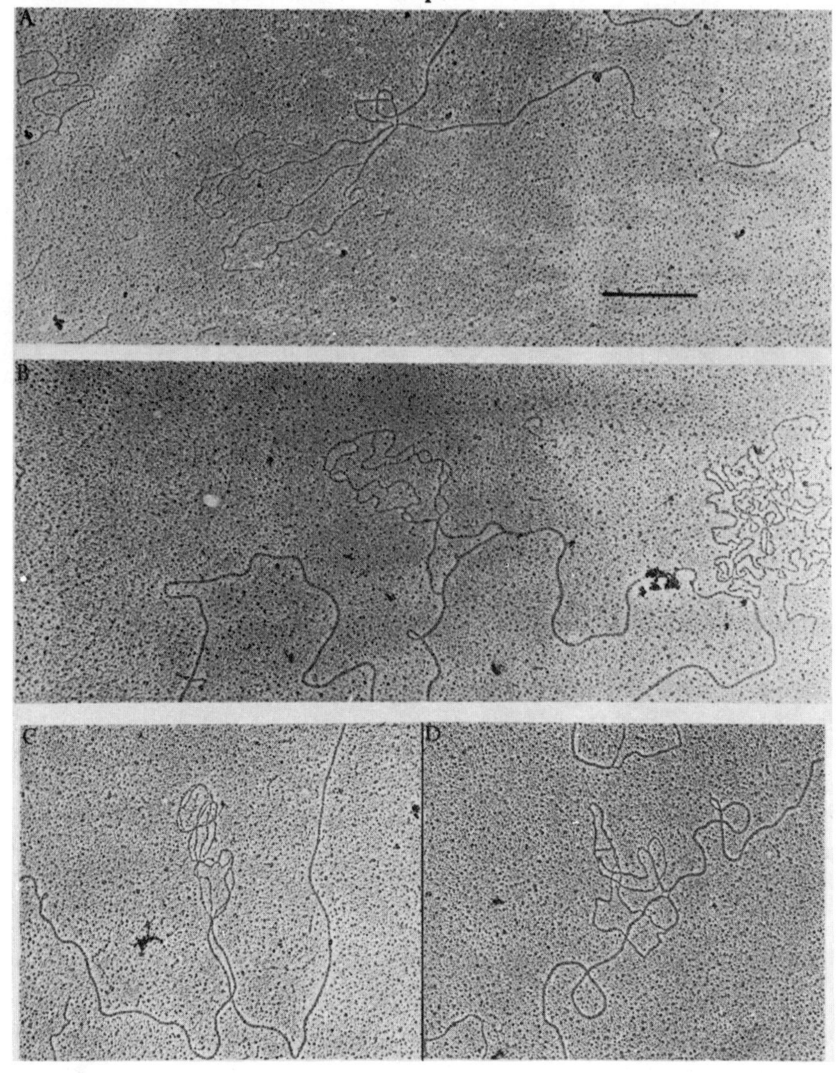

Figure 1

(*I*) Heteroduplexes of Tc10 and Tc106. (*A*) Tc10/Tc106 heteroduplex in which loops have not interacted. (*B*) Tc10/Tc106 heteroduplex in which loops have interacted. (*C*) Homoduplex (Tc10/Tc10 or Tc106/Tc106) in which loops have interacted. (*D*) Homoduplex (Tc10/Tc10 or Tc106/Tc106) in which loops have not interacted but branch migration has occurred around the point of the insertions. (Bar = 0.5 μm.)

The structure seen in *IB* differs from that in *IA* in that there has been partial pairing between the two single-stranded loops of the inserted material. The pattern of paired and unpaired regions is reproducible among the many such structures we have seen. We interpret these structures to mean that the insertions carried by Tc10 and Tc106 are actually totally homologous, and that such

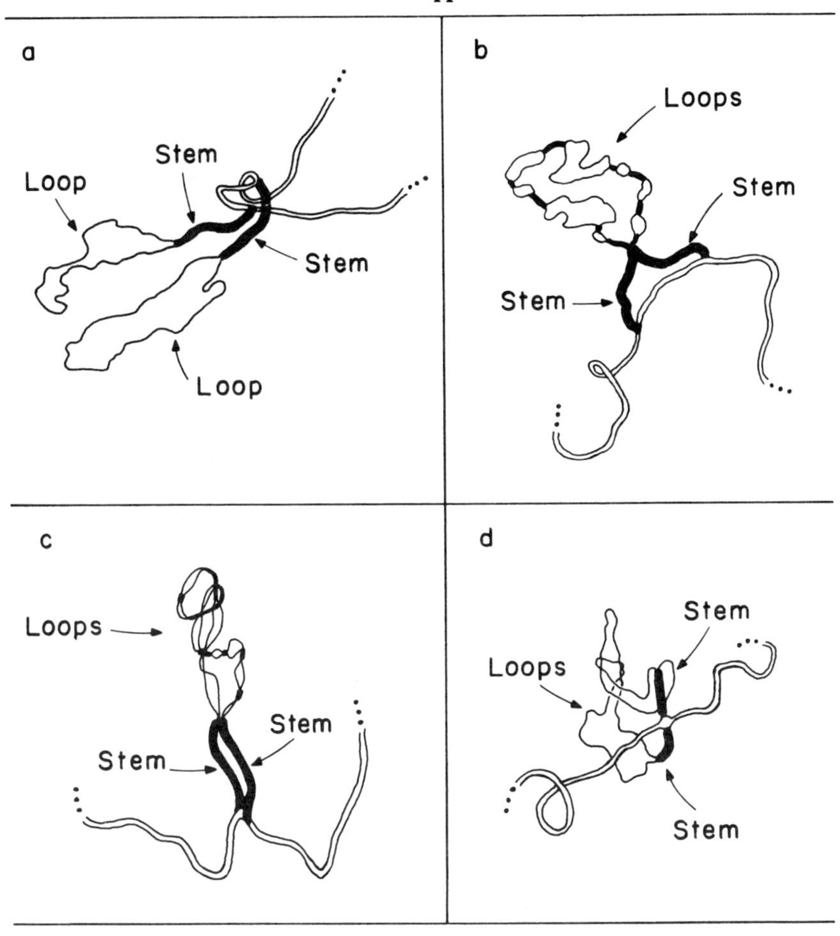

Figure 1 *(continued)*
structures represent instances in which intramolecular pairing between the self-complementary portions of each strand preceded intermolecular pairing between the complementary strands of the loop. Complete pairing in the loop region apparently is obstructed by steric constraints imposed by the prior pairing of the self-complementary regions. Implicit in this interpretation is the idea that the insertions carried by P22Tc10 and P22Tc106 are not only homologous in base sequence but are also oriented in the same direction with respect to the P22 genome. (*I* and *II* are reprinted, with permission, from Kleckner et al. [1975], where further discussion of their interpretation can be found.)

(*II*) Interpretation of heteroduplexes: *a* to *d* correspond to micrographs *A* to *D* in *I*. (▬▬) Paired stems (double-stranded); (——) paired regions within the loop (double-stranded); (——) unpaired regions within the loop (single-stranded); (═══) P22 DNA (double-stranded).

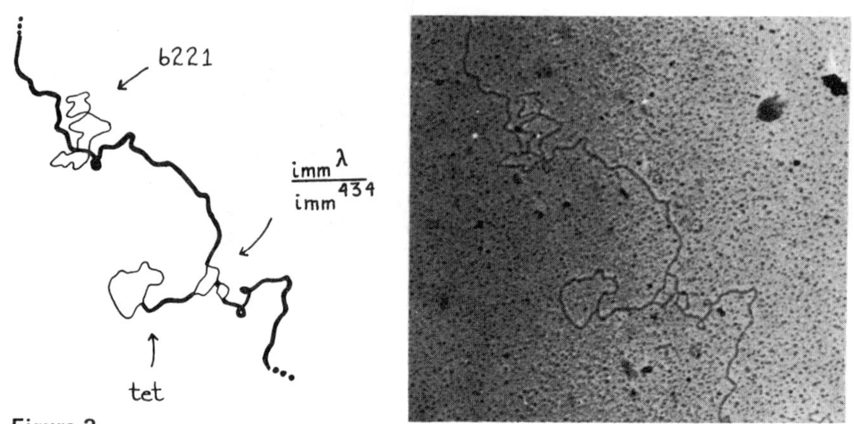

Figure 2
Heteroduplex of λ*b*221 *imm*$^{\lambda}$::Tn*10*/λ*imm*^{434}. This λ*b*221 Tn*10* transducing phage carries the Tn*10* insertion within the region of nonhomology with λ*imm*^{434}. The ratio of loop to stem length for this molecule is about 2.6. The same ratio is observed for Tn*10* insertions carried by P22Tc10 (from which this λ Tn*10* is descended) and by P22Tc106 (see Fig. 1).

crosses among insertion mutations, and some deletion mapping experiments (Kleckner et al. 1975 and unpubl.; J. Roth et al., unpubl.). It is clear from Figure 3 that the distribution of insertion mutations is not uniform over the genetic map. The insertions fall into clusters, and each cluster is defined to a resolution limited only by the reversion rate of the insertion mutations (10^{-7} to 10^{-8}). The results with the insertion mutations isolated in the *recA*$^-$ strain were the same.

The most striking aspect of the data in Figure 3 is the evidence provided for "hot spots" for insertion of Tn*10*; one cluster in *hisG* contains 27 of the 55 independent insertions from the *rec*$^+$ strain and 9 of the 30 from the *recA*$^-$ strain. The resolution to which this site (or cluster of very closely linked sites) is defined is quite high. I. Hoppe and J. Roth (unpubl.) have established a fine-structure deletion map of *hisG*; they have shown that the insertions at this site fall into a deletion interval occupied by none of 70 mapped *hisG* point mutations. Assuming that the point mutations are more or less randomly distributed, they estimate the size of this interval to be no more than 25 base pairs. In the two-point crosses used to define this site, the resolution is finer than that required to detect recombination between each insertion and the nearest deletion end point; therefore, the distance between insertions is likely to be substantially less than the size of the deletion interval. Finally, using deletions generated by one of the insertions at this site (see below), we have carried out

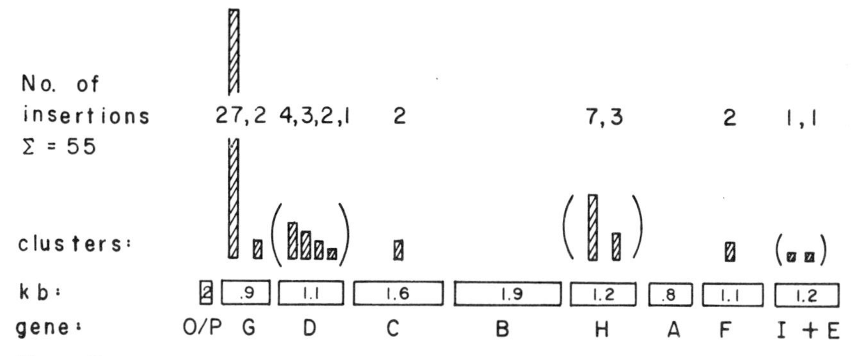

Figure 3
Distribution of Tn*10* insertions in the histidine operon. Each of 55 independent *his*::Tn*10* insertions isolated in *rec*$^+$ bacteria was characterized by complementation tests. Each insertion was then crossed against each of the other Tn*10* insertions in the same gene. A complete set of reciprocal P22-mediated transduction crosses was performed. Relative to the frequency of transductants obtained when *his*$^+$ bacteria are used as donors, transduction frequencies of 10^{-3} to 10^{-4} can be detected. The sensitivity depends on the reversion frequency of the particular *his*::Tn*10* insertion used as recipient. For genes in which the relative order of clusters is not known, the clusters are shown in parentheses. Complementation and transduction experiments were performed as described previously (Kleckner et al. 1975).

crosses with even better resolution and have observed no detectable recombination, suggesting again that all 36 insertions might actually be at exactly the same point.

Physical studies on Tn*10* insertions into coliphage lambda also suggest that two independent insertions in the lambda *c*I gene may lie at the same site (D. Barker and N. Kleckner, in prep.).

Insertion of Tn*10* Is Usually Precise

The overwhelming majority of Tn*10*-generated auxotrophs isolated in *Salmonella* have been found to revert to prototrophy (with concomitant loss of the tetracycline resistance determinant). All of the 85 *his*::Tn*10* insertions discussed above revert to *his*$^+$ at frequencies of 10^{-7} to 10^{-8} (N. Kleckner, unpubl.), and at least 45 of the 50 Tn*10*-generated auxotrophs at other locations were also capable of giving prototrophic tetracycline-sensitive revertants (Kleckner et al. 1975).

From this, we conclude that insertion of Tn*10* is sufficiently precise (with respect to the recipient DNA molecule) that in almost every insertion the damaged gene can be restored to functional form at a detectable frequency (greater than 10^{-8}). We cannot exclude the possibility that small alterations in the nucleotide sequence of the host chromosome may sometimes occur during either integration or excision of Tn*10.* Although fewer Tn*10*-generated auxotrophs in *E. coli* have been examined, most of these revert to prototrophy as well.

Tn*10* Can Be Inserted in Either Orientation With Respect to the Entire *Salmonella* Chromosome

The orientations of various Tn*10* insertions in the *Salmonella* chromosome have been determined genetically by R. Menzel and J. Roth (unpubl.). This is accomplished by determining the direction in which particular F′ episomes, each carrying a known Tn*10* insertion, will mobilize the chromosomes of various *Salmonella* hosts which carry an additional Tn*10* insertion within the bacterial chromosome. Since mobilization will frequently occur by integration of the F′ into the chromosome via the Tn*10* homology, the direction of mobilization of the chromosome will depend on the relative orientations of the two Tn*10* elements in question. Of three insertions located near the *put* locus in *Salmonella,* two are oriented in one direction and one in the other.

Precise Excision

As indicated above, mutations caused by insertion of Tn*10* virtually always revert, indicating that the Tn*10* element can be excised precisely. The spontaneous reversion frequencies of different Tn*10* insertion mutations vary over a 1000-fold range (10^{-6} to 10^{-9} revertants in a saturated culture). Even among strains carrying different insertions within the same gene, the reversion rates can differ by as much as 400-fold. Precise excision of Tn*10* is not dependent on the *recA* function of the host and does not show any marked dependence upon temperature between 30°C and 40°C. These properties are similar to those reported for IS*1* by Reif and Saedler (1975).

Precise excision of Tn*10* does not seem to be associated with reinsertion of the intact element elsewhere in the bacterial chromosome. Independent revertants of several *his*::Tn*10* auxotrophs were tested for tetracycline resistance; fewer than 1/1000 were *tet*^{R}. We are inclined to interpret this result as suggesting that precise excision is not a normal aspect of the mechanism of translocation of Tn*10.*

In considering the relationship of precise excision to the other illegitimate recombination events involving Tn*10,* one must consider the possibility that Tn*10* does not participate actively in precise excision, but instead may be excised passively as the result of some unrelated host repair or recombination process.

Imprecise Excision

Precise excision of Tn*10* (which reconstructs in functional form the structural gene damaged by the original insertion) occurs very rarely. In contrast, *imprecise* excision (loss of the tetracycline resistance determinant *without* concomitant restoration of structural gene function) occurs at a much higher frequency. Revertants of *hisG*::Tn*10* insertions in which the polar effects of the insertion on an adjacent gene (*hisD*) are relieved arise at a frequency of 10^{-4}. The overwhelming majority of these revertants have lost the tetracycline resistance determinant, but none (less than 10^{-3}) are precise excisions, since none regained *hisG* function (Kleckner et al. 1975). The frequency of tetracycline-sensitive derivatives in *hisG*::Tn*10* cultures has also been determined directly by visually scoring colonies on tetracycline indicator plates. The frequency of approximately 10^{-4} for such derivatives is consistent with previous results, and some of them exhibit relief of polarity on *hisD* function.

The products of these imprecise excision events have been analyzed further, and the results will be presented in detail elsewhere (N. Kleckner, K. Reichardt, and D. Botstein, in prep.). In brief, 324 tetracycline-sensitive derivatives from eight independent clones of a *hisG*::Tn*10* insertion at the "hot spot" described above were isolated and characterized by a number of genetic tests. They were scored for the functionality of the nearest known bacterial genes on either side of the Tn*10* insertion: on the left, *phs* (production of H_2S); on the right, *hisD* (growth on histidinol), *morph* (plaque morphology of phage P22), *gnd* (gluconate as source of carbon), and *rfb* (O-antigen biosynthesis and phage resistance). The derivatives were also tested, as both donor and recipient, in transductional crosses with several of the nearest known *hisG* point mutants, some of which are likely to map within 25 base pairs of the Tn*10* insertion site (see above).

The tetracycline-sensitive derivatives fell into a number of discrete categories on the basis of these tests (Table 1). One very surprising feature of these "imprecise excisants" is that over half of them (classes 1-4) not only lost the tetracycline resistance determinant but they also acquired detectable new lesions nearby on the bacterial chromosome. The population of *hisG*::Tn*10* bacteria as a whole (prior to selection for tetracycline-sensitives) exhibits no such lesions at frequencies greater than 10^{-3}. It seems likely, therefore, that the appearance of these nearby lesions is in some way connected with imprecise excision of the Tn*10* element. Our interpretation of this result is that when the Tn*10* element excises imprecisely, it often "sees" or interacts with another stretch of DNA during that process.

Table 1
Behavior of Tetracycline-sensitive Derivatives of *hisG*::Tn*10* with Respect to Markers on Either Side of Original Tn*10* Insertion Point

			Left		Right						
			recombination with *hisG* (left) markers		recombination with *hisG* (right) markers						
Class	Percent of total	*phs*	derivative as donor	derivative as recipient	derivative as donor	derivative as recipient	expression of *hisD*	*morph*	*gnd*	*rfb*	Interpretations
1	13	+ or –*									
a			none	none	normal	normal	–	+	+	+	deletions leftward from Tn*10* site
b		+	normal	normal	none	none	–	+ or –*	+	+	deletions rightward from Tn*10* site
2	17	+ or –*									
a			normal	low	normal	normal	–	+	+	+	inversions from Tn*10* site leftward
b		+	normal	normal	normal	low	–	+ or –*	+	+	inversions from Tn*10* site rightward
3	11	+	P22-resistant		P22-resistant		+	+	+	–	inversion from Tn*10* site to *rfb*

4	18	+	normal	very low	normal	very low	–	+	+	+	nontransducible rearrangement
5	42	+	normal	normal	normal	normal	+ or –*	+	+	+	alteration confined to Tn*10* element

Map Order of Relevant Markers: - *phs* -- *hisG* (left) -- Tn*10* -- *hisG* (right) -- *hisD* -- *morph* -- *gnd* -- *rfb* --

Tetracycline-sensitive derivatives of NK120 (*edd⁻hisG*::Tn*10*) were selected (324 from eight independent clones) by ampicillin treatment in the presence of tetracycline (10 μg/ml). They were scored for *gnd, rfb, phs,* and *morph.* In addition, since the original *hisG*::Tn*10* insertion is polar on *hisD* (histidinol dehydrogenase), they were tested for *hisD* function by testing their ability to grow on minimal medium supplemented with histidinol. After these tests, a number of derivatives chosen at random were characterized further as to reversion to his^+, to tetracycline resistance, and to growth on histidinol (*hisD* function); none revert spontaneously to tetracycline resistance.

Each derivative was then tested for its behavior as recipient and donor in P22-mediated transductional crosses with known *his* mutants. In scoring the behavior of *tet*-sensitive derivatives as recipients, "low" means that the frequency of recombinants was 5- to 20-fold lower than when the parental *hisG*::Tn*10* strain (NK120) was used as recipient; "very low" means that the frequency of recombinants was reduced 100-fold or more compared to NK120.

*+ or – means that some derivatives scored as + and some as – for the indicated phenotype.

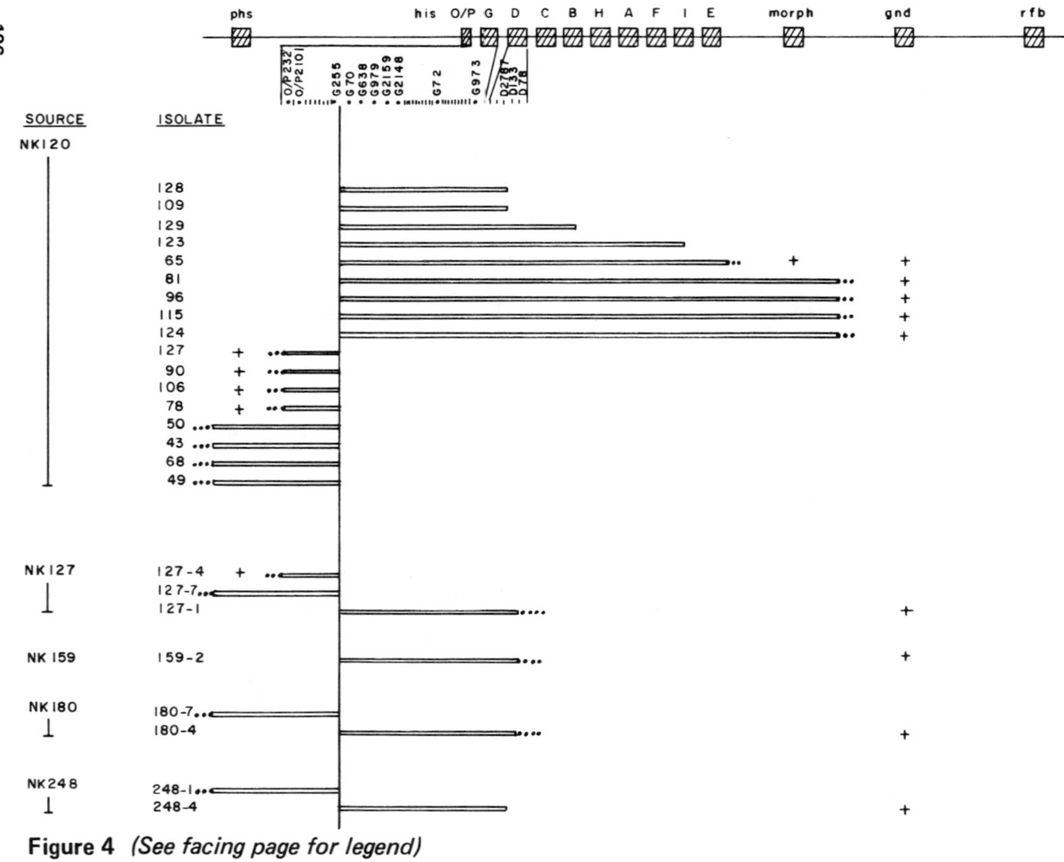

Figure 4 *(See facing page for legend)*

As summarized in Table 1, class-1 derivatives are missing blocks of markers on one side or the other of a common genetic end point; this end point corresponds (to the resolution afforded by the density of *hisG* point mutations) to the site of the original Tn*10* insertion (Fig. 4). These derivatives are therefore deletions apparently formed concomitantly with the imprecise excision of Tn*10*.

Class-2 derivatives, when used as recipients, recombine less frequently than normal with blocks of *hisG* point mutations located to one side or the other of the site of the original insertion. When used as *donors,* all frequencies are normal. As in the case of deletions, these blocks have a common end point which coincides with the original insertion location. Sometimes, the class-2 derivatives have also lost the function of a neighboring gene located on the same side as the point mutations with which recombination is low. We suspect that these tetracycline-sensitive derivatives are inversions formed concomitantly with imprecise excision of Tn*10*. In cases where the distal end point of the putative inversion interrupts a neighboring gene, the gene function is lost. In our view, class 3 is similar, but here the inversion has an end point in *rfb* (causing, among other things, P22 resistance) which makes transductional analysis impossible. However, preliminary results using chromosome mobilization by F′ *his* indicate that derivatives in this class indeed contain inversions. It should be noted that the *rfb* operon is transcribed in the direction opposite to the *his* operon; thus an inversion of the type we propose could restore expression of *hisD* function. Indeed, we find that all the class-3 derivatives have regained *hisD* function. Class-3 derivatives can be selected directly by virtue of their ability to use histidinol as a source of histidine. They arise at comparable frequencies in isogenic rec^+ and $recA^-$ hosts.

Class-4 tetracycline-sensitive derivatives are normal donors for all point mutations in the *hisG* gene and have lost no nearby gene functions. However, as

Figure 4
Tetracycline-sensitive deletions. The top line shows a map of the *his* region and adjacent genes. Below this is an expanded portion spanning *O/P, G,* and *D,* in which known *his* mutations are indicated by numbers adjacent to dots. Each "tick" (vertical mark) represents two additional point mutations mapped by I. Hoppe and J. Roth (unpubl.). The vertical line starting between G255 and G70 and extending downward represents the common end point of the deletions mapped in the lower part of the figure. A plus sign indicates that the gene above (*phs* at left; *morph* or *gnd* at right) is present, showing that the extent of the deletion is less than the distance to the gene. These deletions were obtained among tetracycline-sensitive derivatives of several independent *hisG*::Tn*10* insertions at the apparent hot spot. They were tested, both as donors and as recipients, in P22-mediated transductional crosses with known *his* mutations, as described elsewhere (Kleckner et al. 1975). Each deletion was also tested for other markers in the *his* region (see text). Bacterial strains NK120, NK127, NK159, NK180, and NK248 are edd^- and carry independent *hisG*::Tn*10* insertions (Kleckner et al. 1975).

recipients in transduction, they cannot be transduced efficiently by a single transducing particle (not even by phage grown on wild-type *his*$^+$ hosts). Although we know very little about this class, we suppose that it represents rearrangements (inversions or, possibly, translocations) which span a length of DNA larger than that which fits into a P22 phage head.

Class-5 derivatives are the simplest of all. These have no aberration at all relative to their *his*::Tn*10* parent save their loss of the tetracycline resistance phenotype. These we suppose to be changes which are totally internal to the Tn*10* element. It should be noted that these do not revert spontaneously to tetracycline resistance, suggesting that they may well be deletions or other destructive rearrangements within the DNA of the Tn*10* element.

We have grouped a variety of different recombination events under the heading of "imprecise excision." The relationships among these different events and the relationship of imprecise excision to precise excision and translocation are not clear. For Tn*10,* there are at least 1000 imprecise excision events for each precise one, a result quite different from results obtained with phages Mu and lambda. Following heat-pulse curing of a *trp*::lambda lysogen, Shimada et al. (1973) found that of 34 cured strains all had regained normal *trp* function. Bukhari (1975) found that about 10% of the polarity-relief revertants obtained from a *lacZ*::Mu *cts X* lysogen had regained *lacZ* function, suggesting that excision was precise approximately one in ten times. We do not know whether this large quantitative difference in the ratio of imprecise to precise excision events reflects a fundamental difference in mechanism or is simply a consequence of the fact that Tn*10* carries IS*3* sequences close to one another.

In summary, the genetic consequences associated with imprecise excision of Tn*10* suggest that imprecise excision often involves a rearrangement of DNA within the Tn*10* element, or in its vicinity, or both. Since this sort of illegitimate recombination is so frequent, it seems reasonable to suppose that Tn*10* plays an active role in generating these rearrangements, and that the mechanism of imprecise excision is in some way related to the mechanism of Tn*10* translocation.

Tn*10* Influences Deletion Generation

Several different types of experiments have shown that Tn*10* can in some way influence the production of deletions that remove chromosomal material immediately adjacent to its point of insertion. Some of these deletions remove the tetracycline resistance determinant as well and some do not. We favor the view that Tn*10* plays an active role in generating these deletions because of the frequencies at which they arise and because of the striking analogy between these deletions and the deletions generated by the lambda *int* system (see below); it should also be noted that Reif and Saedler (1975) have come to a similar conclusion for IS*1*-generated deletions. However, there is no direct evidence at the present time to support this view.

Tetracycline-sensitive Deletions

Our investigation of imprecise excision of Tn*10* (described in the preceding section) revealed that approximately 15% of all spontaneous tretracycline-sensitive derivatives of a *hisG*::Tn*10* insertion carried deletions of genetic markers in the histidine operon. These deletions extended leftward or rightward from a point at or closely linked to the site of the original Tn*10* insertion (Fig. 4).

There has been no genetic or physical mapping of the end points of these deletions within the Tn*10* element itself. Since these derivatives do not revert spontaneously to tetracycline resistance at frequencies greater than 10^{-8}, it seems likely that the deletion extends into the resistance determinant. However, in at least one case, the deletion appears not to remove all of the Tn*10* element. Deletion KR109 still carries some portion of the Tn*10* element which is capable of generating further deletions; such deletions have one end point at the site of the original insertion but the other end point is moved further down the histidine operon. The ability of IS-generated deletions to generate further deletions has also been documented by Reif and Saedler (this volume).

Tetracycline-resistant Deletions

In 1972 Chan and Botstein isolated deletions from a P22 phage harboring the Tn*10* element (P22Tc10). The P22Tc10 genome is too large to fit in a single P22 particle. Since P22 packages its DNA by sequential headfuls from a concatemer, the particles made upon induction of a P22Tc10 lysogen do not contain a complete genome and do not have any terminal repetition, and, as a result, they do not produce either phage or lysogens when they singly infect *Salmonella*. One can therefore select deletions of P22Tc10 by selecting tetracycline-resistant lysogens following low-multiplicity infection, since only genomes containing deletions will have regained terminal repetition. A map of the deletions obtained is reproduced in Figure 5. Clearly, the deletions share a common end point on the P22 genome which coincides with the position of the Tn*10* insertion; this common end point suggests that Tn*10* plays a role in the generation of these deletions.

Chan (1974) noticed another unusual feature of these deletions. They were obtained from a lysate of P22 *c2ts*29 Tc10 which had been prepared by UV induction of a lysogen; they were not obtained from a lysate made by heat induction of the same lysogen. It would appear that UV plays a role in the Tn*10*-influenced generation of deletions. Henderson and Weil (1975) have reported recently that *int*-generated deletions of lambda (which have a common end point at the phage attachment site) are also present at much higher frequencies in lysates prepared by UV induction than in lysates prepared by heat induction. In this case, there can be little doubt that UV somehow stimulates the production of deletions by the *int* system; it seems possible that the same is true in the case of Tn*10*.

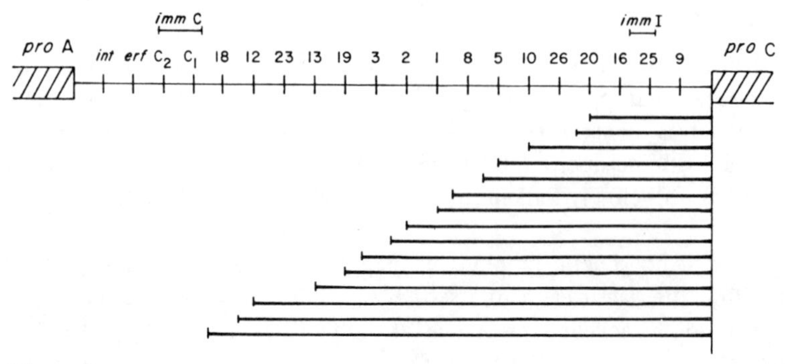

Figure 5
Map of P22 prophage showing extent of tetracycline-resistant prophage deletions isolated from P22Tc10. A P22Tc10 lysogen carries the Tn*10* insertion at the extreme right-hand end of the prophage genome (between gene 9 and the attachment site). Each line represents the material deleted in an individual strain. These deletions are described in more detail by Chan and Botstein (1972) and Chan et al. (1972). (Reprinted, with permission, from Chan et al. 1972.)

Tetracycline-resistant deletions having one end point at the point of insertion of a Tn*10* element have also been obtained in *Salmonella* without the intervention of phage. Deletions of this type were isolated from a strain carrying the Tn*10* element in *gnd* (which is linked to *his*) following penicillin selection for *his*⁻ mutants (N. Kleckner, unpubl.). These results, together with the observation of délétions during imprecise excision, suggest that the influence of Tn*10* on deletion generation does not depend on any P22 function.

None of the tetracycline-resistant deletions have yet been analyzed physically to determine exactly the end points of the deletions at (or within) the Tn*10* element. It is therefore not known whether the entire Tn*10* element still exists intact in any of these deletions. The deletions isolated thus far from P22 have their Tn*10* distal end points scattered throughout the phage genome (Fig. 5).

Tetracycline-sensitive deletion derivatives arise at frequencies of 10^{-3} to 10^{-4} in plate lysates of lambda Tn*10* phages. Three such deletion variants have been analyzed by heteroduplex mapping. Although each carries a deletion of material including portions of the Tn*10* element, no specific sites on either the phage genome or the element itself seem to be involved. One deletion is completely internal to the Tn*10* element itself, and two others remove variable amounts of Tn*10* and lambda material (D. Barker and N. Kleckner, in prep.). The interpretation of these results is subject to the reservation that we do not know whether any lambda function(s) affect either the frequency or the nature (the end points) of these deletions, but clearly these results resemble those found for tetracycline-sensitive deletion derivatives in the *his* operon of *Salmonella.*

Tn*10* appears to be involved in the generation of deletions which have one end point at or very near the site of the Tn*10* insertion. The extent of these deletions within the Tn*10* element itself is not well characterized, but both tetracycline-sensitive and tetracycline-resistant deletions can be obtained. Currently, we favor the view that the Tn*10* element plays an active role in the generation of these deletions.

SUMMARY

Tn*10* has been implicated in several different kinds of illegitimate recombination events: translocation and insertion into many different sites, excision (both precise and imprecise) from these sites, and formation of deletions and other chromosomal aberrations. The behavior of Tn*10* is quite analogous to that observed for the IS sequences IS*2* and IS*1*; deletions are generated, translocation occurs, precise excision is independent of temperature, events are independent of *recA* function, insertions are polar, and insertion shows some degree of specificity. It seems highly probable that the properties of Tn*10* as a translocatable element derive from the properties of the IS*3* sequences at its ends, although it remains to be shown whether either or both of these IS*3* sequences are functionally intact.

Although we do not yet understand the mechanisms involved in any of these illegitimate recombination events, some of the properties of Tn*10* (as listed below) should limit the kinds of models that can be entertained.

1. The insertion part of the translocation process seems to be quite precise both with respect to the element itself and with respect to the "recipient" DNA molecule. A discrete unit of genetically and physically well-defined DNA is inserted at the new location, and a gene that receives such an insertion is *not* damaged in any way which precludes subsequent restoration of gene function.
2. Insertion of Tn*10* shows a definite specificity for particular sites and a preference for some sites over others. This behavior is quite different from that of both phage Mu-1, which does not display any detectable specificity for particular sites in *lacZ* (Bukhari and Zipser 1972), and phage lambda, which shows an overwhelming preference for its normal attachment site and a marked preference for particular sites even when the primary attachment site is deleted (Shimada et al. 1973). Of the other translocatable elements, the Tn*5* (kanamycin resistance) element seems to insert randomly (Berg, this volume), whereas IS*1* and IS*2* seem to display a degree of specificity roughly comparable to Tn*10* (Starlinger and Saedler 1972, 1976). If all of these translocatable entities share a common mechanism, some explanation for this apparent continuum of insertion specificities will have to be found.
3. Virtually nothing is known about excision events which might precede or accompany translocation. Our observation that fewer than 10^{-3} precise

revertants of *his*::Tn*10* auxotrophs retain the resistance determinant does show that precise excision is not inevitably accompanied by reinsertion of the element elsewhere. This observation, plus the fact that precise excision occurs at such very low frequencies, raises the possibility that precise excision might not be involved in the translocation process. Imprecise excision, on the other hand, might well be involved in the translocation mechanism, since the frequency of imprecise excision is high. However, we have no evidence that any excision is necessary for translocation, and the possibility of a nondestructive (possibly replicative) translocation mechanism cannot be overlooked.

Acknowledgments

We are especially grateful to John Roth and his colleagues for productive collaboration, stimulation, and access to unreported strains and information essential to this work. We thank Katherine Reichardt for dedicated and skillful technical assistance; Ahmad Bukhari, Allan Campbell and Robert Weisberg for particularly illuminating points of discussion; and Susan Lucas for preparation of the manuscript. The research was supported by grants from the National Institutes of Health (GM 18973) and the American Cancer Society (VC18D). Nancy Kleckner was supported by an NIH postdoctoral fellowship (AI05109) and David Bostein by an NIH Career Development Award (GM 70325).

REFERENCES

Berg, D. E., J. Davies, B. Allet and J.-D. Rochaix. 1975. Transposition of R factor genes to bacteriophage λ. *Proc. Nat. Acad. Sci.* **72**:3628.

Bukhari, A. 1975. Reversal of mutator phage Mu integration. *J. Mol. Biol.* **96**: 87.

Bukhari, A. and D. Zipser. 1972. Random insertion of Mu-1 DNA within a single gene. *Nature* **236**:240.

Chan, R. K. 1974. Specialized transduction by bacteriophage P22. Ph.D. dissertation, Biology Department, Massachusetts Institute of Technology, Cambridge.

Chan, R. K. and D. Botstein. 1972. Genetics of bacteriophage P22. I. Isolation of prophage deletions which affect immunity to superinfection. *Virology* **49**:257.

——. 1976. Specialized transduction by bacteriophage P22 in *Salmonella typhimurium*: Genetic and physical structure of the transducing genomes and the prophage attachment site. *Genetics* **83**:433.

Chan, R. K., D. Botstein, T. Watanabe and Y. Okada. 1972. Specialized transduction of tetracycline resistance by phage P22 in *Salmonella typhimurium.* II. Properties of a high-frequency transducing lysate. *Virology* **50**:883.

Couturier, M. 1976. The integration and excision of the bacteriophage Mu-1. *Cell* **7**:155.

Gottesman, M. M. and J. L. Rosner. 1975. Acquisition of a determinant for chloramphenicol resistance by coliphage lambda. *Proc. Nat. Acad. Sci.* **72**: 5014.

Hedges, R. W. and A. E. Jacob. 1974. Transposition of ampicillin resistance from RP4 to other replicons. *Mol. Gen. Genet.* **132**:31.

Heffron, F., C. Reubens and S. Falkow. 1975. The translocation of a plasmid DNA sequence which mediates ampicillin resistance: Molecular nature and specificity of insertion. *Proc. Nat. Acad. Sci.* **72**:3623.

Henderson, D. and J. Weil. 1975. Recombination-deficient deletions in bacteriophage λ and their interactions with *chi* mutations. *Genetics* **79**:143.

Hu, S., K. Ptashne, S. N. Cohen and N. Davidson. 1975. The αβ-sequence of F is IS3. *J. Bact.* **123**:687.

Kleckner, N., R. K. Chan, B.-K. Tye and D. Botstein. 1975. Mutagenesis by insertion of a drug-resistance element carrying an inverted repetition. *J. Mol. Biol.* **97**:561.

Kopecko, D. J. and S. N. Cohen. 1975. Site-specific *recA*-independent recombination between bacterial plasmids: Involvement of palindromes at the recombinational loci. *Proc. Nat. Acad. Sci.* **72**:1373.

Malamy, M. H. 1967. Frameshift mutations in the lactose operon of *E. coli. Cold Spring Harbor Symp. Quant. Biol.* **31**:189.

———. 1970. Some properties of insertion mutations in the *lac* operon. In *The lactose operon* (ed. J. R. Beckwith and D. Zipser), p. 359. Cold Spring Harbor Laboratory, Cold Spring Harbor, New York.

Malamy, M. H., M. Fiandt and W. Szybalski. 1972. Electron microscopy of polar insertions in the *lac* operon of *Escherichia coli. Mol. Gen. Genet.* **119**:207.

Ptashne, K. and S. N. Cohen. 1975. Occurrence of insertion sequence (IS) regions on plasmid DNA as direct and inverted nucleotide sequence duplications. *J. Bact.* **122**:776.

Reif, H.-J. and H. Saedler. 1975. IS*1* is involved in deletion formation in the *gal* region of *E. coli* K12. *Mol. Gen. Genet.* **137**:17.

Sharp, P. A., S. N. Cohen and N. Davidson. 1973. Electron microscope heteroduplex studies of sequence relations among plasmids of *Escherichia coli.* II: Structure of drug resistance (R) factors and F factors. *J. Mol. Biol.* **75**:235.

Shimada, K., R. Weisberg and M. E. Gottesman. 1973. Prophage lambda at unusual locations. II. Mutations induced by bacteriophage lambda in *Escherichia coli* K12. *J. Mol. Biol.* **80**:297.

Starlinger, P. and H. Saedler. 1972. Insertion mutations in microorganisms. *Biochimie* **54**:177.

Starlinger, P. and H. Saedler. 1976. IS-elements in microorganisms. In *Current topics in microbiology and immunology,* vol. 75, p. 111. Springer Verlag, Berlin.

Tye, B.-K., R. K. Chan and D. Botstein. 1974. Packaging of an oversize transducing genome by phage P22. *J. Mol. Biol.* **85**:45.

Watanabe, T., Y. Ogata, R. K. Chan and D. Botstein. 1972. Specialized transduction of tetracycline resistance by phage P22 in *Salmonella typhimurium.* I. Transduction of R factor 222 by phage P22. *Virology* **50**:874.

Insertion and Excision of the Transposable Kanamycin Resistance Determinant Tn5

D. E. Berg
Department of Biochemistry
University of Wisconsin
Madison, Wisconsin 53706

Transposable elements are found inserted in the genomes of prokaryotes and eukaryotes and can move from the positions at which they were first detected to other nonhomologous regions. They influence the activity of genes near their sites of insertion and also cause chromosome aberrations.

As discussed in Section I, the first elements identified in bacteria were short (800-1400 base pairs) DNA sequences found as insertions in genes that had mutated spontaneously. They were distinguished on the basis of size and sequence relationships in electron microscope heteroduplex studies and were designated IS*1*, IS*2*, and IS*3*. Multiple copies of these elements are present in the chromosome of *Escherichia coli*. They are also found in plasmids, such as the fertility factor F, and appear to be involved in the insertion of these plasmids into the bacterial chromosome.

Recently, transposable elements that encode resistance to antibiotics have been described (Hedges and Jacob 1974; Berg et al. 1975; Kleckner et al. 1975; Heffron et al. 1975; Gottesman and Rosner 1975; Barth et al. 1976). They are larger than the IS elements mentioned above, and, in several of them, sequences present at one end are repeated at the other end in reverse orientation (Berg et al. 1975; Heffron et al. 1975; Kleckner et al. 1975).

Temperate bacteriophages Mu and lambda resemble both IS elements and transposable resistance determinants in their capacities to insert into and excise from bacterial genes (Shimada et al. 1972; Howe and Bade 1975; Bukhari 1975). The similarities suggest possible evolutionary relationships between the integration-excision systems of these genetic units. In addition, homology between IS elements and parts of transposable resistance determinants has been seen in two cases (IS*3*-Tn*10* [Ptashne and Cohen 1975] and IS*1*-TN*9* [MacHattie and Jackowski, this volume]).

This report deals with a transposable kanamycin resistance element, Tn*5*, discovered during the analysis of lambdoid phages that had acquired genes for resistance during growth on *E. coli* harboring a resistance plasmid. The two phage lines studied carry insertions of the same 5200-base-pair segment of plasmid JR67 at two different sites in the λ genome: one near *c*III and the other in *rex*; 1450 base pairs at one end of the segment are repeated in reverse orientation at the other end. The positions of the two Tn*5* insertions, their sequence identity, and the fact that they are insertions, not substitutions, all indicated that the Tn*5* segment might be a transposable element (Berg et al. 1975; D. E. Berg and B. Allet, in prep.).

The present experiments document transposition of Tn*5* from λ Tn*5* vectors to the *E. coli* chromosome and from site to site on the chromosome. It is shown that Tn*5* inactivates genes by insertion, that Tn*5*-induced mutations can revert by excision of Tn*5*, and that this excision is not associated with transposition of Tn*5* to new locations.

Transposition of Tn*5* from λ to the Bacterial Chromosome

Approximately 10^{-2} of cells infected with λ*b*221 Tn*5* phages were transduced to Kan^r (see Fig. 1). If either of the generalized recombination pathways (*red* or *rec*) was inactive, the transductant frequency was 3-7 $\times$ 10^{-3}, and if both pathways were inactive, the transductant frequency was 10^{-3}.

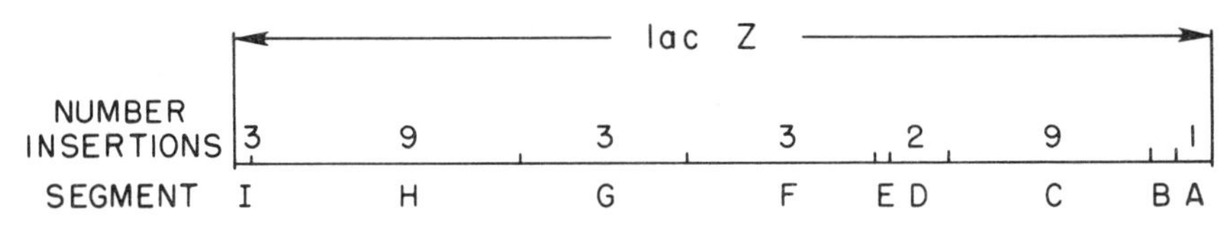

Figure 1
Distribution of Tn*5* insertions in *lacZ*. *E. coli* K12 strain DB1162 (F$^-$ *lac*$^+$ *proB*$^-$ *thy*$^-$ *str*^{r}) was grown to stationary phase in tryptone broth containing 0.2% maltose and infected with phage λ*b*221 *c*I857 Tn*5* (Tn*5* insertion near gene *c*III) at a multiplicity of five phage per cell. After 15 min adsorption at 30°C, the cells were diluted 20-fold in tryptone broth and aerated 30 min at 30°C to permit expression of resistance. Lac$^-$ Kan^r transductants were selected on 1.5% agar plates containing 1% tryptone, 1% NaCl, 0.5% yeast extract, 1% lactose, 30 μg/ml 2,3,5-triphenyltetrazolium chloride, and 30 μg/ml kanamycin sulfate. *LacZ*$^-$ mutants, identified as white colonies when streaked on XGal plates (Miller 1972), comprised approximately one-third of the Lac$^-$ mutants.

The *lac*$^-$ mutations were assigned to different deletion segments, designated "A" through "I," in crosses with F$'$ *proB*$^+$ *lacZ* deletion strains CSH 13 through CSH 20 (Miller 1972). Pro$^+$ exconjugants were selected and replica-plated onto MacConkey lactose agar, and the formation of Lac$^+$ recombinants was scored.

The operator region and the *lacY* gene lie to the right and left of *lacZ*, respectively, as drawn here. The map is drawn to the scale of Zipser et al. (1970).

For mapping alleles within a deletion group, the *lacZ*::Tn*5* alleles were first recombined into an F$'$ *proB*$^+$ *lac*$^+$ episome; these F$'$ *pro*$^+$ *lacZ*::Tn*5* episomes were transferred into an F$^-$ *proB lac* deletion strain and then crossed pairwise with the *lacZ*::Tn*5* derivatives of DB1162.

The *b*221 deletion in λ*b*221 Tn*5* removes the phage integration site, thus rendering the phage genome incapable of integrating by λ-specific mechanisms. Most transductants (⩾ 98%) do not contain a λ prophage and thus must have been formed by the transposition of the Tn*5* segment from its λ vector to the *E. coli* chromosome. The 1-2% of transductant clones that do contain λ prophages are unstable and spontaneously lose λ at frequencies of 10^{-2}-10^{-3}.

One percent of the Kan^r transductants are auxotrophs, and 0.01% are Lac$^-$. Therefore, Tn*5* can serve as a mutagen. Similar observations have been made with a transposable tetracycline resistance determinant, which, however, undergoes transposition only at a frequency of about 10^{-6} (Kleckner et al. 1975).

Specificity of Tn*5* Insertion

The distribution of 30 sites of Tn*5* insertion in the *lacZ* gene was determined by deletion mapping. These *lacZ*::Tn*5* insertion mutations are found in seven different regions of *lacZ* (Fig. 1). Pairwise crosses between 20 mutants picked from deletion groups C, F, G, and H were performed to see if they were located at different sites. Lac$^+$ recombinants were obtained in all crosses involving 19 of the mutations; one mutation in group C behaved like a small deletion in that it failed to recombine with two other insertions in group C. These results show that Tn*5* insertion into *lacZ* is not site-specific.

Localization and Stability of Tn*5* Insertions

To determine the number of Tn*5* elements inserted per chromosome, seven different F$^-$ *lacZ*::Tn*5* *proB*$^-$ *str*^{r} strains were crossed with HfrH (*pro*$^+$ *lac*$^+$ *str*^{s}), and *pro*$^+$ *lac*$^+$ *str*^{r} recombinants were selected. In each of the crosses, ⩾ 99% of the recombinants were Kan^s, indicating that the Tn*5* element causes mutation by insertion into genes and that usually only one Tn*5* element is inserted per chromosome. However, 0.1-1% of the Pro$^+$ Lac$^+$ exconjugants from each cross remained Kan^r, indicating the existence of subclones in the F$^-$ populations in which a Tn*5* element was present elsewhere in the chromosome. This frequency of transposition from one chromosomal site to another is similar to the frequency of transposition during λ*b*221 Tn*5* infection.

Loss of the Tn*5* Element from Its Site of Insertion

Nineteen of 20 *lacZ*::Tn*5* insertion mutations studied to date revert to *lacZ*$^+$ (frequency approximately 10^{-6}). Ninety-nine percent of revertants from the seven insertions studied in more detail were Kan^s. Since few (~1%) Kan^r clones are found among revertants, these sensitive revertants must arise by excision and loss of Tn*5* from the cell.

Test crosses with 15 of the Kan^r Lac$^+$ revertants indicated that the resistance determinant was no longer linked to the *lac* operon. Thus these rare Kan^r Lac$^+$ revertants may have arisen in clones in which Tn*5* had been transposed to a new site prior to the reversion event.

Each of the 30 *lacZ*::Tn*5* insertions obtained is LacY$^-$; thus insertions of Tn*5* in an operon are polar on the expression of distal genes. Revertants of seven insertions to the LacY$^+$ phenotype were selected on MacConkey melibose agar at 41°C. The majority (60-95%, depending on the insertion) remained Lac$^-$. Nearly all of these partial revertants were also Kan^s.

Two classes of LacY$^+$ revertants can be distinguished on the basis of their ability to revert to *lacZ*$^+$. Most of those that were found to be unable to revert behaved like deletions in crosses, failing to give recombinants with one or more closely linked *lac*$^-$ alleles. It is likely that the revertable *lacY*$^+$*Z*$^-$ mutations retain part of the Tn*5* insertion (although the genes for resistance to kanamycin and the determinants of polarity have been deleted).

IMPLICATIONS FOR THE MECHANISM OF TRANSPOSITION

The 5000-base-pair kanamycin resistance determinant Tn*5* can be transposed from phage λ to the *E. coli* chromosome and from one chromosomal site to another at frequencies of 10^{-2}-10^{-3}. It causes mutation by insertion into genes. Many potential sites of Tn*5* insertion exist; at least 19 are found in the *lacZ* gene alone. The revertability of 19 of 20 *lacZ*::Tn*5* mutations indicates that usually no recipient DNA sequences are lost during insertion.

Transposition of the Tn*5* element to new sites on the *E. coli* chromosome occurs at a frequency of 10^{-2}-10^{-3}, whereas excision of the Tn*5* element, detected as reversion, occurs at a frequency of 10^{-5}-10^{-6}. Furthermore, revertants of *lacZ*::Tn*5* insertions are usually Kan^s. There is thus the paradox that the excisions of Tn*5* that we detect do not lead to transposition of Tn*5* to new sites.

How can these observations be explained? The mechanism of transposition proposed in Figure 2 is viewed as a modification of that of λ phage integration (see Campbell 1971; Gottesman and Weisberg 1971). An enzyme complex ("transposase") which recognizes and binds the ends of the Tn*5* element specifically also binds another DNA molecule with little or no sequence specificity. Transposition occurs by (1) cleavage of the DNA duplex at the junctions between the Tn*5* element and the flanking sequences, (2) cleavage of the bound recipient DNA molecule, (3) realignment of the newly created termini, and (4) ligation of Tn*5* to the recipient molecule. After excision of Tn*5*, the remainder of the donor molecule is released, since its ends no longer contain sequences specifically bound by the complex. The freed ends would not be joined by recombination since they are not homologous to each other and therefore would probably be degraded exonucleolytically.

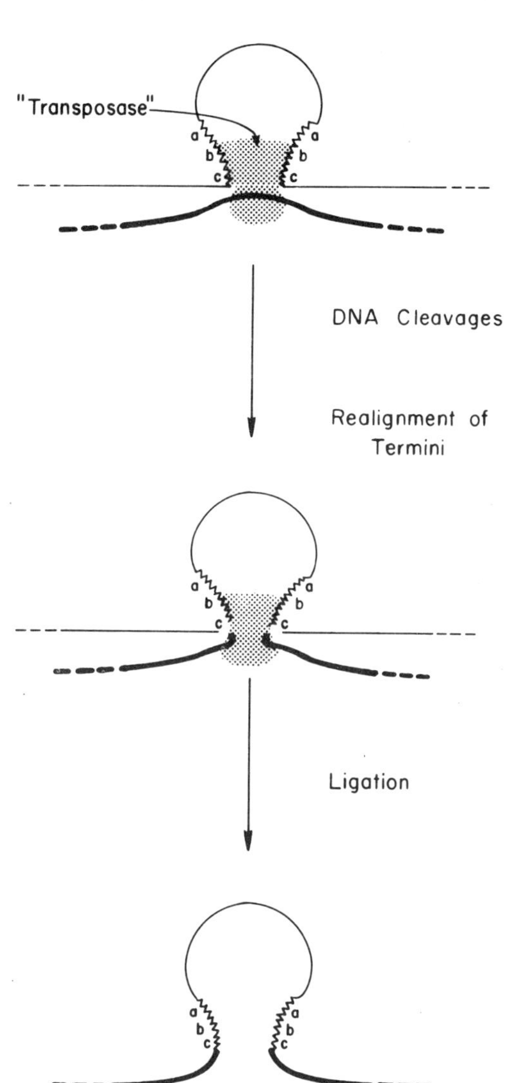

Figure 2
A proposed mechanism of transposition. Transposition of Tn*5* results in insertion of the element (here shown as the segment with ends "cba" and "abc") into a nonhomologous DNA sequence or molecule (bold line). The sequences "cb" and "bc" at the ends of the Tn*5* element are bound specifically by a "transposase" enzyme complex. The complex also binds another DNA molecule without regard to its sequence. Three DNA cleavages occur: one at each junction of Tn*5* with the "vector" DNA and one in the recipient molecule. The ends of the Tn*5* segment and the recipient molecule are aligned and ligated while bound by the complex. The remainder of the vector molecule may be released and degraded exonucleolytically.

The principal departures of this model from that of λ phage integration are the lack both of sequence specificity for the recipient DNA molecule and of a covalently closed circular intermediate form. The lack of specificity results in insertion of Tn*5* at many different sites and the loss of the remainder of the donor molecule. Lack of circularity of the Tn*5* DNA need not impede its transposition to new sites; ends of linear intermediates might be held in proximity to each other by the "transposase" complex, as well as being protected by it from exonucleolytic degradation. The circularity of λ DNA prior to its integration could reflect adaptation of λ's integration mechanism to the use of the products of other pathways, e.g., replication.

How do the detectable products of Tn*5* excision arise? Reversion of *lacZ*::Tn*5* insertion mutations to Lac^+ suggests DNA sequence-specific recognition and cleavage similar to that involved in transposition. Reversion could be viewed as an abortive transposition event in which components from the "transposase" complex bind the segment containing Tn*5*, remove Tn*5*, and restore the ancestral sequence (designated KL MN in Fig. 3). Although some of the partial ($LacY^+Z^-$) revertants might also arise from this process, it is clear that others must arise by the processes generally responsible for spontaneous deletion formation (Franklin 1971).

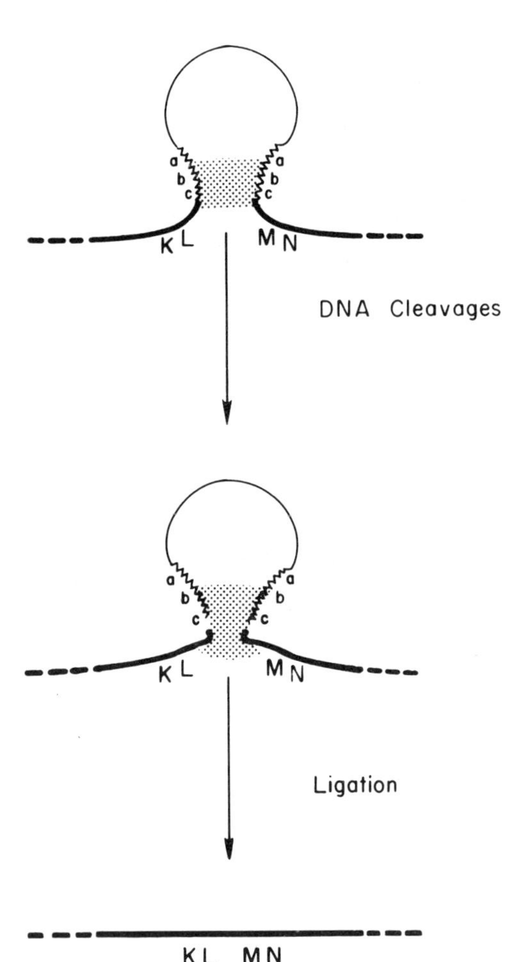

Figure 3
A proposed mechanism of reversion of Tn*5* insertion mutations. An enzyme complex similar to that employed for transposition (Fig. 2) binds specifically the Tn*5* segment, cleaves the DNA at "c," and permits ligation of the ends near sequences "L" and "M" which result from removal of Tn*5*.

SUMMARY

Transposition of the Tn*5* (kanamycin resistance) element from a λ Tn*5* phage to the *E. coli* chromosome or from one chromosomal site to another occurs at frequencies of 10^{-2}-10^{-3}. Tn*5* can insert into many different sites (>19 in *lacZ*). Each of 30 insertions in *lacZ* is polar on expression of the distal *lacY* gene. Nineteen of 20 *lacZ*::Tn*5* insertion mutations revert to Lac$^+$, and also to LacY$^+$Z$^-$, at frequencies of approximately 10^{-4}-10^{-6}. Nearly all revertants are Kan^s, which implies that excision of Tn*5*, detectable as reversion, is not associated with the transposition of Tn*5* to new sites. Models were proposed to account for the lack of specificity in the choice of a Tn*5* insertion site, the apparent loss of the DNA molecule that had contained Tn*5* during transposition of Tn*5* to new sites, and the capacity of Tn*5* insertion mutants to revert to wild type.

Acknowledgments

This work was initiated in the Départment de Biologie Moléculaire, Université de Genève, Switzerland. I thank C. Berg, L. Caro, J. Davies, M. Howe, and G. Kellenberger-Gujer for stimulating discussions. The work was supported by grants from the Fonds National Suisse de la Recherche Scientifique to L. Caro, the National Science Foundation to J. Davies, and by NIH Grant T32 CA090705 from the National Cancer Institute, DHEW.

REFERENCES

Barth, P., N. Datta, R. Hedges and N. Grintner. 1976. Transposition of a DNA sequence encoding trimethoprim and streptomycin resistances from R483 to other replicons. *J. Bact.* **125**:800.

Berg, D., J. Davies, B. Allet, and J.-D. Rochaix. 1975. Transposition of R factor genes to bacteriophage λ. *Proc. Nat. Acad. Sci.* **72**:3628.

Bukhari, A. 1975. Reversal of mutator phage Mu integration. *J. Mol. Biol.* **96**:87.

Campbell, A. 1971. Genetic structure. In *The bacteriophage lambda* (ed. A. D. Hershey), p. 138. Cold Spring Harbor Laboratory, Cold Spring Harbor, New York.

Franklin, N. 1971. Illegitimate recombination. In *The bacteriophage lambda* (ed. A. D. Hershey), p. 175. Cold Spring Harbor Laboratory, Cold Spring Harbor, New York.

Gottesman, M. E. and R. A. Weisberg. 1971. Prophage insertion and excision. In *The bacteriophage lambda* (ed. A. D. Hershey), p. 113. Cold Spring Harbor Laboratory, Cold Spring Harbor, New York.

Gottesman, M. M. and J. L. Rosner. 1975. Acquisition of a determinant for chloramphenicol resistance by coliphage lambda. *Proc. Nat. Acad. Sci.* **72**:5014.

Hedges, R. and A. Jacob. 1974. Transposition of ampicillin resistance from RP4 to other replicons. *Mol. Gen. Genet.* **132**:31.

Heffron, F., C. Rubens and S. Falkow. 1975. Translocation of a plasmid DNA sequence which mediates ampicillin resistance: Molecular nature and specificity of insertion. *Proc. Nat. Acad. Sci.* **72**:3623.

Howe, M. and E. Bade. 1975. Molecular biology of bacteriophage Mu. *Science* **190**:624.

Kleckner, N., R. Chan, B.-K. Tye and D. Botstein. 1975. Mutagenesis by insertion of a drug resistance element carrying an inverted repetition. *J. Mol. Biol.* **97**:344.

Miller, J. 1972. *Experiments in molecular genetics.* Cold Spring Harbor Laboratory, Cold Spring Harbor, New York.

Ptashne, K. and S. Cohen. 1975. Occurrence of insertion sequence (IS) regions on plasmid deoxyribonucleic acid as direct and inverted nucleotide sequence duplications. *J. Bact.* **122**:776.

Shimada, K., R. Weisberg and M. E. Gottesman. 1972. Prophage lambda at unusual chromosomal locations. I. Location of the secondary attachment sites and the properties of the lysogens. *J. Mol. Biol.* **63**:483.

Zipser, D., S. Zabell, J. Rothman, T. Grodzicker and M. Wenk. 1970. Fine structure of the gradient of polarity in the *Z* gene of the *lac* operon of *Escherichia coli. J. Mol. Biol.* **49**:251.

Transposition and Deletion of Tn*9*: A Transposable Element Carrying the Gene for Chloramphenicol Resistance

J. L. Rosner* and M. M. Gottesman†
*Laboratory of Molecular Biology
National Institute of Arthritis, Metabolism and Digestive Diseases
and †Laboratory of Molecular Biology, National Cancer Institute
National Institutes of Health, Bethesda, Maryland 20014

This paper summarizes our work on Tn*9*, a transposable genetic element (or transposon) which carries an R-factor-derived gene for chloramphenicol resistance (*cam*). Transposons may be defined as genetic elements capable of integrating into numerous, apparently nonhomologous, sequences of DNA. For a given transposon, a specific, nonpermuted sequence is observed regardless of the site it occupies. Furthermore, recombination functions of *Escherichia coli* that are required for promoting recombination of homologous DNAs are not required for transposition. These observations suggest that each transposon has an insertion mechanism that is specific for sequences on its own DNA and relatively nonspecific for sequences on the DNA with which it recombines.

The Transposability of *cam*

The transposability of the *cam* gene was first indicated by the accidental derivation of P1*cam* from P1 in the course of experiments involving P1-mediated transduction of R-factor genes (Kondo and Mitsuhashi 1964). Subsequently, Kondo and Mitsuhashi (1966) showed that F could acquire *cam* from P1*cam*. The striking fact that there could be "alternate association of the Cm^r determinant from an R factor with different genetic elements, first with P1 and then with the F factor" (Kondo and Mitsuhashi 1966) was hard to evaluate since little was understood of the molecular biology of R factors, P1, and F.

An opportunity for exploring the transposability of *cam* was provided by the observation that λ could acquire *cam* from a defective derivative of P1*cam* (Scott 1973). Did λ*cam*1 arise by the usual mechanism for formation of specialized transducing λ or was some other mechanism at work? If λ*cam*1 were formed by the usual mechanism for generating specialized transducing

We respectfully dedicate this paper to the memory of the late Professor Theodosius Dobzhansky.

λ phage, the *cam* genes would be located at one of the λ attachment sites (*att*) (Gottesman and Weisberg 1971). However, electron microscopic analysis of heteroduplex DNA from λ*cam*1 and appropriately marked λ standards showed that *cam* was located within the *b*2 region of λ and not at *att* (Gottesman and Rosner 1975). Furthermore, λ*cam* phages could be generated from strains in which *cam* was present on a P1 plasmid and the λ prophage was located at any of three different chromosomal sites. Analysis of heteroduplex DNA made from one of these isolates (called λ*cam*3) and the original λ*cam*1 showed that *cam*1 and *cam*3 were present at different sites in the *b*2 region (Gottesman and Rosner 1975). A third *cam* insertion (*cam*107) into λ has been shown to be at a third site in the *b*2 region (MacHattie and Jackowski, this volume). The ability of *cam* to insert into λ is not restricted to the *b*2 region, however, since λ*b*2 *cam* may be isolated and at frequencies comparable to λ$b2^+$*cam* (J. L. Rosner, unpubl.). Thus *cam* is able to insert into many sites on λ.

The acquisition of *cam* by λ is independent of λ Red and *Escherichia coli* RecA functions, but it was not clear whether a P1 function was required (Gottesman and Rosner 1975). Kleckner et al. (1975) have shown that when *tet* (tetracycline resistance) is transposed from P22*tet* to the *Salmonella* chromosome, about 1% of the resistant cells are auxotrophs due to the insertion of *tet* into genes required for prototrophy. To see whether *cam* transposition could occur in the absence of P1, the transfer of *cam* from λ*cam* to the *E. coli* chromosome was studied. A *recA* or *recA*(λcI857*h*80*att*80) *E. coli* culture was infected with λcI857*bio*69*gam*210*nin*5*cam*1 at 31°C (where λ is repressed), and chloramphenicol-resistant bacteria were selected at 31°C. Under these conditions, the infecting phage cannot lysogenize. About 2×10^{-5} infected cells were chloramphenicol-resistant, and 8 out of 1000 of these were auxotrophs (J. L. Rosner and R. G. Bieser, unpubl.). Similar experiments in which *recA*(*gal-att*λ-*bio*) Δ(λcI857*h*80*att*80) bacteria were infected with either λ*cam*1, λ*cam*1*red*3, λ*bio*69*gam*210*nin*5*cam*1, or λ*bio*11*cam*1 under repressed conditions gave rise to about 2×10^{-5} chloramphenicol-resistant cells. Thus transposition of *cam* to the chromosome appears to be independent of P1, *recA, red,* and other λ functions.

cam Is Part of a Transposon

The *cam* gene is a portion of a discrete sequence of DNA which is transposed as a unit. It encodes the enzyme chloramphenicol acetyl transferase which acetylates chloramphenicol and thereby renders it inactive (Shaw 1967; Kondo et al. 1970; de Crumbrugghe et al. 1973). Chloramphenicol acetyl transferase consists of four identical subunits of about 20,000 daltons each (Shaw and Brodsky 1968; W. Shaw, pers. comm.). To encode this information, about 600 base pairs are required. Since the DNA inserted in λ*cam*1, *cam*3, and *cam*107 is about 2400 to 2600 base pairs, *cam* must be only a part of this insertion. Moreover, the insertions in *cam*1 and *cam*3 are homologous and nonpermuted

(Gottesman and Rosner 1975), suggesting that the entire sequence transposes as a unit. Since *cam* has many insertion sites in nonhomologous DNAs and is part of a larger sequence which transposes as a unit independently of functions required for the recombination of homologous DNAs, we have concluded that *cam* is present on a transposon for which the designation Tn*9* has been given.

*IS*1 *Brackets* cam *in Tn*9

A number of transposons carrying drug resistance genes have been identified recently (Berg et al. 1975; Heffron et al. 1975; Kleckner et al. 1975; Kopecko and Cohen 1975). Several of these have a sequence at one end which is repeated in inverted orientation at the other end. These inverted repeats, readily seen by electron microscopy, focused attention on transposon ends. In the case of Tn*10* (which carries *tet*), the inverted repeats have been identified as IS*3* (Ptashne and Cohen 1975). No inverted repeats have been observed for Tn*9* (Gottesman and Rosner 1975). However, MacHattie and Jackowski (in prep. and this volume) have shown that Tn*9* has homologous sequences of about 800 base pairs at each end and that these sequences are homologous to IS*1*. The same conclusion was reached by Chow and Bukhari (this volume) for Tn*9* inserted in bacteriophage Mu.

The presence of direct repeats of IS*1* on Tn*9* prompts some speculation as to their origin. The IS*1* repeats probably reflect the presence of direct repeats of IS*1* on certain R factors (Hu et al. 1975; Ptashne and Cohen 1975). However, the IS*1* repeats are not closely linked to *cam.* We were unable to isolate λ*cam* from bacteria harboring the R factor R1 and λ (Gottesman and Rosner 1975). IS*1* repeats may have come to be close to *cam* by successive transposition of each IS*1* or by IS*1*-mediated deletion of intervening DNA (Reif and Saedler 1975). In either case, Tn*9* was formed under selective pressure: alternate cycling of P1-mediated transduction of *cam* and F-mediated conjugal transfer of *cam* (Kondo and Mitsuhashi 1964). By applying strong selective pressure, P1 phage carrying *cam* or *kan* have been isolated recently (Mise and Arber 1976; Takano and Ikeda 1976).

IS-mediated transposition is not limited to drug resistance genes. A novel P1-*pro* prophage carries bacterial *proAB* genes and the P1 gene 2 bracketed by two Tn*9* elements (Rosner 1975). Bacteria deleted for *proAB* are efficiently and jointly transduced for *cam pro* gene 2. That this group of genes (R factor, bacterial, and phage in origin) has been transposed is shown by their map locations: between *ctr* and *his* in one case and between *his* and *gal* in another (Rosner 1975 and unpubl.). This shows further how larger transposable elements can be formed from simpler ones.

*Instability of Tn*9

A striking feature of Tn*9* is its instability. Lysates of P1*cam* contain a variable fraction of particles (usually 5-20%) which have irreversibly lost the ability to transduce *cam* (Kondo and Mitsuhashi 1964). Kondo et al. (1970) reported that phage that had lost their ability to transduce *cam* (which will be referred to as *cam*Δ) had a decreased buoyant density in CsCl, indicating that the *cam* DNA had been lost.

We have found that λ*cam* phage also lose their ability to transduce *cam* and that this is due to a loss of a portion of Tn*9* (J. L. Rosner and M. M. Gottesman, in prep.; MacHattie and Jackowski, this volume). When λ*cam*1 lysogens are induced, about 2% of the phage are λ*cam*Δ. However, lysates prepared by multiple cycles of infection (as by the confluent lysis plate method) contain up to 99% λ*cam*Δ. (This is true for several independently isolated λ*cam.*) CsCl density gradient centrifugation indicates that whereas λ*cam*1 has 5% more DNA than λ, λ*cam*1Δ has about 2% more DNA than λ. Electron microscopy of heteroduplex DNA prepared from either of two λ*cam*1Δ phage and λ*imm*434 or λ*cam*1 verified that a portion, but not all, of Tn*9* is lost.

The finding that IS*1* elements bracket *cam* helps explain the loss of *cam.* Homologous recombination between the IS*1* elements would result in the excision of *cam* and one IS*1* while one copy of IS*1* would remain on the phage. Indeed, MacHattie and Jackowski (this volume) have shown that λ*cam*107Δ phage do have one copy of IS*1*. However, several questions concerning the loss of *cam* remain to be answered. None of several thousand λ*cam*1 lysogenic *rec*$^+$ cells were found to have lost *cam* after several generations of growth in the absence of chloramphenicol. Why didn't the *E. coli* Rec system excise *cam*? Various *red* mutations have been crossed into λ*cam*1, yet (with the exception discussed below) they produce *cam*Δ phage even in *recA* or *recB* hosts. Why does the loss of *cam* occur in the absence of systems which promote homologous recombination?

An explanation for *rec*-independent loss of *cam* was first suggested by our observation that a particular λ*bio*69*gam*210*nin*5*cam*1Δ3 isolate had a deletion of *cam* which extended into the λ genome toward *att.* Since IS*1* is present on each side of *cam* and since IS*1* promotes deletions (Reif and Saedler 1975), we suggest that some of the *cam*Δ arise as IS*1*-promoted deletions.

Deletions of λ adjacent to Tn*9* that do not eliminate the *cam* gene have been observed by us and by MacHattie and Jackowski. One such deletion was found after crossing *cam* from λ*cam*1*bio*69*gam*210*nin*5 into λ*red*3. The deletion removes about 6% of the λ genome adjacent to Tn*9,* but it is not known whether any of Tn*9* is missing. This λ*cam*1*red*3 strain shows little if any loss of *cam* even in *rec*$^+$ cells. However, a second λ*cam*1*red*3 strain produced in the same cross does not have a detectable deletion and does lose *cam* frequently. This phenomenon is the subject of continuing investigation.

SUMMARY

A unique genetic element, Tn*9,* carries the *cam* gene which confers upon bacteria high-level resistance to the antibiotic chloramphenicol. Tn*9*, like other recently described transposons, facilitates its own transposition to diverse replicons in the absence of known recombination functions. Unlike some transposons, Tn*9* suffers high-frequency loss of its antibiotic resistance activity. This appears to be due to two different mechanisms: (1) known recombination systems acting on directly repeated sequences that bracket the *cam* gene and (2) Tn*9*-promoted deletions of itself and/or of neighboring genes.

Acknowledgments

We thank S. Bailey and R. Bieser for their assistance with some of the experiments, R. Bird for encouragement, L. MacHattie for communicating his work to us prior to publication, and A. Martinson for typing the manuscript.

REFERENCES

Berg, D. E., J. Davies, B. Allet and J.-D. Rochaix. 1975. Transposition of R factor genes to bacteriophage λ. *Proc. Nat. Acad. Sci.* **72**:3628.

de Crumbrugghe, B., W. V. Shaw, I. Pastan and J. L. Rosner. 1973. Stimulation by cyclic AMP and ppGpp of chloramphenicol acetyl transferase synthesis. *Nature New Biol.* **241**:237.

Gottesman, M. E. and R. A. Weisberg. 1971. Prophage insertion and excision. In *The bacteriophage lambda* (ed. A. D. Hershey), p. 113. Cold Spring Harbor Laboratory, Cold Spring Harbor, New York.

Gottesman, M. M. and J. L. Rosner. 1975. The acquisition of a determinant for chloramphenicol resistance by coliphage lambda. *Proc. Nat. Acad. Sci.* **72**: 5041.

Heffron, F., C. Rubens and S. Falkow. 1975. Translocation of a plasmid DNA sequence which mediates ampicillin resistance: Molecular nature and specificity of insertion. *Proc. Nat. Acad. Sci.* **72**:3623.

Hu, S., E. Ohtsubo, N. Davidson and H. Saedler. 1975. Electron microscope heteroduplex studies of sequence relations among bacterial plasmids: Identification and mapping of the insertion sequences IS1 and IS2 in F and R plasmids. *J. Bact.* **122**:764.

Kleckner, N., R. K. Chan, B.-K. Tye and D. Botstein. 1975. Mutagenesis by insertion of a drug-resistance element carrying an inverted repetition. *J. Mol. Biol.* **97**:561.

Kondo, E. and S. Mitsuhashi. 1964. Drug resistance of enteric bacteria. IV. Active transducing bacteriophage P1CM produced by the combination of R factor with bacteriophage P1. *J. Bact.* **88**:1266.

——. 1966. Drug resistance of enteric bacteria. VI. Introduction of bacteriophage P1CM into *Salmonella typhi* and formation of P1dCM and F-CM elements. *J. Bact.* **91**:1787.

Kondo, E., D. K. Haapala and S. Falkow. 1970. The production of chloramphenicol acetyltransferase by bacteriophage P1CM. *Virology* **40**:431.

Kopecko, D. J. and S. N. Cohen. 1975. Site specific *recA*-independent recombination between bacterial plasmids: Involvement of palindromes at the recombinational loci. *Proc. Nat. Acad. Sci.* **72**:1373.

Mise, K. and W. Arber. 1976. Plaque-forming transducing bacteriophage P1 derivatives and their behaviour in lysogenic conditions. *Virology* **69**:191.

Ptashne, K. and S. N. Cohen. 1975. Occurrence of insertion sequence (IS) regions on plasmid deoxyribonucleic acid as direct and inverted nucleotide sequence duplications. *J. Bact.* **122**:776.

Reif, H.-J. and H. Saedler. 1975. IS1 is involved in deletion formation in the *gal* region of *E. coli* K12. *Mol. Gen. Genet.* **137**:17.

Rosner, J. L. 1975. Specialized transduction of *pro* genes by coliphage P1: Structure of a partly diploid P1-*pro* prophage. *Virology* **67**:42.

Scott, J. R. 1973. Phage P1 cryptic. II. Location and regulation of prophage genes. *Virology* **53**:327.

Shaw, W. V. 1967. The enzymatic acetylation of chloramphenicol by extracts of R factor-resistant *Escherichia coli*. *J. Biol. Chem.* **242**:687.

Shaw, W. V. and R. F. Brodsky. 1968. Characterization of chloramphenicol acetyltransferase from chloramphenicol-resistant *Staphylococcus aureus*. *J. Bact.* **95**:28.

Takano, T. and S. Ikeda. 1976. Phage P1 carrying kanamycin resistance gene of R factor. *Virology* **70**:198.

Physical Structure and Deletion Effects of the Chloramphenicol Resistance Element Tn9 in Phage Lambda

L. A. MacHattie and J. B. Jackowski
Department of Medical Genetics
University of Toronto
Toronto, Canada, M5S 1A8

Lambda phages carrying a chloramphenicol resistance determinant derived from an R factor via phage P1Cm (Kondo and Mitsuhashi 1964) can be obtained at frequencies of 10^{-6} to 10^{-7} by induction of a λ in a host lysogenic for P1Cm (Scott 1973; Gottesman and Rosner 1975). The location of the R-determinant DNA within the *b*2 region of the lambda chromosome (this work, and Gottesman and Rosner 1975) indicates an unusual mode of pickup by lambda. Since this resistance element is translocated as a unit from a drug resistance plasmid into P1 and from P1 into an F episome (Kondo and Mitsuhashi 1967), as well as into lambda, it is designated as a transposon and is now called Tn*9* (Rosner and Gottesman; Campbell et al.; both this volume). In this paper, λ phage particles carrying Tn*9* are referred to as λ*cam*.

Structure of the *cam* Insertions

Insertion End Points within Lambda DNA

Measurements of heteroduplex molecules formed between the DNAs of the λ*cam* phages and those of other λ strains carrying deletions and substitutions of known position and size revealed the pattern of insertions diagrammed in Figure 1. It is apparent that the Tn*9* insertion variants derived from the original isolate, λ*cam*107, have been altered only by deletions extending to the left (*cam*105) or to the right (*cam*104, 108, and 112) from the original insertion site at physical map position 0.498. A similar tendency toward the production of adjacent deletions has been noted in the case of an IS*1* insertion in the *gal* operon of *E. coli* (Reif and Saedler 1975). The frequency of adjacent deletions is such that they have appeared in more than 50% of the chloramphenicol-resistant recombinant products so far isolated from crosses with λ*cam*107. However, the phages already having deletions appear much less likely to give rise to further-deleted progeny. It may be speculated that packaging or stability problems with the oversized genome of λ*cam*107 (105% of $λ^+$) probably exert a selective pressure which amplifies the apparent frequency of deletions in this strain.

Structure of the Inserted DNA Segment

In heteroduplexes between the DNAs of *cam*105 and *cam*104 or *cam*112, the duplex formed by the insertion-segment strands was bounded at both ends by deletion loops (that of 105 on the left and that of 104 or 112 on the right, as shown by the "type-1" configurations of Figure 2 a and b) and hence could be measured accurately. The value obtained, 0.0529 (± 0.0009, 16 measurements), agrees well with the single-stranded measurements of the inserted DNA segment (0.0532 ± 0.0004, 205 measurements).

In many of these heteroduplexes, however, the expected full-length insertion duplex did not form, but instead configurations like those shown in Figure 2 as types 2 and 3 were found. The simplest interpretation of these is that the left and right ends of the *cam* insertion DNA are direct repeats of the same nucleotide sequence, about 0.015 of λ^+ in length.

The derivation of the chloramphenicol resistance determinant from the drug resistance plasmid R14 (Kondo and Mitsuhashi 1964), together with the recent finding of Hu et al. (1975) that the insertion sequence IS*1* occurs, directly repeated, at the two junctions of the R determinant with the RTF DNA in several R plasmids, leads to the supposition that a direct repeat at the two ends of the *cam* insertion could be IS*1*. The generation of adjacent deletions such as

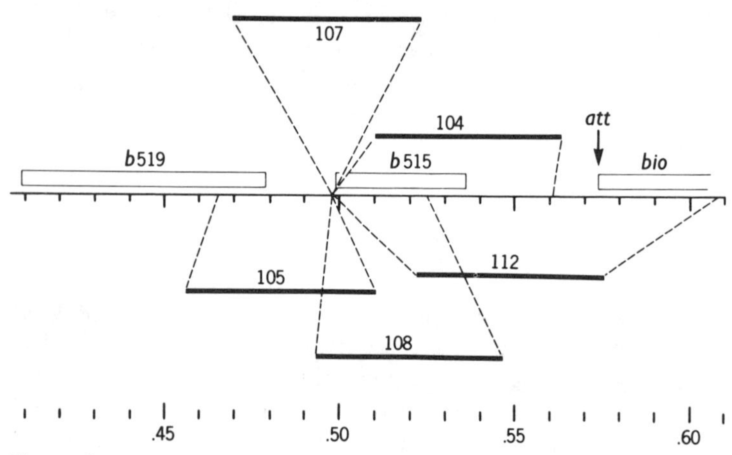

Figure 1
Physical map of *cam* insertions in strains derived from λ*cam*107 as determined by heteroduplex measurements on about 40 different preparations with respect to single-stranded and duplex reference DNA segments of previously known size. Map position given as fraction of the λ^+ DNA length (46,500 base pairs; Davidson and Szybalski 1971). The dark segments labeled with strain numbers represent the inserted DNA, in each case of length 0.0532 of λ^+ DNA, and the dashed lines indicate the points of insertion into the λ DNA sequence. For comparison, the positions of deletions *b*519 and *b*515 (Davis and Parkinson 1971; Fiandt et al. 1971) are shown according to our measurements versus the left end of the *bio*11 insertion, assumed to lie at 0.574.

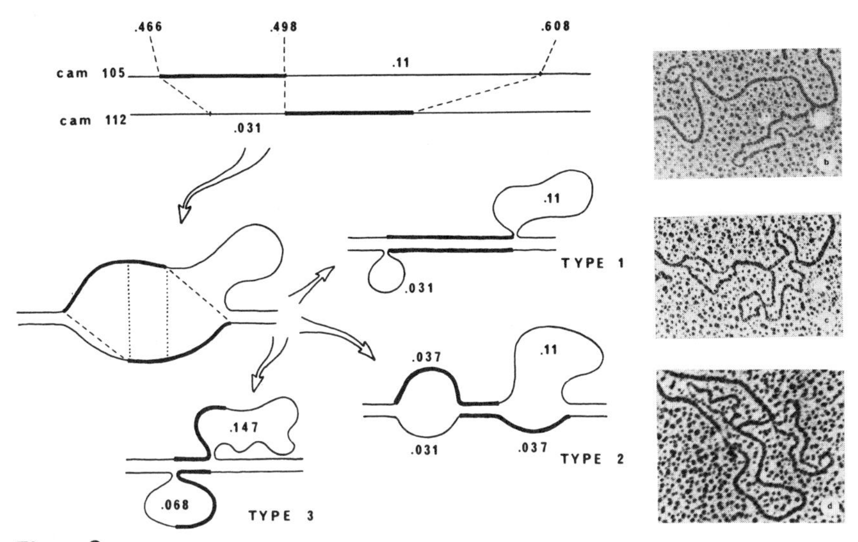

Figure 2
Heteroduplex configurations formed between the *cam* insertion segments of two different λ*cam* strains. (*a*) Diagrammatic representation of the case of *cam*105 vs. *cam*112, showing the three alternative patterns formed. (*b, c, d*) Electron micrographs of the type-1, type-2, and type-3 configurations: (*b, c*) *cam*105:*cam*112 and (*d*) *cam*105:*cam*104 (in which the loop to the right is 0.100 in length instead of 0.147).

we find in this family of λ*cam* phages would also be consistent with this hypothesis since IS*1* has been observed to favor the occurrence of such deletions (Reif and Saedler 1975 and this volume).

This idea was tested by examining heteroduplexes of λ*cam*105i^{21} DNA with those of phages λp*lac*5::IS*1* (MS348) and λp*lac*5::IS*1* (MS520), which carry oppositely oriented IS*1* insertions within the *lacZ* gene (Malamy et al. 1972). The heteroduplexes with λp*lac*5 MS520 showed the formation of the expected duplex of length 0.0170 (± 0.0005 S.E.M., 24 measurements) between the appropriate parts of the *lac* and *cam* strands. Figure 3 shows the configurations observed, in which the IS*1* in *lac* is seen to hybridize with either the left or the right copy of the directly repeated segment in the *cam* strand. Measurements of λp*lac*5 MS520:λp*lac*5 MS348 heteroduplexes confirmed the presence of insertions of approximately IS*1* size at the expected locations in the two phage genomes, and λp*lac*5 MS348:λ*cam*105i^{21} heteroduplexes showed no homologies between the *lac* and *cam* insertion strands, indicating that no inverted copies of IS*1* occur in the *cam* insertion DNA.

Since the *cam* (Tn*9*) insertion segment, of length 0.0532 of λ^+, contains two copies of IS*1*, totaling 0.0340 of λ^+ in length, the structural gene for the chloramphenicol-inactivating enzyme CAT must be contained in the remaining central

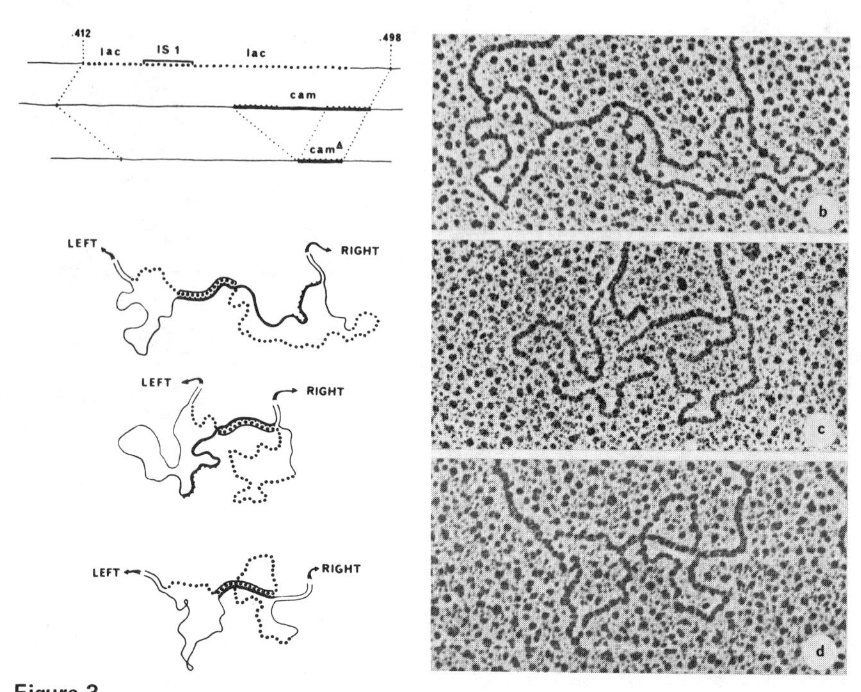

Figure 3
Heteroduplex patterns formed by the hybridization of the IS*1* insertion in λ*plac*5 MS520 DNA with those present at the two ends of the *cam*105 insertion and with the IS*1* remaining in λ*cam*Δ105 DNA. (*a*) Diagrammatic interpretation of the configurations seen in the electron micrographs *b, c,* and *d.*

segment of the insertion DNA, of length 0.0192 of λ^+ or 890 base pairs. This enzyme, reported to have a molecular weight of 78,000 (Shaw and Brodsky 1968), must therefore be supposed to be made up of either four or three identical subunits, with the structural gene for the subunit occupying, respectively, $\frac{2}{3}$ or $\frac{7}{8}$ of the 890-nucleotide-pair segment.

Tandem Duplications

R plasmids selected for increased levels of drug resistance by growing the host cells in high levels of drugs are found to contain amplified amounts of the R-determinant segments of their DNA (Rownd et al. 1971). One form of the amplified DNA consists of tandem duplications of the R-determinant segment within the R factor. Since the Tn*9* insertion in λ appeared to have the structure of a plasmid R-determinant segment carrying only a single resistance element, it was of interest to know whether it, too, might show the capacity to generate tandem duplications.

A preparation of λ*cam*112i^{21} phage particles banded in equilibrium CsCl gradients in the Model E ultracentrifuge was found to form three major peaks,

Figure 4
Bands formed by a $\lambda cam112i^{21}$ phage preparation centrifuged with marker λ^+ particles in an equilibrium CsCl gradient in the Model E ultracentrifuge. (UV densitometer tracing) Density increases to the right. The major bands from left to right are identified as $\lambda cam\Delta112i^{21}$, $\lambda cam112i^{21}$, λcam^2112i^{21}, and λ^+. (There is also a suggestion of a λcam^3112 i^{21} band to the left of λ^+.)

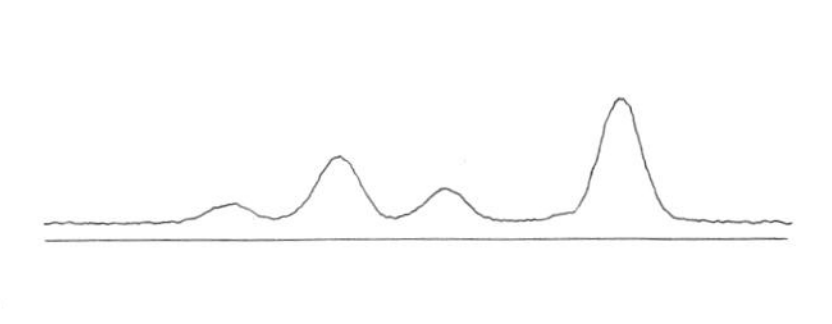

as shown in Figure 4. The corresponding densities were calculated from the results of 15 banding runs to be 0.00748 (± 0.00006 S.E.M.), 0.01200 (± 0.00005), and 0.01672 (± 0.00006) g/ml less than that of marker λ^+ phage included in the same gradient. According to the empirical relations derived by Bellett et al. (1971), these densities correspond to those expected for λ phages carrying shortened DNA molecules, of respective sizes 0.861, 0.899, and 0.936 times the size of λ^+ DNA. These values agree closely with the genome sizes of $\lambda cam\Delta112i^{21}$, $\lambda cam112i^{21}$, and λcam^2112i^{21} (0.860, 0.897, and 0.934, according to our heteroduplex measurement data), where cam^2 stands for a duplicated Tn*9* insertion containing two copies of the CAT gene segment and three copies of IS*1*. As a further test of the origin of the density heterogeneity in this phage preparation, self heteroduplexes of the DNA extracted from it were measured by electron microscopy. Relative insertion/deletion loops of two size classes, 0.036 ± 0.001 and 0.072 ± 0.007 of λ^+ length, were found at map position 0.495 ± 0.020 (S.D.), the approximate position of Tn*9* insertion. These values confirm the DNA size differences estimated from the phage particle densities and indicate that the additional DNA in the cam^2 genome was present as a tandem duplication rather than being a second insertion at a different site.

Tn*9* Excision

All of these λ*cam* phages show a marked tendency to lose the chloramphenicol-resistant phenotype during growth, such that in a typical plate lysate from 10% to 70% of the progeny of a chloramphenicol-resistant phage may be chloramphenicol-sensitive.

Cutout Geometry

Heteroduplex comparisons of the DNA molecule of a chloramphenicol-sensitive derivative with that of its chloramphenicol-resistant parent show that the principal cause of loss of resistance is the deletion of a segment of apparently constant size (measured value 0.0365 ± 0.0006 S.E.M., 78 measurements) from the R-determinant DNA segment. With the aim of detecting whatever heterogeneity might exist in the excision geometry, these determinations were made using chloramphenicol-sensitive phages grown, without cloning, directly from the chloramphenicol-resistant parents.

In a heteroduplex between, for example, the DNAs of the chloramphenicol-sensitive phage λ*cam*Δ105 and resistant λ*cam*104, the structures formed by their insertion segments can be identified by virtue of the deletion loops at either end. The configurations seen indicate that the insertion remaining in λ*cam*Δ105 DNA is a single copy of the terminal repeat segment of Tn*9*, which can hybridize with either the left or the right copy in the intact Tn*9* segment of λ*cam*104. This was confirmed by the examination of heteroduplexes formed between the DNAs of λ*plac*5 MS520 and the deleted phage λ*cam*Δ105i^{21}. An example is shown in Figure 3d. Again, the IS*1* duplex showed a length of 0.0171 (± 0.0005 S.E.M., 26 measurements), indicating a full-sized copy of IS*1* in the deleted DNA. That IS*1* makes up approximately the whole of the insertion in λ*cam*Δ DNA is shown by heteroduplexes between wild-type λ and *cam*Δ DNAs, in which the single-stranded insertion segments have a length of 0.0173 (± 0.0003 S.E.M., 78 measurements). The geometry of *cam* excision therefore resembles that of an homologous recombination event between the two IS*1* copies in Tn*9*. The similarity is further supported by our finding that DNA minicircles of the same size as the excised segment may be recovered from infected cells (L. A. MacHattie and J. B. Jackowski, unpubl.).

Excision Frequency

Assays of the rates at which λ*cam* phages give rise to chloramphenicol-sensitive progeny during growth on plates have yielded some information about the genetic factors involved in *cam* excision. Two methods of detecting the chloramphenicol-resistant phenotype have been employed, both dependent on the production of chloramphenicol-resistant "lysogens" (see note to Table 1) and both yielding comparable results, though with a reproducibility that was less than ideal. When growth took place through 17 to 22 doublings, losses during extraction of progeny pfu (plaque-forming units) had relatively minor effects on the calculated rates. However, judging from the DNA size effect observed (see below), the results may have reflected significant bias due to differential survival rates of *cam*Δ vs. *cam* pfu on the plates, thus necessitating the use of empirical controls for DNA size.

The rates of *cam* excision observed for several different λ*cam* phages grown on rec^+ or rec^- host cells are shown in Table 1 and Figure 5. The graph of excision rate vs. DNA size (Fig. 5) shows the following effects: (1) The apparent excision rate increased fairly strongly with DNA size in the range 99-105% of the λ$^+$ DNA size but was less size-dependent below that range. (2) Right-arm substitutions in the phage chromosome coming from λ *int am*29i^{21} or λ*c*I857*CF*1 resulted in significantly increased excision rates, whereas those carrying *red* β*am*270 or γ*am*210 mutations left the excision rates essentially unchanged. (3) The combination of *recA, recB* mutations in the host and *red* β*am*270 in the phage reduced the excision frequency to the lowest values yet observed

Table 1

Apparent *cam* Excision Frequencies in *rec*⁻ and *rec*⁺ Host Cells

Phage[a]	DNA size[b]	Excision frequency[c]		
		rec⁺	*recA*⁻	*recA*⁻*B*⁻
λ*cam*107	1.053	7.50 ± 0.98 (5)		
λ*cam*107i^{21}	1.007	3.50 ± 0.56 (5)		
λ*cam*105	1.021	2.40 ± 0.18 (17)	5.33 ± 0.61 (2)	
λ*cam*105i^{21}	0.975	2.25 ± 0.49 (2)		
λ*cam*105*CF*1	0.961	1.87 ± 0.21 (3)		
λ*cam*105 γ*am*210	1.021	2.86 ± 0.15 (6)	4.17 ± 0.40 (3)	
λ*cam*104	0.990	1.02 ± 0.12 (12)		0.82 ± 0.18 (4)
λ*cam*104 β*am*270	0.990	0.75 ± 0.20 (4)		0.36 ± 0.14 (4)
λ*cam*112	0.943	0.77 ± 0.10 (4)		
λ*cam*112i^{21}	0.897	1.33 ± 0.08 (3)		
λ*cam*112 β*am*270	0.943	0.62 ± 0.09 (10)		0.23 ± 0.04 (3)

Two methods were used to detect the chloramphenicol-resistant phenotype in individual plaques: (1) by turbid plaque morphology on plates containing 4 μg/ml of chloramphenicol, seeded with ten times the usual number of host cells (*E. coli* W3110), and incubated at 32°C for 36 to 48 hr, and (2) by the growth of chloramphenicol-resistant colonies introduced by stabs from the centers of plaques (on standard plates) into plates containing 20 μg/ml of chloramphenicol, which were then incubated at 32°C for 18 to 24 hr. *Cam* excision rates per doubling of phage number were calculated by the relation $E = 1 - (F/C)^{1/n}$, where F and C are the fractions of chloramphenicol-resistant plaque-forming units (pfu) in the progeny and parent stocks, respectively, and n is the number of doublings of phage number, $= \log_2 A$, A being the growth amplification ratio, (total progeny pfu)/(total parental pfu).

[a] The right-arm markers were crossed into the original λ*cam* phages from the following: i^{21} from λ*b*519*b*515*int*29i^{21}*ts* (obtained indirectly from E. Signer), *CF*1, a *nin*5-like deletion extending from 0.834 to 0.895 (our measurements), from λ*c*I857*CF*1 (obtained from C. R. Fuerst), γ*am*210 from λ γ*am*210*c*I857 (obtained from M. Pearson), and β*am*270 from λ β*am*270*c*I857 (obtained from M. Pearson).

[b] Expressed in terms of the chromosome of wild-type lambda as unit size. Values are derived from our heteroduplex measurements plus the figures for shortening of the genome due to the i^{21} substitution (–0.046) from Davis and Parkinson (1971) and due to the *CF*1 deletion (–0.0604) from our heteroduplex measurements.

[c] Excision frequencies are given as percent per doubling of phage number. The *rec*⁺, *recA*⁻, and *recA*⁻*B*⁻ *E. coli* host strains used were W3110 (obtained from S. Kumar), DM455, and JC5495 (both obtained from A. J. Becker), respectively.

(0.2% to 0.5% per doubling), whereas the single *recA* mutation increased the excision frequency somewhat, both for λ*cam*105 and for λ*cam*105 γ*am*210.

The first observation above indicates a possible second reason for the relatively high incidence of deleted phage types among cross-products of λ*cam*107 that were selected for chloramphenicol resistance transducing ability: the deleted phage types have a higher probability of retaining their chloramphenicol resistance.

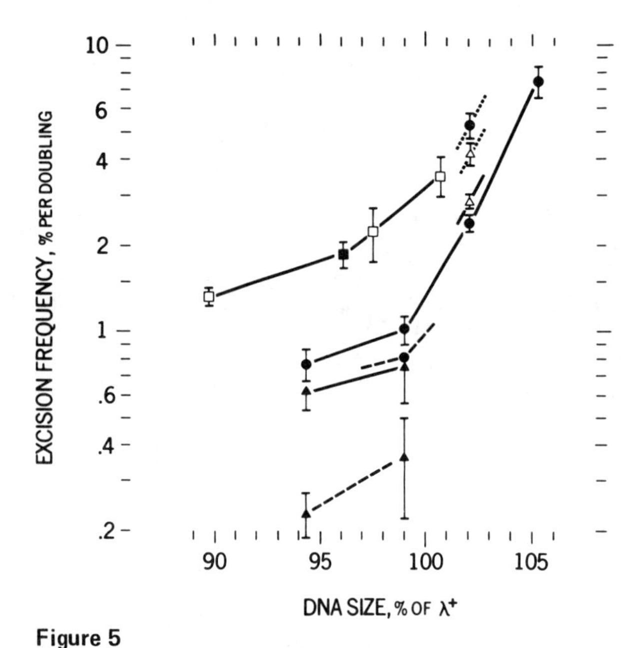

Figure 5
Apparent rates of loss of chloramphenicol resistance during growth on plates plotted against λ*cam* DNA size (data from Table 1). The *E. coli* host cell used is indicated by the type of line: (——) W3110 (rec^+),(– – – –) JC495 (*recA*–*B*–), and (• • • • •) DM455 (*recA*–). The λ*cam* phage type is shown by the symbols: (●) original isolates; recombinants carrying substituted chromosomal right arms (□) i^{21}, (■) *CF*1, Δ γ*am*210, and (▲) β*am*270. To identify individual phage types, refer to the DNA size scale (DNA sizes given in Table 1, column 2). Standard errors of the mean are indicated by vertical bars.

The second observation above would suggest that our original λ*cam* isolates may carry a defect in the right arm of the lambda chromosome that depresses the excision frequency to the same extent as does a *red* β^- or γ^- mutation. However, this putative defect differs in phenotype from the *red* β^- or γ^- mutations in that the original λ*cam* isolates were found to plaque almost normally on a $polA^-$ host, on which *red* or γ mutants plated under the same conditions do not form plaques.

In speculating about the possible genetic origins of the residual *cam* excision activity in the absence of *red* and *rec* functions or of the additional activity shown by derivatives carrying a right-arm replacement as compared to the original λ*cam* isolates, it may be noted that the λ*cam*112 data suggest that lambda genes *int* and *xis* are not involved. The deletion in *cam*112, extending to map position 0.608, should have deleted both of these genes, which Enquist and Weisberg (this volume) have determined to occupy the region 0.574 to 0.603 approximately. Thus λ*cam*112 β^-, which shows the residual excision activity, and λ*cam*112i^{21}, which shows the additional activity, should both be *int*⁻, *xis*⁻. (However, definitive tests for *int* and *xis* functions in these phages have not been carried out.)

SUMMARY

We have investigated the DNA structures of a λ*cam* phage (Scott 1973) and of a number of its derivatives and have made preliminary studies of the "*cam* excision" phenomenon by which Tn*9* is spontaneously lost during phage growth. The Tn*9* insertion is shown to promote deletions of lambda sequences adjacent to the insertion site. The structural gene for chloramphenicol acetyl transferase (CAT), the enzyme responsible for the drug-resistant phenotype (Shaw 1967); Kondo et al. 1970), is contained in an 890-base-pair segment at the center of Tn*9*, bracketed on either side by a copy of the insertion sequence IS*1* (Hirsch et al. 1972; Malamy et al. 1972; Finadt et al. 1972). Tandem duplications of the drug resistance determinant have been observed in λ*cam* phages carrying a shortened genome, which allows for accommodation of the additional Tn*9* DNA within the normal λ head structure. Tn*9* excision normally takes place by reciprocal recombination between the two copies of IS*1*, deleting part of the inserted DNA segment as a minicircle and leaving behind one copy of IS*1*. This event appears to be promoted by both the *red* and *rec* generalized recombination systems but, under *red*⁻, *rec*⁻ conditions, it still occurs at fairly high frequencies, of the order of 0.3% per doubling of phage number.

Acknowledgments

The authors are indebted to Andrew J. Becker, Michael H. Malamy and Mark L. Pearson for numerous phage and host strains and for creative discussions; to J. L. Rosner and M. M. Gottesman for communicating and discussing results prior to publication; and to the Medical Research Council of Canada for funding this research.

REFERENCES

Bellett, A. J. D., H. G. Busse and R. L. Baldwin. 1971. Tandem genetic duplications in a derivative of phage lambda. In *The bacteriophage lambda* (ed. A. D. Hershey), p. 501. Cold Spring Harbor Laboratory, Cold Spring Harbor, New York.

Davidson, N. and W. Szybalski. 1971. Physical and chemical characteristics of lambda DNA. In *The bacteriophage lambda* (ed. A. D. Hershey), p. 45. Cold Spring Harbor Laboratory, Cold Spring Harbor, New York.

Davis, R. W. and J. S. Parkinson. 1971. Deletion mutants of bacteriophage lambda. III. Physical structures of *att*. *J. Mol. Biol.* **56**:403.

Fiandt, M., W. Szybalski and M. H. Malamy. 1972. Polar mutations in *lac, gal* and phage λ consist of a few IS-DNA sequences inserted with either orientation. *Mol. Gen. Genet.* **119**:223.

Fiandt, M., Z. Hradecna, H. A. Lozeron and W. Szybalski. 1971. Electron micrographic mapping of deletions, insertions, inversions and homologies in the DNAs of coliphages lambda and phi 80. In *The bacteriophage lambda* (ed. A. D. Hershey), p. 329. Cold Spring Harbor Laboratory, Cold Spring Harbor, New York.

Gottesman, M. M. and J. L. Rosner. 1975. Acquisition of a determinant for chloramphenicol resistance by coliphage lambda. *Proc. Nat. Acad. Sci.* **72**: 5041.

Hirsch, H.-J., P. Starlinger and P. Brachet. 1972. Two kinds of insertions in bacterial genes. *Mol. Gen. Genet.* **119**:191.

Hu, S., E. Ohtsubo, N. Davidson and H. Saedler. 1975. Electron microscope heteroduplex studies of sequence relations among bacterial plasmids: Identification and mapping of the insertion sequences IS1 and IS2 in F and R plasmids. *J. Bact.* **122**:764.

Kondo, E. and S. Mitsuhashi. 1964. Drug resistance of enteric bacteria. IV. Active bacteriophage P1CM produced by the combination of R factor with bacteriophage P1. *J. Bact.* **88**:1266.

——. 1967. Drug resistance of enteric bacteria. VI. Introduction of bacteriophage P1CM into *Salmonella typhi* and formation of P1dCM and F-CM elements. *J. Bact.* **91**:1787.

Kondo, E., D. K. Haapala and S. Falkow. 1970. The production of chloramphenicol acetyltransferase by bacteriophage P1CM. *Virology* **40**:431.

Malamy, M. H., M. Fiandt and W. Szybalski. 1972. Electron microscopy of polar insertions in the *lac* operon of *Escherichia coli. Mol. Gen. Genet.* **119**:207.

Reif, H. J. and H. Saedler. 1975. IS1 is involved in deletion formation in the *gal* region of *E. coli* K12. *Mol. Gen. Genet.* **137**:17.

Rownd, R., H. Kasamatsu and S. Mickel. 1971. The molecular nature and replication of drug resistance factors of *Enterobacteriaceae. Ann. N.Y. Acad. Sci.* **182**:188.

Scott, J. R. 1973. Phage P1 cryptic. II. Location and regulation of prophage genes. *Virology* **53**:327.

Shaw, W. V. 1967. The enzymatic acetylation of chloramphenicol by extracts of R factor-resistant *E. coli. J. Biol. Chem.* **242**:687.

Shaw, W. V. and R. F. Brodsky. 1968. Characterization of chloramphenicol acetyltransferase from chloramphenicol-resistant *Staphylococcus aureus. J. Bact.* **95**:28.

Structure and Location of Antibiotic Resistance Determinants in Bacteriophages P1Cm and P7 (ϕ Amp)

T. Yun and D. Vapnek
Program in Genetics and
Departments of Microbiology and Biochemistry
University of Georgia
Athens, Georgia 30602

Insertions of antibiotic resistance determinants into the P1 plasmid prophage can occur in P1 lysogens harboring R factors. Following induction of these lysogens, nondefective P1 phages able to transduce antibiotic resistance markers at a high frequency have been isolated. These include P1Cm (chloramphenicol; Kondo and Mitsuhashi 1964), P1Km (kanamycin; Takano and Ikeda 1976), P1TcCm (tetracycline, chloramphenicol), and P1SmSu (streptomycin, sulfonamide; Mise and Arber 1975). In addition, bacteriophage P7 (formerly called ϕAmp), a nondefective transducing phage which carries the determinant for ampicillin resistance (Ap^r) (Smith 1972), has been shown to be closely related to P1. Both phages are indistinguishable morphologically, and their high frequency of recombination with each other suggests a high degree of sequence homology (Chesney and Scott 1975; Walker and Walker 1976).

Since P1 DNA has a large terminal redundancy and appears to be packaged by a headful mechanism (Ikeda and Tomizawa 1969), one possible explanation for the origin of these nondefective transducing phages is that they result from an insertion of the antibiotic determinant, with a concomitant decrease in the extent of terminal redundancy of the phage DNA. To test this hypothesis, the extent of the terminal redundancies of phages P1, P1Cm, and P7 have been determined and compared with the size of the chloramphenicol and ampicillin insertions. These experiments have also allowed a correlation to be made between the genetic and physical maps of phages P1 and P7.

Determination of Terminal Redundancy

Since P1 circularizes during lysogenization by recombination between its terminally redundant ends, the difference in molecular weights of the phage and prophage DNAs is a measure of the terminal redundancy (Ikeda and Tomizawa 1969). To determine these molecular weights, phage and prophage DNAs were purified and analyzed by electron microscopy. The results of this analysis (Table 1) demonstrate that all three phage DNAs have very similar molecular

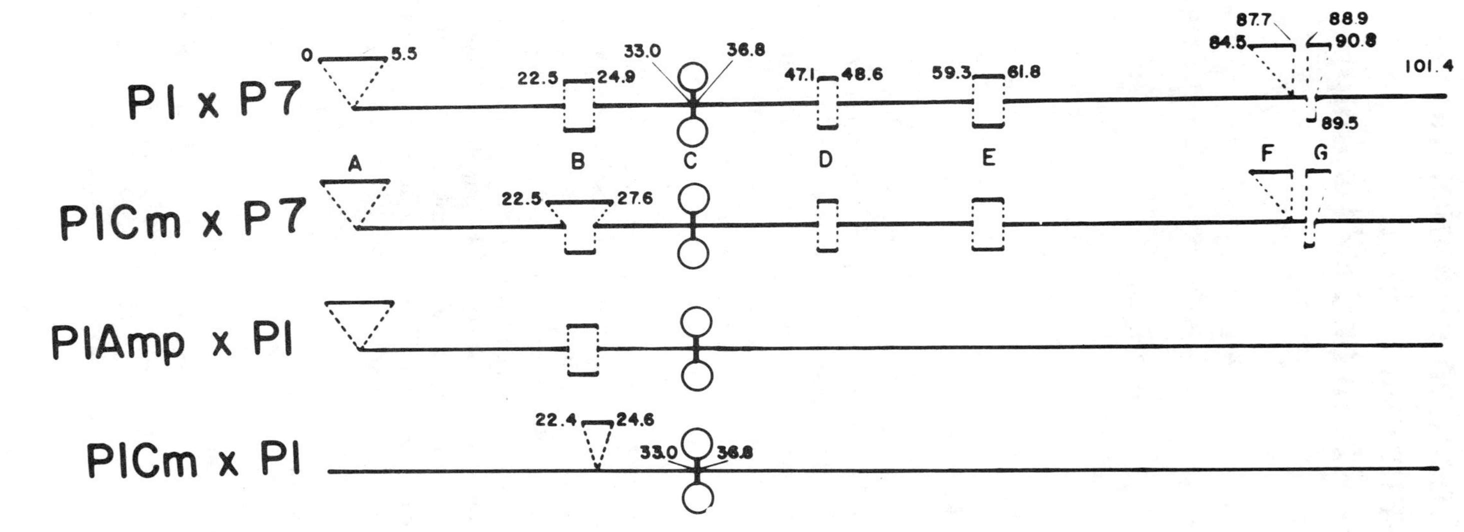

Figure 1
Heteroduplex analysis was carried out as described by Sharp et al. (1972). The final spreading solution was adjusted to 60% formamide, 0.1 M Tris, 10 mM EDTA, 0.1 mg/ml cytochrome c and spread onto a hypophase of 30% formamide, 10 mM Tris, and 1 mM EDTA. These conditions are equivalent to a T_m of $-13°C$ (Davis and Hyman 1971). ϕX174 and ColE1 DNA molecules, used as single-strand and double-strand standards for length measurements, were added to each heteroduplex preparation. Electron micrographs were projected using a point-light-source enlarger and measured with an electronic planimeter (Numonics Corporation). The circular heteroduplexes are presented in linear order to conform to the gentic map of Chesney and Scott (1975) and D. Walker and J. Walker (pers. comm.). Coordinates given are in kilobases. One kilobase is equal to 1000 nucleotides or nucleotide pairs for single- or double-strand DNA, respectively.

Table 1
Molecular Weight Determination of Phage and Prophage DNA

	Phage			Prophage		
	number measured	m.w. ($\times 10^{-6}$)	kilobases	number measured	m.w. ($\times 10^{-6}$)	kilobases
P1	30	66.1 ± 1.5	99.7	18	60.9 ± 1.7	91.8
P1Cm	23	65.5 ± 2.2	98.8	35	62.5 ± 1.1	94.3
P7	18	65.4 ± 1.2	98.6	39	65.0 ± 2.4	98.0

Phage and prophage DNAs were mounted by the aqueous technique and shadowed using the procedure of Davis et al. (1971). ColE1 DNA, molecular weight 4.2×10^6 (Bazaral and Helinski 1968), was added, as a molecular weight standard, to samples before spreading. Molecular weights in kilobases (kb) were calculated by using the relationship 1×10^6 daltons duplex DNA = 1508 base pairs (Sharp et al. 1972).

weights, but there are significant differences in the sizes of the prophage DNAs. From these measurements, terminal redundancies are calculated as 7.9 kb for P1, 4.5 kb for P1Cm, and 0.6 kb for P7.

Size and Location of the Cm^r Determinant

Lee et al. (1974) showed that P1 carries an inverted repeat of about 600 bases separated by about 3 kb. This region, which is homologous to the G loop of phage Mu (Chow and Bukhari, this volume), provided a convenient physical marker on the circular P1 genome. By heteroduplex analysis of P1/P1Cm, and utilizing the inverted repeat region as a marker, the size and location of the Cm^r determinant were determined. Results of these heteroduplex studies demonstrated one region of nonhomology, an insertion of 2.2 kb about 8.4 kb from the inverted repeat (Fig. 1). This suggests that the insertion includes the Cm^r determinant.

Homology between P1 and P7

Heteroduplexes between phage P1 and P7 DNAs showed six regions of nonhomology (Fig. 1). Two of these (Fig. 1, A and F) were insertions or deletions, and four (Fig. 1, B, D, E, and G) were substitutions (or inversions). The inverted repeat (Fig. 1, C) present in P1 was also present in P7. In a majority of the heteroduplex molecules, the DNA between the inverted repeat was in either the lariat or the melted-loop form, but, in some cases, this DNA was double-stranded, indicating that the two phages were homologous in this region. Since the sum of the homologous regions was 84 kb and the sizes of the P1 and P7 prophages were 92 and 98 kb, respectively, the extent of sequence homology between these two phages was 85% with respect to P7 and about 90% with respect to P1.

Heteroduplexes between P7 and P1Cm were formed in order to determine the relationship of the Cm^r determinant to the regions of nonhomology between P1 and P7. These heteroduplexes (Fig. 1) demonstrated no new regions of nonhomology but did show a different pattern in the B loop. In heteroduplexes between P1 and P7, both strands of the B loop were of approximately equal size, whereas in the P1Cm/P7 heteroduplex, one strand of the B loop was 2.6 kb larger than the other. This result confirmed that the Cm^r determinant was in the B-loop region, about 8.5 kb from the inverted repeat.

Identification of the Ap^r Determinant in P7

The location of the Ap^r determinant was identified by analysis of P1Ap/P7 heteroduplexes. P1Ap was derived by recombination between P1 and P7 and carries the immunity region (*C4*) of P1 and Ap^r from P7 (J. Scott, pers. comm.). These heteroduplexes (Fig. 1) showed only two regions of nonhomology, the A and B loops. This indicated that one of these two regions carried the Ap^r determinant.

Chesney and Scott (1975) showed that Cm^r and Ap^r are not close on the genetic map. Since we demonstrated above that the Cm^r determinant was inserted in the B loop (Fig. 1), we conclude that the A loop (Fig. 1) contains the Ap^r determinant. The size and structure of this region is similar to the transposable Ap^r determinant studied by Heffron et al. (1975).

DISCUSSION

Results of the heteroduplex analyses demonstrate that the Cm^r determinant of P1Cm and the Ap^r determinant of P7 exist as insertions of 2.2 kb and 5.5 kb, respectively. Molecular weight measurements of P1Cm and P7 phages and prophages support the hypothesis that this additional DNA is accommodated by decreasing the terminal redundancy of the phage DNA. For example, the 2.2-kb insertion of the Cm^r determinant in P1Cm leads to a decrease in terminal redundancy of 3.4 kb, compared to the terminal redundancy of P1, whereas P7, with a 5.5-kb Ap^r determinant, has a terminal redundancy of only 0.6 kb compared to the 7.9-kb terminal redundancy of P1.

Both the Cm^r determinant and the Ap^r determinant have been shown to be transposable (Gottesman and Rosner 1975; Heffron et al. 1975). Recent experiments with the Cm^r determinant in phages λ and Mu have demonstrated that this determinant is bounded by directly repeated IS*1* insertion sequences (MacHattie and Jackowski; Rosner and Gottesman; Bukhari et al.; all this volume). Recombination between IS*1* sequences could delete this region of the genome and may account for the observed instability of the Cm^r determinant in P1 (Kondo and Mitsuhashi 1964; Rosner 1972).

Since the Ap^r determinant in P7 codes for the TEM-1 β-lactamase (Matthew and Hedges 1976) and has a structure similar to the translocatable Ap^r determinant (Heffron et al. 1975), it seems likely that this determinant arose in P7 by a translocation event.

Results of the heteroduplex analysis are presented in Figure 1 so as to conform to the map order of Chesney and Scott (1975) and D. Walker and J. Walker (pers. comm.). Based on their maps, the leftmost insertion loop represents the Ap^r determinant, the first region of nonhomology, Cm^r, followed by the inverted repeat region. Although the function of this last region in phage P1 is not known, it has been located between Cm^r and the *C4* immunity gene (R. Mural, T. Yun and D. Vapnek, unpubl.). Further analysis of other P1 × P7 hybrid phages to correlate the rightward portions of the genetic and physical maps is currently in progress.

SUMMARY

Heteroduplex analysis has been used to determine the structure and location of the chloramphenicol resistance determinant (Cm^r) in bacteriophage P1Cm and the ampicillin resistance determinant (Ap^r) in bacteriophage P7. The results demonstrate that the Cm^r determinant exists as an insertion of 2.2 kb in P1Cm and that the Ap^r determinant is an insertion of 5.5 kb in P7. A comparison of the genome sizes of phages P1Cm and P7 with their corresponding plasmid prophages suggests that the additional DNA found in P1Cm and P7 is accommodated by reducing the terminal redundancy of the phage DNA.

Acknowledgments

We thank Dr. June Scott for phage strains, and Rick Mural and June Scott for helpful discussions during the course of this work. This research was supported by U.S. Public Health Service Grant GM 20160. One of us (D.V.) is the recipient of a Research Career Development Award GM 00090 from the National Institute of General Medical Sciences, and the other (T.Y.) is supported by a NIGMS predoctoral training grant GM 7103.

REFERENCES

Brazaral, M. and D. R. Helinski. 1968. Circular DNA forms of colicinogenic factors E1, E2 and E3 from *Escherichia coli*. *J. Mol. Biol.* **36**:185.

Chesney, R. H. and J. R. Scott. 1975. Superinfection immunity and prophage repression in phage P1. II. Mapping of the immunity-difference and ampicillin-resistance loci of P1 and ϕAmp. *Virology* **67**:375.

Davis, R. W. and R. W. Hyman. 1971. A study in evolution: The DNA base sequence homology between coliphages T7 and T3. *J. Mol. Biol.* **62**:287.

Davis, R. W., M. Simon and N. Davidson. 1971. Electron microscope heteroduplex methods for mapping regions of base sequence homology in nucleic acids. In *Methods in enzymology* (ed. L. Grossman and K. Moldave), vol. 21D, p. 413. Academic Press, New York.

Gottesman, M. M. and J. L. Rosner. 1975. Acquisition of a determinant for chloramphenicol resistance by coliphage lambda. *Proc. Nat. Acad. Sci.* **72**: 5041.

Heffron, F., C. Rubens and S. Falkow. 1975. Translocation of a plasmid DNA sequence which mediates ampicillin resistance. Molecular nature and specificity of insertion. *Proc. Nat. Acad. Sci.* **72**:3623.

Ikeda, H. and J. Tomizawa. 1969. Prophage P1, an extrachromosomal replication unit. *Cold Spring Harbor Symp. Quant. Biol.* **33**:791.

Kondo, E. and S. Mitsuhashi. 1964. Drug resistance of enteric bacteria. IV. Active transducing bacteriophage P1Cm produced by the combination of R factor with bacteriophage P1. *J. Bact.* **88**:1266.

Lee, H. J., E. Ohtsubo, R. C. Deonier and N. Davidson. 1974. Electron microscope heteroduplex studies of sequence relations among plasmids of *Escherichia coli. J. Mol. Biol.* **89**:585.

Matthew, M. and R. W. Hedges. 1976. Analytical isolectric focusing of R factor-determined β-lactamases: Correlation with plasmid compatibility. *J. Bact.* **125**:713.

Mise, K. and W. Arber. 1975. Plaque-forming transducing bacteriophage P1 derivatives and their behavior in lysogenic conditions. *Virology* **69**:191.

Rosner, J. L. 1972. Formation, induction, and curing of bacteriophage P1 lysogens. *Virology* **48**:679.

Sharp, P. A., M.-T. Hsu, E. Ohtsubo and N. Davidson. 1972. Electron microscope heteroduplex studies of sequence relations among plasmids of *E. coli. J. Mol. Biol.* **71**:471.

Smith, H. W. 1972. Ampicillin resistance in *Escherichia coli* by phage infection. *Nature New Biol.* **238**:205.

Takano, T. and S. Ikeda. 1976. Phage P1 carrying kanamycin resistance gene of R factor. *Virology* **70**:198.

Walker, D. H. and J. T. Walker. 1976. Genetic studies of coliphage P1. II. Relatedness to ϕAmp as shown by marker rescue tests. *J. Virol.* **19**:271.

Amplification of the Tetracycline Resistance Determinant on Plasmid pAMα1 in *Streptococcus faecalis*

D. B. Clewell and Y. Yagi
The Dental Research Institute, and
Departments of Oral Biology and Microbiology
Schools of Dentistry and Medicine
Ann Arbor, Michigan 48109

Plasmid pAMα1 is a 6.0-megadalton, nonconjugative genetic element that determines tetracycline resistance (Tc^r) in *Streptococcus faecalis* (Clewell et al. 1975). It is one of three plasmids originally identified in *S. faecalis* strain DS-5, a multiple-drug-resistant, hemolytic, clinical isolate (Clewell et al. 1974). It is present in the cell to the extent of about ten copies per chromosomal genome equivalent. The cultivation of bacteria harboring pAMα1 for a number of generations (40–50) in the presence of subinhibitory concentrations of tetracycline (Tc) results in the development of higher levels of resistance to the drug. This phenomenon is accompanied by an increase in size of pAMα1, originally detected by sucrose density gradient analyses, and we have suggested that this reflects an increased gene dosage of the Tc^r determinant (Clewell et al. 1975). By conjugal transfer to a plasmid-free strain of *S. faecalis* (strain JH2-2 [Jacob and Hobbs 1974]), we have constructed stains that harbor only pAMα1 (Dunny and Clewell 1975), such as strain DT-11.

The minimum inhibitory concentration (MIC) of tetracycline in strain DT-11 is about 25 μg/ml. Tc-sensitive variants of DT-11 appear spontaneously at a frequency of about 1% during growth at 37°C, and such variants consistently have a plasmid that sediments at 22S, in contrast to the 28S value of the parent plasmid (Yagi and Clewell 1976). This corresponds to a decrease in size of about 2.65 megadaltons of DNA. Heteroduplex structures consisting of plasmid DNA from Tc-sensitive variants and the parent plasmid (see Fig. 1) clearly indicate the deletion of a single segment of DNA (Yagi and Clewell 1976).

The following experiment was carried out to determine what happens when strain DT-11 is grown in the presence of drug. Overnight cultures were diluted in fresh medium containing 20 μg Tc/ml, grown to late log phase, subcultured successively into medium with 30, 40, and 50 μg Tc/ml, and then subcultured into drug-free broth (see Fig. 2). At various times, aliquots were plated on Penassay agar containing different concentrations of Tc to determine colony-forming units. Cells were not allowed to go into stationary phase. The results are illustrated in Figure 2. When DT-11 cells were grown in the presence of Tc, the percentage of cells capable of growing on higher levels of drug increased. When these cells were subsequently cultured in the absence of drug, the level of resistance dropped.

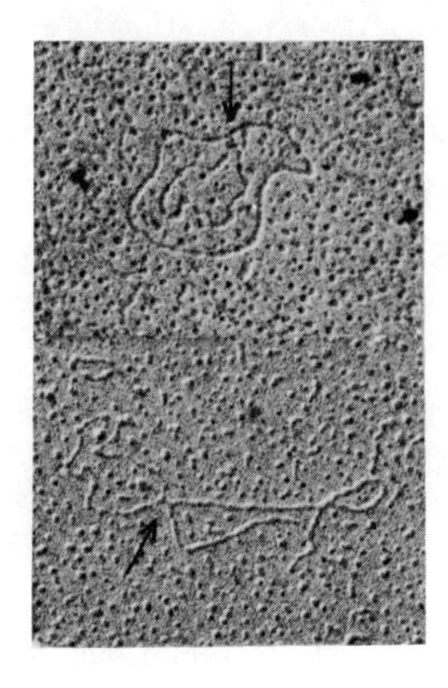

Figure 1
Heteroduplex structures consisting of plasmid DNA from strain DT-11 and a tetracycline-sensitive variant DT-11C1. Arrows indicate the points where the single-stranded deletion loops originate.

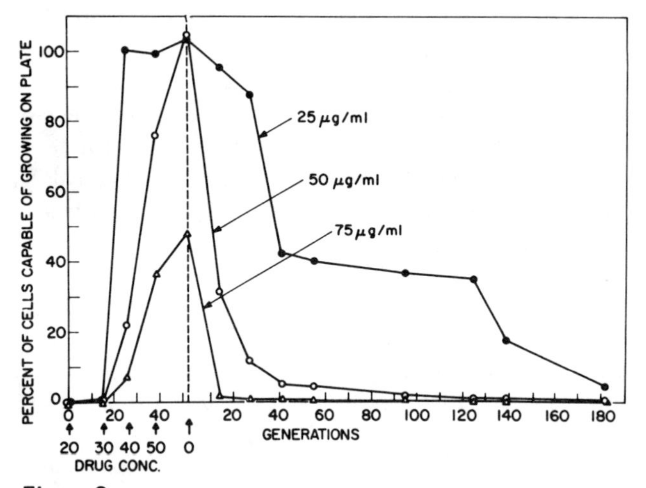

Figure 2
Level of tetracycline resistance of strain DT-11 during growth in the presence of drug and subsequent growth in the absence of drug. Approximately 3×10^6 cells of a fresh, overnight culture of DT-11 were inoculated into 100 ml of M9-YE medium containing 20 μg Tc/ml. When the turbidity reached approximately 85 Klett-units (log phase), 0.01 ml of a culture was transferred into fresh broth (prewarmed) containing 30 μg Tc/ml. After a similar period of cultivation, transfers were made in the same manner to fresh medium containing 40 and finally 50 μg Tc/ml, after which transfers were subsequently made to drug-free broth. Immediately prior to each transfer, appropriate dilutions were plated onto Difco Penassay Broth agar plates containing 0, 25, 50, and 75 μg Tc/ml (see arrows in figure). The percentage of cells capable of growing on the drug plates relative to the total count on the drug-free plates was determined and plotted in each case. The estimate of the number of generations of growth was based on the growth rate and time required after each transfer to attain the final turbidity. (Reprinted, with permission, from Yagi and Clewell 1976.)

During the course of the above experiment, the growth rate of the cells varied with the concentration of Tc in the medium and the length of time the cells had been growing in the presence of the drug. The doubling time on initial exposure was about 6 hours compared to 100 minutes during the last four generations in the highest concentration of drug used. Subsequent growth in the absence of drug gave a doubling time of 45 minutes, similar to that for unexposed cells. As previously seen with strain DS-5C1 (Clewell et al. 1975), there is no strong selection against cells previously grown in the presence of drug.

We examined sedimentation properties of pAMα1 molecules isolated from the cells that were grown as described in the above experiment in the presence and subsequently in the absence of Tc. A portion of each culture was withdrawn just prior to each subculture and labeled with [^{3}H]thymidine for approximately three generations. A control culture of DT-11 cells without Tc was grown with [^{14}C]thymidine. The cultures were centrifuged, mixed, lysed, and sedimented to equilibrium in ethidium bromide-cesium chloride density gradients. Plasmid DNA was isolated from the satellite peaks and analyzed in 5% to 20% neutral sucrose gradients (Fig. 3). After a number of generations in the presence of Tc, we observed the appearance of plasmid molecules which sediment faster than the normal 28S pAMα1 molecule (Fig. 3a–d), and the relative amount of 28S molecules decreased. We also observed that the heavier molecules gave rise to discrete peaks sedimenting in the regions of 34, 38, and 41S. In Figure 3, e and f show that when the cells are subsequently grown in the absence of Tc there is a gradual return to the normal state.

Plasmid DNA isolated from cells grown in the presence of Tc was examined by electron microscopy and found to consist of a heterogeneous population (Yagi and Clewell 1976). Contour-length measurements indicated that the molecules differ in size by 2.65-megadalton units of DNA (see Fig. 4). Figure 4C corresponds to a normal pAMα1, and Figure 4D–J correspond to molecules with one to seven additional 2.65-megadalton units. We have also detected a few small circles of 2.65 megadaltons (Fig. 4A) and structures corresponding to pAMα1 with a 2.65-megadalton deletion (Fig. 4B).

The amplification phenomenon thus appears to involve the generation of repeated, 2.65-megadalton segments of DNA on which the Tc^r determinant resides. Agarose gel analysis of *Eco*RI restriction fragments corroborates this view and shows that the repeats are arranged in tandem (Yagi and Clewell 1976).

It is noteworthy, with respect to the regulation of pAMα1 replication, that amplification experiments consistently indicate that the total mass of covalently closed circular plasmid DNA as a fraction of chromosomal mass remains constant regardless of the extent of amplification (Clewell et al. 1975; Yagi and Clewell 1976). Thus cells harboring larger molecules have fewer copies of these replicons than cells with smaller molecules. Insofar as the redundant segment on the amplified structures is less than half the size of pAMα1, an overall increase in gene dosage would still be generated.

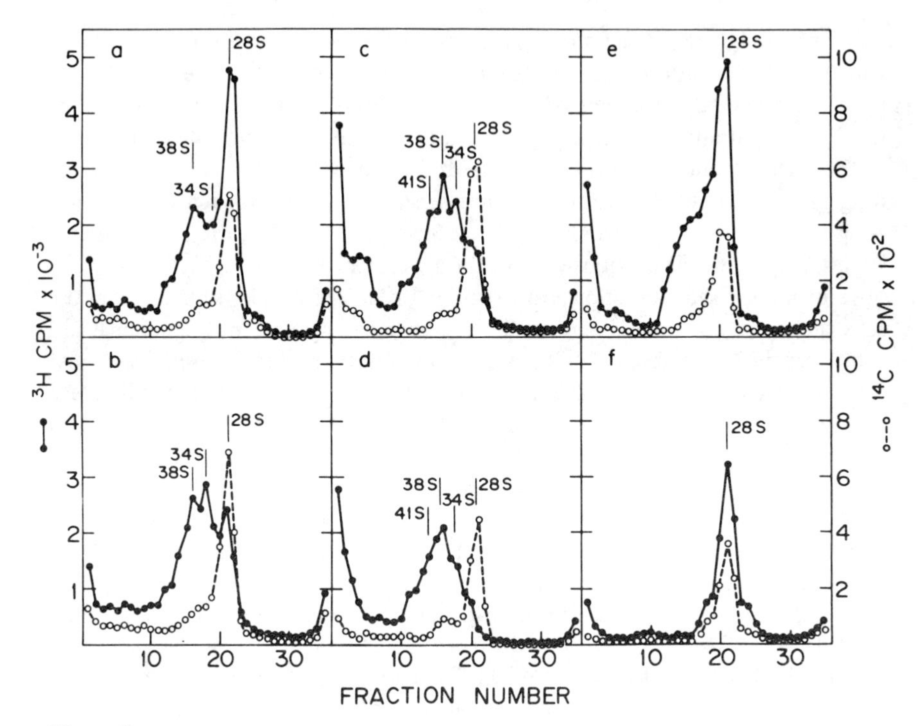

Figure 3
Sedimentation analyses of covalently closed circular DNA molecules isolated from DT-11 cells after growth in the presence of tetracycline and subsequent growth in the absence of the drug. The data presented here relate to the experiment shown in Fig. 2. In certain cases just prior to the transfer of cells to fresh medium, a 1-ml portion was removed and inoculated into 15 ml of fresh medium containing an identical concentration of tetracycline as well as [^{3}H]thymidine. After three generations, the cells were collected by centrifugation and then mixed with cells of a parallel culture of ^{14}C-labeled cells never previously exposed to Tc. Samples of lysates were centrifuged to equilibrium in EtBr-CsCl buoyant density gradients, and fractions containing the covalently closed circular DNA were pooled, dialyzed, and sedimented through 5% to 20% neutral sucrose density gradients (Beckman SW50.1, 48,000 rpm for 60 min at 15°C). (*a*) Cells after 15 generations in the presence of 20 μg Tc/ml; (*b*) cells after 11 more generations now in 30 μg Tc/ml; (*c*) cells after 14 more generations now in 40 μg Tc/ml; (*d*) cells after 13 more generations now in 50 μg Tc/ml; (*e*) cells after 27 generations now in the absence of Tc; and (*f*) cells after 154 more generations still in the absence of Tc. (Reprinted, with permission, from Yagi and Clewell 1976.)

The amplification data support the notion that there is a specific sequence on pAMα1 that is located at two sites on the molecule defining a 2.65-megadalton segment containing the Tc^r determinant (Fig. 5) (Clewell et al. 1975; Yagi and Clewell 1976). The generation of the structures found in Tc-sensitive variants,

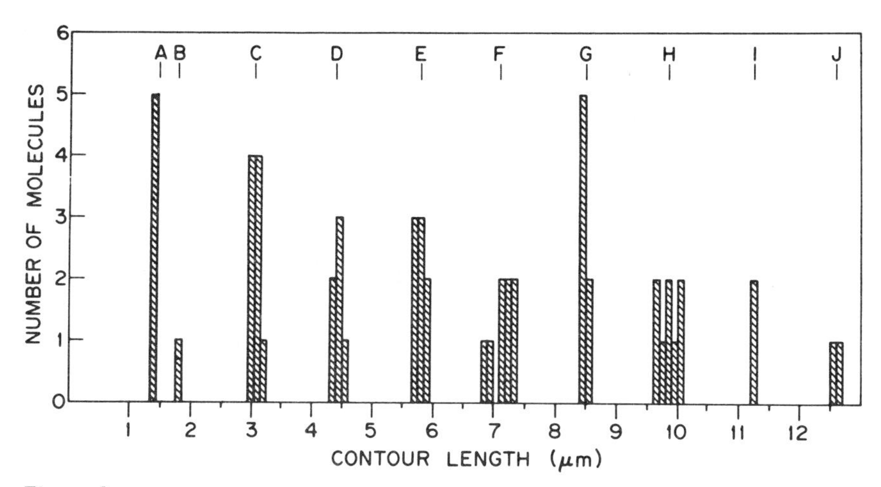

Figure 4
Histogram of contour lengths of circular plasmid molecules isolated from DT-11 cells grown in the presence of tetracycline. Contour lengths of a number of circles were measured and were found to fall in specific size classes as depicted above. (Reprinted, with permission, from Yagi and Clewell 1976.)

as well as those found in an "amplified" state, can be explained on the basis of reciprocal recombinational events between these sequences. We call the sequence RS for recombination sequence.

We have demonstrated the presence of the RS sequences by restriction enzyme analysis. Treatment of "amplified" forms of pAMα1 with the *Eco*RI nuclease produces three fragments (α_1, α_2, and α_3), as compared to two fragments (α_1 and α_2) from "normal" pAMα1 (Yagi and Clewell 1976). A summary of the cleavage patterns of the various forms of pAMα1 is given in Figure 6. The plasmid found in Tc-sensitive variants, αd, representing a pAMα1 with a 2.65-megadalton deletion is not sensitive to *Eco*RI. This indicates that the two

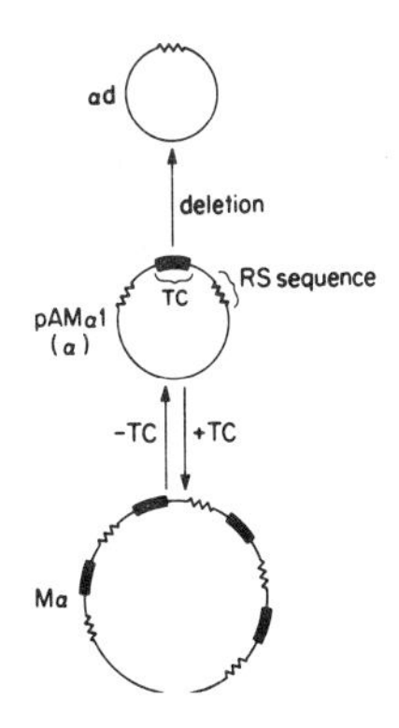

Figure 5
An illustration of the generation of "deleted" and "amplified" forms from pAMα1. RS corresponds to recombination sequence. Such sequences are proposed to occur on both sides of the Tc^r determinant. As a result of recombinational events between these sequences, a deletion or amplification arises. Growth in the presence of Tc results in the selection of "amplified" forms. αd refers to the "deleted" structure; Mα refers to the "amplified" ("modified") structure.

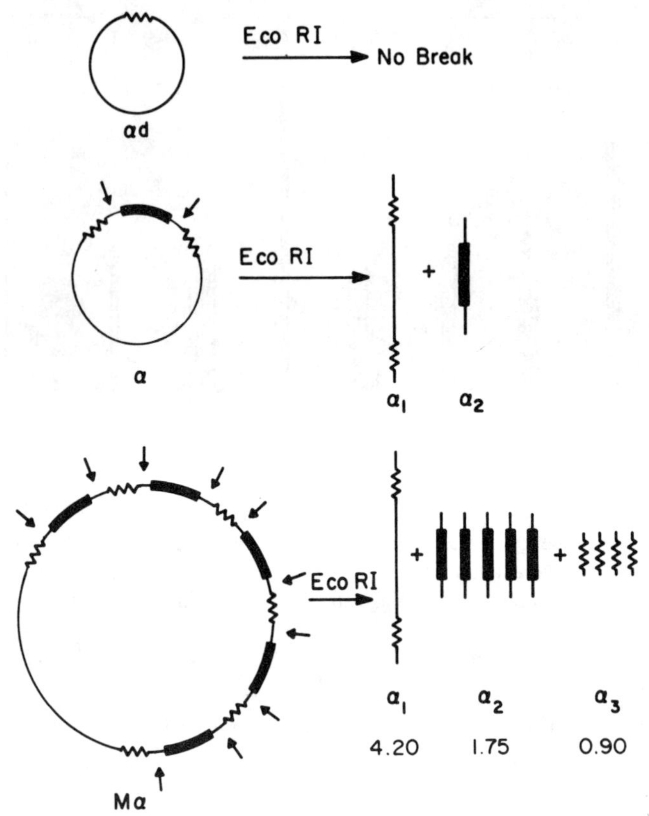

Figure 6
Summary of the fragmentation pattern resulting from the treatment of various forms of pAMα1 with the restriction endonuclease *Eco*RI and an interpretation of where the cleavage sites are located with respect to the RS sequences (zig-zags) and the Tc^r determinant (heavily dark segment). The corresponding molecular weights (in millions) are indicated for each fragment: α_1, α_2, and α_3. α corresponds to pAMα1; αd corresponds to the structure found in Tc-sensitive variants; Mα corresponds to "amplified" ("modified") structures of pAMα1. The arrows indicate the approximate cleavage sites. (Reprinted, with permission, from Yagi and Clewell 1976.)

cleavages on pAMα1 (abbreviated α) occur between the proposed RS sequences (i.e., that segment that is deleted in αd). The third fragment (α_3), which appears upon digestion of "amplified" structures (abbreviated Mα for "modified"), must serve to connect the α_2 segments in the intact circles. On the basis of recombination models that have been suggested for the generation of Mα structures, the α_3 fragment should contain a single copy of the RS sequence. Thus heteroduplex experiments between fragments and certain forms of pAMα1

(α and αd) should be particularly revealing with respect to the existence of the proposed RS sequences.

The three fragments α_1, α_2, and α_3, whose corresponding molecular weights are 3.35×10^6, 1.75×10^6, and 0.90×10^6, respectively, can be easily separated by sucrose density gradient centrifugation. A preparation of α_3 was heteroduplexed with a preparation of αd, giving rise to the structures shown in Figure 7a (Yagi and Clewell 1977). The presence of only one tail projecting from the circular αd structure indicates that annealing has occurred on one end of the α_3 fragment. Assuming that annealing involves the putative RS sequence (with the exception of the RS sequence, there is no basis for αd and α_3 to have homologous sequences), the latter must exist very close to one of the *Eco*RI cleavage sites. (The insensitivity of αd to *Eco*RI digestion indicates that cleavage does not occur within an RS sequence.) The annealed portion represents about one-fourth the length of the α_3 fragment. On this basis, a more accurate interpretation can now be made of the *Eco*RI cleavage pattern of Mα DNA, as illustrated in Figure 8. The regions between the RS sequences are marked arbitrarily (a, b, c, etc.).

When α_3 DNA was heteroduplexed with α, only structures with one tail were observed. This would be expected even if α_3 anneals at two sites on α, since at one of the sites α_3 would be expected to anneal completely. Our model assumes that the two RS sequences have the same polarity. The presence of two RS sequences in α is more easily demonstrated by heteroduplexing α_1 and α_3.

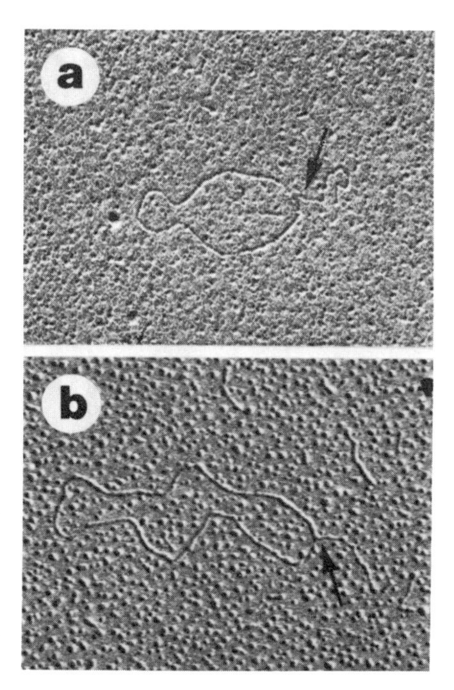

Figure 7
Heteroduplex structures involving the α_3 fragment, α, and αd. (*a*) α_3 heteroduplexed with αd. (*b*) α_3 heteroduplexed with α (pAMα1). Arrows point to the origins of the single-stranded tails.

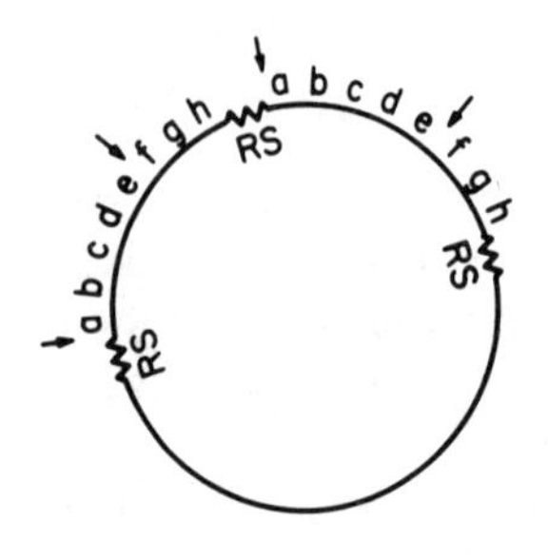

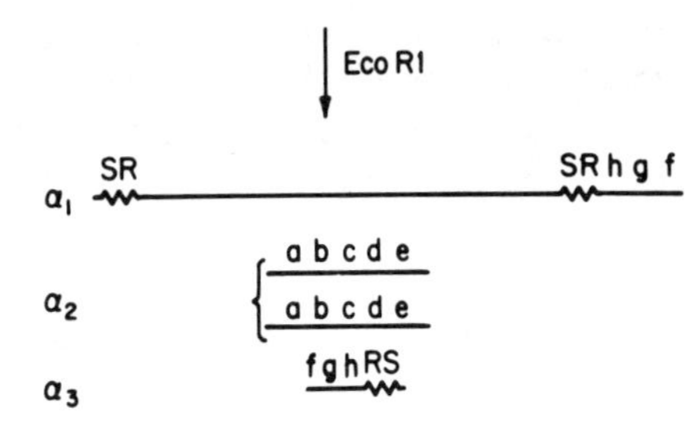

Figure 8

A refined map of the *Eco*RI cleavage pattern on "amplified" molecules. The sequences between the two RS sequences are arbitrarily indicated (a, b, c, etc.), as are the corresponding sequences on the resulting fragments. (Reprinted, with permission, from Yagi and Clewell 1977.)

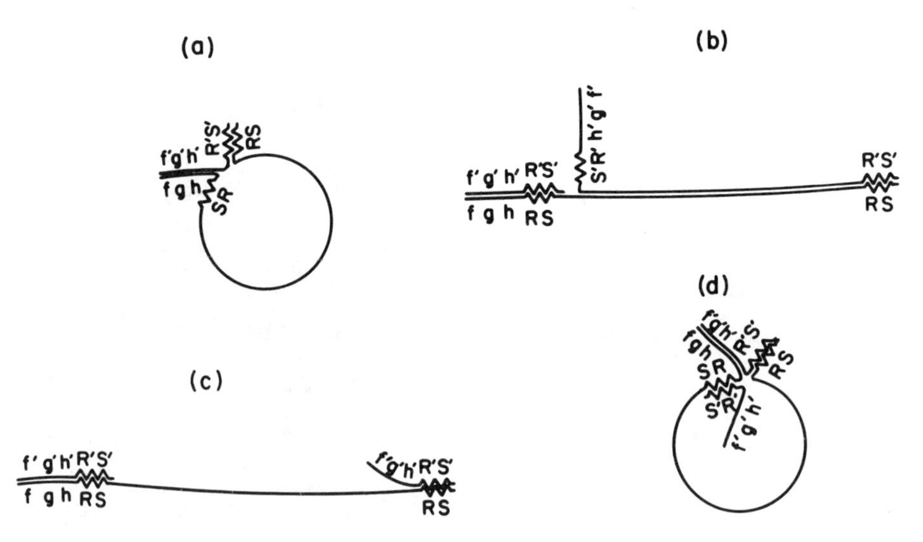

Figure 9

Predicted heteroduplex structures involving the *Eco*RI-generated fragments α_1 and α_3. (*a–d*) Four structures that should be easily found in heteroduplex experiments. (Reprinted, with permission, from Yagi and Clewell 1977.)

Figure 9 shows some of the heteroduplex structures that would be expected in such an experiment. Structures resembling all of the structures illustrated in Figure 9 were in fact observed, and examples of these are shown in Figure 10. On the basis of the length of the shorter tail in Figure 10a, the size of the RS sequence has been estimated to correspond to 0.25 megadalton or close to 380 nucleotide pairs. It can also be deduced from calculations based on measurements of these heteroduplex structures that corresponding points on the two RS sequences of pAMα1 are spaced by 2.65 megadaltons of DNA (Yagi and Clewell 1977). That all of the predicted structures were easily detected in actual heteroduplex experiments supports our model of repeated RS sequences in one orientation. Self annealing experiments using α_1 or pAMα1 (separately) failed to reveal any "stem" structures, thus indicating an absence of repeated sequences with opposite polarity.

The demonstration of this sequence, which will henceforth be designated RS1, lends strong support for models of the amplification of Tc resistance which are based on recombinational events. Three such models are shown in Figure 11.

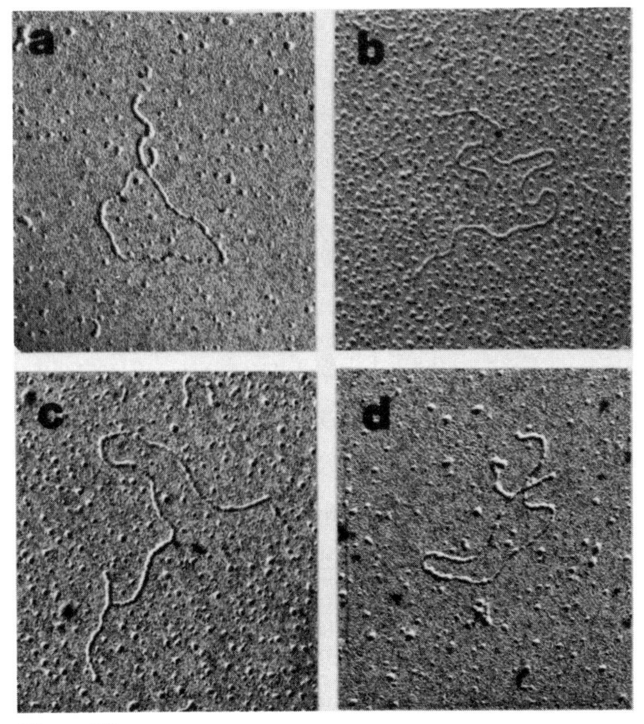

Figure 10
Electron micrographs of heteroduplex structures consisting of α_1 and α_3. (*a–d*) Predicted structures shown in Fig. 9,a–d, respectively.

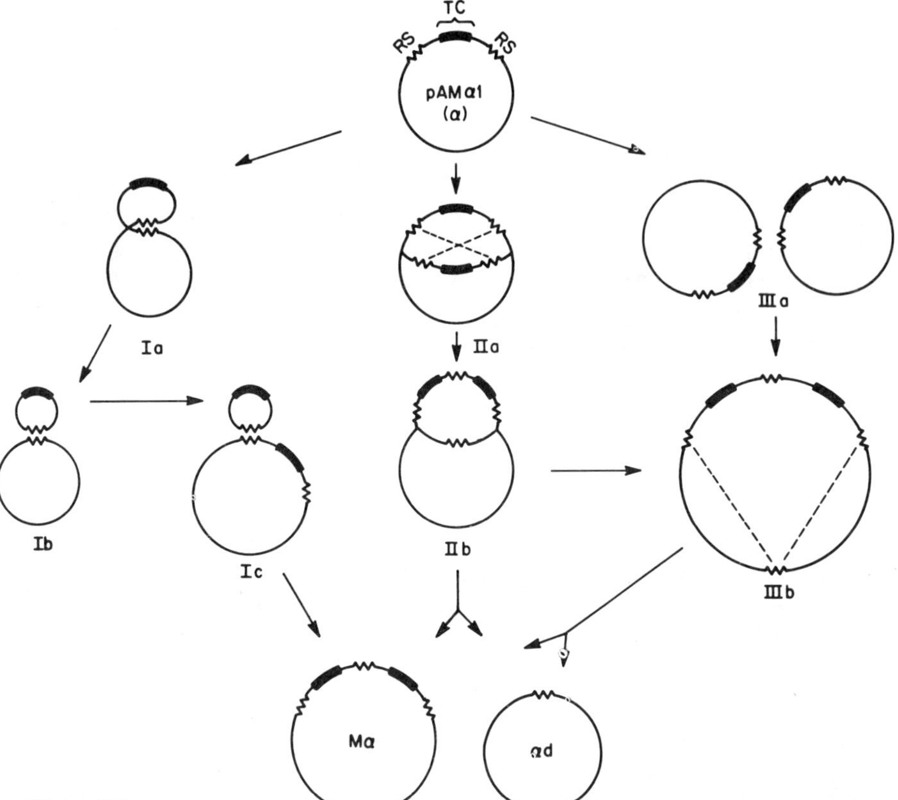

Figure 11

Models for the generation of plasmid DNA molecules with a deletion or with repeats of the Tc^r determinant. In Ia through Ic, an intramolecular recombinational event results in the looping out of a small, 2.65-megadalton, circle which then recombines with an intact molecule of pAMα1 resulting in a duplication. IIa through IIb represents an uneven recombinational event between RS sequences on the daughter strands of a partially replicated plasmid. Depending on the nature of the events that follow, this structure could give rise to a dimeric structure (IIIb) or two independent molecules, one of which contains a duplication, whereas the other contains a deletion. In the case where a dimeric structure is generated, a subsequent recombinational event between two RS sequences (see dashed lines in IIIb) would be required to generate two independent molecules. The third model (IIIa through IIIb) involves an uneven recombinational event between the RS sequences of two independent molecules, giving rise to a dimeric structure. A subsequent recombination would be required for generation of two independent molecules, one with a duplication, the other with a deletion. (Reprinted, with permission, from Yagi and Clewell 1977.)

In all cases, higher degrees of amplification may arise from repeats of basic recombinational processes, as described in Figure 11. The reversibility of the phenomenon is also accounted for in these models. However, reversibility in all cases may represent "loop-out" deletions of redundant segments of DNA.

We assume that the recombinational events occur at relatively low frequency (probably on the order of 10^{-2} to 10^{-3} per generation based on the frequency of spontaneous appearance of Tc-sensitive variants), and that recombinant structures are selected under appropriate conditions (e.g., during presence of Tc). We have no evidence favoring one model over another. Indeed, it is conceivable that all are operating to some extent. Model II is attractive because recombinational events are confined within a single molecule. Intramolecular encounters of RS1 sequences would seem likely to occur more frequently than intermolecular encounters.

Models similar to Model I have been suggested recently by others in relation to certain R plasmids which amplify in *Proteus mirabilis* (Rownd et al. 1975; Ptashne and Cohen 1975; Hu et al. 1975). R1, R6, and R100 all contain amplifiable segments bordered by two homologous sequences having the same orientation. In these plasmids, the sequence corresponds to IS*1* (Ptashne and Cohen 1975; Hu et al. 1975).

The pAMα1 plasmid might serve as an ideal vehicle for the amplification of heterologous DNA inserted using in vitro recombinant DNA techniques. Since the *Eco*RI cleavage sites are both within the segment bordered by the RS1 sequences, it should be possible to insert heterologous *Eco*RI-generated DNA fragments, which could then be amplified passively by selecting for amplification of Tc resistance. A test of the feasibility of such a system awaits the development of a transformation method for inserting recombinant DNA into *S. faecalis*. Efforts to transform plasmid-free strains of *S. faecalis* using pAMα1 DNA have thus far been unsuccessful (G. M. Dunny, Y. Yagi and D. B Clewell, unpubl.).

SUMMARY

The 6.0-megadalton *Streptococcus faecalis* plasmid pAMα1 determines resistance to tetracycline (Tc). Studies in our laboratory involving electron microscopy, sedimentation, and the restriction endonuclease *Eco*RI have shown that when cells harboring pAMα1 are grown in the presence of Tc for a number of generations, a reversible amplification phenomenon occurs, generating tandem repeats of a 2.65-megadalton segment of the plasmid. On the basis of heteroduplex studies between various forms of pAMα1 and fragments generated by *Eco*RI, we have obtained direct evidence for the presence of a small (380 base pairs) sequence, designated RS1, located on both sides of the Tc determinant. The RS1 sequences have the same orientation, and corresponding points on the two sequences are separated by 2.65 megadaltons of DNA. The demonstration of these sequences, which had been predicted from the characteristics of the amplification phenomenon, strongly supports the notion that the amplification process is the result of a series of recombinational events.

Acknowledgments

This work was supported by U.S. Public Health Service research grants AI10318 (National Institute of Allergy and Infectious Diseases) and DE02731 (National Institute of Dental Research). One of the authors (D.B.C.) is the recipient of a Research Career Development Award, K04 AI00061 (National Institute of Allergy and Infectious Diseases). We are grateful to Drs. H. Whitfield, W. Folk and R. Helling for their help in learning the techniques involved in the *Eco*RI endonuclease analyses. We also thank Gary Dunny, Arthur Franke and Margaret Mikus for their help with certain technical aspects of this work.

REFERENCES

Clewell, D. B., Y. Yagi and B. Bauer. 1975. Plasmid-determined tetracycline resistance in *Streptococcus faecalis*: Evidence for gene amplification during growth in presence of tetracycline. *Proc. Nat. Acad. Sci.* **72**:1720.

Clewell, D. B., Y. Yagi, G. M. Dunny and S. K. Shultz. 1974. Characterization of three plasmid deoxyribonucleic acid molecules in a strain of *Streptococcus faecalis*: Identification of a plasmid determining erythromycin resistance. *J. Bact.* **117**:283.

Dunny, G. M. and D. B. Clewell. 1975. Transmissible toxin (hemolysin) plasmid in *Streptococcus faecalis* and its mobilization of a noninfectious drug resistance plasmid. *J. Bact.* **124**:784.

Hu, S., E. Ohtsubo, N. Davidson and H. Saedler. 1975. Electron microscope heteroduplex studies of sequence relations among bacterial plasmids: Identification and mapping of the insertion sequences IS1 and IS2 in F and R plasmids. *J. Bact.* **122**:764.

Jacob, A. and S. Hobbs. 1974. Conjugal transfer of plasmid-borne multiple antibiotic resistance in *Streptococcus faecalis* var *zymogenes*. *J. Bact.* **117**:360.

Ptashne, K. and S. N. Cohen. 1975. Occurrence of insertion sequence (IS) region on plasmid deoxyribonucleic acid as direct and inverted nucleotide sequence duplications. *J. Bact.* **122**:776.

Rownd, R., D. Perlman and N. Goto. 1975. Structure and replication of R-factor deoxyribonucleic acid in *Proteus mirabilis*. In *Microbology–1974* (ed. D. Schlessinger), p. 76. American Society of Microbiology, Washington, D. C.

Yagi, Y. and D. B. Clewell. 1976. Plasmid-determined tetracycline resistance in *Streptococcus faecalis*: Tandemly repeated resistance determinants in amplified forms of pAMα1. *J. Mol. Biol.* **102**:583.

——. 1977. Identification of a small sequence located at two sites on the amplifiable tetracycline resistance plasmid pAMα1 in *Streptococcus faecalis*. *J. Bact.* **129**:400.

SECTION III
Bacteriophage MU: A Transposable Element

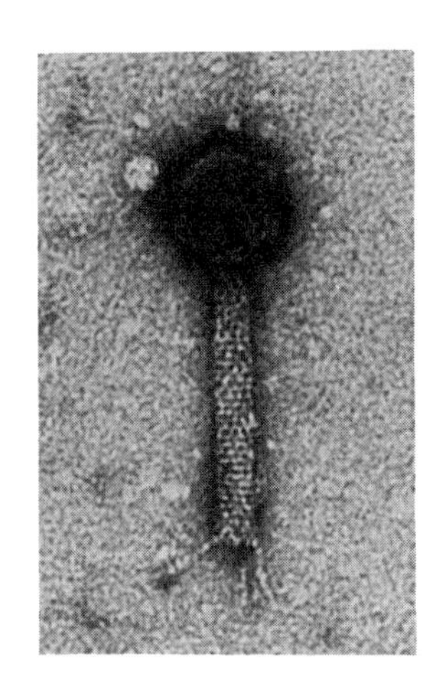

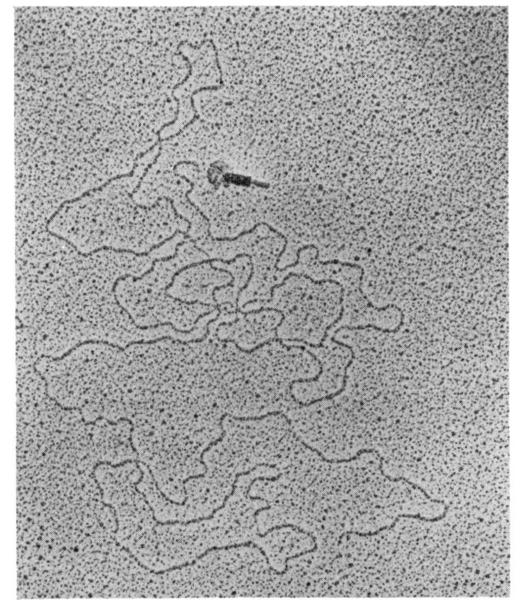

The discovery of the temperate bacteriophage Mu preceded the studies that established the phenomenon of IS elements. In 1963, A. L. Taylor (*Proc. Nat. Acad. Sci.* 50:1043) described the isolation of a bacteriophage that caused mutations in *E. coli*. By a series of genetic linkage tests, Taylor confirmed that the mutations resulted from the insertion of the phage genome into the host genes. In the late 1960s, Taylor and his associates, mainly J. Martuscelli from the University of Mexico and also A. I. Bukhari, then a graduate student, developed some of the basic features of the Mu system. Much of this work remains unpublished. Martuscelli showed that, in an Hfr strain, insertion of Mu DNA caused a 1-minute delay in the time of transfer of a marker that followed the prophage during conjugation. This genetic experiment thus indicated that Mu DNA was linearly inserted into host DNA and was equivalent to about 1% of the *E. coli* chromosome, which can be transferred completely in about 100 minutes. J. Abelson and his coworkers at the University of California, San Diego, independently also characterized some of the important properties of the Mu system. The second phase of Mu work dates from the early 1970s. The initial findings resulting from this work were discussed at the Mu workshop held at the Cold Spring Harbor Laboratory in 1972. The genetic results presented at the workshop can be found in Abelson et al. (*Virology* 54:90 [1973]).

Mu resembles other temperate phages in its morphology. The section-title figure shows a Mu particle at top. The Mu virion is composed of an icosahedral head, 540 Å in diameter; a contractile tail sheath, 1000 Å long and 180 Å wide; a base plate; and at least three tail spikes (To et al., *J. Ultrastruct. Res.* 14:441 [1966]; Martuscelli et al., *J. Virol.* 8:551 [1971]). The electron micrograph below shows a Mu DNA molecule ejected from a virion. Mu DNA is a simple linear duplex of approximately 36–38-thousand nucleotide pairs. However, as some of the papers in this section describe, when Mu DNA is denatured and renatured, the renatured molecules exhibit some highly unusual features. (The electron micrographs in the section-title figure were provided by A. L. Taylor and L. Canedo, University of Colorado Medical Center.)

The Mechanism of Bacteriophage Mu Integration

A. I. Bukhari, E. Ljungquist, F. de Bruijn and H. Khatoon
Cold Spring Harbor Laboratory
Cold Spring Harbor, New York 11724

The temperate bacteriophage Mu presents a unique mode of integrative recombination. It inserts its DNA at randomly distributed sites on the chromosome of its host bacterium *Escherichia coli*. The mechanism by which this remarkably promiscuous integrative recombination occurs remains largely obscure. Recent work from our laboratory suggests that Mu integration is a highly complex process in which replication of viral DNA might be intimately linked to its assimilation into host DNA. This process may be a prototype for the integration of other transposition elements found in prokaryotes.

We discuss here the integration system of Mu and compare the phenomenon of Mu insertion with other systems of integrative recombination. For a detailed background on the Mu integration problem, the reader is referred to Bukhari (1976).

The Integration System

Insertion into the Host Genome

The conclusion that Mu DNA is inserted at random sites is based on two lines of evidence. First, randomly picked lysogens carry prophage Mu in different parts of the host chromosome; about 2% of the lysogens carry prophage-linked auxotrophic mutations (Taylor 1963). The genes inactivated by Mu are not localized in any particular segment of the chromosome. Second, fine genetic mapping of Mu insertions within a single gene shows that there are no notable "hot spots" for Mu insertion within the gene. Bukhari and Zipser (1972) and Daniell et al. (1972) examined a large number of Mu insertions in the *lacZ* gene and concluded that the insertion sites were distributed throughout the gene. The number of different insertion sites (definitely more than 50) indicates that if a sequence is recognized for Mu integration, it must not be larger than two or three nucleotide pairs. On the strength of the available evidence, it is reasonable to assume that Mu does not require a specific DNA sequence for insertion. Recently it has been found that Mu can insert with apparently normal efficiency into plasmids and bacterial chromosomes that have G+C contents quite different from that of *E. coli* (Dénarié et al.; Faelen et al.; both this volume). These observations provide further support for the conclusion that Mu engineers the insertion of its DNA irrespective of the host sequences encountered.

Attachment Region of the Mu Genome

Although Mu insertion is nonspecific with respect to the host sequences, Mu DNA is always inserted in a specific manner into the host genome. All prophages have the same gene order regardless of their site of insertion (Abelson et al. 1973). Thus a specific part of Mu DNA must be recognized for integration. As shown in Figure 1, the phage gene order, determined by vegetative crosses, is identical to the prophage gene order, although the prophage can be in opposite orientation with respect to the host markers (see Abelson et al. 1973 and references therein). Electron microscope heteroduplex studies have also shown that no discernable rearrangement of phage DNA occurs upon integration (Hsu and Davidson 1972, 1974). The collinearity of the phage and prophage genomes indicates that the specific "attachment" region of Mu DNA is defined by the ends of Mu DNA or sequences near the ends.

Reversion of Mu-induced Mutations

Mutations caused by the insertion of Mu DNA are very stable and do not revert at a detectable frequency. Even when the mutations are caused by Mu *cts* (temperature-inducible Mu mutants which have mutations in the immunity gene *c* and presumably make a thermosensitive repressor), revertants cannot be obtained by a simple heat-treatment of the mutant cells. However, revertants can be obtained if the Mu *cts* prophages carry the *X* mutations (Bukhari 1975). The Mu *X* mutants are defective prophages which cannot express their killing functions. These mutants are found among heat-resistant survivors of Mu *cts* lysogens. Mu *cts X* DNA can be lost spontaneously from the host cells. This excision generally is imprecise and occurs at a frequency of 10^{-5}-10^{-7}. Precise excision can be detected at a frequency of 10^{-6}-10^{-8}. Precise excision leads to restoration of the wild-type activity of the gene into which Mu was inserted.

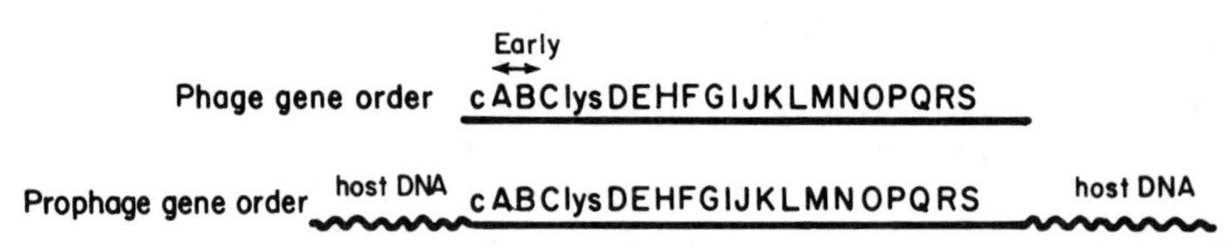

Figure 1

The genetic map of bacteriophage Mu. Both the phage and prophage have the same gene order. The immunity gene *c*, which apparently encodes Mu repressor, is the first marker at the left end of the map. The last gene at the right end is the *S* gene. All of these genes are located in the α segment of Mu, to the left of the G segment (see Chow and Bukhari, this volume). The left end is called the *c* end and the right end the *S* end. At least some of the genes expressed early in the lytic cycle are located near the *c* end. The *X* mutations have roughly been mapped close to the *B* gene. The map as shown here has been slightly modified from the 1973 Cold Spring Harbor map of Mu (see Appendix C7).

The excision event is at least partly controlled by Mu-coded functions (Bukhari 1975). Reversion to wild type via the Mu *X* pathway can be detected in most but not all mutants induced by Mu *cts*. In those cases where revertants cannot be isolated, the mutants have been shown to contain deletions at the site of Mu insertion (M. M. Howe and A. I. Bukhari, unpubl.). The finding that most Mu-induced mutants can be forced to revert to wild type leads to two important conclusions: (1) In most cases, Mu insertion occurs without any damage to the host sequences. (2) The ends of prophage Mu are specific, and the junction of Mu DNA to host DNA can be recognized.

In view of the above considerations, the integration system of Mu can be summarized as follows:

1. Integration of Mu involves linear insertion of viral DNA (25×10^6 m.w. or 36-38 kb).
2. Specific sequences in Mu DNA react with nonspecific sequences in host DNA during the integration process.
3. Specific sequences of Mu are defined by the prophage ends.
4. The sequences at prophage ends are at or very close to the phage DNA ends.
5. The phage DNA ends (actually sequences close to the ends, as discussed below) generally must act in a coordinated fashion so that Mu insertion occurs at a point and does not cause any deletion of the host sequences.

The Integrative Precursor

An important step central to understanding the mechanism of Mu integration would be the elucidation of the form of Mu DNA that reacts with host DNA during insertion. We define this form as the integrative precursor. It is clear that in temperate phage systems such as λ and P2, the inserting viral DNA has a circular form. The prophage gene order in these cases is permuted with respect to the phage gene order because the attachment sites on the phage genomes are located away from the phage DNA ends (Campbell 1971; see also Calender and Six; Nash et al.; both this volume). In Mu, however, the prophage gene order is not permuted, and no inferences concerning the form of the integrative precursor can be made. However, we can assume that the ends (or sequences close to the ends) of phage DNA interact during the integration process. Therefore it is imperative that we review briefly the structure of Mu DNA ends before discussing the possible nature of the Mu integrative precursor.

Unlike other well-known temperate phage systems, such as λ, P1, P2, and P22, Mu DNA has neither cohesive ends nor terminally repetitious sequences. Instead, it has heterogeneous ends which differ in sequence from molecule to molecule (Daniell et al. 1975; Bukhari et al. 1976; for reviews, see Howe and Bade 1975; Bukhari 1976; Chow and Bukhari, this volume). The heterogeneous sequences consist of host DNA. The left end (*c* end) of Mu DNA contains about 100 host base pairs, whereas the right end (*S* end) contains about 1500 base pairs of host DNA (see Fig. 2). Because of the heterogeneous host DNA at its ends, Mu does

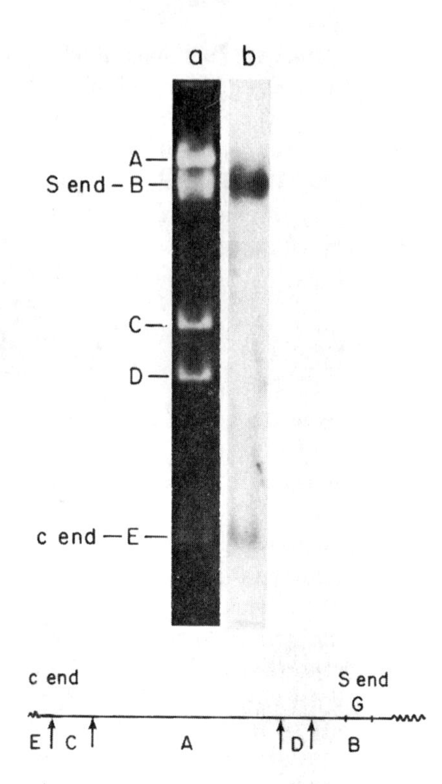

Figure 2
Demonstration of the presence of host DNA at the ends of Mu DNA. Mu DNA, extracted from phage particles, was cut with a combination of endonucleases *Eco*RI and *Hin*dIII. These enzymes cleave Mu DNA into five fragments (Allet and Bukhari 1975), as shown in the cleavage map. The fragments were separated by electrophoresis in 1% agarose gels, stained with ethidium bromide, and photographed in ultraviolet light. The fragments were then denatured in gels, blotted onto a nitrocellulose paper, and hybridized to ^{32}P-labeled denatured *E. coli* DNA (Bukhari et al. 1976). The ethidium bromide stain pattern is shown to the left, and the autoradiograph of the same fragments after hybridization is shown to the right. It can be seen that the two end fragments (labeled B and E) hybridize to *E. coli* DNA, whereas the internal fragments of Mu do not.

not have any obvious means of fusing its ends to form circular DNA molecules. The host sequences are acquired by Mu as a consequence of headful packaging of Mu DNA from maturation precursors in which Mu DNA is flanked by host sequences (Bukhari and Taylor 1975; Chow and Bukhari, this volume). A Mu lysate prepared from a single plaque still has a population of phage particles containing different end sequences. This must mean that the sequences at the ends are randomized during Mu growth; that is, Mu sheds old host sequences and picks up new host sequences. Thus the old host sequences must be removed from the phage DNA when it is integrated in order to generate maturation precursors with new sequences. Hsu and Davidson (1974) have provided evidence by electron microscope heteroduplex techniques that at least the *S*-end host sequences are lost upon Mu integration.

The unusual structure of Mu DNA leads to some interesting questions concerning the form of the integrative precursor: How does Mu lose the host sequences at its ends during the integration process? Does the integrative precursor have a form free of host DNA? Does Mu DNA assume a circular form, as is the case with other temperate phages? To answer these questions and to analyze the structure of the integrative precursor, we have examined the behavior of prophage Mu DNA upon induction and the fate of phage DNA after infection of host cells.

Behavior of Prophage Mu DNA upon Induction

So far, it has not been possible to detect Mu DNA free of host DNA. It is known that many copies of Mu DNA are integrated into host DNA during the lytic cycle of the phage (Razzaki and Bukhari 1975). The formation of heterogeneous circles carrying Mu DNA and varying amounts of host DNA also indicates that Mu DNA undergoes multiple rounds of integration into the host genome during Mu growth (Waggoner et al. 1974; Schroeder et al. 1974; Waggoner et al., this volume). This multiple integration of Mu apparently is responsible for the formation of maturation precursors from which Mu DNA is packaged during morphogenesis. Therefore integrative precursors continuously must be formed during the process of Mu growth. An analysis of the process by which these integrative precursors are generated after heat induction of a Mu *cts* lysogen, carrying only a single Mu *cts* prophage, is critical to our understanding of the form of Mu DNA required for integration. For example, the Mu prophage may be excised cleanly after induction, giving rise to a form free of host DNA. We have experimentally tested whether Mu DNA is excised efficiently after induction.

To examine the possibility of Mu excision, we first located prophage Mu DNA in the host DNA. We extracted the total DNA from lysogenic cells, cleaved it with restriction endonucleases, and separated the fragments by electrophoresis in agarose gels. The DNA fragments were denatured in gels and transferred to a nitrocellulose paper, as described by Southern (1975). Hybridization with denatured ^{32}P-labeled Mu DNA caused the fragments containing Mu DNA to become labeled, and these fragments could be visualized after autoradiography. We were thus able to locate both the internal Mu fragments and the end fragments arising from fusion of host DNA to the right and to the left of Mu DNA. With this technique, the fate of the junction fragments after induction was followed throughout the Mu lytic cycle (Ljungquist and Bukhari 1977). Figure 3 shows that, under the normal conditions for the lytic cycle, the original junction fragments remain intact after induction and replication of Mu DNA. In similar experiments with bacteriophage λ lysogens, we could readily monitor the excision of λ DNA, as shown by the disappearance of junction fragments and the appearance of a phage fragment containing the λ attachment site. These observations indicate that, unlike λ, Mu is not excised upon induction. Thus if a free form of Mu DNA exists, it does not arise from excision of Mu DNA. This must mean that Mu DNA is replicated in situ and that replicas (copies) of Mu DNA are integrated at different sites. The structure of the replication products of Mu DNA has yet to be unraveled. Recent experiments by B. Waggoner and M. Pato (pers. comm.) and by us indicate that upon induction of a Mu prophage, host sequences adjacent to Mu DNA are not amplified; that is, if Mu DNA is replicated in situ, the replication does not proceed into the host DNA.

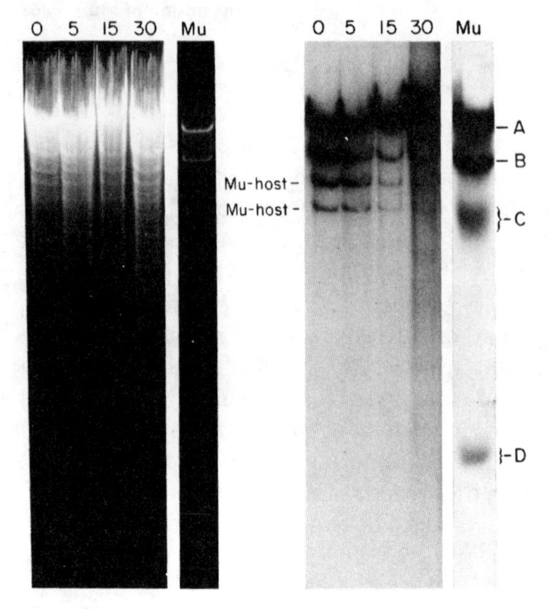

Figure 3
Persistence of prophage Mu at the original site after induction. A culture of strain BU563, lysogenic for Mu *cts*62, was heat-induced, and samples were withdrawn just before induction (min 0) and at different times after shift to 43°C. The total cellular DNA in each case was digested with *Bal*I endonuclease and subjected to electrophoresis in 1% agarose gels. The digestions gave the complex patterns shown on the left. To unmask the fragment corresponding to the Mu prophage, the fragments were transferred to nitrocellulose papers and hybridized with ^{32}P-labeled Mu DNA as described in Fig. 2. *Bal*I cuts Mu DNA at three sites, generating four fragments (see Kahmann et al., this volume). The two end fragments are "fuzzy" because of the heterogeneity of the ends (the marker Mu DNA in the figure). In a lysogen, the two Mu-host fusion fragments give sharp bands and have mobilities different from those of the end fragments of mature Mu DNA. These fragments are marked Mu-host in the autoradiographs. It can be seen that these two fragments, one representing the right-end fusion fragment and the other the left-end fusion fragment, are not altered after prophage induction. If there were excision of Mu DNA, these junction fragments would be expected to be cut and to give rise to new fragments from the excision products. No new fragments are seen. In similar experiments with bacteriophage λ lysogens, disappearance of junction fragments and appearance of the λ fragment containing the λ attachment site could be readily observed. Thus this technique can be used for in vivo assay for viral excision. In λ, excision occurred even if DNA replication was blocked with nalidixic acid. In Mu, no change was seen for long periods (more than 2 hr after induction) if DNA synthesis was blocked. Under normal conditions when Mu DNA is replicating, new Mu-host junction fragments are continually being generated as many copies of Mu DNA are integrated at different sites. As shown in the autoradiograph in the figure, the generation of new fusion fragments shows up as a background of hybridization smear at later times after induction.

Behavior of Mu DNA upon Infection

One approach to answering the question of whether the integrative precursor of Mu assumes a circular form would be to infect cells with ^{32}P-labeled Mu particles and then follow the fate of the labeled DNA. We have examined in detail the fate of ^{32}P-labeled Mu DNA after infection, using sucrose gradient centrifugation, ethidium bromide-CsCl density centrifugation, and restriction endonuclease cleavage analysis of the DNA (E. Ljungquist and A. I. Bukhari, in prep.). No covalently closed circles of the parental Mu DNA can be detected after infection of either sensitive or immune (lysogenic) cells. A prominent form of parental Mu DNA which moves about twice as fast as the linear Mu DNA can be seen in neutral sucrose gradients after infection of immune cells but not after infection of sensitive cells. This peak is absent, however, from alkaline sucrose gradients. Furthermore, the bulk of parental (^{32}P-labeled) Mu DNA can be shown to preserve its free ends up to 40 minutes after infection. Careful examination of parental DNA after various physiological manipulations, such as infection of sensitive cells blocked in DNA synthesis but not in protein synthesis, has failed to reveal any trace of covalently closed circles. We conclude, therefore, that Mu does not have an efficient mechanism for fusing its ends to form circular molecules. It should be noted that covalently closed circular forms of DNA are found in the cells beginning approximately in the middle of the lytic cycle. These circles are not pure Mu circles, since they are of variable size and contain both Mu DNA and host DNA (Waggoner et al., this volume). No bona fide Mu circles (without host DNA) have so far been demonstrated in a population of heterogeneous circles.

A surprising result of the experiments described above was that not only is the infecting parental DNA not converted to a covalently closed circular form, but it is also not integrated efficiently into the host genome. This is indicated by three observations: (1) Most of the parental DNA preserves its free ends late into the lytic cycle. (2) ^{32}P-labeled parental DNA does not show up in the heterogeneous circles. (3) Not more than 10% of the infecting Mu DNA can be shown to be taken up by the host DNA. This experiment involves separation of host DNA from Mu DNA by electrophoresis in 0.3% agarose gels (E. Ljungquist and A. I. Bukhari, in prep.), as shown in Figure 4. It can be seen that after infection of cells with ^{32}P-labeled Mu, most of the ^{32}P-label does not enter the host DNA. Poor incorporation of the parental Mu DNA into the host DNA is highly intriguing since we know that Mu DNA is integrated efficiently at different sites during Mu growth. This result therefore forces the conclusions that the copies (the replication products) of Mu DNA take part in the integration process and that the parental DNA itself cannot be integrated efficiently.

Replication and Integration

In trying to answer the questions outlined above, we have encountered an interesting situation. Both the excision experiments and the experiments on

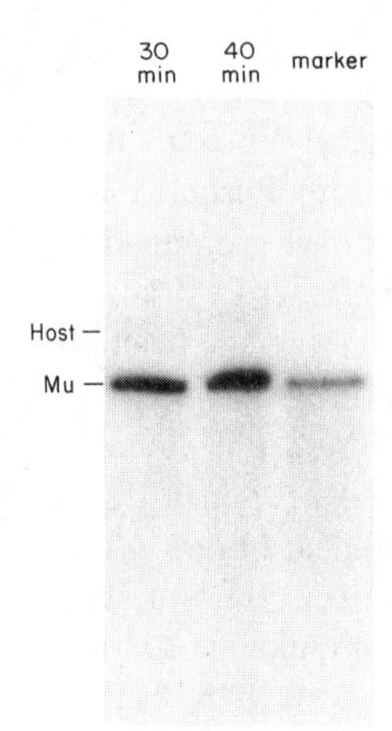

Figure 4
Noninsertion of parental Mu DNA into host DNA after infection. *E. coli* cells were infected with ^{32}P-labeled Mu particles, the unadsorbed phage were removed by washing, and the cells were lysed at different times to extract the DNA. The DNA was subjected to electrophoresis in 0.3% agarose horizontal slab gels. Mu DNA can be separated from host DNA under these conditions. In the experiment shown in the figure, the multiplicity of infection was 5, and cell lysis was complete about 60 min after infection. Only a small amount of the parental label is found incorporated in the host DNA by 40 min. The DNA appearing in the phage bands at different times apparently contains linear Mu DNA molecules since the endonuclease digestion yields Mu end fragments.

the fate of the parental DNA suggest that replicas of Mu DNA are the molecules active in integration. Gene-*A* mutants of Mu are known to be defective in DNA replication as well as integration (Wijffelman et al. 1974; Faelen et al. 1975). Our experiments with Mu *tsA* mutants have further documented that the *A*-gene mutants are deficient in DNA replication even when they are already inserted in the host genome. It is quite possible that the gene-*A* function of Mu is involved in integration as well as replication of Mu DNA. Thus replication of Mu DNA appears to hold the key to the problem of Mu integration.

If Mu DNA replication does not proceed into the host sequences flanking the phage DNA, it might help to explain how Mu sheds old host sequences during its integration. Mu DNA replication might generate a form free of host DNA. This form may be the active precursor for integration. Whether this form has sufficient half-life to be detectable would depend upon the mechanics of replication and integration. If replication and integration are in some manner coupled, the integrative precursor may not be detectable. However, if replication is completed first and is then followed by integration, it might be possible to detect the integrative precursor. We would like to emphasize at this point that failure of the infecting parental Mu DNA to be converted to a circular form does not necessarily reflect on the structure of the integrative precursor. No statement on the structure of the replication products of Mu DNA can be made at this time.

Mu and Other Integration Systems

Although Mu is a temperate bacteriophage and has the required functions for lytic and lysogenic cycles, it differs from other temperate phages in its integration properties and the intracellular behavior of its DNA. In many of its properties, it resembles the transposable IS elements and transposons (see Bukhari 1976).

Phages That Follow the Campbell Model

Lambda is a well-known representative of the temperate phages that follow the Campbell model of integrative recombination. In these phages, the integrative precursor is known to be circular. The phages have linear DNA molecules, but the ends of their DNA are either cohesive (single-stranded, complementary sequences at 5′ ends) or terminally repetitious (allowing the ends to recombine). The linear molecules can therefore be converted into circles. The "attachment site" (the site for integrative recombination) is located away from the molecular ends of the phage DNA, and thus the prophage gene order is permuted with respect to the phage gene order. No extensive replication of DNA molecules appears to be required for integration. In Mu, the prophage gene order is not permuted with respect to the phage order, and phage DNA replication appears to be a prerequisite for integration.

ϕ105

The only other well-known phage in which the prophage gene order and the phage gene order are identical is ϕ105 of *Bacillus subtilis* (Armentrout and Rutberg 1970). Interest in this phage has been further stimulated by evidence that, during the reversal of lysogeny, the derepressed ϕ105 is not excised from the cell chromosome prior to replication but apparently replicates as a complex of phage and bacterial DNA (Armentrout and Rutberg 1971; Rutberg 1973; Shapiro et al. 1974). The ϕ105 lysogens can be induced by mitomycin C, a property exhibited by the lambdoid phages but not by P2-like phages or Mu.

Despite the two strikingly similar features (i.e., the collinearity of phage and prophage gene orders and the nonexcision of prophage DNA upon induction), ϕ105 and Mu probably have significant differences in their mechanisms of integration. First, unlike Mu, ϕ105 has a single specific attachment site for integration on the host chromosome. Second, Garro and coworkers have recently shown that a 24–26 $\times$ 10^6-dalton DNA duplex of ϕ105 exists primarily as a nicked circle upon extraction of DNA from phage particles (B. M. Scher, D. H. Dean and A. J. Garro, pers. comm.). They have inferred from these results that ϕ105 DNA has cohesive ends (complementary, single-stranded DNA) and thus is likely to behave as a circle upon infection of host cells. Third, when prophage ϕ105 DNA replicates in situ, the replication proceeds into the host DNA, and host genes adjacent to the prophage apparently are amplified. Preliminary evidence on Mu indicates that, upon prophage Mu induction, replication does not proceed into the host DNA.

Thus ϕ105 is an interesting case of site-specific integrative recombination in which reversal of integration deviates from the classic excision mode of bacteriophage λ. One point worth noting is that the sequences at the cohesive ends of

ϕ105 appear to define both a site for attachment to the host chromosome and a site for packaging DNA into phage particles.

Transposition Elements

Transposition of IS elements and transposons is discussed extensively in other papers in this volume. Insertion of these elements is specific with respect to their DNA (a specific sequence is recognized for integration), but the level of specificity for host DNA is not clear. For some IS elements, there are definite "hot spots" of integration in genes. Thus, at least in these cases, a small, specific host sequence is preferred for integration.

A striking property of IS elements is that they can be excised from host DNA at a low frequency. Mutations induced by IS elements can revert at a frequency of 10^{-6}-10^{-7}. However, this excision frequency is clearly lower than the apparent transposition frequency. Berg (this volume) has found that kanamycin transposon Tn*5* has an overall transposition frequency of 10^{-2} but that its excision from a given site in the host genome occurs at a much lower frequency. This conclusion can also be reached for IS elements by simply calculating the transposition frequency. The frequency at which a specific IS-induced mutation can be isolated is about 10^{-6}. If there are 10^{4} sites in host DNA at which IS can be integrated (assuming two sites per gene for about 5000 gene equivalents in the chromosome of *E. coli*), the overall frequency of transposition must not be less than 10^{-2}. Furthermore, excision of transposons does not lead to their reintegration. For example, when mutations caused by the insertion of tetracycline and kanamycin resistance transposons revert, the revertants almost always lose the transposon. We infer from these arguments that excision of IS elements and transposons cannot account for the efficiency of their transposition.

The discrepancy between excision and integration is also evident in Mu. As we have discussed above, the Mu *X* mutants can be excised from the host genes at a frequency of 10^{-5}-10^{-7}, and this excision does not lead to reintegration of Mu DNA (Bukhari 1975; Bukhari and Froshauer 1977). Very little excision of prophage DNA after induction can be seen, although many copies of Mu DNA are "transposed" to different sites during the lytic cycle. It can be postulated, therefore, that, in both Mu and transposition elements, the integrative precursor is generated by replication, and at least a copy of the resident element is retained at the original site. It should be noted that, as in Mu, the DNA of transposition elements free of host DNA has not been detected so far.

The excision behavior of Mu *X* mutants is similar to that of the IS elements and transposons (see Berg; Botstein and Kleckner; both this volume). Excision can be precise, but, more frequently, it is imprecise, leaving behind deletions and probably other types of rearrangements in the gene affected by excision. In the case of Mu *X* mutants, the Mu DNA appears to be completely analogous to that of an insertion element. We have found by DNA-DNA hybridization studies that the *X* mutations are caused by the insertion of IS*1* into Mu DNA (see also Chow and Bukhari, this volume). So far, all Mu mutants that can be excised (*X* mutants) appear to have insertion of IS*1* in Mu. It is open to question whether or not the IS*1* element and Mu DNA interact in any way during the excision process. The possible relationship between Mu and the IS elements has yet to be explored.

Mu Integration in Perspective

Mu might represent a class of elements in which the mode of integrative recombination involves replication of DNA before the formation of the integrative precursor. This mode is distinct from that generally exhibited by temperate phages such as λ, P2, and P22. In the classic model of λ, integration proceeds by recombination of the circular integrative precursor into the host DNA. The recombination occurs at specific sites and is mediated by specific proteins. Rescue of viral DNA from host DNA is an exact reversal of the integration process, such that resident viral DNA is physically excised from host DNA. In Mu, however, rescue of viral DNA does not appear to proceed via physical excision of resident viral DNA. Resident viral DNA seems to remain unaltered, and copies of DNA are synthesized and translocated to different host sites. The transposition elements might behave in a manner similar to that of Mu.

SUMMARY

We have studied the problem of Mu DNA integration and excision by genetic as well as biochemical techniques, involving ultracentrifugation, gel electrophoresis, and DNA-DNA hybridization. The parental Mu DNA molecules, after infection of host cells, are not converted to circular molecules. Furthermore, only a very small amount of parental DNA is assimilated into host DNA. Upon induction of a prophage, Mu DNA is not excised efficiently from the host chromosome, and yet many copies of Mu DNA are found to integrate rapidly at different sites in the host chromosome during Mu growth. We propose that, both after induction of a prophage and after infection of cells by Mu particles, Mu DNA undergoes replication such that the replication leads to the formation of integrative precursors of Mu. The structure of the integrative precursor remains obscure. The replication-integration model for Mu insertion might also apply to the insertion of transposition elements.

Acknowledgments

We thank Mike Botchan and Tom Maniatis for discussions, and Linda Ambrosio and Susan Froshauer for help with many of the experiments. Original work in this paper was supported by grants from the National Science Foundation and the National Cystic Fibrosis Foundation, and by a Career Development Award to A. I. B. from the National Institutes of Health.

REFERENCES

Abelson, J., W. Boram, A. I. Bukhari, M. Faelen, M. Howe, M. Metlay, A. L. Taylor, A. Toussaint, P. van de Putte, G. C. Westmaas and C. A. Wijffelman. 1973. Summary of the genetic mapping of prophage Mu. *Virology* **54**:90.

Allet, B. and A. I. Bukhari. 1975. Analysis of Mu and λ-Mu hybrid DNAs by specific endonucleases. *J. Mol. Biol.* **92**:529.

Armentrout, R. W. and L. Rutberg. 1970. Mapping of prophage and mature deoxyribonucleic acid from temperate *Bacillus* bacteriophage ϕ105 by marker rescue. *J. Virol.* **6**:760.

———. 1971. Heat induction of prophage ϕ105 in *Bacillus subtilis*: Replication of the bacterial and bacteriophage genomes. *J. Virol.* **8**:455.

Bukhari, A. I. 1975. Reversal of mutator phage Mu integration. *J. Mol. Biol.* **96**:87.

———. 1976. Bacteriophage Mu as a transposition element. *Annu. Rev. Genet.* **10**:389.

Bukhari, A. I. and S. Froshauer. 1977. Insertion of a chloramphenicol resistance determinant in bacteriophage Mu. *Gene* (in press).

Bukhari, A. I. and A. L. Taylor. 1975. Influence of insertions on the packaging of host sequences covalently linked to mutator phage Mu DNA. *Proc. Nat. Acad. Sci.* **72**:4399.

Bukhari, A. I. and D. Zipser. 1972. Random insertion of Mu-1 DNA within a single gene. *Nature New Biol.* **236**:240.

Bukhari, A. I., S. Froshauer and M. Botchan. 1976. The ends of bacteriophage Mu DNA fragments separated by gel electrophoresis. *Nature* **264**:580.

Campbell, A. 1971. Genetic structure. In *The bacteriophage lambda* (ed. A. D. Hershey), p. 13. Cold Spring Harbor Laboratory, Cold Spring Harbor, New York.

Daniell, E., E. E. Kohne and J. Abelson. 1975. Characterization of the inhomogeneous DNA in virions of bacteriophage Mu by DNA reannealing kinetics. *J. Virol.* **15**:237.

Daniell, E., R. Roberts and J. Abelson. 1972. Mutations in the lactose operon caused by bacteriophage Mu. *J. Mol. Biol.* **69**:1.

Faelen, M., A. Toussaint and J. de Lafonteyne. 1975. Model for the enhancement of λ-*gal* integration into partially induced Mu-1 lysogens. *J. Bact.* **121**:873.

Howe, M. and E. Bade. 1975. Molecular biology of bacteriophage Mu. *Science* **190**:624.

Hsu, M.-T. and N. Davidson. 1972. Structure of bacteriophage Mu-1 DNA and physical mapping of bacterial genes by Mu-1 DNA insertion. *Proc. Nat. Acad. Sci.* **69**:2823.

———. 1974. Electron microscope heteroduplex study of the heterogeneity of Mu phage and prophage DNA. *Virology* **58**:229.

Ljungquist, E. and A. I. Bukhari. 1977. State of prophage Mu DNA upon induction. *Proc. Nat. Acad. Sci.* (in press).

Razzaki, T. and A. I. Bukhari. 1975. Events following prophage Mu induction. *J. Bact.* **122**:437.

Rutberg, L. 1973. Heat induction of prophage ϕ105 in *Bacillus subtilis*: Bacteriophage induced bidirectional replication of the bacterial chromosome. *J. Virol.* **12**:9.

Schroeder, W., E. G. Bade and H. Delius. 1974. Participation of *Escherichia coli* DNA in the replication of temperate bacteriophage Mu-1. *Virology* **60**:534.

Shapiro, J., D. H. Dean and H. O. Halvorson. 1974. Low frequency specialized transduction with *Bacillus* phage ϕ105. *Virology* **62**:393.

Southern, E. M. 1975. Detection of specific sequences among DNA fragments. *J. Mol. Biol.* **98**:503.

Taylor, A. L. 1963. Bacteriophage-induced mutation in *Escherichia coli. Proc. Nat. Acad. Sci.* **50**:1043.

Waggoner, B. T., M. S. Gonzalez and A. L. Taylor. 1974. Isolation of heterogeneous circular DNA from induced lysogens of bacteriophage Mu-1. *Proc. Nat. Acad. Sci.* **71**:1255.

Wijffelman, C., M. Gassler, W. F. Stevens and P. van de Putte. 1974. On the control of transcription of bacteriophage Mu. *Mol. Gen. Genet.* **131**:85.

Characterization of Covalently Closed Circular DNA Molecules Isolated after Bacteriophage Mu Induction

B. T. Waggoner,* M. L. Pato*† and A. L. Taylor†
*Division of Molecular and Cellular Biology
National Jewish Hospital and Research Center, and
†Department of Microbiology
University of Colorado Medical Center
Denver, Colorado 80206

Mu-1, a temperate bacteriophage of *Escherichia coli* K12, is capable of generating stable, random mutations in its host (Taylor 1963). These Mu-induced mutations are generated by the linear insertion of the Mu genome into the inactivated host gene (Martuscelli et al. 1971; Boram and Abelson 1971; Bukhari and Zipser 1972; Daniell et al. 1972). However, at this time, there is no clear understanding of the steps preceding the insertion of the phage genome.

The models for Mu integration have been patterned after those for integration in other, better known temperate phage systems. Campbell proposed a model for bacteriophage lambda integration which involves a covalently closed circular form of the virus DNA as an intermediate (Campbell 1962). In the Campbell model, a single recombination event between the host chromosome and the circular viral DNA results in the linear insertion of the phage genome. Since this model was proposed, it has been shown that not only is the lambda genome circularized prior to integration, but that P2 and P22 similarly utilize a circular form of their genomes for integration (Gottesman and Weisberg 1971; Bertani and Bertani 1971; Levine 1972). Several lines of evidence suggest that integration of Mu DNA, as well, might be facilitated by the formation of a circular intermediate. For example, Mu DNA can integrate in either orientation with respect to the host chromosome by a recombination event involving the recognition of the ends of the phage genome. Involvement of Mu DNA ends is inferred from the finding that the gene order in the prophage is identical to that of the vegetative phage genome (Hsu and Davidson 1972; Abelson et al. 1973). The insertion of Mu normally occurs without simultaneous deletions of host DNA (Martuscelli et al. 1971; Hsu and Davidson 1972; Bukhari and Zipser 1972; Bukhari 1975). These findings are best explained if the Mu integrative precursor has a circular form, and we therefore sought such a precursor. We report here the characterization of a circular form containing Mu DNA that appears during the lytic cycle of Mu.

Isolation of Covalently Closed Circular, Supercoiled DNA during Mu Development

Excision is postulated to be essentially the reverse process of integration (Campbell 1962). Therefore, we considered that the initial excision product, or possibly its progeny, isolated after Mu induction might be the circular form of Mu DNA analogous to the circular molecules isolated after lambda induction (Lipton and Weissbach 1966).

In a typical experiment, a lysogen of Mu *cts*4 (a heat-inducible mutant of Mu; Waggonner et al. 1974) carrying a single prophage was labeled with [^{3}H] thymidine for several generations and then induced at 42°C in the continued presence of the label. At various times after induction, the DNA was isolated and purified by centrifugation in a cesium chloride-ethidium bromide (CsCl-EtBr) density gradient (Radloff et al. 1967). By 14 minutes after induction, a small peak of DNA banding at a higher density than the bulk chromosomal DNA was observed (Fig. 1). This material bands in a position corresponding to the location of

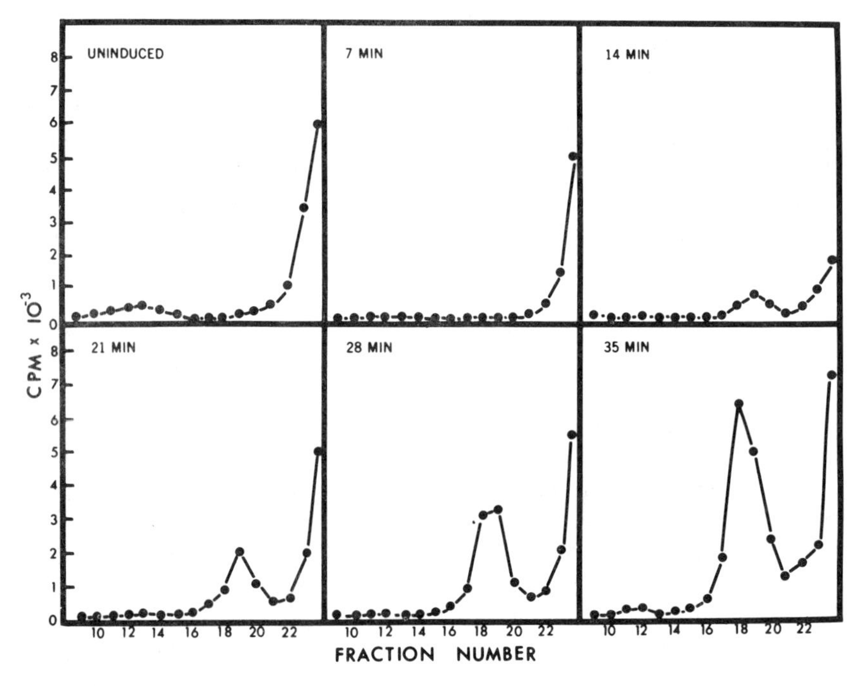

Figure 1

Kinetics of appearance of CCC-DNA after induction of a Mu *cts*4 lysogen. Samples were taken prior to induction and at 7, 14, 21, 28, and 35 min after induction. Extracts were centrifuged in CsCl-EtBr density gradients, collected, fractionated, and counted as described by Martuscelli et al. (1971). Only the denser fractions (1-24) are plotted. More than 70% of the input material was recovered in all gradients.

covalently closed circular (CCC) DNA molecules. The formation of the CCC-DNA molecules occurs only when Mu development is normal; the CCC-DNA is absent when a Mu *cts* lysogen is maintained at 30°C or when the induced Mu *cts* lysogen is defective in phage replication due to amber mutations in genes *A* or *B* (data not shown). It is clear from Figure 1 that not only does the CCC-DNA first appear by 14 minutes, but that it continues to accumulate in amount and reaches a maximum level of 2.0–2.5% of the bulk chromosomal DNA shortly before lysis (Fig. 2). Numerous experiments were done to determine the time of appearance of the CCC-DNA molecules after induction; two such experiments are shown in Figure 2. In one case, CCC-DNA first appeared 12–14 minutes after induction when the latent period was 40–42 minutes; in the same experiment, cell viability began to decline between 5 and 10 minutes after induction, and Mu-specific DNA synthesis began between 6 and 8 minutes (Fig. 3). The rate of synthesis of Mu DNA was such that by 9 minutes after induction there were two prophage copies per cell; by 12 minutes, four copies; and by 15 minutes, six copies. From these data, it is clear that at least one round of phage DNA synthesis was completed before CCC-DNA molecules were detected in the CsCl-EtBr density gradients. The molecular mechanisms involved in the decline in cell viability, phage DNA replication, or the formation of the CCC-DNA are not well

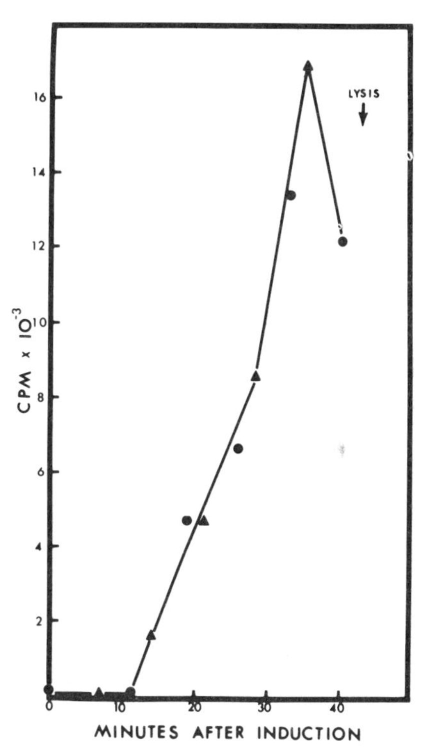

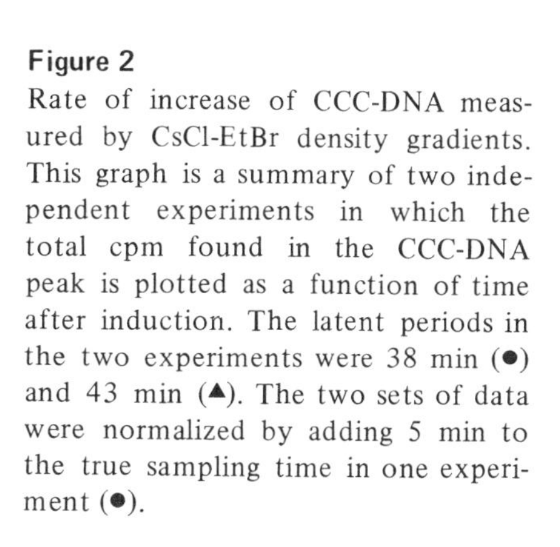
Figure 2
Rate of increase of CCC-DNA measured by CsCl-EtBr density gradients. This graph is a summary of two independent experiments in which the total cpm found in the CCC-DNA peak is plotted as a function of time after induction. The latent periods in the two experiments were 38 min (●) and 43 min (▲). The two sets of data were normalized by adding 5 min to the true sampling time in one experiment (●).

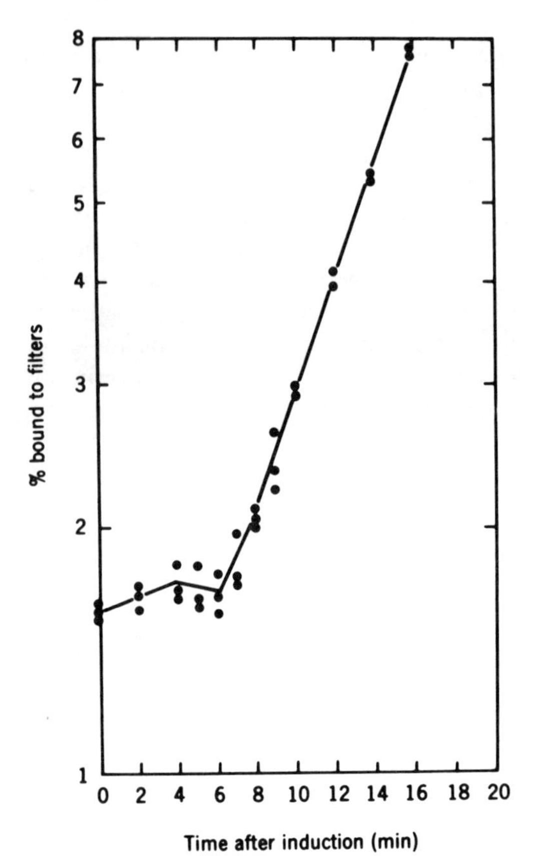

Figure 3

Synthesis of Mu-specific DNA after induction of Mu *cts*4. A monolysogen of Mu *cts*4 was grown (30°C) and induced (42°C) in the presence of [^{3}H] thymidine (50 μCi/ml), and samples were prepared for DNA-DNA annealing as described by Kuempel (1972). The specific activity of the [^{3}H]DNA was 2.4 × 10^6 μCi/ml. The [^{3}H]DNA samples were diluted appropriately, mixed with purified ^{32}P-labeled Mu DNA and unlabeled *E. coli* DNA, and boiled for 15 min in 0.5 N NaOH. Annealing mixtures contained < 0.08 μg [^{3}H]DNA, < 0.005 μg ^{32}P-labeled Mu DNA, 20 μg *E. coli* DNA, and Denhardt polymer mixture (Denhardt 1966). A nitrocellulose filter with 5 μg unlabeled Mu DNA attached by the procedure of Denhardt (1966) was added to each annealing mixture, and annealing was carried out in a total volume of 1 ml for 24–36 hr at 65°C. Binding to filters without DNA was 0.2–0.5% of the input TCA-precipitable [^{3}H]-DNA; binding of ^{32}P-labeled Mu DNA to filters containing Mu DNA was 85–95% of the input TCA-precipitable ^{32}P-labeled Mu DNA. Each sample was run in triplicate, and each point on the figure represents the average of three determinations. Random host sequences from the SE regions of Mu DNA comprise 3–5% of the DNA on the filters and account for the binding of approximately 0.65% of the TCA-precipitable input [^{3}H]DNA. All values were corrected for nonspecific binding.

understood at this time. Furthermore, the cause and effect relationships (if any) between these three early events in Mu development are just now being investigated (for reviews, see Howe and Bade 1975; Couturier 1976).

Molecular Lengths of Covalently Closed Circular DNA Molecules

The circular form of lambda DNA isolated after induction is approximately a monomer in length, although dimers and trimers have been isolated (Tomizawa and Ogawa 1969). Thus it was surprising to find that the sizes of the CCC-DNA molecules induced by Mu development were exceedingly variable, and that most of the molecular lengths did not correspond to either a Mu monomer size or an integral multiple of the length of the mature Mu genome (Table 1). Due to the heterogeneous nature of the sizes of these CCC-DNA forms, they will be referred to as Hc-DNA throughout this report. The data in Table 1 also show that the overall length distribution of the Hc-DNA molecules changes during the developmental cycle of Mu. The average contour lengths in kilobases (kb) of Hc-DNA isolated 20, 30, and 40 minutes after induction are 85.5 kb, 79.9 kb, and 71.5 kb, respectively. The range of observed molecular lengths was 22–201 kb at 20 minutes after induction, 36–156 kb at 30 minutes, and 38–120 kb at 40 minutes. The molecular events involved in the reduction in size of the Hc-DNA are not known, but the host recombination (*recA*) system does influence this overall size distribution of the Hc-DNA. As shown in Table 1-D, when Hc-DNA is isolated from a Mu *cts recA* lysogen at 30 minutes after induction, a greater proportion of the population consists of larger molecules than when Hc-DNA is isolated from a Mu *cts* rec^+ lysogen (Table 1-B). Therefore, it seems likely that the effect of the *recA* gene product is to reduce the larger circles to smaller ones during the developmental cycle of Mu. This is not an essential function, however, since the phage yields in *recA* and rec^+ lysogens are similar.

Mu DNA in the Hc-DNA

To obtain evidence for the presence of Mu DNA in the Hc-DNA molecules, heteroduplexes were formed between mature Mu *cts* DNA (from Mu phage grown by induction) and Hc-DNA molecules (isolated 30 minutes after induction). A mixture of Hc-DNA and Mu DNA was denatured, reannealed, spread by the formamide method, and analyzed in the electron microscope (Sharp et al. 1972).

A total of 32 interpretable DNA structures were measured. Several of the types observed are schematically diagrammed in Figure 4A. Two types of typical homoduplex structures formed between mature Mu DNA are illustrated in Figure 4A, 1 and 2 (Bade 1972; Daniell et al. 1973b). Sixteen homoduplexes were measured which contained four distinct regions, designated α, G, β, and

Table 1

Measured Length of Individual Covalently Closed Circular DNA Molecules Obtained after Induction of a Mu *cts*4 Lysogen

		1 x Mu 36.9		2 x Mu 73.8		3 x Mu 110.7			4 x Mu 147.6	
A				82.0	98.2					
				80.6	94.6					
				78.5	91.5					
				78.4	90.6					
			65.8	78.0	90.2	114.4				
			64.2	77.9	88.1	114.1				
			61.1	76.9	87.7	113.5				
			59.2	76.5	86.6	110.7				
			59.1	74.9	86.5	107.8				
		49.2	58.6	74.0	85.4	106.2				
		48.1	57.0	72.4	85.1	104.5				
	34.4	41.5	56.6	71.5	84.8	103.7	129.8	142.0		
	26.8	38.2	52.0	69.2	83.6	102.9	128.4	137.5	161.5	
	22.1	37.9	51.2	68.5	82.7	100.5	117.3	133.1	149.3	201.5
B			65.5							
			65.3	82.0						
			63.0	81.4	98.2					
		50.0	61.0	77.3	93.5					
		46.7	59.2	75.8	93.2	113.6				
		44.7	58.3	75.2	92.4	112.1				
		40.9	57.6	73.3	88.5	110.9				
		40.3	57.0	72.7	88.5	110.0				
		39.2	56.4	71.7	88.0	107.6				
		37.6	55.2	71.1	88.0	101.4	123.3			
		37.1	53.6	69.0	85.3	100.5	120.9	136.8		
		37.0	51.7	67.7	85.3	99.4	119.7	136.1		
		36.5	51.2	67.0	83.3	98.5	116.2	133.8	156.7	
C			66.4							
		50.4	66.2							
		50.1	65.6							
		49.9	65.1							
		49.6	65.1							
		45.7	61.3	81.5	89.0					
		45.5	60.9	78.0	88.0					
		42.3	58.8	72.3	88.0	107.7				
		41.2	58.1	71.9	87.6	106.8				
		41.0	57.9	70.9	87.4	106.7				
		39.4	56.9	70.0	86.5	103.2	120.8			
		39.1	53.9	68.8	84.1	102.4	116.1			
		38.6	52.7	68.5	83.0	99.0	115.2			
D			65.1							
			64.0							
			62.5							
			60.6							
			59.8							
			59.6							
			58.6							
		50.3	58.3							
		49.2	57.3	82.1						
		48.6	55.5	77.4						
		46.2	55.0	76.2						
		43.6	55.0	70.8			126.2		162.4	285.8
		43.5	54.4	69.0	98.3	110.4	120.6		154.2	184.7
		40.0	53.5	67.7	93.4	104.3	117.0	144.0	153.9	174.0
		39.7	52.1	67.3	89.6	103.3	115.4	136.7	153.0	166.7
	0–34	35–50	51–66	67–82	83–98	99–114	115–130	131–146	147–162	>163

Individual Circular Molecules

Molecular Length Range, kb

Molecules were isolated at 20 min (*A*), 30 min (*B*), and 40 min (*C*) after induction of a Mu *cts*4 lysogen, and 30 min (*D*) after induction of a *recA* Mu *cts*4 monolysogen. The lengths predicted for circles containing integral multiples (1–4) of a Mu genome are designated above. Length measurements were made as described by Waggoner et al. (1974). Measurements are given in kilobases (kb), 1000 base pairs for double-stranded DNA or 1000 bases for single-stranded DNA, and are calibrated against circular, double-stranded phage PM2 DNA (9.1 kb) or single-stranded phage ϕX174 DNA (5.2 kb) (Waggoner 1975).

split end (SE) (Fig. 4A, 1); six were measured which had two distinct regions, termed $\alpha G\beta$ and SE (Fig. 4A, 2). The SE segment is present in all Mu DNA homoduplexes and consists of randomly derived *E. coli* DNA (Daniell et al. 1973a, 1975). The G region can undergo spontaneous inversion (Hsu and Davidson 1972). Renaturation of Mu DNA strands containing G regions in opposite orientation produces the G bubble. The control homoduplexes were not photographed at random since it was desirable to obtain a larger number of measurements of the G bubble. Normally, the G bubble is present in 30-50% of the Mu DNA homoduplexes formed when Mu is grown by induction (Daniell et al. 1973a; Hsu and Davidson 1974). However, for unknown reasons, this frequency is reduced to $< 0.5\%$ when Mu is obtained from an infective cycle, even when the infecting phage is obtained by induction (Daniell et al. 1973a).

The average lengths of the segments unique to Mu are shown in Table 2. The lengths of these characteristic Mu regions are compared to analogous regions found in the experimental heteroduplex DNA structures (Fig. 4A, 3 and 4). Because the SE segment is lost during Mu integration (Hsu and Davidson 1974), the length of the $\alpha G\beta$ region represents the entire length of the Mu prophage. The mean lengths for the $\alpha G\beta$, G, and β regions (Table 2) in the heteroduplexes between Mu DNA and Hc-DNA correspond so closely to the analogous regions in mature Mu DNA homoduplexes that we conclude that Mu DNA is present in the Hc-DNA molecules. In addition, the Mu DNA is covalently linked to non-Mu sequences of variable lengths. DNA-DNA annealing experiments of Schroeder et al. (1974) have indicated that the non-Mu sequences are derived from the *E. coli* host. These data show that Mu DNA is covalently attached to *E. coli* sequences of variable lengths to form Hc-DNA molecules.

Sequence Heterogeneity in Hc-DNA

Since the host DNA sequences associated with Mu DNA in the Hc-DNA vary in length, we wanted to test whether their nucleotide sequences also vary. To determine the sequence complexity of the Hc-DNA, circles isolated 30 minutes after induction were self-annealed. Some of the heteroduplex structures observed in the electron microscope are shown in Figure 4B. From measurements of 16 heteroduplexes (Table 2), it is clear that Mu DNA sequences in different Hc-DNA molecules have reannealed. The length of Mu DNA in these structures is defined. This indicates that the sequences adjacent to the ends of the Mu DNA are different and hence do not reanneal. We infer that Mu DNA derived from the primary prophage site or its progeny becomes covalently linked to many different host segments during the course of Mu development. This result is consistent with the earlier observation that Mu DNA continues to be integrated at different sites throughout the vegetative cycle (Schroeder and van de Putte 1974; Razzaki and Bukhari 1975).

In addition to heteroduplexes between different Hc-DNA molecules, nine heteroduplexes between inverted repeat sequences in single Hc-DNA molecules

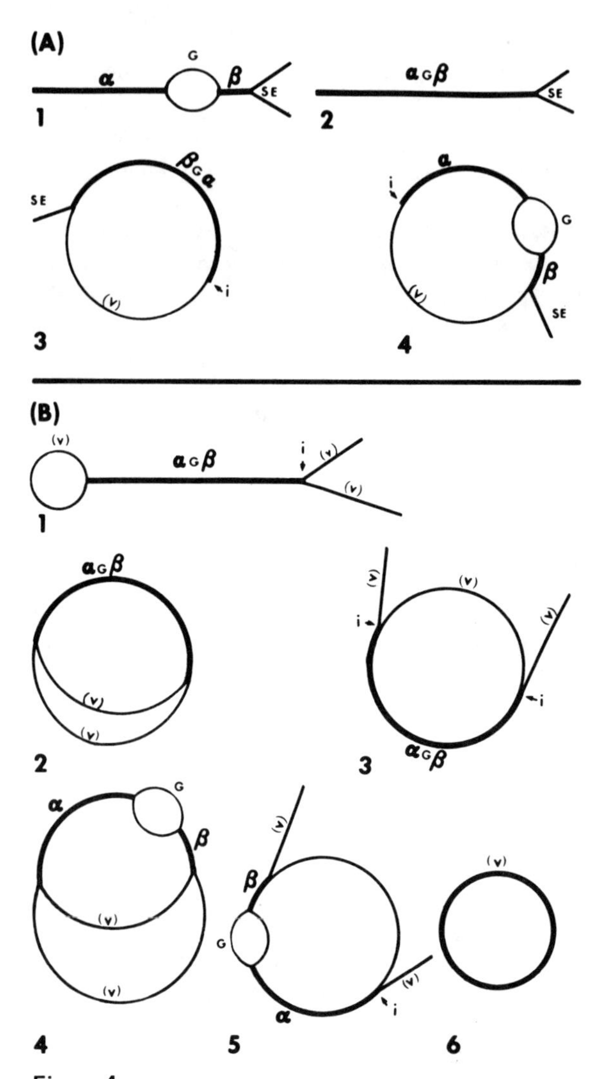

Figure 4

An interpretative schematic representation of relevant DNA structures. The heavy lines are duplex DNA; light lines are single-stranded DNA. An "i" is placed in the schematic diagram in regions which are indeterminate due to the difficulty in accurately pinpointing the junction of double-stranded and single-stranded DNA. Regions of DNA that are variable in length are indicated by the letter "v."

(*A*) Structures formed following reannealing of Hc-DNA with mature Mu *cts* DNA: (*1*) A control homoduplex of mature Mu DNA containing the four distinct regions characteristic of Mu DNA–α, G, β, and SE. (*2*) A control homoduplex of mature Mu DNA in which the G bubble has reannealed, showing the SE and the duplex region $\alpha G\beta$. (*3*) A heteroduplex structure formed between a

Table 2
Summary of Molecular Length Measurements of Heteroduplex DNA Structures

Type of heteroduplex	No. of examples	Mean length, kilobases		
		$\alpha G\beta$	G	β
Self-annealed mature Mu DNA containing a G bubble	16	34.7 ± 2.0	2.9 ± 0.5	1.6 ± 0.1
Self-annealed mature Mu DNA without a G bubble	6	35.1 ± 1.6		
Mature Mu DNA annealed to Hc-DNA	10	35.9 ± 1.6	3.2 ± 0.8	1.7 ± 0.7
Heteroduplexes between different Hc-DNA molecules	16	35.0 ± 2.9	3.2 ± 0.8	1.7 ± 0.2
Heteroduplexes between inverted repeat sequences in single Hc-DNA molecules	9	33.9 ± 4.2		

Figure 4 *(continued)*
complete Hc-DNA circle and mature Mu DNA containing the SE and $\alpha G\beta$ segments. (*4*) A heteroduplex structure formed between a complete Hc-DNA circle and mature Mu DNA showing the α, G, β, and SE segments.

(*B*) Structures formed by self-annealing Hc-DNA: (*1*) An intrastrand heteroduplex structure formed by reannealing of an inverted repeat sequence present in a single-stranded linear molecule derived from Hc-DNA. The length of the double-stranded DNA between the terminal forks corresponds to the $\alpha G\beta$ length of Mu. (*2*) A heteroduplex structure formed between two complete Hc-DNA molecules of different sizes containing the $\alpha G\beta$ region. (*3*) A heteroduplex structure formed between one complete Hc-DNA molecule and a linear strand derived from an Hc-DNA molecule showing the duplex region $\alpha G\beta$. (*4*) A heteroduplex structure formed between two complete Hc-DNA molecules of different sizes containing the α, G, and β regions. (*5*) A heteroduplex structure formed between one complete Hc-DNA molecule and a broken strand derived from an Hc-DNA molecule. (*6*) A completely duplex molecule probably due to incomplete denaturation; more rigorous denaturation before renaturation reduced the amount of this form.

were measured (Fig. 4B, 1; Table 2). The length of the duplex region is consistent with the interpretation that these structures represent a reannealing of two Mu genomes inserted in opposite orientation in a single Hc-DNA molecule. It is of interest to note that the G region is in the same orientation in all such molecules measured. The absence of the G bubble suggests that the orientation of G may be fixed within a single cell during vegetative growth, but that this fixed orientation may be different in other cells, thus giving rise to the G bubbles observed in heteroduplexes between different Hc-DNA molecules. It appears, therefore, that the heterogeneity in length of the Hc-DNA molecules is due both to the amount of *E. coli* sequences present and to the number of Mu copies present. The observation that the amount of Mu DNA present in the population of Hc-DNA molecules continues to increase throughout the infective cycle of Mu is consistent with this hypothesis (Schroeder et al. 1974).

SUMMARY

Bacteriophage Mu development induces the formation of covalently closed circular DNA molecules. These circular molecules appear approximately 12–14 minutes after induction of a heat-inducible prophage. It is possible that phage DNA replication is a prerequisite for their formation since their appearance is preceded by phage-specific DNA replication during normal phage development and they are absent from induced Mu lysogens defective in phage DNA replication. The most distinctive property of these circular forms is that they are heterogeneous in length (Hc-DNA) and reveal no consistent correlation to the size of a Mu genome or any integral multiple of the unit genome size. In addition, the overall average length of the Hc-DNA decreases during the lytic cycle. Heteroduplex analysis has demonstrated that Mu DNA sequences are present in individual circular molecules covalently attached to host DNA. From this study, it is postulated that the Hc-DNA owes its length heterogeneity to the number of Mu copies present as well as to the length of the host sequences present. The host DNA segments vary not only in length but also in nucleotide sequence. It is proposed that these noncomplementary host sequences are acquired by repeated integration events during the vegetative cycle of Mu. The role of Hc-DNA in Mu development remains to be determined. (For a discussion of the possible involvement of Hc-DNA in the packaging of mature Mu DNA, see Bukhari and Taylor [1975].)

Acknowledgments

This work was supported by U.S. Public Health Service Research Grants GM-13712 to A. L. T. and AI-11075 to M. L. P.

REFERENCES

Abelson, J., W. Boram, A. I. Bukhari, M. Faelen, M. Howe, M. Metlay, A. L. Taylor, A. Toussaint, P. van de Putte, G. C. Westmaas and C. A. Wijffelman.

1973. Summary of the genetic mapping of prophage Mu. *Virology* **54**:90.
Bade, E. G. 1972. Asymmetric transcription of bacteriophage Mu-1. *J. Virol.* **10**:1205.
Bertani, E. and G. Bertani. 1971. Genetics of P2 and related phages. *Adv. Genet.* **16**:199.
Boram, W. and J. Abelson. 1971. Bacteriophage Mu integration: On the mechanism of Mu-induced mutations. *J. Mol. Biol.* **62**:171.
Bukhari, A. I. 1975. Reversal of mutator phage Mu integration. *J. Mol. Biol.* **96**:87.
Bukhari, A. and D. Zipser. 1972. Random insertion of Mu-1 DNA within a single gene. *Nature New Biol.* **236**:240.
Bukhari, A. and A. L. Taylor. 1975. Influence of insertions on packaging of host sequences covalently linked to bacteriophage Mu DNA. *Proc. Nat. Acad. Sci.* **72**:4399.
Campbell, A. M. 1962. Episomes. *Adv. Genet.* **11**:101.
Couturier, M. 1976. The integration and excision of the bacteriophage Mu-1. *Cell* **7**:155.
Daniell, E., W. Boram and J. Abelson. 1973a. Genetic mapping of the inversion loop in bacteriophage Mu DNA. *Proc. Nat. Acad. Sci.* **70**:2153.
Daniell, E., D. E. Kohne and J. Abelson. 1975. Characterization of the inhomogeneous DNA in virions of bacteriophage Mu by DNA reannealing kinetics. *J. Virol.* **15**:739.
Daniell, E., R. Roberts and J. Abelson. 1972. Mutations in the lactose operon caused by bacteriophage Mu. *J. Mol. Biol.* **69**:1.
Daniell, E., J. Abelson, J. S. Kim and N. Davidson. 1973b. Heteroduplex structures of bacteriophage Mu DNA. *Virology* **51**:237.
Denhardt, D. T. 1966. A membrane-filter technique for the detection of complementary DNA. *Biochem. Biophys. Res. Comm.* **23**:641.
Gottesman, M. E. and R. A. Weisberg. 1971. Prophage insertion and excision. In *The bacteriophage lambda* (ed. A. D. Hershey), p. 113. Cold Spring Harbor Laboratory, Cold Spring Harbor, New York.
Howe, M. M. and E. G. Bade. 1975. Molecular biology of bacteriophage Mu. *Science* **190**:624.
Hsu, M.-T. and N. Davidson. 1972. Structure of inserted bacteriophage Mu-1 DNA and physical mapping of bacterial genes by Mu-1 DNA insertion. *Proc. Nat. Acad. Sci.* **69**:2823.
——. 1974. Electron microscope heteroduplex study of the heterogeneity of Mu phage and prophage DNA. *Virology* **58**:229.
Kuempel, P. L. 1972. Deoxyribonucleic acid-deoxyribonucleic acid hybridization assay for replication-origin deoxyribonucleic acid of *Escherichia coli*. *J. Bact.* **110**:917.
Levine, M. 1972. Replication and lysogeny with phage P22 in *Salmonella typhimurium*. *Curr. Topics Microbiol. Immunol.* **58**:138.
Lipton, A. and A. Weissbach. 1966. The appearance of circular DNA after lysogenic induction in *E. coli* CR34 (λ). *J. Mol. Biol.* **21**:517.
Martuscelli, J., A. L. Taylor, D. J. Cummings, V. A. Chapman, S. S. Delong and L. Cañedo. 1971. Electron microscopic evidence for linear insertion of bacteriophage Mu-1 in lysogenic bacteria. *J. Virol.* **8**:551.
Radloff, R., W. Bauer and J. Vinograd. 1967. A dye-buoyant density method for

the detection and isolation of closed circular duplex DNA: The closed circular DNA in HeLa cells. *Proc. Nat. Acad. Sci.* **57**:1514.

Razzaki, T. and A. I. Bukhari. 1975. Events following prophage Mu induction. *J. Bact.* **122**:437.

Schroeder, W. and P. van de Putte. 1974. Genetic study of prophage excision with a temperature inducible mutant of Mu-1. *Mol. Gen. Genet.* **130**:99.

Schroeder, W., E. G. Bade and H. Delius. 1974. Participation of *Escherichia coli* DNA in the replication of temperate bacteriophage Mu-1. *Virology* **60**:534.

Sharp, P. A., M.-T. Hsu, E. Ohtsubo and N. Davidson. 1972. Electron microscope heteroduplex studies of sequence relations among plasmids of *Escherichia coli. J. Mol. Biol.* **71**:471.

Taylor, A. L. 1963. Bacteriophage-induced mutation in *Escherichia coli. Proc. Nat. Acad. Sci.* **50**:1043.

Tomizawa, J.-I. and T. Ogawa. 1969. Replication of phage lambda DNA. *Cold Spring Harbor Symp. Quant. Biol.* **33**:533.

Waggoner, B. T. 1975. Heterogeneous circular DNA isolated from *E. coli* after bacteriophage Mu-1 induction. Ph.D. dissertation, University of Colorado, Denver.

Waggoner, B., N. Gonzalez and A. L. Taylor, 1974. Isolation of heterogeneous circular DNA from induced lysogens of bacteriophage Mu-1. *Proc. Nat. Acad. Sci.* **71**:1255.

Mu-mediated Illegitimate Recombination as an Integral Part of the Mu Life Cycle

A. Toussaint and M. Faelen
Laboratoire de Génétique
Université libré de Bruxelles
Rhode St. Genèse, Belgium

A. I. Bukhari
Cold Spring Harbor Laboratory
Cold Spring Harbor, New York 11724

The temperate bacteriophage Mu can integrate its DNA at randomly distributed sites on the chromosome, the plasmids, and the prophages of its host bacterium *Escherichia coli* (Taylor 1963; Toussaint 1969; Bukhari and Zipser 1972; Daniell et al. 1972). The life cycle of Mu has been studied by several groups (see, for review, Howe and Bade 1975; Couturier 1976; Bukhari 1976; and several articles in this volume). One of the main attributes of Mu is its ability to engineer illegitimate recombination. Mu is known to mediate insertion of circular DNA molecules, to promote transposition of host genes, and to cause deletions in the host genome. Chromosomal rearrangements appear to occur continuously during Mu growth. These rearrangements, however, are not brought about merely by Mu-coded enzymes; direct participation of Mu DNA is a prerequisite for the occurrence of these events. We discuss here the Mu-mediated recombinational events with particular reference to the lytic cycle of Mu.

The Lytic Cycle of Mu

Three characteristic features of bacteriophage Mu have been recognized:

1. Mu DNA, as extracted from purified phage particles, has host DNA at both ends.
2. During phage development (the lytic cycle), Mu DNA is integrated continuously at different sites on the host genome.
3. Approximately midway during the lytic cycle, covalently closed circles containing Mu DNA as well as host DNA begin to appear.

Host DNA at Mu Ends

As discussed elsewhere in this volume, a striking property of Mu DNA is the presence of host sequences at its ends (see Bukhari et al.; Chow and Bukhari; Bade et al.; all this volume). These sequences are not inserted into the host DNA and apparently are lost during the integration process (Hsu and Davidson 1974). There is strong evidence that these host sequences arise from headful packaging of phage DNA from maturation precusors that contain covalently linked Mu DNA and host DNA (Bukhari and Taylor 1975; Chow and Bukhari, this volume). This evidence is based on observations that insertions in Mu DNA cause a decrease in the length of host sequences at the *S* end. Since Mu acquires a wide assortment of host sequences (a plaque-purified lysate contains Mu particles with different host sequences), the integration of the Mu genome must occur at many different sites to generate maturation precursors with different host sequences.

Integration during the Lytic Cycle

Genetic experiments have shown that integration of Mu DNA occurs at a high frequency during the phage lytic cycle. Integration of Mu during the lytic cycle was detected by mating the F′-episome-carrying strains after prophage Mu induction or infection with Mu. Some of the F′ episomes transferred to an F⁻ strain were shown to have acquired a Mu prophage (Schroeder and van de Putte 1974; Razzaki and Bukhari 1975). At least ten copies of Mu were estimated to be inserted per *E. coli* chromosome (Razzaki and Bukhari 1975). Association of Mu DNA with host DNA has also been suggested by the formation of heterogeneous circles during the lytic cycle (Waggoner et al. 1974; Schroeder et al. 1974). It has now been confirmed that these circles contain Mu DNA linked to different segments of host DNA (Waggoner et al., this volume).

Mode of Heterogeneous Circle Formation

Multiple integration of Mu at different sites, starting with a lysogen carrying one copy of prophage Mu at a given site, is formally analogous to transposition of Mu DNA. Heterogeneous circular DNA molecules containing different segments of host DNA and Mu DNA presumably arise as a result of such "transposition events." Electron microscope heteroduplex studies have so far shown that all of the circles contain at least one copy of Mu DNA (see Waggoner et al., this volume). Parker and Bukhari (1976) have presented genetic evidence that circles generated during the Mu lytic cycle always contain Mu DNA. In their experiment, an Hfr strain lysogenic for a temperature-inducible (*cts*) Mu located at a site distant from the integrated F factor was heat-induced, and the heterogeneous circles carrying the F factor were recovered by mating with F⁻ strains. Almost all of the F′ episomes thus formed were shown to have a Mu prophage. It was inferred, therefore, that direct participation of Mu DNA is a required step in the formation of heterogeneous circles.

The covalently closed, heterogeneous circles vary in length, ranging from one Mu genome length to 6–8 Mu lengths, and can be detected after Mu DNA has undergone five to six cycles of replication (Waggoner et al., this volume). Replication of Mu DNA is such that there are two copies of Mu DNA at 9 minutes and about six copies at 15 minutes after prophage induction. The heterogeneous circles can be seen at approximately 14-15 minutes after induction. The exact role of these circles in the Mu life cycle is not understood. They might be used as maturation precursors, or progenitors of maturation precursors, since they begin to decline in number at the end of the Mu lytic cycle, at the time when Mu particles begin to mature.

Mu-mediated Illegitimate Recombination

The propensity of Mu for illegitimate recombination is indicated by the formation of heterogeneous circles during the lytic cycle. The illegitimate recombinational events also lead to a variety of genetic rearrangements which can be detected at different stages during the Mu life cycle.

Mu-mediated Insertion of Non-Mu DNA

Circular DNA molecules, such as the circular forms of the bacteriophage λ genome and F′ episomes, can be inserted into host DNA with Mu as a vector for integration (Faelen et al. 1971; van de Putte and Gruijthuijsen 1972). This integration of circular non-Mu DNA occurs in a *recA*$^-$ host and is dependent upon Mu-specific functions. The inserted DNA is always flanked on each side by a copy of Mu DNA, and both Mu prophages have the same orientation. If Mu-mediated insertion is obtained after infection of cells with Mu, the Mu-DNA-Mu structure is located at random points on the host chromosome (Toussaint and Faelen 1973). However, if Mu-mediated insertion is examined after partially inducing Mu *cts* lysogens, a surprising result is obtained. In monolysogens, the Mu-DNA-Mu structure is located at the original site of prophage insertion (Faelen et al. 1975). That is, instead of one prophage, there are two copies of the Mu genome at the original site, and the newly inserted DNA is sandwiched between the two copies. If the induced Mu lysogen contains two Mu prophages at different locations, then in about 50% of the cases the non-Mu DNA is integrated at the site of one of the prophages and is surrounded by two Mu genomes which have the same genotype and orientation as the original prophage. In the other 50% of the cases, the non-Mu DNA is located at the site of the second prophage and surrounded by two Mu prophages which have the genotype of the second prophage. This finding strongly suggests that excision of Mu DNA does not occur during integration of non-Mu DNA and that the second copy of the prophage arises from in situ duplication of the prophage DNA.

Mu-mediated Transposition

One of the most interesting genetic rearrangements caused by Mu is the transposition of genes. This transposition can be readily observed in strains carrying transmissible plasmids whether the cells are *recA*$^-$ or *recA*$^+$. During Mu growth, various segments of the chromosome are transposed onto plasmids, which can be recovered by mating with F⁻ cells if suitable markers are available for selection (Faelen and Toussaint 1976; Faelen et al., this volume). As is the case with Mu-mediated insertion, the transposed DNA is surrounded by two Mu prophages which have the same orientation. All *E. coli* markers can be transposed to a plasmid at a relatively high frequency (2×10^{-4} per recipient cell receiving the plasmid). Transposition appears to require replication of Mu DNA since Mu mutants blocked in DNA replication (*X* mutants; see Bukhari 1975) are unable to carry out transposition of the host genes.

Mu-mediated Inversion

There is some preliminary evidence that Mu can cause inversions of host sequences located adjacent to a Mu prophage (M. Faelen, unpubl.). However, this phenomenon has yet to be explored in detail.

Mu-promoted Deletions

Mu can generate deletions of host sequences. These deletions occur at the site of Mu insertion. The deletions apparently can be generated by Mu in three different ways. One is the production of deletions during the process of Mu lysogenization. Although most mutations caused by Mu insertion behave as point mutations, in some cases deletions are found to be present at the site of prophage insertion (Cabezon et al. 1975; M. M. Howe, pers. comm.). Out of ten Mu *cts*62 insertions in the *lacZ* gene, three were found to have caused deletion of at least a part of the *lac* operon (see Bukhari 1975; M. M. Howe and A. I. Bukhari, unpubl.). The second mode of deletion formation is detected after partial induction of a Mu *cts* lysogen. In some of the survivors of partial induction, deletions of host DNA linked to the prophage can be detected (M. Faelen, unpubl.). It is possible that both of the above processes of deletion formation are the same, and that deletions seen after lysogenization of cells by Mu actually result immediately after the phage DNA has been inserted.

The third process by which Mu can cause deletions involves the excision of the *X* mutants of prophage Mu (Bukhari 1975). Normally, Mu DNA is not excised from the host chromosome. Mu-induced mutants do not revert to wild type even if no deletion of host sequences is present at the site of Mu insertion. However, it has been shown that if a Mu *cts* prophage carries *X* mutations, which eliminate the phage killing functions and render the prophage defective, the prophage DNA can be excised at a low frequency. The excision sometimes is precise (causing the host mutant to revert to wild type) but mostly it is imprecise. Precise excision can be detected at a frequency of 10^{-6}-10^{-8}, whereas imprecise excision is observed at a frequency of 10^{-5}-10^{-7}. Imprecise excision generally involves removal of Mu DNA accompanied by deletion of host sequences. The deletions left after Mu excision are of various sizes and frequently span both sides of the prophage. Some deletions appear to be very small, covering not more than 100 base pairs (H. Khatoon and A. I. Bukhari, unpubl.). It should be noted that this process of deletion formation produces a different end result than the two modes outlined above. Mu DNA is either completely or partially deleted as a result of Mu *X* excision, whereas a complete copy of a Mu prophage is retained at the site of a deletion generated during lysogenization of cells with Mu or after induction of a prophage.

A Model for Interaction of Mu DNA and Host DNA

In discussing the mechanism of Mu DNA-host DNA interaction, the following points deserve particular attention: (1) Mu DNA has always been found to be associated with host DNA. No intracellular form of Mu DNA free of host DNA has been detected, and no excision of prophage DNA after induction has been demonstrated. (2) Mu DNA is directly involved in heterogeneous circle formation and in transpositions and insertions of non-Mu DNA that occur during the lytic cycle or during lysogenization of cells by Mu. The transposed or inserted DNA is always linked to the ends of Mu DNA. (3) Mu-mediated integration of non-Mu DNA by a partially derepressed prophage involves duplication of the prophage and insertion of non-Mu DNA between the two Mu copies. (4) Since Mu DNA is packaged from maturation precursors that contain host DNA linked to Mu DNA, integration of Mu is a necessary step in the lytic cycle of Mu. No plaque-forming mutants defective in integration have been isolated (M. Howe, pers. comm.). Only the *A*-gene mutants have been found defective in functions related to integration, and these mutants are also defective in DNA replication (Wijffelman et al. 1974; Bukhari 1975; Faelen et al. 1975).

The above considerations imply that Mu DNA is replicated in situ and that the copies react with host DNA to produce various illegitimate recombinational events. These inferences are consistent with the findings of E. Ljungquist and A. I. Bukhari (see Bukhari et al., this volume). Faelen and coworkers (Faelen et al. 1975; Faelen and Toussaint 1976) have proposed a model for Mu DNA-host DNA interaction which is based on the duplication of prophage DNA, followed by reaction between the ends of the Mu genome and host DNA. This model, with a slight modification, is shown in Figure 1. The Mu DNA ends that react with each other and with host DNA are shown by solid triangles, and the host sites that react with Mu DNA are shown by solid squares. The model is elaborated below.

The mechanism by which linear phage DNA is inserted into host DNA (steps 1 and 2) is not known. Bukhari et al. (this volume) have postulated that this insertion in some manner involves replication of phage DNA. Once the phage DNA is inserted, it can behave as any other prophage DNA. The primary difference between Mu infection of cells and induction of a Mu lysogen is that, in infection, every infected bacterium in the culture would have Mu inserted at a different location, whereas, in a monolysogen, the original prophage has a unique location. Step 3 in Figure 1 shows duplication of the prophage DNA. Duplication of phage DNA as shown in the figure can result from replication

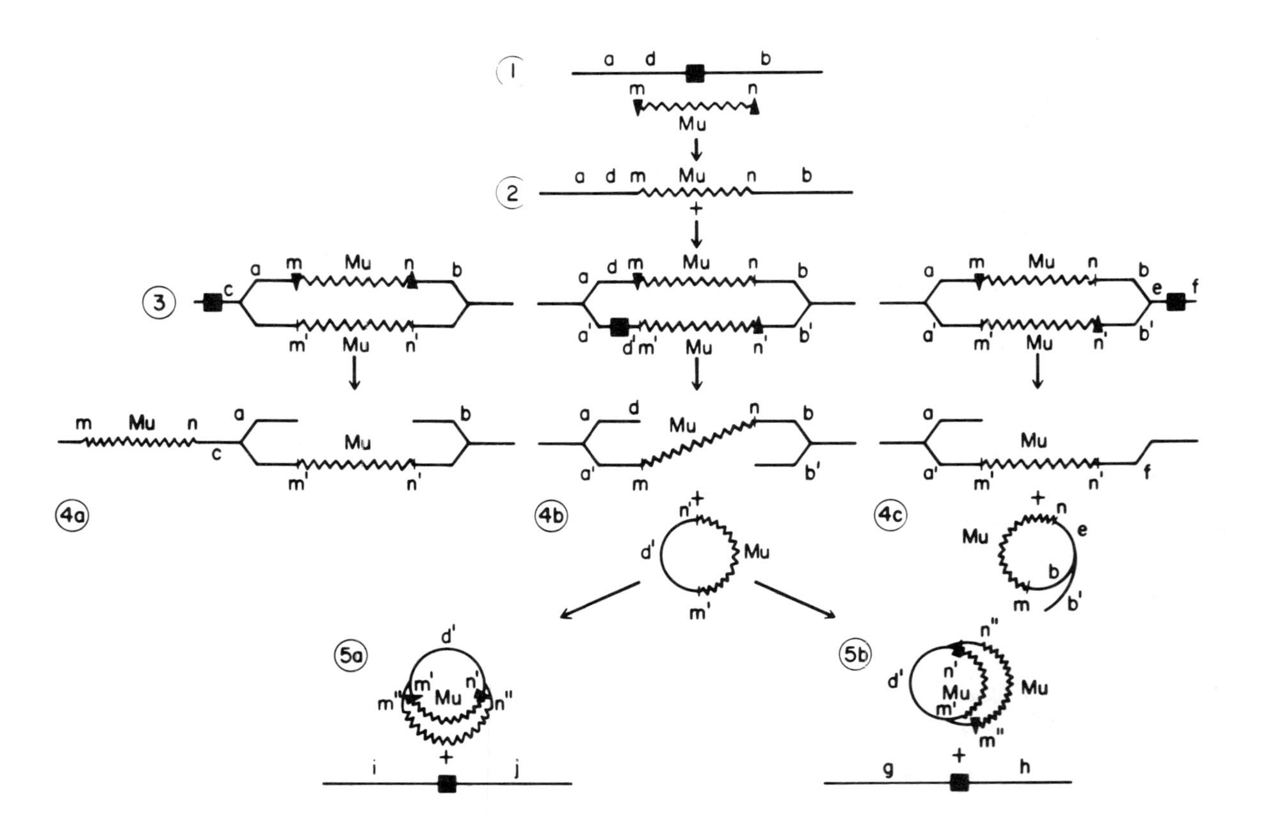
1
2
3
4a
4b
4c
5a
5b
Mu

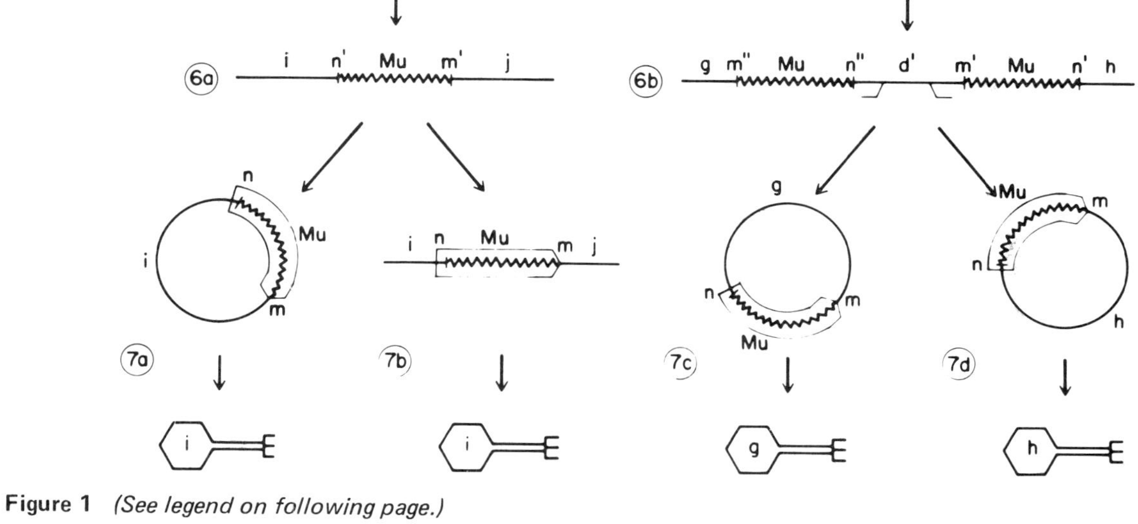

Figure 1 *(See legend on following page.)*

Figure 1 *(See preceding pages for figure.)*

A model for interaction of Mu DNA and host DNA. (*a–j*) Sequences of the bacterial chromosome; (*m, n*) specific sequences at the extremities of Mu DNA. In this scheme, *m* and *n* lie at the *c* end and the *S* end of Mu DNA, respectively. Recombination sites at Mu ends are shown by solid triangles (▲) and those in the host DNA by solid squares (■).

Steps:

(*1–2*) Mu DNA is inserted into host DNA after infection through unknown intermediate steps.

(*3*) The DNA which contains the Mu genome has been duplicated. In this figure, the host sequences adjacent to the prophage are also shown to be duplicated. There is no evidence that replication initiated in Mu DNA can proceed into host DNA. The duplication of host sequences might occur because of replication of the bacterial chromosome. However, the recombinational events postulated here may be related to the Mu-specific duplication of the Mu genome, and bacterial chromosome replication might not be directly involved in these events. Thus the replication of host sequences is not assumed to be mandatory for the recombinational events shown in the figure. The only requirement assumed is generation of "active" *m* and *n* sequences. The *m* and *n* sequences, normally activated by Mu DNA replication, might in some cases be activated by host DNA replication spanning prophage ends.

(*4*) One Mu copy has transposed near the *c* sequence of the bacterial chromosome (*a*). The other copy remains at the original position. (*b*) The *m* and n' Mu sequences and a point of the host chromosome located between a' and d' undergo recombination. This recombination generates a ring containing one Mu and the host sequence d'. The bacterial chromosome still carries one whole Mu. If it is properly repaired, it contains a deletion of d' adjacent to the prophage. (*c*) The m, n', and a point of the host DNA between the *e* and *f* sequences undergo recombination. This generates a rolling circle which contains one Mu, *b*, and b'. If properly repaired, the chromosome contains a deletion of b' and *e* adjacent to the retained Mu prophage.

(*5–6*) Mu DNA in the circle is duplicated (*5a*). This step should require a functional Mu replication system. One copy of Mu is transposed between *i* and *j* bacterial sequences as shown in *6a*. (*5b*) One end of each Mu genome, shown as m″ and n′, and a point of the host DNA located between *g* and *h* undergo an integrative recombination (a process similar to that shown in *4b*). As shown in *6b*, the d' fragment is thus transposed, surrounded by two Mu genomes similarly oriented, between *g* and *h*. The two Mu prophages can further generate a ring with one Mu and *g* or *h* or a circle containing two Mu's, d', and *g* or *h* (not shown).

(*7*) A few circles which can be generated by the steps described above. Maturation of phage DNA from these circles or from molecules carrying Mu inserted in different host sequences would result in mature virions with different bacterial sequences (indicated by letters in phage heads).

of the bacterial chromosome. Mu-specific replication will result in duplication of prophage DNA, but there is no evidence that replication can proceed into host DNA. Although the figure shows replication of host sequences along with phage DNA, only the phage DNA replication is the main requirement of the model. Once the phage DNA is replicated, it can react in various ways. Step 4a in the figure shows that one copy of Mu can be transposed to a different site. One copy of the prophage, however, remains at the original location. Step 4b shows that the ends of two different prophages react with a site in the host DNA releasing a circle carrying host DNA as well as one copy of Mu DNA. This type of reaction may result in a "rolling circle" type of structure with a host DNA tail (step 4c). One copy of Mu remains on the chromosome and is linked at one side to a deletion of the host sequence, now carried in a circle. The Mu DNA in a circle shown at step 4b may again replicate, one copy of which might be reinserted (transposition of Mu DNA–steps 5a-6a), or both copies of Mu DNA may react with a site in host DNA in such a manner that both Mu genomes are reinserted with the bacterial DNA in between (Mu-mediated transposition–steps 5b-6b). Note that the transposed bacterial DNA is flanked by Mu prophages in the same orientation. The inserted Mu DNA at steps 6a and 6b may be packaged into virions or may again be replicated and give rise to heterogeneous circles that might act as maturation precursors (steps 7a-7d).

In the scheme presented in Figure 1, Mu-mediated integration of circular non-Mu DNA would occur at a stage where Mu DNA has been duplicated. This is shown in Figure 2. Interaction of one end of one Mu copy, the opposite end of the second copy, and non-Mu DNA would result in the integration of non-Mu DNA between two Mu prophages similarly oriented. The Mu-DNA-Mu structure is located at the site previously occupied by the single Mu genome.

The model discussed above was designed primarily to explain Mu-mediated transpositions and insertions and the formation of heterogeneous circles. It is consistent with the general notion that the integrative precursor of Mu is generated by replication of Mu DNA (Bukhari et al., this volume). However, the model is based mainly on the reactivity of the ends and not on Mu DNA replication per se. The exact role of Mu DNA replication in the processes outlined above remains to be determined.

SUMMARY

Integration of Mu DNA into host DNA occurs continuously during the lytic cycle of bacteriophage Mu. The Mu integration system promotes a series of illegitimate recombinational events during Mu growth. Heterogeneous circles containing Mu DNA and different segments of host DNA are formed, and the host genes are translocated from one site to another. Mu can also mediate insertion of circular non-Mu DNA molecules which are unable to insert by themselves into the host chromosome. Mu DNA is an active participant in

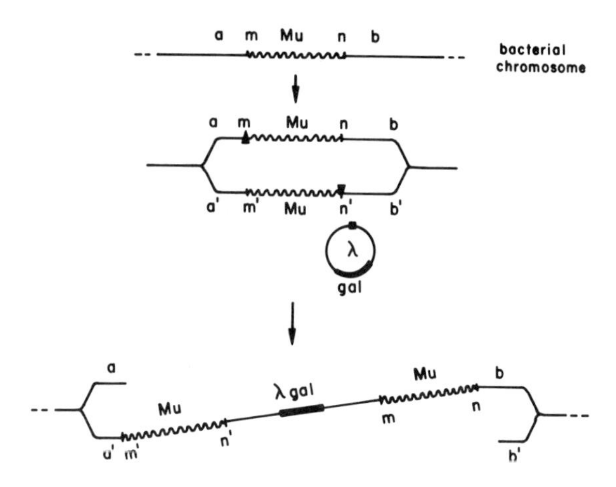

Figure 2
Mu-mediated integration of circular non-Mu DNA. As in Fig. 1, a Mu prophage has been duplicated. Mu sequences *m* and *n′* interact with a random point of the circular non-Mu DNA (solid square [■] in λ*gal* circle) for integrative recombination. The non-Mu DNA is integrated between two Mu prophages which have the genotype and the orientation of the original single Mu. The broken host sequences might be repaired or degraded by host enzymes.

these recombinational events. Heterogeneous circles apparently always contain Mu DNA, and the DNA segment transposed or inserted by Mu is always flanked by two Mu prophages in the same orientation. A model has been proposed to explain the role of Mu DNA in these events. The model postulates that Mu DNA, inserted in the host chromosome, is first replicated in situ, giving rise to duplicated Mu genomes. The ends of the phage genomes can then react with each other and with host DNA to generate various genetic rearrangements.

Acknowledgments

We thank J. de Lafonteyne, O. Huisman, M. Couturier and F. van Vliet for stimulating discussions.

REFERENCES

Bukhari, A. I. 1975. Reversal of mutator phage Mu integration. *J. Mol. Biol.* **96**:37.

——. 1976. Bacteriophage Mu as a transposition element. *Annu. Rev. Genet.* **10**:389.

Bukhari, A. I. and A. L. Taylor. 1975. Influence of insertions on packaging of host sequences covalently linked to bacteriophage Mu DNA. *Proc. Nat. Acad. Sci.* **72**:4399.

Bukhari, A. I. and D. Zipser. 1972. Random insertion of Mu-1 DNA within a single gene. *Nature New Biol.* **236**:240.

Cabezon, T., M. Faelen, M. De Wilde, A. Bollen and R. Thomas. 1975. Expression of ribosomal protein genes in *E. coli. Mol. Gen. Genet.* **137**:125.

Couturier, M. 1976. The integration and excision of bacteriophage Mu-1. *Cell* **1**:155.

Daniell, E., R. Roberts and J. Abelson. 1972. Mutations in the lactose operon caused by bacteriophage Mu. *J. Mol. Biol.* **69**:1.

Faelen, M. and A. Toussaint. 1976. Mu-1: A tool to transpose and to localize bacterial genes. *J. Mol. Biol.* **104**:525.

Faelen, M., A. Toussaint and M. Couturier. 1971. Mu-1 promoted integration of a λ-*gal* phage in the chromosome of *E. coli* K12. *Mol. Gen. Genet.* **113**:367.

Faelen, M., A. Toussaint and J. de Lafonteyne. 1975. A model for the enhancement of λ-*gal* integration into partially induced Mu-1 lysogens. *J. Bact.* **121**: 873.

Howe, M. and E. G. Bade. 1975. Molecular biology of bacteriophage Mu. *Science* **190**:624.

Hsu, M.-T. and N. Davidson. 1974. Electron microscope heteroduplex study of the heterogeneity of Mu phage and prophage DNA. *Virology* **58**:229.

Parker, V. and A. I. Bukhari. 1976. Genetic analysis of heterogeneous circles formed after prophage Mu induction. *J. Virol.* **19**:756.

Razzaki, T. and A. I. Bukhari. 1975. Events following prophage Mu induction. *J. Bact.* **122**:437.

Schroeder, W. and P. van de Putte. 1974. Genetic study of prophage excision with a temperature inducible mutant of Mu-1. *Mol. Gen. Gent.* **130**:99.

Schroeder, W., E. G. Bade and H. Delius. 1974. Participation of *E. coli* DNA in the replication of temperate bacteriophage Mu-1. *Virology* **60**:534.

Taylor, A. L. 1963. Bacteriophage induced mutations in *E. coli. Proc. Nat. Acad. Sci.* **50**:1043.

Toussaint, A. 1969. Insertion of phage Mu-1 within prophage λ: A new approach for studying the control of the late functions in bacteriophage λ. *Mol. Gen. Genet.* **106**:89.

Toussaint, A. and M. Faelen. 1973. Connecting two unrelated DNA sequences with a Mu dimer. *Nature New Biol.* **242**:1.

van de Putte, P. and M. Gruijthuijsen. 1972. Chromosome mobilization and integration of F factors in the chromosome of a *recA* strain of *E. coli* under the influence of bacteriophage Mu-1. *Mol. Gen. Genet.* **118**:173.

Waggoner, D. T., N. S. Gonzales and A. L. Taylor. 1974. Isolation of heterogeneous circular DNA from induced lysogens of bacteriophage Mu-1. *Proc. Nat. Acad. Sci.* **71**:1255.

Wijffelman, C., M. Gassler, W. Stevens and P. van de Putte. 1974. On the control of transcription of bacteriophage Mu-1. *Mol. Gen. Genet.* **131**:85.

On the *kil* Gene of Bacteriophage Mu

P. van de Putte, G. Westmaas, M. Giphart and C. Wijffelman
Department of Biochemistry
University of Leiden
Leiden, The Netherlands

Induction of *E. coli* cells lysogenic for a thermoinducible Mu prophage leads to the death of the cells as a result of replication and maturation of the phage. However, even in the absence of replication and expression of late genes, as is the case upon induction of an *A* amber (*Aam*) or *B* amber (*Bam*) lysogen, the host cells can be killed (Wijffelman et al. 1974). After induction of a defective Mu prophage in which the β end is deleted, only early Mu functions are expressed, but the cells are killed (Wijffelman et al. 1974; Wijffelman and van de Putte 1974; Westmaas et al. 1976). Mu-specific DNA synthesis does not take place after induction of such defective prophages or after induction of prophages carrying *Aam* or *Bam* mutations. These observations imply that an early Mu gene encodes a product that can kill the host cells. This gene is designated the *kil* gene.

Although Mu *Aam* or *Bam* lysogens are killed upon prophage induction, the *Aam* and *Bam* phage mutants fail to kill the sensitive cells after infection. This indicates that the *kil* gene product acts on a prophage. There is some evidence suggesting that the killing of host cells might be due to single-strand damage near the left end, the *c* end, of Mu prophage. There is a difference in the kinetics of killing of $PolA^+$ and $PolA^-$ cells by thermoinducible defective prophages in which the right prophage end is deleted. As shown in Figure 1, in the Pol^+ cells the damage, presumably caused by the *kil* gene product, can be repaired during the first 15 minutes after induction, whereas in $PolA^-$ cells the damage is irreversible. To analyze further the *kil* gene of Mu, we have mapped the *kil* gene and isolated Kil^- mutants.

Localization of the *kil* Gene

With Defective Prophages

Two thermoinducible defective prophages, deleted from the β end, have been described. Both can complement *A* and *B* amber mutants after induction but

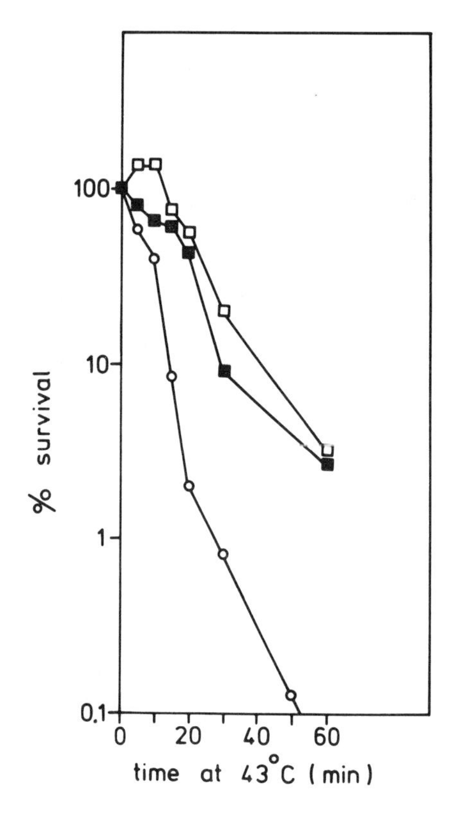

Figure 1
Survival of different Mu lysogens in which the *kil* gene is expressed at 43°C. Cells were grown to a density of 3 X 10^8 cells/ml at 32°C; at time zero, the cells were transferred to 43°C. Samples were withdrawn after different times of incubation at 43°C, chilled in ice, and survivors measured at 32°C on tryptone agar plates.
(○—○) KMBL 1617, a strain which contains a complete Mu *cts* prophage in the *trp* operon; (□—□) KMBL 1672, isogenic with 1617, except that the Mu prophage is deleted from the β end by a *tonB* deletion; the α segment (from the left end) to the *S* gene is still present; (■—■) PP316 isogenic PolA$^-$ derivative of KMBL 1672.

they lack the genes *C*[1] to *S*. One type kills the cell after induction (Westmaas et al. 1976), whereas the other does not (Howe 1973). This strongly suggests that the *kil* gene is located between genes *B* and *C*.

With λ-Mu Hybrids

The defective λ-Mu hybrids isolated by Boram and Abelson (1973) contain the bacterial gene *pgl* and different portions of the early region of Mu. To determine whether the *kil* gene was located on the λ*pgl*-Mu hybrids, they were used for quantitative Pgl$^+$ transduction to a Mu-sensitive and a Mu-immune recipient. The hybrid phages fell into two classes: in one class, the transduction frequency was a factor of 20 lower in the Mu-sensitive recipient than in the Mu-immune recipient. Phages in this last class were considered to possess the Mu *kil* gene. The *kil* gene was mapped between genes *B* and *C*. The plaque-forming λ-Mu hybrids were also used in mapping experiments (Bukhari and Allet 1975). With these phages, it was more difficult to determine whether or not the *kil* gene of Mu was present since λpMu's do not carry a bacterial gene that can be used in transduction experiments. Therefore, the presence or absence of the *kil* gene

[1] Gene *C* lies to the right of gene *B* and should not be confused with the immunity gene *c* located at the left end of the Mu genome.

was determined in the following way: Mutants have been isolated which have an insertion in the early region of Mu and which survive temperature induction at 43°C (Bukhari 1975). In these so-called *X* mutants, the expression of the *kil* gene is blocked. A strain lysogenic for Mu *cts X* and for bacteriophage λ was induced at 43°C and simultaneously infected with different λpMu hybrids. In these λ-immune cells, only the Mu part of the hybrid is expressed. After the infection, the survival of the cells was measured. The λpMu's that did kill the cells were considered to be Kil^+. With this method also, the *kil* gene was mapped between genes *B* and *C*, confirming the mapping data obtained with the defective prophages and the defective λ-Mu hybrids.

The Function of the *kil* Gene

Some information on the function of the *kil* gene was derived from transfection experiments with Mu DNA using Ca^{++}-treated *E. coli* cells (Cosloy and Oishi 1973). Even in $RecBC^-$, $SbcB^-$ cells, which are transformed with linear coli DNA at a higher efficiency than normal cells, the transfection with linear Mu DNA is extremely low. One possibility is that, on normal Mu infection, a protein is injected with Mu DNA, and that this protein is essential for the protection of linear Mu DNA. Assuming that such a protein is an early protein, we performed the following experiment: $RecBC^-$, $SbcB^-$ cells lysogenic for a thermoinducible defective Mu prophage were induced for 20 minutes at 43°C and subsequently subjected to the normal transfection procedure with Mu DNA. At the time of transfection, the cells would contain early Mu proteins. Under these conditions, a strong stimulation of transfection (factor of 100) was found (Table 1). To test whether the products of the early genes *A* or *B* of Mu were responsible for the stimulation of transfection, a nonpermissive *recBC,sbcB* strain was made lysogenic for Mu *cts Aam* or *Bam*. These strains gave nearly the same stimulation of transfection, indicating that the *A* and *B* gene products are not involved. Since amber mutants of other early Mu functions are not available, the transfection-stimulating factor was mapped with the λ-Mu hybrids, which have por-

Table 1
Influence of Early Mu Proteins on the Frequency of Mu Transfection

Strain	Transfection frequency (infective centers/no. of bacteria)
JC7620 (*recBC,sbcB*)	4.0×10^{-7}
JC7620, *trp*::(Mu *cts*62 $\Delta D \ldots \beta$)	3.5×10^{-5}
JC7620, *trp*::(Mu *cts*62 *Bam*)	3.0×10^{-5}
JC7620, *trp*::(Mu *cts*62 *Aam*)	2.0×10^{-5}

Experimental conditions: Cells were induced for 20 min at 43°C and treated with a high concentration of Ca^{++} (100 mmoles). Mu DNA (15 μg/ml) was taken up during 10 min at 42°C. Infective centers were measured on an *E. coli* K12 indicator strain. $\Delta D \ldots \beta$ means a Mu deletion which has removed the phage DNA from gene *D* to the right end, the β end of the prophage.

tions of different lengths of the early region of Mu. The transfection procedure was changed as follows: The RecBC⁻, SbcB⁻ cells, which were made lysogenic for λ, were infected with the different λ-Mu hybrids prior to the Ca^{++} treatment. Under these conditions, only the Mu functions are expressed. The results (Table 2) show that all of the λ-Mu hybrids that contain the *kil* gene stimulate transfection by Mu DNA, whereas hybrids without the *kil* gene either cause no stimulation or do so to a much lesser extent. This strong correlation between the presence of the *kil* gene product and the stimulation of transfection with Mu DNA was further confirmed by experiments with *kil* mutants (see below, "Isolation of Kil^- Mutants"). The stimulation of transfection appears to be specific for Mu DNA. Transformation with coli DNA is not affected by the *kil* gene product.

In transfection experiments, the Mu DNA used carried a kil^+ gene. Since the kil^+ gene of a defective prophage is required for high-level transfection, it means that the presence of the *kil* gene product is needed at an early step of Mu development before the kil^+ gene of the transfecting DNA is expressed. The *kil* gene may produce a product that protects the linear Mu DNA against nucleolytic breakdown or a product that plays a role in the attachment of the Mu DNA to the bacterial membrane. For this reason, we hypothesize that the *kil* gene product is an internal Mu protein which is injected with the DNA upon normal Mu infection.

Identification of the *kil* Gene Product

The λ-Mu hybrids used in transfection experiments were also used to characterize the *kil* gene product. The *kil* protein was identified by gel electrophoresis,

Table 2

Localization on the Mu Genome of Transfection-stimulating Factor with λ-Mu Hybrids

Helper phage	Transfection frequency (infective centers/no. of bacteria)
λd*pgl*Mu*c*	4.0×10^{-7}
λd*pgl*Mu*cA*	4.0×10^{-7}
λd*pgl*Mu*cAB*	1.0×10^{-6}
λd*pgl*Mu*cABkil*	2.0×10^{-5}
λd*pgl*Mu*cABkilC*	2.5×10^{-5}
λpMu*ctsA*(96–49)	2.0×10^{-6}
λpMu*ctsAB*(98–96)	5.0×10^{-6}
λpMu*ctsABkil*(102)	6.0×10^{-5}
λpMu*ctsABkilClys*(504)	7.0×10^{-5}

Experimental conditions: cells of JC7620 (λ) were infected with different helper phages for 20 min and subsequently treated with a high concentration of Ca^{++} (100 mmoles). Mu DNA (15 μg/ml) was taken up during 10 min at 42°C. Infective centers of Mu were measured on an *E. coli* K12 (λ) indicator strain.

as described by Hendrix (1971) for several λ proteins. The cells used in these experiments were lysogenic for λ*ind* to prevent expression of λ genes after irradiation and infection. The cells were UV-irradiated with a dose of 4000 erg/mm^2 and divided in two portions. One portion was infected with a λ-Mu hybrid containing the *kil* gene and labeled with [^{14}C] leucine, the other was infected with the hybrid lacking the *kil* gene and labeled with [^{3}H] leucine for 20 minutes. The cells were chilled and the two different portions mixed. The proteins were then analyzed by electrophoresis in a sodium dodecyl sulfate-polyacrylamide gel.

An example of such a dual-labeling experiment is shown in Figure 2. One polypeptide is made by the λ-Mu hybrid containing the *kil* gene, whereas it is absent in the λ-Mu Kil$^-$ hybrid. This was found both for the plaque-forming and for the defective λ-Mu hybrids. We infer from these experiments that the *kil* gene product is a polypeptide of approximately 23,000 daltons molecular weight.

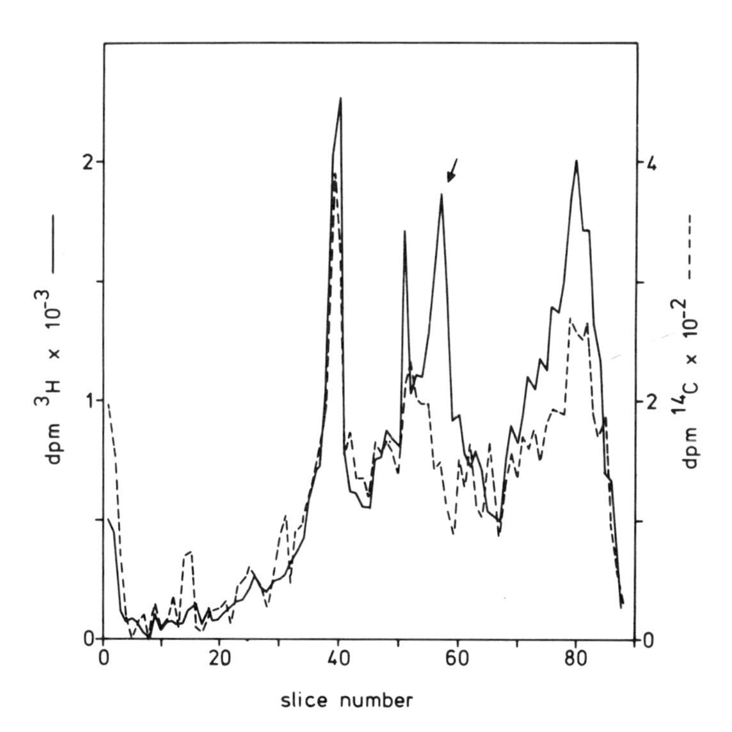

Figure 2

Dual-labeling experiment with λpMu hybrids containing and lacking the *kil* gene. The position of the *kil* gene product is indicated by the arrow.
(------) λpMu*ctsAB* (98-96); (———) λpMu*ctsABkil* (102).

Isolation of Kil⁻ Mutants

For the isolation of Kil$^-$ mutants, the simplest procedure would be to examine the temperature-resistant survivors of a Mu *cts* lysogen; however, the *kil* gene product is probably not the only killing function of Mu. The replication of Mu DNA and the expression of some of the late genes might also lead to cell death. An indication that, for survival at 43°C, both the *kil* function and replication have to be blocked comes from the isolation of *X* mutants, which are found among the temperature-resistant survivors of Mu *cts* lysogens (Bukhari 1975). In this type of mutant, the insertion of an IS*1* in the early region of Mu prevents the expression not only of the *kil* gene but also of gene *B*. The *B* gene apparently is involved in integration, replication, and late RNA synthesis (van de Putte and Gruijthuijsen 1972; Faelen and Toussaint 1973; Wijffelman et al. 1974). To reduce the effects of Mu DNA replication, Kil$^-$ mutants were sought from a strain lysogenic for a thermoinducible defective prophage located in the *trp* operon. The prophage was deleted from the *β* end and lacked gene *S*. After induction of this strain, no Mu-specific replication or late RNA synthesis occurs. By plating this defective lysogen at 43°C, several mutants with different characteristics were found. Among these were Kil$^-$ mutants having the following properties: (1) they had retained a temperature-sensitive immunity for Mu; (2) they were able to complement *A* and *B* mutants; and (3) they had lost the ability to help λFec$^-$ mutants to grow in RecA$^-$ cells.

The last point stems from the observation that induction of Mu *cts* prophage enables λFec$^-$ phages (Manly et al. 1969) to grow in a RecA$^-$ background (F. van Vliet, pers. comm.). To establish which Mu function is involved, we tested a number of Mu mutants. Strains lysogenic for Mu *cts*62 Δ(*C*-*β*), Mu *cts Aam*, and Mu *cts Bam* were all able to rescue λFec$^-$ phages, indicating that an early Mu function is involved but not the genes *A* and *B*. Only a Mu *cts X* lysogen failed to help the λFec$^-$ phages to grow. Since, in this strain, the expression of the *kil* gene as well as that of the *B* gene is blocked, it is likely that the *kil* gene product is responsible for the growth of λFec$^-$ phages in a RecA$^-$ background.

These strains were used for transfection experiments. The defective prophages with the *kil* mutations were transferred to a RecBC$^-$, SbcB$^-$ strain by P1 transduction, and it was found that the ability to stimulate the transfection with Mu DNA was reduced by a factor of 20 to 30. Therefore, we infer that the *kil* gene product has the following properties: (1) it is responsible for the killing of the host; (2) it can stimulate transfection with Mu DNA; and (3) it helps λFec$^-$ phages to grow in RecA$^-$ strains.

SUMMARY

One early Mu gene can cause killing of the host cells. This gene is located between genes *B* and *C* of Mu, as shown by mapping experiments with defective Mu prophages and with λ-Mu hybrid phages.

The presence of the *kil* gene product in the cell stimulates strongly the frequency of transfection with Mu DNA. Moreover, the presence of the *kil* gene product allows the propagation of λFec^- phages in $RecA^-$ strains.

Lambda-Mu hybrids carrying the *kil* gene produce a polypeptide with a molecular weight of 23,000 daltons; this polypeptide is not made by λ-Mu hybrids lacking the *kil* gene.

Several Kil^- mutants have been isolated.

Note Added in Proof

Recent experiments suggest that the three functions which are attributed in this paper to the *kil* gene product, namely (1) killing of the host cell, (2) allowing the growth of λFec^- mutants in a RecA background, and (3) stimulation of transfection of Mu DNA, might be due to the activity of a cluster of genes located between the genes *B* and *C*. This is inferred from analyses with restriction enzymes of the Kil^- mutants lacking these functions which show that they all contain polar insertions (IS*1* type).

REFERENCES

Boram, W. and J. Abelson. 1973. Bacteriophage Mu integration: On the orientation of the prophage. *Virology* **54**:102.

Bukhari, A. I. 1975. Reversal of mutator phage Mu integration. *J. Mol. Biol.* **96**:87.

Bukhari, A. I. and B. Allet. 1975. Plaque-forming λ-Mu hybrids. *Virology* **63**:30.

Cosloy, S. D. and M. Oishi. 1973. Genetic transformation in *Escherichia coli* K12. *Proc. Nat. Acad. Sci.* **70**:84.

Faelen, M. and A. Toussaint. 1973. Isolation of conditional defective mutants of temperate phage Mu-1 and deletion mapping of the Mu-1 prophage. *Virology* **54**:117.

Hendrix, R. W. 1971. Identification of proteins coded in phage lambda. In *The bacteriophage lambda* (ed. A. D. Hershey), p. 355. Cold Spring Harbor Laboratory, Cold Spring Harbor, New York.

Howe, M. M. 1973. Prophage deletion mapping of bacteriophage Mu-1. *Virology* **54**:93.

Manly, K. F., E. R. Signer and C. M. Radding. 1969. Non-essential functions in bacteriophage λ. *Virology* **37**:177.

van de Putte, P. and M. Gruijthuijsen. 1972. Chromosome-mobilization and integration of F-factors in the chromosome of RecA strains of *E. coli* under the influence of bacteriophage Mu-1. *Mol. Gen. Genet.* **118**:173.

Westmaas, G. C., W. L. van der Maas and P. van de Putte. 1976. Defective prophages of bacteriophage Mu. *Mol. Gen. Genet.* **145**:81.

Wijffelman, C. and P. van de Putte. 1974. Transcription of bacteriophage Mu. *Mol. Gen. Genet.* **135**:327.

Wijffelman, C., M. Gassler, W. F. Stevens and P. van de Putte. 1974. On the control of transcription of bacteriophage Mu. *Mol. Gen. Genet.* **131**:85.

Bacteriophage Mu Genome: Structural Studies on Mu DNA and Mu Mutants Carrying Insertions

L. T. Chow and A. I. Bukhari
Cold Spring Harbor Laboratory
Cold Spring Harbor, New York 11724

The genome of the temperate bacteriophage Mu appears as a linear, double-stranded DNA molecule upon disruption of mature phage particles. It has been estimated from electron microscope measurements that the Mu DNA duplex is about 36-38 kilobases (kb) in length, corresponding to a molecular weight of about 25×10^6 (Martuscelli et al. 1971; Daniell et al. 1973b). Structural analyses in several laboratories have revealed that Mu DNA has some highly interesting features. In this paper, we will review briefly the structural organization of Mu DNA and describe electron microscope studies on the invertible segment of the Mu genome and on insertions in the *X* mutants of Mu.

Main Features of Mu DNA

Both ends of Mu DNA have heterogeneous DNA of variable length. When a population of Mu DNA molecules is denatured and the separated strands are allowed to reanneal, the simple linear duplexes are not regenerated. Instead, one end of almost every molecule is split into single-stranded tails, as visible in the electron microscope (Daniell et al. 1973b; the discovery was also made independently by Delius and Bade, see Bade 1972). The so-called split ends in renatured molecules, diagrammed in Figure 1, are always located at the right end, the *S*-gene end, of Mu DNA (Daniell et al. 1973a). The lengths of the single-stranded tails vary, ranging from 0.5 kb to 3.2 kb, with a mean of 1.5 kb (Daniell et al. 1973b; Bukhari and Taylor 1975). The split ends are seen whether Mu particles are grown from a single plaque or from an induced lysogen carrying a single prophage. Thus the heterogeneous sequences represented by split ends appear to be randomized during the growth cycle of Mu. Figure 1 also shows a nonrenaturable region located internally in Mu DNA, near the split ends in some of the renatured molecules. The noncomplementary region, visible in the electron microscope, is found primarily in renatured DNA molecules of Mu particles prepared by induction. Mu particles prepared by infection show the bubble at a frequency of less than 1%. The DNA sequences comprising the bubble have been termed the G segment (Hsu and Davidson 1974).

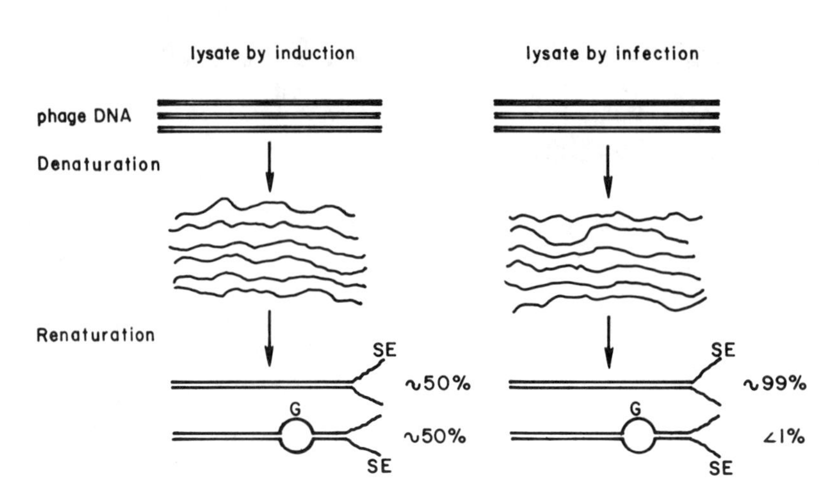

Figure 1
Structures observed in the electron microscope after denaturation and renaturation of Mu DNA extracted from mature particles. The relative frequencies of molecules with and without the G bubble are shown next to the typical structures. (Adapted from Bukhari 1976.)

In contrast to the right end (*S* end) of Mu DNA, the left end (*c* end) shows no prominent split ends in renatured Mu DNA molecules. However, Allet and Bukhari (1975) found that this end also contains heterogeneous DNA and varies in length by about 100 base pairs. If T4 gene-32 protein is used for binding to single-stranded DNA, very small split ends at the *c* end can be seen in renatured Mu DNA (H. Delius, pers. comm.).

Heterogeneity of Mu ends results from the presence of host sequences that differ from molecule to molecule. Renaturation kinetics of Mu DNA in the presence of *E. coli* DNA indicated that *E. coli* sequences were present in Mu DNA (Daniell et al. 1975). The presence of heterogeneous sequences at Mu ends has now been confirmed by direct DNA-DNA hybridization experiments (Bukhari et al. 1976; see Bukhari et al., this volume). Figure 2 presents a composite picture of the Mu genome as developed from the studies described above. The DNA has been divided into the α, G, and β segments. The α segment has a mean length of 30.7 ± 1.1 kb, G has a mean length of 2.9 ± 0.4 kb, and β has a mean length of 1.6 ± 0.1 kb (Bukhari and Taylor 1975). The left end of α (the *c* end of Mu DNA) has about 100 host base pairs, and the right end of β (the *S* end of Mu DNA) has, on the average, approximately 1.5 kb of host DNA. The host DNA at the *S* end is sometimes referred to as the SE segment. All the known essential genes of Mu are located in the α segment.

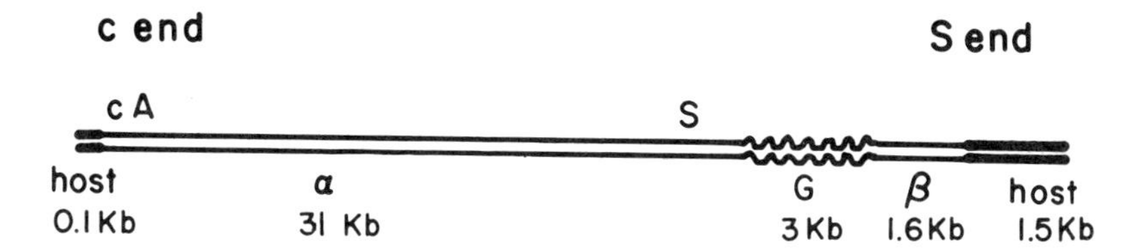

Figure 2
Structure of Mu DNA. The approximate sizes of the different segments are given in kilobases. The segments are not drawn to scale. Average values are given for host sequences (which vary in length) at the ends. The known essential genes of Mu (*A* through *S*) are located in the α segment. The *c* gene at the end is the immunity gene.

The G Segment

Structure of G

The occurrence of the G bubble in some molecules that have been denatured and renatured is attributed to the presence of the G segment in opposite orientations with respect to α and β (Hsu and Davidson 1974). When Mu DNA strands carrying G in opposite orientations renature, the G segment cannot pair and is seen as a noncomplementary inversion loop. The G segment is flanked by inverted repetitious sequences that are probably less than 50 nucleotide pairs in length (Hsu and Davidson 1974). These inverted repetitions are observed as a reproducible strand crossover when the single-stranded DNA is mounted for electron microscopy under moderately denaturing conditions. Within the middle 1 kb of G, there is another pair of inverted repetitions of lower stability. These inverted sequences give rise to inter- and intrastrand secondary structures when the single-stranded or self-renatured DNA is mounted on EM grids under mildly denaturing conditions. The structure of G is diagrammed in Figure 3.

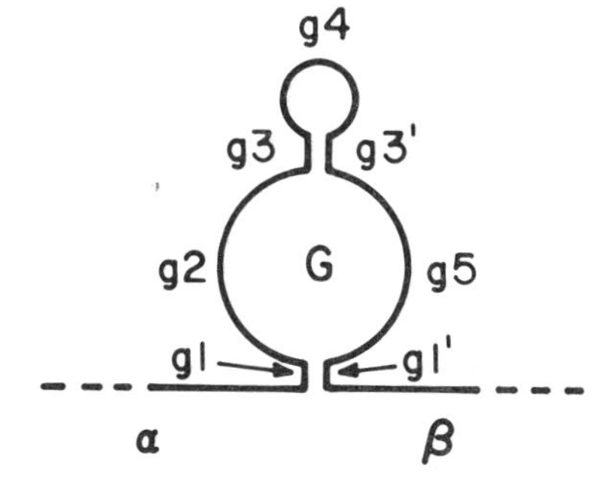

Figure 3
Structure of the G segment. A single-stranded structure generated by the part of Mu DNA containing G is shown. g1 and g1′ are the inverted repetitions at the ends of G. g3, g3′, and g4 represent the internal stem-loop structure. The sum of g3, g4, and g3′ is about 1 kb, or one-third the entire G sequence. The individual lengths of g3 (g3′) and g4 are variable, depending upon the spreading conditions employed for electron microscopy. g2 and g5 are each about 1 kb in length. (Adapted from Chow and Bukhari 1976.)

Inversion of G

The inversion of the G segment is thought to occur by recombination between inverted repetitions at the ends of G. The recombination occurs in Mu prophages and is independent of the *E. coli recA, B,* and *C* systems (Hsu and Davidson 1974; Chow et al. 1977 and this volume). Experiments with plaque-forming λ-Mu hybrids carrying the G segment of Mu have indicated that the gene for inversion function is located within or close to G (Allet and Bukhari 1975). Chow et al. (this volume) have described mutants in which parts of the G and β regions of Mu are deleted. In one of the mutants, only the β segment is affected by the deletion, but the mutant, called the *gin*$^-$ mutant, is unable to invert the G segment. Thus the β region either encodes or controls the G inversion function.

As shown in Figure 1, the inversion is detected primarily in Mu particles grown by induction of Mu lysogens. The reason why Mu particles grown by infection contain primarily the same G orientation is not clearly understood. D. Kamp and R. Kahmann (pers. comm.) have found that lysogens of Mu *cts* gram-negative mutants yield defective phage particles, if the G segment is frozen into the orientation opposite to the lytic orientation. These particles do not form plaques on Mu-sensitive hosts. Thus, although prophages with either G orientation appear to replicate upon induction, only the particles with the lytic orientation of G might be able to infect the sensitive cells productively. This selection of G orientation after infection would result in accumulation of Mu particles with the lytic orientation, assuming that the process of G inversion is not efficient during Mu growth.

Presence of the G Segment in Bacteriophage P1

The temperate bacteriophage P1 has an invertible segment which is about 3 kb long (Lee et al. 1974). This segment of P1 DNA is flanked by inverted repetitious sequences of 0.62 kb. It has been shown by both electron microscopy (Chow and Bukhari 1977) and DNA-DNA hybridization experiments (F. de Bruijn and A. I. Bukhari, in prep.) that the 3-kb invertible segment of P1 is homologous to the G segment of Mu. A heteroduplex between the G segment of Mu and P1 is shown in Figure 4. When Mu DNA is cleaved with restriction endonucleases, the fragments containing the G sequences specifically hybridize to the P1 DNA fragment containing the invertible segment. These experiments show that G in Mu is identical in sequence to the P1 invertible segment, although small differences in sequence cannot be ruled out. It cannot be ascertained whether the less than 50-base-pair inverted repetitions at the ends of Mu G are derived from the much larger 0.6-kb inverted repetitions of P1. However, the G segment of Mu can be inverted by the inversion system of P1 (L. Chow, R. Kahmann and D. Kamp, unpubl.).

The presence of G sequence in both Mu and P1 is surprising since the two phages differ in many respects. P1 DNA is terminally repetitious, circularly permuted, and more than twice as long as Mu DNA (Ikeda and Tomizawa 1969).

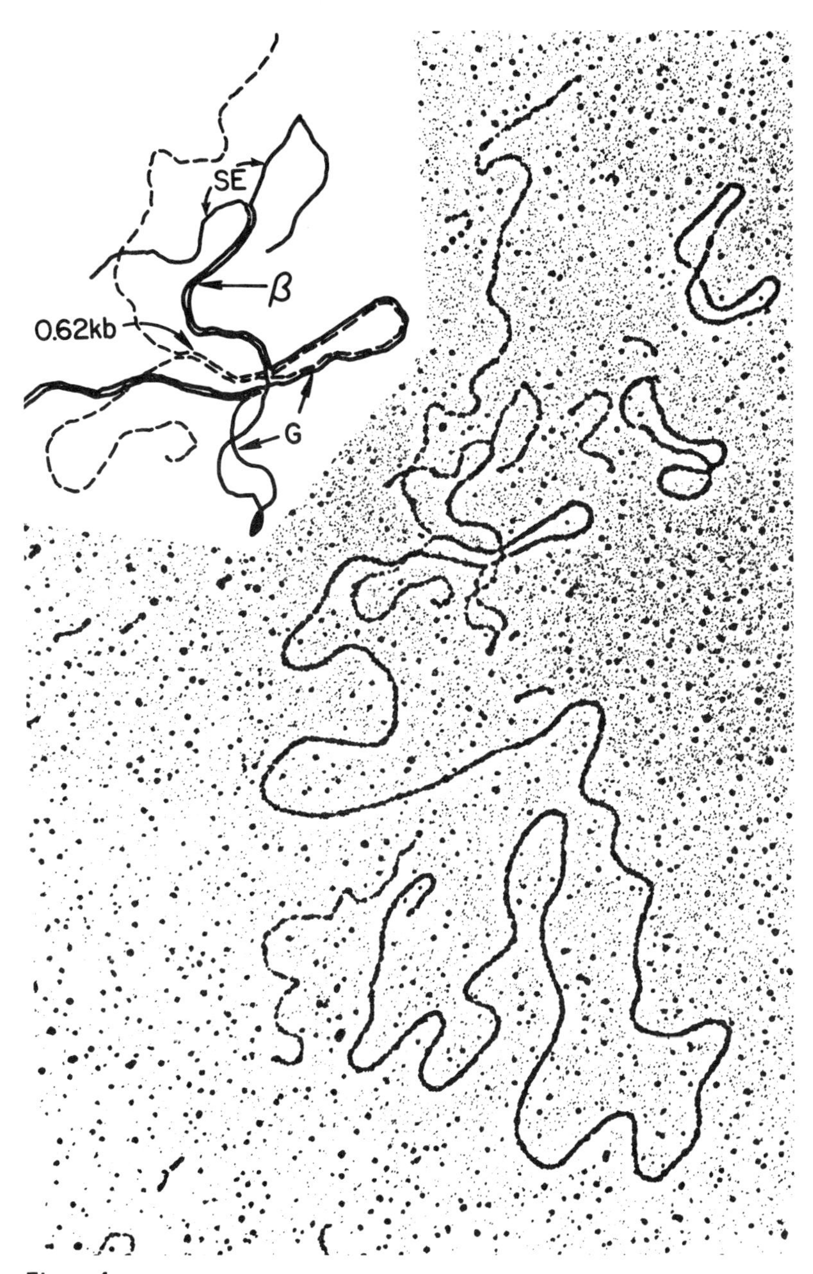

Figure 4

A heteroduplex formed between the P1 invertible segment and one arm of the noncomplementary G loop of a Mu/Mu heteroduplex. The P1 DNA has been digested with the *Eco*RI endonuclease. The inset is an enlarged tracing of the region of interest. (———) Mu DNA; (– – – –) P1 DNA. Circular molecules are ϕX174 RF II.

It is not integrated into the host chromosome and remains as a plasmid in the lysogenic cells. It is possible that the G segment is either a transposon or a remnant of a transposon and that the two phages have acquired the sequence from another source merely by translocation.

Analogy between G Inversions in Mu and Phase Variation in Salmonella

It should be noted that an invertible DNA sequence, such as G in Mu, has the capability of functioning as an on-off switch for gene expression, depending, of course, upon the information encoded in the invertible segment. Thus if an invertible segment is located adjacent to a promoter, the genes in the invertible segment might be expressed only in one orientation or the invertible segment might promote the expression of nearby genes only in one orientation. Zweig et al. (1977) have proposed that this type of mechanism is responsible for the ability of *Salmonella* strains to switch from one flagellar antigen (phase) to another. This switch in phases (phase variation) occurs with a small probability which is two to three times higher than the frequency of mutation. The structure of the flagellar element appears to be specified by two genes, *H1* and *H2*. The ability to switch from the expression of one gene to another is linked to the *H2* gene; when the *H2* locus is turned on, *H1* is inactive, and when *H2* is inactive, *H1* is expressed (Lederberg and Lino 1956; Enomoto and Stocker 1975). Zweig et al. (1977) have cloned a DNA segment carrying the *H2* gene of *Salmonella typhimurium* in a ColE1 plasmid vector and a phage λ vector. They have presented evidence that the *H2* gene in the cloned segment can be turned off and on and that this is related to the inversion of a segment of approximately 750 base pairs close to the *H2* gene. Their results indicate that phase-2 off and phase-2 on are associated with specific orientations of the segment adjacent to the *H2* gene. The process of phase transition was found to be asymmetric; the rate of transition from phase-2 on to phase-2 off appeared to be at least tenfold higher relative to the rate in the reverse direction. In Mu, it has been noted that the DNA of phage particles grown by induction is modified as compared to the phage particles grown by infection (Allet and Bukhari 1975; Toussaint 1976). It remains to be determined whether the G orientation influences the expression of the DNA modification function (the *mom* gene), or some other function, of Mu.

Generation of Host Sequences at Mu Ends

Bukhari and Taylor (1975) have proposed that Mu DNA is encapsidated by a headful mechanism from maturation precursors that contain host DNA covalently linked to both ends of Mu DNA. Packaging is presumed to start at the *c* end of Mu, cutting the adjacent host sequences, and to terminate in host sequences at the other end, resulting in the packaging of host sequences covalently linked to both ends of Mu DNA. This proposal was based on the observations that insertions in Mu DNA cause a reduction in the *S*-end host sequences. Insertions of about 750 base pairs in Mu DNA were found to reduce, by about 600 pairs, the length of host sequences at the *S* end (Bukhari and Taylor 1975). The presence of a small amount of host DNA at the *c* end apparently is unrelated to the size of Mu DNA being packaged (Bukhari et al. 1976).

The headful packaging model has been further strengthened by analysis of the Mu mutants (Mu *Xcam*) that carry insertions of Tn*9*, the 2.8-kb transposon for chloramphenicol resistance. Since Tn*9* is larger than the average length of the host sequences at the *S* end, headful packaging terminates within Mu DNA instead of within the bacterial DNA. Figure 5 depicts the effect of Tn*9* insertions on the packaging of host sequences. In heteroduplexes between Mu *Xcam* and Mu *cts*62, no loss of DNA was observed at the *c* end. However, more than 99.9% of the Mu *Xcam* DNA molecules had lost the bacterial DNA and a variable fraction of the adjoining β sequence. Figure 6 shows Mu *Xcam*/Mu and Mu *Xcam*/Mu *Xcam* heteroduplexes. These heteroduplexes do not have two single strands comprising the split ends. In the Mu *Xcam*/Mu heteroduplex (Fig. 6A), there is only one single strand at the end because host DNA is linked only to the normal Mu DNA strand. In the Mu *Xcam*/Mu *Xcam* heteroduplex (Fig. 6B), no bacterial DNA can be seen at the end. Since the amount of DNA packaged is not absolutely precise, different lengths of the β sequence are lost along with the *S*-end bacterial DNA in the Mu *Xcam* mutants. Additional evidence for the headful packaging of Mu DNA has been obtained by Chow et al. (1977 and this volume), who have found that deletions in Mu DNA cause a corresponding increase in length of the host sequences at the *S* end.

Characterization of Mu Mutants Carrying Insertions

Isolation of Insertion Mutants

Insertions in Mu DNA have been very useful in studies on the Mu genome. The simplest way to obtain such insertions is to select for *X* mutants of Mu

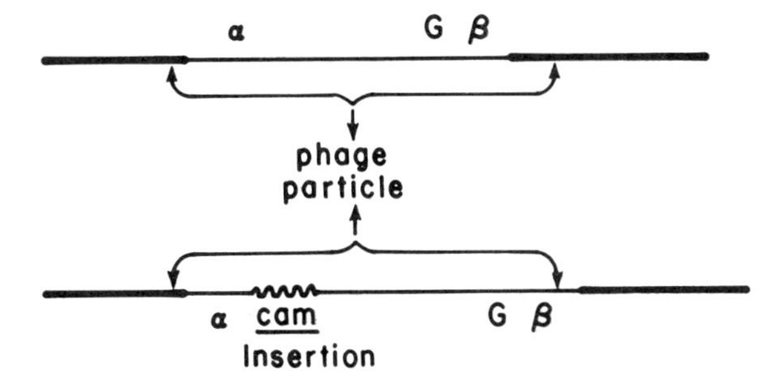

Figure 5
Headful packaging of Mu DNA. Encapsidation of Mu DNA involves packaging of approximately 37 kilobases beginning from the *c* end. When *cam* (Tn*9*) insertions are present, the packaging reaction is completed before host DNA attached at the *S* end is packaged. In addition, a variable portion of the β segment is also clipped since the packaging reaction is not absolutely precise.

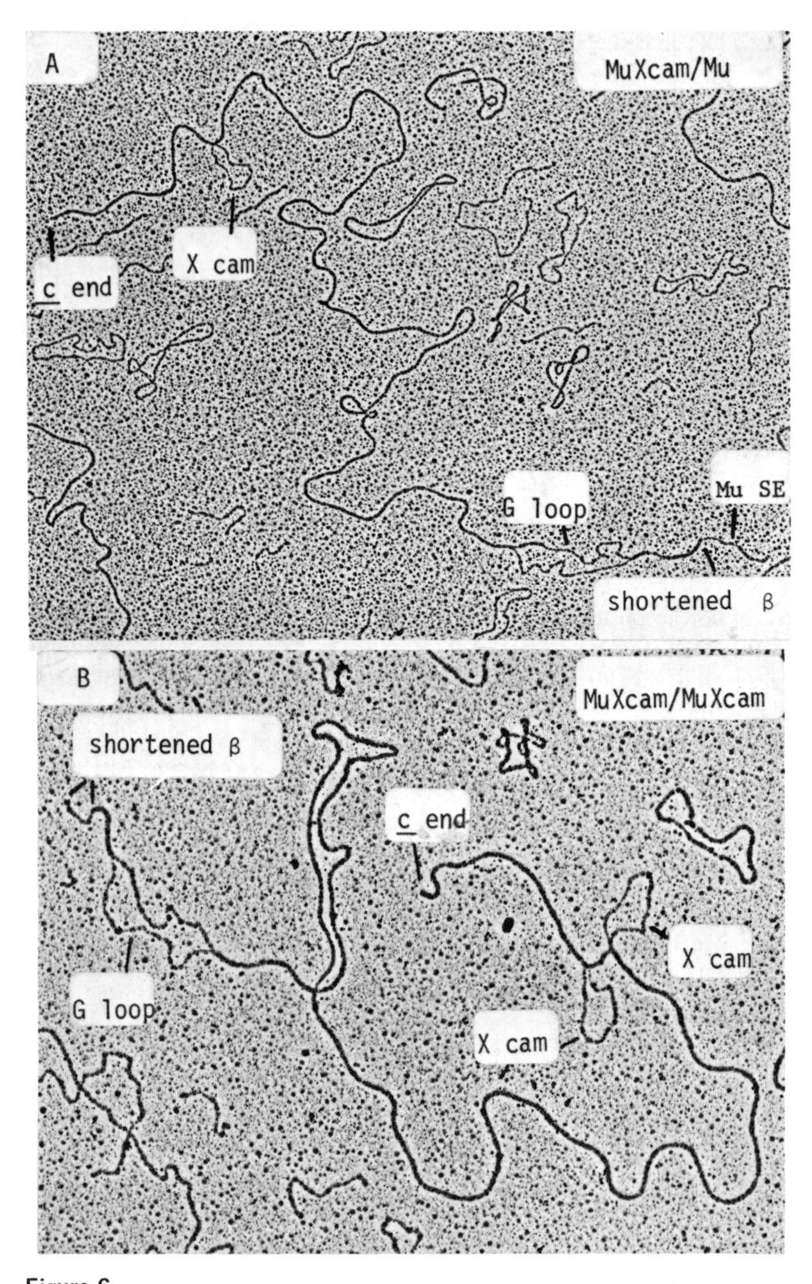

Figure 6

(*A*) A heteroduplex of Mu *Xcam*/Mu showing the *cam* insertion, a shortened β segment of Mu *Xcam*, and the SE from the Mu strand. (*B*) A heteroduplex of Mu *Xcam*/Mu *Xcam* showing two *cam* insertions of opposite orientations at different locations and two β sequences shortened to different extents.

(Bukhari 1975). The *X* mutants, defined by their ability to be excised at a low frequency, always carry insertions and can be selected by plating Mu *cts* lysogens at 42-44°C. At these temperatures, the Mu *cts* prophage is induced, killing the host cells. Some of the surviving colonies carry the defective Mu prophages called the *X* mutants. The *X* mutants contain insertions which eliminate the killing functions of Mu. These mutants can be grown by induction with a Mu *cts* prophage helper. The spontaneously isolated *X* mutants have been shown to contain insertions of about 750 base pairs near the *c* end of Mu (Bukhari and Taylor 1975).

The procedure for isolating *X* mutants has also been used to insert Tn*9*, the chloramphenicol resistance transposon, in the killing genes of Mu. When Mu *cts*, P1*cam* (P1 carrying Tn*9*) dilysogens are plated at high temperatures, about 2% of the heat-resistant survivors are found to have defective Mu prophages (Mu *Xcam* mutants) in which mutations result from the insertion of Tn*9* (Bukhari and Froshauer 1977).

Characterization of Mu Xcam *Mutants*

The Mu *Xcam* mutants contain an insertion of about 2.8 ± 0.2 kb. No deletion of Mu sequences can be observed at the site of insertion (Fig. 6). The insertion can occur in either orientation at one of the two preferred sites (3.3 kb or 3.9 kb from the *c* end). These results confirm the conclusions of Rosner and Gottesman and MacHattie and Jackowski (both this volume) that the chloramphenicol resistance gene moves as a discrete DNA unit and is thus a transposon. We have found in Mu *Xcam* mutants that the Tn*9* insertion can be at either of the two sites (3.3 kb or 3.9 kb) even if the lysates are prepared by inducing cells (Mu *Xcam*, Mu *cts* dilysogens) grown from a single colony (Chow and Bukhari 1977). This observation implies that the Tn*9* transposon is translocated at high efficiency from one site to another during Mu growth. However, the observation could merely reflect the presence of two Tn*9* copies in the Mu *Xcam* prophages such that segregation of one of the copies occurs continuously during Mu growth. Another interesting observation on the Mu *Xcam* mutants is that they can revert to wild-type Mu. This result confirms that Tn*9* insertion occurs without causing a deletion in Mu sequences and also shows that Tn*9* can be excised precisely. The excision frequency of Tn*9*, however, differs widely in different Mu *Xcam* mutants. It can be as low as 10^{-6}-10^{-7} or as high as 10^{-3}-10^{-4}.

Mu *Xcam* particles do not transduce chloramphenicol resistance at a high frequency whether or not a helper phage is present. The transduction frequencies vary from 10^{-5} to 10^{-7} per Mu *Xcam* particle in the mixed lysates (the number of Mu *Xcam* particles is estimated by the relative amounts of Mu *ctsX* and Mu *cts* DNA). The chloramphenicol resistance transductants that do arise at a low frequency generally have the *cam* determinant (Tn*9*) translocated from Mu DNA into the host DNA. The failure of the Mu *Xcam* particles to be integrated, assuming that they are injected normally, could be ascribed to the loss of the right end (part of the β segment), which, along with the Mu-specific sequence on the *c* end, is necessary for Mu integration.

The Insertion in Spontaneous X *Mutants Is IS*1

The insertion of about 750 base pairs in the Mu *X* mutants has been identified as the IS*1* element. The identification was based on DNA-DNA hybridization between an IS*1*-containing plasmid and the Mu *X* mutants (F. de Bruijn and A. I. Bukhari, unpubl.) and independently by electron microscopic examination of heteroduplexes of the plasmid DNA containing IS*1* and Mu *X* DNA (Chow and Bukhari 1977).

*Tn9 Is Flanked by Two IS*1 *Elements*

L. MacHattie (pers. comm.) first suggested that the *cam* insertion (Tn*9*) in λ*pcam* transducing phages might contain two IS*1* sequences in direct order. We have obtained evidence that this is indeed the case. Two single strands of pSM3 (a 7.2-kb plasmid containing one IS*1*; see Ohtsubo and Ohtsubo, this volume) were found to hybridize with a single strand of Mu *Xcam* at positions about 1100 nucleotides apart at the proper map position (Chow and Bukhari 1977). They were also observed to hybridize at the two flanking 0.8-kb portions of the *cam* insertion loop in Mu *Xcam*/Mu *cts* heteroduplexes. Restriction endonuclease cleavage and DNA-DNA hybridization studies also show the presence of two IS*1* copies (with one cut by the enzyme *Bal*I in each) and a cleavage site for the enzyme *Eco*RI in the *cam* determinant or the Tn*9* transposon (Bukhari and Froshauer 1977; F. de Bruijn and A. I. Bukhari, unpubl.).

SUMMARY AND GENERAL COMMENTS

An interesting picture of the bacteriophage Mu genome has developed from electron microscope and DNA-DNA hybridization studies of Mu DNA. Furthermore, Mu mutants with insertions have been useful not only in understanding some of the characteristics of the Mu genome but also in revealing some properties of the insertion elements.

There are two major characteristics of wild-type Mu DNA. One is the presence of the 3-kb-long invertible sequence, called the G segment, which is flanked by inverted repetitious sequences of less than 50 nucleotide pairs. The role of G in the Mu life cycle is not known. The G segment of Mu is the same as the 3-kb invertible region of the bacteriophage P1 genome. In P1, however, G is flanked by inverted repetitions 0.6 kb in length. The other characteristic feature of wild-type Mu DNA is the presence at the ends of host sequences that differ in size and sequence from molecule to molecule. The left end (*c* end) of Mu contains, on the average, 100 host base pairs, whereas the right end (*S* end) contains about 500 to 3200 base pairs of host DNA.

The host sequences at Mu ends result from oriented headful packaging of Mu DNA from maturation precursors in which Mu DNA is covalently linked to host DNA. Packaging is initiated from the *c* end of Mu DNA, cutting to the left of the *c* end of Mu, and terminates in host sequences linked to the *S* end. Insertions in Mu DNA reduce the length of host sequences at the *S* end. Insertion of the Tn*9* transposon (2.8 kb in length) completely eliminates host sequences at the *S* end and also causes a variable loss of the β segment of Mu. Presumably because of this loss of a part of β, Mu mutants carrying Tn*9* appear to be defective in integration.

The insertions in the spontaneously isolated *X* mutants of Mu have been identified as IS*1*. The *X* mutations can also be caused by the insertion of Tn*9*, which has two flanking copies of IS*1* arranged in direct order.

Acknowledgments

We are thankful to Drs. T. Broker, R. Kahmann and D. Kamp for discussions. We wish to acknowledge support from the National Cancer Institute (Cancer Center Grant to Cold Spring Harbor Laboratory), the National Science Foundation, and the National Cystic Fibrosis Foundation. A. I. B. holds a Career Development Award of the National Institutes of Health.

REFERENCES

Allet, B. and A. I. Bukhari. 1975. Analysis of bacteriophage Mu and lambda-Mu hybrid DNAs by specific endonucleases. *J. Mol. Biol.* **92**:529.

Bade, E. 1972. Asymmetric transcription of temperate bacteriophage Mu-1. *J. Virol.* **10**:1205.

Bukhari, A. I. 1975. Reversal of mutator phage Mu integration. *J. Mol. Biol.* **96**:87.

——. 1976. Bacteriophage Mu as a transposition element. *Annu. Rev. Genet.* **10**:389.

Bukhari, A. I. and S. Froshauer. 1977. Insertion of a chloramphenicol resistance determinant in bacteriophage Mu. *Gene* (in press).

Bukhari, A. I. and A. L. Taylor. 1975. Influence of insertions on packaging of host sequences covalently linked to bacteriophage Mu DNA. *Proc. Nat. Acad. Sci.* **72**:4399.

Bukhari, A. I., S. Froshauer and M. Botchan. 1976. Ends of bacteriophage Mu DNA. *Nature* **264**:580.

Chow, L. and A. I. Bukhari. 1976. The invertible DNA segments of coliphages Mu and P1 are identical. *Virology* **74**:242.

——. 1977. Electron microscopic heteroduplex characterization of bacteriophage Mu mutants carrying IS*1* or the chloramphenicol resistance transposon Tn*9*. *Gene* (in press).

Daniell, E., W. Boram and J. Abelson. 1973a. Genetic mapping of the inversion loop in bacteriophage Mu DNA. *Proc. Nat. Acad. Sci.* **70**:2153.

Daniell, E., D. E. Kohne and J. Abelson. 1975. Characterization of the inhomogeneous DNA in virions of bacteriophage Mu by DNA reannealing kinetics. *J. Virol.* **15**:739.

Daniell, E., J. Abelson, J. S. Kim and N. Davidson. 1973b. Heteroduplex structures of bacteriophage Mu DNA. *Virology* **51**:237.

Enomoto, M. and B. A. D. Stocker. 1975. Integration, at *hag* or elsewhere, of H2 (phage-2 flagellin) genes transduced from *Salmonella* to *Escherichia coli*. *Genetics* **81**:595.

Hsu, M.-T. and N. Davidson. 1974. Electron microscope heteroduplex study of the heterogeneity of Mu phage and prophage DNA. *Virology* **58**:229.

Ikeda, H. and H.-I. Tomizawa. 1969. Prophage P1, an extrachromosomal replication unit. *Cold Spring Harbor Symp. Quant. Biol.* **33**:791.

Lederberg, J. and T. Lino. 1956. Phase variation in *Salmonella*. *Genetics* **46**: 1475.

Lee, H. J., E. Ohtsubo, R. C. Deonier and N. Davidson. 1974. Electron microscope heteroduplex studies of sequence relations among plasmids of *Escherichia coli*. V. *ilv*$^+$ deletion mutants of F14. *J. Mol. Biol.* **89**:585.

Martuscelli, J., A. L. Taylor, D. Cummings, V. Chapman, S. Delong and L. Cañedo. 1971. Electron microscopic evidence for linear insertion of bacteriophage Mu-1 in lysogenic bacteria. *J. Virol.* **8**:551.

Toussaint, A. 1976. The DNA modification function of temperate phage Mu-1. *Virology* **70**:17.

Zweig, J., M. Silverman, M. Hilmen and M. Simon. 1977. Recombinational switch for gene expression. *Science* **196**:170.

Electron Microscope Studies of Nondefective Bacteriophage Mu Mutants Containing Deletions or Substitutions

L. T. Chow, R. Kahmann and D. Kamp
Cold Spring Harbor Laboratory
Cold Spring Harbor, New York 11724

Bacteriophage Mu DNA is a linear duplex 36-38 kilobase pairs (kb) in length. Various regions of the Mu chromosome have been distinguished on the basis of heteroduplex structures visualized in the electron microscope after the DNA is denatured and self-annealed. They have been designated, from left to right, α, G, β, and SE (split end) (Daniell et al. 1973; Hsu and Davidson 1974; see Fig. 2 in Chow and Bukhari, this volume). In electron microscope studies of heteroduplexes prepared between Mu *vir* DNA and total bacterial DNA from strain DK445, a dilysogen of lambda *cts*857 *Sam*7 *lac*5::Mu cts^+, we have discovered that the Mu prophage (in the bacterial chromosome) contains an as yet to be identified segment of DNA 2.6 kb long inserted in the middle of the β region (Fig. 1). This prophage proved to be an ideal source for the isolation of nondefective mutants containing deletions or substitutions in apparently nonessential regions of the Mu chromosome. These mutants were anticipated because the headful packaging of DNA into Mu phage particles requires that the unknown insertion be compensated for by deletions (Bukhari and Taylor 1975; Chow and Bukhari, this volume). The insertion is sufficiently long so that viable Mu phage containing the entire insertion without compensating deletions have not yet been isolated. Seven independent mutants have been obtained and studied by electron microscope heteroduplex methods. The results are summarized in Figure 2 and are discussed below.

All mutants have growth rate, burst size, and plaque morphology indistinguishable from wild-type Mu. Mu 445-3 is isolated as a clear plaque mutant; all the others can lysogenize and form normal turbid plaques.

Three of the mutants (445-1, 2, and 6) turn out to be identical. They contain the entire insertion found in the prophage. To compensate for the increase in genome size, the mutants have deleted the β sequence to the left of the insertion (0.85 kb) plus 1 kb of the adjoining G sequence. Since the three mutants were isolated independently, the two end points of the deletion are "hot spots" for recombination. One of the end points is approximately the junction between the G internal inverted repetition and the right unique portion of G in the lytic orientation (see Fig. 3 g3′-g5 of Chow and Bukhari, this volume). The other

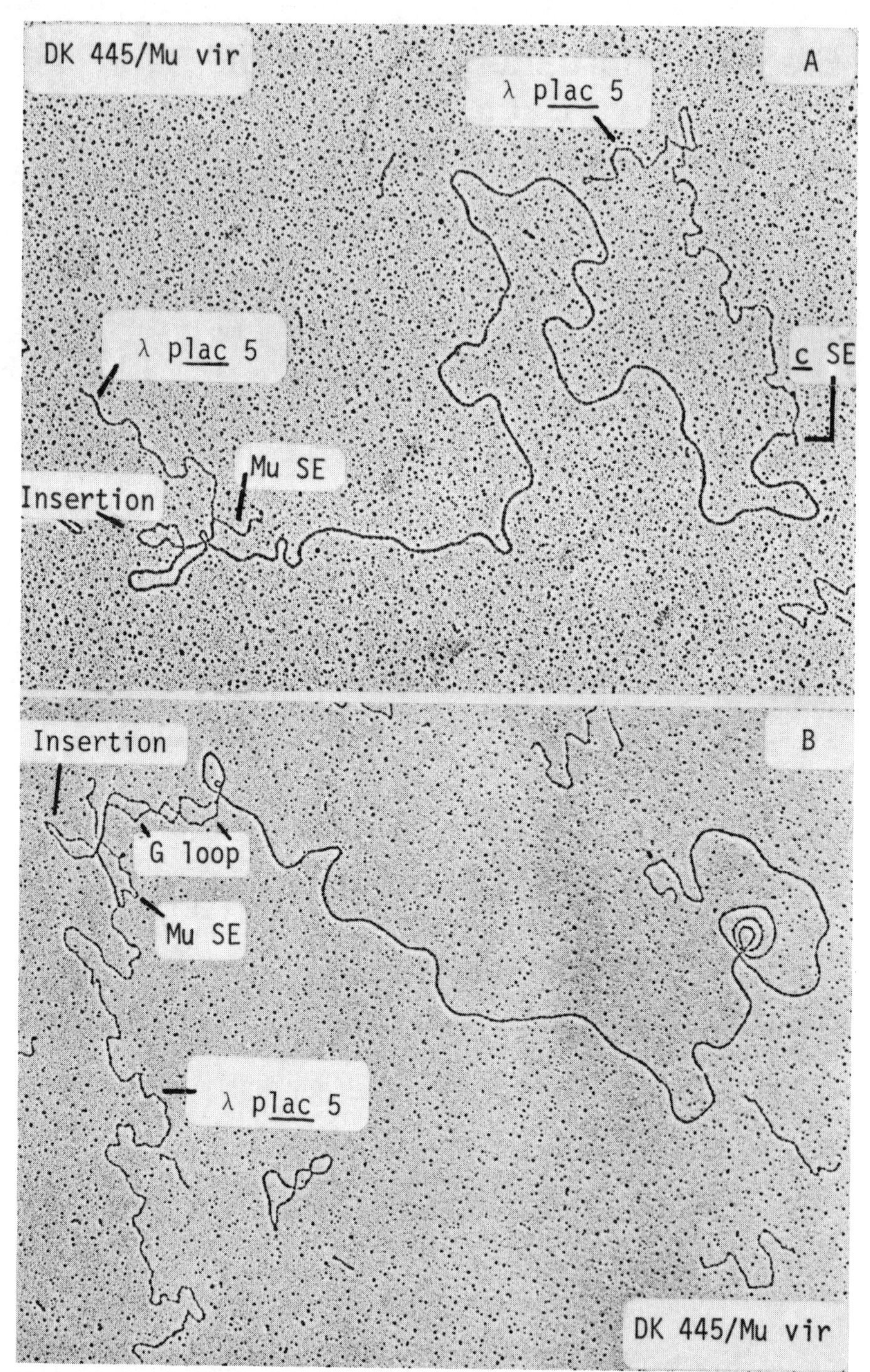

Figure 1 *(See facing page for legend)*

end point is the left junction between β and the insertion DNA. A fourth mutant (445-7) is similar to these three except that the right 1.3 kb of the G sequence have been deleted so that it no longer has the internal inverted repeat stem-loop structure in the single-stranded DNA.

One mutant (445-3) has a deletion 2.3 kb in length spanning most of the β sequence (the entire insertion DNA) and 0.8 kb of the adjoining G sequence. The only β sequence remaining is the rightmost 0.24 kb next to the SE. In this mutant, the encapsidated bacterial DNA comprising the SE is 4–5 kb or longer; thus the split end is increased by the length of the deletion (Fig. 3). Combined with our finding that the bacterial DNA (SE) is lost in the insertion mutant Mu *Xcam* (Chow and Bukhari, this volume), this observation further confirms the headful packaging model.

Two mutants have intact G DNA segments. Mu 445-8 has a substitution in the right half of β by a portion of the original insertion in the prophage; it still retains the extreme right 0.15 kb of the β DNA segment. Mutant 445-5 (subsequently converted to *cts*62), with a substitution in the left half of β by a different portion of the original insertion in the prophage (Fig. 2B), no longer inverts its G sequence (Fig. 4). This result suggests that the gene coding for the enzyme responsible for the G inversion or its promoter is located in this portion of the β region between the point of insertion and the G-β junction. This mutant does not modify its DNA to the same extent as wild-type Mu, again conforming to the suggestions by other workers that there is a correlation between the orientation of the G sequence and the expression of the Mu modification system (*mom*) (Allet and Bukhari 1975; Toussaint 1976). The concurrent loss of the *mom* function and the invertability of the G sequence suggest that the control of gene expressions could be mediated through inversions of the DNA segments encoding those genes.

In the *prophage* that contains the 2.6-kb insertion DNA and that has given rise to these viable mutants, the G sequence can invert. This conclusion is drawn from observing a noncomplementary G loop at a frequency of 40% in heteroduplexes between the prophage DNA and Mu *vir* DNA (Fig. 1B). (Mu *vir* DNA has its G loop fixed in the lytic orientation and therefore can be used to determine the invertability of other G segments.) Nevertheless, all seven mutants have their G sequence in the same orientation as that found in Mu grown by infection.

Figure 1
Heteroduplexes of the prophage Mu DNA in dilysogen DK445 (in which Mu has integrated in the *lac* region of a λ*lac*5 prophage) and Mu *vir* DNA, showing the unknown insertion of 2.6 kb in the middle of β (0.85 kb from both the G-β and β-SE junctions). (*A*) The two G sequences are of the same orientation. The heteroduplex also shows the short bacterial sequence at the *c* end of the Mu *vir* DNA (*c* SE). (*B*) The two G sequences are of opposite orientations. For schematic representation of the Mu prophage in DK445, see Fig. 2B.

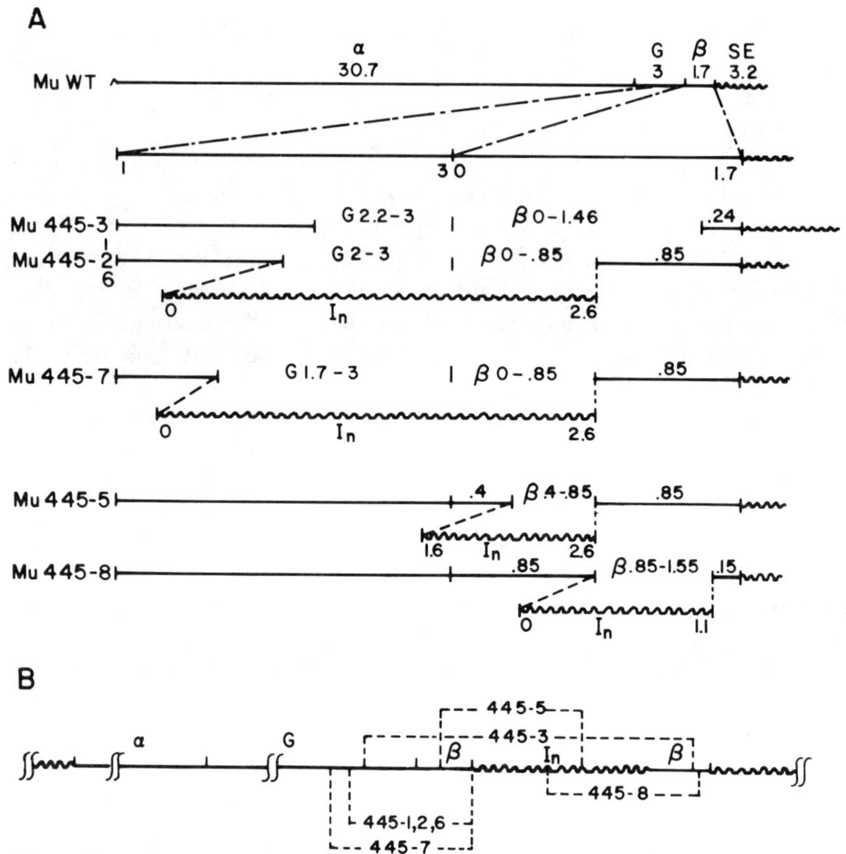

Figure 2

(A) Physical maps of wild-type Mu and nondefective Mu mutants. The region covering β and two-thirds of the G sequence is expanded in scale to allow more detailed illustration. The G segment of DNA is drawn in the lytic orientation. The lengths of various segments are shown in kilobases. The sequences of G, β, and the insertion (In) have been assigned coordinates G 0–3 in the lytic orientation, β 0–1.7, and In 0–2.6. All mutants have α and G 0–1. The SE DNA in Mu 445-3 is lengthened, whereas that in Mu 445-1, 2, or 6 is slightly but noticeably shortened. Gaps represent deletions; (——) Mu DNA; (〰) insertion DNA and other bacterial DNA; (– – – –) construction lines.

(B) Physical map of the Mu prophage in dilysogen DK445. (– – – –) Deletion events generating the various mutants above; (〰) *lac* DNA of lambda *cts*857 *Sam*7 *lac*5.

Figure 3

(A) A heteroduplex of Mu 445-3/Mu showing the deletion loop (G 2.3–3 + β 0–1.46), a very short β remaining, $\beta\star$ (β 1.46–1.7), and a long SE DNA from Mu 445-3. (B) Heteroduplex of Mu 445-3/Mu 445-3 showing two long SE DNA.

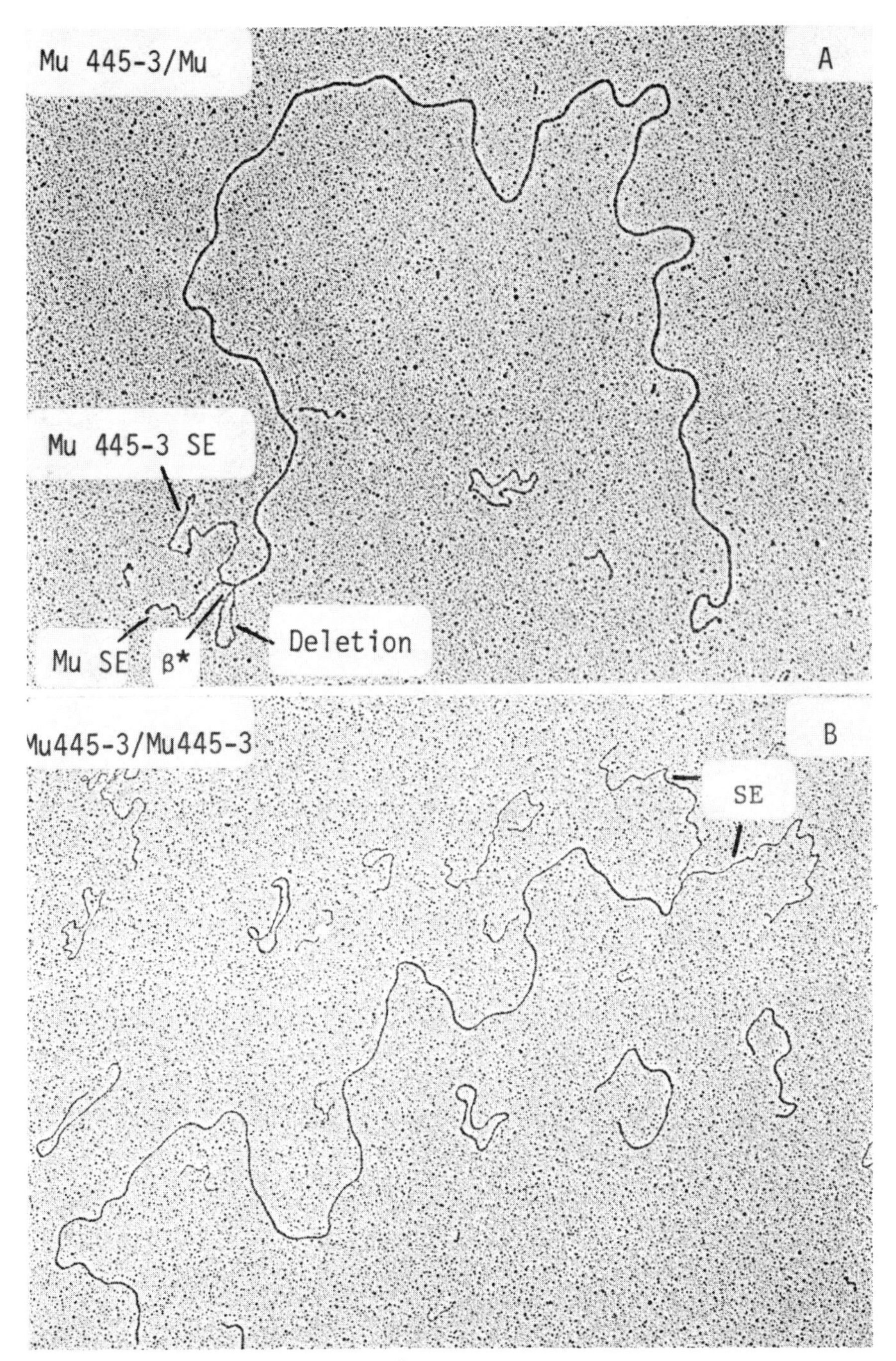

Figure 3 *(See facing page for legend)*

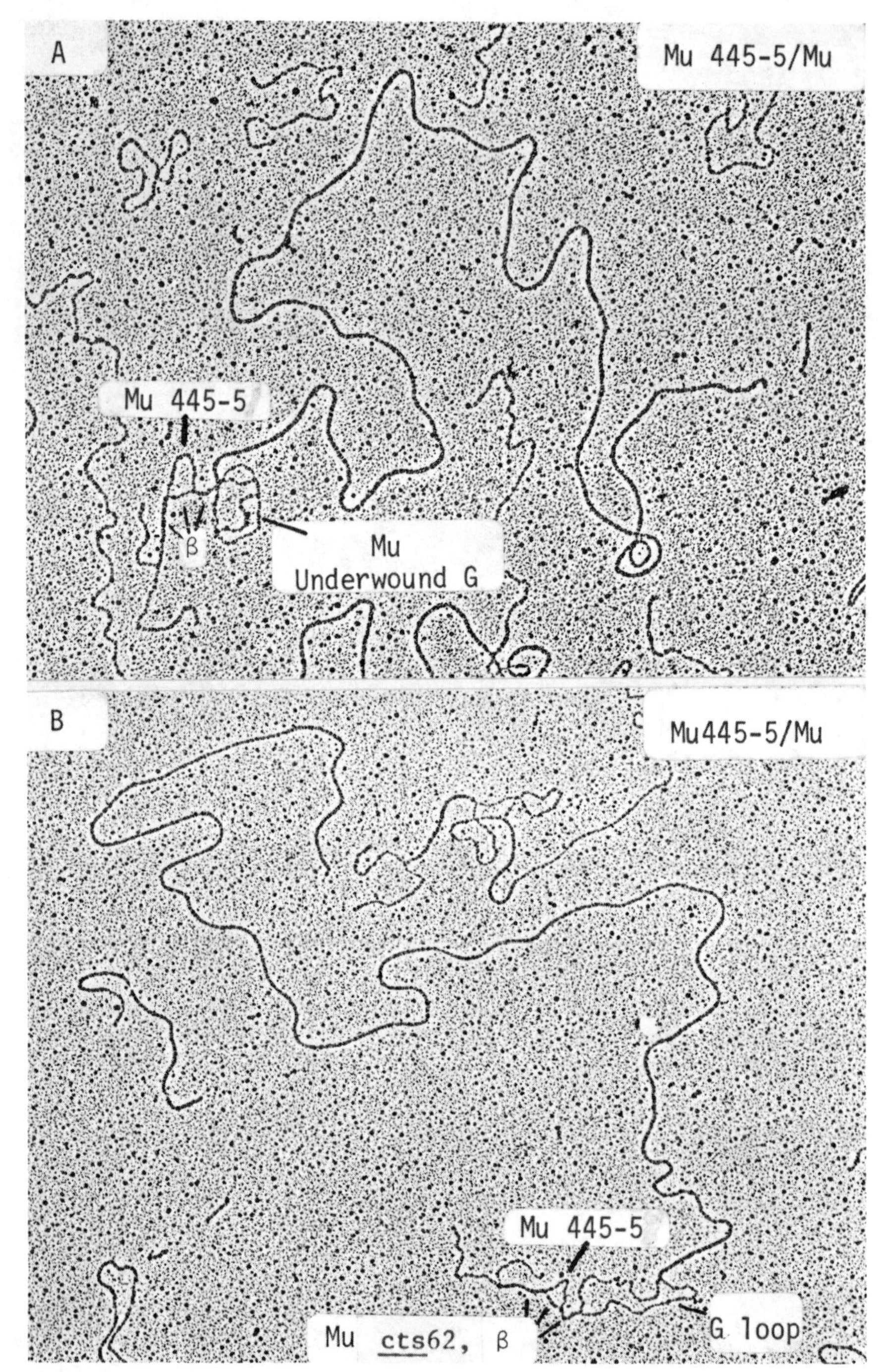

Figure 4 *(See facing page for legend)*

This raises the possibility that there is an essential function located within the remaining G sequence which would have been lost had the deletion occurred while the G sequence was in the opposite orientation. Such deletions would thus be lethal, and mutants would not be recovered as plaque formers.

SUMMARY

Seven nondefective Mu mutants have been isolated from a prophage that contains an insertion of unknown nature in the middle of the β sequence. All of them have growth rate, burst size, and plaque morphology indistinguishable from wild-type Mu. They have been characterized by electron microscope heteroduplex methods and were found to have deletions or substitutions in the G and β regions.

Acknowledgments

We thank Dr. Thomas Broker for critical reading of this manuscript and Dr. David Zipser for his support of this work. We are also pleased to acknowledge the assistance of Ms. Marie Moschitta and Mr. Robert Yaffee in the preparation of the manuscript.

This work has been funded by Cancer Center Grant CA13106 to Cold Spring Harbor Laboratory and by grants to Dr. David Zipser from the National Science Foundation (GB-32041 XI) and the U.S. Public Health Service (5 ROI GM17612-06). R. K. and D. K. are recipients of fellowships of the Deutsche Forschungsgemeinschaft.

REFERENCES

Allet, B. and A. I. Bukhari. 1975. Analysis of bacteriophage Mu and lambda-Mu hybrid DNAs by specific endonucleases. *J. Mol. Biol.* **92**:529.

Bukhari, A. I. and A. L. Taylor. 1975. Influence of insertions on packaging of host sequences covalently linked to bacteriophage Mu DNA. *Proc. Nat. Acad. Sci.* **72**:4399.

Figure 4

Heteroduplexes of Mu 445-5/Mu *cts*62 (grown by induction) showing the substitution in β. The three segments of β of Mu *cts*62 have coordinates β 0–0.4, β 0.4–0.85, and β 0.85–1.7. The long arm of the substitution loops is In 1.6–2.6 of Mu 445-5. (*A*) The G sequences are of the same orientation and form an underwound loop (see Broker et al., this volume). (*B*) The G sequences are of opposite orientations. No such heteroduplex is seen if Mu *vir,* instead of Mu *cts*62 DNA, is used. For nomenclature and schematic representation of sequences in Mu 445-5, see Fig. 2A.

Daniell, E., J. Abelson, J. S. Kim and N. Davidson. 1973. Heteroduplex structures of bacteriophage Mu DNA. *Virology* **51**:237.

Hsu, M.-T. and N. Davidson. 1974. Electron microscope heteroduplex study of the heterogeneity of Mu phage and prophage DNA. *Virology* **58**:229.

Toussaint, A. 1976. The DNA modification function of temperate phage Mu-1. *Virology* **70**:17.

Structure and Packaging of Mu DNA

E. Bade
Fachbereich Biologie
Universität Konstanz
D-7750 Konstanz, Germany

H. Delius
EMBO Laboratory
D-6900 Heidelberg, Germany

B. Allet
Département de Biologie Moléculaire
Université de Genève
CH-1211 Genève, Switzerland

Much of the biochemical and biophysical work on phage Mu was stimulated by the discovery of the unusual homoduplex structure of phage DNA (Bade 1972; Daniell et al. 1973) and the observation that Mu DNA contains covalently linked *E. coli* DNA sequences at its variable end (Bade et al. 1974; Daniell et al. 1975; Howe and Bade 1975).

The 25×10^6 daltons of Mu DNA have an average base composition of 50% G + C (Martuscelli et al. 1971). The bases are distributed in a nonrandom fashion along the phage chromosome, thus allowing the separation of Mu DNA strands by poly(U,G) binding and equilibrium centrifugation (Bade 1972).

Mu DNA, obtained from phage grown either by heat induction of lysogens or by lytic infection, was measured both before and after cleavage with *Eco*RI and *Hin*dIII restriction endonucleases. The DNA was also partially denatured. The length distributions are shown in Tables 1 and 2. The partial denaturation pattern of *Hin*dIII fragments is shown in Figure 1. The most striking feature of the partial denaturation maps is the clustering of readily denaturable (AT-rich) segments close to the ends of the phage DNA. This typical partial denaturation pattern should prove useful for the physical analysis of DNA structures containing complete or partial Mu genomes (e.g., partially deleted Mu prophages).

The length measurements of the restriction fragments show further that the ends of Mu DNA (fragments C and B for both restriction endonucleases used here) are heterogeneous in size. This result is well in agreement with previous reports on heterogeneity of the variable end. Recently, the left (immunity) end

Table 1
*Eco*RI Fragments of Mu DNA

Fragment	Length range (μ)	No. of fragments measured	Av. length (μ)	S.D. (μ)	S.D. (%)	m.w. (× 10^6 daltons)
A	5.50–6.50	29	6.071	± 0.102	1.69	11.578
B	4.00–5.50	35	4.905	± 0.266	5.44	9.354
C	1.55–1.85	58	1.679	± 0.055	3.31	3.202
					Total	24.134
DNA of phage PM2			3.356	± 0.091	2.70	6.4

Size (contour length measurements) of *Eco*RI restriction fragments of Mu DNA.

Table 2
*Hin*dIII Fragments of Mu DNA

Fragment	Length range (μ)	No. of fragments measured	Av. length (μ)	S.D. (μ)	S.D. (%)	m.w. (× 10^6 daltons)
A	8.20–8.80	13	8.485	± 0.095	1.13	16.181
B	3.65–4.50	26	4.022	± 0.202	5.03	7.670
C	0.30–0.43	30	0.349	± 0.029	8.32	0.666
					Total	24.517
DNA of phage PM2		9	3.356	± 0.056	1.66	6.4

Size (contour length measurements) of *Hin*dIII fragments of Mu DNA.

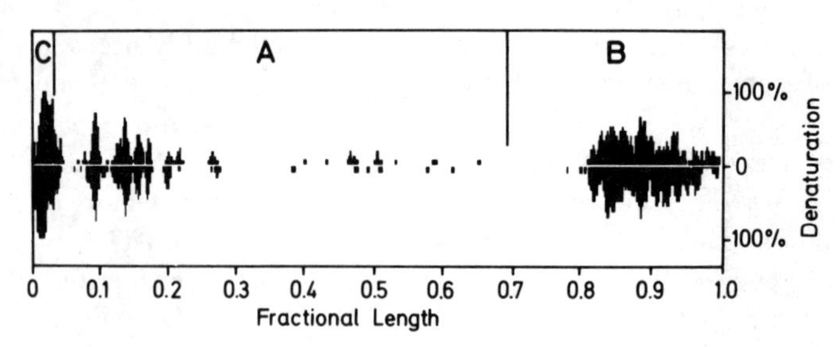

Figure 1
Partial denaturation map of whole Mu DNA (lower half of figure) and its *Hin*dIII restriction fragments (upper half of figure). Denaturation was achieved by a 60-min incubation of 6 M sodium perchlorate, pH 7.5, and 2.5% formaldehyde, at 30°C.

of the phage DNA has also been shown to be heterogeneous in size both by gel electrophoresis (Allet and Bukhari 1975) and by electron microscopy of Mu DNA homoduplexes after treatment with gene-32 protein of phage T4 (H. Delius, as reported at the EMBO Workshop on Mu, Brussels, 1975). We find that this left end of Mu DNA also contains *E. coli* DNA sequences (E. Bade, as reported at the EMBO Workshop on Mu, Brussels, 1975; see Table 3). These *E. coli* sequences are indeed heterogeneous since they are eliminated from homoduplex molecules by the single-strand-specific nuclease S1 (Table 3). The presence of *E. coli* DNA at both ends of Mu DNA has also been shown by Bukhari et al. (1976 and this volume).

We have previously postulated (Shroeder et al. 1974) that the DNA of phage Mu is replicated in association with *E. coli* DNA. We showed that covalently closed ring molecules containing both host and phage DNA are formed during replication. To explain the presence of *E. coli* DNA at the variable end of the DNA molecule and the production of generalized transducing particles of Mu, we proposed that oriented packaging of phage Mu DNA occurs from hybrid *E. coli*-Mu replication intermediates beginning from the *c* end of Mu (Schroeder et al. 1974). Bukhari and Taylor (1975) have presented strong evidence that the packaging of Mu involves headful packaging of DNA. The maturation precursors presumably contain *E. coli* DNA linked to both ends of Mu DNA.

These inferences are consistent with observations that no "tandem" structures of Mu DNA have ever been detected, either genetically or physically, and that in every stage of the phage life cycle studied so far, Mu DNA has always been shown to be intimately associated with (integrated into) host DNA.

Table 3
E. coli DNA Sequences on *Eco*RI Fragments of Mu DNA

	DNA on filter (^{32}P)		DNA hybridized (cpm)	
Fragment	treatment	cpm	*E. coli* [^{3}H]	Mu [^{3}H]
A	denat./renat.	608	10	1126
	denat./renat./S1	502	7	961
B	denat./renat.	576	477	1199
	denat./renat./S1	689	16	852
C	denat./renat.	357	58	405
	denat./renat./S1	311	6	307

Homology of *Eco*RI restriction fragments with *E. coli* DNA. ^{32}P-labeled restriction fragments of Mu DNA eluted from 0.5% agarose gels were used. Comparable amounts of DNA from denatured and reannealed fragments were bound to filters (Bade 1972) after hydroxylapatite chromatography, with and without prior treatment with single-strand-specific nuclease S1. Hybridizations between sonicated [^{3}H]thymidine-labeled DNA from *E. coli* and Mu were continued for 22 hr in 50% of formamide–4 × SSC.

It remains open to further investigation whether replication, integration, and packaging of Mu DNA share common steps.

SUMMARY

The DNA of phage Mu and the fragments generated from it by *Eco*RI and *Hin*dIII were analyzed by DNA hybridization and partial denaturation mapping. The phage's chromosome is flanked by covalently linked host DNA sequences. The outer segments of Mu DNA consist of AT-rich sequences. The oriented packaging of Mu DNA (Schroeder et al. 1974) is proposed to occur through scissions just outside of these AT-rich segments.

Acknowledgments

Part of this cooperative work was performed at Cold Spring Harbor Laboratory. We thank J. D. Watson and R. F. Gesteland for support.

REFERENCES

Allet, B. and A. I. Bukhari. 1975. Analysis of bacteriophage Mu λ-Mu hybrid DNAs by specific endonucleases. *J. Mol. Biol.* **92**:529.

Bade, E. G. 1972. Asymmetric transcription of bacteriophage Mu-1. *J. Virol.* **10**:1205.

Bade, E. G., H. Delius and W. Schroeder. 1974. Structure and replication of Mu DNA. *Fed. Proc.* **33**:1487.

Bukhari, A. I., S. Froshauer and M. Botchan. 1976. Ends of bacteriophage Mu DNA. *Nature* **264**:580.

Daniell, E., D. E. Kohne and J. Abelson. 1975. Characterization of the inhomogeneous DNA in virions of bacteriophage Mu by DNA reannealing kinetics. *J. Virol.* **15**:739.

Daniell, E., J. Abelson, J. S. Kim and N. Davidson. 1973. Heteroduplex structures of bacteriophage Mu DNA. *Virology* **51**:237.

Howe, M. M. and E. G. Bade. 1975. The molecular biology of bacteriophage Mu. *Science* **190**:624.

Martuscelli, J., A. L. Taylor, D. J. Cummings, V. A. Chapman, S. S. DeLong and L. Cañedo. 1971. Electron microscopic evidence for linear insertion of bacteriophage Mu-1 in lysogenic bacteria. *J. Virol.* **8**:551.

Schroeder, W., E. G. Bade and H. Delius. 1974. Participation of *E. coli* DNA in the replication of temperate bacteriophage Mu. *Virology* **60**:534.

DNA Partial Denaturation Mapping Studies of Packaging of Bacteriophage Mu DNA

M. M. Howe,* M. Schnös† and R. B. Inman†
Departments of *Bacteriology and †Biochemistry, and †Biophysics Laboratory
University of Wisconsin
Madison, Wisconsin 53706

Information concerning the mechanism of packaging of bacteriophage DNA can be obtained by determining the spatial organization of the DNA within the bacteriophage particle. When phages such as λ, P2, 186, and P4 are subjected to mild formaldehyde cross-linking conditions, either before or during protein film spreading for electron microscopy, the phage heads lyse and much of the released DNA is found attached to the proximal end of the phage tail. Partial denaturation mappings of DNA-tail complexes show that phage tails are always joined to unique DNA ends (Chattoraj and Inman 1974; Thomas 1974). This attachment results from a phage structure in which a specific end of the DNA molecule is situated at the phage head-tail junction. Presumably, it is this specific DNA end which is the last to be packaged during maturation and the first to be injected during infection.

Bacteriophage Mu DNA is unusual in that it contains at one end, the variable end, a large segment of host DNA which is heterogeneous in both length and DNA base sequence (Daniell et al. 1973b; Bade et al. 1974). This variable end gives rise to nonhybridizing split ends (SE) in renatured Mu DNA molecules (Daniell et al. 1973b). Bukhari and Taylor (1975) have shown that the presence of *X* mutations, insertions near the immunity end of Mu, cause a shortening in the length of the variable ends. They postulate that this relationship results from packaging of Mu DNA from an integrated form by a mechanism which proceeds from the immunity end to the variable end.

In the present investigation, we have studied Mu DNA packaging by partial denaturation mapping analysis of Mu DNA-tail complexes generated by limited formaldehyde treatment of Mu particles.

Generation of Mu DNA-Tail Complexes

Particles of bacteriophage Mu have rigid tails of uniform diameter which attach to the head via a short, constricted region. In particles with contracted tails, the tails exhibit thick and thin regions. The thick region, which is the contracted sheath, is always found next to the constriction adjacent to the head and therefore serves to mark the head-proximal end of the tail (Inman et al. 1976).

When Mu particles are spread for electron microscopy by the protein film technique in a buffer containing the cross-linking agent formaldehyde at pH 10 (Chattoraj and Inman 1974), the phage heads are usually disrupted and unrecognizable. The phage tails are contracted but otherwise appear to be intact and are often attached to Mu DNA ends. The attachment of DNA to the end of the tail containing the contracted sheath indicates that it is the proximal end of the tail that is involved in DNA attachment.

Partial Denaturation Mapping of Mu DNA-Tail Complexes

When phage Mu is spread under the conditions described above but at pH 11.0–11.2, some of the DNA is still found attached to phage tails but in addition exhibits regions of partial denaturation. These partially denatured regions occur in characteristic locations (AT-rich regions) on the DNA and can be used to distinguish the different ends of the molecule.

The locations of denatured sites in 43 Mu DNA-tail complexes have been mapped. First, the DNA molecules were oriented using denatured sites at 2.8, 5.4, 8.0, and 11.2 μm. Since this set of molecules exhibited a rather wide total length distribution (13.9 ± 0.3 μm), they were not normalized to identical length but rather were individually lengthened or shortened to give best fit of all sites (the changes in length were small and resulted in a length distribution [13.9 ± 0.3 μm] unchanged from the original). Molecules were also shifted either to the right or left to give best fit of denatured sites (again the magnitude of these shifts was small [0.00 ± 0.14 μm]). The histogram average of the 43 molecules, shown in Figure 1, exhibits many unique melting regions characteristic of Mu DNA at 11.5% denaturation. The denaturation map has many denatured sites near the ends of the DNA, fewer denatured sites in the middle, and is remarkably symmetrical. In all the DNA-tail complexes observed, the phage tails are attached to a unique end of the DNA which corresponds to the right end of the denaturation pattern.

Partial Denaturation Mapping of Mu DNA Heteroduplexes

In order to relate the partial denaturation pattern to the known physical and genetic maps of Mu (Howe and Bade 1975), we have taken advantage of the fact that Mu DNA contains two characteristic regions of heterogeneity, the variable end and the G segment. The variable end is composed of a segment of bacterial DNA which differs from molecule to molecule and thus gives rise to nonhybridizing SE in reannealed DNA (Daniell et al. 1973b; Bade et al. 1974; Daniell

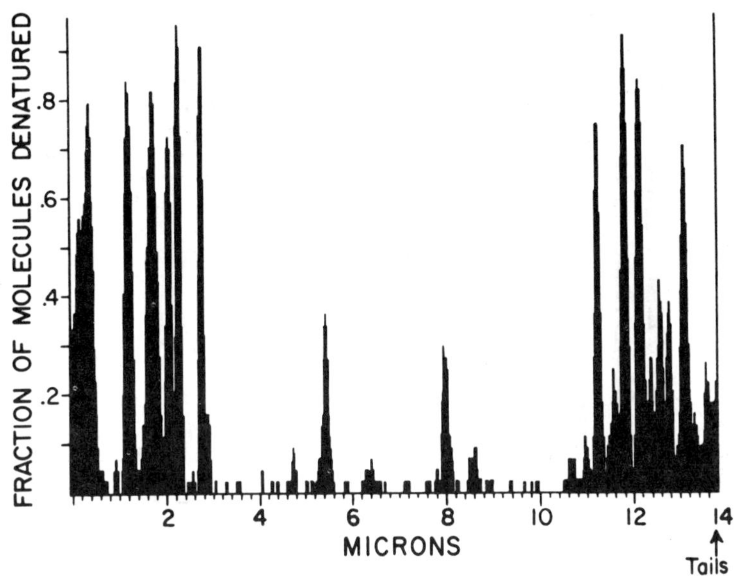

Figure 1
Histogram average of aligned denaturation maps of Mu DNA-tail complexes.

Preparation and purification of Mu (Mu *cts*62 *gov*3261 *momA*) lysates are described in Howe (1973a,b) and in Inman et al. (1976). Methods used for spreading phage and DNA are reported in Chattoraj and Inman (1974). High pH buffer used for partial denaturation mapping is given in Schnös and Inman (1970).

Samples were examined and photographed on 7-mm roll film in an EM-300 electron microscope. Molecules were measured using a projection desk (Physical Sciences Laboratory, Stoughton, Wis.) and a digitizer (Numonics Corporation, North Wales, Pa.) interfaced to a Hewlett-Packard 9820 calculator-plotter system. Methods for orienting and normalizing the lengths of the molecules are described in the text.

The histogram average of denatured sites from 43 DNA-tail complexes is given in the shaded regions of the figure.

et al. 1975). The G segment is a region that can be inverted with respect to the rest of the DNA (Hsu and Davidson 1974). These inversions result in single-stranded G bubbles in some proportion of the heteroduplexes formed when Mu DNA is denatured and reannealed. An investigation of the relationship between the regions of heterogeneity and known genes has shown that both heterogeneous regions occur at the right end of the genetic map (Daniell et al. 1973a).

We have therefore derived a denaturation map for partially denatured Mu heteroduplex molecules containing SE and G bubbles. The heteroduplex molecules were oriented with the G bubbles and SE at the right. They were not normalized individually but were only shifted slightly to the right or left to give best alignment of denatured sites. The histogram average of 35 heteroduplexes (Fig. 2) clearly shows that the denatured sites observed at the left end

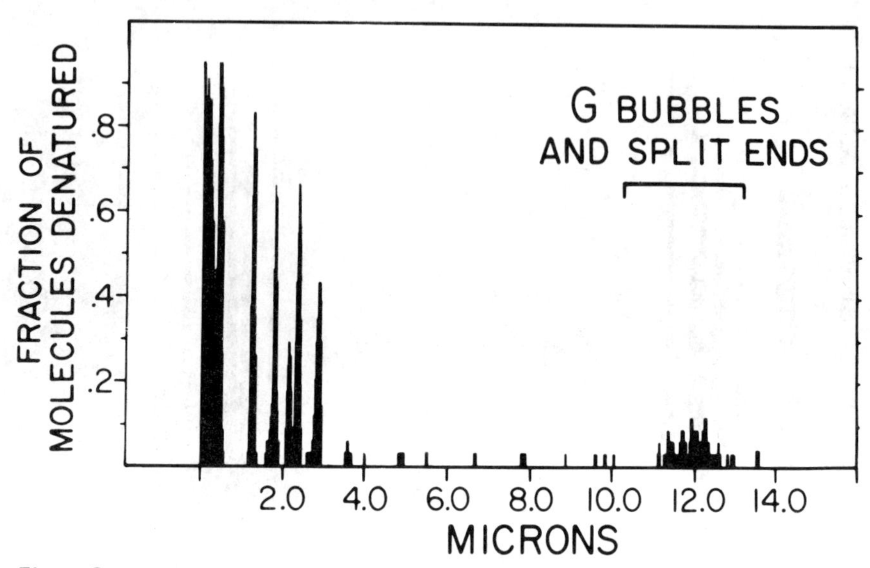

Figure 2

Denaturation pattern of Mu heteroduplexes. The heteroduplexes were oriented with G bubbles and SE on the right in the position indicated. The shaded areas of the histogram denote denatured sites. Techniques for heteroduplex preparation and subsequent partial denaturation mapping are reported in Chattoraj and Inman (1972).

of Mu heteroduplexes correspond very closely to those observed at the left end of DNA-tail complexes. Thus the tail is attached at the genetic right end, the variable end of the Mu genome.

Mapping of Mu DNA at Different Degrees of Denaturation

The DNA molecules in DNA-tail complexes depicted in Figure 1 were oriented using four characteristic denatured sites. Due to the symmetry of the partial denaturation pattern and to the existence of molecules lacking one or more of these denatured sites, the orientation of up to 25% of these molecules might be considered ambiguous. Therefore the attachment of the tail to the right end of the denaturation pattern was considered conclusive for only 75% of these molecules.

In an attempt to remove this ambiguity, the experiment was repeated using higher and lower degrees of denaturation. The histogram averages of 35 complexes at 9.4% denaturation and 39 complexes at 31.6% denaturation are shown in Figure 3 a and b. At low denaturation, the DNA molecules are denatured at several sites near each end, but the middle region of the DNA is primarily double-stranded. At higher denaturation, there are many denatured sites within the middle of the molecule, and often the sites at the right ends coalesce to form

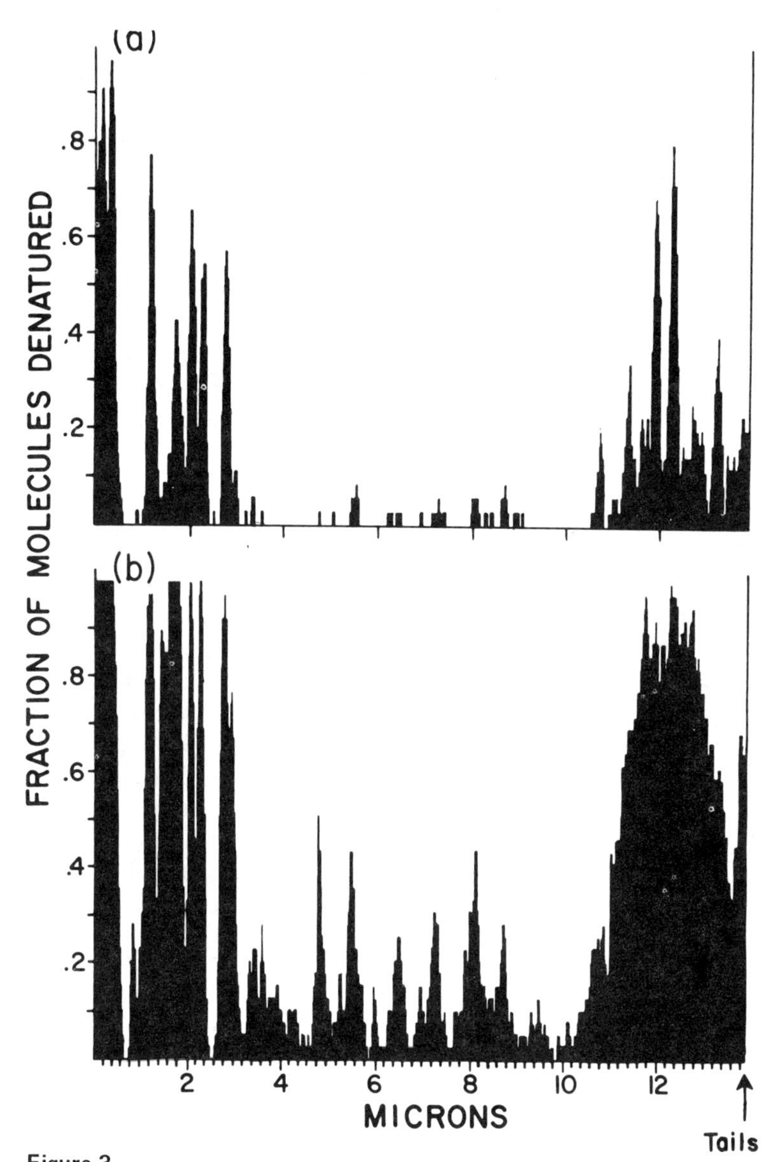

Figure 3

Denaturation patterns of Mu DNA-tail complexes at different degrees of denaturation.

The histograms represent partial denaturation patterns of Mu *cts*62 *gov*3261 *momA* species from a mixed lysate which contained in addition Mu *cts*62 *X*5004 in the proportion 55% Mu *gov* and 45% Mu *X*. The lysates were prepared as described in Inman et al. (1976). Further procedures are described in the legend to Fig. 1.

The histogram average for 35 complexes at 9.4% denaturation is given in *a* and that for 39 complexes at 31.6% denaturation is given in *b*. In all cases, the phage tail was situated at the right end of the denaturation map.

large bubbles. Although the denaturation patterns observed differ in the number of denatured sites and the number of molecules exhibiting each site, they remain rather symmetrical and still do not allow the unambiguous orientation of all the molecules.

Denaturation Mapping of Mu *X* DNA-Tail Complexes

Mu *X* DNA contains an insertion of 720 base pairs located 4000 base pairs from the immunity end of Mu DNA (Bukhari and Taylor 1975). We hypothesized that partial denaturation mapping of Mu *X* DNA-tail complexes should help resolve the orientation ambiguity in two ways: (1) the insertion should shift the denaturation pattern and make it less symmetrical; and (2) since the insertion maps near the immunity end of Mu, the DNA end exhibiting the shift must be the immunity end.

A mixed lysate of Mu and Mu *X* was examined under partially denaturing conditons similar to those used for Mu in Figure 1. As before, the DNA was attached to the proximal end of phage tails and exhibited partially denatured regions. On the basis of the denaturation pattern, the molecules could be divided into two classes: (1) those with a denaturation pattern similar to that of Mu observed previously and in which the tails were always at the right end of the denaturation map, and (2) those with a denaturation pattern expected for Mu *X*, where the Mu denaturation pattern was shifted 0.3 μm to the right from a position 11% from the left end. The relative frequencies of the two types of molecules were 45% and 55% for Mu *X* and Mu, respectively, values in agreement with those of Bukhari and Taylor (1975). The molecules were normalized to an average length of 13.9 μm in the way described for Figure 1 and were aligned at 0.0 μm. The histogram average denaturation patterns are given in Figure 4. As can be seen, the two leftmost denatured sites for Mu and Mu *X* coincide exactly, whereas the subsequent sites coincide only if the Mu pattern is shifted to the right by 0.3 μm, an amount equivalent to the size of the *X* insertion (compare Fig. 4 a and b).

In all the molecules thus characterized, the tails are unambiguously attached to the right end of the denaturation map.

Orientation of DNA-Tail Attachment in Aberrant Complexes

Out of 351 partially denatured DNA-tail complexes examined from Mu and Mu/Mu *X* lysates, 94.6% have the DNA attached to the proximal end of the tail, as described above. In 3.7% of the complexes, the DNA is attached to the *distal* end of the tail, and, in 1.7%, the tail is found at an internal position on the DNA molecule. When the DNA denaturation patterns of 12 such aberrant complexes were compared to the pattern observed for the normal complexes, it was clear

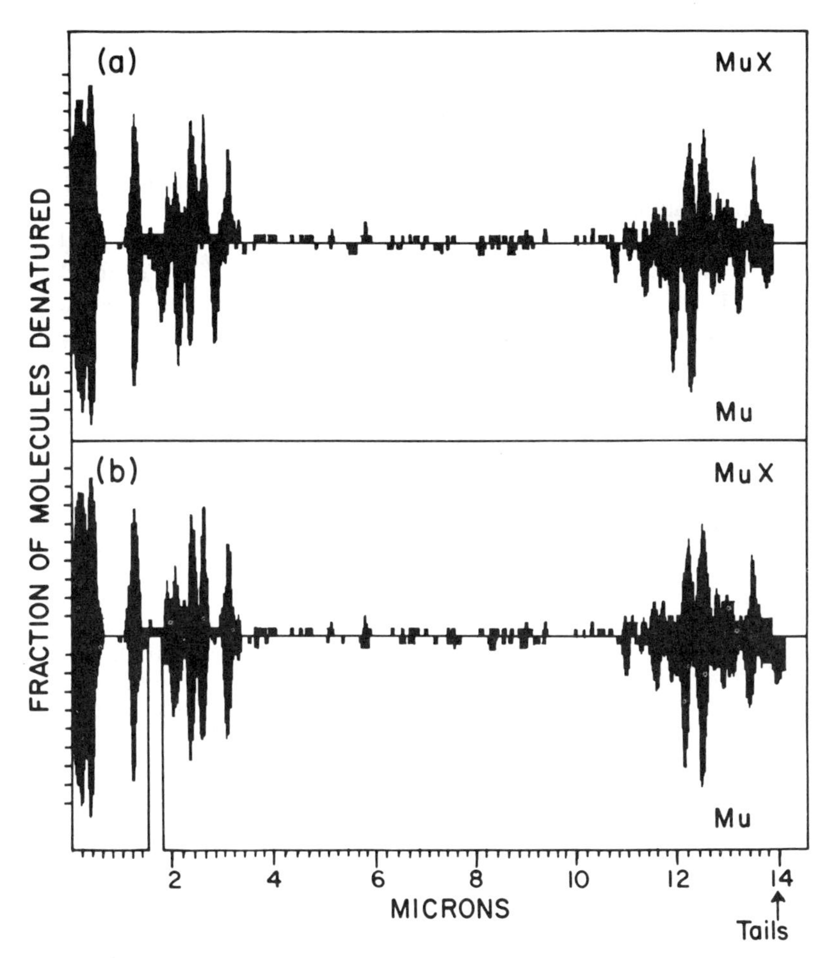

Figure 4
Denaturation map of Mu *X* DNA-tail complexes. (*a*) Histogram average denaturation pattern of Mu *X* (above) and Mu (below) DNA-tail complexes aligned at their left ends. (*b*) The same data presented in *a*, but with the Mu histogram shifted 0.3 μm to the right at the 1.5 μm position to allow for the insertion in Mu *X*.

that the DNA should be oriented such that the tails were at or near the left end of the denaturation pattern. Therefore, the aberrant DNA-tail association is due to partial ejection of the DNA through the phage tail. This observation supports the hypothesis that the unique end of Mu DNA packaged in association with the phage tail is also the end that leaves the phage particle first upon injection.

SUMMARY

When bacteriophage Mu is subjected to mild formaldehyde cross-linking conditions and spread for electron microscopy, the phage heads lyse, and one end of the released DNA is found attached to the proximal end of the phage tail. If spreading is performed at pH 11.0–11.2, the DNA exhibits characteristic regions of partial denaturation located predominantly near the ends of the Mu DNA molecule. Examination of the partial denaturation patterns of Mu DNA-tail complexes for wild-type Mu and for the insertion mutant Mu *X* reveals that the tail is attached to a unique end of the DNA. Comparison of denaturation maps of DNA-tail complexes with those of Mu DNA heteroduplexes containing G bubbles and SE indicates that the tail is attached to the variable end of Mu DNA. Similar analysis of denaturation maps of Mu DNA partially ejected through the tail demonstrates that the variable end is also the first end to be ejected from the phage particle. These results support the hypothesis that the variable end is the last section of Mu DNA to be packaged and the first to be released upon injection.

Acknowledgments

This research was supported by the College of Agriculture and Life Sciences, University of Wisconsin, Madison, and by National Science Foundation and National Institutes of Health grants to M. M. H., and National Institutes of Health and American Cancer Society grants to R. B. I.

REFERENCES

Bade, E. G., H. Delius and W. Schroeder. 1974. Structure and replication of Mu1 DNA. *Fed. Proc.* **33**:1487.

Bukhari, A. I. and A. L. Taylor. 1975. Influence of insertions on packaging of host sequences covalently linked to bacteriophage Mu DNA. *Proc. Nat. Acad. Sci.* **72**:4399.

Chattoraj, D. K. and R. B. Inman. 1972. Position of two deletion mutations on the physical map of bacteriophage P2. *J. Mol. Biol.* **66**:423.

——. 1974. Location of DNA ends in P2, 186, P4 and lambda bacteriophage heads. *J. Mol. Biol.* **87**:11.

Daniell, E., W. Boram and J. Abelson 1973a. Genetic mapping of the inversion loop in bacteriophage Mu DNA. *Proc. Nat. Acad. Sci.* **70**:2153.

Daniell, E., D. E. Kohne and J. Abelson. 1975. Characterization of the inhomogeneous DNA in virions of bacteriophage Mu by DNA reannealing kinetics. *J. Virol.* **15**:739.

Daniell, E., J. Abelson, J. S. Kim and N. Davidson. 1973b. Heteroduplex structures of bacteriophage Mu DNA. *Virology* **51**:237.

Howe, M. M. 1973a. Prophage deletion mapping of bacteriophage Mu-1. *Virology* **54**:93.

——. 1973b. Transduction by bacteriophage Mu-1. *Virology* **55**:103.

Howe, M. M. and E. G. Bade. 1975. Molecular biology of bacteriophage Mu. *Science* **190**:624.
Hsu, M.-T. and N. Davidson. 1974. Electron microscope heteroduplex study of the heterogeneity of Mu phage and prophage DNA. *Virology* **58**:229.
Inman, R. B., M. Schnös and M. M. Howe. 1976. Location of the "variable end" of Mu DNA within the bacteriophage particle. *Virology* **72**:393.
Schnös, M. and R. B. Inman. 1970. Partial denaturation of thymine- and 5-bromouracil-containing λ DNA in alkali. *J. Mol. Biol.* **49**:93.
Thomas, J. O. 1974. Chemical linkage of the tail to the right-hand end of bacteriophage lambda DNA. *J. Mol. Biol.* **87**:1.

Asymmetric Hybridization of Mu Strands with Short Fragments Synthesized during Mu DNA Replication

C. Wijffelman and P. van de Putte*
Laboratory of Genetics and *Laboratory of Molecular Genetics
State University of Leiden
Leiden, The Netherlands

The temperate *E. coli* phages such as λ and P2 have circular DNA replication intermediates which correspond in length to the size of the phage genome (MacHattie and Thomas 1964; Schnös and Inman 1971). Such molecules are not found during the development of bacteriophage Mu, although covalently closed circles are found. These are very heterogeneous in size, ranging from 36-156 kilobases, and contain both bacterial and phage DNA (Waggoner et al. 1974; Schroeder et al. 1974). Mu DNA synthesis starts at about 8 minutes after induction of a Mu lysogen (C. Wijffelman, as reported at the EMBO Workshop on Mu, Brussels, 1975), and the first covalently closed circles are detected about 15 minutes after induction (Waggoner et al. 1974; Schroeder et al. 1974). Thus heterogeneous circles arise after Mu DNA replication, and perhaps these circles represent structures which arise as a consequence of Mu replication. However, it is not clear whether these circles play a direct role in Mu DNA replication, and, until now, attempts to isolate intermediates in the replication of Mu have not been successful. Such intermediates were essential in the determination of the direction and origin of replication of λ, P2, and SV40 by electron microscopy and denaturation mapping (Schnös and Inman 1970, 1971; Fareed et al. 1972). Moreover, denaturation mapping would give complicated results in Mu because of the more or less symmetric denaturation map of Mu DNA (Breepoel et al. 1976). We have therefore sought to determine the direction of Mu replication by a method based on DNA-DNA hybridization.

EXPERIMENTAL RATIONALE

Semiconservative replication starts at a fixed point and proceeds sequentially along the DNA. Since the two strands of DNA have opposite polarity, the overall direction of chain growth on one strand is $5' \rightarrow 3'$ and on the other $3' \rightarrow 5'$. At least one of the two strands is replicated discontinuously, involving $5' \rightarrow 3'$ synthesis of short fragments (Okazaki fragments) and their joining by ligase

(Okazaki et al. 1969). Recently, it was proposed that the other strand is also replicated discontinuously but that the rate of joining of the fragments is faster (Kurosawa and Okazaki 1975). It has been shown that the short fragments always result from synthesis in the direction opposite to the movement of the replication fork (Louarn and Bird 1974). In P2, where by means of other methods replication had been proved to be unidirectional (Schnös and Inman 1971), it has been demonstrated that most of the short fragments anneal with one strand (Kurosawa and Okazaki 1975; Kainuma-Kuroda and Okazaki 1975). If in Mu the replication is unidirectional, a similar result might be obtained. Therefore we examined the pattern of hybridization of Mu fragments to the separated strands of Mu DNA.

RESULTS

Our experiment to determine whether Mu replication is bidirectional or unidirectional was as follows: A strain lysogenic for Mu *cts*62 was induced at 43°C. After 35 minutes, at which time most of the DNA synthesis is apparently Mu DNA synthesis, the DNA was labeled by a short pulse of [^{3}H]thymidine and subsequently fractionated as a function of size by sedimentation in an alkaline sucrose gradient with a cushion of saturated CsCl. The radioactivity was found in two positions in the gradient. The bulk of the DNA ("long DNA") sedimented on the CsCl cushion, whereas the "short DNA" (8-12S) sedimented in a broad peak in the alkaline sucrose. The long and short DNAs were pooled separately, purified by isopycnic CsCl centrifugation, sonicated, and hybridized to filters loaded with the *r* strand of Mu DNA, the *l* strand of Mu DNA, and denatured calf thymus DNA. The results are shown in Table 1, lines 1, 2, and 5. The labeled short DNA anneals mainly with the *l* strand of Mu DNA, whereas the long DNA anneals equally well with both strands of Mu. The interpretation of these results, however, is complicated by the fact that the G segment of Mu can undergo inversion (Daniell et al. 1973). The G segment in a prophage can occur in opposite orientations. When such an inverted G segment is replicated, short DNA will be synthesized, which would anneal with the "wrong" strand of the Mu DNA. This would lead to an extensive background of hybridization with the *r* strand of Mu DNA. Therefore, we also hybridized the fragments with separated strands of a λpMu hybrid phage which lacks the G region. This plaque-forming λ phage contains the *cABClys* part of Mu DNA (Bukhari and Allet 1975). By hybridization with uniformly labeled ^{3}H-Mu DNA, we first determined that the length of the Mu part of this phage is about 28% that of a Mu DNA molecule. Furthermore, we found by hybridization with purified Mu RNA, specific to the *r* strand, that the *r* strand of the hybrid phage also contained the *r* strand of the Mu DNA. The results of the hybridizations of the separated strands of this hybrid phage with the long and short DNAs obtained after induction of Mu are shown in Table 1, lines 3 and 4. Again, the labeled

Table 1
Strand Specificity of Pulse-labeled Mu DNA after Induction of a Mu Lysogen

	Radioactive DNA cpm bound to filter		Percent strand-specific Mu DNA	
DNA on filter	long DNA	short DNA	long DNA	short DNA
r-Mu	7335	2768	48.5	15.8
l-Mu	7786	13,292	51.5	84.2
r-λMu	2239	840	50.0	10.3
l-λMu	2229	4803	50.0	89.7
calf thymus	158	328	–	–

A culture (100 ml) of KMBL 1614, *trp*::(+Mu-1 *cts*62) was induced at 43°C. After 35 min, the culture was labeled for 15 sec with [^{3}H]thymidine (0.5 mCi, sp. act. 49 Ci/mM), lysed, and centrifuged in an alkaline sucrose gradient on a cushion of saturated CsCl. Fractions were assayed for radioactivity, and the long DNA on the CsCl cushions and the short DNA (8–12S) were pooled separately; these were purified by banding in CsCl. The DNA-containing fractions were pooled, dialyzed, sonicated, and hybridized to separated DNA strands bound to filters. The percentage of strand-specific Mu DNA in one fraction is given by the ratio: (radioactivity bound to one strand/radioactivity bound to both strands) × 100, after subtracting the background (DNA bound to calf thymus DNA on the filter).

short DNA anneals mainly with the *l* strand; however, the percentage of Mu DNA hybridizing with the *r* strand of this phage is lower than that with the *r* strand of Mu (column 5, lines 1 and 3). The difference may reflect the influence of the inverted G area.

INTERPRETATION

In bacteriophage P2, replication is unidirectional (Schnös and Inman 1971). Recently, under experimental conditions comparable to those used here, it was found that about 90% of the labeled P2 short DNA anneals with one strand (Kurosawa and Okazaki 1975; Kainuma-Kuroda and Okazaki 1975). By analogy, our interpretation of these results is that the replication of Mu DNA is completely or almost completely unidirectional.

The direction of replication can be inferred from these results by taking into account the previously determined direction of transcription of Mu (Wijffelman and van de Putte 1974). Since mRNA synthesis on Mu DNA is asymmetric and proceeds from left to right in the 5′ → 3′ direction on the *r* strand, the polarity of the Mu DNA strands is known (Fig. 1). Because the discontinuous synthesis of short fragments is on the *l* strand (from 5′ → 3′), the overall direction of Mu DNA replication is from left to right (Fig. 1). If the direction of replication is unidirectional throughout the life cycle of Mu, it is very likely that the start of Mu DNA synthesis takes place at the immunity end, the left end of Mu DNA.

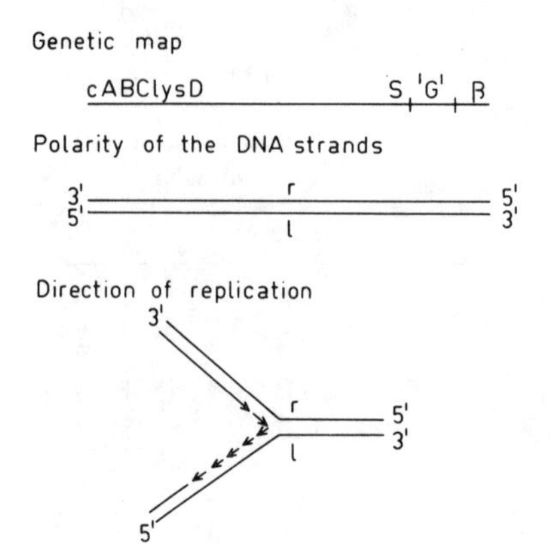

Figure 1
Polarity of Mu DNA showing the direction of replication fork.

SUMMARY

Short pieces of DNA synthesized at late stages of the lytic cycle of Mu hybridize preferentially to the *l* strand of Mu DNA. By analogy with the P2 system, this result is interpreted to mean that replication is unidirectional; the direction of replication is determined to be from left to right on the Mu DNA (Fig. 1).

Acknowledgments

We wish to thank M. Giphart and T. Goosen for many fruitful discussions and their criticism on the manuscript.

REFERENCES

Breepoel, H., J. Hoogendorp, J. E. Mellema and C. Wijffelman. 1976. Linkage of the variable end of the bacteriophage Mu DNA to the tail. *Virology* **74**:279.

Bukhari, A. I. and B. Allet. 1975. Plaque-forming λ-Mu hybrids. *Virology* **63**:30.

Daniell, E., W. Boram and J. Abelson. 1973. Genetic mapping of the inversion loop in bacteriophage Mu DNA. *Proc. Nat. Acad. Sci.* **70**:2153.

Fareed, G. C., C. F. Garon and N. P. Salzman. 1972. Origin and direction of simian virus 40 deoxyribonucleic acid replication. *J. Virol.* **10**:484.

Kainuma-Kuroda, R. and R. Okazaki. 1975. Mechanism of DNA chain growth. XII. Asymmetry of replication of P2 phage DNA. *J. Mol. Biol.* **94**:213.

Kurosawa, Y. and R. Okasaki. 1975. Mechanism of DNA chain growth. XIII. Evidence for discontinuous replication of both strands of P2 phage DNA. *J. Mol. Biol.* **94**:229.

Louarn, J. M. and R. E. Bird. 1974. Size distribution and molecular polarity of newly replicated DNA in *E. coli. Proc. Nat. Acad. Sci.* **71**:329.

MacHattie, L. A. and C. A. Thomas. 1964. DNA from bacteriophage lambda: Molecular length and conformation. *Science* **144**:1142.

Okazaki, R., T. Okazaki, K. Sakabe, K. Sugimoto, R. Kainuma, A. Sugino and N. Iwatsuki. 1969. In vivo mechanism of DNA chain growth. *Cold Spring Harbor Symp. Quant. Biol.* **33**:129.

Schnös, M. and R. B. Inman. 1970. Position of branch points in replicating λ DNA. *J. Mol. Biol.* **51**:61.
——. 1971. Starting point and direction of replication in P2 DNA. *J. Mol. Biol.* **55**:31.
Schroeder, W., E. G. Bade and H. Delius. 1974. Participation of *E. coli* DNA in the replication of temperate bacteriophage Mu-1. *Virology* **60**:534.
Waggoner, B. T., N. S. Gonzalez and A. L. Taylor. 1974. Isolation of heterogeneous circular DNA from induced lysogens of bacteriophage Mu-1. *Proc. Nat. Acad. Sci.* **71**:1255.
Wijffelman, C. and P. van de Putte. 1974. Transcription of bacteriophage Mu: An analysis of the transcription pattern in the early stage of phage development. *Mol. Gen. Genet.* **135**:327.

Mapping of Restriction Sites in Mu DNA

R. Kahmann, D. Kamp and D. Zipser
Cold Spring Harbor Laboratory
Cold Spring Harbor, New York 11724

Purification or cloning of specific restriction fragments would be helpful for a more detailed understanding of functions encoded in specific parts of the Mu genome. Furthermore, ligation of appropriate restriction fragments could be used to generate deletions, insertions, or inversions in Mu DNA itself. *Eco*RI and *Hin*dIII maps of Mu, and a partial *Hin*dII map, have been established previously by Allet and Bukhari (1975) and by Bade et al. (this volume). This paper reports our extension of the restriction map of Mu using 13 new enzymes. We will also describe preliminary fine-structure mapping of the left-terminal *Hin*dIII fragment with ten additional restriction enzymes and the purification of a fragment that contains the left end of Mu.

Mapping of Restriction Sites in Mu DNA

Mu *cts*62 *mom*$^-$ DNA was digested with various restriction enzymes and analyzed on agarose slab gels, as described by Sugden et al. (1975). To identify the origin of certain fragments, we made use of some of the characteristic features of Mu DNA. Both ends of Mu DNA are covalently linked to different amounts of heterogeneous bacterial DNA (Daniell et al. 1973; Allet and Bukhari 1975; Bukhari et al. 1976; Bade et al., this volume). The end fragments appear on gels as diffuse bands. The size variation of the left end is 50 to 100 base pairs (bp). This fragment appears diffuse only if its size is less than 2000 bp (Fig. 1, slot 1). Fragments originating from the variable end, the right end, contain a few hundred to 3200 bp of apparently random bacterial DNA and can be identified as very diffuse bands if they are larger than 4000 bp but smaller than 10,000 bp (Fig. 1, slot 11). If they are smaller than 4000 bp, they appear as broad smears on overexposed films. Sites in the G segment give rise to weaker bands owing to the two G orientations in Mu *cts*62 *mom*$^-$ DNA (Fig. 1, slot 10). The order of restriction sites was determined by using mixed digests in all possible combinations of enzymes (a few examples are shown in Fig. 1, slots 1-11). In a first approximation, we determined which particular restriction fragment was cleaved

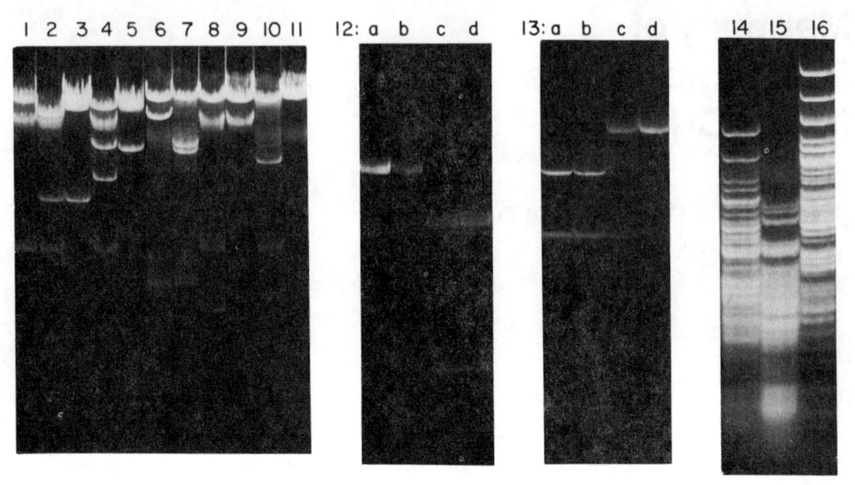

Figure 1
Agarose gel analysis of Mu DNA cleaved with various restriction enzymes. Mu *cts*62 *mom*⁻ DNA isolated from phage grown by heat induction was cleaved with *Pst*I (slot 1), with a mixture of *Pst*I and *Bam*HI (slot 2), with *Bam*HI (slot 3), with a mixture of *Pst*I and *Eco*RI (slot 4), with *Eco*RI (slot 5), with *Bal*I (slot 6), with a mixture of *Pst*I and *Bal*I (slot 7), with a mixture of *Pst*I and *Sal*I (slot 8), with *Sal*I (slot 9), with a mixture of *Pst*I and *Kpn*I (slot 10), and with *Kpn*I (slot 11). The fragments were analyzed on 1.2% agarose slab gels containing 0.5 μg ethidium bromide per ml and run under a constant voltage of 100 V for 3 hr. The purified left-terminal *Hin*dIII fragment of Mu DNA was cleaved with increasing amounts of *Hha*I (slots 12a–d) or with increasing amounts of *Hpa*II (slots 13d–a). Mu *cts*62 DNA was cleaved with *Hha*I (slot 14), a mixture of *Hha*I and *Hae*III (slot 15), and *Hae*III (slot 16). The cleavage products were analyzed on 2% agarose slab gels containing 0.5 μg ethidium bromide/ml and run for 2 hr under a constant voltage of 140 V. The DNA was visualized directly by fluorescence under short-wave UV light and photographed on Polaroid P/N 55 film, using Kodak no. 23 red filter.

by other enzymes. This, together with size estimations made with an *Eco*RI digest of λ*b*522 as a size marker, allowed us to map restriction sites of *Bal*I, *Bam*HI, *Bgl*II, *Kpn*I, *Pst*I, and *Sal*I in whole Mu DNA (Fig. 2). *Blu*I, *Sac*I, *Sac*II, *Sma*I, and *Xba*I do not cut Mu DNA, as was shown in mixed digests with *Eco*RI. The rightmost *Eco*RI site position was relocated 22,500 bp from the left end, refining the 25,000-bp estimate of Allet and Bukhari (1975). This was done by heteroduplex methods, using in-vitro-constructed λ-Mu hybrid phages (L. T. Chow, R. Kahmann and D. Kamp, in prep.). The position of the *Sal*I site was confirmed by L. T. Chow with heteroduplex methods. The mapping allows us to dissect Mu DNA into fragments averaging about 3000 bp, the largest being 6000 bp long. In the left *Eco*RI end fragment, the two *Bgl*I sites and the one

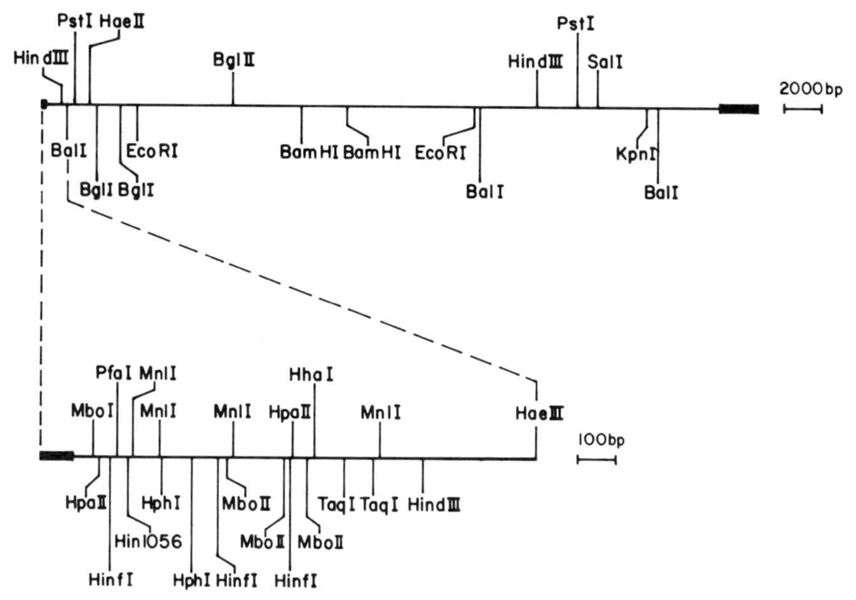

Figure 2
Restriction map of Mu DNA. The size of Mu DNA has been estimated to be 37,000 bp. This comprises the α segment (30,700 bp), the G segment (3000 bp), the β region (1700 bp), and the variable end with an average length of 1450 bp (Daniell et al. 1973; Bukhari and Taylor 1975; Chow and Bukhari, this volume). The light horizontal lines represent Mu; the heavy horizontal bars, heterogeneous bacterial DNA. The upper line represents whole Mu DNA, the lower one an enlargement of the left-terminal *Bal*I fragment. All cleavage sites are marked with vertical lines and lettered with the name of the corresponding restriction enzyme. (For a detailed description of the enzymes, see Roberts 1976.)

*Hae*II site have been mapped, so that we can generate even smaller pieces of this fragment. This result is of special interest because the early functions of Mu map in this region.

Mapping of Restriction Sites in the Left-terminal *Hin*dIII Fragment

For mapping of restriction sites in the 1000-bp-long left-terminal *Hin*dIII fragment of Mu, we could again make use of the heterogeneity of the left end of Mu DNA. Upon digestion of the purified *Hin*dIII fragment with other enzymes, two types of gel pattern could be observed. One pattern revealed only sharp bands (Fig. 1, slot 13a). Our interpretation is that one site is close to the left end and that the resulting end fragment is too diffuse to be visible. This could be proved with incomplete digests of the same fragment. A diffuse end fragment appears now as a partial product (Fig. 1, slot 13b, third band from the top). The other gel pattern observed shows one diffuse band, the end fragment, and, depending on the number of other sites, one or more sharp bands (Fig. 1, slot 12d).

To discover the order of the fragments, we performed either partial-digest (Fig. 1, slots 12 and 13) or mixed-digest experiments and took size measurements. In addition, smaller or larger end fragments of Mu were purified, cleaved, and compared with one another. The preliminary restriction map of the left-terminal *Hin*dIII fragment resulting from experiments with *Hpa*II, *Hha*I, *Hin*1056, *Hin*fI, *Hph*I, *Mbo*I, *Mbo*II, *Mnl*I, *Pfa*I, and *Taq*I is presented in the lower part of Figure 2. *Hae*III does not cut the left-terminal *Hin*dIII fragment. The first site was mapped at 1300 bp from the left end. The first *Hha*I site in Mu DNA is located at a distance of 600 bp from the left end. *Hha*I makes more than 30 cuts in Mu DNA. The diffuse end fragment, however, migrates with other fragments and cannot be recovered in pure form starting with whole Mu DNA (Fig. 1, slot 14). Since *Hae*III does not cleave the left-terminal *Hha*I fragment, we used a mixed digest of both enzymes. All comigrating and larger *Hha*I fragments are cut by *Hae*III (Fig. 1, slot 15). The largest remaining fragment is the diffuse, left-terminal *Hha*I fragment, which can now be isolated easily without going through several purification steps.

SUMMARY

Restriction enzyme cleavage maps of Mu DNA for the enzymes *Bal*I, *Bam*HI, *Bgl*II, *Kpn*I, *Pst*I, and *Sal*I have been constructed. No cleavage sites of the enzymes *Blu*I, *Sac*I, *Sac*II, *Sma*I, and *Xba*I have been detected. In addition, the results of a fine-structure analysis of the left-terminal *Hin*III fragment of Mu DNA using ten other restriction enzymes are presented.

Acknowledgments

We wish to thank R. J. Roberts and P. A. Myers for advice and for a generous supply of restriction enzymes. This work was supported by fellowships from the Deutsche Forschungsgemeinschaft to R. K. and D. K., and by research grants to D. Z. from the National Science Foundation (PCM74-00016 A01) and the U.S. Public Health Service (GM17612).

REFERENCES

Allet, B. and A. I. Bukhari. 1975. Analysis of bacteriophage Mu and λ-Mu hybrid DNA's by specific endonucleases. *J. Mol. Biol.* **92**:529.

Bukhari, A. I. and A. L. Taylor. 1975. Influence of insertions on packaging of host sequences covalently linked to bacteriophage Mu DNA. *Proc. Nat. Acad. Sci.* **72**:4399.

Bukhari, A. I., S. Froshauer and M. Botchan. 1976. Ends of bacteriophage Mu DNA. *Nature* **264**:580.

Daniell, E., J. Abelson, J. S. Kim and N. Davidson. 1973. Heteroduplex structures of bacteriophage Mu DNA. *Virology* **51**:237.

Roberts, R. J. 1976. Restriction endonucleases. *CRC Crit. Rev. Biochem.* **4**: 123.

Sugden, B., B. DeTroy, R. J. Roberts and J. Sambrook. 1975. Agarose slab gel electrophoresis equipment. *Anal. Biochem.* **68**:36.

SECTION IV
Nonhomologous Recombination and the λ Paradigm

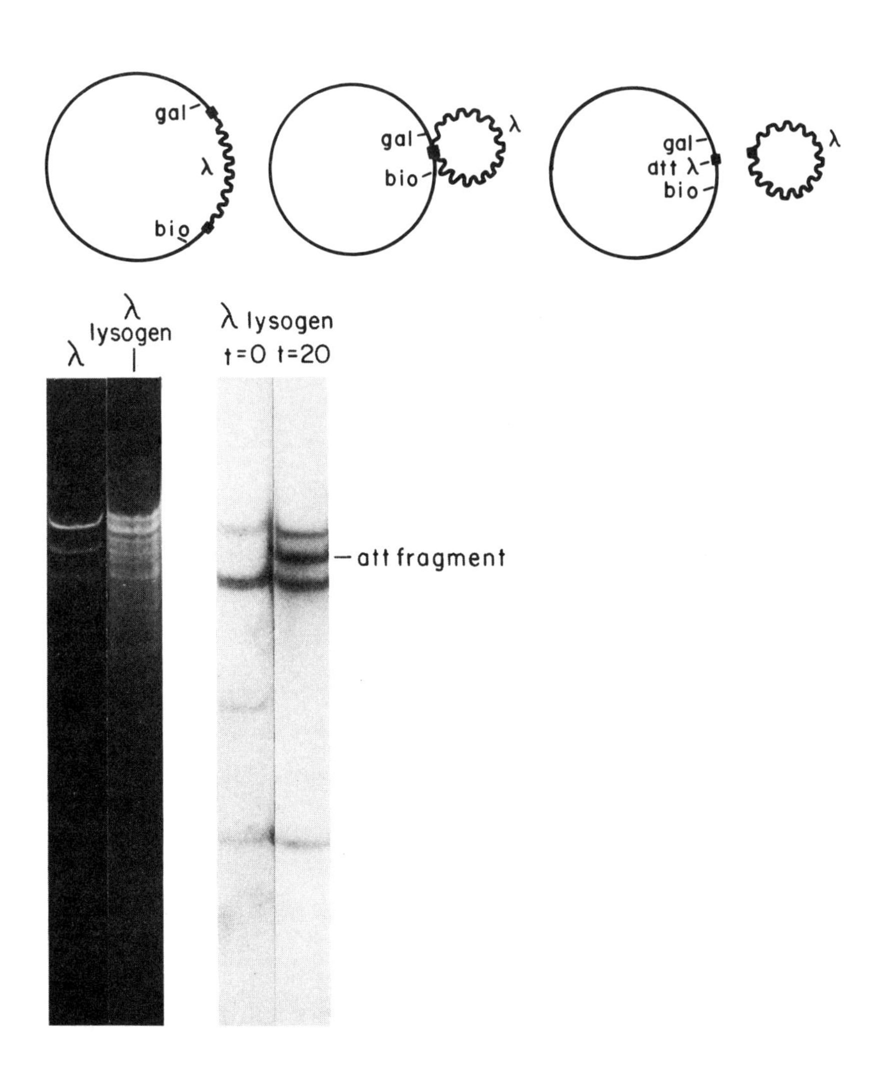

Insertion of the bacteriophage λ genome into the chromosome of *E. coli* has been a primary system for studies on integrative recombination. According to the classical model of Campbell, integration of λ involves recombination between the host DNA and a circular form of λ DNA. Specific sequences (attachment or *att* sites) in the host DNA and the λ DNA are recognized for recombination such that the process culminates in complete linear insertion of the circles. Rescue of the inserted DNA from the host is visualized as physical excision of the DNA in a manner that is a reversal of the insertion process. The *att* site for λ is located between the *gal* and the *bio* operons on the linkage group of *E. coli*. In the section-title figure, the diagram depicts integration-excision of λ from the chromosome of *E. coli*.

The gel pictures at the bottom show an in vivo assay for excision of prophage λ (E. Ljungquist and A. I. Bukhari, Cold Spring Harbor Laboratory). In this assay, DNA of the lysogenic cells is cleaved with a restriction endonuclease and fractionated in agarose gels. The DNA fragments are denatured in gels, transferred to nitrocellulose paper (Southern, *J. Mol. Biol.* 98:503 [1975]), and hybridized with ^{32}P-labeled phage DNA. The fragments containing the prophage DNA become visible after autoradiography. The gel pictures at the left show λ DNA and DNA from a λ lysogen cleaved with *Bgl*II and stained with ethidium bromide after electrophoresis in agarose. At the right is the lysogen DNA after hybridization with ^{32}P-labeled λ DNA. At $t = 0$, before prophage induction, the fragment containing the λ*att* site is absent. The arrow shows a λ-host junction fragment. The second junction fragment migrates close to another λ fragment and is not clearly resolved here. At $t = 20$ (20 min after prophage induction), the disappearance of the junction fragment and the appearance of the *att* fragment can be seen. This reaction is observable 10–15 minutes after prophage induction.

Flexibility in Attachment-site Recognition by λ Integrase

L. Enquist and R. Weisberg
Laboratory of Molecular Genetics
National Institute of Child Health and Human Development
National Institutes of Health
Bethesda, Maryland 20014

Insertion of phage λ DNA into the *Escherichia coli* chromosome, which occurs with high frequency after infection, adds a prophage to the host genome (Campbell 1962; see also Gottesman and Weisberg 1971). Prophage excision, which occurs shortly after lysogenic induction, reverses insertion: it regenerates exact copies of the original phage and host chromosomes. Insertion and excision require enzymes that are encoded by the *int* and *xis* genes of the phage, as shown in the following diagram:

$$\text{phage DNA} + \text{host DNA} \underset{\text{p}int,\ \text{p}xis}{\overset{\text{p}int}{\rightleftharpoons}} \text{inserted prophage.}$$

The *int* and *xis* proteins act by promoting recombination at specific sequences called attachment sites, one located in the phage DNA, one in the host DNA, and one at each terminus of an inserted prophage. To a first approximation, they do not promote recombination in regions lacking attachment sites nor do they promote insertion or excision of λ-related phages that carry structurally different but functionally analogous attachment sites (see Gottesman and Weisberg 1971). However, Shimada et al. (1972, 1973), armed with a selective method that allowed them to detect rare events, observed insertions of λ DNA at locations other than the normal host attachment site. These abnormal insertions resembled normal insertions in that the phage attachment site was used, *int* was required for their formation, and *int* and *xis* were required for their excision. Abnormal insertions have thus far been found at more than 25 sites with frequencies that vary markedly from one site to another. The cumulative frequency of abnormal insertion at all loci is between 0.1% and 1% that of insertion at the normal host attachment site.

The strong site preference of normal and abnormal λ insertion distinguishes λ from phage Mu, which inserts its DNA with approximately equal frequency at a great many loci on the host chromosome (see Howe and Bade 1975). The transposable drug resistance elements appear to resemble Mu more than λ in their insertion-site preference (this volume). Lambda also differs from Mu and the transposable drug resistance elements in excision: excision of prophage λ, when promoted by *int* and *xis,* always, or nearly always, regenerates the original nucleotide sequence of the host DNA (Shimada et al. 1972, 1973); excision of Mu or transposable drug resistance elements rarely does so (Bukhari 1975; Kleckner et al. 1975; this volume). These differences may reflect a fundamental dissimilarity in the mechanism of insertion or they may be less profound. Lambda integrase promotes recombination between different attachment-site pairs with frequencies that depend on the structure of the pair, the growth temperature, and the activity of *xis* protein (Guerrini 1969; Guarneros and Echols 1973). Phage Mu and the insertion elements probably produce proteins that, like λ integrase, promote insertion and excision. We suggest that all of these "*int*" proteins act by promoting recombination between a site in the inserting DNA and another in the receptor DNA and that differences between insertion of λ and that of the other DNAs reflect differences in the site preference of their *int* proteins. Site preference, however, is very likely to be mutable, and we surmise that mutations in *int,* for example, might change the frequency of abnormal relative to normal λ insertion. In fact, wild-type phage P2 already behaves like a λ mutant with reduced site specificity, and P2 variants with more stringent specificity have been obtained (Six 1966; see also Bertani and Bertani 1971). We also know that the prophage attachment sites are less promiscuous than the phage attachment site in participating in recombination with other sites (see Gottesman and Weisberg 1971). Again, *int* mutations might increase the promiscuity of the prophage attachment sites and thus contribute to imprecise excision.

In the remainder of this paper, we shall describe our methods for finding unusual *int* mutants and present evidence showing that some *int* mutations impair recombination between one pair of attachment sites but not another. This demonstrates that small changes in *int* protein structure can alter its site preference.

Wild-type λ*int* Protein Promotes Insertion and Excision at Abnormal Sites

The high relative frequency of abnormal λ insertions suggests that they are promoted by wild-type *int* protein interacting with an abnormal substrate rather than a mutant protein interacting with its normal substrate. To confirm this, we induced six abnormal lysogens and measured qualitatively the ability of the resulting phage to integrate in a wild-type and an attachment-site-deleted host (Table 1). In all cases, at least 99% of the phage resembled wild-type λ in that they integrated well in the wild-type and poorly in the attachment-site-deleted host. A small proportion of the phage failed to integrate in either host; these phage, unlike the majority class, were also *int*$^-$ and *xis*$^-$ by the test

Table 1
Insertion of Phage Released from Abnormal λ Lysogens

Lysogen	Prophage location	Burst size (phage/cell)	Number of plaques tested	Number integrating into *att*BOB′	Number integrating into (*att*BOB′)$^{\Delta}$
RW347	*proA/B*	5	350	350	0
RW363	*tsx*	0.001	360	360	0
RW460	near *lys*	0.8	375	375	0
KS507	*trpC*	0.006	263	261[a]	0
RW594	*malB*	0.02	401	401	0
RW614	*galT*	0.03	263	262	0

All lysogens carried single prophages as determined by sensitivity to λ*c*I*c*17 (Shimada et al. 1972). Lysogens were induced as described by Shimada et al. (1972), and aliquots giving about 100 plaques were plated on strain KS2 (*att*λ)$^+$ and on KS2 (*att*λ)$^{\Delta}$. The plaques were tested for lysogens as described by Gottesman and Yarmolinsky (1968). The burst of a normal lysogen is about 100 phage per cell.

[a]The two strains not integrating were analyzed and found to be *int*$^-$*xis*$^-$ and resistant to EDTA.

described in the next section and are presumably deletion or substitution mutants of the *int-xis* region. We conclude that an aberrant activity of wild-type *int* protein promotes insertion and excision at abnormal bacterial sites. We suspect, however, that a more powerful selection would uncover phage mutants that insert and excise well at abnormal host sites and we are currently seeking such mutants.

Selection of *int* Mutants with Changed Site Preference

We have devised a method for selecting a large number of excision-defective mutants. It is based on the observation that reversion to wild type of a λ insertion within a host gene does not occur without *int* and *xis* function. This is shown for λ insertions within the *galT* gene (Table 2). Heteroimmune superinfection with λ*int*$^+$*xis*$^+$ increases the frequency of *gal*$^+$ revertants by several orders of magnitude. The intermediate level of reversion of the λ*int*$^+$*xis*$^+$ insertion without heteroimmune superinfection probably reflects basal expression of the two genes in the presence of prophage repressor. Similar results have been obtained for λ insertions in *proA/B* and *trpC* (Shimada and Campbell 1974; L. Enquist and R. Weisberg, unpubl.). We conclude that precise excision of intragenic insertions of λ DNA is not promoted by host functions acting alone but requires the phage-encoded excision functions. This has also been inferred for insertions of phage Mu DNA (Taylor 1963; Bukhari 1975). We have therefore used the absence of reversion of an intragenic λ insertion after superinfection as a screen for deficiency in *int* or *xis* function of the superinfecting phage (Enquist and Weisberg 1976).

Table 2
Reversion of λ Insertions in *galT*

Prophage genotype	*gal*$^+$ revertants per cell plated	
	no superinfection	superinfection
int$^-$*xis*$^+$	$\leqslant 10^{-8}$	4×10^{-3}
int$^+$*xis*$^-$	$\leqslant 10^{-8}$	4×10^{-3}
int$^+$*xis*$^+$	0.3×10^{-3}	1×10^{-3}

The λ and the λ*xis am*6 insertions in *galT* were isolated in a *galE* (*att*λ)$^\Delta$ host as described by Shimada et al. (1972). The λ*int am*29 insertion was isolated in the same way except that the host was coinfected with wild-type λ. The normal bacterial attachment site and the *galE*$^+$ allele were replaced in each strain by P1 transduction, as described by Enquist and Weisberg (1976). The cells were grown to saturation in tryptone broth and plated on Penassay-galactose-TTC agar (Shimada et al. 1972) with or without superinfection with two particles per cell of λ*imm*^{434} as indicated. The number of cells plated was always greater than 10^8 per plate; where necessary, an aliquot of strain 594 (λ*imm*^{434}*int am*29) was added. This *gal* mutant does not revert to *gal*$^+$. Red *gal*$^+$ papillae were counted after 40 to 48 hr incubation at 32°C. The number of cells plated was determined by spreading an approximately diluted aliquot on a nutrient plate.

The number of revertants per cell plated gives a high estimate of the true reversion frequency since *gal*$^-$ cells can grow and be reinfected after plating on Penassay-galactose-TTC agar. To obtain a better estimate, uninfected cells were spread on galactose-minimal plates. The frequency of *gal*$^+$ revertants was $\leqslant 10^{-10}$ for the λ*int* and λ*xis* insertions and 5×10^{-8} for the wild-type insertion.

To find mutants with a changed site preference, we examined two collections of *int* mutants, one selected for defective excision as described above, the other for defective insertion (Gottesman and Yarmolinsky 1968). For each set, we checked the unselected phenotype. We also examined three conditional *int* mutants in semipermissive conditions. We found seven unusual mutants: six (including two tested by Guarneros and Echols 1970) were less deficient for excision than for insertion (Table 3, class I), and one was more defective for excision than for insertion (class II). Class-II mutants were indeed atypical: only 1 of 143 excision-defective *int* mutants was insertion-proficient. Class-I mutants were easier to find (see Table 3 legend). We conclude that simple mutational alterations of *int* protein can change its site preference. The structural basis of these changes is unknown, and we postpone further discussion of this to a later report (L. Enquist and R. Weisberg, in prep.).

DISCUSSION AND CONCLUSIONS

We find that wild-type integrase is sufficiently flexible to act at abnormal host sites with low efficiency and that mutation can change its ability to distinguish between different pairs of normal attachment sites. Shulman and Gottesman (1973) and Shimada et al. (1975) have proposed that λ insertions in the host

Table 3
Lambda Integrase Mutants with Altered Site Preference

Mutants	Recombination frequency (% of *int*$^+$) integrative	excisive
Class I		
*int ts*2001 (36°C)	0.1[a]	100[a]
*int ts*2004 (36°C)	0.1[a]	100[a]
*int am*29 (*supD*)	1.0[a]	50[a]
*int*35[b]	low[c]	high[c]
*int*5[d]	low[d]	high[d]
*int*7[d]	low[d]	high[d]
Class II		
*int*2268[e]	50[a]	1.0[a]

[a] Integrative and excisive recombination were measured in λ*att*^{2} *int2red*114*imm*^{21}*c* carrying either *att*P-*att*B (Nash 1974) or *att*L-*att*R, respectively (Shulman and Gottesman 1971). For *int ts*2001, *int ts*2004, and *int am*29, the extent of recombination was determined by measuring EDTA-resistant phages after infection of *recA* strains lysogenic for λ*int-c*226 derivatives of each mutant. The *int ts* mutants gave the results shown at the intermediate temperature of 36°C. The *int am*29 mutant exhibited the effect shown when suppressed by *supD*. All three conditional mutants were integration- and excision-defective in nonpermissive conditions.

[b] Lambda *int*35 was one of the 14 integration-defective mutants isolated using the EMBO test (Gottesman and Yarmolinsky 1968).

[c] Integrative recombination was measured with the EMBO test and excisive recombination with the red plaque test (Enquist and Weisberg 1976). "Low" means no detectable integration and "high" means a normal or near normal red plaque.

[d] Lambda *int*5 and λ*int*7 were tested by Guarneros and Echols (1970).

[e] Lambda *int*2268 was isolated because it made a colorless plaque. It integrated with normal efficiency by the EMBO test. The numbers given in the table were obtained as described in footnote *a* above.

chromosome occur only at sites containing a specific nucleotide sequence, called the common core. The common core, which is also present in the phage attachment site, is thought to have a dual function (Shulman and Gottesman 1973): it interacts specifically with *int* protein and it allows base pairing between complementary single strands of the phage and host attachment sites. Shulman et al. (1976) have recently suggested that sites of abnormal λ insertion, in contrast to the normal bacterial site, contain only a segment of the common core. If the minimum required segment length is small, the number of potential insertion sites may be very large indeed. The less frequently utilized ones may be detectable only when the relative frequency of insertions at the preferred site(s) is decreased by phage mutations. If such mutations are found, the insertion of λ, Mu, and the transposable drug resistance elements may be seen as variations on a common theme.

REFERENCES

Bertani, L. and G. Bertani. 1971. Genetics of P2 and related phages. *Adv. Genet.* **16**:199.

Bukhari, A. 1975. Reversal of mutator phage Mu integration. *J. Mol. Biol.* **96**:87.

Campbell, A. 1962. The episomes. *Adv. Genet.* **11**:101.

Enquist, L. and R. Weisberg. 1976. The red plaque test: A rapid method for identification of excision defective variants of bacteriophage lambda. *Virology* **72**:147.

Gottesman, M. E. and R. Weisberg. 1971. Prophage insertion and excision. In *The bacteriphage lambda* (ed. A. D. Hershey), p. 113. Cold Spring Harbor Laboratory, Cold Spring Harbor, New York.

Gottesman, M. E. and M. Yarmolinsky. 1968. Integration negative mutants of bacteriophage lambda. *J. Mol. Biol.* **31**:487.

Guarneros, G. and H. Echols. 1970. New mutants of bacteriophage λ with a specific defect in excision from the host chromosome. *J. Mol. Biol.* **47**:565.

——. 1973. Thermal asymmetry of site-specific recombination of bacteriophage λ. *Virology* **52**:30.

Guerrini, F. 1969. On the asymmetry of λ integration sites. *J. Mol. Biol.* **46**:523.

Howe, M. and E. Bade. 1975. Molecular biology of bacteriophage Mu. *Science* **190**:624.

Kleckner, N., R. Chan, B.-K. Tye and D. Botstein. 1975. Mutagenesis by insertion of a drug-resistance element carrying an inverted repetition. *J. Mol. Biol.* **97**:561.

Nash, H. 1974. λ*att*B-*att*P, a λ derivative containing both sites involved in integrative recombination. *Virology* **57**:207.

Shimada, K. and A. Campbell. 1974. Lysogenization and curing by Int-constitutive mutants of phage λ. *Virology* **60**:157.

Shimada, K., R. Weisberg and M. E. Gottesman. 1972. Prophage lambda at unusual chromosomal locations. I. Location of the secondary attachment sites and the properties of the lysogens. *J. Mol. Biol.* **63**:483.

——. 1973. Prophage lambda at unusual chromosomal locations. II. Mutations induced by bacteriophage lambda in *Escherichia coli* K12. *J. Mol. Biol.* **80**:297.

——. 1975. Prophage lambda at unusual chromosomal locations. III. The components of the secondary attachment sites. *J. Mol. Biol.* **93**:415.

Shulman, M. and M. E. Gottesman. 1971. Lambda att^2 : A transducing phage capable of intramolecular *int-xis* promoted recombination. In *The bacteriophage lambda* (ed. A. D. Hershey), p. 477. Cold Spring Harbor Laboratory, Cold Spring Harbor, New York.

——. 1973. Attachment site mutants of bacteriophage lambda. *J. Mol. Biol.* **81**:461.

Shulman, M. K. Mizuuchi and M. E. Gottesman. 1976. New *att* mutants of phage λ. *Virology* **72**:13.

Six, E. 1966. Specificity of P2 for prophage site I on the chromosome of *Escherichia coli* strain C. *Virology* **29**:106.

Taylor, A. L. 1963. Bacteriophage-induced mutation in *Escherichia coli. Proc. Nat. Acad. Sci.* **50**:1043.

Isolation of *Escherichia coli* Mutants Unable to Support Lambda Integrative Recombination

H. I. Miller and D. I. Friedman
Department of Microbiology
University of Michigan Medical School
Ann Arbor, Michigan 48109

Integration of λ into the *Escherichia coli* chromosome, according to the Campbell model (1962), requires both unique sites on the phage (*att*P) and bacterial (*att*B) chromosomes as well as the product of the phage *int* gene (Gottesman and Weisberg 1971) (Fig. 1). There is no direct evidence that this site-specific recombination process requires any bacterial products. However, recent evidence based on an in vitro study of λ integrative recombination suggests a role for host functions (H. Nash, pers. comm.). Excision of λ from the *E. coli* chromosome, the reverse of integration, requires two unique sites, *att*L and *att*R (see Fig. 1), and the products of the phage *int* and *xis* genes (Gottesman and Weisberg 1971).

Recombination between the various phage, bacterial, and hybrid attachment sites appears to occur between partially nonhomologous regions of DNA and does not require the participation of the host generalized recombination system (Rec) (Gottesman and Weisberg 1971). Other elements able to insert into the *E. coli* chromosome, such as phage Mu (Couturier 1976), IS elements (see Starlinger and Saedler 1972 for review), and translocatable antibiotic resistance factors (Kleckner et al. 1975; Gottesman and Rosner 1975; Berg et al. 1975), also appear to insert by a Rec-independent mechanism involving recombination between nonhomologous regions of DNA. It is likely, therefore, that host factors involved in λ integration may also play a role in translocation of other elements that appear to depend on exchanges of nonhomologous DNA for their insertion into the *E. coli* chromosome.

In order to determine if any host factors are needed in the λ integration reaction (and if so how many), we designed a procedure to select mutants of *E. coli* unable to support λ integrative recombination. Reported here is the preliminary characterization of two mutants isolated by this procedure. These mutants are designated *him* for *h*ost *i*ntegration *m*ediator.

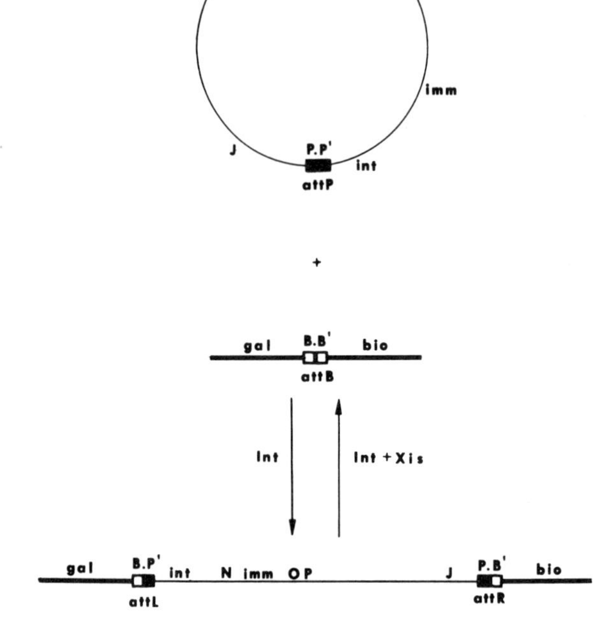

Figure 1 Integration and excision of λ. (Reprinted, with permission, from Gottesman and Weisberg 1971.)

The Selection

The procedure to select *him* mutants uses a λcI857*cam*1*int-cNam*7*Nam*53 multiple mutant as the selective agent and is based on the following considerations: In the absence of *N* activity, λ expresses few functions: replication functions *O* and *P* are expressed at a low level (Ogawa and Tomizawa 1968), and integration and repression functions are not expressed to any significant degree (Gottesman and Weisberg 1971). Although the N^- phage cannot grow normally, it can maintain itself as an autonomously replicating plasmid (Signer 1969; Court and Sato 1969; Lieb 1970). The presence of the *int-c* mutation (Int-constitutive) permits constitutive and high-level production of integrase, the product of the *int* gene (Shimada and Campbell 1974). When integrated into the bacterial chromosome, the unrepressed λN^- prophage is lethal to the host—a lethality due to replication in situ by the N^- phage (Eisen et al. 1969). Thus the infecting *int-c* phage should always integrate and kill the host bacterium unless λ integration is blocked. The presence of the chloramphenicol resistance determinant (*cam*1) (Gottesman and Rosner 1975) in the phage genome and of chloramphenicol in the selective media insure that only infected cells can survive. Thus infected bacteria in which the λN^-*cam int-c* cannot integrate harbor the phage as a plasmid and form chloramphenicol-resistant colonies. All host and phage strains used in this study are listed in Table 1.

Table 1
Bacterial and Phage Strains

	Relevant genotype	Source
Bacteria		
K-37	Su°	M. Yarmolinsky
K-294	Su°, Δ(*gal-att*λ-*bio*)	R. Weisberg
K-125	Su°, $(\lambda^{++})_n$	this laboratory
K-594	*him*57	this laboratory
K-568	*him*79	this laboratory
Phage		
λN^-*cam int-c*	*cam*1*int-c*226*Nam*7,53*c*I857 derived from *cam*1*red*3*c*I857*S*7	J. L. Rosner
λ*att*B-*att*P	*att*B-*att*P*int*2*xis*1*red*114*imm*^{21}*c*	H. Nash
λ*att*L-*att*R	*att*L-*att*R*int*2*xis*1*red*114*imm*^{21}*c*	H. Nash
λ*xis*$^-$*red*$^-$	*xis*1*red*3*c*I857	H. Nash
λ*int-c red*$^-$	*h int-c*226*red*3*c*I857	L. Enquist
λ*red*$^-$	*red*3*c*I857*S*7	this laboratory
λ*att*λ	*c*I857	M. Yarmolinsky
λ*att*80	*att*80*int*80*c*I857	R. Weisberg

The results shown in Table 2 demonstrate the effectiveness of this selection procedure. K-37, a Su° host, when infected with λN^-*cam int-c* survives on chloramphenicol media at a frequency of about 10^{-5}. In contrast, K-294, a Su° host unable to integrate λ because of the attachment-site deletion, shows nearly complete survival.

In addition to *him* and bacterial attachment-site mutants, a number of other classes of bacterial and phage mutants might be selected by this procedure: these include bacterial mutants interfering with λ DNA replication and phage mutants that eliminate expression of the replication genes, inactivate the *int-c* phenotype, or change the attachment site. To increase the probability of obtaining host mutants and thus reduce the effect of phage mutants of all types, the bacterial strain K-37 was mutagenized with nitrosoguanidine (Adelberg et al. 1965) prior to selection. This treatment results in a 100-fold increase in bacteria surviving the infection.[1] Using the method of Gottesman and Yarmolinsky (1968), survivors of this selection were then tested for their ability to support λ integration. Because one obvious class of survivors would be expected to carry mutations or deletions of the λ attachment site, we also tested for integration of a phage carrying the attachment specificity of φ80. The attachment site for φ80 lies distant from that for λ (Gottesman and Weisberg 1971).

[1] An alternative means of reducing the effect of phage mutants, namely increasing the multiplicity of infection, is not feasible in view of the protective effect of multiple copies of λN^- plasmids on bacteria carrying unrepressed λN^- prophage (Friedman and Yarmolinsky 1972).

Table 2
Survival of Bacteria Infected with λ*N*⁻*cam int-c* in the Presence of Chloramphenicol

Bacteria	Survival frequency
K-37	10^{-5}
K-294 (Δ*att*λ)	> 0.9
K-568 (*him*79)	> 0.9

Bacteria were grown to a concentration of 5×10^8/ml in tryptone broth containing 0.2% maltose 10 mM $MgSO_4$ (TAB) and then infected with λN^-*cam int-c* at a multiplicity of infection of 0.5. Adsorption was for 60 min at 31°C. Aliquots of dilutions were spread on LB agar plates containing 12.5 μg/ml chloramphenicol and 10 mM EGTA (Sigma). Plates were incubated for 18 hr at 34°C. Survival frequency is the number of survivors on chloramphenicol plates/the number of infecting *cam*1 phage (~ 60% of the phage in the lysate carried *cam*1). For details on all media, see Friedman and Yarmolinsky (1972).

Survivors unable to integrate both λ*att*λ and λ*att*80 were subsequently cured of the λ plasmid by streaking on agar plates (Antibiotic II Media–Difco) containing 25 μg/ml acridine orange (Eastman) (Friedman and Yarmolinsky 1972) and tested for chloramphenicol sensitivity.

Integrative Recombination of λ in *him* Mutants

Methods for quantitatively measuring integrative and excisive recombination have been published (Shulman and Gottesman 1971; Nash 1974, 1975). For integrative recombination, the λ variant λ*att*B-*att*P is used. This phage carries bacterial genes flanked by the normal substrates for the Int system. The larger genome of λ*att*B-*att*P results in phage particles that are sensitive to the heavy metal chelating agent EDTA (Parkinson and Huskey 1971). During growth in the presence of an active Int system, the bacterial genes are deleted by intramolecular recombination, resulting in phage particles insensitive to EDTA treatment.

In the experiments outlined in Table 3, we determined the extent to which *him* mutants could support integrative recombination by comparing the frequency of formation of EDTA-resistant recombinant phage following single-cycle growth of λ*att*B-*att*P in both him^- and him^+ bacteria. Since the λ*att*B-*att*P phage is int^-, Int function must be supplied in *trans*. The efficient formation of recombinant phage in K-37 (him^+) does not differ significantly when either $\lambda xis^- red^-$ or λ*int-c* red^- is used to supply *int* gene product at any of the three temperatures tested. At 37°C, recombinant formation in *him*57 is reduced slightly. However, at 38.5°C, *him*57 shows a 20-fold reduction in the formation of EDTA-resistant progeny. In contrast, *him*79 exhibits a 40–100-fold reduction in recombinant formation at either temperature tested when $\lambda xis^- red^-$ is used to supply *int* gene product. The effect of the *him*79 mutation is partially overcome if λ*int-c* red^- is used to supply *int* gene product, probably a consequence of the

Table 3
Integrative Recombination in λ*att*B-*att*P

Bacteria	Temperature °C	Percent EDTA-resistant progeny using Int supplied by	
		λ*xis*⁻*red*⁻	λ*int-c red*⁻
K-37	31	48.2	66.6
	37	68.8	52.3
	38.5	30.0	52.0
K-594 (*him*57)	37	26.9	–
	38.5	1.9	–
K-568 (*him*79)	31	1.3	10.0
	37	0.5	3.7

Bacteria were grown to a density of ~2×10^8/ml in TAB, pelleted, and resuspended at the same concentration in 10 mM $MgSO_4$. The cells were then infected with either λ*xis*⁻*red*⁻ or λ*int-c red*⁻ at a multiplicity of 5 and λ*imm*^{21}*att*B-*att*P at a multiplicity of 1. Following a 20-min adsorption at room temperature, the infected cultures were diluted 1000-fold into prewarmed TAB and incubated for 90 min at the indicated temperatures. Chloroform was added, samples were incubated an additional 20 min, and the lysates were titrated for total *imm*^{21} phage on tryptone agar plates using a λ lysogen (K-125) as the bacterial lawn. Recombinant phage were assayed on the same lawn after the following additional treatment: A 100-fold dilution of the lysate was incubated in 10 mM Tris-HCl (pH 7.4) and 10 mM EDTA at 41°C for 15 min. The $MgSO_4$ concentration was adjusted to 20 mM prior to titration. Phage adsorption in all cases was >90%, and the burst sizes of all phages in both *him*⁻ mutants were equivalent to that in the *him*⁺ host. Lysates of λ*att*B-*att*P initially contained 0.2% EDTA-resistant phage. The *him*57 mutant was selected at 40°C, whereas the *him*79 mutant was selected at 34°C.

higher levels of integrase expressed from the constitutive promoter (Shimada and Campbell 1974).

Excisive Recombination of λ in *him* Mutants

The technique for measuring excisive recombination employs another variant of λ, λ*att*L-*att*R (Shulman and Gottesman 1971). Since the added bacterial genes are carried between the left and right prophage attachment sites, *att*L and *att*R, efficient formation of EDTA-resistant recombinant phage progeny is dependent on both *int* and *xis* gene products.

Table 4 shows the results of single-cycle growth experiments of λ*att*L-*att*R performed in *him*⁺ and *him*⁻ hosts. Under conditions where integrative recombination is severely reduced in the *him*⁻ mutants, excisive recombination shows at most a two- to threefold reduction.

Generalized Recombination in *him* Mutants

Generalized recombination was tested by studying recombination between λ mutants. The effect of the phage generalized recombination system (Red) was

Table 4
Excisive Recombination in λ*att*L-*att*R

Bacteria	Temperature °C	Percent EDTA-resistant progeny
K-37	31	69.4
	37	68.5
	39	47.6
K-594 (*him*57)	37	36.8
	39	13.9
K-568 (*him*79)	31	22.2
	37	24.4

Same procedure as given in the legend to Table 3, except that λ*red*$^-$ was used at a multiplicity of 5 and λ*imm*^{21}*att*L-*att*R was used at a multiplicity of 1.

eliminated by the use of *red*$^-$ phage (Signer 1971). The effect of Int-promoted recombination was eliminated by the use of markers located on the same side of the *att*P site (Gottesman and Weisberg 1971). The frequency of formation of amber$^+$ recombinant phage from crosses of two λ amber mutants in both *him*57 and *him*79 is equivalent to that in K-37 (data not shown). We conclude that generalized recombination is not affected by *him* mutations. Confirming this conclusion is the observation that a λ derivative carrying the *bio*11 substitution forms plaques on both *him*57 and *him*79. This phage is unable to form plaques on *recA*$^-$ hosts (Zissler et al. 1971).

Establishment of Repression in *him* Mutants

Phage carrying either the wild-type *c*I repressor or the temperature-sensitive *c*I857 repressor form turbid plaques on both *him* mutants (the latter at 32°C). This result indicates that the failure of λ to form immune lysogens in the *him* mutants is due to a defect in integration per se and not to a defect in the establishment of repression. Although failure to establish repression could account for a negative test for lysogeny by the method of Gottesman and Yarmolinsky, it could not explain the failure of λ*att*B-*att*P to recombine in the *him* mutants, a process in which repressor should play no role.

Physiological Effects of the *him* Mutations

Both *him* mutants exhibit depressed growth rates in tryptone broth. We have no evidence at present as to whether this effect is a result of secondary mutations induced by nitrosoguanidine mutagenesis or whether it is associated with the failure to support integrative recombination. Experiments are now in progress to distinguish between these two alternatives.

DISCUSSION

Reconstruction experiments outlined in Table 2 indicate that mutants of *E. coli* unable to integrate λ, due either to a structural defect in the bacterial attachment site or to a mutation in a host factor required for λ integration, survive the selection at a frequency approximately five orders of magnitude higher than wild-type bacteria. In addition, we were able to isolate a mutant temperature-sensitive for integration (*him*57) by performing the selection at high temperature (see Table 3). Thus these selection and screening procedures should facilitate the isolation of a variety of bacterial mutants defective in λ integration.

Although the *him* mutants were selected for their inability to integrate a phage with the attachment specificity of λ, the integration of a phage with a different specificity (*att*80) was also reduced. Since both the attachment specificity and Int proteins of λ and ϕ80 are different, the fact that the same factor(s) is required for their integration may indicate that the Him host factor may not be involved in the site-specific component of integration but rather may be a factor involved directly in the recombinational event.

Both *int* and *xis* gene products are required for lambda excision. It is surprising, therefore, that excisive recombination assayed with λ*att*L-*att*R is not greatly affected in *him* mutants under conditions where integrative recombination is severely reduced. We have not yet measured directly the frequency of excision of an integrated prophage in *him* mutants. However, the existence of mutations mapping within the *int* gene of λ that allow excisive recombination but not integrative recombination also suggests that excision requires less Int activity than integration (Enquist and Weisberg, this volume).

SUMMARY

We have reported here both a method for isolating mutants of *E. coli* which inhibit integration of the bacteriophage λ genome into the host chromosome (*him*) and the initial characterization of two such *him* mutants.

Acknowledgments

The authors thank Howard Nash for phage strains and helpful suggestions and Justine Posner for technical excellence in preparing the manuscript. This work was supported by grants from the National Institute of Allergy and Infectious Diseases (1RO 1 A111459-01) and from the National Science Foundation (GB-41719).

REFERENCES

Adelberg, E. A., M. Mandel and G. C. C. Chen. 1965. Optimal conditions for mutagenesis by *N*-methyl-*N'*-nitro-*N*-nitrosoguanidine in *E. coli* K12. *Biochem. Biophys. Res. Comm.* 18:788.

Berg, D., J. Davies, B. Allet and J. D. Rochaix. 1975. Transposition of R factor genes to bacteriophage λ. *Proc. Nat. Acad. Sci.* **72**:3628.

Campbell, A. 1962. The episomes. *Adv. Genet.* **11**:101.

Court, D. and K. Sato. 1969. Studies of novel transducing variants of λ: Dispensibility of genes N and Q. *Virology* **39**:348.

Couturier, M. 1976. The integration and excision of the bacteriophage Mu-1. *Cell* **7**:155.

Eisen, H., L. H. Pereira da Silva and F. Jacob. 1969. The regulation and mechanism of DNA synthesis in bacteriophage lambda. *Cold Spring Harbor Symp. Quant. Biol.* **33**:755.

Friedman, D. I. and M. B. Yarmolinsky. 1972. Prevention of the lethality of induced λ prophage by an isogenic λ plasmid. *Virology* **50**:472.

Gottesman, M. E. and R. A. Weisberg. 1971. Prophage insertion and excision. In *The bacteriophage lambda* (ed. A. D. Hershey), p. 113. Cold Spring Harbor Laboratory, Cold Spring Harbor, New York.

Gottesman, M. E. and M. Yarmolinsky. 1968. Integration-negative mutants of bacteriophage lambda. *J. Mol. Biol.* **31**:487.

Gottesman, M. M. and L. Rosner. 1975. Acquisition of a determinant for chloramphenicol resistance by coliphage lambda. *Proc. Nat. Acad. Sci.* **72**:5041.

Kleckner, N., R. Chan, B.-K. Tye and D. Botstein. 1975. Mutagenesis by insertion of a drug-resistance element carrying an inverted repetition. *J. Mol. Biol.* **97**:561.

Lieb, M. 1970. λ Mutants which persist as plasmids. *J. Virol.* **6**:218.

Nash, H. 1974. λ*att*B-*att*P, a λ derivative containing both sites involved in integrative recombination. *Virology* **57**:207.

——. 1975. Integrative recombination in bacteriophage lambda: Analysis of recombinant DNA. *J. Mol. Biol.* **91**:501.

Ogawa, T. and J. Tomizawa. 1968. Replication of bacteriophage DNA. I. Replication of DNA of lambda phage defective in early functions. *J. Mol. Biol.* **38**:217.

Parkinson, J. and R. Huskey, 1971. Deletion mutants of bacteriophage lambda. I. Isolation and initial characterization. *J. Mol. Biol.* **56**:369.

Shimada, K. and A. Campbell. 1974. Int-constitutive mutants of bacteriophage lambda. *Proc. Nat. Acad. Sci.* **71**:237.

Shulman, M. and M. E. Gottesman. 1971. Lambda att^2: A transducing phage capable of intramolecular *int-xis* promoted recombination. In *The bacteriophage lambda* (ed. A. D. Hershey), p. 477. Cold Spring Harbor Laboratory, Cold Spring Harbor, New York.

Signer, E. R. 1969. Plasmid formation: A new mode of lysogeny of phage λ. *Nature* **223**:158.

——. 1971. General recombination. In *The bacteriophage lambda* (ed. A. D. Hershey), p. 139. Cold Spring Harbor Laboratory, Cold Spring Harbor, New York.

Starlinger, P. and H. Saedler. 1972. Insertion mutations in microorganisms. *Biochimie* **54**:177.

Zissler, J., E. Signer and F. Schaefer. 1971. The role of recombination in growth of bacteriophage lambda. I. The gamma gene. In *The bacteriophage lambda* (ed. A. D. Hershey), p. 455. Cold Spring Harbor Laboratory, Cold Spring Harbor, New York.

A Mutant of *Escherichia coli* Deficient in a Host Function Required for Phage Lambda Integration and Excision

J. G. K. Williams* and D. L. Wulff
Department of Molecular Biology and Biochemistry
University of California
Irvine, California 92717

H. A. Nash
National Institute of Mental Health
National Institutes of Health
Bethesda, Maryland 20014

The integration of bacteriophage λ into the *Escherichia coli* chromosome requires the product of the λ*int* gene (Gottesman and Weisberg 1971). Biochemical studies have shown that an activity present in uninfected *E. coli* is also required for the recombination (Nash et al., this volume). In this paper, we report the isolation and characterization of an *E. coli* mutant in which λ integration is defective.

Isolation and Genetic Mapping of the Mutant

A log-phase culture of *E. coli* strain UC3163 (W3110 $leu^-trp^-lys^-ser^-ilv^-str^-hfl^-$), following treatment with the mutagen *N*-methyl-*N*′-nitro-*N*-nitrosoguanidine (NG) according to the method of Adelberg et al. (1965), was plated on tryptone agar. Several hundred individual colonies were selected and grown up in tryptone broth at 37°C to stationary phase and then used as indicator bacteria with several λ phage strains. One isolate, designated UC3171, gave no plaques with phage strain λ*c*I71*cin*-1 but behaved normally with all other phages tested. (We had included λ*c*I71*cin*-1 among the test phages because of its unusual properties. On most host strains, such as C600 and UC3163, λ*c*I71*cin*-1 forms turbid plaques at 37°C and clear plaques at 41°C. Quantitative studies of lysogenization of strain C600 have shown that most bacteria surviving infection at 37°C carry the λ*c*I71*cin*-1 prophage in an unstable plasmid state. A small fraction of the survivors are stable lysogens [Wulff 1976; D. A. Hamel and D. L. Wulff, unpubl.].)

*Present address: Department of Molecular Biophysics and Biochemistry, Yale University, New Haven, Connecticut 06510.

In determinations of lysogeny by the method of Gottesman and Yarmolinsky (1968), it was found, unexpectedly, that λ^+ does not form stable lysogens of the mutant strain UC3171, even though λ^+ plaques are turbid on this strain. These observations suggest an integration defect, and the responsible host mutation has been named *hid*, for *h*ost *i*ntegration *d*efective.

We located the *hid* mutation on the *E. coli* chromosome by mating UC3171 with several *hid*$^+$ Hfr strains, selecting for various nutritional markers, and testing these recombinants for the Hid character. Hfr strains KL96 and KL99 were found to transfer the *hid*$^+$ marker at a high frequency, whereas Hfr strains KL208 and PK191 were unable to transfer *hid*$^+$ at all during a 60-minute mating period. This places the *hid* mutation between 33 minutes and 43 minutes on the *E. coli* genetic map (Low 1973; Bachmann et al. 1976). The *hid* mutation has been located more accurately by P1 cotransduction experiments. Using the *hid*$^-$ strain UC3171 as donor and the *aroD*$^-$ strain AB1360 (*thi*$^-$*argE*$^-$*his*$^-$*proA*$^-$*aroD*$^-$ *lacY*$^-$*galK*$^-$*mtl*$^-$*xyl*$^-$*tsx*$^-$*su*$^+$) as recipient (Wallace and Pittard 1967), we found Aro$^+$ transductants which were phenotypically Hid$^-$. (The *hid* marker was most easily scored on the basis of plaque morphology with λ*c*I71*cin*-1. Lambda *c*I71*cin*-1 forms large plaques with turbid centers on Hid$^+$Aro$^+$ transductants of strain AB1360 and somewhat smaller plaques having uniform deep turbidity and indistinct fuzzy edges on Hid$^-$Aro$^+$ transductants of this strain. We do not understand why λ*c*I71*cin*-1 forms plaques on the Hid$^-$ transductants but not on the original mutant strain, nor can we explain the differences in plaque morphology on Hid$^-$Aro$^+$ and Hid$^+$Aro$^+$ transductants). Of 20 Aro$^+$ transductants scored by these methods, 13 were also *hid*$^-$, indicating that *hid* lies close to *aroD*, at about 37-38 minutes on the *E. coli* genetic map (Bachmann et al. 1976).

Properties of the Mutant

Since the original *hid*$^-$ strain grows poorly, we suspected the presence of other, uncharacterized mutations. Therefore, in all further experiments, we used isogenic Hid$^+$Aro$^+$ and Hid$^-$Aro$^+$ transductants of *E. coli* strain AB1360. With these strains, designated UC2198 (*hid*$^+$) and UC2195 (*hid*$^-$), it was established that the *hid* mutation has no significant effect on cell growth, sensitivity to UV light, burst size following λ^+ infection, plaque morphology with λ^+, or *rec*-mediated recombination in λ. The latter was determined by using the isogenic *hid*$^+$ and *hid*$^-$ strains as hosts for crosses of λ*bio*10*imm*^{21}*S*7 with λ*bio*10 and tabulating the fraction of λ*bio*10*imm*^{21}*S*$^+$ recombinants. These crosses test for *rec*-mediated recombination because *bio*10 derivatives are Red$^-$ and Int$^-$ (Signer et al. 1969; Signer 1971).

Integrative recombination in isogenic *hid*$^+$ and *hid*$^-$ host strains was determined using the test phage λ*att*B-*att*P (Nash 1974), which contains both attachment sites involved in integration. Recombination between these sites produces a deleted phage genome; consequently, recombinant phage are resistant to chelating agents, whereas parental phage are sensitive. Integrative recombination of a λ*att*B-*att*P *int am*29 is reduced about 50-fold in the *hid* strain UC2195 compared to the hid^+ strain UC2198 (Table 1, lines 1 and 2). The recombination defect is not due to poor suppression of the *int* amber mutation: UC2195 remains defective as a host for λ*att*B-*att*P recombination in the presence of superinfecting λint^+ (Table 1, lines 3 and 4).

Excisive recombination was determined in a manner similar to that described above for integrative recombination, except that the test phage was λ*att*R-*att*L (Shulman and Gottesman 1971), which contains both attachment sites involved in excision. Excisive recombination is greatly reduced in the hid^- strain UC2195 compared to the isogenic hid^+ strain UC2198 (Table 1, lines 7 and 8).

Studies on Integrative Recombination In Vitro

The defect in hid^- strains was investigated further in an in vitro system for measuring integrative recombination. Two kinds of extracts are used in these experiments: extracts from a nonlysogen to provide host-coded functions and extracts from an *int*-constitutive lysogen to provide *int* gene product. Extracts from hid^- strains are defective in promoting integrative recombination in this system (Table 2, lines 1 and 3). This could be due to reduced *int* gene expression in hid^- strains or to a defective host component. Lambda *int* gene activities in extracts of hid^+ and hid^- *int*-constitutive lysogens were compared after mixing with extracts of a hid^+ nonlysogen (the latter providing the required host components). The same activity is observed in either case (Table 2, lines 1 and 2).

Table 1
Site-specific Recombination In Vivo

Test phage	Helper phage	Host	Percent recombination
λ*att*B-*att*P *int am*29		UC2198 (hid^+)	28
λ*att*B-*att*P *int am*29		UC2195 (hid^-)	<0.5
λ*att*B-*att*P *int*2	λint^+	UC2198 (hid^+)	39
λ*att*B-*att*P *int*2	λint^+	UC2195 (hid^-)	1
λ*att*B-*att*P *int*2		UC2198 (hid^+)	<0.1
λ*att*B-*att*P *int*2		UC2195 (hid^-)	<0.1
λ*att*R-*att*L *int am*29		UC2198 (hid^+)	77
λ*att*R-*att*L *int am*29		UC2195 (hid^-)	3

The host cells were grown at 31°C in tryptone maltose broth, infected (5–8 particles/cell) with test phage and helper phage as indicated, diluted 100-fold into broth, shaken at 31°C for 90 min, and treated with chloroform. An aliquot of the lysate was diluted 100-fold into 0.01 M Tris-HCl (pH 7.4) containing 0.01 M Na_2 EDTA and heated at 41°C for 15 min. The proportion of phage in the lysate surviving this treatment is given in the last column. The λ*att*B-*att*P test phage were *xis*1*red*114imm^{434} clear derivatives; the λ*att*R-*att*L test phage were imm^{21} clear derivatives. The λint^+ helper phage was λ*xis*1*red*3*c*I857. The *int am*29 mutation is suppressed and provides *int* function in these experiments; *int*2 is a missense mutation.

Table 2
Integrative Recombination In Vitro

Source of extract						
1. nonlysogen		2. lysogen		3. nonlysogen		Activity
genotype	μg	genotype	μg	genotype	μg	(% recombinant phage)
hid$^+$	114	*hid*$^+$	40			15
hid$^+$	114	*hid*$^-$	39			20
hid$^-$	94	*hid*$^-$	39			2
hid$^-$	94	*hid*$^-$	39	*hid*$^+$	49	17
hid$^-$	94	*hid*$^-$	39	*hid*$^-$	41	4

Reactions were performed as described by Nash et al. (this volume). Briefly, hydrogen-bonded circular λ*att*B-*att*P *int*2*xis*1*imm*^{434} clear *Sam*7 DNA is incubated at 25°C with buffer, spermidine, ATP, and the indicated amounts of enzyme preparation (listed as μg of protein). At the end of the reaction, the nucleic acid is extracted from the mixture and used to transfect spheroplasts. The proportion of transfected phage resistant to chelating agents is taken to represent the fraction of DNA recombined in the in vitro reaction, and this is shown in the last column. Crude extracts prepared as described by Nash (1975) were used for 2 and 3; high-speed supernatants derived from crude extracts were used for 1. The enzyme fractions of 1 and 2 were dialyzed against 0.05 M Tris (pH 7.4) for 2 hr. The nonlysogen strains used were UC2198 (*hid*$^+$) and UC2195 (*hid*$^-$); the corresponding lysogens were constructed by first lysogenizing the *hid*$^+$*aroD*$^-$ strain AB1360 with λ*h int*-*c*226*c*I857 and then transducing to Aro$^+$ with a *hid*$^-$*aro*$^+$ donor. To test the transductants for the Hid phenotype, the plating characteristics of λ*c*I71*cin*-1 were determined on non-immune derivatives. The in vivo integration defect of a *hid*$^-$ transductant relative to a *hid*$^+$ transductant was essentially confirmed by a method similar to that described in Table 1.

indicating that the *hid* mutation does not affect *int* gene expression. The activities of host components in extracts of *hid*$^+$ and *hid*$^-$ nonlysogens were compared after mixing with extracts of a *hid*$^-$ *int*-constitutive lysogen (the latter providing *int* gene product). Extracts from the *hid*$^+$ nonlysogen are tenfold more active than those from the corresponding *hid*$^-$ strain (Table 2, lines 2 and 3). We conclude that *hid*$^-$ strains are defective in a host component required for integrative recombination. In agreement with observations in vivo, the defect in recombination of the *hid*$^-$ strain in vitro is also not complete.

The defect in *hid*$^-$ extracts can be overcome by the addition of extract from a *hid*$^+$ nonlysogen (Table 2, lines 3 and 4). This indicates that *hid*$^-$ strains are not producing an inhibitor of recombination but are missing an active component. As yet, we have no additional information concerning either the product(s) under the control of the *hid*$^-$ gene or the step(s) in recombination it affects. It also remains to be seen whether other DNA insertions, deletions, and translocations are affected by this mutation.

SUMMARY

Bacteriophage lambda does not lysogenize *hid⁻* strains of *E. coli* efficiently. The *hid* mutation is located at 37-38 minutes on the *E. coli* genetic map, in close proximity to *aroD*. Both integrative and excisive recombination of lambda are greatly reduced in infections of *hid⁻* host strains. Cell-free extracts of *hid⁻* strains are deficient in a host component required for in vitro integrative recombination.

Acknowledgments

This work was supported by National Science Foundation Grant No. GB 43726X to D. W. The authors thank Michael Mahoney and Carol Anne Robertson for technical assistance. J. W. would like to thank Charles Radding for his helpful discussions and for the use of his laboratory during the final stages of this project.

REFERENCES

Adelberg, E. A., M. Mandel and G. C. C. Chen. 1965. Optimal conditions for mutagenesis by *N*-methyl-*N*′-nitro-*N*-nitrosoguanidine in *Escherichia coli* K12. *Biochem. Biophys. Res. Comm.* **18**:788.

Bachmann, B. J., K. B. Low and A. L. Taylor. 1976. Recalibrated linkage map of *Escherichia coli* K-12. *Bact. Rev.* **40**:116.

Gottesman, M. E. and R. A. Weisberg. 1971. Prophage insertion and excision. In *The bacteriophage lambda* (ed. A. D. Hershey), p. 113. Cold Spring Harbor Laboratory, Cold Spring Harbor, New York.

Gottesman, M. and M. Yarmolinsky. 1968. Integration negative mutants of lambda. *J. Mol. Biol.* **31**:487.

Low, B. 1973. Rapid mapping of conditional and auxotrophic mutations in *Escherichia coli* K12. *J. Bact.* **113**:798.

Nash, H. A. 1974. *att*B-*att*P, a λ derivative containing both sites involved in integrative recombination. *Virology* **57**:207.

———. 1975. Integrative recombination of bacteriophage lambda DNA *in vitro*. *Proc. Nat. Acad. Sci.* **72**:1072.

Shulman, M. and M. Gottesman. 1971. Lambda att^2: A transducing phage capable of intramolecular *int-xis* promoted recombination. In *The bacteriophage lambda* (ed. A. D. Hershey), p. 477. Cold Spring Harbor Laboratory, Cold Spring Harbor, New York.

Signer, E. 1971. General recombination. In *The bacteriophage lambda* (ed. A. D. Hershey), p. 139. Cold Spring Harbor Laboratory, Cold Spring Harbor, New York.

Signer, E. R., K. Manly and M. Brunstetter. 1969. Deletion mapping of the *cIII-N* region of bacteriophage lambda. *Virology* **39**:137.

Wallace, B. J. and J. Pittard. 1967. Genetic and biochemical analysis of the isoenzymes concerned in the first reaction of aromatic biosynthesis in *Escherichia coli*. *J. Bact.* **93**:237.

Wulff, D. L. 1976. Lambda *cin*-1, a new mutation which enhances lysogenization by bacteriophage lambda, and the genetic structure of lambda *cy* region. *Genetics* **82**:401.

Integrative Recombination of Bacteriophage Lambda—The Biochemical Approach to DNA Insertions

H. A. Nash,* K. Mizuuchi,† R. A. Weisberg,‡
Y. Kikuchi* and M. Gellert†
*National Institute of Mental Health
†National Institute of Arthritis, Metabolism, and Digestive Diseases
‡National Institute of Child Health and Human Development
National Institutes of Health
Bethesda, Maryland 20014

The recombination that normally leads to the insertion of the λ genome into the chromosome of its *E. coli* host has been studied in vitro (Nash 1975; Mizuuchi and Nash 1976). This work may be useful in at least two ways for investigating other DNA insertions: First, as described elsewhere in this volume, the mechanism of insertion of other DNAs may be thought of as analogous to the insertion of bacteriophage lambda DNA. Therefore, information on the nature of the substrate, cofactors, enzymes, and reaction mechanism of lambda insertion may shed light on DNA insertions of other kinds. Second, and perhaps of more immediate value, we have developed several assays for lambda integrative recombination in vitro that should prove applicable to the in vitro study of various other DNA insertions. In this report, we will describe briefly three assays based on different aspects of integrative recombination and will indicate some ways in which each assay has been used.

Intramolecular Recombination

All three assays have utilized variants of the same substrate, the chromosome of λ*att*B-*att*P, a single phage DNA molecule carrying both sites used in integrative recombination (Nash 1974). The two attachment sites on the same molecule can recombine with one another. Indeed, in all the in vitro studies we have done, it appears that intramolecular recombination is the only mode that occurs at a detectable frequency (Nash 1975; K. Mizuuchi, unpubl.). This probably reflects the enhanced local concentration of attachment sites that are connected to each other via the phosphodiester backbone. A similar hypothesis can account for the enhanced joining of λ cohesive ends belonging to a single molecule compared to the joining of ends on different molecules (Wang and Davidson 1966). Applying the formulation of Wang and Davidson, the effective local concentration of attachment sites of λ*att*B-*att*P is found to be 5 × 10^{12} sites/ml. This is 50- to 100-fold greater than the concentration of DNA substrate molecules usually employed in the assay, which of course is the effective concentration for intermolecular events.

If collision between attachment sites is a rate-limiting step in recombination and is determined by the concentration of sites, intermolecular recombination should occur when the substrate DNA is sufficiently concentrated. This is certainly true for infected cells. The intracellular concentration of phage DNA is calculated to be more than 10^{12} molecules/ml, and, in such infected cells, λ*att*B-*att*P undergoes both intermolecular and intramolecular site-specific recombination (Nash 1974). Furthermore, Syvanen (1974) has shown that site-specific intermolecular recombination between two molecules carrying *att*P does occur in vitro. Significantly, the methods used by Syvanen are expected to produce extracts containing about 100-fold higher concentrations of DNA substrate than is used in our reactions. It remains to be seen whether intermolecular integrative recombination in vitro can occur when the substrate is at an elevated concentration.

Transfection Assay

The assay used in the initial studies of integrative recombination in vitro depended on conversion of recombinant DNA to mature phage (Nash 1975). Gottesman and Gottesman (1975) used a similar assay to study in vitro excisive recombination of phage λ. In a typical assay, the DNA is extracted from the reaction mixture and used to transfect spheroplasts. Since recombination of the attachment sites removes the DNA segment lying between them, the transfected recombinant phage have a smaller genome than parental phage and can be identified by their resistance to chelating agents. The proportion of recombinant transfected phage is taken as the proportion of recombinant DNA produced by the reaction. This assay requires that no recombination occur during the transfection; this has been achieved by performing the transfections in the absence of an active phage *int* gene. The sensitivity of this assay is limited primarily by the efficiency of transfection. Under usual conditions, several hundred DNA molecules are successfully transfected, and recombinant proportions of 1% or greater are easily detected.

Using the transfection assay, we have made a preliminary biochemical characterization of the integrative recombination reaction. Cell-free extracts are derived from cells lysogenic for a derivative of phage lambda which constitutively expresses the *int* gene (Nash 1975). High-speed centrifugation of the extracts yields an inactive pellet fraction that regains activity on mixing with a soluble fraction prepared from a nonlysogen (Table 1, lines 1-3). Therefore, integrative recombination requires one or more components made by the bacterial host. Mutants of *E. coli* which may affect host-coded function(s) are described

Table 1
Phage-specific and Host-specific Enzyme Fractions

Source of enzyme fractions		
lysogen (μg)	nonlysogen (μg)	Activity (% recombinant phage)
Extract (210)	–	26.1
Pellet (106)	–	<0.1
Pellet (106)	supernatant (84)	19.8
–	pellet (100) + supernatant (84)	<0.1
Pellet (106)	heated supernatant (84)	<0.1
Heated pellet (106)	supernatant (84)	<0.1

For each reaction, a mixture containing 36 mM Tris-HCl (pH 7.4), 10 mM spermidine, 33 mM potassium phosphate buffer, 3 mM $MgCl_2$, 2.65 mM ATP, 0.165 mg/ml *E. coli* soluble RNA, and 0.5 μg λ*att*B-*att*P *int*2*xis*1*imm*^{434} clear *Sam*7 hydrogen-bonded circular DNA in a total volume of 75 μl was incubated 10 min at 25°C. To this was added 50 μl 0.05 M Tris-HCl (pH 7.4) containing the indicated amounts of various enzyme fractions and sufficient bovine serum albumin to make a total protein content of 0.5 to 1.0 mg. Incubation was continued at 25°C for 30 min.

Crude extracts were made as before (Nash 1975) from nonlysogen N99 *recB*21 or from AB2470 *recB*21 lysogenic for λ*int-c*, a derivative of λ that expresses the *int* gene constitutively (Shimada and Campbell 1974). Supernatant and pellet fractions were separated by centrifugation of crude extracts at 240,000*g* for 3 hr. All enzyme fractions were dialyzed against 0.05 M Tris-HCl (pH 7.4) for 2 hr before use. Where indicated, enzyme fractions were heated to 45°C for 10 min and chilled prior to addition to reaction mixtures.

The reactions were quenched by addition of EDTA, sodium dodecyl sulfate (SDS), and $NaClO_4$. The DNA was extracted with chloroform-isoamyl alcohol, dialyzed, and used to transfect spheroplasts of strain KL176 *recArecBsu*$^-$λ^r. The resulting phage were plated on strain N720 *su*III using nonselective and pyrophosphate containing plates. The ratio of pfu on these two plates is the measure of in vitro recombination activity. The titer of transfected phage on nonselective plates was similar for all combinations of enzyme fractions tested.

elsewhere in this volume (Miller and Friedman; Williams et al.). Fractions derived entirely from a nonlysogen are not active (Table 1, line 4). Thus a phage-specific component is also required for integrative recombination; this must be present in the pellet fraction derived from lysogenic cells. It has also been shown that the phage-specific component includes the product of the *int* gene, since phage-specific activity displays increased thermolability in vitro when extracted from lysogens containing temperature-sensitive mutations of the *int* gene (L. Enquist, R. A. Weisberg and H. A. Nash, in prep.). As yet, we have no biochemical evidence for other phage-coded components. The supernatant and pellet fractions are both completely inactivated by brief heating (Table 1, lines 5 and 6). This thermolability, as well as the retention of activity on dialysis, suggests further that the activity of both host- and phage-coded fractions is protein-dependent.

Restriction Fragment Assay

Recently, we have developed an assay in which recombinant DNA is identified by its novel restriction fragments (Mizuuchi and Nash 1976). As seen in the map of Figure 1, the removal of the DNA between the *att* sites by integrative recombination results in the loss of two restriction fragments (B and E) of the parental DNA. The product DNA, λ*att*L, contains a new fragment (Rb) which is formed by fusion of the DNA previously separated by the removed region. The parental and recombinant fragments can be separated by agarose gel electrophoresis (Fig. 1). After in vitro recombination reactions, we find that, in addition to fragment Rb, the DNA which lies between *att*B and *att*P and which is removed by recombination can also be detected as a new fragment (Ra) of 4.2×10^6 daltons. Under optimal conditions, recombination of as little as 1% to 2% of the substrate DNA can be detected by loss of fragments B and E and production of fragments Ra and Rb.

Studies with the restriction fragment assay show that closed circular DNA is the only efficient form of the substrate (Mizuuchi and Nash 1976). Linear *att*B-*att*P DNA is a poor substrate; hydrogen-bonded circular DNA is an effective substrate only if it is converted to the closed circular form. As shown in Figure 2, the recombination proficiency of hydrogen-bonded circular DNA is abolished if

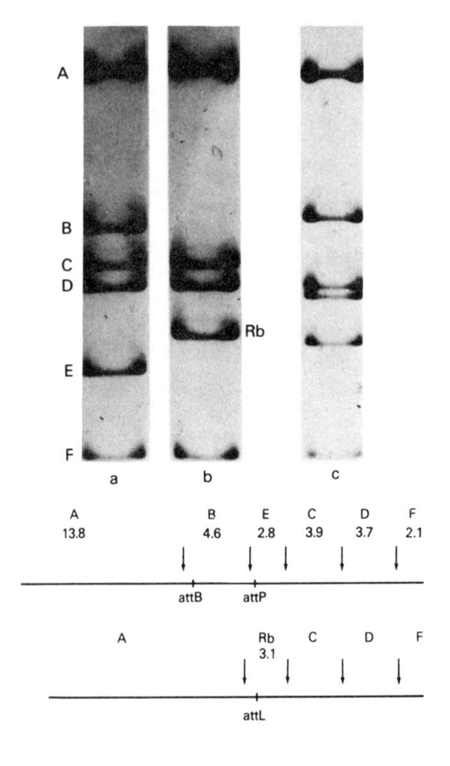

Figure 1

Restriction map of substrate and recombinant DNA. DNA (0.5 μg) from purified phage stocks of (*a*) λ*att*B-*att*P, (*b*) λ*att*L, and (*c*) λ were treated with *Eco*RI restriction endonuclease, separated by electrophoresis in 0.7% agarose, and stained with ethidium bromide. The fluorescent DNA bands are shown in the upper portion of the figure; the molecular lengths (in daltons $\times 10^{-6}$) and map positions of the fragments are shown in the lower portion. λ*att*B-*att*P and λ*att*L were *int*2*xis*1*red*114*imm*[434]*S*7 derivatives; λ standard was λ*c*I857*S*7. The DNA samples were heated to 75°C for 5 min before electrophoresis to produce the linear form of DNA. In circular forms of the DNA where the molecular ends are covalently joined, fragments A and F are fused to form a larger fragment, AF (see Fig. 2, top).

E. coli ligase in the extract is inhibited by nicotinamide mononucleotide (NMN). By contrast, the efficiency of closed circular DNA is only slightly reduced by the identical treatment (Fig. 2c,d). In addition, circles with interruptions introduced by pancreatic DNase or X rays are also ineffective substrates if the interruptions are prevented from being resealed by NMN.

Closed circles are required for the recombination reaction because only they can be converted to a supertwisted form. We have discovered a host-coded enzyme, DNA gyrase, which introduces negative superhelical turns into relaxed closed circular DNA (Gellert et al. 1976). The enzyme and its cofactor ATP are not required for integrative recombination when a negatively supertwisted circular DNA substrate is used but are absolute requirements with relaxed substrate DNA. The substrate for integrative recombination must therefore either be provided in a supertwisted form or be made supertwisted by components of the extract (K. Mizuuchi, M. Gellert and H. A. Nash, in prep.). As yet, we do not know why the supertwisted DNA substrate is preferentially utilized by the extracts.

Circular Product

The in vitro reaction carries out a complete breakage and rejoining cycle of recombination. Fragments Ra and Rb have the expected molecular lengths, and no fragments arising from broken recombinant ends or chiasmata are found. Since the substrate for integrative recombination is a circle and breakage and reunion occur, it follows that the products of the in vitro reaction will be circles. For intramolecular recombination, two product circles should be formed; they should have molecular weights of 27 and 4 million daltons and contain recombinant attachment sites *att*L and *att*R, respectively. Moreover, if the DNA substrate is supercoiled when breakage and reunion occur and if recombination is reciprocal, the product circles should be wound around each other in a catenate (see Fig. 3). Such interwound circles are indeed the principal product of in vitro integrative recombination (in prep.). Figure 4 shows an electron micrograph of one such catenate; more highly interwound catenates have also been found.

Filter Assay

The demonstration of circular products derived from a circular substrate suggested the assay described in Figure 5. A DNA substrate with a single restriction endonuclease target site forms one product circle devoid of restriction targets. Treatment of a reaction mixture with restriction endonuclease yields linear unreacted substrate DNA and both linear and circular product DNA. Circular DNA product is then resolved from linear DNA by preferential trapping on membrane filters (Saucier and Wang 1973). Table 2 shows that the assays of recombination by the filter and restriction fragment methods are in good

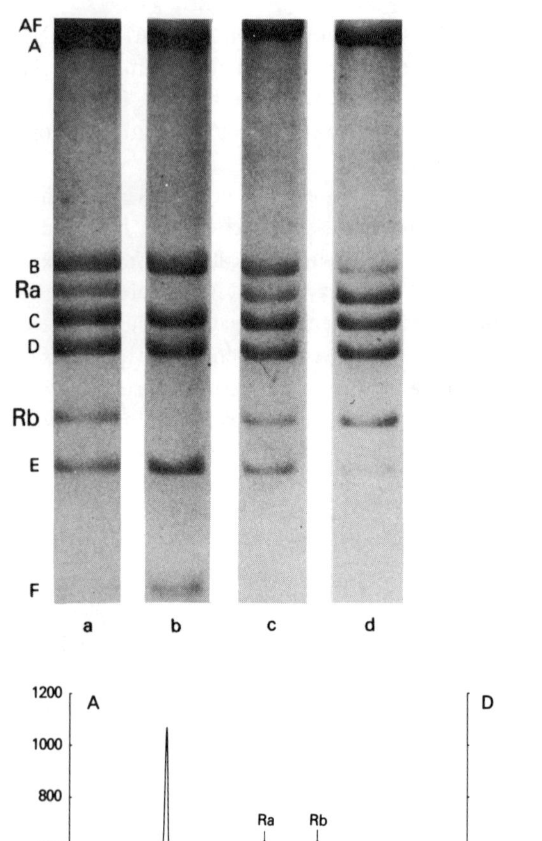

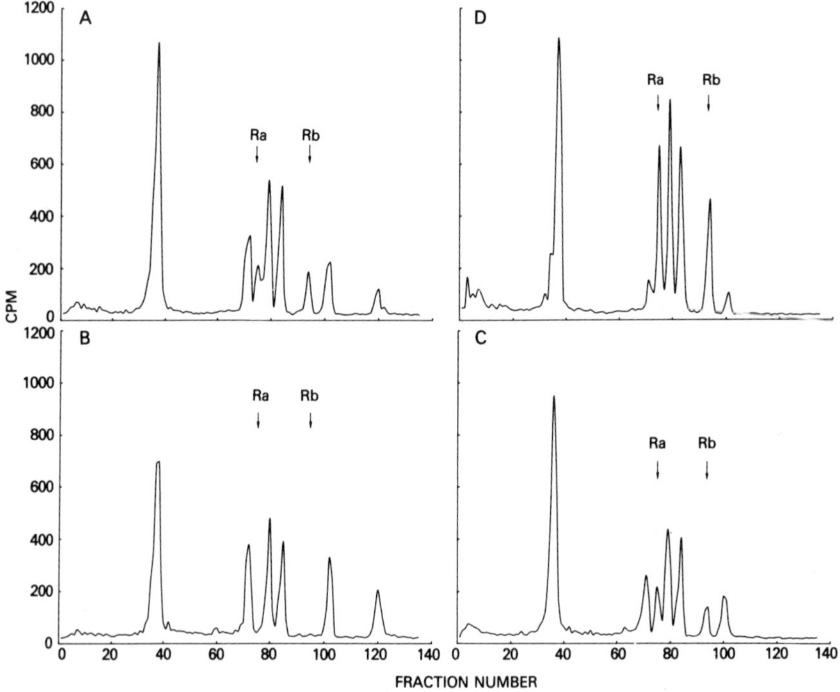

Figure 2 *(see facing page for legend)*

Figure 3
The origin of catenated products. The DNA of λ*att*B-*att*P is indicated as a continuous ribbon; attachment sites are at the junction of the shaded and unshaded regions. The diagram on the left shows the pathway of integrative recombination in a relaxed circular molecule. The attachment sites are aligned as for a conventional genetic exchange; subsequent breakage and reunion produces an equal mixture of free and catenated products. The diagram on the right shows the pathway of integrative recombination as a substrate with a few interwound supertwists. Following alignment of the attachment sites as before, genetic exchange produces only catenated products.

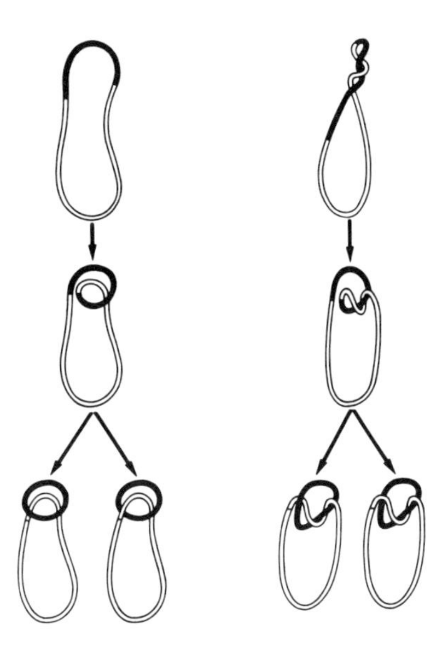

Figure 2
Restriction fragment assay with closed circular and hydrogen-bonded circular substrates. Recombination reactions were performed as described in Table 1, using either hydrogen-bonded circular DNA (*a*, *b*) or closed circular DNA (*c*, *d*). Enzymatic ligation of hydrogen-bonded circles was used to prepare closed circular DNA; this DNA contained no superhelical turns. In reactions *b* and *c*, 10 μl of 40 mM NMN was added to the reaction prior to the addition of enzyme. The enzymes used were similar to those described in Table 1, line 3. The reactions were quenched by addition of *N*-ethylmaleamide (NEM); after 10 min, dithiothreitol (DDT) was added, and the mixture was adjusted to 50 mM NaCl and 6 mM $MgCl_2$ and treated with 1 μl endonuclease *Eco*RI (Miles Laboratories) for 10 min at 37°C. EDTA was then added to 18 mM, and the DNA was extracted with chloroform-isoamyl alcohol (recovery >90%) and subjected to electrophoresis in 0.7% agarose. After electrophoresis, the slab was stained with ethidium bromide. The fluorescent bands are shown in the upper portion of the figure. Each lane of the agarose slab was cut out and sectioned into 1-mm pieces; the pieces were made soluble in $NaClO_4$, dissolved, and the radioactivity counted. A plot of the radioactivity in each fraction is shown in the lower portion of the figure. The expected positions of recombinant fragments Ra and Rb are indicated by arrows.

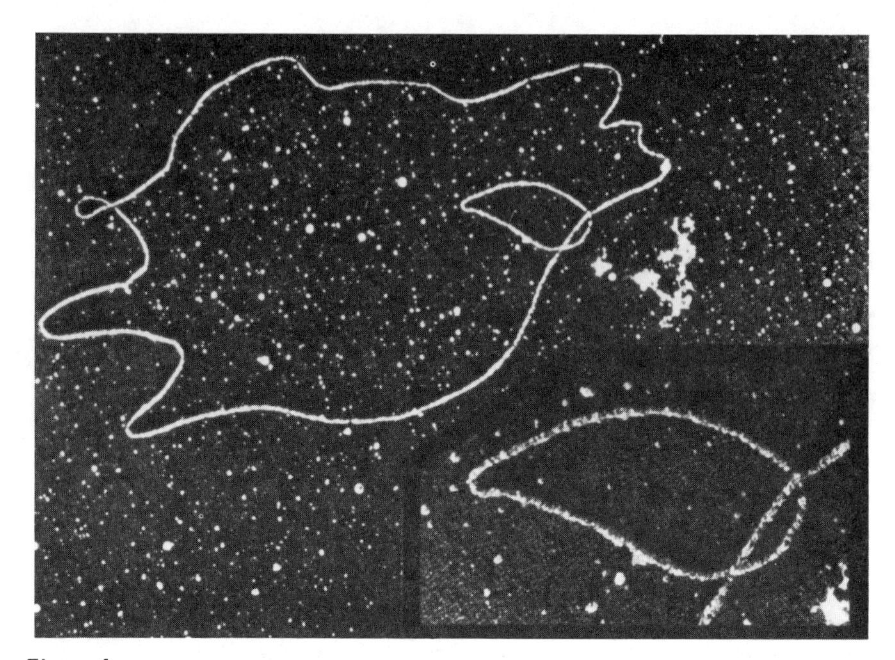

Figure 4
Catenated product DNA. Recombination in vitro using closed circular substrate DNA was carried out as described for Figure 2. The reaction was quenched with EDTA and centrifuged to equilibrium in a cesium chloride-ethidium bromide gradient. The lower band, containing closed circular DNA, was collected, extracted with butanol to remove the ethidium bromide, and dialyzed. The circular DNA was X-irradiated to introduce nicks and spread by the aqueous Kleinschmidt technique as described by Davis et al. (1971). Ethidium bromide was added to the spreading solution at a concentration of 100 μg/ml to stiffen the DNA (Freifelder 1971). The grid was stained with uranyl acetate and shadowed with Pt:Pd (80:20) at an approximately 5° angle from two directions differing by 90°. In 42 separate observations, the ratio of lengths of the two catenated circles was in agreement with that expected for recombinant DNA.

Table 2
Comparison of Restriction Fragment and Circular Product Assays

Incubation time (min)	Extent of recombination	
	restriction fragment assay	circular product assay
5	0.32	0.22
10	0.40	0.33
20	0.59	0.63
30	0.70	0.69

Reactions were performed as described in Table 1, the enzyme fractions being similar to those in line 3 of Table 1. In both assays, at the indicated times, the reactions were quenched with NEM, treated with DTT, and adjusted with NaCl and $MgCl_2$ as described in Fig. 2. (*Notes continued on facing page.*)

Figure 5

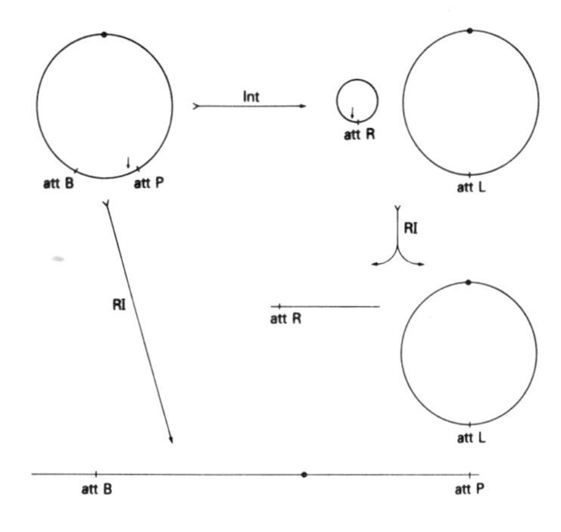

Filter assay for circular product. A derivative of λ*att*B-*att*P in which four of the five *Eco*RI restriction target sites are removed by mutation is used as substrate. The remaining site is indicated by an arrow. Mutations in the three sites mapping between *att*P and the molecular ends (dot) have been described by Murray and Murray (1974); a spontaneous mutant in the target site in the bacterial substitution mapping to the left of *att*B was isolated in a similar manner. The figure shows the conversion of circular substrate DNA to circular product DNA by in vitro recombination. Catenation of the product DNA has been omitted for clarity since the subsequent breakage of one of the product circles by *Eco*RI restriction endonuclease leads to a free product circle.

For the restriction fragment assay, the substrate was negatively supertwisted, closed circular DNA, which was produced by infection of *recA* cells with λ*att*B-*att*P *int*2*xis*1*red am*270*gam am*210cI857*Sam*7 (Reuben et al. 1974). Restriction endonuclease digestion and electrophoresis were carried out as described for Fig. 2, except that tube gels (5 × 110 mm) were used. Fragments Rb and E were visualized by ethidium bromide fluorescence, cut from the gel, made soluble in $NaClO_4$, dissolved, and the radioactivity counted. The fractional extent of recombination, x, was calculated from

$$\frac{X}{1-x} = \frac{\text{cpm in fragment R}_b}{\text{cpm in fragment E}} \times \frac{\text{m.w. fragment E}}{\text{m.w. fragment R}_b}.$$

The last factor in this equation is taken as 0.90 (Fig. 1).

For the circular product assay, the substrate was negatively supertwisted, closed circular DNA produced by enzymatic ligation of hydrogen-bonded circles in the presence of ethidium bromide (Wang 1971). The DNA was derived from λ*att*B-*att*P *int*2*xis*1cI *am* carrying the mutant restriction target sites as indicated in Figure 5. Digestion of the quenched reaction mixtures was performed for 15 min at 33°C with 1 μl *Eco*RI restriction endonuclease (Miles Laboratories) and 1 ng pancreatic DNase. The pancreatic DNase digestion introduced 1-5 nicks per product circle, which relaxed the circles and thereby enhanced their retention on filters. The digestion was stopped by the addition of 0.9 ml of 1% SDS containing 0.1 M NaCl and 0.01 M Na_2 EDTA. The mixture was filtered at 2-4 ml/min through nitrocellulose membrane filters (Schleicher and Schuell, 0.45 μ pore size, 25 mm diameter) that had just been washed with 4 ml of 0.1 M NaCl containing 0.01 M Na_2 EDTA. The filters were then dried and counted. The fractional extent of recombination, x, was calculated as

$$x = \frac{1}{0.87} \times \frac{\text{cpm on filter}}{\text{retention factor}}.$$

The retention factor was experimentally determined as the cpm trapped on a filter from a reaction mixture carried through the entire procedure described above but omitting the restriction endonuclease from the digestion. The factor of 1/0.87 corrects for the size of the circular product compared to the substrate DNA. All data are corrected for a zero-time blank value of $x = 0.06$.

quantitative agreement. Because of its rapidity and ease, the filter assay should greatly expedite the purification of the various host- and phage-coded enzymes involved in integrative recombination.

SUMMARY

Cell-free extracts containing phage- and bacterial-encoded proteins carry out site-specific genetic recombination of coliphage lambda. Recombinant DNA made in vitro may be detected by assaying recombinant phage produced after transfection or by direct examination of the DNA with one of several different techniques. Our studies have shown that recombination in vitro requires a negatively supertwisted DNA substrate which can be generated from relaxed circles by the action of an *E. coli* enzyme, DNA gyrase. Recombination in vitro is reciprocal, yielding two recombinant DNA circles linked together in a catenate.

Acknowledgments

We are grateful to Dr. J. V. Maizel for his advice concerning the electron microscopy. We thank Dr. R. Yuan for suggesting the filter method for trapping of circular DNA. Carol Anne Robertson and Mary O'Dea provided expert technical assistance.

REFERENCES

Davis, R. W., M. Simon and N. Davidson. 1971. Electron microscope heteroduplex methods for mapping regions of base sequence homology in nucleic acids. In *Methods in enzymology* (ed. L. Grossman and K. Moldave), vol. 21, p. 413. Academic Press, New York.

Freifelder, D. 1971. Electron microscopic study of the ethidium bromide-DNA complex. *J. Mol. Biol.* **60**:401.

Gellert, M., K. Mizuuchi, M. H. O'Dea and H. A. Nash. 1976. DNA gyrase: An enzyme that introduces superhelical turns into DNA. *Proc. Nat. Acad. Sci.* **73**:3872.

Gottesman, S. and M. Gottesman. 1975. Excision of prophage λ in a cell-free system. *Proc. Nat. Acad. Sci.* **72**:2188.

Mizuuchi, K. and H. Nash. 1976. A restriction assay for integrative recombination of bacteriophage λ DNA *in vitro.* Requirement for closed circular DNA substrate. *Proc. Nat. Acad. Sci.* **73**:3524.

Murray, N. E. and K. Murray. 1974. Manipulation of restriction targets in phage λ to form receptor chromosomes for DNA fragments. *Nature* **251**:476.

Nash, H. A. 1974. λ*att*B-*att*P, a λ derivative containing both sites involved in integrative recombination. *Virology* **57**:207.

———. 1975. Integrative recombination of bacteriophage lambda DNA *in vitro.* *Proc. Nat. Acad. Sci.* **72**:1072.

Reuben, R., M. Gefter, L. Enquist and A. Skalka. 1974. New method for large-scale preparation of covalently closed λ DNA molecules. *J. Virol.* **14**:1104.

Saucier, J.-M. and J. C. Wang. 1973. Selective retention of double-stranded circular deoxyribonucleic acid by membrane filtration. *Biochemistry* **12**:2755.

Shimada, K. and A. Campbell. 1974. Int-constitutive mutants of bacteriophage lambda. *Proc. Nat. Acad. Sci.* **71**:237.

Syvanen, M. 1974. *In vitro* genetic recombination of bacteriophage λ. *Proc. Nat. Acad. Sci.* **71**:2496.

Wang, J. C. 1971. Use of intercalating dyes in the study of superhelical DNAs. In *Procedures in nucleic acid research* (ed. G. L. Cantoni and D. R. Davies), vol. 2, p. 407. Harper and Row, New York.

Wang, J. C. and N. Davidson. 1966. On the probability of ring closure of lambda DNA. *J. Mol. Biol.* **19**:469.

The Integrase Promoter of Bacteriophage Lambda

A. Campbell and L. Heffernan
Department of Biological Sciences
Stanford University
Stanford, California 94305

S.-L. Hu and W. Szybalski
McArdle Laboratory for Cancer Research
University of Wisconsin
Madison, Wisconsin 53706

Bacteriophage λ was the first genetic element shown to be inserted into the linear continuity of a preexisting chromosome (see Campbell 1972). The λ genome (DNA length 46.5 kb) is much longer than a simple insertion sequence. However, that segment of λ specifically involved in insertion and excision is only about 1.5 kb. This segment contains the actual site of insertion, *att*P, plus two genes, *int* and *xis*. These genes determine the products, integrase and excisionase, respectively, whose biochemical actions in breaking and joining DNA are beginning to be understood (Nash et al., this volume). The λ genome differs from known IS elements in its very strong preference for insertion at one specific site of the host chromosome.

The integrase protein causes insertion of λ DNA into the host chromosome. The *xis* gene product is not required for insertion but, together with integrase, is needed to excise viral DNA from the chromosomes of lysogenic cells. The reaction can thus be represented as

$$\text{phage} \underset{int\,+\,xis}{\overset{int}{\rightleftarrows}} \text{prophage.}$$

The existence of different catalytic requirements for insertion and excision raises the possibility that the direction of the reaction may be so controlled that insertion is favored at some stages of the viral life cycle and excision at others.

For some years, differential control at the transcriptional level seemed unlikely because *int* and *xis* were both known to be transcribed from the major leftward promoter p_L (Gottesman and Weisberg 1971). In lysogenic cells, this promoter is completely repressed by the product of the *c*I gene. When derepressed, p_L originates a transcript that includes genes for regulation (*N* and *c*III) and recombination (*gam* and *red*). These genes are arranged in the order *int xis redX redB gam* *c*III *N* p_L *c*I. More recently, Shimada and Campbell (1974a) demonstrated the presence of a promoter (p_I), somewhere between the *int* and *red* genes, which can serve as an origin of transcription for *int,* even in prophages from which p_L and surrounding DNA have been deleted. The *xis* gene, on the other hand, is efficiently expressed only when transcribed from p_L.

The *c*II and *c*III products stimulate transcription of the repressor gene *c*I (Eisen and Ptashne 1971). Recent evidence (Katzir et al. 1976; Court et al., this volume) shows that integrase formation following infection also depends strongly on these products. When *c*II and *c*III are supplied, integrase is formed even when p_L is repressed, indicating that the *c*II or *c*III product activates *int* transcription from some other promoter, probably p_I. To establish a stably lysogenic condition, both repression of genes leading to phage production and insertion of viral DNA into the host chromosome are needed. Therefore it makes sense that both repressor and integrase are formed in response to the same effectors, *c*II and *c*III.

Shimada and Campbell (1974a) isolated a number of mutants (*int-c*) that form elevated amounts of integrase even in the prophage state where repressor is present. One way this phenotype can arise is by a change in the sensitivity of p_L to repressor. Other possible mechanisms include changes in the properties of p_I or other relevant sites or the creation or insertion of new promoters. Operator constitutive mutations at p_L do occur and are easily distinguishable from changes at p_I. The remaining *int-c* isolates may include examples of other types.

Table 1 shows the properties of several *int-c* mutants. There are clearly quantitative differences among them. Some specific facts seem noteworthy:

(1) Among the three *xis*1*int-c* lysogens, the correlation between messenger level and *trpB* enzyme is excellent. The *int-c*57 promoter seems to be about three times as strong as the other two. However, this correlation does not extend to *int-c*226, where the *trpB* enzyme level in an *int-c*226 lysogen is threefold lower than would be expected from the amount of message. The reason for these discrepancies is unknown. Because these three *xis*1*int-c* lysogens were made originally from the same lysogen of λ*xis*1 inserted into the *trp* operon, we worried that an additional mutation might have occurred in the process of replacing λ*xis*$^+$ by λ*xis*1 in *trp* or that an extraneous mutation might have arisen within the stock. However, the *trpB* level with *int-c*226 remains the same when the *trp* region of a λ*int-c*226 lysogen is transduced into the parent strain KS507 by phage P1. Also, all the λ*int-c* lysogens in Table 1 can be cured with similar efficiency by heteroimmune superinfection. Thus any extraneous mutation must lie within the λ-*trp* region and not in the insertion site itself.

(2) The integrase activity of λ*int-c*548 is much lower than would be expected from the transcription and *trpB* results. This difference is observed consistently, both with the λ*trpBint-c* *c*I857 phages obtained by induction of the original *int-c* lysogens and with λ*h int-c* *c*I857 recombinants derived from them (preparation described by Shimada and Campbell 1974b).

Table 1
Properties of λ*int-c* Mutants

Phage	*trpB* Enzyme[a]	*int* Message[b]	Int activity[c]
λ*c*I857	0.1	7	<0.0001
λ*int-c*226*c*I857	2.4	192	0.05
λ*xis*1*int-c*548*c*I857	2.0	73	0.002
λ*xis*1*int-c*57*c*I857	5.9	234	0.03
λ*xis*1*int-c*60*c*I857	1.6	72	0.009

[a] Assay was according to the method of Smith and Yanofsky (1962) in lysogens carrying the indicated prophage inserted into the *trp* operon. The *trpB* gene is transcribed by read-through from the prophage into adjacent bacterial DNA.

[b] ^{3}H-labeled mRNA (1-min pulse at 30°C) extracted from noninduced lysogens (the same as used for *trpB* enzyme assay) was prehybridized to *l* strands of λ*b*2 DNA, eluted, and then hybridized analytically to *l* strands of λ, λ*bio*16A, λ*bio*386, and λ*bio*7-20 (for methods, see Bøvre et al. 1971). *Int* transcription was measured as percent of total ^{3}H-labeled input RNA that hybridized to the *att-bio*386 region (in λ*bio*386, the *int* and *xis* genes of λ are replaced by bacterial DNA), and results are expressed as the percentages of the p_{rm}-*c*I-*rex* (Imm) transcription normalized per 1% of λ length, assuming that the assayed lengths of the Imm and Int mRNAs are 4 %λ and 3 %λ units, respectively.

[c] Assayed by homoimmune curing of strain RW240, followed by superinfection with the phages listed in column 1. Strain RW240 is *recA*$^-$*galK*$^-$(λ*gal int*29*c*I857)(λ*int*29*c*I857). It is Gal$^+$ because of the λ*gal* prophage, which, having *att*B-*att*P boundaries, can be excised by integrase action. Results are expressed as number of Gal$^-$ cells produced per infecting phage particle.

The observed differences among *int-c*548, *int-c*57, and *int-c*60 are of interest because these mutants carry an IS*2* element in the antipolar, promoter-generating orientation, within or near the *xis* gene (Zissler et al., this volume). On the other hand, λ*int-c*226 has no IS insertion. The *int-c* phenotype of the three IS*2*-carrying mutants thus presumably derives from the inserted IS*2* promoter and therefore should be the same promoter in all cases. This is compatible with the very similar amounts of *int* message and *trpB* enzyme in *int-c*548 and *int-c*60. The poor integrase activity in *int-c*548 might then mean either (a) the *int-c*548 insertion is actually within the *int* gene, very close to the *N* terminus, or (b) the *int* gene copies in the *int-c*548 and *int-c*60 messages are translated with different efficiencies. Actually, the IS*2* insertion in λ*int-c*548 is about 150 base pairs to the left of *int-c*60 and might well affect the *int* gene or the *int* ribosomal binding site (Zissler et al., this volume). The latter possibility is interesting because of evidence (Court et al., this volume) that integrase production is largely *c*II-dependent even when transcription from p_L is derepressed. This finding suggests that the natural message originating at p_L, though containing the entire *int* gene, is translated inefficiently compared to that originating at p_I. The *int-c*57 mutation resembles *int-c*60 in the ratio of *trpB* enzyme:*int* message:Int activity, but the amounts are elevated about threefold. The various possible bases for this difference (altered promoter within IS*2*, polar fusions, effects of genetic background, etc.) have not yet been distinguished. The *int-c57* and *int-c60* insertions are physically indistinguishable (M. Fiandt and W. Szybalski, unpubl.).

Characteristically, λ*xis*$^{+}$*int-c* mutants are deficient in (but not devoid of) excisionase activity following derepression but do exhibit some excisionase activity even when p_L is repressed (Shimada and Campbell 1974b). Also, most of them map to the right of the *b*538 deletion, which dissects the *xis* gene, but to the left of the *bio*386 end point (L. Enquist, pers. comm.). We might have interpreted this as indicating that the normal promoter p_I lies within *xis*. However, recent results of A. Honigman, S.-L. Hu and W. Szybalski (in prep.) on the isolation of the short leader sequence for the Int mRNA, which crosses the *bio*386 deletion end point, suggested to them that the *int-c* mutations, because of their location to the left of the *bio*386 end point, probably do not define the p_I promoter. Rather, they might either inactivate the t_I terminator for the Int leader (e.g., *int-c*226) or are the promoter-carrying IS*2* insertion in orientation II (Saedler et al. 1974), which is located downstream of the t_I site. (The IS*2* insertions described here were isolated from an *xis*1 mutant. Thus it is not known whether they abolish excisionase completely or partially.) Direct genetic mapping of the normal promoter p_I relative to known *xis* and *int-c* mutations and to the end points of the *b*538 and *bio*386 deletions (located at 60.2 %λ and 60.3 %λ, respectively) (Szybalski and Szybalski 1974a; M. Fiandt and W. Szybalski, unpubl.) has not yet been accomplished.

The location and properties of p_I may be viewed in two contexts. On the one hand, we may try to relate them to the overall biology of λ, as we have already discussed with respect to the response to *c*II and *c*III. On the other hand, we may consider how the integrase promoter may relate to the evolutionary origin of the 1.5-kb segment of λ that determines its insertion functions. Much of the existing variation among phages and plasmids can be comprehended on the basis of the idea that such elements frequently have come about by the association of smaller elements derived from diverse sources (Campbell 1972; Sybalski and Szybalski 1974b).

According to this viewpoint, the segment of λ that functions in prophage insertion would represent the lineal descendent of something related to present-day insertion sequences. The p_I promoter might then very well be the direct descendent of the promoter that controlled the insertion and excision functions of that element, which has secondarily modified its control properties to become well integrated into the λ life cycle.

Acknowledgments

We thank L. Enquist, A. Honigman and J. Zissler for communication of unpublished results. The work at Stanford University was supported by National Institutes of Health Grants AI08573 and 5T01-GM 158. The work at the University of Wisconsin was supported by National Institutes of Health Grant CA-07175.

REFERENCES

Bøvre, K., H. A. Lozeron and W. Szybalski. 1971. Techniques of RNA-DNA hydridization in solution for the study of viral transcription. In *Methods in virology* (ed. K. Maramorosch and H. Koprowski), vol. 5, p. 271. Academic Press, New York.

Campbell, A. 1972. Episomes in evolution. *Brookhaven Symp. Biol.* **23**:534.

Eisen, H. and M. Ptashne. 1971. Regulation of repressor synthesis. In *The bacteriophage lambda* (ed. A. D. Hershey), p. 239. Cold Spring Harbor Laboratory, Cold Spring Harbor, New York.

Gottesman, M. E. and R. A. Weisberg. 1971. Prophage insertion and excision. In *The bacteriophage lambda* (ed. A. D. Hershey), p. 113. Cold Spring Harbor Laboratory, Cold Spring Harbor, New York.

Katzir, N., A. Oppenheim, M. Belfort and A. B. Oppenheim. 1976. Activation of the lambda *int* gene by the *c*II and *c*III gene products. *Virology* **74**:324.

Saedler, H., H.-J. Reif, S. Hu and N. Davidson. 1974. IS*2*, a genetic element for turn-off and turn-on of gene activity in *E. coli. Mol. Gen. Genet.* **132**: 265.

Shimada, K. and A. Campbell. 1974a. *Int*-constitutive mutants of bacteriophage lambda. *Proc. Nat. Acad. Sci.* **71**:237.

——. 1974b. Lysogenization and curing by *int*-constitutive mutants of phage λ. *Virology* **60**:157.

Smith, H. O. and C. Yanofsky. 1962. Enzymes involved in the biosynthesis of tryptophan. In *Methods in enzymology* (ed. S. P. Colowick and N. O. Kaplan), vol. 5, p. 794. Academic Press, New York.

Szybalski, E. H. and W. Szybalski. 1974a. Physical mapping of the *att-N* region of coliphage lambda: Apparent oversaturation of coding capacity in the *gam-ral* segment. *Biochimie* **56**:1497.

——. 1974b. Visualization of the evolution of viral genomes. In *Viruses, evolution and cancer* (ed. E. Kurstak and K. Maramorosch), p. 563. Academic Press, New York.

Position Effects of Insertion Sequences IS2 near the Genes for Prophage λ Insertion and Excision

J. Zissler, E. Mosharrafa and W. Pilacinski
Department of Microbiology
University of Minnesota
Minneapolis, Minnesota 55455

M. Fiandt and W. Szybalski
McArdle Laboratory for Cancer Research
University of Wisconsin
Madison, Wisconsin 53706

Many so-called polar mutations, including *gal3* which was isolated over 20 years ago by Morse et al. (1956), were later found to be insertions of short sequences of DNA. These were named IS (for insertion sequences), and most were shown to comprise only a few classes, being either about 800 base pairs long (IS*1*) or 1100 to 1400 base pairs long (IS*2*, IS*3*, IS*4*) (Fiandt et al. 1972; Hirsch et al. 1972; Malamy 1972). The IS*1* element appeared in both orientations with respect to the disrupted operon, and both orientations were polar. The IS*2* element, however, appeared in only one orientation if selected as a polar mutation. It was shown later that, in the opposite orientation, called orientation II, the IS*2* element exerts a positive effect on gene expression (Saedler et al. 1974). Thus the IS*2* insertion can "turn on" or "turn off" particular genes depending on its orientation within an operon.

The IS*2* insertion appears fortuitously in several strains of bacteriophage λ, and the earliest one characterized was the so-called *b*1 insertion between genes *P* and *Q* (Fiandt et al. 1971). Especially interesting to us at present is the occurrence of IS*2* near the λ genes for integration and excision. In particular, the presence of IS*2* could explain the phenotype of several λ variants.

We show here that λ*crg* (Fischer-Fantuzzi and Calef 1964) has an IS*2* insertion in orientation I at 60.7 %λ, which is just upstream from the *xis* gene. In this orientation, IS*2* may exert a polar effect on *xis,* since λ*crg* is partially defective in *xis* function (Adhya and Campbell 1970; L. Enquist, pers. comm.). The IS*2* insertion may also influence excision directly and cause the formation of cryptic (*cry*) prophage.

The IS*2* element resides in other λ phages, including the spontaneous mutant λ*bi*2 (Mosharrafa et al. 1976), and in some mutants selected as carrying up-promoter mutations (Shimada and Campbell 1974). In λ*bi*2 and λ*int-c*, the IS*2* insertion is in orientation II, at 61.6 %λ and 59.9-60.2 %λ, respectively. In this orientation, IS*2* is in position to turn on expression of Int function. Whereas *int-c* does cause constitutive Int expression, the *bi*2 insertion surprisingly does not. These effects on gene expression may be explained by considering both the orientation of the element and its position relative to other control sequences in the operon.

IS*2* INSERTIONS

The *bi*2 Mutation

Our present interest in insertion sequences began with a variant of λ that has an increased density and which we called λ*bi*2 (Mosharrafa et al. 1976). The density increase corresponds to approximately 2.8 %λ units of DNA. This mutant arose serendipitously from a λ*c*I857*S*7 strain and appeared to carry an insertion because it showed a new pattern of DNA fragments after digestion with the restriction endonuclease *Eco*RI. Specifically, the DNA fragment $C_{3.5}$ (m. w. ~ 3.5 × 10^6 daltons) observed for normal λ (Thomas and Davis 1975) was replaced by a larger fragment (m. w. ~ 4.4 × 10^6 daltons), which we designated $C_{4.4}$ (Fig. 1). Only the strain carried in the laboratory of J. Zissler, which we renamed λ*bi*2*c*I857*S*7, had acquired this alteration, whereas four other isolates of λ*c*I857*S*7, obtained from the laboratories of R. W. Davis, D. I. Friedman, E. Signer, and W. Szybalski contained the normal $C_{3.5}$ fragment.

The original λ*c*I857*S*7 and its spontaneous λ*bi*2*c*I857*S*7 variant were studied further by electron microscopy of appropriate heteroduplexes, as described by Mosharrafa et al. (1976). The increased size of the C fragment is due to the presence of an IS*2* element, at 61.6 %λ, which is upstream from the gene *xis* (see Fig. 2). The position of 61.6 %λ is measured in relation to the *att* site (57.3 %λ; Westmoreland et al. 1969) and is measured as the duplex DNA in the interval between the *att* site (marked by the *gal* substitution or the *b*2 deletion) and the *bi*2 loop in heteroduplexes λ*gal*8-490/λ*bi*2*c*I857*S*7 (interval = 4.24 ± 0.24 %λ; Mosharrafa et al. 1976) and λ*b*2/λ*bi*2*c*I857*S*7 (interval = 4.34 ± 0.40 %λ; present data).

The IS*2* insertion in λ*bi*2 must lie in orientation II since it interacts with another IS*2* (*r*32) in *l*/*r* heteroduplexes with λ*r*32 (Mosharrafa et al. 1976). In orientation II, the *bi*2 insertion might turn on constitutively the *int* and *xis* genes located downstream (Saedler et al. 1974). However, the λ*bi*2 prophage does not appear to produce the *int* and *xis* gene products at an elevated constitutive level, as measured by breakdown of λatt^2 (see Shimada and Campbell 1974) upon superinfection of the λ*bi*2 lysogen (W. Pilacinski, unpubl.).

The identification of an IS*2* insertion near the *xis* gene in λ*bi*2 led us to wonder whether previously described mutations in this region of λ might be

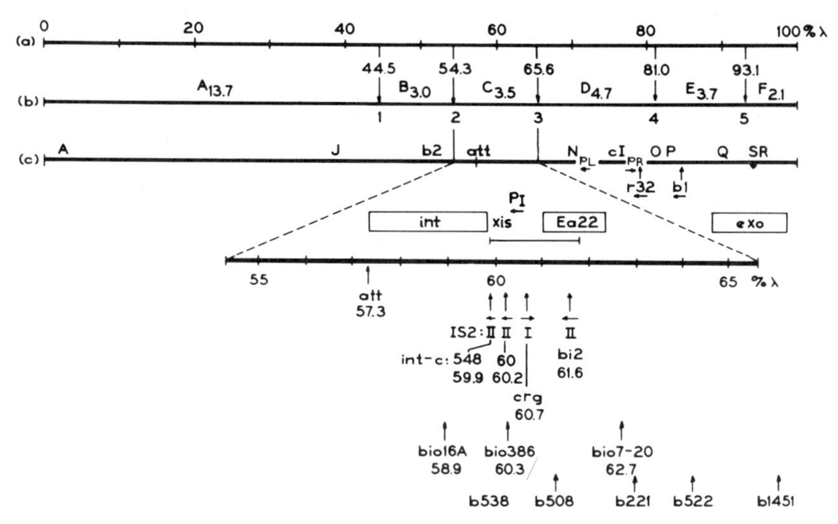

Figure 1
Physical and genetic maps of bacteriophage λ. (*a*) Physical map coordinates expressed in %λ units (%λ = 465 base pairs). (*b*) The six fragments generated by *Eco*RI restriction endonuclease (see Fig. 2a). (*c*) Genetic map of λ, including selected λ genes and the IS*2* insertions at the *r*32 and *b*1 sites (Fiandt et al. 1971, 1972). The $C_{3.5}$ fragment is expanded and shows positions of IS*2* insertions (*bi*2, *crg, int-c*60, and *int-c*548). Insertion *int-c*57 (now shown) is indistinguishable for *int-c*60.

The orientations of transcription from the promoter in IS*2* insertions and from the p_L, p_R, and p_I promoters are indicated by the horizontal arrows. (For further details, see Mosharrafa et al. 1976.) (The position of the right *b*538 end point is here shown revised, based on unpublished measurements of E. H. Szybalski. When measured versus the *crg* insertion, the positions of the *b*538 and *bio*386 end points are 60.2 %λ and 60.3 %λ, respectively [M. Fiandt and W. Szybalski, unpubl.].)

due to similar insertions. In particular, one primary suspect was the mutation *crg*, which results in abnormal excision leading to the formation of *cry* prophage (Fischer-Fantuzzi and Calef 1964; Adhya and Campbell 1970; Marchelli et al. 1968). Another suspect was the mutation *int-c*, which results in constitutive expression of the *int* gene (Shimada and Campbell 1974). In fact, as shown in Figure 2, this proved to be the case, since both *crg* and some of the *int-c* mutations are associated with IS*2* insertions.

The *crg* Mutation

The λ*crg* phage has an IS*2* insertion in polar orientation I; it forms a duplex with the IS*2* insertion (490) of λ*gal*8-490 (Fiandt et al. 1972) in the λ*crg*842/λ*gal*8-490 heteroduplex. As with λ*bi2*, this insertion probably occurred spontaneously,

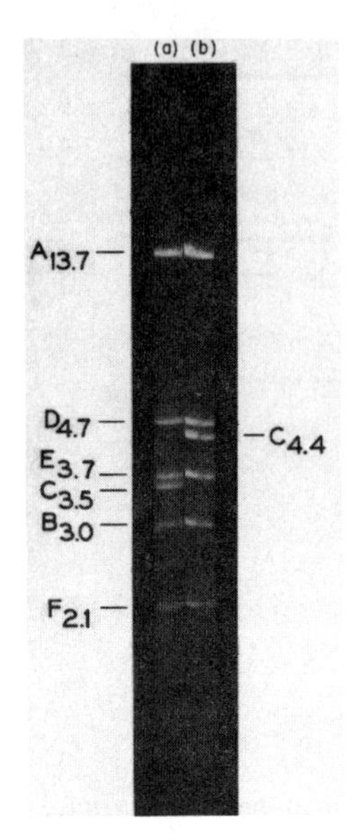

Figure 2
Agarose gel electrophoresis of λ DNA fragmented by *Eco*RI endonuclease. (*a*) λcI857*S*7 (normal pattern of λ DNA fragments). (*b*) Pattern for the IS*2* insertion variants λ*bi*2, λ*crg,* and λ*int-c*60. (For details, see Mosharaffa et al. 1976.)

presumably during the time when the strain was propagated and stored in the early 1960's in E. Calef's laboratory.

IS*2* in λ*crg* is located at 60.7 %λ, being separated from *bi*2 by 0.92 ± 0.08 %λ units in the heteroduplex λ*crg*842/λ*bi*2cI857*S*7. This places the *crg* insertion just upstream from the *xis* gene. Adhya and Campbell (1970) observed that λ*crg* is partially defective in *xis* function, and this has been supported by additional tests (L. Enquist, pers. comm.). We conclude that λ*crg* may be partially *xis*-deficient since the polarity of IS*2* in orientation I may decrease the expression of *xis* function. Polarity could occur even in the presence of the λ gene-*N* product, since relief of polarity by *N* is probably incomplete (Adhya et al. 1974).

The *int-c* Mutations

The *int-c*60 mutation (Campbell et al., this volume), which we studied first, involves an IS*2* insertion in orientation II, since it forms a duplex with the IS*2* insertion of λ*r*32 (Fiandt et al. 1972) in the λ*int-c*60/λ*r*32 heteroduplex. The IS*2* element in λ*int-c*60 is located at 60.2 %λ, being separated from the *crg* insertion by 0.47 ± 0.08 %λ units in the heteroduplex λ*int-c*60/λ*crg*842. This mapping places the *int-c* insertion probably within the *xis* gene. Since the insertion lies in orientation II, its promoter may be responsible for the constitutive expression of gene *int,* or the *trp* genes when λ*int-c* is inserted into the *trp* operon (Shimada and Campbell 1974).

The order of the *int-c*60 and *crg* insertions is further confirmed by heteroduplexes between them and λ*bio*386 (end point of 60.3 %λ; Fig. 1; Szybalski and Szybalski 1974), which showed that the *crg* insertion is 0.40 ± 0.11 %λ units to the right of the *bio*386 deletion end point, whereas the *int-c*60 insertion is within this deletion.

We have found that λ*int-c*548 (Campbell et al., this volume) also carries an IS*2* insertion. The insertion element lies in orientation II, at a site about 150 base pairs to the left of *int-c*60. In addition, our strain of λ*int-c*548 has a large *b*2-type deletion, which explains why bands B and C are missing in an electropherogram of the *Eco*RI fragments of λ*int-c*548 (not shown in Fig. 2). We do not know when the *b*2-type deletion occurred.

Not all of the λ*int-c* mutants carry the IS insertions as we did not observe any single-stranded DNA loop in the *xis* region in heteroduplexes between λimm^{434} and λ*int-c*570 or λ*int-c*226.

DISCUSSION

The insertion sequence IS*2* seems to occur rather frequently in phage λ near or within the *xis* gene. We do not understand the significance of the spontaneous and unselected occurrence of at least two insertions (*bi*2 and *crg*), which indicates a high frequency of this kind of mutation in that part of the λ genome which alone governs the integration and excision of λ DNA. We also do not as yet understand whether IS*2* is "foreign" to phage λ, or whether IS or similar sequences might act in normal chromosomal mechanics. In particular, we suspect that IS*2* may share some similar insertion mechanism with the *att-int-xis* region of λ.

The effects of IS*2* insertion on gene expression might be better understood by considering the normal regulation of the λ genes in this region. The transcription of the *int* gene could originate either at the p_L or p_I promoter (Fig. 1). However, the expression of the Int function from a transcript originating at p_L is apparently inefficient (Court et al., this volume). Moreover, Int expression from the wild-type p_I promoter requires the functions *c*II and *c*III (Katzir et al. 1976; Court et al., this volume). These functions may be required to antiterminate a very short RNA leader sequence identified by A. Honigman, S.-L. Hu and W. Szybalski (in prep.). Interestingly, they have shown that this leader RNA initiates just to the right of the *bio*386 boundary and terminates at a site, t_I, just to the left of this point (Fig. 1). Thus we might speculate that the two *int-c* mutations (*int-c*60 and *int-c*548) are constitutive for the Int function because they represent an IS*2* insertion which, in orientation II, carries a new promoter for constitutive *int* transcription. We believe that the p_I promoter is located to the right of the *bio*386 boundary and that the *int-c* insertions have to be located downstream from the t_I terminator for the Int leader RNA. We might also speculate that some or all *int-c* mutations of the 226 or 570 type, which we found not to carry IS insertions, may inactivate the t_I terminator and thus render the p_I-initiated *int* expression independent of the *c*II and *c*III functions. It might be of interest that there is a significant difference in Int expression, but not transcription, between the IS*2* mutants, with λ*int-c*548 expressing several times less Int function than λ*int-c*57 and 60, the latter two probably inserted at the same site (Fig. 1; M. Fiandt and W. Szybalski, unpubl.; Campbell et al., this volume). The simplest interpretation is that the IS*2* insertion in λ*int-c*548 interferes with the translation of the *int* message, since, as seen in Figure 1, it is located extremely close to the beginning of gene *int* and might affect the ribosome binding site, or another translation-controlling site, or even the *int* protein itself.

The observation that *int-c* mutations, selected for the medium level of constitutive Int expression, are all caused by the IS*2* insertions in orientation II provides strong evidence that the IS*2* element carries a constitutive promoter with orientation II.

There are several other observations, some of which could be explained by this simplified model. For example, the *bi*2 mutation does not show constitutive Int expression, although the IS*2* insertion in λ*bi*2 resides in the promoter-generating orientation II (Fig. 1). Possibly any transcription promoted by the *bi*2 insertion is terminated by terminator t_I. However, to provide a better understanding of this and other controls, a more detailed analysis of the *b*2-*att-int-xis* region is required.

SUMMARY

Three kinds of mutations in bacteriophage λ were found to be associated with the IS*2* insertion element. Two of these (λ*bi*2 and λ*crg*) appeared without any known selective pressure, and the third (λ*int-c*) was selected as an up-promoter for the Int function. In λ*bi*2 and three λ*int-c* phages, the IS*2* element was inserted with a promoter-generating orientation at the 61.6 %λ (*bi*2), 59.9 %λ (*int-c*548), and 60.2 %λ (*int-c*57 and 60) sites, respectively. The *bi*2 insertion has no obvious effect on *xis* or *int* expression, whereas the *int-c* insertions promote an elevated constitutive expression of the gene *int*. The *crg* insertion at 60.7 %λ has a polar orientation and appears to cause aberrant excisions leading to the formation of λ*cry* lysogens. The positions and orientations of these insertions could explain the phenotypes of several λ variants, which should prove valuable for studies on the λ integration and excision functions.

Acknowledgments

Work at the University of Minnesota was supported, in part, by a grant from the National Science Foundation (BMS 7412225) and by training grant GM966 to the Department of Microbiology from the National Institute of General Medical Sciences, National Institutes of Health. Work at the University of Wisconsin was supported by the Program-Project Grant from the National Cancer Institute (CA-07175). We thank Dr. A. Campbell for suggesting that an IS insertion might be responsible for the *crg* character, Dr. L. W. Enquist for sending us his unpublished data on the mapping of the *xis* mutations, Dr. M. L. Pearson and C. Epp for communicating their results on the sizes and locations of the presumed *xis* and Ea22 proteins, and Dr. E. H. Szybalski for the physical data on the *b*538 deletion.

REFERENCES

Adhya, S. and A. Campbell. 1970. Crypticogenicity of bacteriophage λ. *J. Mol. Biol.* **50**:481.

Adhya, S., M. Gottesman and B. de Crombrugghe. 1974. Release of polarity in *Escherichia coli* by gene *N* of phage λ: Termination and antitermination of transcription. *Proc. Nat. Acad. Sci.* **71**:2534.

Fiandt, M., W. Szybalski and M. H. Malamy. 1972. Polar mutations in *lac, gal* and phage λ consist of a few IS-DNA sequences inserted with either orientation. *Mol. Gen. Genet.* **119**:223.

Fiandt, M., Z. Hradecna, H. A. Lozeron and W. Szybalski. 1971. Electron micrographic mapping of deletions, insertions, inversions, and homologies in the DNA of coliphage lambda and phi 80. In *The bacteriophage lambda* (ed. A. D. Hershey), p. 329. Cold Spring Harbor Laboratory, Cold Spring Harbor, New York.

Fischer-Fantuzzi, L. and E. Calef. 1964. A type of λ prophage unable to confer immunity. *Virology* **23**:209.

Hirsch, H. J., P. Starlinger and P. Brachet. 1972. Two kinds of insertions in bacterial genes. *Mol. Gen. Genet.* **119**:191.

Katzir, N., A. Oppenheim, M. Belfort and A. B. Oppenheim. 1976. Activation of the lambda *int* gene by the *c*II and *c*III gene products. *Virology* **74**:324.

Malamy, M. H., M. Fiandt and W. Szybalski. 1972. Electron microscopy of polar insertions in the *lac* operon of *Escherichia coli. Mol. Gen. Genet.* **119**:207.

Marchelli, C., L. Pica and A. Soller. 1968. The cryptogenic factor in λ. *Virology* **34**:650.

Morse, M. L., E. M. Lederberg and J. Lederberg. 1956. Transduction in *Escherichia coli* K-12. *Genetics* **41**:142.

Mosharrafa, E., W. Pilacinski, J. Zissler, M. Fiandt and W. Szybalski. 1976. Insertion sequences IS*2* near the gene for prophage λ excision. *Mol. Gen. Genet.* **147**:103

Saedler, H., H.-J. Reif, S. Hu and N. Davidson. 1974. IS*2*, a genetic element for turn-off and turn-on of gene activity in *E. coli. Mol. Gen. Genet.* **132**:265.

Shimada, K. and A. Campbell. 1974. Int-constitutive mutants of bacteriophage lambda. *Proc. Nat. Acad. Sci.* **71**:237.

Szybalski, E. H. and W. Szybalski. 1974. Physical mapping of the *att*-N region of coliphage lambda: Apparent oversaturation of coding capacity in the *gam-ral* segment. *Biochimie* **56**:1497.

Thomas, M. and R. W. Davis. 1975. Studies on the cleavage of bacteriophage lambda DNA with *Eco*RI restriction endonuclease. *J. Mol. Biol.* **91**:315.

Westmoreland, B. C., W. Szybalski and H. Ris. 1969. Mapping of deletions and substitutions in heteroduplex DNA molecules of bacteriophage lambda by electron microscopy. *Science* **163**:1343.

The Phage λ Integration Protein (Int) Is Subject to Control by the *c*II and *c*III Gene Products

D. Court,* S. Adhya,* H. Nash† and L. Enquist‡
*National Cancer Institute
†National Institute of Mental Health and
‡National Institute of Child Health and Human Development
National Institutes of Health
Bethesda, Maryland 20014

The choice between lytic and lysogenic growth of phage λ appears to rest primarily on the expression of the *c*II and *c*III gene products (see Fig. 1). This was first inferred from the fact that *c*II and *c*III mutants formed clear plaques and lysogenized poorly (Kaiser 1957). Later work showed that *c*II and *c*III are positive regulators of the synthesis of *c*I repressor (Reichardt and Kaiser 1971; Echols and Green 1971). An additional effect of these functions is the inhibition of lytic gene expression (McMacken et al. 1970; Court et al. 1975).

Repression by *c*I product is only a part of the lysogenic response. In a stable lysogen, the phage DNA is inserted into the *Escherichia coli* genome by a site-specific recombination event. A phage protein, Int, is required for this recombination. We report here that the regulatory products *c*II and *c*III which govern repressor synthesis also control the production of Int.

Several years ago, one of us (D.C.) partially purified a protein that is synthesized when *E. coli* is infected with $\lambda c\text{II}^+ c\text{III}^+$ but not with $\lambda c\text{II}^- c\text{III}^-$ (Fig. 2). The purification scheme used is similar to one developed recently by H. Nash (unpubl.) to isolate the *int* gene product, as assayed by an in vitro integrative recombination system (Nash 1975). Furthermore, the *c*II,*c*III-dependent protein has a subunit molecular weight (40,000) the same as that reported for the *int* gene product (Hendrix 1971). We postulated that the *c*II,*c*III-dependent protein and Int are the same and have tested this hypothesis by examining the effect of mutations in *c*II or *c*III on integrative recombination in vivo. Our conclusion is that *c*II and *c*III control integration by modulating *int* gene expression. H. Echols and colleagues (pers. comm.) and Katzir et al. (1976) have demonstrated independently that *c*II and *c*III are required for Int protein synthesis.

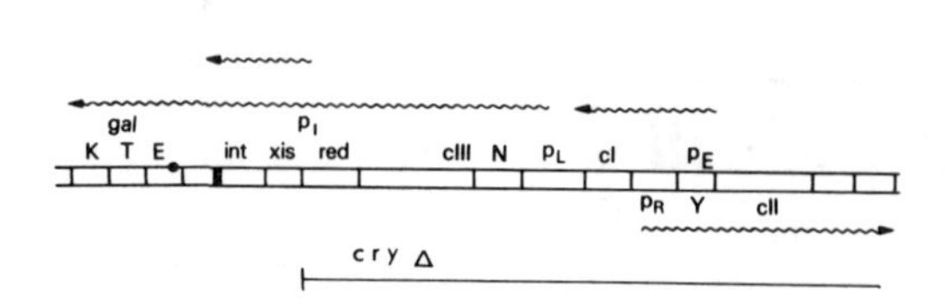

Figure 1
Gene organization and transcription of the "early region" of bacteriophage λ. Transcripts are shown as wavy lines; arrows indicate direction. Gene symbols are described in the text. *cry* Δ is a deletion.

Reduction of Integrative Recombination by *c*II and *c*III Mutants

Table 1 shows that a defect of the *c*II gene, *c*II28, lowers integrative recombination fivefold. Other *c*II mutants (*c*II123, *c*II2002) reduce recombination to about the same degree (data not shown). We also see a smaller but reproducible reduction with a *c*III mutation. When both *c*II and *c*III are inactive, integrative

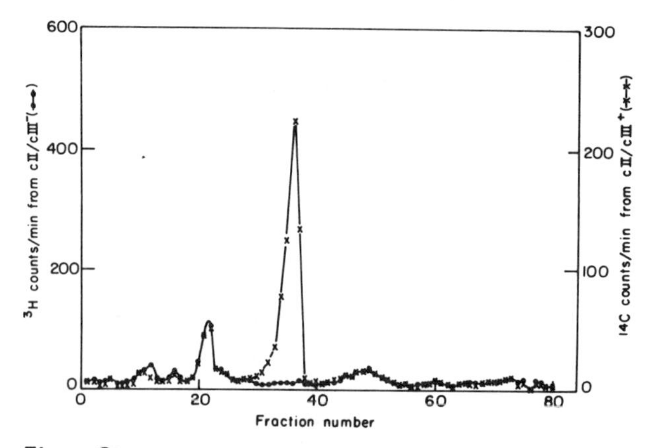

Figure 2
Gel profile of a *c*II,*c*III-dependent protein. Cells are infected with either λ*c*I14*cY*42*Qam*21 or λ*c*I14*c*I128*c*III611*Qam*21 (*c*II$^+$*c*III$^+$ or *c*II$^-$*c*III$^-$) at a multiplicity of ten. Growth conditions were as described by Court et al. (1975). [^{14}C]leucine and [^{3}H]leucine, respectively, are added for 10 min after infection. Growth is stopped by quickly chilling the culture. Cells from the two cultures are mixed and extracted together. The preparation of the crude extract has been described by Echols and Green (1971). The crude extract was centrifuged 60 min at 100,000*g*. The pellet was saved and eluted in buffer with 0.5 M NH_4Cl. By analysis of the eluate on SDS-polyacrylamide gels, a protein (^{14}C-labeled) of ~40,000 daltons is found associated only with the *c*II$^+$*c*III$^+$ infection. When this eluate is dialyzed against a low-salt buffer, a precipitate forms. This precipitate contains the 40,000-dalton protein. This figure shows the gel profile of the precipitate fraction which was resolubilized in high salt and centrifuged in a 10–30% gylcerol gradient in 0.5 M NH_4Cl.

Table 1
Int Activity of Clear Mutants of Phage λ

Genotype	Relative Int activity
cI^-	100
cI^-cIII^-	57
cI^-cII^-	20
$cI^-cII^-cIII^-$	15
cI^-cY^-	115
$cI^-cY^-int^-$	0.3

Int activity is measured by a site-specific recombination event between the phage attachment site *att*P and the bacterial attachment site *att*B. Both of these *att* sites are located on the same phage DNA, i.e., λ*att*B-*att*P (λatt^2). This phage, as well as the conditions used to determine Int activity, have been described (Nash 1974). In this experiment, $\lambda imm^{434}att^2$ carried the mutations $int2xis1red114imm^{434}$cI-1cII28cIII611 to eliminate any *int*, cII, or cIII expression. A λ helper phage present in all infections supplied the *int* function; its genotype is given in the first column. The clear mutations used are cI14, *c*Y42, cII28, and cIII611. The *int* mutation is *int*6. The bacteria were infected with $\lambda imm^{434}att^2$ phage at low multiplicity (<1) and the λ helper at high multiplicity (~8). The yield of λimm^{434} phage was counted as plaques formed on a λ lysogen, N116, which does not permit the helper λ to grow. Relative Int activity is expressed as percent $\lambda imm^{434}att^2$ phage recombined among total λimm^{434} phage progeny, normalized to 110% for cI^-.

recombination is further reduced. These results suggest that the *c*II,*c*III-dependent protein described in Figure 2 is the product of the *int* gene and that *c*II and *c*III affect recombination by modulating the expression of the *int* gene.

How might *c*II and *c*III affect the modulation? Both are known regulatory elements that establish the synthesis of repressor by an effect upon transcription from a region near *c*II (Spiegelman et al. 1972). The effect could therefore be an indirect result of their regulation of *c*I. To dissociate any such indirect effects of *c*I on *int*, we have included a cI^- mutation in all the strains used. Table 1 also shows that the *cY* region, which is essential for *c*II,*c*III-mediated repressor synthesis and lytic gene inhibition, is unimportant for *int* activity, i.e., a *cY* mutation did not reduce integrative recombination.

Dispensability of Immunity Region, Which Includes the p_L Promoter, for *c*II, *c*III Activation of *int* Gene

Is the p_L promoter required for *c*II,*c*III-dependent Int synthesis? We eliminated p_L transcription either by repression with *c*I repressor or by deletion of the promoter. In both cases, *c*II,*c*III activation of Int synthesis occurred (Table 2).

The site of the Int-stimulating action of *c*II,*c*III must still be present in the cryptic prophage listed in Table 2. The *cry* deletion removes *red,* but the *int-xis* region is still present. Shimada and Campbell (1974) have demonstrated the existence of a low-level constitutive promoter, p_I, that is located in the *int-xis-*

Table 2
Int Activity from the Prophage p_I Promoter

Infecting phage	Int activity from prophage genotype *int*$^+$	*int*$^+$ (*cry* deletion) Δ	*int-c*
λ*c*I$^-$*c*II$^+$*c*III$^+$	17.1	33.0	56.8
λ*c*I$^-$*c*II$^-$*c*III$^-$	<0.1	<0.1	41.0

Int activity is measured, as in Table 1, by recombination between *att*P and *att*B in a λ*att*^{2} phage. The data are the average from two experiments. In these experiments, the infecting phage was always λ*att*^{2}*int*$^-$, which was infected at an m.o.i. of 10. The host carries an *int*$^+$ prophage. No activity is seen with *int*$^-$ prophage.

Prophage genotypes: In column 1, the prophage is wild-type immunity λ, and p_L is repressed. In column 2, the prophage carries a deletion that has removed prophage genes from *red* through *R*, including p_L and the immunity region; *int* and at least part of *xis* is not deleted (L. Enquist, unpubl.). This prophage is called λ*cry* and was derived from λ*crg* prophage (Adhya and Campbell 1970). In column 3, the prophage is also immunity λ and repressed but it carries a mutation (*int-c*) that allows constitutive synthesis of *int* even when repressed (Shimada and Campbell 1974).

Infecting phage genotype: In line 1, λ*att*^{2}*int*2*xis*1*red*114*imm*^{434}cI-1; in line 2, λ*att*^{2}*int*2-*xis*1*red*114*imm*^{434}cI-1cII28cIII611.

The λ*att*^{2} phage and their recombinants in the various infections were counted as plaques formed on bacterial strain N99.

red region and is capable of transcribing the *int* gene. An attractive hypothesis is that *c*II,*c*III stimulates *int* expression by acting at the level of transcription from this promoter. Alternatively, *c*II,*c*III could stimulate translation of message made from this promoter.

It is possible that *c*II and *c*III may in addition affect Int synthesis through events originating at p_L. However, since the *c*II,*c*III-dependent activity for p_I is high and since no mutants exist which inactivate p_I, this possibility is difficult to test.

Shimada and Campbell (1974) have isolated high-level constitutive promoter mutants, *int-c*, that occur in the same region as p_I. When *int* gene is expressed from such an *int-c* promoter, integrative recombination is independent of *c*II and *c*III (Table 2). This result is in agreement with our hypothesis that *c*II and *c*III affect *int* expression rather than some aspect of the recombination event.

Effect of Multiplicity of Infection on Integration

The number of λ particles infecting a cell greatly influences the alternative between a lytic or lysogenic infection cycle (Fry 1959). At high multiplicities of phage per cell (>5), the lysogenic response is enhanced. The *c*II and *c*III genes have been shown to be responsible for this mode of control (Kourilsky 1974; Court et al. 1975; Reichardt 1975). The activation of *int* gene by *c*II and *c*III is also multiplicity-dependent (see Table 3).

Table 3
*c*II,*c*III Multiplicity Dependence

Infecting phage	M.o.i.	Int activity from a repressed prophage *int*$^+$
λ*c*I$^-$*cY*$^-$*c*II$^+$*c*III$^+$	0.5	0.3
λ*c*I$^-$*cY*$^-$*c*II$^+$*c*III$^+$	2.5	5.0
λ*c*I$^-$*cY*$^-$*c*II$^+$*c*III$^+$	10.0	76.5
λ*c*I$^-$*c*II$^-$	0.5	0.5
λ*c*I$^-$*c*II$^-$	2.5	0.3
λ*c*I$^-$*c*II$^-$	10.0	0.5

The design of this experiment is the same as that in Table 2. The prophage used is λ*imm*^{434} in which p_L is repressed. Infecting phage genotypes: In lines 1, 2, and 3, λ*att*^{2}*int*2 *xis*1*red*114*imm*λ*c*I14*cY*42; in lines 4, 5, and 6, λ*att*^{2}*int*2*xis*1*red*144*imm*λ*c*I14*c*II28.

Interpretation of *c*II,*c*III Control

Messenger RNA originating at p_L can extend through *int* into bacterial genes. This extended message does not require a functional *c*II gene (Adhya et al. 1974), nor is *c*II required for expression of functions located before or beyond the *int* gene. Lambda *c*II$^-$ is not defective in Xis, the product of a gene located before *int* (L. Enquist, unpubl.). Moreover, galactokinase, the product of a bacterial gene located beyond *int* and also transcribed from the p_L promoter, is unaffected by the presence or absence of *c*II (Adhya et al. 1974). Since p_L is one of the most active promoters in *E. coli* (S. Adhya, unpubl.), why is the expression of Int from p_L insufficient in the absence of *c*II and *c*III (Table 1)? This question has not yet been resolved and is under study.

Control of Int synthesis is very much like control of repressor synthesis. For both, the *c*II requirement is much greater than that of *c*III (Reichardt and Kaiser 1971; Echols and Green 1971). In addition, there is an effect of multiplicity associated with all *c*II,*c*III effects (Court et al. 1975; Reichardt 1975). The actual mechanism of *c*II,*c*III action has not been determined; however, it is likely that the same mechanism will hold true for both the *int* and *c*I regions.

The regulation of both integration and repression by the same regulatory genes, *c*II and *c*III, would ensure that both processes are coordinated. This coordination could be advantageous to the bacteriophage since integration prior to cessation of lytic activities would result in death of the bacterial host.

Acknowledgments

We thank A. Campbell, R. Weisberg and M. Gottesman for helpful discussions. H. Echols provided laboratory space and advice to D. Court in the initial purification of the *c*II,*c*III-dependent protein. H. Echols and A. Oppenheim provided information prior to publication. We acknowledge the technical help of Carol Robertson.

REFERENCES

Adhya, S. and A. Campbell. 1970. Crypticogenicity of bacteriophage λ. *J. Mol. Biol.* **50**:481.

Adhya, S., M. Gottesman and B. de Crombrugghe. 1974. Release of polarity in *Escherichia coli* by gene N of phage λ: Termination and antitermination of transcription. *Proc. Nat. Acad. Sci.* **71**:2534.

Court, D., L. Green and H. Echols. 1975. Positive and negative regulation by the *c*II and *c*III gene products of bacteriophage λ. *Virology* **63**:484.

Echols, H. and L. Green. 1971. Establishment and maintenance of repression by bacteriophage lambda. *Proc. Nat. Acad. Sci.* **68**:2190.

Fry, B. A. 1959. Conditions for the infection of *Escherichia coli* with lambda phage and for the establishment of lysogeny. *J. Gen. Microbiol.* **21**:676.

Hendrix, R. W. 1971. Identification of proteins in phage lambda. In *The bacteriophage lambda* (ed. A. D. Hershey), p. 355. Cold Spring Harbor Laboratory, Cold Spring Harbor, New York.

Kaiser, A. D. 1957. Mutations in a temperate bacteriophage affecting its ability to lysogenize *E. coli. Virology* **3**:42.

Katzir, N., A. Oppenheim, M. Belfort and A. Oppenheim. 1976. Activation of the lambda *int* gene by the *c*II and *c*III gene products. *Virology* **74**:324.

Kourilsky, P. 1974. Lysogenization by bacteriophage lambda. II. Identification of genes involved in the multiplicity dependent processes. *Biochimie* **56**:1517.

McMacken, R., N. Mantei, B. Butler, A. Joyner and H. Echols. 1970. Effect of mutations on the *c*II and *c*III genes of bacteriophage λ on the macromolecular synthesis in infected cells. *J. Mol. Biol.* **49**:639.

Nash, H. A. 1974. λ*att*B-*att*P, a λ derivative containing both sites involved in integrative recombination. *Virology* **57**:207.

——. 1975. Integrative recombination of bacteriophage lambda DNA *in vitro. Proc. Nat. Acad. Sci.* **72**:1072.

Reichardt, L. F. 1975. Control of bacteriophage lambda repressor synthesis after phage infection: The role of the *N*, *c*II, *c*III, and *cro* products. *J. Mol. Biol.* **93**:267.

Reichardt, L. and A. D. Kaiser. 1971. Control of λ repressor synthesis. *Proc. Nat. Acad. Sci.* **68**:2185.

Shimada, K. and A. Campbell. 1974. Int-constitutive mutants of bacteriophage lambda. *Proc. Nat. Acad. Sci.* **71**:237.

Spiegelman, W. G., L. F. Reichardt, M. Yaniv, S. F. Heinemann, A. D. Kaiser and H. Eisen. 1972. Bidirectional transcription and the regulation of phage λ repressor synthesis. *Proc. Nat. Acad. Sci.* **69**:3156.

Temperate Coliphage P2 as an Insertion Element

R. Calendar
Molecular Biology Department
University of California
Berkeley, California 94720

E. W. Six
Microbiology Department
University of Iowa
Iowa City, Iowa 52240

F. Kahn
Genetics Department
University of Lund
Lund, Sweden

P2 is a temperate coliphage with a linear DNA genome of 33 kilobase pairs (Inman and Bertani 1969). When wild-type P2 infects *Escherichia coli,* 15-20% of the infected cells survive and become lysogenic (Bertani 1957). Such cells are immune to superinfection by P2 (Bertani 1954), due to the presence of the P2 repressor protein encoded by gene *C* (Bertani 1968) (Fig. 1).

Integration Sites

It has been shown by conjugation and P1 transduction that there are at least ten distinct sites on the *E. coli* genome at which the P2 genome can be integrated (Bertani and Six 1958; Six 1960, 1961, 1963, 1966; Kelly 1963). By determining the gene order of P2 prophages at three of these sites, Calendar and Lindahl (1969) have demonstrated that the P2 genome contains a specific attachment site (*att*) which is located between tail gene *D* and immunity gene *C* (Figs. 1 and 2). Int-promoted recombination between infecting P2 phages also occurs at this site (Lindahl 1969b). Since the attachment site is not located at the ends, the P2 genome must circularize and integrate, via a single reciprocal crossover event, into each of its chromosomal sites. Circularization occurs because of the complementary single-stranded ends of P2 DNA (Mandel 1967). A list of the known P2 integration sites in the host genome is given in Table 1.

In *E. coli* strain C, P2 shows strong preference for integration at one site. Hence this site (I) is virtually always occupied first. Doubly lysogenic strains can be isolated which carry their second prophage at a variety of secondary sites (Bertani and Six 1958; Six 1960, 1961, 1963, 1966). *E. coli* strain K lacks the

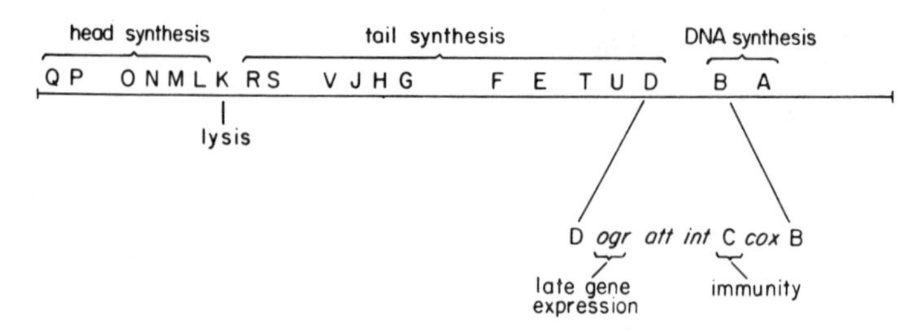

Figure 1
Genetic map of phage P2, based on Lindahl (1969a, 1970, 1971) and Sunshine et al. (1971), for essential genes and immunity gene *C*. The *att* site was mapped by Calendar and Lindahl (1969), the *int* (integration) gene by Lindahl (1969a), the *ogr* gene by Sunshine and Sauer (1975), and the *cox* mutation by Lindahl and Sunshine (1972). Map distances correspond to physical distances in so far as they are known (see Appendix C4).

preferred site I, and singly lysogenic strains of *E. coli* K can be isolated which carry prophages at a variety of chromosomal sites, among which II and H are most commonly occupied (Kelly 1963). Prophage maps of P2 at sites I, II, and H are depicted in Figure 2. It was possible to construct such prophage maps simply by P1 transduction because P2 lysogenizes repeatedly at these three sites. Thus P2 shows a greater degree of integration specificity than phage Mu, which integrates at random (Bukhari and Zipser 1972; Daniell et al. 1972). However, P2 integration is less specific than that of phage λ. Lambda integrates at only one site (at 17 min) on the *E. coli* genetic map, unless that site has been deleted (Shimada et al. 1972), in which case integration of λ prophage at second sites occurs 200-fold less frequently. Cells doubly lysogenic for λ are found to carry both prophages at the same site "in tandem" (Calef et al. 1965). But P2 does not normally form prophage tandems. Rather, it is found that doubly lysogenic *E. coli* C strains carry one prophage at the preferred site I and the other at a secondary site. Occupation of secondary sites appears to occur quite frequently; about 2% of cells lysogenized with P2 carry more than one P2 (Bertani 1962). This conclusion is supported by the similar frequencies found for the establishment of P2 as a second prophage in cells carrying the heteroimmune prophage P2 Hy *dis* at site I (Six 1963, 1966).

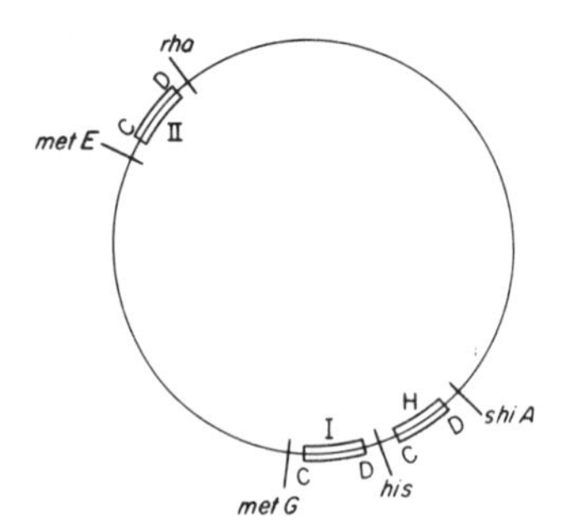

Figure 2
Position and orientation of P2 prophages, based on Calendar and Lindahl (1969). Only the prophage genes *C* and *D* are shown, to indicate orientation, because of their proximity to the ends of the prophage map. Prophage location I occurs in *E. coli* C but not in *E. coli* K, whereas location H is found in *E. coli* K but not in *E. coli* C (Kelly 1963).

Table 1
Known Chromosomal Locations for P2 Prophage

Designation	Map location, min[a]	Notes
I	48	found in *E. coli* strain C, where it is the preferred site (Bertani and Six 1958; Calendar and Lindahl 1969).
H	44	found in *E. coli* strain K; occupied in 3/13 of the lysogenic strains studied by Kelly (1963)
II	85	most common site for a second prophage in *E. coli* C; occupied in 8/13 of the *E. coli* K lysogenic strains studied by Kelly (1963)
III	32	unusual attachment site in *E. coli* C (Bertani and Six 1958; Six 1966; Wiman et al. 1970)
E	61-62	unusual attachment site in *E. coli* C (E. W. Six and F. Kahn, unpubl.)
IV-IX	?[b]	unusual attachment sites in *E. coli* C (Six 1960, 1966, 1968, and unpubl.)

More detailed information is given by Bertani and Bertani (1971).
[a] One-hundred-minute map of *E. coli* C (Wiman et al. 1970) or *E. coli* K (Bachmann et al. 1976). Additional evidence for the locations of sites I–III can be found in Wiman et al. (1970).
[b] Not precisely located.

Integration specificity in P2 must be determined in part by the precise sequence of the attachment site in the host genome since transduction of the histidine region from *E. coli* K into *E. coli* C can create a hybrid strain which has lost the preferred attachment site I characteristic of *E. coli* C (Sunshine and Kelly 1967).

Modification of the P2 Site Affinity

A P2 prophage carried at site II generates spontaneously two kinds of phage, now termed P2*saf*$^+$ and P2*saf*. The former corresponds to the P2 that has established the prophage at site II. P2*saf* shows a greater tendency than P2*saf*$^+$ to establish itself as second prophage in cells already carrying P2 Hy *dis* at site I. Genetic analysis indicates that P2*saf* originates from an interaction between P2*saf*$^+$ and site II, and it appears likely that P2*saf* contains in the attachment site on its chromosome a part of site II (Six 1963, 1966). The *saf* character apparently increases the affinity of P2 for site II. Unexpectedly, *saf* was also found to enable P2 to integrate next to a P2 Hy *dis* prophage already occupying site I (Six 1968 and unpubl.).

In one case, a P2 Hy *dis saf* phage, in superinfection of a lysogen (strain C-231 of Six 1966), established itself as the second prophage at a new site, E, closely linked to *arg*1 (absence of prophage at E was 67% cotransducible with that marker by P1 when the donor strain was nonlysogenic and the recipient strain carried P2 prophage at site E) (E. W. Six and F. Kahn, unpubl.).

Integration-deficient Mutants

Choe (1969) and Lindahl (1969b) isolated mutants of P2 that are unable to integrate at any chromosomal site, analogous to the *int* mutants of phages P22 and λ (Smith and Levine 1967; Zissler 1967). The most interesting property of the P2*int* gene is the control of its expression, which differs sharply from that observed for phage λ. The λ repressor controls the expression of the λ*int* gene; the only double lysogens isolated after homoimmune superinfection of λ lysogenic cells carry two λ genomes in tandem, formed by *recA*-promoted insertion of the second prophage into the first (Calef et al. 1965; Gottesman and Yarmolinsky 1969). In contrast, in P2 lyogens superinfected with P2, second P2 prophages are established at sites separate from that of the first prophage. As shown by Bertani (1970), the insertion of a second P2 prophage following homoimmune superinfection is dependent on the P2*int* gene product. In light of this finding, one would conclude that the P2*int* gene cannot be under strict negative regulation by P2 repressor. Bertani (1970) has proposed that the P2*int* gene belongs to a constitutive operon which is physically disrupted by the act of integration in such a way that the *int* gene is separated from its main promoter. Consistent with this idea is the finding that tandem P2 prophages can exist in stable form only when at least one of them is *int*$^-$: the one whose transcription unit is intact (Bertani 1971). If IS elements also carry *int* genes, it will be interesting to see whether their expression is also under this "split-operon control."

P2 Prophage Excision

P2*int*$^-$ lysogens can be formed by complementation with P2*int*$^+$. These lysogenic strains do not release phage spontaneously (Choe 1969; Lindahl 1969b). Thus the *int* gene product is also needed for excision. However, derepression of a P2*int*$^+$ prophage producing a temperature-sensitive repressor does not lead to efficient excision, unless P2*int* product is supplied by superinfecting P2*int*$^+$ (Bertani 1970). This finding further supports the concept of split-operon control for *int* function. It remains unclear, however, as to how normal prophage excision and spontaneous phage release take place. It appears that generalized recombination is not needed for spontaneous phage release since *E. coli recA* (P2) exhibits normal levels of such phage production (R. Calendar, unpubl.; T. Laffler and S. Luria, pers. comm.). The inability of P2 to be produced efficiently after derepression has been overcome by isolation of a mutation termed *nip*$^-$ (*non*inducible *p*rophage), which is mapped very near the repressor gene and which may exhibit altered regulation of *int* gene expression (Calendar et al. 1972). Lindahl and Sunshine (1972) have found another genetic element which is needed for P2 prophage excision. They isolated P2*cox* (*c*ontrol *o*f e*x*cision) mutants, which can integrate but cannot be excised and hence, in this regard, are analogous to λ*xis* mutants (Guarneros and Echols 1970). But *cox* mutations differ from *xis* mutations in that they increase the frequency of *int*-promoted recombination between two coinfecting phages. Furthermore, the only *cox* mutation mapped is located to the right of gene *C* (Fig. 1).

Aberrant Excision: Deletion of the *his* Region

Kelly and Sunshine (1967) noticed that *E. coli* K strains lysogenic for P2 at site H segregate His⁻ mutants at low frequency. His⁻ mutants are also observed at low frequency among the survivors of an infection by P2. These His⁻ mutants are created by deletion of the entire P2 prophage and histidine operon, as well as the *xonA, gnd,* and *rfb* genes, which span a region of almost 2% of the *E. coli* genetic map (Sunshine and Kelly 1971). This kind of specialized deletion, called "eduction," probably occurs by an aberrant excision whereby one end of the prophage pairs with a segment of the host chromosome about 2 minutes distant.

The eduction phenomenon depends on the P2*int* gene product and upon prophage attachment site H (Sunshine 1972). The *cox* mutations tested do not affect the formation of His⁻ eductants. The frequency of His⁻ eductants can be raised 100-fold by use of a P2*nip*⁻ prophage or by superinfection of *E. coli* K $(P2)_H$ by P2*int*⁺. Both these approaches allow the *int* gene to be expressed constitutively (Sunshine 1972).

Phages Related to P2

P2 represents a large group of temperate coliphages which share similar morphologies and DNA replication patterns (Bertani and Bertani 1971) and are most easily recognized by their inability to grow on *E. coli rep* mutants (Calendar et al. 1970). Among these, phages PK and 299 must carry an integration system similar or identical to P2 since they cause eduction of the histidine region (Sunshine 1972) and, like P2, prefer site I for integration (E. W. Six, unpubl.). Phage 186 has a quite different integration specificity and attaches at 57 minutes on the *E. coli* linkage group (Woods and Egan 1972). Satellite phage P4, which depends upon P2 as a helper, has yet another specificity, attaching to the host chromosome at 94 minutes (Six and Klug 1973; Usher 1975). Unlike for P2 and its relatives, P4 growth is *rep*-independent (Lindqvist and Six 1971).

SUMMARY AND CONCLUSIONS

P2 is an insertion element that has less attachment specificity than phages λ and P22 but more specificity than phage Mu or the IS elements. The P2 integration gene is not regulated by P2 repressor and may be under split-operon control. Aberrant excision of P2 prophage in site H causes deletion of 2 minutes of the *E. coli* chromosome in the *his* region. This phenomenon may be a special example of excisive deletion caused at many sites by a variety of transposons, such as Tn*10* and Tn*9* (Botstein and Kleckner; MacHattie and Jackowski; both this volume).

Acknowledgments

The research reported here was supported by U.S. Public Health Service Grants AI08722 and AI04043 from the National Institute of Allergy and Infectious Diseases; by National Cancer Institute Grant CA14097, National Science Foundation Grant BMS74-19607, and American Cancer Society Grant VC-188; and by funds from the Swedish Medical Research Council. The advice of G. Bertani is greatly appreciated.

REFERENCES

Bachman, B., K. B. Low and A. L. Taylor. 1976. Recalibrated linkage map of *Escherichia coli* K-12. *Bact. Rev.* **40**:116.

Bertani, G. 1954. Lysogenic versus lytic cycle of phage multiplication. *Cold Spring Harbor Symp. Quant. Biol.* **18**:65.

——. 1962. Multiple lysogeny from a single infection. *Virology* **18**:131.

Bertani, G. and E. W. Six. 1958. Inheritance of prophage P2 in bacterial crosses. *Virology* **6**:357.

Bertani, L. E. 1957. The effect of the inhibition of protein synthesis on the establishment of lysogeny. *Virology* **4**:53.

——. 1968. Abortive induction of bacteriophage P2. *Virology* **36**:87.

——. 1970. Split operon control of a prophage genome. *Proc. Nat. Acad. Sci.* **65**:331.

——. 1971. Stabilization of P2 tandem double lysogens by *int* mutations in the prophage. *Virology* **46**:426.

Bertani, L. E. and G. Bertani. 1971. Genetics of P2 and related phages. *Adv. Genet.* **16**:200.

Bukhari, A. I. and D. Zipser. 1972. Random insertion of Mu-1 DNA within a single gene. *Nature New Biol.* **236**:240.

Calef, E., C. Marchelli and F. Guerrini. 1965. The formation of superinfection double lysogens of phage λ in *Escherichia coli* K-12. *Virology* **27**:1.

Calendar, R. and G. Lindahl. 1969. Attachment of prophage P2: Gene order at different chromosomal sites. *Virology* **39**:867.

Calendar, R., G. Lindahl, M. Marsh and M. G. Sunshine. 1972. Temperature-inducible mutants of P2 phage. *Virology* **47**:68.

Calendar, R., B. Lindqvist, G. Sironi and A. J. Clark. 1970. Characterization of Rep$^-$ mutants and their interaction with P2 phage. *Virology* **40**:72.

Choe, B. K. 1969. Integration-defective mutants of bacteriophage P2. *Mol. Gen. Genet.* **105**:275.

Daniell, E., R. Roberts and J. Abelson. 1972. Mutations in the lactose operon caused by bacteriophage Mu. *J. Mol. Biol.* **69**:1.

Gottesman, M. E. and M. B. Yarmolinsky. 1969. The integration and excision of the bacteriophage lambda genome. *Cold Spring Harbor Symp. Quant. Biol.* **33**: 735.

Guarneros, G. and H. Echols. 1970. New mutants of bacteriophage λ with a specific defect in excision from the host chromosome. *J. Mol. Biol.* **47**:565.

Inman, R. B. and G. Bertani. 1969. Heat denaturation of P2 phage DNA: Compositional heterogeneity. *J. Mol. Biol.* **44**:533.

Kelly, B. 1963. Localization of P2 prophage in two strains of *Escherichia coli*. *Virology* **19**:32.

Kelly, B. L. and M. G. Sunshine. 1967. Association of temperate phage P2 with the production of histidine negative segregants by *Escherichi coli*. *Biochem. Biophys. Res. Comm.* **28**:237.

Lindahl, G. 1969a. Genetic map of bacteriophage P2. *Virology* **39**:839.

———. 1969b. Multiple recombination mechanisms in bacteriophage P2. *Virology* **39**:1861.

———. 1970. Characterization of *rep* mutants and their interaction with P2 phage. *Virology* **40**:72.

———. 1971. On the control of transcription in bacteriophage P2. *Virology* **46**: 620.

Lindahl, G. and M. G. Sunshine. 1972. Excision-deficient mutants of bacteriophage P2. *Virology* **49**:180.

Lindqvist, B. H. and E. W. Six. 1971. Replication of bacteriophage P4 DNA in a nonlysogenic host. *Virology* **43**:1.

Mandel, M. 1967. Infectivity of phage P2 DNA in the presence of helper phage. *Mol. Gen. Genet.* **99**:88.

Shimada, K., R. A. Weisberg and M. E. Gottesman. 1972. Prophage lambda at unusual chromosomal locations. I. Location of the secondary attachment sites and the properties of the lysogens. *J. Mol. Biol.* **63**:483.

Six, E. W. 1960. Prophage substitution and curing in lysogenic cells superinfected with heteroimmune phage. *J. Bact.* **80**:728.

———. 1961. Inheritance of prophage P2 in superinfection experiments. *Virology* **14**:220.

———. 1963. Affinity of P2 *rd 1* for prophage sites on the chromosome of *Escherichia coli* strain C. *Virology* **10**:375.

———. 1966. Specificity of P2 for prophage site I on the chromosome of *Escherichia coli* strain C. *Virology* **29**:106.

———. 1968. Prophage site specificities of P2 phages. *Bact. Proceed.* p. 159.

Six, E. W. and C. A. C. Klug. 1973. Bacteriophage P4: A satellite virus depending on a helper such as prophage P2. *Virology* **51**:327.

Smith, H. O. and M. Levine. 1967. A phage P22 gene controlling integration of prophage. *Virology* **31**:207.

Sunshine, M. G. 1972. Dependence of eduction on P2 *int* product. *Virology* **47**: 61.

Sunshine, M. G. and B. Kelly. 1967. Studies on P2 prophage-host relationships. I. Alteration of P2 prophage localization patterns in *Escherichia coli* by interstrain transduction. *Virology* **32**:644.

——. 1971. Extent of host deletions associated with bacteriophage P2-mediated eduction. *J. Bact.* **108**:695.

Sunshine, M. G. and B. Sauer. 1975. A bacterial mutation blocking P2 phage late gene expression. *Proc. Nat. Acad. Sci.* 72:2770.

Sunshine, M. G., M. Thorn, W. Gibbs, R. Calendar and B. Kelly. 1971. P2 phage amber mutants: Characterization by use of a polarity suppressor. *Virology* **46**: 691.

Usher, D. 1975. The effect of P4 of the fertility of male *Escherichia coli.* Ph.D. thesis, University of Iowa, Iowa City.

Wiman, M., G. Bertani, B. Kelly and I. Sasaki. 1970. Genetic map of *E. coli* strain C. *Mol. Gen. Genet.* **107**:1.

Woods, W. H. and J. B. Egan. 1972. Integration site of noninducible coliphage 186. *J. Bact.* **111**:303.

Zissler, J. 1967. Integration-negative (*int*) mutants of phage λ. *Virology* **31**:189.

Recombination Models for the Inverted DNA Sequences of the Gamma-Delta Segment of *E. coli* and the G Segments of Phages Mu and P1

T. R. Broker
Cold Spring Harbor Laboratory
Cold Spring Harbor, New York 11724

The 3000-base-pair invertible DNA segments of bacteriophages Mu and P1 are flanked by sequence duplications inverted with respect to one another (Hsu and Davidson 1974; Lee et al. 1974). These "G" segments have been shown by electron microscope heteroduplex methods to be identical (Chow and Bukhari, this volume), but the inverted repetitions of Mu are about 20 base pairs long, whereas those of P1 are about 620 base pairs long. Reciprocal intramolecular recombination between inverted duplications enables the Mu or P1 G segments to reverse their orientation with respect to the rest of the chromosome (Hsu and Davidson 1974; Lee et al. 1974). The function of the DNA inversion is not yet clear.

The 5700-base-pair $\gamma\delta$ segment found in the chromosome of *E. coli* and in the *E. coli* fertility factor F is a sequence that is involved in recombination between F factor and bacterial DNA (Ohtsubo et al. 1974a; Davidson et al. 1975; Palchaudhuri et al. 1976). F′14 and the fused F′KLF5 each contain two copies of $\gamma\delta$, arranged in direct order at the two junctions of F and bacterial DNA (see Fig. 3, C_1) (Ohtsubo et al. 1974a; Palchaudhuri et al. 1976). Intramolecular recombination between the $\gamma\delta$ sequences is frequent and results in the instability of these F′s in $recA^+$ or $recA^-$ hosts. Isolation of the $\gamma\delta$ sequence on the defective transducing phage $\phi 80d_3 ilv^+ rrn^+ su^+ 7\gamma\delta$ by L. Soll (Ohtsubo et al. 1974b) permitted our discovery that about 20 nucleotides at one end of the $\gamma\delta$ are duplicated, in inverted order, at the other end (Broker et al., this volume).

Recombination of the R segment and of the $\gamma\delta$ segment appear to be quite different, even though both segments are flanked by inverted duplications. Recombination of the $\gamma\delta$ segment is characterized by the following properties:

1. Legitimate reciprocal recombination utilizes two complete copies of the $\gamma\delta$ sequence (hence four flanking duplications) located on the same chromosome or on independent chromosomes. (Illegitimate recombinations involving either end of $\gamma\delta$ have also been observed [Davidson et al. 1975; Palchaudhuri et al. 1976].) Integration of F factor is known or suspected to occur at, among other insertion sequences, a $\gamma\delta$ segment located at any of four different

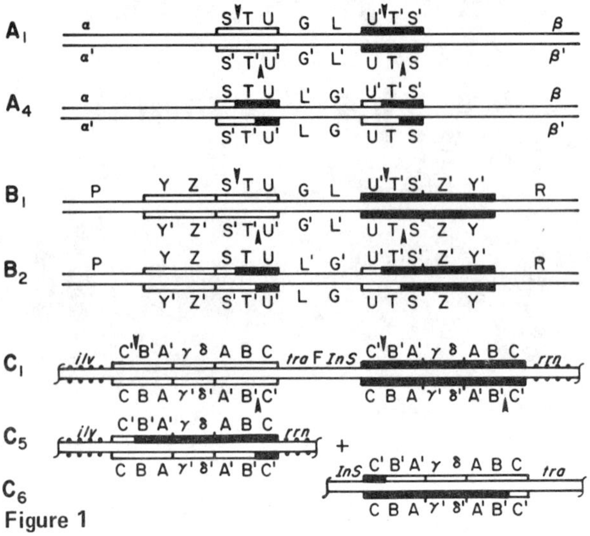

Figure 1
Comparison of proposed recombinations at inverted duplications in Mu, P1, and F′14.

A_1 and A_4: Mu G segment before and after inversion, respectively; B_1 and B_2: P1 G segment before and after inversion, respectively; and C_1 and $C_{5,6}$: F′14 $\gamma\delta$ segments before and after segregation, respectively.

Segments STU in $A_{1,4}$ and $B_{1,2}$ are about 20 nucleotides long (but are not necessarily the same in Mu and P1); segment YZ in $B_{1,2}$ is about 600 nucleotides long; segment ABC in $C_{1,5,6}$ is about 20 nucleotides long; and segment $\gamma\delta$ in $C_{1,5,6}$ is about 5660 nucleotides long.

Each reciprocal recombination requires four single-strand nicks (arrows), two on each strand. For Mu and P1 G-segment inversion, the two pairs of staggered nicks are close together within each sequence STU. Only short heteroduplex overlaps are established. In $\gamma\delta$, the proposed single-strand nicking activity cuts only once in each terminal duplication, between C′ and B′. Two $\gamma\delta$ sequences (each with flanking inverted duplications) are therefore needed for recombination. Heteroduplex overlaps are established through the entire $\gamma\delta$ sequence.

E. coli sites (27 min, Hfr EC8; 83 min, Hfr AB313; 86 min, RA-2; and 61 min, RA-1) (Low 1967; Ohtsubo et al. 1974a; Palchaudhuri et al. 1976). F-factor integration at any particular host $\gamma\delta$ sequence is of unique and reproducible polarity. Thus the identical orientation of the two $\gamma\delta$ sequences located in both F′14 and F′KLF5 indicates parallel alignment of the $\gamma\delta$ sequences during Hfr formation.

2. The inverted duplications flanking $\gamma\delta$ do not promote segment inversion. If they did, inversion would have led to two configurations of $\phi 80d_3$ DNA (as in Mu and P1) and to as many as eight configurations of F′14 and of KLF5 in which the $\gamma\delta$ sequences individually, together, or combined with the F segment would be inverted with respect to the bacterial segment. No such

variability has been reported from EM heteroduplex studies (see Ohtsubo et al. 1974a,b; Palchaudhuri et al. 1976; Broker et al., this volume).

I present here a model to account for these properties of $\gamma\delta$ recombination. The model is compared with that proposed for the inversion of the G segment of Mu (Hsu and Davidson 1974).

Mu and P1 inversion must require two pairs of closely spaced, possibly staggered, single-strand nicks, one pair on each terminal duplication (Fig. 1, A_1 and B_2). In phage P1 DNA, the paired nicks could, in principle, be placed at any symmetrical position within the long inverted repetitions (Fig. 1, B_1), but it seems likely they occur adjacent to the G segment, as in Mu, because the Mu G segment can be inverted in *trans* by the inversion system of P1 prophage (L. Chow, R. Kahmann and D. Kamp, pers. comm.). This suggests that the inverted sequences of Mu (STU, Fig. 1) are included in the inverted sequences of P1 (YZSTU, Fig. 1). Recombination probably occurs by some form of direct exchange of the short cohesive ends generated by the nicks (Fig. 1, A_4 and B_2) (Hsu and Davidson 1974). An unlikely alternative to direct exchange is recombination promoted by branch migration (Broker and Lehman 1971). The latter requires concurrent axial rotation of the duplexes to relieve and to reestablish helical interwinding of the strands (Platt 1955; Broker 1973). But for intrachromosomal recombination of inverted duplications, the interacting sequences would necessarily twist in counter-rotation rather than in the synchronous axial rotation required for branch migration (see below and Fig. 2, A_2) (Broker and Doermann 1975). Therefore, without an additional assumption of strand nicks (swivels) (e.g., in sequence GL of Fig. 2), the participation of axial rotation and branch migration in G-segment inversion are unlikely.

Integrative or excisive recombination of $\gamma\delta$ requires four single-strand nicks, just as does inversion in Mu and P1. Similarly, the nicks are probably staggered

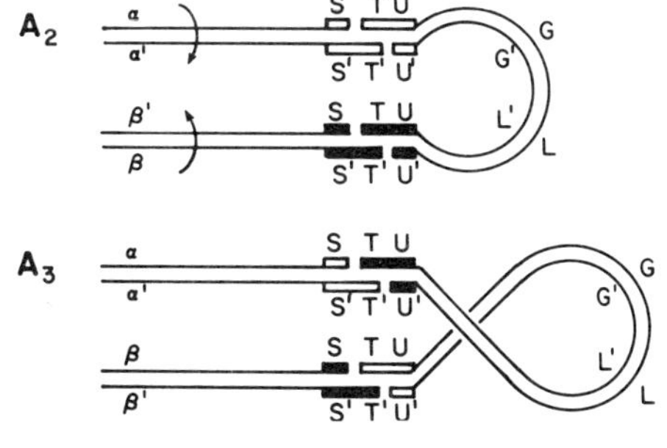

Figure 2
DNA segment inversion. Structures A_2 and A_3 are recombination intermediates in the pathway between structures A_1 and A_4 in Fig. 1. Recombination is proposed to occur by direct exchange of cohesive termini generated by a sequence-specific, double-strand endonuclease. Therefore, heteroduplex overlaps are limited to the sequence between the staggered nicks within the inverted duplications. Branch migration with coordinate axial rotation can probably be ruled out because recombining segments in inverted order on the same DNA duplex would be in counter-rotation rather than in the synchronous rotation necessary for strand exchange (arrows) (cf. Fig. 3, C_3).

on opposite strands to allow the formation of base-paired heteroduplex joints between the recombining segments. To account for the differences between $\gamma\delta$ integration-excision and G inversion, I propose that the pairs of nicks for integrative or excisive recombination of the $\gamma\delta$ sequence are placed 5700 base pairs apart, at opposite ends of each $\gamma\delta$ segment, whereas the pairs of nicks for G inversion are located within 20 bases, at each end of the G segment. This would result in one pair of nicks per $\gamma\delta$ segment and two pairs of nicks per G segment. To achieve this arrangement, the endonuclease for $\gamma\delta$ recombination would cut only a single strand at each of four identical base sequences of the two copies of $\gamma\delta$, e.g., between C′ and B′ (Fig. 1, C_1). Because the flanking duplications are inverted, scissions would be introduced on opposite strands.

Pairing between $\gamma\delta$ segments must include interaction of the sequences between the flanking duplications to assure proper orientation of the synapse. If pairing involved only the duplicated sequences, but not the internal unique sequences, $\gamma\delta$ recombination would be expected to proceed with either orientation, since the duplicated regions are inverted with respect to one another. To provide this orientation, I propose that stabilization of the $\gamma\delta$ synapse (Fig. 3, C_2) requires concerted and reciprocal double-strand exchange (Broker and Lehman 1971) to proceed from the terminal duplications into the adjoining unique sequences of $\gamma\delta$. This could occur only if the two internal sequences were aligned in the same orientation. Strand exchange should then continue throughout the entire sequence (Fig. 3) until the distal pair of single-strand nicks between B′ and C′ (on the strands opposite those with the initiating nicks) result in termination of exchange (Fig. 3, C_4). The F-specific and bacterial sequences, each with a single heteroduplex copy of $\gamma\delta$, would then segregate (Fig. 3, $C_{5,6}$). In principle, the initiation and termination of strand exchange could take place at either end of the segment. However, the polarity of axial rotation dictates the direction of branch migration in the recombination intermediate and should thus determine which end of the $\gamma\delta$ sequence is utilized for initiation of strand exchange. Little is known about the driving forces for axial rotation, except that they must be associated with DNA replication and transcription as well as with recombination (Broker and Doermann 1975).

The concerted strand exchange imposes a stringent requirement for complete homology of the two interacting $\gamma\delta$ sequences. Should either sequence become altered to the extent that reciprocal strand exchange could not pass the region of divergence, its capacity for legitimate recombination would be lost. It is possible that at least part of the enzyme system responsible for $\gamma\delta$ recombination is encoded within the $\gamma\delta$ sequence. If that were the case, then the demand for perfect homology during strand exchange through the recombination gene(s) would preserve the physical and functional linkage between the gene for the nuclease and the cleavage sites in the flanking inverted duplications. Such linkage should select against mutational alteration of both the gene and the cleavage sites.

Figure 3

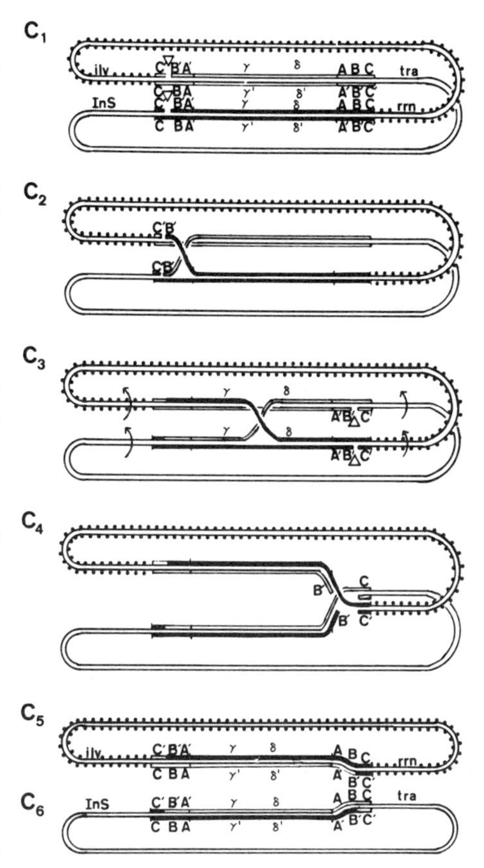

Model of $\gamma\delta$ sequence recombination. The diagrams illustrate segregation of F factor and *E. coli* DNA from F′14. The $\gamma\delta$ sequences are distinguished by open and solid wide lines. Each $\gamma\delta$ is about 5700 nucleotide pairs long. The flanking inverted duplications of ABC are about 20 nucleotide pairs long. Bacterial sequences *ilv* and *rrn* (⁝⁝⁝⁝) and F-factor sequences *tra* and InS (insertion sequences) (═══) are indicated. C_1: Circular F′14 with two $\gamma\delta$ sequences arranged in direct order. Single-strand nicks (▽) are imposed between C′ and B′. C_2: Initiation of crossover by exchange of B′ sequences (Holliday [1964] structure). C_3: Concerted double-strand exchange. The polarity of branch migration from γ to δ is dictated by the synchronous axial rotations of the double helices, as indicated by arrows. All four strands at the exchange site are topologically equivalent, as can be better appreciated by referring to the helical (Broker and Doermann 1975) or the isomeric open box (Broker and Lehman 1971) representations of the same structure (also see Emerson 1969). Single-strand nicks (Δ) are imposed between B′ and C′. C_4: Termination of crossover. Migration of the branch past the single-strand nicks causes the recombination intermediate to dissociate. C_5 and C_6: Segregated bacterial and F-factor DNA each contain a complete but hybrid $\gamma\delta$ sequence. Integration is proposed to proceed via a similar strand exchange through the entire 5700-base-pair region.

SUMMARY

The main features of a model proposed for integration and excision of the *E. coli* $\gamma\delta$ recombination sequence are that (a) the inverted duplications flanking the $\gamma\delta$ segment serve as recognition sites for single-stranded DNA endonucleases, and (b) recombination proceeds by a concerted double-strand exchange that initiates at the nick at one end of each $\gamma\delta$ sequence, continues through the entire sequence, and terminates at the nick at the opposite end, with the establishment of reciprocal 5700-base-pair heteroduplex overlaps in each $\gamma\delta$ segment.

Acknowledgments

I wish to thank Dr. Louise T. Chow for stimulating discussions and Ms. Marie Moschitta and Mr. Robert Yaffe for assistance during the preparation of the manuscript. I was supported by a National Cancer Institute Cancer Center Contract to Cold Spring Harbor Laboratory.

REFERENCES

Broker, T. R. 1973. An electron microscopic analysis of pathways for bacteriophage T4 DNA recombination. *J. Mol. Biol.* **81**:1.

Broker, T. R. and A. H. Doermann. 1975. Molecular and genetic recombination of bacteriophage T4. *Annu. Rev. Genet.* **9**:213.

Broker, T. R. and I. R. Lehman. 1971. Branched DNA molecules: Intermediates in T4 recombination. *J. Mol. Biol.* **60**:131.

Davidson, N., R. C. Deonier, S. Hu and E. Ohtsubo. 1975. Electron microscope heteroduplex studies of sequence relations among plasmids of *Escherichia coli.* X. Deoxyribonucleic acid sequence organization of F and F-primes and the sequences involved in Hfr formation. In *Microbiology–1974* (ed. D. Schlessinger), p. 56. American Society for Microbiology, Washington, D.C.

Emerson, S. 1969. Linkage and recombination at the chromosome level. In *Genetic organization,* (ed. E. W. Caspari and A. W. Ravin), vol. I, p. 267. Academic Press, New York.

Holliday, R. 1964. A mechanism for gene conversion in fungi. *Genet. Res.* (Camb.) **5**:282.

Hsu, M.-T. and N. Davidson. 1974. Electron microscope heteroduplex study of the heterogeneity of Mu phage and prophage DNA. *Virology* **58**:229.

Lee, H.-J., E. Ohtsubo, R. C. Deonier and N. Davidson. 1974. Electron microscope heteroduplex studies of sequence relations among plasmids of *Escherichia coli.* V. ilv^+ deletion mutants of F14. *J. Mol. Biol.* **89**:585.

Low, B. 1967. Inversion of transfer modes and sex factor-chromosome interactions in conjugation in *Escherichia coli. J. Bact.* **93**:98.

Ohtsubo, E., R. C. Deonier, H.-J. Lee and N. Davidson. 1974a. Electron microscope heteroduplex studies of sequence relations among plasmids of *Escherichia coli.* IV. The F sequences in F14. *J. Mol. Biol.* **89**:565.

Ohtsubo, E., L. Soll, R. C. Deonier, H.-J. Lee and N. Davidson. 1974b. Electron microscope heteroduplex studies of sequence relations among plasmids of *Escherichia coli.* VIII. The structure of bacteriophage ϕ80 d_3 ilv^+ su^+7, including the mapping of the ribosomal RNA genes. *J. Mol. Biol.* **89**:631.

Palchaudhuri, S., W. K. Maas and E. Ohtsubo. 1976. Fusion of two F-prime factors in *Escherichia coli* studied by electron microscope heteroduplex analysis. *Mol. Gen. Genet.* **146**:215.

Platt, J. R. 1955. Possible separation of intertwined nucleic acid chains by transfer twist. *Proc. Nat. Acad. Sci.* **41**:181.

An Electron Microscope Study of Actively Recombining Plasmid DNA Molecules

H. Potter and D. Dressler
Department of Biochemistry and Molecular Biology
Harvard University
Cambridge, Massachusetts 02138

The experimental systems used to study genetic recombination have been as varied as the process itself is interesting. In this paper, we will present the results of a recombination study which employed plasmid DNA as an experimental system. An immediate benefit that arises from using such a system to study recombination is that the number of plasmid genomes can be increased from about 20 to 1000 per cell by culturing the cells in the bacteriostatic agent chloramphenicol (Clewell 1972). One may take advantage of this specific amplification process to fill the cell with homologous DNA molecules that might engage in recombination. It is then possible to search among the isolated plasmid DNA for molecules that are actively recombining.

Our data are based on the electron microscopic observation of over 800 DNA molecules that appear to be in the process of genetic recombination. These molecules are judged to represent recombination intermediates because (1) they contain two genome-length elements; (2) these elements are connected at a region of homologous DNA; (3) the DNA strand substructure in the region of the connection can often be seen in the electron microscope and is consistent with the fine structure expected for a genetic crossover; and (4) these molecules are absent in recombination-deficient ($RecA^-$) bacteria.

Our results provide evidence supporting the physical existence of a recombination intermediate postulated over a decade ago on genetic grounds by Robin Holliday (1964, 1968, 1974). This intermediate has served as the focal point for a class of recombination models that have evolved over the last several years incorporating contributions from several sources (Fogel and Hurst 1967; Cross and Lieb 1967; Emerson 1969; Hotchkiss 1971, 1974; Sigal and Alberts 1972; Meselson 1972; Meselson and Radding 1975). As a starting point, we will first describe the Holliday recombination intermediate.

The Holliday Recombination Intermediate

Recombination involving the Holliday intermediate may be pictured as shown in Figure 1. Two homologous double helices are aligned, and, in each, the

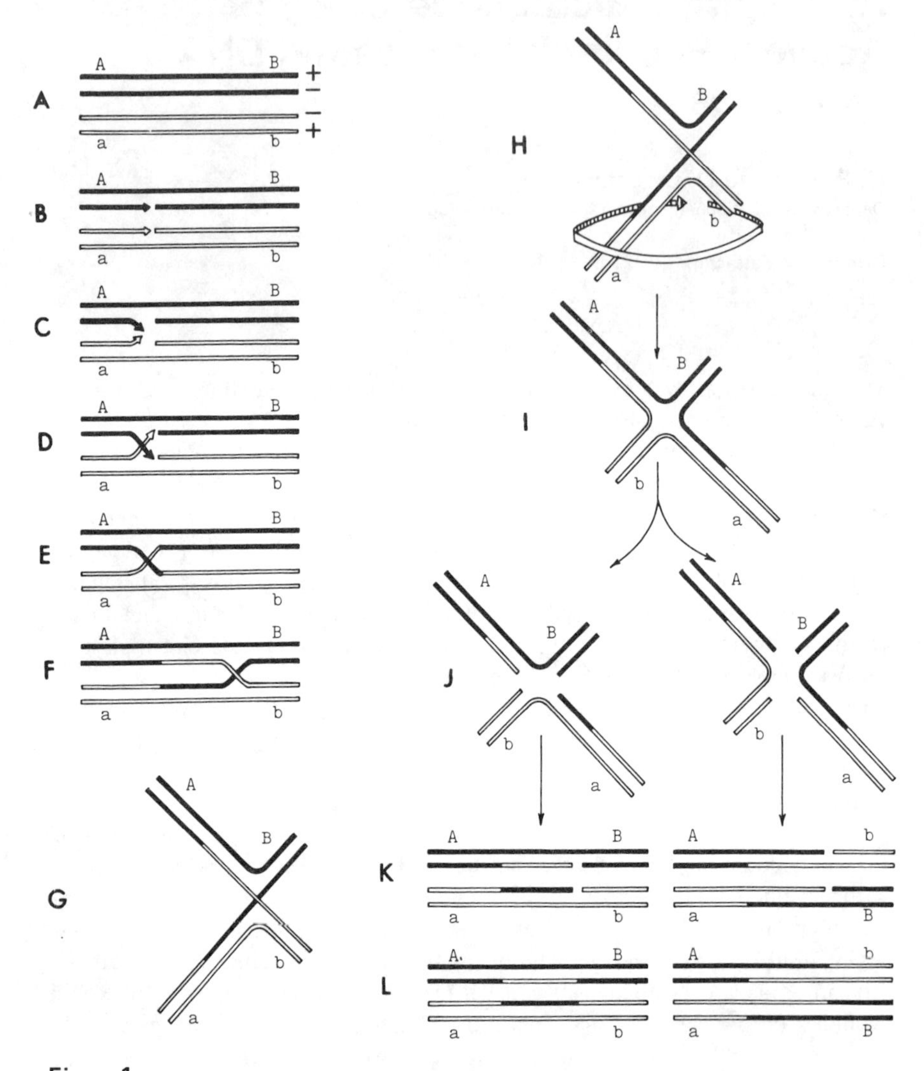

Figure 1
A drawing of the prototype Holliday model for genetic recombination. Details of the model and evidence in support of the central recombination intermediate are presented in the text. The electron micrographs shown in Figs. 3 and 5 correspond to the recombination intermediate diagramed in G and I.

positive strands (or, alternatively, the negative strands) are nicked open in a given region. The free ends thus created breathe away from the complementary strands to which they had been hydrogen-bonded and become associated instead with the complementary strands in the homologous double helix (Fig. 1A–D). The result of this reciprocal strand invasion is to establish a tentative physical connection between the two DNA molecules. This linkage can be made stable through a process of DNA repair, which in this case can be as simple as the formation of two phosphodiester bonds by means of the enzyme ligase (Fig. 1-E).

The recombination intermediate thus formed is the Holliday structure. By building a space-filling model of DNA, Sigal and Alberts (1972) have shown that, in this intermediate, steric hindrance is minimal and virtually all of the bases can be paired.

Once formed, the intermediate need not be static. A continuing strand transfer by the two polynucleotide chains involved in the crossover can occur, moving the point of linkage between the two DNA molecules to the right or to the left (Fig. 1-F). This dynamic property of the Holliday structure can lead to the development of regions of heterozygous DNA during recombination. In fact, such regions of heterozygosity are frequently detected in recombinant chromosomes and they provided the impetus for the original proposal of this model by Holliday (Kitani et al. 1962; Hurst et al. 1972; see also Russo 1973; White and Fox 1974; Enea and Zinder 1976). The rate of bridge migration has been discussed by Meselson (1972) in terms of the rotary diffusion of DNA double helices; he has calculated that this rate is high enough to allow the rapid formation of hybrid regions under physiological conditions.

The maturation of the Holliday intermediate can occur in either of two reciprocal ways (Emerson 1969; Sigal and Alberts 1972; Holliday 1974). This key property of the Holliday structure is most easily appreciated if one draws the intermediate in another planar form (Fig. 1-I). Then, cutting on an east-west axis or a north-south axis allows the release of recombinant DNA molecules in which, on either side of the potentially heterozygous region, the parental alleles are either conserved in their original linkage or reciprocally exchanged (Fig. 1-L). Genetic distances are measured among the 50% of the cases in which the flanking markers are exchanged.

When genetically heterozygous regions are created during recombination, they must eventually be rectified. Thus the overall recombination process is completed by repair of the heterozygous regions with enzymes that recognize base-pair mismatches or, alternatively, by the conversion of these regions to a homozygous state through subsequent rounds of DNA replication.

The experiments to be described here provide direct physical evidence in support of the recombination intermediate proposed by Holliday on genetic grounds.

Observation of Figure Eights

To search for recombination intermediates, cells containing the colicin E1-derived plasmid pMB9 were cultured in growth medium containing chloramphenicol, harvested, and made into a lysate, as described in the legend to Figure 2. The closed circular plasmid DNA molecules were then recovered from the cell lysate by equilibrium centrifugation in CsCl-EthBr density gradients.

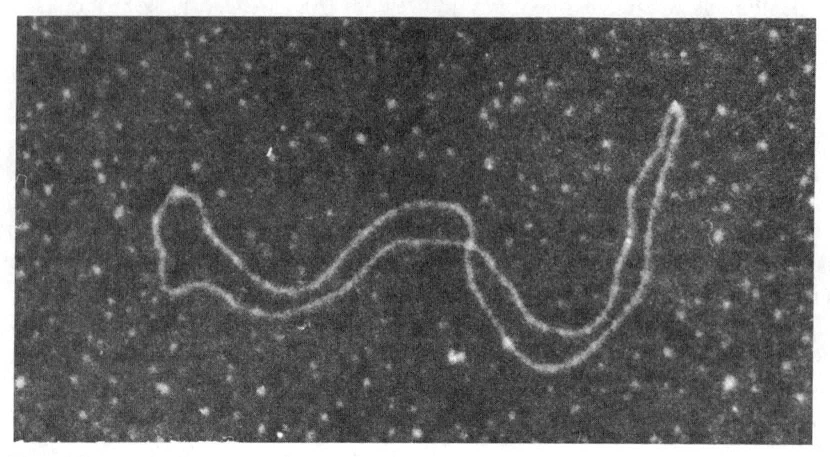

Figure 2
A double-size colicin DNA molecule shaped like a figure eight. The involvement of such molecules in recombination is discussed in the text.

The specific colicin plasmid we studied is pMB9 (obtained from Dr. H. Boyer). This is a 3.6-million-dalton double-stranded DNA circle related to ColE1.

The DNA was prepared by a modification of the method of Clewell (1972). Plasmid-containing cells (HB-129) totaling 5×10^{10} were lysed at a concentration of 1.5×10^{10} cells/ml, as described by Clewell (1972). Lysis was completed by adding Sarkosyl to a concentration of 0.5% and heating to 60°C until the lysate became clear and viscous (about 2 min). The solution was then chilled, diluted threefold with 50 mM Tris, pH 8, and spun at 50,000*g* for 30 min at 2°C. CsCl (0.93 g/ml) and EthBr (300 μg/ml) were added to the supernatant from the clearing spin, and the solution was centrifuged to equilibrium at 36,000 rpm for 60 hr at 15°C in a Beckman number 40 angle rotor.

The material from the lower (supercoil) band was recovered and immediately extracted four times with isopropanol (previously equilibrated with a saturated CsCl solution). The final aqueous phase was dialyzed for three periods of 1 hr each, at 20°C, against 10 mM Tris, 1 mM EDTA, pH 8. The DNA was precipitated with ethanol, lyophilized, and resuspended at about 500 μg/ml in 10 mM Tris, 1 mM EDTA, pH 8, in deionized water.

The DNA was examined in the electron microscope by the Davis et al. (1971) modification of the Kleinschmidt and Zahn protein monolayer technique (Kleinschmidt 1968), as described previously by Wolfson et al. (1972).

When the plasmid DNA molecules were treated lightly with DNase to remove superhelical twists and examined in the electron microscope, they were found to consist of a set of monomer-size DNA rings and related multimers. A small percentage of these DNA rings appeared to be touching each other and thus were candidates for intermediates in recombination. Such "figure-eight" structures have been observed in cells containing the DNA circles of the viruses S13 (Doniger et al. 1973; Thompson et al. 1975), $\phi\chi$ (Benbow et al. 1974, 1975), and lambda (Valenzuela and Inman 1975) and have been considered in terms of recombination (for a discussion, see Potter and Dressler 1976).

Figure 2 shows a colicin figure-eight molecule. To define the structure of such molecules, we have opened them with the restriction enzyme *Eco*RI. This enzyme cleaves monomeric colicin DNA rings once, at a unique site, generating unit-size rods. If the image in Figure 2 reflects pairs of monomer circles interlocked like two links in a chain, or double-length circles accidentally crossing themselves in the middle, then the enzyme would cleave the interlocked monomer rings or the dimer circles twice, releasing separated monomer rods. On the other hand, should they be two plasmid circles, covalently connected at a point of DNA homology, the enzyme would convert figure-eight structures into bilaterally symmetrical dimers, shaped like the Greek letter chi. These "chi forms" are the subject of this paper.

Observation of Chi Forms

Figure 3 is an electron micrograph of a chi-shaped molecule. We have observed more than 800 such molecules among colicin DNA after linearization with *Eco*RI. These dimeric forms have been observed in a simple background of about 80,000 unit-length monomer rods. The frequency of chi forms in 25 different preparations ranged from 0.5% to 3%.

Could these crossed molecules have arisen from an accidental overlap between two monomer rods? This is unlikely for two reasons: First, the DNA was spread for electron microscopy at a low concentration so that accidental overlaps would be virtually nonexistent. Second, the crossed molecules always have a special symmetry. The point of contact between the unit-size colicin genomes occurs so as to divide the structure into two pairs of equal-length arms (Figs. 3 and 4). Since the point of contact between the two genomes creates this symmetry with respect to a defined base sequence (the RI-cut ends), it must almost certainly occur at a region of DNA homology. Thus, in these dimeric colicin DNA forms, two plasmid genomes appear to be held together by the interaction of homologous DNA.

The data in Figure 4 provide evidence that the point of contact between the plasmid genomes can occur at numerous locations along the plasmid DNA molecule—that is, at various distances from the RI cutting site.

The Crossover Point

In addition to observing plasmid genomes touching at a point of DNA homology, one can often see the nature of the DNA strand substructure in the crossover region. Figure 5A shows an example of a molecule that has become locally denatured in the crossover region during spreading for electron microscopy. The single strands connecting the four arms of the recombining molecules are visible and allow the structure to be correlated exactly with the intermediate shown in Figure 1-I. This is one of the two representations of the Holliday

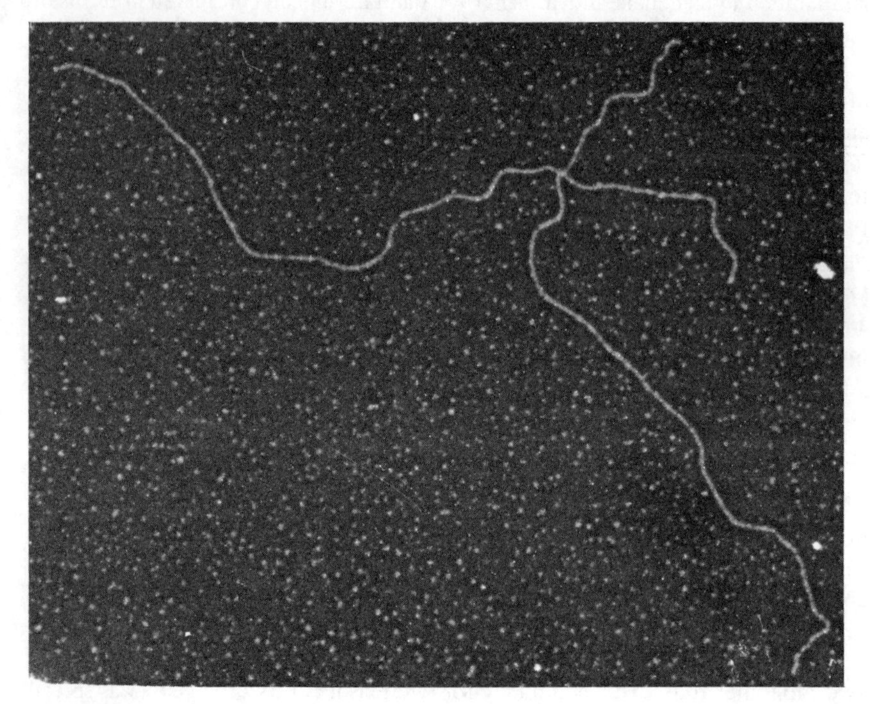

Figure 3
An example of a chi form, consisting of two unit-size plasmid DNA molecules held together at a region of homology. These colicin molecules are observed after linearization of colicin DNA with the restriction enzyme *Eco*RI.

For the RI cutting reaction, the DNA was used at a concentration of 20 μg/ml in the following buffer: 100 mM Tris, pH 7.5, 50 mM NaCl, 5 mM $MgCl_2$, 1 mM dithiothreitol, and 0.2% Triton X-100. The reaction was allowed to proceed at 37°C until about 75% of the molecules were cut, which required about 30 sec.

After RI cutting, the DNA was examined in the electron microscope as described in the legend to Fig. 2. The chi forms are stable for a period of days when maintained at 0°C. A few minutes at 37°C, however, is sufficient to allow almost all of them to roll apart into two separated monomer rods.

Almost all (>90%) of the crossed molecules were chi forms, meaning that, upon measurement, they consisted of two pairs of equal-length arms (± 10%) (representative data are shown in Fig. 4).

In different experiments, the cells were allowed to replicate colicin DNA molecules in the presence of chloramphenicol for varying periods of time ranging from 10 to 30 hr. We found that a greater percentage yield of chi forms was obtained after longer periods of incubation in chloramphenicol. This is consistent with the idea that the increase in the concentration of plasmid DNA rings over time promotes the bimolecular formation of recombinants.

Figure 4
An analysis of the lengths of the arms in 25 randomly chosen chi forms. The lengths of the four arms were measured, summed, and divided by two to obtain the unit genome length. The proportional lengths of the two shorter arms were then plotted, one as the abscissa and the other as the ordinate of a single point. Similarly, the two longer arms were used to produce a single point for the curve. The fact that the points generate essentially a straight line of slope 1 establishes that the chi forms contain pairs of equal-length arms. Furthermore, the finding that pairs of arms have different lengths indicates that the point of contact between the two genomes can occur at many locations, perhaps randomly. All of the other chi forms have been similarly photographed, traced, measured, and analyzed, yielding the same result.

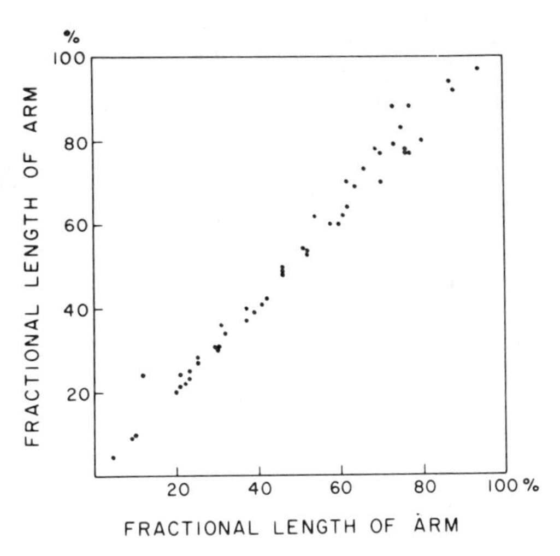

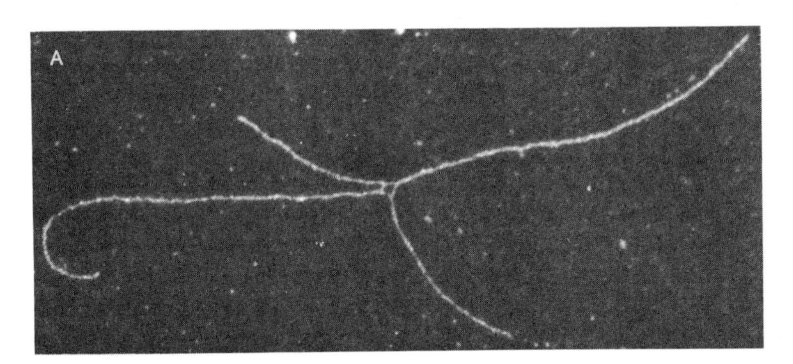

Figure 5
Figure 5A shows an example of one of 80 chi forms in which the covalent strand connections in the region of the crossover can be seen. Apparently the molecule has been strained during spreading for electron microscopy, and the single strands in the crossover have been pulled apart. The short arms of this molecule differ in length by 10%; the long arms differ in length by 8%.

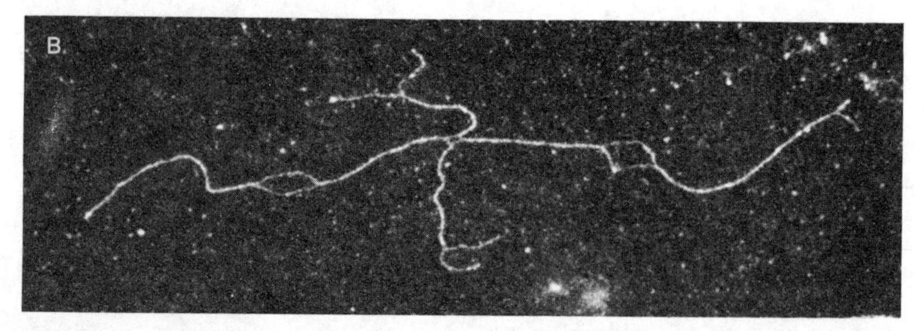

Figure 5 *(continued)*
Figure 5B shows a chi form which has been prepared for electron microscopy in the presence of a high concentration of formamide (Wolfson et al. 1972). Under these conditions, the DNA double helix is stressed, and those regions particularly rich in AT base pairs undergo a localized denaturation. This sequence-specific denaturation allows the homologous arms in the molecule to be identified. Furthermore, the covalent strand connections in the region of the crossover can be seen.

In these and the other 80 open molecules, the homologous arms appear in a *trans* configuration. This geometry is expected, as has been discussed in connection with Fig. 1 (compare Fig. 5 with Fig. 1-I).

structure (visualized in a plane) and is the one expected to be most easily observed in an open state in the electron microscope. When the structure assumes this planar form, it is predicted that the equal-length arms will always appear in the *trans* configuration (as in Fig. 1-I). This prediction has been fulfilled in all of the 80 open molecules we have observed.

Figure 5B shows another example of a molecule that has become partially denatured during spreading for electron microscopy, in this case because the DNA was prepared in a high concentration of formamide. It is again possible to see the covalent connections in the region of the crossover. Furthermore, the homologous arms can be identified by their characteristic denaturation patterns.

In sum, we believe that Figures 3 and 5 correspond to the two planar representations of the Holliday intermediate as shown in Figure 1, G and I.

Effect of the *recA* Locus

In 25 different preparations of colicin DNA from wild-type cells, the frequency of chi forms ranged from 0.5% to 3%. Similar percentages of chi forms were observed in material from a RecB-C strain (MM486 from M. Meselson). However, when we examined material obtained from three independent recombination-deficient ($RecA^-$) strains (MM152 from M. Meselson, HB101 from H. Boyer, and RM201J from F. Stahl), few structures larger than monomers and no chi forms were found among 8000 molecules. The fact that the chi forms were absent in these recombination-deficient cells is the basis for our belief that they are intermediates in recombination.[1]

Recombination between Circular Chromosomes

Thus far, we have discussed intermediates in recombination in terms of linear DNA molecules. However, it must be remembered that the experimentally observed chi form actually represents an RI-digested figure-eight structure. Whereas the *formation* of the recombination intermediate does not, as far as is known, depend upon whether the two input chromosomes are linear or circular, the *maturation* of the intermediate is crucially influenced by the structural constraints imposed by circularity.

Within a circular recombination intermediate, the maturation at the crossover point can still proceed essentially as shown in Figure 1-I. However, as can be verified by forming circles from the rods shown in Figure 1, one type of maturation cut (the east-west cut) will produce the parental circles with their original genes still in linkage; the only genetic change will be the possible existence of heterozygous regions developed by rotary diffusion. The other (north-south) maturation cut generates one double-length DNA circle, not two independent monomers. This dimeric form must be further processed if two recombinant monomers are to be obtained. In principle, this could involve a second recombination event, mechanistically identical to the first, making a new figure eight with a new crossover position. This new figure eight then has a 50% chance of being matured into two independent monomers with a reciprocally recombinant arrangement of genes.

It is expected, therefore, that circular dimers are a product of the maturation of two monomer circles engaged in the crossover stage of genetic recombination. The data in Table 1 support this expectation. Here it is seen that transfection of purified monomer rings into *rec*$^{+}$ cells leads to the development of a population containing both monomer and multimer plasmid genomes. Upon transfection into *recA*$^{-}$ cells, however, the monomers can only replicate to form more monomers; in the absence of a functional *recA* recombination system, very few multimeric plasmid forms arise. Similarly, dimers transfected into *recA*$^{-}$ cells remain dimers.

[1] Could the chi forms be the result of some process other than recombination? The devil's advocate position here would be that the figure-eight molecules and related chi forms are not truly intermediates in recombination but instead could be near-terminal products of DNA replication. Conceivably, at the end of a cycle of Cairns' replication, the two almost completed daughter circles might undergo a set of strand-nicking events (like those discussed in connection with Fig. 1 in terms of recombination). In principle, such a set of nicking events could yield figure eights, which, after bridge migration and cleavage with *Eco*RI, would become the chi-shaped structures we have observed. In view of the *recA* data, there is no reason to interpret the chi forms in terms of replication. However, should the *recA* gene ultimately prove to have pleiotropic effects extending into replication, then the possibility that the figure eights and chi forms are products of DNA synthesis would have to be reconsidered.

Table 1
Fate of Purified Monomers and Dimers Transfected into Rec$^+$ and RecA$^-$ Cells

	Resulting plasmid population			
	monomers	dimers	higher multimers	Total
Monomers transfected into				
Rec$^+$	635	265	100	1000
RecA$^-$	989	11	0	1000
Dimers transfected into				
Rec$^+$	16	863	121	1000
RecA$^-$	2	985	13	1000

Monomer- and dimer-size plasmid DNA molecules were recovered from a sucrose velocity gradient and used to transfect Rec$^+$ and RecA$^-$ cells (strains 294 and 152, respectively, from M. Meselson). Cells that received plasmids were selected by virtue of their conversion to tetracycline resistance. Individual cells were recovered from each transfection and grown into cultures from which plasmid DNA was purified (see legend to Fig. 2). Electron microscopy was then used to determine the number of monomer and multimer plasmid species.

The data in Table 1 also provide an additional and surprising result. Transfection of dimers into *recA*$^+$ cells does not result in the rapid regneration of a spectrum of monomeric and multimeric species. The dimers engage in recombination to form tetramers and higher multimers, but they do not readily undergo the intramolecular recombination event that would lead to the production of monomers. How can this apparent problem be explained? The answer is unknown, but several possibilities may be mentioned. Perhaps the initiation of recombination is normally an asymmetric event (see, e.g., Fox 1966; Dressler and Wolfson 1972; Meselson and Radding 1975; Holloman et al. 1975). One molecule might be supercoiled and the other relaxed–the former (by giving up its superhelical twists and acquiring a region of local denaturation) could then serve as a recipient for a strand invasion from the latter. Thus, because it is entirely supercoiled or entirely relaxed, a circular dimer would be slow to initiate recombination. An alternative idea would be that recombination can occur between two supercoiled elements, but that the intramolecular crossover event cannot readily take place in a small circle because there are too few superhelical twists to provide a basis for the extensive local denaturations that would necessarily precede reciprocal strand invasion. Our current experiments are designed to choose between these two possibilities.

SUMMARY

In this paper we have presented evidence about the nature of recombination intermediates, using the DNA of the plasmid ColE1 as an experimental system. The evidence consists of the electron microscopic analysis of more than 800 molecules that appear to represent intermediates in the process of recombination. Specifically, we find (after isolating colicin DNA and linearizing it with the restriction enzyme *Eco*RI to remove problems associated with circularity) crossed molecules with twice the normal colicin DNA content. These molecules consist of two genome-length elements held together at a region of DNA homology (Figs. 3 and 5). The molecules can be recovered at a frequency of about 1% from wild-type cells but are not present among the colicin DNA forms isolated from recombination-deficient (*recA*$^-$) cells. We have termed the experimentally observed molecules "chi forms" and believe that they represent the recombination intermediate proposed by Holliday on genetic grounds.

It will be a major source of satisfaction to learn the nature of the recombination intermediates involved in the other types of DNA rearrangements discussed in this book—the integration and excision of prophage genomes and the movement of insertion sequences—and to compare these structures with the *recA*-mediated intermediate that has been discussed here in the context of general recombination.

Note Added in Proof

With respect to the data in Table 1, which show that multimers are a product of genetic recombination, we have the following additional observation: it is possible to observe the flow of plasmid DNA rings from the monomer to the multimer state *and back again* under the guidance of the *recA* recombination system (see Potter and Dressler 1977). However, the second, intramolecular recombination event (which divides the multimer circle into two smaller DNA rings) builds up monomers rather slowly.

Acknowledgments

The research of our laboratory is made possible by grants from the American Cancer Society (NP-57) and from the National Institutes of Health (GM-17088). H.P. is supported by a National Institutes of Health Training Grant (5T01 GM-00138) and D.D. by a Public Health Service Research Career Development Award (GM-70440).

REFERENCES

Benbow, R., A. Zuccarelli and R. Sinsheimer. 1975. Recombinant DNA molecules of bacteriophage $\phi\chi$ 174. *Proc. Nat. Acad. Sci.* **72**:235.

Benbow, R., A. Zuccarelli, A. Shafer and R. Sinsheimer. 1974. Exchange of parental DNA during genetic recombination in bacteriophage $\phi\chi$ 174. In *Mechanisms in recombination* (ed. R. Grell), p. 3. Plenum Press, New York.

Clewell, D. 1972. Nature of Col E1 plasmid replication in *Escherichia coli* in the presence of chloramphenicol. *J. Bact.* **110**:667.

Cross, R. and M. Lieb. 1967. Heat-inducible λ phage. V. Induction of mutations in genes *O, P,* and *R*. *Genetics* **57**:549.

Davis, R., M. Simon, and N. Davidson. 1971. Electron microscope heteroduplex methods for mapping regions of base sequence homology in nucleic acids. In *Methods in enzymology* (ed. L. Grossman and K. Moldave), vol. 21, p. 413. Academic Press, New York.

Doniger, J., R. C. Warner and I. Tessman. 1973. Role of circular dimer DNA in the primary recombination mechanism of bacteriophage S-13. *Nature New Biol.* **242**:9.

Dressler, D. and J. Wolfson. 1972. Discussion. In *DNA synthesis in vitro* (ed. R. Wells and R. Inman), p. 448. University Park Press, Baltimore.

Emerson, S. 1969. Linkage and recombination at the chromosome level. In *Genetic organization* (ed. E. Caspari and A. Ravin), vol. 1, p. 267. Academic Press, New York.

Enea, V. and N. Zinder. 1976. Heteroduplex DNA: A recombinational intermediate in bacteriophage F1. *J. Mol. Biol.* **101**:25.

Fogel, S. and D. Hurst. 1967. Meiotic gene conversion in yeast tetrads and the theory of recombination. *Genetics* **57**:455.

Fox, M. 1966. On the mechanism of integration of transforming deoxyribonucleate. *J. Gen. Physiol.* (Suppl.) **49**:183.

Holliday, R. 1964. A mechanism for gene conversion. *Genet. Res.* **5**:282.

——. 1968. Genetic recombination in fungi. In *Replication and recombination of genetic material* (ed. W. Peacock and R. Brock), p. 157. Australian Academy of Science, Canberra.

——. 1974. Molecular aspects of genetic exchange and gene conversion. *Genetics* **78**:273.

Holloman, W., R. Wiegand, C. Hoessli and C. Radding. 1975. Uptake of homologous single-stranded fragments by superhelical DNA: A possible mechanism for initiation of genetic combination. *Proc. Nat. Acad. Sci.* **72**:2394.

Hotchkiss, R. 1974. Models of genetic recombination. *Annu. Rev. Microbiol.* **28**:445.

——. 1971. Toward a general theory of genetic recombination in DNA. *Adv. Genet.* **16**:325.

Hurst, D., S. Fogel and R. Mortimer. 1972. Conversion-associated recombination in yeast. *Proc. Nat. Acad. Sci.* **69**:101.

Kitani, Y., L. Olive and A. El-ani. 1962. Genetics of *Sordaria fimicola*. V. Aberrant segregation at the *G* locus. *Amer. J. Bot.* **49**:697.

Kleinschmidt, A. 1968. Monolayer techniques in electron microscopy of nucleic acid molecules. In *Methods in enzymology* (ed. S. Colowick and N. Kaplan), vol. 12B, p. 125. Academic Press, New York.

Meselson, M. 1972. Formation of hybrid DNA by rotary diffusion during genetic recombination. *J. Mol. Biol.* **71**:795.

Meselson, M. and C. Radding. 1975. A general model for genetic recombination. *Proc. Nat. Acad. Sci.* **72**:358.

Potter, H. and D. Dressler. 1976. On the mechanism of genetic recombination: Electron microscopic observation of recombination intermediates. *Proc. Nat. Acad. Sci.* **73**:3000.

——. 1977. On the mechanism of genetic recombination: The maturation of recombination intermediates. *Proc. Nat. Acad. Sci.* (in press).

Russo, V. 1973. On the physical structure of λ recombinant DNA. *Mol. Gen. Genet.* **122**:353.

Sigal, N. and B. Alberts. 1972. Genetic recombination: The nature of a crossed strand-exchange between two homologous DNA molecules. *J. Mol. Biol.* **71**:789.

Thompson, B., C. Excarmis, B. Parker, W. Slater, J. Doniger, I. Tessman and R. Warner. 1975. Figure-8 configuration of dimers of S13 and $\phi\chi$ 174 replicative form DNA. *J. Mol. Biol.* **91**:409.

Valenzuela, M. and R. Inman. 1975. Visualization of a novel junction in bacteriophage λ DNA. *Proc. Nat. Acad. Sci.* **72**:3024.

White, R. and M. Fox. 1974. On the molecular basis of high negative interference. *Proc. Nat. Acad. Sci.* **71**:1544.

Wolfson, J., D. Dressler and M. Magazin. 1972. Bacteriophage T7 DNA replication: A linear replicating intermediate. *Proc. Nat. Acad. Sci.* **69**:499.

SECTION V
Eukaryotic Systems

This section focuses on some interesting manifestations of eukaryotic phenomena in which insertion sequences might be involved. The section is not intended to be comprehensive. It leaves out some well-known phenomena, such as joining of the genes for variable and constant portions of antibody molecules. Of course, the most clear-cut cases of insertions in eukaryotic cells are those involving integration of viral genomes into the host DNA. The integration of tumor viruses, such as simian virus (SV) 40, adenoviruses, and RNA viruses (in which a DNA copy of the RNA gemome is made), is currently being studied in depth. In particular, the reader is referred to recent studies on the arrangement of SV40 sequences in transformed cells (Botchan et al., *Cell* 9:269 [1976]; Ketner and Kelly, *Proc. Nat. Acad. Sci.* 73:1102 [1976]).

The most famous transposition elements in eukaryotic systems are the controlling elements in maize. The section-title figure shows two maize ears which exhibit differences in expression of a gene required for anthocyanin pigment formation. The phenotype of each kernel reflects the action of a two-element control system, both elements of which are potentially transposable. One element resides at the locus of the gene and modulates its action. The second element resides elsewhere in the chromosome complement. The gene-associated element responds to the second element only when this second element is in its active phase. On each ear, the kernels with variegated patterns exhibit this interaction. When the second element is inactive, the gene-associated element provides the phenotypes shown by the faintly pigmented kernels on the ear to the right and by the uniformly more darkly pigmented kernels on the ear to the left. (Photo is from a study by B. McClintock, Carnegie Institution of Washington, Cold Spring Harbor, New York.)

An Introductory Note on Controlling Elements in Maize

Controlling elements in *Zea mays* have been the subject of much discussion among geneticists since the pioneering work of McClintock (1950, 1952, 1957, 1961). When bacteriophage Mu and the insertion sequences in *Escherichia coli* were discovered in the sixties, immediate analogies were drawn between the maize and *E. coli* elements. Currently, every discussion of prokaryotic insertion sequences includes a reference to the maize elements. The intricacies of the maize elements, however, have remained largely obscure to the molecular and bacterial geneticists. The primary reason is a lack of familiarity with the *Zea mays* genetic system. It is not possible in this book to give the necessary background required for a critical discussion of the evidence for controlling elements. For detailed information on the controlling elements, the reader is referred to an excellent review by Fincham and Sastry (1974)[1] and references therein. The following excerpt from that review outlines the properties of the controlling elements.

> The term "controlling element" was coined by McClintock (1957) to describe transposable elements of apparently sporadic occurrence, which make themselves visible through their abnormal control of the activities of standard genes. Most simply, a controlling element may inhibit activity of a gene through becoming integrated in, or close to, that gene. From time to time, either in germinal or somatic tissue, it may be excised from this site and, as a result, the activity of the gene is often more or less restored, while the element may become reintegrated elsewhere in the genome where it may affect the activity of another gene.
>
> In the simplest examples (the autonomous or one-element systems), the only element that needs to be considered is the one that resides in or close to the affected gene, apparently acting autonomously with regard to its inhibition of gene action and its occasional transposition. Frequently, however, the element inhibiting gene action is itself controlled by another element, in some way complementary to it, located elsewhere in the genome. In such cases the excision of the first element, with consequent release of gene activity, does not occur except in the presence of the second element, which appears to supply some missing excision function. A situation of this kind is called a *two-element* system, and McClintock (1961) has termed the element at the affected locus the *operator* and that acting on it from a distance the *regulator* element. The analogy with bacterial regulator-operator systems should not be pushed too far, and we propose to use the term *receptor,* rather than operator which has a precise meaning in molecular biology that may not be appropriate. . . .

[1]Extract is reprinted with permission from Fincham and Sastry (1974).

In the autonomous systems, the regulator and receptor components are integral parts of the same element, both residing at the locus under control. An extremely important conclusion . . . is that a nonautonomous (two-element) system can originate from an originally autonomous one through loss of the regulator function; following such an event the receptor component may continue to respond to a regulator elsewhere in the genome. Regulators are, in fact, more often than not identified through their effects on nonautonomous receptors. An active regulator promotes not only excision of a responsive receptor but (at least in many cases) its own excision and transposition as well.

The receptor-regulator relationship is a highly specific one and three classes of elements have been recognized on the basis of this specificity. These have been given the names *Dotted (Dt)* (Rhoades 1938, 1942), *Activator (Ac)* (McClintock 1950, 1952), and *Suppressor-mutator (Spm).* Peterson's *Enchancer (En)* appears to be the same as *Spm,* while Brink's *Modulator (Mp)* is homologous with, if not identical to, *Ac.* Receptors of each class respond only to their own regulator and are quite unaffected by the other two. While its status is not altogether clear, the determinant of the instability of *R-stippled* may represent a fourth class.

The *Spm* element differs from *Dt* and *Ac* in being commonly capable of regulating the activity of a gene even in the absence of mutation-like (excision) events. The receptor component of *Spm* may, in the absence of the regulator, bring about only a partial or even quite slight inhibition of the activity of the associated gene. In such cases the regulator may act at a distance on the receptor to suppress the gene activity in most cells as well as to release full gene activity, presumably by receptor excision, in some of them. The *suppressor* and activity-releasing (*mutator*) functions of the *Spm* regulator are separable, in as much as derivatives of *Spm* are known that have lost the second while retaining the first.

The three classes of element differ strikingly in their dosage effects on the frequency of excision. In the case of *Ac* this frequency is maximal with one copy of the element, is strikingly reduced by a second copy, and is much lower again with a third (note that the endosperm is triploid). *Dt* shows the opposite dosage effect; release of activity of the susceptible allele a_1 increases in frequency at least in proportion to the number of *Dt* copies present (Rhoades 1942). *Spm* shows no dosage effect, at least for mutation frequency of the susceptible allele, and behaves in this respect as a simple dominant.

Finally, it must be mentioned that both receptor and regulator components of controlling elements frequently change their properties. These changes, though they may be transmitted through many cell divisions like genetic mutations, are often referred to as changes in *state* or *phase,* in token of their high frequency, their strong tendency to revert, and (in some cases) their high degree of predictability at certain stages of plant development.

REFERENCES

Fincham, J. R. S. and G. R. F. Sastry. 1974. Controlling elements in maize. *Annu. Rev. Genet.* **8**:15.

McClintock, B. 1950. The origin and behavior of mutable loci in maize. *Proc. Nat. Acad. Sci.* **36**:344.

——. 1952. Chromosome organization and gene expression. *Cold Spring Harbor Symp. Quant. Biol.* **16**:13.
——. 1957. Controlling elements and the gene. *Cold Spring Harbor Symp. Quant. Biol.* **21**:197.
——. 1961. Some parallels between gene control systems in maize and in bacteria. *Am. Natur.* **95**:265.
Rhoades, M. M. 1938. Effect of the *Dt* gene in the mutability of the a_1 allele in maize. *Genetics* **23**:377.
——. 1942. The genetic control of mutability in maize. *Cold Spring Harbor Symp. Quant. Biol.* **9**:138.

The Position Hypothesis for Controlling Elements in Maize

P. A. Peterson
Department of Agronomy
Iowa State University
Ames, Iowa 50011

Controlling elements in maize were first recognized by McClintock (1952), who identified transposable chromosome elements that affect the functioning of various loci and behave as discrete genetic determinants. In the ensuing years, additional studies on elements in maize which show transposability have been reported (Brink and Nilan 1952; Peterson 1953; Doerschug 1973; Gonella and Peterson 1975).

A characteristic feature of controlling elements is their effect on the mutability of various loci (Fig. 1). Controlling elements are considered the causal agents for the higher mutation rates, but thus far the empirical definition of controlling elements has been based mainly on genetic studies, and it has not been possible to elucidate their exact nature.

Controlling-element Systems and Their Components

Controlling elements are recognized by their effect on functioning genes: their association with a gene alters its activity, and subsequent excision of these elements might lead to the restoration of normal gene activity. Excision might be followed by the reintegration of the transposable element at a new location in the genome.

There are two types of systems associated with controlling elements: the two-element system and the autonomous system. A *two-element* controlling system consists of a *receptor* (Fincham and Sastry 1974) and a *regulatory* element. The identifiable receptor element, such as the Dissociation element, *Ds* (McClintock 1952), or the Inhibitor element, *I* (Peterson 1960), which is adjacent to the locus and in the *cis* position with the affected locus alters or inhibits gene activity. This receptor element responds to a second element, the regulatory element, which excises or alters the receptor element such that the gene activity might be restored. This receptor–regulatory-element interaction is highly specific and describes a controlling-element system. Thus far, four controlling-element systems have been described in maize. The receptor-regulatory elements are: *a-dt-Dt* (Rhoades 1936, 1938, 1942; Nuffer 1961; Doerschug 1973), *Ds-Ac* (McClintock 1952, 1965), *I-(En-Spm)* (Peterson 1953, 1960, 1961, 1965;

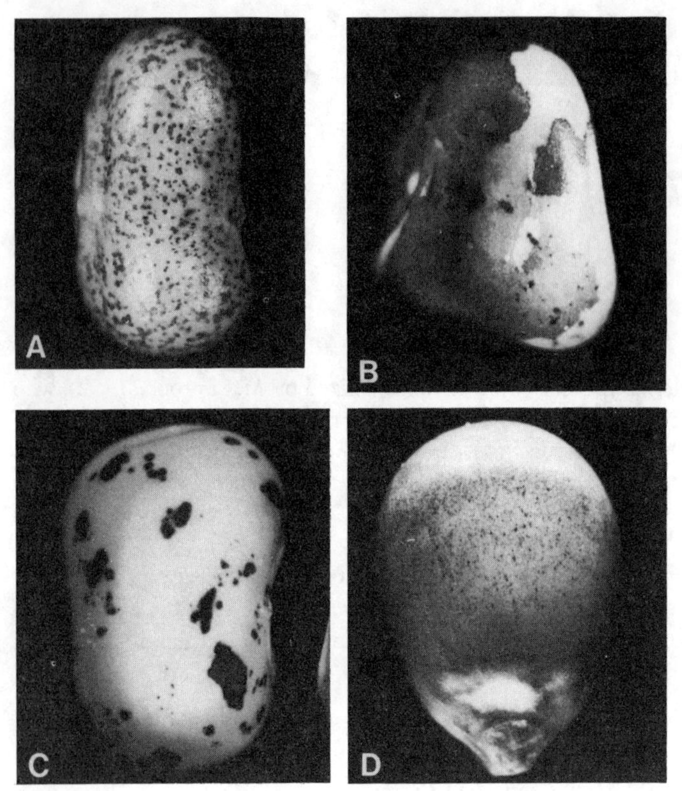

Figure 1
Four kernels expressing different types of mutability. Each of the *a2-m* alleles arose in a stock containing an *En*. (*A*) The original *En* (as expressed on a standard receptor allele, *a-m(r)102b*) that gave rise to the *a2-m* alleles in *B, C,* and *D*. (*B*) *a2-m 1 1511*–colored spots on a colorless background. (*C*) *a2-m 7 8018*–coarse, large (early occurring) sectors on a colorless background. (*D*) *a2-m 6 8140*–very fine (late occurring) spots on a relatively pale background.

McClintock (1956, 1957), and *Icu-Fcu* (Gonella and Peterson 1974, 1975). These systems are mutually exclusive (Table 1) and can be identified only when a "controlled allele" (a receptor element at an affected locus) responds to the presence of a specific active regulatory element by exhibiting a phenotypic change. Mutability effects on different alleles have been recognized with one or another of these systems. One example, such as mutable pericarp (*P-vv=P-rrMp*), was established as part of the *Ds-Ac* system because *Ds* responds to *Mp* (Barclay and Brink 1954) by expressing *Ds*-type chromosome breaks (Brink and Williams 1973). This makes *Mp* homologous with *Ac* in its effect on *Ds*. In a similar manner, the *En* and *Spm* elements were identified as part of the same system by means of their similar effect on a common receptor element. These systems and their component elements are listed in Table 1.

Table 1
Responses of Receptor Elements to Regulatory Elements

Regulatory elements	Receptor elements				
	Ds	*I*	*a-m-1*	*a-dt*	*r-cu*
Ac (Mp)	+	–	–	–	–
En (Spm)	–	+	+	–	–
Dt	–	–	–	+	–
Fcu[a]	–	–	–	–	+

Mutability, +; no mutability, –.
[a] Data of J. Gonella and P. A. Peterson (unpubl.).

In the one-element or *autonomous system,* a regulatory element can be identified at the affected locus; it alters gene activity (Peterson 1970b). Release of the inhibition of gene activity is associated with excision of the regulatory element from its locus position and, at least sometimes, reintegration of the element at a new position (McClintock 1952; Brink and Nilan 1952; Peterson 1970a,b, 1976). In the autonomous system, the regulatory element can inhibit gene activity and self-excise. If the new integration site is the site of a functioning gene, the same cycle of inhibition and excision can occur and be recognized. A receptor element of a two-element system can be used to identify the autonomous element of a particular system (Barclay and Brink 1954; Peterson 1970a,b).

Relation between the Receptor and Regulatory Elements

In the origin of a mutable allele, in the *En* system for example (Peterson 1968, 1976), mutability arises when the regulatory element, Enhancer (*En*), becomes inserted at a locus such as the *A2* locus, which affects color in the aleurone layer of the corn kernel (Fig. 2). The resulting phenotypic change is from a colored kernel to a colorless kernel with spots of color (Figs. 1 and 2). The colored spots apparently result from excision events at the *A2* locus (Fig. 2C) which release the inhibiting effects of the *cis*-located regulatory element. This *A2*-mutable allele (whose phenotype shows spots of color on a colorless background) is represented by an *A2-En* linkage, and this linkage is unbroken until *En* undergoes transposition and the autonomously controlled allele is freed from suppression by the regulatory element. Excision of the regulatory element usually leads to several phenotypic classes of derivatives (Peterson 1970a).

One class of derivative may have a phenotype indistinguishable from that of the normal allele. This indicates that excision of the regulatory element from the locus can leave the locus unimpaired (Fig. 2C). Another derivative may have a phenotype of an intermediate level of expression, indicating that the

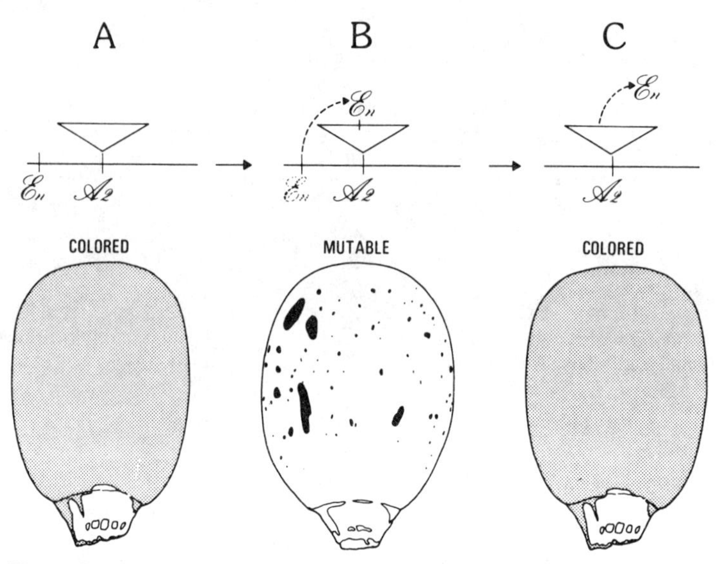

Figure 2
Diagrammatic representation of the result of *En* transpositions leading to changes in *A2*. The associated phenotypes representing the gene configurations are illustrated. A fully active *A2* allele (*A*) becomes mutable (*B*) following implantation of *En* at the *A2* locus. Each of the spots in the kernel in *B* represents an emission of *En* as illustrated in *C*.

expression of the allele has been modified (Fowler and Peterson 1974; Reddy and Peterson 1976). A third type, a colorless phenotype, is a nonfunctioning allele. There are two types of nonfunctioning alleles. The more frequent is the type that does not have the capacity to respond to a regulatory element, as described for two-element systems. These alleles are called nonresponders, [*a2-m(r)*]. The other type of nonfunctioning derivative is that which responds to regulatory elements by expressing mutability. It is this latter type that shows a relationship between the two elements of a controlling-element system. Such derivatives, and in this case *a2-m(r)* is an example, are considered receptive alleles with a receptor element at the affected locus.

Therefore, in the origin of a mutable allele, the implanted regulatory element generates a number of derivatives. One is the functionless type that responds. Tests show that this functionless responding allele includes a receptor element which presumably arose as a consequence of the transposition of the regulatory element from the locus site. This receptor element is responsive only to the regulatory element that was responsible for the excision event. Thus an element remaining at the locus after excision of the regulatory element responds specifically to the excised regulatory element. Hence, systems of interacting receptor-regulatory elements arise (Table 1).

The Position Hypothesis as a Basis for States of Controlling Elements

One of the more striking features of controlling-element expression is the diversity of distinct patterns of mutability (Fig. 1B,C,D). These patterns are a function of two events, namely, timing and frequency of mutation. Large, colored sectors which contribute to a greater amount of tissue during ontogeny of the aleurone (Fig. 1B,C) result from mutations occurring early in aleurone development. Mutations occurring later contribute to a smaller amount of tissue expressing the mutated gene (Fig. 1A,D). Frequency is related to the number of mutation events. Spot frequency can vary from the very frequent number evident in Fig. 1A and D to cases where only one or two spots are evident (Peterson 1961). An exact frequency of mutation cannot be gauged from phenotype changes. Many mutation events probably cause no detectable changes.

The combined effect of timing and frequency results in specific phenotypic patterns of mutability, which are referred to as *states* of the receptor or regulatory elements. When a responsive allele, such as *a2-m(r),* colorless in the absence of the regulatory element *En* but showing spots of color in the presence of *En* (indicating that a receptor allele is present), is tested against an assortment of *En* types, a wide range of patterns is observed. These patterns are consistent and, with a common receptor allele such as *a2-m(r),* are indications of the different states of *En.* Conversely, different states of a receptor element can be differentiated in tests with a common regulatory element. The final pattern is the result of the interaction between the states of the regulatory element and the states of the *cis*-located receptor element.

When transpositions of the regulatory element *En* occur, an assortment of patterns is observed. In Figure 1, three kernels illustrate the original patterns of three independently isolated mutable alleles at the *A2* locus, *a2-m 1 1511, a2-m 6 8140,* and *a2-m 7 8018.* When the regulatory element *En* that was present in each of the three plants giving rise to the three alleles is tested for its effect on a standard *a-m(r)* tester, each of these parental-source *En*'s gives a similar pattern (Fig. 1A). (These three particular alleles were chosen as an example because *En* was of the same state.) This indicates that a particular state of *En,* as expressed on a standard *a-m(r)* tester, can give rise to strikingly different patterns at a target locus. It may be hypothesized that the resultant patterns are a consequence of different sites of implantation of *En* at the *A2* locus and, therefore, that the position of the controlling element determines the state of mutability.

McClintock has described similar states in the *Spm-En* and *Ds-Ac* systems. In addition, she has documented phenotypically deviant derivatives of original states. For a pictorial presentation of these states, refer to McClintock (1965), especially Figures 2 and 3.

SUMMARY

In maize, the insertion of controlling elements generates a broad spectrum of diverse derivatives at most loci examined in most mutable gene systems. This diversity includes distinguishable patterns that are determined by the timing and frequency of mutation events, as well as an array of heritable, distinct, stable expressions that, among the loci governing the genes of the anthocyanin pathway expressed in the kernel, range from colorless to full-colored types. The hypothesis is proposed that this diversity is not caused by a change in the genetic information of the inserted element, but rather by its position at the locus.

Acknowledgments

Journal paper no. J-8535 of the Iowa Agriculture and Home Economics Experiment Station, Ames, Iowa. Project no. 1884. This research was supported by National Science Foundation Grant GB38328.

REFERENCES

Barclay, P. C. and R. A. Brink. 1954. The relation between modulator and activator in maize. *Proc. Nat. Acad. Sci.* **40**:1118.

Brink, R. A. and R. A. Nilan. 1952. The relation between light variegated and medium variegated pericarp in maize. *Genetics* **37**:519.

Brink, R. A. and E. Williams. 1973. Mutable R-Navajo alleles of cyclic origin in maize. *Genetics* **73**:273.

Doerschug, E. B. 1973. Studies of *dotted*, a regulatory element in maize. I. Inductions of *dotted* by chromatid breaks. II. Phase variation of *dotted*. *Theoret. Appl. Genet.* **43**:182.

Fincham, J. R. S. and G. R. K. Sastry. 1974. Controlling elements in maize. *Annu. Rev. Genet.* **8**:12.

Fowler, R. G. and P. A. Peterson. 1974. The *a2-m(r-pa-pu)* allele of the *En* controlling element system in maize. *Genetics* **76**:433.

Gonella, J. and P. A. Peterson. 1974. Variegation associated with controlling element systems in tribal maize from Colombia. *Maize Genet. Coop. Newsl.* **48**:66.

——. 1977. The *F-cu* 2-unit controlling element system. In *Genetics and breeding of maize* (ed. D. B. Walden). Wiley-Interscience, New York. (In press.)

McClintock, B. 1952. Chromosome organization and genic expression. *Cold Spring Harbor Symp. Quant. Biol.* **16**:13.

——. 1954. Mutations in maize and chromosomal aberrations in *Neurospora*. *Carnegie Inst. Wash. Year Book* **53**:254.

——. 1956. Intranuclear systems controlling gene action and mutation. *Brookhaven Symp. Biol.* **8**:58.

——. 1957. Controlling elements and the gene. *Cold Spring Harbor Symp. Quant. Biol.* **21**:197.

——. 1965. The control of gene action in maize. *Brookhaven Symp. Biol.* **18**: 162.

Nuffer, M. G. 1961. Mutation studies at the A_1 locus in maize. I. A mutable allele controlled by *Dt*. *Genetics* **46**:625.
Peterson, P. A. 1953. A mutable pale green locus in maize. *Genetics* **38**:682.
——. 1960. The pale green mutable system in maize. *Genetics* **45**:115.
——. 1961. Mutable a_1 of the *En* system in maize. *Genetics* **46**:759.
——. 1965. A relationship between the *Spm* and *En* control systems in maize. *Am. Natur.* **99**:391.
——. 1968. The origin of an unstable locus in maize. *Genetics* **59**:391.
——. 1970a. Controlling elements and mutable loci in maize: Their relationship to bacterial episomes. *Genetica* **41**:33.
——. 1970b. The *En* mutable system in maize. III. Transposition associated with mutational events. *Theoret. Appl. Genet.* **40**:367.
——. 1977. Controlling elements: The induction of mutability at the *A2* and *C* loci in maize. In *Genetics and breeding of maize* (ed. D. B. Walden). Wiley-Interscience, New York. (In press.)
Reddy, A. R. and P. A. Peterson. 1976. Germinal derivatives of the *En* controlling-element system in maize: Characterization of colored, pale, and colorless derivatives of *a2-m*. *Theoret. Appl. Genet.* **48**:269.
Rhoades, M. M. 1936. The effect of varying gene dosage on aleurone colour in maize. *J. Genet.* **33**:347.
——. 1938. Effect of the *Dt* gene in the mutability of the a_1 allele in maize. *Genetics* **23**:377.
——. 1942. The genetic control of mutability in maize. *Cold Spring Harbor Symp. Quant. Biol.* **9**:138.

The Case for DNA Insertion Mutations in *Drosophila*

M. M. Green
Department of Genetics
University of California
Davis, California 95616

Evidence for the existence of DNA insertion mutations in *Drosophila melanogaster* is of an entirely circumstantial nature. The evidence comes from a limited number of *D. melanogaster* mutants which exhibit genetic properties analogous in many details to the genetic properties of established insertion mutants in *Escherichia coli.*

Those genetic properties that originate from *Drosophila* experiments and that characterize presumptive insertion mutants in *D. melanogaster* can be summarized as follows:

1. The mutant reverts spontaneously to wild type at a measurable frequency. In this case, two general classes of mutants can be recognized: reasonably stable mutants which revert at a frequency on the order of 10^{-5} to 10^{-6} gamete and unstable mutants or mutable genes which revert at inordinately high frequencies of 10^{-3} to 10^{-4} gamete.
2. The spontaneous revertability is significantly increased following exposure to agents known to produce deletions; e.g., X rays or specific mutator genes.
3. The mutant may be causally associated with the production of specific deletions at the chromosome site to which it is mapped.
4. The mutant is associated with a significant reduction in interallelic recombination.

A brief description of specific putative insertion mutants follows. These mutants have been selected because they illustrate a number of the genetic properties listed above.

Stable Putative Insertion Mutants

The mutant white-ivory (w^i) is spontaneous in origin and is a functional allele at the *white* locus on the X chromosome. Spontaneous reversion of w^i assayed in homozygous females occurs at a frequency of 5×10^{-5} gamete (Bowman 1965). Following X irradiation, the reversion frequency can be increased significantly (Lewis 1959); an applied dose of 4000 rad, for example, results in a 50-fold increase in reversions (Bowman and Green 1966). A second property of w^i is its effect on interallelic crossing over. Whereas the frequency of crossing over between the *w* alleles w^a and w^e is approximately 1×10^{-4}, in the presence of w^i (as a heterozygote) this frequency is strikingly reduced (Lewis 1959; Bowman 1965; Bowman and Green 1966) to a frequency of less than 1×10^{-5}. Yet reversions of w^i to wild type exhibit normal interallelic recombination (Bowman 1965; Bowman and Green 1966). Both the reversion and recombination properties are consistent with the interpretation that w^i is causally associated with a DNA insertion whose presence depresses recombination and whose loss leads to a restoration of the wild-type condition. Presumably, the inserted DNA is "foreign" to the site of integration. Being foreign, or nonhomologous, the insertion adversely influences local chromosome pairing, resulting in a decrease in crossing over. Excision simultaneously restores normal pairing and normal crossing over.

The mutational properties of w^i are not unique, and at least three other independent, nonallelic X-chromosome mutants exhibit comparable behavior. These mutants are yellow-2 (y^2) body color, scute (*sc*) bristles missing, and forked-3N (f^{3N}) bristles. Spontaneous reversions to wild type of y^2 and *sc*, although rare, have been recovered, occurring at a frequency of less than 2×10^{-5} chromosome scored (Green 1961). On the other hand, f^{3N} reverts spontaneously to wild type at a frequency of around 2×10^{-5} chromosome (Woodruff 1975). Following X irradiation of homozygous females with a dose of 5000 rad, the reversion frequency of each mutant has been significantly increased, both y^2 and *sc* reverting at a frequency of 6×10^{-5} chromosome (Green 1961). In a comparable study in which homozygous f^{3N} females were X-irradiated with 4200 rad, the f^{3N} reversion frequency increased 30-fold, to 6×10^{-4} (Lefevre and Green 1959). Consistent with the X-ray reversion results is the additional observation that the reversion frequencies of both y^2 and f^{3N} are significantly increased in the presence of a mutator gene (Green 1970) that functions primarily, and probably exclusively, to make deletions (Green and Lefevre 1972). These high reversion frequencies associated with agents that make deletions suggest that y^2, *sc*, and f^{3N}, like w^i, are causally associated with DNA insertions whose loss is concomitant with the reversion to wild type.

Unstable Putative Insertion Mutants

Among the several mutable genes described in *D. melanogaster*, more genetic information is presently available on *white-crimson* (w^c) than on any other. Therefore, its genetic behavior will be considered here in some detail. Mutable w^c arose from w^i following X irradiation of a homozygous w^i female (Green 1967). As its name indicates, w^c is easily separable from w^i on the basis of eye color. It is also separable on the basis of its mutation spectrum, both quantitative and qualitative. Because w^c mutates premeiotically at an inordinate rate, its precise mutation frequency cannot be determined readily. It will suffice to note here that when homozygous w^c females are tested for mutability, one or more newly arisen mutations will occur among the progeny of about one in every four females. Among w^c males, approximately one in eight yield new mutants among their progeny. On the basis of eye-color phenotype, a number of different classes of new mutants can be identified. At the extremes are the wild-type (w^+) and white-eyed (w) individuals. Other exceptions manifest eye-color phenotypes readily separable from w^c, w^+, and w. These include some whose eye color is intermediate between w^+ and w^c, some whose eye color is intermediate between w^c and w, and some inseparable from w^i from which w^c originally arose. The new mutants recovered are most frequently w^+. In addition to phenotypic differences, the new mutants also differ in mutability. Thus, to date, all w^+ exceptions are mutationally stable. Those new mutants phenotypically inseparable from w^i behave mutationally the same as w^i, reverting to w^+ infrequently. New mutants intermediate in phenotype are mutationally unstable and produce a spectrum of mutants not unlike that described for w^c, ranging in phenotype from w^+ to w. Among the white-eyed mutants, two classes can be recognized. One group is mutationally unstable, reverting to wild type and to revertants with pigmented-eye phenotypes. A second group is mutationally stable and turns out to include deletions. The deletions will be considered in more detail below. Mapping experiments establish a clear-cut affinity between w^i and w^c: both map to the same site, and w^c depresses interallelic recombination just as w^i does.

The regular production of deletions is a characteristic feature of w^c and its mutable derivatives. The deletions produced differ in length. By and large, most are small and difficult to define cytologically. Others are longer, easily defined by polytene chromosome cytology, and may include the loss of as many as 15 polytene chromosome bands. Since these deletions occur in both females and males, it would appear that, whatever their mode of origin, meiotic crossing over is not involved. Consistent with this notion is the observation that many identical deletions occur as clusters, indicating that they occur before meiosis. Whatever their mode of origin, the deletions apparently begin at the site of w^c, extend either to the left or to the right, and do not overlap. Two lines of evidence support this inference. First, larger deletions can be recovered from small deletions. This means that whatever is inserted at w^c that is responsible for the small deletions is not lost in the deletion process and, consequently, is responsible for induction of the larger deletions. Second, transpositions have been obtained from w^c-induced deletions in which the w^c phenotype is now "turned on." Such phenotypic expression could occur only if whatever is responsible for the w^c process is not lost in the course of deletion production. These facts illustrate the overall analogous behavior in deletion induction of the w^c-associated inserted DNA and the IS*1* and IS*2* DNA elements of *E. coli.*

Transposition of Integrated DNA Elements

To date, no authenticated case of the transposition of a specific DNA element from one genetic site to another has been reported in *D. melanogaster.* Such transpositions have been imputed to account for the occurrence of an X-chromosome, mutable, miniature-wing mutant (Green 1975) and for the polar mutants at the *w* locus (Rasmuson et al. 1974); the latter is considered in more detail in the next section. DNA-element-associated transpositions have been found in which the element plus the genic DNA at the site of integration have been excised from their X-chromosome location and transposed to and integrated into the third chromosome (Green 1969). The fact that all transpositions occurred spontaneously and in both males and females implies that excision and subsequent integration do not involve a meiotic crossover event. The transpositions mapped to discrete, independent sites on the third chromosome. That the transpositions were associated with a DNA element derives from the observation that, in their new position, they retained their high mutability.

A quite separate transposition, derived from an independent *w*-locus mutable system, has been reported (Judd 1975). A remarkable series of transpositions have been summarized by Ising and Ramel (1973). In the latter case, a chromosome segment including a minimum of two, linked, contiguous genes (*white* and *roughest*) and associated with at least three polytene chromosome bands transposed from its site in the X chromosome to the second chromosome. Transpositions were subsequently found to occur from the second chromosome to other chromosomes.

The question of whether or not transpositions are invariably mediated by insertion elements is raised by the transposition of the "sex realizer" in the phorid fly *Megaselia scarlaris* (Mainx 1964). The sex realizer or "maleness determiner" is mapped at the end of a specific chromosome. Spontaneous transpositions of the sex realizer from one chromosome end to the end of a nonhomologous chromosome occur frequently. Are these transpositions mediated by insertion elements or is the sex factor itself an insertion element? These questions are at present unanswered.

Polar Mutations

The most compelling case reported thus far for an insertion mutation in *D. melanogaster* are the two independent occurrences alluded to above, and these have been analyzed in some detail (Rasmuson et al. 1974). Since in all essential details these mutants mimic the polar insertion of the *gal* operon in *E. coli* (Starlinger and Saedler 1972), their origin and genetic behavior will be summarized here.

Both mutants stem from a direct tandem duplication of the entire *w* region. The two segments of the duplication are marked with two different *w* alleles, w^{sp} in the left segment and w^{17g} in the right segment. The two mutants complement one another, producing a near wild-type eye color. From this tandem duplication, two independent mutants were found which no longer exhibited the complementary eye color but were white-eyed. Genetic analysis of both mutants established conclusively that the tandem duplication was intact and that the right segment retained the w^{17g} marker. The left segment of each exception was separated and isolated from the tandem duplication by unequal crossing over, and each segment proved to manifest the genetic characteristics of a *w* deletion. Each *w* deletion depressed interallelic crossing over and each mutated spontaneously, without marker exchange, to w^{sp}, the marker gene originally present in the tandem duplication. Mutation to w^{sp} was accompanied by a restoration of normal interallelic crossing over. These findings mean that the presumptive *w* deletions are, in reality, not deletions but functionally "turned-off" *w* regions. Presumably, an integrated DNA element turned off the region, resulting in a pseudodeletion phenotype. When the element that depresses interallelic crossing over is excised, the lost function is restored.

Sites of DNA Insertion

The majority of DNA insertions are recognized by their association with a specific mutation and its phenotypic effects. However, at least two types of mutables in which the presumed inserted DNA cannot be tied to a specific phenotypic alteration have been associated with the *w* locus.

In one type, the presence of inserted DNA is recognized through the frequent production of small *w* deletions (Gethmann 1971). The *w* deficiencies, all of which appear to be identical, arise from wild type premeiotically both in males and females. The genetic instability has been mapped to a specific region within the *w* region.

A second class of mutables associated with the *w* locus has been uncovered by virtue of the influence of the *w* locus on the phenotypic expression of the X-linked eye-color mutant zeste (*z*). Independently, the same mutable system at the w^+ locus has been monitored by a graded series of phenotypic changes at the *z* locus, from z^+ to *z* phenotype (Judd 1969, 1975; Kalisch and Becker 1970). Although involvement of an inserted element has been considered (Judd 1969, 1975), it is not clear precisely what mutational changes occur at the *w* locus to evince changes in the *z* phenotype. It is clear, however, that discernable phenotypic changes occur at the *w* locus and that, in the absence of the *z* mutant, the recorded mutational events would not be detected.

An analogous situation involving the *w* locus has been described (Rasmuson and Green 1974) in which there is an alternate "switching on" and "switching off" of the zeste eye-color phenotype. Control of the "on-off" switch occurs at the w^+ locus, and, in the absence of the *z* mutant, no changes can be detected.

The association of the switch mechanism with an inserted element at the *w* locus is suggested by the regular, frequent recovery of *w* deletions in this mutable system.

Changes in the Orientation of Insertions

One mutable system discussed above involves the switching on and switching off of the zeste eye color. In the genetic analysis of this system (Rasmuson and Green 1974), no structural difference could be discerned in the w^+ locus when it was switched on (evoking a zeste eye phenotype) or when it was switched off (evoking a wild-type eye color). The inserted element, postulated from the induction of deletions, mapped to the same place in the w^+ region in both states. Therefore, it was suggested that the difference between switched on and switched off lies in the orientation of the inserted element. Thus "on" represents one orientation and "off" represents the inverted orientation, similar to what has been established for IS*2* in *E. coli* (Saedler et al. 1974).

A parallel situation has been found in the case of certain mutable, recessive, X-linked, singed (*sn*) bristle mutants in *D. melanogaster.* A series of *sn* mutants have been extracted from flies caught in the wild in the USSR and Israel (R. L. Berg, M. D. Golubovsky, Yu. N. Ivanov and M. M. Green, unpubl.). Some of the *sn* mutants proved to be mutable, and one, designated $sn^{77\text{-}27}$, has been thoroughly studied in Novosibirsk and Davis (Ivanov 1975; M. D. Golubovsky, Yu. N. Ivanov and M. M. Green, unpubl.) and will be annotated here. Among *sn* mutants, the bristle phenotype varies from allele to allele; in some, the gnarling of the bristles is slight, in others moderate, and in still others extreme. In addition, some *sn* mutants are female-sterile, others female-fertile; all are male-fertile. The phenotype of $sn^{77\text{-}27}$ is a moderate bristle alteration and sterility of homozygous females. Mutationally, $sn^{77\text{-}27}$ is unstable: somatic reversion to wild type results in mosaic flies with both $sn^{77\text{-}27}$ and sn^+ (wild-type) bristles, and reversion premeiotically in the germ line gives rise to a cluster of progeny with the sn^+ phenotype. In a typical experiment in which individual $sn^{77\text{-}27}$ males were crossed with homozygous sn^3 (a stable functional allele) females, the following results were obtained: Of a total of 25 $sn^{77\text{-}27}$ males tested, each produced, on the average, 500 female progeny who received one $sn^{77\text{-}27}$ chromosome from their father. The female progeny of 20 males were exclusively *sn* in phenotype, but those of 5 males included both *sn* and sn^+ females. Among the 5 males, the number of sn^+ exceptional females differed; one produced 4 sn^+ females, a second produced 5, a third produced 7, a fourth 13, and a fifth 118 exceptions. (The different numbers of exceptions per male probably reflects the time prior to meiosis when the mutation to sn^+ occurred.) In contrast to most mutable systems studied, the sn^+ exceptions derived from $sn^{77\text{-}27}$ are also mutationally unstable. In a typical experiment in which individual revertant sn^+ males were tested by crossing to sn^3 females, the following results were obtained: Among 23 males tested, each producing on the average

600 female progeny, 15 produced only sn^+ female progeny, and 8 produced both phenotypically sn^+ and *sn* female progeny. Among the latter males, the numbers of *sn* female exceptions recovered per male were: two males produced 1 exception each, one male produced 2 exceptions, two males produced 4 exceptions each, one male produced 14 exceptions, one male produced 15 exceptions, and one male produced 34 exceptions. Progeny tests of the exceptional females showed that two classes of *sn* mutants arose from the mutable sn^+. One class, characteristic of six independently recovered *sn* mutants, was in phenotype inseparable from that of the original $sn^{77\text{-}27}$ mutant; the bristles were a moderate singed and homozygous females were sterile. On subsequent testing, all six reversions to sn^+ were once more recovered. A second class, characteristic of two separately recovered *sn* mutants, produced an extreme singed phenotype in males and females and was female-sterile. On subsequent testing, both proved to be mutationally stable.

A probable explanation for these results is that $sn^{77\text{-}27}$ is an insertion mutant. Reversions to sn^+ are associated with a changed (inverted) orientation of the inserted DNA, whereas mutations to the phenotypically extreme and mutationally stable mutant are the result of deletions of the *sn* gene modulated by insertion elements. This explanation also predicts that some reversions to sn^+ involve the loss of the insertion element and should be mutationally stable. A more detailed study should establish the validity of the proposed explanation.

CONCLUDING REMARKS

In the foregoing narrative an attempt has been made to make a case for the occurrence of inserted DNA elements in *D. melanogaster.* Mutable genes in *D. melanogaster* and insertion mutants in *E. coli* have parallel genetic behavior. Argumentation by analogy, however, has its pitfalls. Unequivocal proof of DNA insertions awaits the day when specific short stretches of eukaryote chromosomes can be routinely transferred to an episome or plasmid and then biochemically assayed (Glover et al. 1975). Clearly this day is in the foreseeable future.

Even though unequivocal proof of DNA insertions is lacking, I would like to pose a few relevant questions. Are the presumed insertions of the same size? If we assume that DNA is inserted, where did it come from? There is some suggestive genetic evidence that not all *Drosophila* insertions are of the same size. This evidence stems from the influence of insertions on interallelic crossing over. Thus some presumed insertions clearly decrease interallelic crossing over (e.g., w^i), whereas others apparently do not (e.g., the deletion-producing mutable of Gethmann [1971]). The influence of interallelic crossing over could be a function of the size (length) of the inserted nonhomologous DNA; longer stretches might decrease crossing over, shorter stretches might be without discernible influence. Possibly orientation as well as site of insertion of the inserted element influences interallelic crossing over. Whatever their size, inserted elements are sufficiently small that they cannot be seen by the present methods of polytene chromosome cytology.

As to the source of inserted elements, nothing concrete can be stated at this time.

SUMMARY

On the basis of cytogenetic criteria, two classes of insertion mutants can be recognized in *D. melanogaster.* One class, the stable insertion mutants, revert spontaneously at a low frequency but under the influence of agents which induce deletions revert at a significantly elevated frequency. A second class, the unstable insertion mutants, revert spontaneously at inordinately high frequencies and generate deletions at the genetic site of insertion. Examples of "polar" mutants and "changes in polarity" can be inferred from the genetic behavior of the insertion mutants. Taken together, the genetic properties attributed to the *Drosophila* insertion mutants are analogous to those described for proved insertion mutants of *E. coli.*

Acknowledgment

This work was supported by National Institutes of Health Grant GM 22221.

REFERENCES

Bowman, J. T. 1965. Spontaneous reversion of the white-ivory mutant of *Drosophila melanogaster. Genetics* **52**:1069.

Bowman, J. T. and M. M. Green. 1966. X-ray induced reversion of the white-ivory mutant of *Drosophila melanogaster. Genetica* **37**:7.

Gethmann, R. C. 1971. The genetics of a new mutable allele at the *white* locus in *Drosophila melanogaster. Mol. Gen. Genet.* **114**:144.

Glover, D. M., R. L. White, D. J. Finnegan, and D. S. Hogness. 1975. Characterization of six cloned DNAs from *Drosophila melanogaster,* including one that contains genes for rRNA. *Cell* **5**:149.

Green, M. M. 1961. Back mutation in *Drosophila melanogaster.* I. X-ray induced back mutations at the *yellow, scute,* and *white* loci. *Genetics* **46**:671.

——. 1967. The genetics of a mutable gene at the *white* locus of *Drosophila melanogaster. Genetics* **56**:467.

——. 1969. Controlling element mediated transpositions of the *white* gene in *Drosophila melanogaster. Genetics* **61**:429.

——. 1970. The genetics of a mutator gene in *Drosophila melanogaster. Mutat. Res.* **10**:353.

——. 1975. Genetic instability in *Drosophila melanogaster*: Mutable miniature (m^u). *Mutat. Res.* **29**:77.

Green, M. M. and G. Lefevre. 1972. The cytogenetics of mutator induced X-linked lethals in *Drosophila melanogaster. Mutat. Res.* **16**:59.

Ising, G. and C. Ramel. 1973. The behavior of a transposing element in *Drosophila melanogaster. Genetics* **73**:s123.

Ivanov, Y. F. 1975. Unstable conditions of singed mutation in *Drosophila melanogaster. Drosophila Inf. Serv.* **51**:71.

Judd, B. H. 1969. Evidence for a transposable element which causes reversible gene inactivation in *Drosophila melanogaster. Genetics* **62**:s29.

——. 1975. Genes and chromomeres of *Drosophila.* In *The eukaryote chromosome* (ed. W. J. Peacock and R. Brock), p. 169. Australian National University Press, Canberra.

Kalisch, W.-E. and H. J. Becker. 1970. Ueber eine Reihe mutabler Allele des *white*-locus bei *Drosophila melanogaster. Mol. Gen. Genet.* **107**:321.

Lefevre, G. and M. M. Green. 1959. Reverse mutation studies on the *forked* locus in *Drosophila melanogaster. Genetics.* **44**:769.

Lewis, E. B. 1959. Germinal and somatic reversion of the ivory mutant in *Drosophila melanogaster. Genetics* **44**:522.

Mainx, F. 1964. The genetics of *Megaselia scalaris* Loew (Phoridae): A new type of sex determination in Diptera. *Am. Natur.* **98**:415.

Rasmuson, B. and M. M. Green. 1974. Genetic instability in *Drosophila melanogaster.* A mutable tandem duplication. *Mol. Gen. Genet.* **133**:249.

Rasmuson, B., M. M. Green and B.-M. Karlsson. 1974. Genetic instability in *Drosophila melanogaster.* Evidence for insertion mutations. *Mol. Gen. Genet.* **133**:237.

Saedler, H., H.-J. Reif, S. Hu and N. Davidson. 1974. IS2, a genetic element for turn-off and turn-on of gene activity in *E. coli. Mol. Gen. Genet.* **132**:265.

Starlinger, P. and H. Saedler. 1972. Insertion mutations in microorganisms. *Biochimie* **54**:177.

Woodruff, R. C. 1975. The control of mutational instability by a new mutator gene of *Drosophila melanogaster. Genet. Res. Camb.* **25**:163.

"Flip-Flop" Control and Transposition of Mating-type Genes in Fission Yeast

R. Egel
Institut für Biologie III der Universität
D-7800 Freiburg, West Germany

Mating Types and Mating-type Genes

Can yeasts be used profitably to study development? Cells of the fission yeast *Schizosaccharomyces pombe* remain unicellular throughout their life cycle and have but a single mode of differentiation. They can terminate vegetative cell division (when circumstances cease to favor growth) and resort to sexual reproduction: two cells fuse to form a zygote, the diploid nucleus undergoes meiosis, four ascospores are produced, and these remain dormant until they meet a more favorable environment.

Cells of two different "mating types," termed ⊕ and ⊖ in *S. pombe,* are always required, although such cells are indistinguishable by morphological criteria. The basic genetics of the mating types have been worked out by Leupold (1959) and Gutz et al. (1974). Essentially, there are two genes controlling these mating types: $mat1^M$ is responsible for ⊖ activity and $mat2^P$ for ⊕ activity in both mating and meiosis. These genes are not strictly allelic and can be carried on the same chromosome, separated by about 1% meiotic recombination. A haploid strain carrying both genes is termed "homothallic"; cells of a clonal culture (which are identical in genotype) can form zygotes and ascospores. Other strains are "heterothallic," only sporulating in crosses with partner cells that express the opposite mating type.

Mating-type Instabilities

Genetic analyses of the mating-type locus in *S. pombe* are motivated not only by the regulatory role to be attributed to those genes, but also by the various genetic changes that are known to happen there. Three main aspects of the latter category are discussed in this paper: (1) transposition of ⊕ genes, (2) "change of state" of ⊕ genes, and (3) a "flip-flop" control of gene expression in the homothallic configuration. Above all, the mating-type locus is known to be a "hot spot" of localized mitotic recombination, which runs as high as meiotic recombination if $mat1^M$ is present homozygously in diploid cells (Gutz et al. 1974). This localized recombination appears to be instrumental in most, if not all, of the instabilities mentioned.

Transposition

Homothallic strains of *S. pombe* can mutate to ⊖ strains, apparently by loss of the *mat2^P* function. Since there is now evidence that three independent functions are encoded by *mat2^P* and that most of the ⊖ mutants have lost these functions simultaneously (Meade and Gutz 1976), the underlying mutations are probably deletions. Homothallic strains can also mutate to ⊕ strains (in fact, 10 times as frequently), but these mutations cannot be caused by *mat1^M* deletions.

Leupold (1959) recognized that normal ⊕ strains carry two linked genes, each of which is sufficient to specify the ⊕ mating type. These genes can be separated by recombination, and only the one proximal to the centromere can mutate spontaneously to *mat1^M*. Since the ⊕ strains derived from homothallic ones are normal in every respect, *mat1^M* should have acquired ⊕ function by mutation (explicitly demonstrated by Meade and Gutz 1976), and this acquisition should be reversible. Yet this kind of mutability is not intrinsic to *mat1^M* itself, pure ⊖ strains being perfectly stable. Instead, a functional *mat2^P* gene is mandatory for such a mutation to occur.

Therefore it was postulated that those mutations were actually duplications of *mat2^P*, and evidence of faithful copying was in fact obtained (Egel 1976a,b) when a partially defective mutant *mat2^{Pm}* (impaired in meiosis but not in conjugation) was used for analogous mutation analyses (see Table 1 for a summary of the strains involved in those studies). In that experiment, the mutant *Pm* allele was duplicated, and the copy could be separated from its original by recombination.

A duplication might, of course, result from a single displaced crossover (Fig. 1). Such an event would produce a tandem repetition in "direct" order (same orientation) but would not affect the ⊖ gene to the left of that repeat. As it happens, however, *mat1^M* is totally inactivated as a result of the duplication, and so this simple explanation is unsatisfactory. Instead, the duplication is thought to arise as an insertion, dismembering *mat1^M* itself or its controlling region. Since that duplication is readily reversible (all these changes take place on the order of 10^{-4} to 10^{-3} per cell division), fully restoring the ⊖ function, the transposed sequence should be subject to precise excision. (A schematic interpretation of the four standard *S. pombe* strains is given in Fig. 2. These strains are also summarized in Table 1.) The transposed *mat2^P* probably originated from the sister strand of the receptor chromatid after DNA replication. Whether the donor strand is aborted or survives as a different mutant is not known.

Changes of State

Activity of *mat2^P* at either position can be greatly reduced as a result of other mutational events, which are significantly correlated with recombination (Egel 1976b); of four colonies isolated from a certain diploid strain (*Pm Pm* / *M P*), which were chosen for their reduced ⊕ activity, all had exchanged their mating-

Table 1
Mating Characteristics and Mutations of the Various Strains

	Mating reaction ↓			
	with ⊕ cells	with ⊖ cells	with ↓ self	Mutations to ↓
M P	+	+	+	*P P; M o*
M o	+	–	–	none
P P	–	+	–	*M P; M*⌊*P*⌉
P o	–	+	–	*M o*
M⌊*P*⌉	+	–	–	*M P; P P*
M Pm	+	x	x	*Pm Pm; M o*
Pm Pm	–	x	–	*M Pm; M* ⌊*Pm*⌉
Pm o	–	x	–	*M o*
M ⌊*Pm*⌉	+	–	–	*M Pm; Pm Pm*
P Pm	–	+	–	*M Pm;* ⌊*P*⌉ *Pm*
⌊*P*⌉ *Pm*	–	x	–	*M Pm; P Pm*
Pm ⌊*P*⌉	–	x	–	*M P; Pm P*
Pm P	–	+	–	not tested

Symbols: +, conjugation, meiosis, and sporulation normal; –, no conjugation, no meiosis (except for segregants); x, conjugation normal, no meiosis, diploid zygotic fission inducible. ⌊*P*⌉, ⌊*Pm*⌉: genes still present but not expressed (except for some residual activity in diploid cells); and *M*⌊*P*⌉: unstable ⊖ strain, first described by Gutz and Doe (1973) and reinterpreted by Meade and Gutz (1976).

type genes (resulting in configuration *Pm*⌊*P*⌉ of Table 1 and in *M Pm* as the reciprocal recombinant), as well as a flanking marker, distal with respect to the centromere. These events seem to require at least one transposed *mat2* copy to be present, since they have never been observed in haploid (*M P*) strains (this feature is still poorly understood). The inactivated $mat2^P$ segments can revert spontaneously and also serve to provide a reactivated copy for transposition. (All the hitherto known strains carrying inactivated segments ⌊*P*⌉ or ⌊*Pm*⌉ are also summarized in Table 1.) Some residual activity is still observable, especially

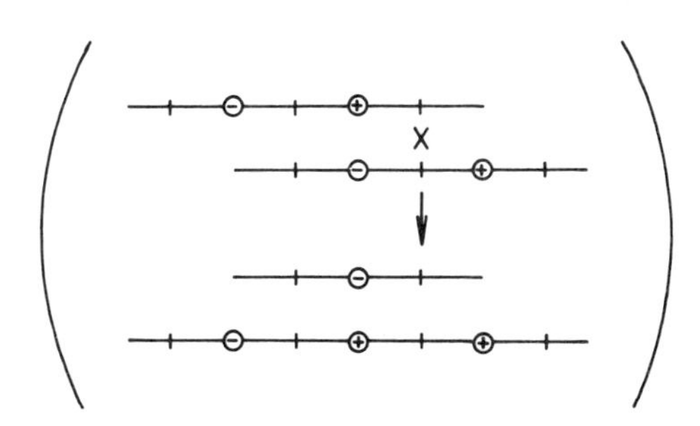

Figure 1
Hypothetic generation of a duplication of the ⊕ gene by a displaced crossover. Since the ⊖ gene to the left is unaffected by such an event, but known to be blocked whenever the ⊕ gene is duplicated, this explanation is refuted.

Figure 2

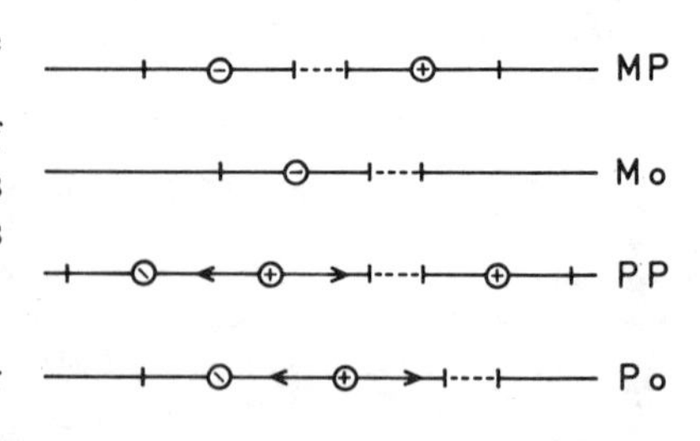

Genetic interpretation of the mating-type locus for the four standard strains of *S. pompe*. The mating-type locus is part of linkage group II; the centromere is located to the left. The abbreviations stand for the following genotypes: *M P*, for $mat1^M$ $mat2^P$; *P P*, for $mat1^M$::$mat2^P$ $mat2^P$; *M o*, for $mat1^M$ $mat2^o$; *P o*, for $mat1^M$::$mat2^P$ $mat2^o$.

$mat1^M$ *and* $mat2^P$ represent complementary mating-type genes; $mat2^o$ is devoid of ⊕ activity and believed to be derived by deletion; $mat1^M$::$mat2^P$ represents the insertion of a transposed copy of $mat2^P$ into $mat1^M$. After the transposition, all the information for the expression of ⊖ activity is still present in the genome, although it is disrupted by the inserted ⊕ gene; it can be fully restored when the inserted sequence is lost by precise excision. The first strain (*M P*) is homothallic; the others are heterothallic of one or the other mating type. The last strain (*P o*) can be obtained from (*P P*) and (*M o*) by recombination.

The mating-type segments M and P are drawn as being separated by a spacer sequence, devoid of functions related to mating type, to account for the 1% recombination observed between these genes--according to conventional genetic mapping. However, if all the recombination events of that region were initiated at a single specific site, the need for such a spacer would disappear. The more detailed model of Fig. 4 is based on this latter view.

in diploid strains, which can produce rare asci that still pass on the inactivated alleles to their progeny. Such changed alleles (as well as other mating-type mutants) can be detected by a distinctive iodine reaction that effectively stains sporulating areas of agar-plate cultures (Fig. 3).

Flip-Flop Control

Formally speaking, homothallic strains of *S. pombe* can produce spores in clonal cultures because they contain the genes for both mating-type activities. Yet single cells, even of homothallic strains, can only express either one or the other mating type. That notion has been substantiated with the aforementioned mutant $mat2^{Pm}$ in crosses with homothallic wild-type strains (Egel 1976a). Two classes of zygotes are produced by such crosses: sporulating and nonsporulating ones. That is exactly as should be expected if cells of both strains can activate either their *M* or their *P* gene (activity designated by "!"). A zygote produced by the configuration (*M P!*)/(*M! Pm*) is then able to sporulate, since only wild-type alleles are activated; conversely, in the combination (*M! P*)/(*M Pm!*), the activation of the mutated *Pm* allele impairs meiosis at such an early stage that these zygotes can be induced to resume diploid fission, but they do not sporulate. The fact that the second class exists at all actually shows that

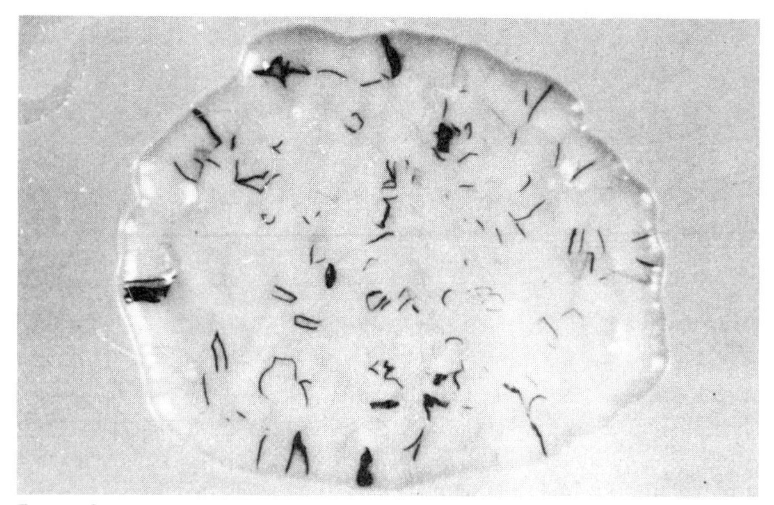

Figure 3
Detection of mating-type mutants. If colonies of *S. pombe* are exposed to vapors of iodine, sporulating areas are conspicuously stained (Gutz et al. 1974). The culture depicted here has grown up from a mixture of ⊖ cells and cells that had been selected for the recovery of ⊕ activity from a virtually inactivated state. Reaction lines, in different shades from faint to strong, developed wherever colonies of opposite mating activity had coalesced. Multiple degrees of reactivation are indicated by the different staining intensities.

generally, once the choice is made, it cannot be reversed in zygotes. Otherwise, the presence of the wild-type *P* allele would have allowed sporulation. During vegetative divisions in the diploid phase, the switch can be reset, since diploid colonies originating from zygotes of the second type are good sporulators again.

The postulated flip-flop has to operate fairly frequently in order to explain the readily observed conjugation of sister cells in homothallic strains (say more than once per ten cell divisions, although an accurate means of measuring this rate has not yet been developed).

The mutually exclusive activation of either $mat1^M$ or $mat2^P$ occurs only in *cis* position. In *trans* position (in diploid strains heterozygous for both mating-type genes), there is no repression in either direction. Hence the switch seems to be set at the level of DNA sequences. The flip-flop switch could no longer be demonstrated after a $mat2^P$ copy had been transposed and inserted into $mat1^M$ (apparently a crucial sequence is dismembered by the insertion). Such a conclusion, of course, could not be derived from a newly arisen *P P* segregant, since both of its *P* segments would be identical. Yet the strains *P Pm* and *Pm P* (see Table 1), which had been constructed as recombinants, did not show any signs of frequent switching. However, it is still an open question whether strains actually do exist which carry two simultaneously active *P* segments. (Since wild-type *P* is epistatic to the mutant *Pm*, the combinations mentioned are inappropriate for settling this issue unequivocally.)

These results indicate that the mating-type locus of fission yeast is indeed uniquely organized. Its regulation seems to be based on recombinational alterations of its sequence, and these rearrangements are precisely localized. The different kinds of genetic instabilities manifested at this locus are probably all interrelated. How such an interrelation might be envisioned is set forth in the following discussion.

Models and Speculations

The tidelands between model-building and unrestrained speculation are unsettled grounds. Yet sounding the foreseeable limitations of a given plot may at least serve to sharpen our conceptions and could even guide the design of decisive experiments.

The transposition of a $mat2^P$ copy to $mat1^M$, as suggested for *S. pombe*, results in a gene duplication. The repeats can be arranged in direct or inverse order, but both possibilities are not equally likely. Repeated sequences are inherently unstable, but differently so for the alternative orientations. A series of direct repeats can be extended or contracted by displaced recombination (see Fig. 1). Because this has not yet been observed for the duplicated ⊕ strain, insertion in reverse order is considered to be more likely (*P* and *P′* in Fig. 4). On the other hand, the spontaneous excision of $mat2^P$ suggests that this segment is bounded by direct repeats (c_1 and c_2 in Fig. 4).

An attractive feature of certain IS elements in bacteria (discussed extensively in other contributions to this book) is that they might possess a promoter signal only in one direction. An enzyme system capable of inverting such an element back and forth would provide a means of effecting a mutually exclusive expression of flanking genes: a type of "flip-flop" switch such as has long been postulated for various systems of alternately active genes.

The aforementioned *cis*-acting exclusion of $mat1^M$ activity when $mat2^P$ is expressed, and vice versa, strongly calls for structural alterations on the DNA level controlling homothallism in *S. pombe*. The flip-around of a regulatory element carrying a promoter signal (R and R′ in Fig. 4) would be an elegant mechanism for the production of the observed effects. Such a mode would require a site-specific recombination mechanism, probably acting within one strand. The well-defined hot spot of interstrand recombination in that region (Gutz et al. 1974) may be just a by-product of the promoter-inverting system. To be inverted, such a promoter element ought to be bounded by inversely repeated sequences (b and b′ in Fig. 4).

The available evidence suggests that both the flip-flop control of gene activity and the rate of localized mitotic recombination suffer from the transposition by which a $mat2^P$ segment is duplicated. Apparently, the specific site normally triggering these events is disrupted by the transposition. As depicted in Figure 4,

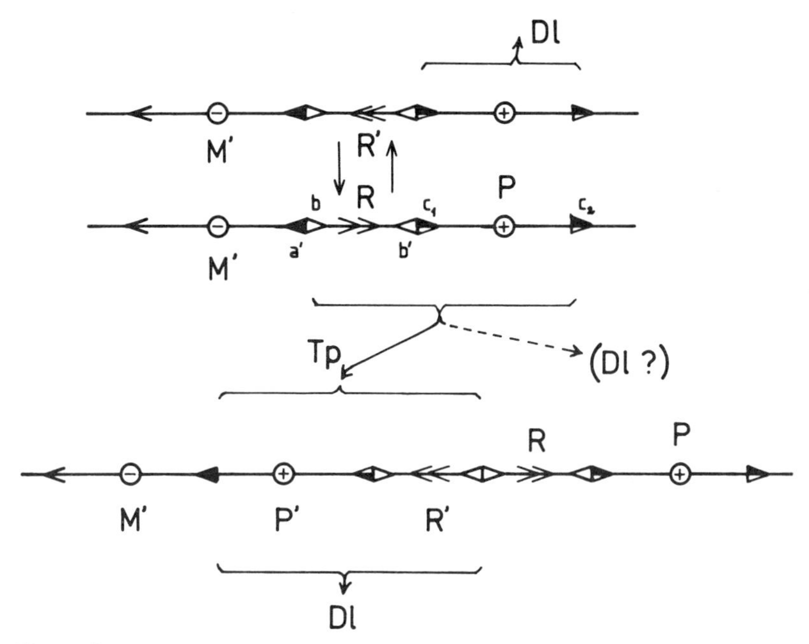

Figure 4
Conjectural view of flip-flop control and transposition of mating-type genes of *S. pombe* in terms of particular DNA segments. *M*, structural gene for ⊖ activity; *P*, structural gene for ⊕ activity; R, a regulatory element carrying a promoter signal; a, b, and c, segments of partial homology, involved in site-specific recombination. All segments with leftward polarity are indicated by a prime added to the original symbol. Dl, deletions to explain known mating-type mutations, or (Dl ?) presumed to be sterile if occurring; Tp, transposition to explain a ⊕-gene duplication that prohibits ⊖ expression by disconnecting the structural gene from its promoter element (of course, the ⊕ segment should carry a terminator signal to prevent read-through). Site-specific recombination should be induced most efficiently by the configuration $\overline{a'b}$, which is disrupted by the transposition.

the promoter element R might be duplicated along with a *P* transposition. This assumption was made in order to account for the observation that wild-type *P* segments could be expressed at either position (in strains *P Pm* and *Pm P* of Table 1).

Admittedly, Figure 4 is a speculative attempt to integrate the features mentioned into one coherent view. Presumably, the sequences a, b, and c are not identical, yet they could share sufficient homology to evoke occasional recombination (a′, b′, and c′ symbolize the same sequences in opposite orientation). The most efficient trigger of the proposed site-specific recombination mechanism should be configuration $\overline{a'b}$, which is adjacent to *mat1*^{M} when no insertion is present. Various deletions would occur because of partially homologous direct repeats. Some of these ought to be sterile if all promoters are deleted. Actually, sterile mutations are the most frequent class of mating-type change (cf. Bresch et al. 1968), although these have so far escaped genetic analyses.

Will this scheme also have room for the briefly mentioned "changes of state," where *mat2*^{P} segments are inactivated almost completely, but reversibly ⌊*P*⌉ or ⌊*Pm*⌉ of Table 1)? It might, if one of the promoter elements becomes inversely locked to the adjacent *mat2*^{P} gene, but in order to explain the residual ⊕ activity, a weak promoter, acting in the reverse directions, must be postulated.

CONCLUDING REMARKS

There is indeed strong evidence that rearrangements of DNA segments are involved in all the mating-type changes of *S. pombe,* and a final theory will probably not be entirely different from the views presented here. The details will of course remain open to discussion until more powerful techniques are available. Heteroduplex analyses (so extensively documented in this volume for bacterial systems) and cloning experiments aimed at isolating DNA fragments spanning the mating-locus will be needed if we are to further our understanding of mating-type changes.

The genetic data presently available suggest, in essence, that the mating-type locus in fission yeast (and budding yeast may not be too much different) contains a controlling element which can be studied in the context of its natural settings. This element still does its normal job and has not gone astray like its apparent counterparts in maize, which appear to have been set free by events so disruptive that they occasionally interfere with differentiation in a somewhat haphazard manner. Such a strictly local action as exemplified in the yeast system may be the prototype of naturally placed controlling elements in general.

SUMMARY

The mating-type locus in the fission yeast *S. pombe* is interpreted as containing a controlling element which directs a mutually exclusive expression of *cis*-positioned complementary mating-type genes in homothallic strains. A site-specific recombination system is postulated to effect the frequent alternation (flip-flop) between both possible modes: the expression of gene *mat1*^{M} to produce ⊖ activity in a particular cell or the expression of gene *mat2*^{P} to produce ⊕ activity. These functions, needed in conjugation and in meiosis, are the two mating types in this yeast. Both mating types coexist in homothallic cell lines, though not in individual haploid cells, and are separate in heterothallic strains. Three classes of interrelated phenomena, observed at lower frequencies than the flip-flop switching itself, are interpreted as by-products of the envisaged locally acting recombination system: mating-type mutations as excision or transposition events, changes of state that result in virtually complete but reversible inactivations, and a conspicuous hot spot of mitotic recombination.

Acknowledgments

Research support from the Deutsche Forschungsgemeinschaft (SFB 46) as well as a travel grant for attending the meeting at which this paper was presented (DFG 477/43/76) are both gratefully acknowledged.

REFERENCES

Bresch, C., G. Müller and R. Egel. 1968. Genes involved in meiosis and sporulation of a yeast. *Mol. Gen. Genet.* **102**:301.

Egel, R. 1976a. The genetic instabilities of the mating-type locus in fission yeast. *Mol. Gen. Genet.* **145**:281.

——. 1976b. Rearrangements at the mating-type locus in fission yeast. *Mol. Gen. Genet.* **148**:149.

Gutz, H. and F. J. Doe. 1973. Two different h^- mating types in *Schizosaccharomyces pombe. Genetics* **74**:563.

Gutz, H., H. Heslot, U. Leupold and N. Loprieno. 1974. *Schizosaccharomyces pombe.* In *Handbook of genetics* (ed. R. C. King), vol. 1, p. 395. Plenum Press, New York.

Leupold, U. 1959. Studies on recombination in *Schizosaccharomyces pombe. Cold Spring Harbor Symp. Quant. Biol.* **23**:161.

Meade, J. H. and H. Gutz. 1976. Mating-type mutations in *Schizosaccharomyces pombe*: Isolation of mutants and analysis of strains with an h^- or h^+ phenotype. *Genetics* **83**:259.

The Cassette Model of Mating-type Interconversion

J. B. Hicks,* J. N. Strathern and I. Herskowitz
Department of Biology, Institute of Molecular Biology
University of Oregon
Eugene, Oregon 97403

Studies of mating-type interconversion in the yeast *Saccharomyces cerevisiae* have led us to propose a new mechanism of gene control involving mobile genes. In this hypothesis, the "cassette model" of mating-type interconversion, cell type is determined by which of two blocs ("cassettes") of regulatory information is inserted into the mating-type locus. These blocs are ordinarily silent because they lack an essential controlling site (such as a promoter or ribosome binding site) and become expressed by placement next to a functional site. The cassettes themselves may be analogous to temperate prophages or transposable drug resistance elements, in which structural genes are flanked by sites governing their mobility. An interesting feature of the mating-type interconversion system is that a gene has been identified which controls the frequency with which these cassettes are activated. In this paper, we shall review briefly the aspects of mating and mating-type interconversion in *S. cerevisiae* which lead to the cassette model.

Mating Types, Sporulation, and the Mating-type Locus

The mating-type locus of *S. cerevisiae* exists in two states, *a* or α, which control the ability of yeast cells to mate and sporulate (for a review, see Mortimer and Hawthorne 1969). Cells that are *a* can mate efficiently with cells that are α, whereas cells of the same mating type mate with each other only rarely (see below). These two types of cells differ from each other in a number of respects; for example, each produces a pheromone (*a* factor by *a* cells and α factor by α cells) which acts specifically on cells of opposite mating type (Duntze et al. 1970; Wilkinson and Pringle 1974; V. L. MacKay, pers. comm.). Diploid *a*/α cells are a third cell type determined by the mating-type locus. These cells, but not *a*/*a* or α/α cells, can be induced to undergo meiosis and sporulation. Furthermore, *a*/α cells are unable to mate and do not produce or respond to the mating pheromones, unlike *a*/*a* and α/α diploids which behave like their respective haploids. MacKay and Manney (1974) have proposed that the mating-type locus

*Present address: Section of Genetics, Development, and Physiology, Cornell University, Ithaca, New York 14853.

specifies regulators of mating and sporulation genes located elsewhere in the genome.

It is clear that *a* and α alleles are distinct entities, as they are codominant–an *a*/α diploid differs from an *a*/*a* or α/α diploid. The *a* allele thus is not simply the absence of α, and the α allele is not simply the absence of *a.* Analysis of mutants with defects at the α mating-type locus indicates that the α mating-type locus specifies at least two functions not produced by the *a* mating-type locus (unpubl.). We view the mating-type loci as nonhomologous blocs of regulatory information, analogous in some respects to the imm^{434} and imm^{λ} regions of temperate phages. An alternative view (see discussion of the "flip-flop" model below) is that the determinants to be *a* or α are both present at the mating-type locus but only one is expressed.

Stable and Unstable Mating Types

Two kinds of *S. cerevisiae* strains can be distinguished on the basis of the stability of the mating-type behavior (Hawthorne 1963b; Mortimer and Hawthorne 1969; Takano and Oshima 1970; Hicks and Herskowitz 1976). In heterothallic strains, a change to opposite mating type is found at a frequency of approximately 10^{-6}. These cells are able to switch to the opposite mating type, again at a frequency of approximately 10^{-6}, and are indistinguishable from conventional laboratory heterothallic strains with respect to all phenotypes tested (see, for example, Hicks and Herskowitz 1977). In homothallic strains, changes to opposite mating types occur frequently, as often as every generation (Hicks and Herskowitz 1976 and unpubl.). These strains carry a dominant nuclear gene (*HO*), unlinked to the mating-type locus, which was introduced into *S. cerevisiae* by genetic crosses with *S. chevalieri* (Winge and Roberts 1949). *HO* cells that have sustained a change in mating type are fully capable of continuing to change mating type. However, when the *HO* gene is removed by genetic crosses, the new mating type is stable and indistinguishable from that in ordinary heterothallic strains. The continued presence of the *HO* gene is therefore not needed for maintenance of the new mating type. In addition, the determinant for the new mating type is allelic with the mating-type locus of heterothallic cells. Thus the *HO* gene acts to increase the frequency of mating-type interconversion and does so by changing the mating-type locus itself (Hawthorne 1963b; Takano and Oshima 1970; Hicks and Herskowitz 1976).

The "Flip-Flop" Model

The existence of mating-type interconversion indicates that all cells contain the information to be both *a* and α but that only one or the other information is expressed. D. C. Hawthorne (pers. comm. cited in Holliday and Pugh 1975) has proposed that the mating-type locus contains the regulators for both *a* and α and that DNA modification of a control site between these regulators determines which one is expressed. In a variation of this scheme, the control site–which, for example, directs RNA polymerase rightwards or leftwards–is flanked by reverse repeated DNA sequences (Fig. 1) (Brown 1976; Hicks and Herskowitz 1977). Intramolecular recombination at these sites would result in a change in the direction of transcription and thus of mating type. In this kind of flip-flop scheme, *HO* might control the synthesis of a DNA-modifying enzyme or a recombination enzyme.

The Recovery Experiment and the Cassette Model

Additional observations by D. C. Hawthorne (pers. comm.) and by us (Hicks and Herskowitz 1977) are not readily explained by the flip-flop hypothesis and have led us to propose a new model for mating-type interconversion. Both homothallic and heterothallic cells with a defect at the α mating-type locus can be converted to functional *a* cells–a result consistent with the flip-flop model. However, these *a* cells are then observed to switch to become functional α cells. In other words, a functional α mating-type locus can be restored through the mating-type interconversion process. We explain this recovery by proposing that yeast cells contain an additional copy (or copies) of the mating-type-locus information. Specifically, we propose that yeast cells contain a silent (unexpressed) copy of α information and a silent copy of *a* information and that the *HO* gene activates this information by inserting it (or a copy) into the mating-type locus (Fig. 2).

The Silent Copies

Genetic studies have revealed the existence of two loci in addition to *HO* which are necessary for mating-type interconversion (Naumov and Tolstorukov 1973; Harashima et al. 1974) and which we propose are the silent *a* and α information. *HMa* is necessary for conversion from *a* to α, and a separate locus, *HM*α, is necessary for conversion from α to *a*. The "inactive" alleles of *HMa* and *HM*α are denoted as *hma* and *hm*α, respectively. Thus *a HO hma HM*α and α *HO HMa hm*α have stable mating types, whereas *a HO HMa hm*α and α *HO hma HM*α are

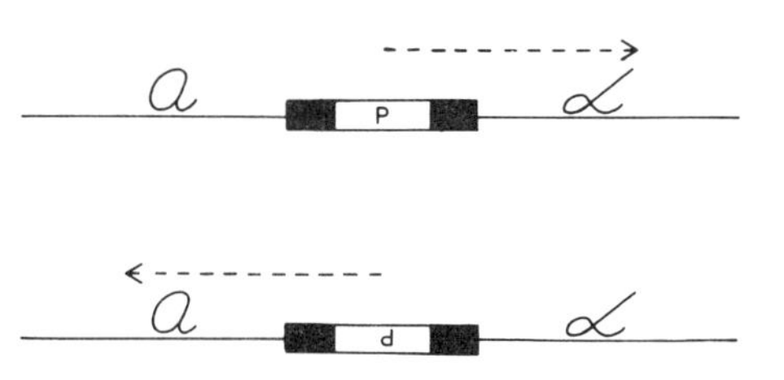

Figure 1
The "flip-flop" model. Diagram of the mating-type locus, which contains regulators for both *a*- and α-mating-type behavior and a regulatory site between these regulators. Black boxes are reverse repeated DNA sequences, which can recombine with each other to invert the promoter (*P*). Dashed lines indicate transcription.

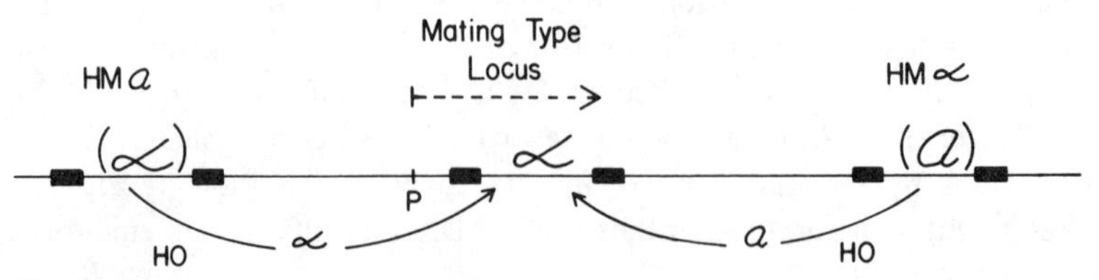

Figure 2
The "cassette" model. Diagram of chromosome III (not drawn to physical or genetic scale) showing the mating-type locus and *HMa* and *HMα* (which are loosely linked to the mating-type locus [Harashima and Oshima 1976]). Boxes represent sequences involved in the recombination event mediated by the *HO* gene. Action of *HO* results in insertion of *a* or α cassettes into the mating-type locus. The mating-type locus contains an essential site (shown here as a promoter, *P*) which allows expression of adjacent mating-type information.

observed to diploidize, that is, to switch mating type. The observation that *a HO hma hmα* and α *HO hma hmα* are both able to switch mating types suggests that *hma* and *hmα* are not simply "inactive" alleles of *HMa* and *HMα* but rather are equivalent to *HMα* and *HMa*, respectively (Naumov and Tolstorukov 1973; Harashima et al. 1974). These relationships are summarized in Table 1. We propose that *HMa* and *hmα* are α information and that *HMα* and *hma* are *a* information. Thus strain *a HO hma HMα* is equivalent to *a HO (a) (a),* and α *HO HMa hmα* is equivalent to α *HO* (α) (α). These strains have a stable mating-type behavior, not because cassette insertion fails to occur, but because cassette insertion does not change the information at the mating-type locus. Strains such as *a HO HMa hmα* [equivalent to *a HO* (α) (α)] contain silent information different from that at the mating-type locus. Insertion of this information into the mating-type locus leads to a change in mating type. In other words, we view *HMa, HMα, hma,* and *hmα* loci as being the substrates upon which the *HO* gene product (or a function under its control) acts.

Although a variety of predictions based on the cassette model have not yet been tested, this model provides a new explanation for an enigmatic observation of Hawthorne (1963a): i.e., heterothallic α cells are capable of switching to the opposite mating type by a deletion on the right arm of chromosome III. Since *HMα* is loosely linked to the mating-type locus on the right arm of chromosome III (Harashima and Oshima 1976; see Fig. 2 above), we propose that the deletion activates the silent *a* information by fusion to the mating-type-locus promoter.

Nonreciprocal Insertion

Before leaving the explicit statement of the model, we would like to point out an additional requirement for the model as it pertains to mating-type interconversion. Specifically, when mating types are interconverted in heterothallic or homothallic strains, the *HMa* and *HMα* loci are not disturbed. In other words,

Table 1
Role of *HMa, hma, HMα,* and *hmα* in Mating-type Interconversion

Genotype	Switch	Cassette model
a HMa HMα HO	yes	*a* (α) (a) *HO*
α HMa HMα HO	yes	α (α) (a) *HO*
a HMa hmα HO	yes	*a* (α) (α) *HO*
α HMa hmα HO	no	α (α) (α) *HO*
a hma HMα HO	no	*a* (a) (a) *HO*
α hma HMα HO	yes	α (a) (a) *HO*
a hma hmα HO	yes	*a* (a) (α) *HO*
α hma hmα HO	yes	α (a) (α) *HO*

Columns 1 and 2 are taken from Harashima et al. (1974) and from Naumov and Tolstorukov (1973). Column 3 translates *HM* and *hm* loci according to the rules of the cassette model: *HMa* and *hmα* are silent copies of α information, (α). *HMα* and *hma* are silent copies of *a* information, (a).

conversion from *a HO HMa HMα* to an α results in a cell of genotype *α HO HMa HMα* and not, for example, *α HO hma HMα*. The activation of silent α information of the *HMa* locus thus must not involve a trade of information between the mating-type locus and *HMa.* Rather, the information at *HMa* is passed unilaterally to the mating-type locus through a nonreciprocal recombination event or by copying of *HMa* followed by insertion.

To summarize, we propose that cell type in *S. cerevisiae* is regulated by a locus into which various blocs of information can be inserted. The mating-type locus is viewed as analogous to a playback head of a tape recorder which can give expression to whatever cassette of information is plugged into it. The means by which the silent information is moved to the mating-type locus seems likely to employ mechanisms discussed elsewhere in this volume (see, in particular, Botstein and Kleckner). One can readily speculate that determination of cell type in higher eukaryotic cells may also proceed by activation of genes via recombination as we have proposed for yeast.

SUMMARY

Haploid cells of the yeast *S. cerevisiae* can be of one of two cell types—either mating-type *a* or mating-type α. In some strains, mating-type behavior is stable, although changes occur at low frequency. In strains carrying the *HO* gene, mating-type behavior is unstable, changing as often as every cell division cycle. We propose that yeast cells contain a silent (unexpressed) copy of α information and a silent copy of *a* information, and that the *HO* gene activates this information by inserting it (or a copy) into the mating-type locus. These silent copies ("cassettes") may be analogous to temperate prophages or transposable drug resistance elements.

Acknowledgments

This work has been supported by Research, Program Project, and Training Grants from the National Institutes of Health. We give special thanks to Vivian MacKay and Thomas Manney for generously providing strains and to Donald Hawthorne for communicating unpublished results.

REFERENCES

Brown, S. W. 1976. A cross-over shunt model for alternate potentiation of yeast mating-type alleles. *J. Genet.* **62**:81.

Duntze, W., V. MacKay and T. R. Manney. 1970. *Saccharomyces cerevisiae*: A diffusible sex factor. *Science* **168**:1472.

Harashima, S. and Y. Oshima. 1976. Mapping of the homothallism genes, *HM* and *HMa,* in *Saccharomyces* yeasts. *Genetics* **84**:437.

Harashima, S., Y. Nogi and Y. Oshima. 1974. The genetic system controlling homothallism in *Saccharomyces* yeasts. *Genetics* **77**:639.

Hawthorne, D. C. 1963a. A deletion in yeast and its bearing on the structure of the mating type locus. *Genetics* **48**:1727.

——. 1963b. Directed mutation of the mating type alleles as an explanation of homothallism in yeast. *Proc. 11th Intern. Congr. Genet.* **1**:34. (Abstr.)

Hicks, J. B. and I. Herskowitz. 1976. Interconversion of yeast mating types. I. Direct observation of the action of the homothallism (*HO*) gene. *Genetics* **83**:245.

——. 1977. Interconversion of yeast mating types. II. Restoration of mating ability to sterile mutants in homothallic and heterothallic strains. *Genetics* (in press).

Holliday, R. and J. E. Pugh. 1975. DNA modification mechanisms and gene activity during development. *Science* **187**:226.

MacKay, V. and T. R. Manney. 1974. Mutations affecting sexual conjugation and related processes in *Saccharomyces cerevisiae.* II. Genetic analysis of nonmating mutants. *Genetics* **76**:273.

Mortimer, R. K. and D. C. Hawthorne. 1969. Yeast genetics. In *The yeasts* (ed. A. H. Rose and J. S. Harrison), vol. 1, p. 385. Academic Press, New York.

Naumov, G. I. and I. I. Tolstorukov. 1973. Comparative genetics of yeast. X. Reidentification of mutators of mating types in *Saccharomyces. Genetika* **9**:82.

Takano, I. and Y. Oshima. 1970. Mutational nature of an allele-specific conversion of the mating type of the homothallic gene *HOα* in *Saccharomyces. Genetics* **65**:421.

Wilkinson, L. E. and J. R. Pringle. 1974. Transient G1 arrest of *S. cerevisiae* cells of mating type α by a factor produced by cells of mating type *a. Exp. Cell Res.* **89**:175.

Winge, Ö. and C. Roberts. 1949. A gene for diploidization in yeast. *Compt. Rend. Trav. Lab. Carlsberg, Ser. Physiol.* **24**:341.

The Origin and Complexity of Inverted Repeat DNA Sequences in *Drosophila*

R. F. Baker and C. A. Thomas, Jr.
Department of Biological Chemistry
Harvard Medical School
Boston, Massachusetts 02115

Inverted repeat sequences have been found to exist in the DNA of all eukaryotes studied thus far, including *Drosophila* (Wilson and Thomas 1974; Cech and Hearst 1975; Schmid et al. 1975). Beyond physical studies showing the existence and size of hairpin DNA resulting from denaturation and concentration-independent renaturation of eukaryotic DNA, little is known about the genomic location, nucleotide sequences, and renaturation complexity of inverted repeat sequences. We report here on studies of these sequences in DNA from cultured Schneider II *D. melanogaster* cells.

During the course of isolation, the DNA was sheared to an average length of about 2200 base pairs. After denaturation in alkali, renaturation at C_0t below 10^{-3}, isolation of double-strand fragments by means of hydroxylapatite, and treatment with the single-strand-specific S1 nuclease, the percentage of DNA resistant to the enzyme was about 3%, and this was independent of the concentration of the renaturing DNA (Table 1). This feature is consistent with the idea that single strands reassociate with themselves by a unimolecular reaction.

Table 2 (line 1) shows that when the 2200-base-pair pieces of DNA were renatured at a maximum C_0t of 10^{-4}, approximately 6.5% to 7% of the starting DNA was at least partially double-stranded, as measured by retention on hydroxylapatite (HA). Second and third rounds (Table 2, line 2) of renaturation caused only a moderate reduction in this percentage. Table 2 (line 3) shows that when the HA fraction of DNA experiencing one round of denaturation was digested to limit with S1 nuclease, about 3% of the starting DNA remained trichloroacetic acid (TCA)-insoluble. When an additional round of fast renaturation was allowed prior to S1 nuclease treatment (Table 2, line 4), the average percentage of TCA-insoluble label was reduced from 3.26% to 2.36%. A third round of renaturation prior to S1 treatment (Table 2, line 5) reduced the recovery percentage by only a few tenths of a percent. Table 2 (line 6) shows that after limit digestion with S1 nuclease, about 85% of the S1-resistant DNA fails to renature rapidly, whereas approximately 90% of the DNA isolated by HA after two rounds of low C_0t renaturation, if not digested, will again renature rapidly. Thus it appears that S1 nuclease, under the conditions used here,

Table 1
Concentration Independence of S1-resistant Fraction

μg/ml	Time (min)	Maximum C_0t	Cpm input	Cpm × 20 resistant	Percent resistant
0.055	1	10^{-5}	6435	2760	2.15
0.22	1	4×10^{-5}	25,740	17,280	3.36
0.11	5	10^{-4}	12,870	7960	3.08
0.22	10	4×10^{-4}	25,740	15,160	2.94
0.55	10	10^{-3}	64,350	41,900	3.26

^{3}H-labeled DNA (sp. act. 1.17×10^5 cpm/μg) was isolated from cultured Schneider II cells (Hamer and Thomas 1975). This DNA was sheared to an average size of 2200 base pairs, denatured in 0.3 M NaOH and renatured in 0.15 M Na phosphate, pH 6.8, at 25°C. After batch mixing with hydroxylapatite (BioRad HTP), the DNA bound at 0.15 M and eluted at 0.4 M Na phosphate was collected, dialyzed against 0.1 M NaCl, precipitated at -20°C with 2 vol. ethanol, and digested to the limit with S1 nuclease. Enzyme treatment with S1 nuclease was in 0.25 M NaCl, 10^{-3} M $ZnSO_4$, 5×10^{-3} M Na acetate, pH 4.8, at 37°C. After S1 treatment, samples were precipitated in 3 ml 10% TCA along with 250 μg each of yeast RNA (Sigma) and bovine serum albumin. After 30 min at 0°C, the samples were centrifuged; the pellets were resuspended in cold 10% TCA and then recentrifuged. The pellets were dissolved in 1 ml of 1 M NaOH, mixed with 15 ml Aquasol (New England Nuclear), 1 ml H_2O, 0.5 ml 3 M acetic acid, and counted.

Table 2
Yield of Rapidly Renaturing DNA

Procedure[a]	Percentage of starting DNA	
	exp. 1	exp. 2
$(\text{Denature/renature–HA})_{1\times}$	6.51	6.95
$(\text{Denature/renature–HA})_{2\times}$	5.84	5.61
$(\text{Denature/renature–HA})_{1\times}$ S1	3.12	3.41
$(\text{Denature/renature–HA})_{2\times}$ S1	2.32	2.40
$(\text{Denature/renature–HA})_{2\times}$ $(\text{denature/renature})_{1\times}$ S1	2.07	2.16
$(\text{Denature/renature–HA})_{2\times}$ S1 $(\text{denature/renature})_{1\times}$ S1	0.34	0.28

[a]The procedure denoted within parentheses consists of denaturation, renaturation at low concentration of DNA, and fractional elution from hydroxylapatite to recover double-strand DNA. This procedure was carried out once (1×) or twice (2×) and was followed by digestion with S1 nuclease where indicated. The last two pairs of experiments involve a third renaturation test preceded or not by S1 digestion. The average size of the starting ^{3}H-labeled DNA was 2200 base pairs. Denaturations were in 0.3 M NaOH for 5 min at room temperature. Renaturations were in 0.15 M Na phosphate, pH 6.8, $C_0t \leqslant 10^{-4}$, at 25°C. S1-nuclease treatments and counting were as described in the notes to Table 1.

breaks the "turnaround" between the inverted repeats. This finding also demonstrates that the formation of the original S1-resistant structures could not have been a bimolecular event.

Bulk *Drosophila* DNA was fractionated on CsCl-actinomycin density gradients as shown in Figure 1. Fractions corresponding to nine zones across the gradient were pooled and tested for the fraction of S1-resistant and rapidly renaturing DNA. The results show that the fraction of S1-resistant sequences in main-band DNA was similar to that in unfractionated DNA. However, the zones containing the various satellite species were greatly reduced in S1-resistant DNA. We interpret these results to mean that most of the adjacent inverted repeat sequences in the *Drosophila* genome are found in euchromatic DNA, not in satellite DNA.

When S1-resistant DNA was electrophoresed on 2.4% agarose gels, most of the DNA migrated as a broad zone with an average length of approximately 100 base pairs. This length is independent of the extent of S1 digestion (Fig. 2).

Since the majority of the turnarounds have been broken by treating hairpin DNA with S1, we have been able to denature the S1-resistant DNA and perform

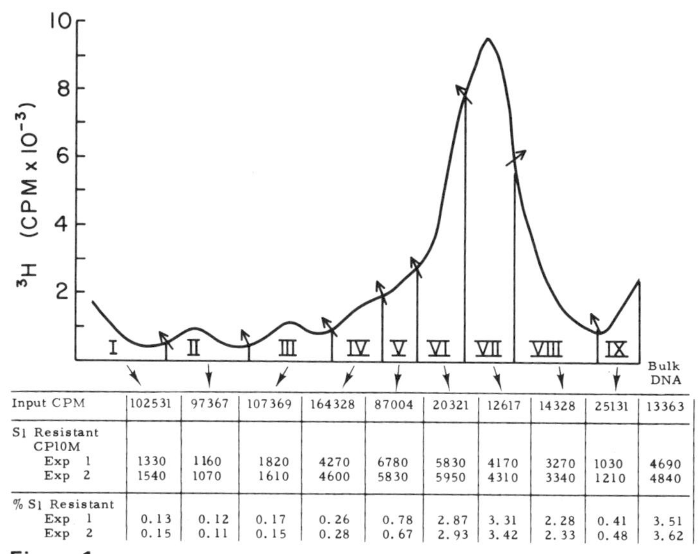

	I	II	III	IV	V	VI	VII	VIII	IX	Bulk DNA
Input CPM	102531	97367	107369	164328	87004	20321	12617	14328	25131	13363
S1 Resistant CP10M										
Exp 1	1330	1160	1820	4270	6780	5830	4170	3270	1030	4690
Exp 2	1540	1070	1610	4600	5830	5950	4310	3340	1210	4840
% S1 Resistant										
Exp 1	0.13	0.12	0.17	0.26	0.78	2.87	3.31	2.28	0.41	3.51
Exp 2	0.15	0.11	0.15	0.28	0.67	2.93	3.42	2.33	0.48	3.62

Figure 1

The percent of rapidly renaturing S1-resistant DNA in various CsCl-actinomycin fractions of *Drosophila* DNA. Bulk ^{3}H-labeled DNA was separated into 34 fractions after centrifugation in CsCl-actinomycin and pooled as shown (Peacock et al. 1974; Hamer and Thomas 1975). Each pool was tested for the fraction of rapidly renaturing S1-resistant DNA as described in the notes to Table 1. The results for two separate experiments are shown beneath the profile, along with the same test for unfractionated DNA.

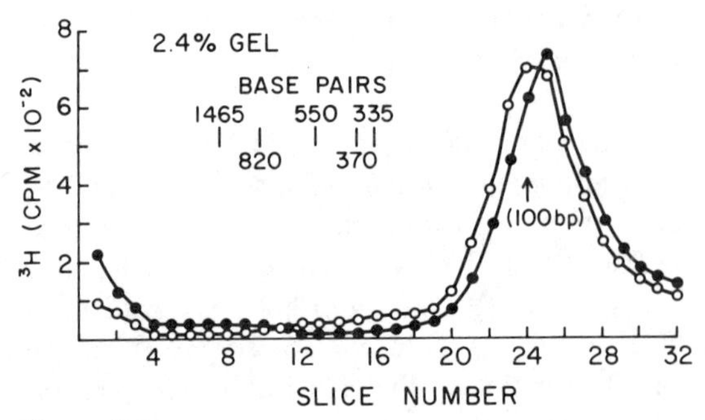

Figure 2
The size distribution of rapidly renaturing S1-resistant *Drosophila* DNA. S1-resistant ^{3}H-labeled DNA was prepared as described in the notes to Table 1. After S1 treatment either to 37% TCA-soluble or to the limit of digestion (95% TCA-soluble), the reaction mixture was phenol-extracted and dialyzed against electrophoresis buffer. Electrophoresis was performed in cyclindrical (8.5 X 0.6 cm) 2.4% agarose gels using the buffer and conditions described by Hamer and Thomas (1975). Length markers were prepared by cleaving SV40 DNA with restriction enzyme *Hae*III. (–○–) 37% TCA-soluble; (–●–) limit of S1 digestion.

renaturation experiments. In these experiments, a background of S1-resistant DNA (approximately 0.3% of bulk, unfractionated DNA, see Table 2), which will not remain denatured at a low C_0t, has been subtracted. This background could represent hairpin DNA whose turnarounds were not broken by S1 nuclease.

Self-driven renaturation of these DNA pieces occurs with a $C_0t_{1/2}$ of 2.0 (Fig. 3A), as compared to *Escherichia coli* DNA which renatures under the same conditions with a $C_0t_{1/2}$ of 8.3. If *E. coli* has a sequence complexity of 4.5×10^6 base pairs, the complexity of the *Drosophila* S1-resistant sequences is calculated to be 4.8×10^5 base pairs. This calculation includes a correction of $(100/500)^{1/2}$ since the sonicated fragments are longer than the S1-resistant segments (Wetmur and Davidson 1968). Figure 3B shows the degree of renaturation at different C_0t's for labeled S1-resistant DNA in the presence of various concentrations of unlabeled S1-resistant DNA. There is an inverse relationship between the rate at which a population of DNA molecules will renature and the sequence complexity of the population (Wetmur and Davidson 1968). Taking *E. coli* DNA as our standard, we estimate from its initial rate of renaturation (Fig. 3B) that the sequence complexity of the S1-resistant *Drosophila* sequences is 5.1×10^5 base pairs.

Schmid et al. (1975) have measured the kinetic composition of *Drosophila* foldback DNA collected on HA. They found all of the frequency components (fractions with different renaturation rates) that were present in unfractionated *Drosophila* DNA, with a slight enrichment for middle-repetitive sequences.

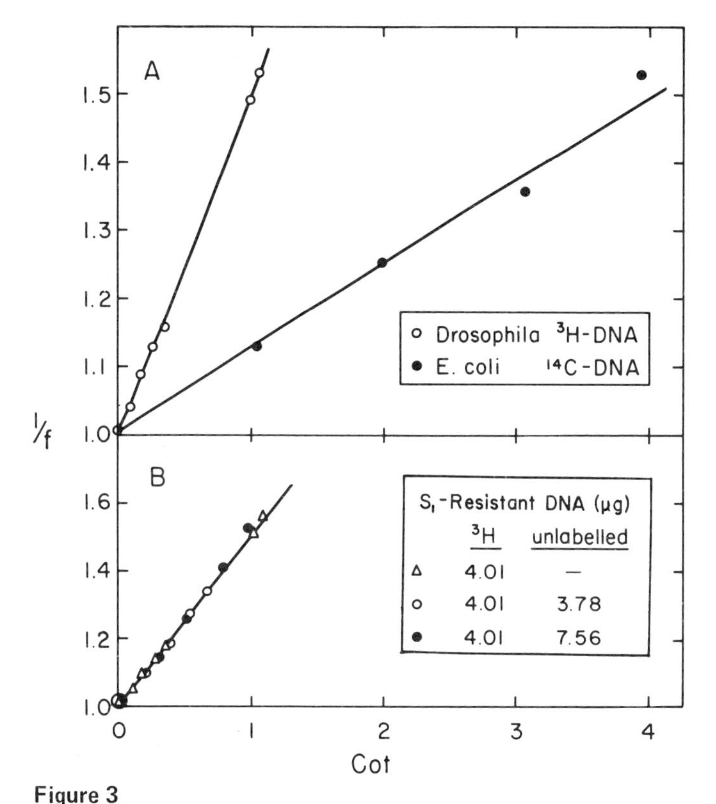

Figure 3
Renaturation of rapidly renaturing S1-resistant *Drosophila* DNA. S1-resistant labeled and unlabeled DNA were prepared as described in the notes to Table 1. After S1 treatment to the limit of digestion, the reaction mixture was phenol-extracted and then dialyzed to 0.18 M NaCl, 10^{-3} M Tris, pH 7.2. Renaturations were at 68°C. (*A*) Renaturation of S1-resistant *Drosophila* ^{3}H-labeled DNA mixed with sonicated (500-base-pair) *E. coli* ^{14}C-labeled DNA as an internal renaturing standard. (*B*) Renaturation of ^{3}H-labeled S1-resistant DNA in the presence of increasing quantities of S1-resistant DNA. The ordinate is the reciprocal of the fraction of DNA that remains S1-sensitive.

After treatment of the HA-derived foldback DNA with single-strand-specific mung bean endonuclease, they found a considerable enrichment for middle-repetitive components. Their results are consistent with our findings that, over a partial renaturation range (Fig. 3), S1-resistant pieces renature as middle-repetitive sequences.

The S1-resistant DNA pieces are about 100 base pairs long. If their complexity is 5×10^5 base pairs, then there are 5000 different kinds of 100-base-pair-long segments. The fraction of S1-resistant DNA in the main-band portion of the CsCl-actinomycin gradient is about 3.4% of the total. If the total euchromatic *C* value is 1.5×10^8 base pairs, this would correspond to 5.1×10^6 base pairs, or ten times the kinetic complexity. Thus each of the 5000 different 100-base-pair-long sequences must be represented ten times.

How are these sequences arranged with respect to each other? Schmid et al. (1975) have presented evidence that the hairpins are very heterogeneous in length, with a weight average of 5000 base pairs. Dividing this number into the amount of S1-resistant DNA (5.1×10^6) means that there are only about 1000 such sequences in the *Drosophila* genome, spaced at distances of $1.5 \times 10^8/10^3$ or 1.5×10^5 base pairs, or 50 μ, or one every five bands in the polytene chromosome. This model is in accord with the findings of Wilson and Thomas (1974). To accommodate the present data, one must assume that each hairpin seen by Schmid et al. (1975) must consist of 50 S1-resistant segments and that this number is very broadly distributed, even though segment length (100 base pairs) is not.

In contrast, consider the model based on the assumption that 100-base-pair hairpins are distributed uniformly throughout the euchromatic DNA to make a total of 5.1×10^6 base pairs. There must be 5×10^4 of them spaced at intervals of $1.5 \times 10^8/5 \times 10^4$ or 3000 base pairs. Clearly, this would conflict with Figure 2 of Wilson and Thomas (1974) and would not agree with the results of Schmid et al. (1975).

Thus it appears likely that the 100-base-pair-long segments, which are only of 5000 different types, are derived from hairpins that are 10 to 50 times longer than they.

SUMMARY

Adjacent inverted repeat DNA sequences from cultured Schneider II *D. melanogaster* cells were isolated as rapidly renaturing DNA hairpins. S1-nuclease-resistant pieces of hairpins constitute about 3.4% of the euchromatic genome and have an average size of about 100 base pairs. Self-driven renaturation of these S1-resistant pieces occurs with a $C_0t_{1/2}$ of 2.0, corresponding to a kinetic complexity of 5×10^5 base pairs. This indicates that there are about ten copies of each of 5000 different sequences in the S1-resistant hairpin DNA of *Drosophila.* The arrangement of S1-resistant segments within adjacent inverted repeat DNA sequences is discussed.

REFERENCES

Cech, T. R. and J. E. Hearst. 1975. An electron microscopic study of mouse foldback DNA. *Cell* 5:429.

Hamer, D. H. and C. A. Thomas, Jr. 1975. The cleavage of *Drosophila melanogaster* DNA by restriction endonucleases. *Chromosoma* 49:243.

Peacock, W. J., D. Brutlag, E. Goldring, R. Appels, C. W. Hinton and D. L. Lindsley. 1974. The organization of highly repeated DNA sequences in *Drosophila melanogaster* chromosomes. *Cold Spring Harbor Symp. Quant. Biol.* **38**:405.

Schmid, C. W., J. E. Manning and N. Davidson. 1975. Inverted repeat sequences in the *Drosophila* genome. *Cell* **5**:159.

Wetmur, J. G. and N. Davidson. 1968. Kinetics of renaturation of DNA. *J. Mol. Biol.* **31**:349.

Wilson, D. A. and C. A. Thomas, Jr. 1974. Palindromes in chromosomes. *J. Mol. Biol.* **84**:115.

Nucleotide Sequence Arrangements in the Genome of Herpes Simplex Virus and Their Relation to Insertion Elements

W. C. Summers and J. Skare
Departments of Therapeutic Radiology and Molecular Biophysics
and Biochemistry and Human Genetics
Yale University School of Medicine
New Haven, Connecticut 06510

Herpes simplex virus is a common virus that infects a high proportion of the human population and gives rise, in many cases, to persistent latent infections. Recent studies on the structure of the double-stranded DNA of the genome of herpes simplex virus type 1 (HSV-1) suggest that this genome has certain striking similarities to insertion elements as known in prokaryotic organisms (Sheldrick and Berthelot 1975; Skare 1976). The results of some of these studies which have been carried out in our laboratory will be summarized here.

The most unusual feature of the HSV-1 genome is the presence of two pairs of inverted repeat sequences. This fact was first reported by Sheldrick and Berthelot (1975) when they examined self-annealed single strands of HSV-1 DNA. These workers observed that both ends of the DNA could anneal with an internal region at a position about 80% of the length of the DNA molecule from one end. This result is shown schematically in Figure 1. They recognized the possibility of intramolecular or intermolecular recombination and noted that the order of genes in the long segment (L) could become inverted relative to the short segment (S) by recombination between the inverted repeat regions.

Both work from our laboratory (Skare et al. 1975; Wagner et al. 1975) and that of others (Hayward et al. 1975; Clements et al. 1976) showed that restriction endonuclease cleavage of HSV-1 DNA generated DNA fragment patterns of unexpected complexity. First, limit digests contain fragments present in less than molar yield. Second, the sum of the molecular weights of the fragments is greater than the well-established molecular weight of the intact genome (about 145 kilobases). These two observations, taken together with the data of Sheldrick and Berthelot (1975), suggested to us and to others that the restriction cleavage patterns could be interpreted in terms of molecular heterogeneity resulting from independent inversions of the L and S segments. Thus a DNA preparation from cloned virus stocks contains approximately equimolar amounts of four sequence isomers: L–S, inverted L–S, L–inverted S, and inverted L–inverted S. Although this model was consistent with the molar yield data and

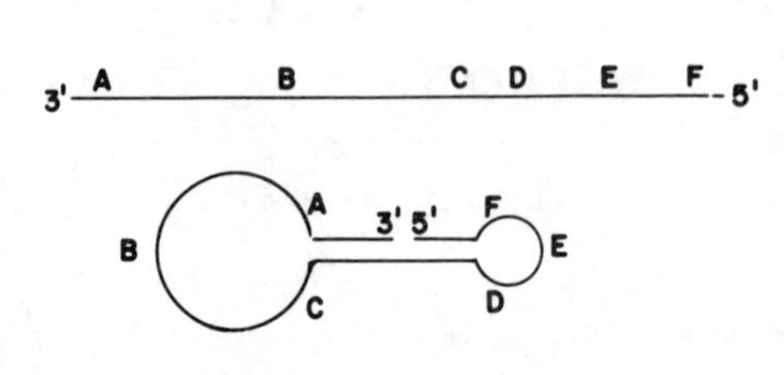

Figure 1
Diagrammatic representation of self-annealed single strand of HSV-1 DNA (after Sheldrick and Berthelot 1975). Both ends anneal at adjacent internal sites to create duplex DNA, as seen by electron microscopy. The repeated region is about 9 kb on the long (L) end and 6.5 kb on the short (S) end (Wadsworth et al. 1975).

with the tentative identification of the terminal fragments, definitive proof for the exact nature of the sequence heterogeneity was lacking. Rigorous confirmation of the model just described has recently been obtained in our laboratory and elsewhere through the complete mapping of restriction endonuclease cleavage sites on HSV-1 DNA.

Endonucleases *Eco*RI (Greene et al. 1974), *Hin*dIII (Smith 1974), and *Xba*I (S. Zain et al., pers. comm.) have been used to cleave the DNA of the KOS strain of HSV-1. The fragments were separated by electrophoresis in 0.3% and 0.5% agarose gels as described previously (Skare et al. 1975; J. Skare and W. C. Summers, in prep.). Individual fragments produced by *Eco*RI were redigested with *Hin*dIII after recovery from the agarose gel, and *Hin*dIII fragments were likewise cleaved by *Eco*RI. DNA fragments produced by *Xba*I were also cleaved with *Eco*RI and *Hin*dIII in separate experiments. In some cases, incomplete digestion products of *Eco*RI were isolated and recleaved with *Eco*RI or *Hin*dIII. The logic of the construction of the cleavage-site maps shown in Figure 2 has been presented elsewhere (Skare 1976; J. Skare and W. C. Summers, in prep.). These maps were deduced from knowledge of primary and secondary cleavage data and from the fact that the length of the genome is about 145 kb (Kieff et al. 1971). There was no reference to the model to decide between possible alternative maps, nor were the observed molar yields employed. Thus the maps were deduced in a way that was logically independent of the model and therefore constitute evidence for the validity of the segment inversion model.

Note that the observed heterogeneity predicts that some fragments should be present in only some of the molecules in the population (e.g., X1a, X3a, etc.). Also, the sum of all fragments in the digest will be greater than the length of the intact genome.

In addition to the gross heterogeneity resulting from the L–S inversions, a microheterogeneity has been observed (Skare et al. 1975) which occurs at the termini of the L segment. When DNA fragments that include the left (L) end of the molecule or the region spanning the L–S joint are examined by high-resolution electrophoresis, a heterogeneity of about 0.3 kb is seen. This heterogeneity is usually manifest as a "fuzzy" band on the standard electropherogram.

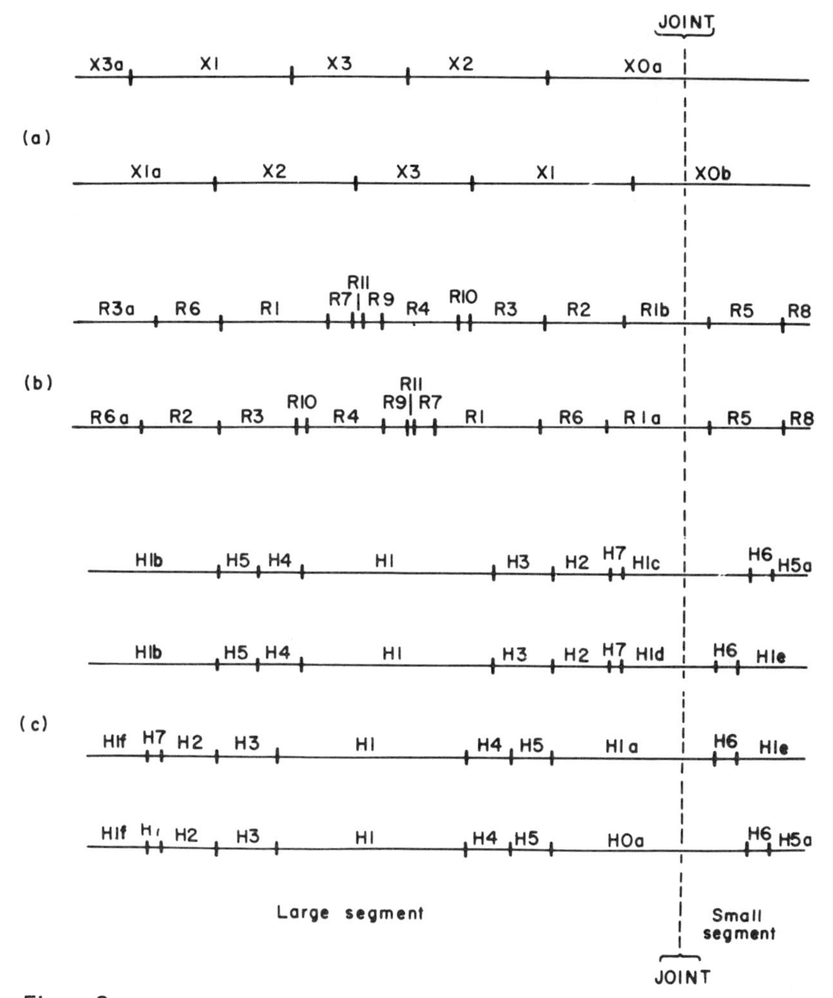

Figure 2

Restriction nuclease cleavage maps of DNA from the KOS strain of HSV-1. Nucleases *Xba*I, *Eco*RI, and *Hin*dIII were used to generate fragments designated X (*a*), R (*b*), and H (*c*). Joint refers to the fusion point of the large and small segments. Fragments are numbered in order of increasing electrophoretic mobility, with the added condition that only fragments present in molar yield receive a single number; submolar fragments are designated by adding a letter to the name of the next fastest molar fragment. All four sequence isomers are distinguishable by *Hin*dIII cleavage, whereas only two maps are distinguishable by *Eco*RI and *Xba*I. This results from the symmetrical *Eco*RI cleavage in the inverted repeat of the S segment and from the absence of *Xba*I cleavage sites in the S segment. Maps are drawn to molecular weight scale.

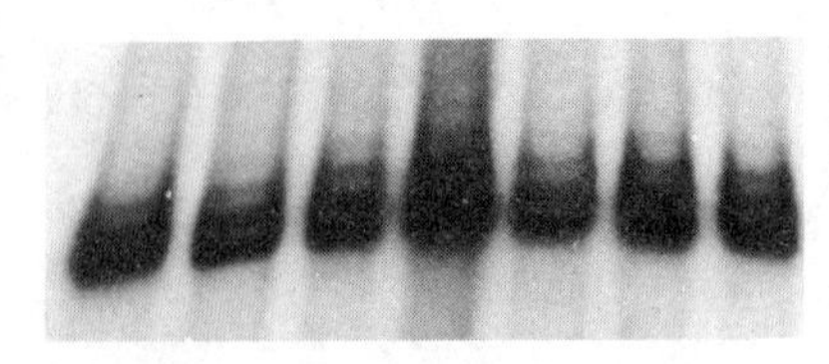

Figure 3
Gel electropherogram of band containing HSV-1 DNA fragment X3a (10.5 kb) and homologous fragments of increasing molecular weight. Increments are approximately 0.3 kb. Results from seven different clones of strain KOS are shown.

The electropherogram of fragment X3a in Figure 3 clearly shows that this band contains a collection of discrete subclasses of DNA differing by the addition of about 0.3 kb. Recleavage of DNA from such bands with other restriction nucleases shows that all of the molecules in the subclasses are related, rather than resulting from the coincidental superposition of two quite unrelated fragments. The subclasses with no addition and with one addition seem to be almost equal and account for more than 90% of the total DNA in each heterogeneous group. The observation that such heterogeneity occurs both at the left end and in the internal region at the juncture of the L and S segments is entirely compatible with, and predicted by, the segment inversion model. If heterogeneity is generated in one end of the L segment, it should appear at the other end after randomization by segment inversion.

The origin of the subterminal microheterogeneity is unclear, but it may be related to the observation of Hyman et al. (1976) that very small inverted repeats of the terminal sequence exist within about 500 base pairs of at least one end of the molecule. Such a structure could result in insertions of discrete DNA segments to yield a homologous series of molecules such as we have observed.

The existence of terminal inverted repeat regions and of segment inversion in the DNA of HSV-1 present clear analogies to other systems discussed in this volume, such as transposable elements and the G-region inversions in bacteriophage Mu. Table 1 lists several properties of HSV-1 and certain other herpesviruses which may be indicative of this analogy.

The origin of the sequence isomers of HSV is not clear. Since the DNA populations that have been analyzed are the product of at least four or five growth cycles after starting with a single virion, the generation of the four isomers from one isomer must take place within that period of growth. Simple recombinational mechanisms may be sufficient or a special site or sequence-specific mechanism may be involved. Several replicational models have also been proposed (Skare 1976).

SUMMARY

As shown by restriction endonuclease cleavage-site mapping, the genome of HSV-1 consists of two unique sequences flanked by inverted repetitions. The orientation of the two unique regions relative to each other is randomized in

Table 1
IS-like Characteristics of Herpesviruses

Characteristic	Organism	Reference
Inverted repetitions (IR) (5–10 kb)	HSV	Sheldrick and Berthelot (1975)
	?EBV	Sugden et al. (1976)
Two invertable segments bounded by IRs	HSV	this work
Symmetric RNAs suggestive of inverted control elements	HSV	Jacquemont and Roizman (1975)
Integration into host DNA	EBV	Adams and Lindahl (1974)
Terminal redundancy (0.7 kb)	HSV	Grafstrom et al. (1975)
Interrupted palindrome (?) at one end (0.5 kb)	HSV	Hyman et al. (1976)
Multiple insertions in terminal IR region	HSV	this work

stocks of virus grown from a single parental virion. Thus viral DNA stocks contain equal amounts of the four possible sequence isomers. Additional heterogeneity occurs at the ends of one of the unique segments and may be related to the reported existence of small inverted repeats within the large inverted repetitions.

Acknowledgments

We thank Dr. Wilma P. Summers and Ms. Liu-Mei Shih for their generous help with cell and virus culture and Dr. S. Zain for communication of her results on endonuclease *Xba*I. This work was supported by grants CA 06519 and CA 16038 from the National Cancer Institute. W.C.S. is a Faculty Research Awardee of the American Cancer Society (PRA 97).

REFERENCES

Adams, A. and T. Lindahl. 1975. Epstein-Barr virus genomes with properties of circular DNA molecules in carrier cells. *Proc. Nat. Acad. Sci.* **72**:1477.

Clements, J. B., R. Cortini and N. M. Wilkie. 1976. Analysis of herpesvirus substructure by means of restriction endonucleases. *J. Gen. Virol.* **30**:243.

Grafstrom, R. A., J. C. Alwine, W. L. Steinhart and C. W. Hill. 1975. Terminal repetitions in herpes simplex virus type 1 DNA. *Cold Spring Harbor Symp. Quant. Biol.* **39**:676.

Greene, P. J., M. C. Betlach, H. W. Boyer and H. M. Goodman. 1974. The *Eco*RI restriction endonuclease. In *DNA replication* (ed. R. B. Wickner), p. 87. Marcel Dekker, New York.

Hayward, G. S., N. Frenkel and B. Roizman. 1975. Anatomy of herpes simplex virus DNA: Strain differences and heterogeneity in the location of restriction endonuclease cleavage sites. *Proc. Nat. Acad. Sci.* **72**:1768.

Hyman, R. W., S. Burke and L. Kudler. 1976. A nearby inverted repeat of the terminal sequence of herpes simplex virus DNA. *Biochem. Biophys. Res. Comm.* **68**:609.

Jacquemont, B. and B. Roizman. 1975. RNA synthesis in cells infected with herpes simplex virus. X. Properties of viral symmetric transcripts and of double-stranded RNA prepared from them. *J. Virol.* **15**:707.

Kieff, E. D., S. L. Bachenheimer and B. Roizman. 1971. Size, composition and structure of the DNA of subtypes 1 and 2 of herpes simplex viruses. *J. Virol.* **8**:125.

Sheldrick, P. and N. Berthelot. 1975. Inverted repetitions in the chromosome of herpes simplex virus. *Cold Spring Harbor Symp. Quant. Biol.* **39**:667.

Skare, J. 1976. Endonuclease *Eco*RI, *Xba*I and *Hin*dIII cleavage sites on herpes simplex virus type 1 DNA. Ph.D. thesis, Yale University, New Haven, Conn.

Skare, J., W. P. Summers and W. C. Summers. 1975. Structure and function of herpesvirus genomes. I. Comparison of five HSV-1 and two HSV-2 strains by cleavage of their DNA with *Eco*RI restriction endonuclease. *J. Virol.* **15**:726.

Smith, H. O. 1974. Restriction endonuclease from *Hemophilus influenzae* Rd. In *DNA replication* (ed. R. B. Wickner), p. 71. Marcel Dekker, New York.

Sugden, B., W. C. Summers and G. Klein. 1976. Nucleic acid renaturation and restriction endonuclease cleavage analyses show that the DNAs of a transforming and a nontransforming strain of Epstein-Barr virus share approximately 90% of their nucleotide sequences. *J. Virol.* **18**:765.

Wadsworth, S. C., R. J. Jacob and B. Roizman. 1975. Anatomy of herpes simplex virus DNA. II. Size, composition and arrangement of inverted terminal repetitions. *J. Virol.* **15**:1487.

Wagner, M., J. Skare and W. C. Summers. 1975. Analysis of DNA of defective herpes simplex virus type 1 by restriction endonuclease cleavage and nucleic acid hybridization. *Cold Spring Harbor Symp. Quant. Biol.* **39**:683.

The Structure of the Adeno-associated Virus Genome

B. J. Carter, L. M. de la Maza and F. T. Jay
Laboratory of Experimental Pathology
National Institute of Arthritis, Metabolism, and Digestive Diseases
National Institutes of Health
Bethesda, Maryland 20014

The adeno-associated viruses (AAV) are defective parvoviruses that grow only in host cells in which additional helper functions are supplied by a coinfecting adenovirus. The precise mechanism of AAV defectiveness is not known.

AAV Genome Orientation

Single AAV DNA strands approximately 4200 nucleotides long (1.4×10^6 daltons) of both plus and minus polarity are separately encapsidated into individual virions in approximately equal amounts (Mayor et al. 1969; Rose et al. 1969). Following purification from virions, the plus and minus AAV DNA strands can anneal to form a variety of duplex monomer and oligomer molecules having either a linear or circular configuration (Koczot et al. 1973; Gerry et al. 1973). Cleavage of AAV-2 DNA duplexes with several restriction endonucleases yields specific fragments that have been mapped on the linear AAV genome (Carter and Khoury 1975; Berns et al. 1975b; Carter et al. 1976) (Fig. 1). These maps were oriented relative to the DNA strand polarity by labeling the 5′ termini of duplex AAV-2 DNA with ^{32}P (Carter et al. 1975) and then cleaving with R·*Eco*RI. The terminal fragments A and B containing the 5′-terminal label were then separated into component strands in CsCl density gradients. The resultant strand polarity determined is shown in Figure 1. The *Eco*RI B fragment was arbitrarily defined as the left molecular end and the RI A fragment as the right molecular end.

Terminal Repetitions in AAV DNA

The nucleotide sequences at the termini of AAV DNA strands exhibit several intriguing properties. Hydroxylapatite chromatography of separated single strands of AAV DNA revealed the presence of self-complementary regions (Carter et al. 1972). Electron microscopy showed that these regions were inverted repetitions at the 5′ and 3′ termini (Koczot et al. 1973). Single AAV

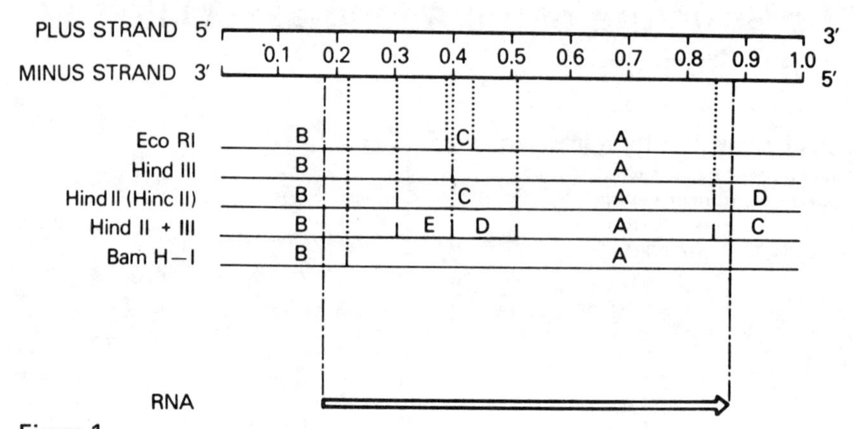

Figure 1
Schematic representation of restriction endonuclease fragment maps and stable RNA transcription map of AAV-2 DNA. The scale represents 1 map unit (genome length) equivalent to approximately 4200 base pairs. The arrow indicates the direction and extent of transcription of the single stable AAV mRNA from the DNA minus strand. This mRNA probably specifies the AAV capsid protein. (Reprinted, with permission, from Carter et al. 1976.)

DNA strands can circularize by annealing of these terminal sequences to form a duplex, hydrogen-bonded "panhandle" equivalent in size to 1% to 2.5% of the total strand length (Berns and Kelly 1974).

Linear duplex AAV-2 DNA that was degraded to a very limited extent by exonuclease III could anneal to form circles by virtue of cohesive ends created by the nuclease digestion (Gerry et al. 1973). This reveals the presence of regular (noninverted) repetitious sequences at or near the duplex termini.

Permutation of AAV DNA Strand Sequence

Duplex molecules of AAV DNA consist of 15S linear monomers and linear and circular oligomers (Koczot et al. 1973; Gerry et al. 1973). The oligomeric molecules can be produced by denaturing and reannealing linear monomers. Also, the oligomeric molecules can be melted to monomer molecules by heating under conditions expected to dissociate cohesive ends (Gerry et al. 1973; Carter and Khoury 1975).

These observations have been interpreted as reflecting the presence of a permutation in the AAV DNA strand sequence. This permutation is limited to a small region of the genome because: (1) the limit product obtained from exonuclease III digestion of AAV DNA duplex does not reassociate (Gerry et al. 1973), and (2) cleavage of AAV DNA with endonuclease *Eco*RI yields three unique fragments of defined size (Carter and Khoury 1975).

Properties of Terminal Fragments of AAV-2 DNA Produced by Cleavage with Restriction Endonucleases

Cleavage of AAV DNA duplex with *Hin*dII+III (Berns et al. 1975b) yielded terminal fragments B and C (Fig. 1) which could each be resolved into two components (Fig. 2). For each fragment, the apparent difference in size of the two components was equivalent to approximately 1% of the AAV genome (i.e., about 40 nucleotides), which was interpreted as evidence for two permutations in AAV DNA strands (Berns et al. 1975b). The appearance of the terminal fragments as doublets is a consistent property of fragments obtained with additional restriction enzymes, such as *Hae*III (Denhardt et al. 1976), *Bam*HI (Carter et al. 1976), *Hae*II, *Pst*I, and *Kpn*I as summarized in Figure 2.

A restriction enzyme which cuts within the terminal inverted repetitions should yield the same terminal fragments from each end of the duplex. Cleavage of AAV duplex DNA, containing a 5′-terminal ^{32}P-label, with *Hae*III endonuclease yields the same two ^{32}P-labeled fragments, designated α and γ (see Fig. 2), from both the left and right molecular termini (Denhardt et al. 1976). These two fragments correspond to the terminal doublet seen with other enzymes and further indicate that *Hae*III cleaves within the inverted repetitions. The ratio of 5′ label in components α and γ is usually about 1:2. In some end-labeled DNA preparations digested with *Hae*III (Denhardt et al. 1976), a third minor component (β) is observed, but this appears to arise partly from denaturation of the α component and partly from incomplete cleavage (L. de la Maza and B. Carter, unpubl.). Since *Hae*III cleaves within the terminal inverted repetitions of AAV DNA, it should also cleave the duplex panhandle regions of purified single strands. *Hae*III cleavage of 5′-^{32}P end-labeled, purified, single AAV strands yields the same terminally labeled fragments as from the left and right ends of linear duplexes (Denhardt et al. 1976). These experiments reveal two important features of AAV DNA. (1) If there are two permutations of AAV strands, then each apparently contains an inverted terminal repetition. (2) The duplex panhandles of single strands contain the same terminal duplex structures as those present at both ends of linear duplexes.

The apparent molecular weights of the γ and β components were consistently about 40 and 75 nucleotides, respectively, as measured on acrylamide gels of varying concentrations. The *Hae*III α component has an unusual property (Denhardt et al. 1976) in that its apparent molecular weight preferentially increases as the acrylamide concentration of the gel is increased. It was proposed, therefore, that the α component may have a branched, nonlinear duplex structure such as a "rabbit-ear" (Fig. 3C). Formation of a "rabbit-ear" structure implies that the population of single strands which exhibits this feature must contain terminal sequences that are not complete inverted repetitions (i.e., that are not fully complementary to each other), as indicated in Figure 3C. The variation in sequence within the terminal repetitions would be required to stabilize a "rabbit-ear" structure, which otherwise would rapidly revert to a normal linear duplex (Fig. 3A).

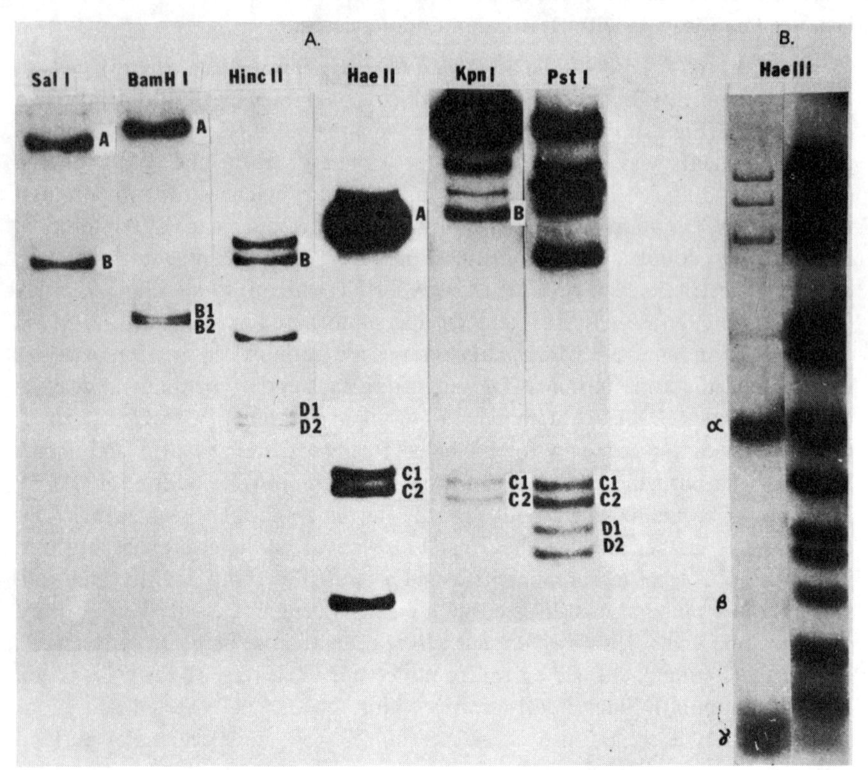

Figure 2
Acrylamide gel electrophoresis of AAV DNA restriction endonuclease cleavage products.

(*A*) Samples of uniformly ^{32}P-labeled AAV DNA were digested with the indicated endonuclease and electrophoresed on a single 3% acrylamide gel slab. Terminal fragments only are indicated by the appropriate letters, and each component of a doublet is indicated by the numerical suffix, e.g., *Hae*II C1 and C2. Terminal fragments larger than 30% of the genome (e.g., *Sal*I B) were not resolved into doublets in these conditions. *Sal*I B, *Hae*II C1 and C2, and *Pst*I C1 and C2 are left-hand and *Sal*I A, *Hae*II A, and *Pst*I D1 and D2 are right-hand terminal fragments. Cleavage with *Kpn*I (generously provided by G. S. Hayward) is incomplete because of the instability of the enzyme, but the cleavage sites can be mapped: *Kpn*I B contains the left-hand 42% and *Kpn*I C1 and C2 the right-hand 9% of the AAV genome.

(*B*) *Hae*III cleavage of AAV-2 DNA containing either a 5′-terminal ^{32}P-label (left-hand track) or a uniform ^{32}P-label (right-hand track). The samples were electrophoresed on a 3-10% acrylamide gradient gel and subsequently autoradiographed. The 5′-terminal label is mainly in components α and γ. On the original autoradiograph, a very small amount of ^{32}P-label was observed in the position of β.

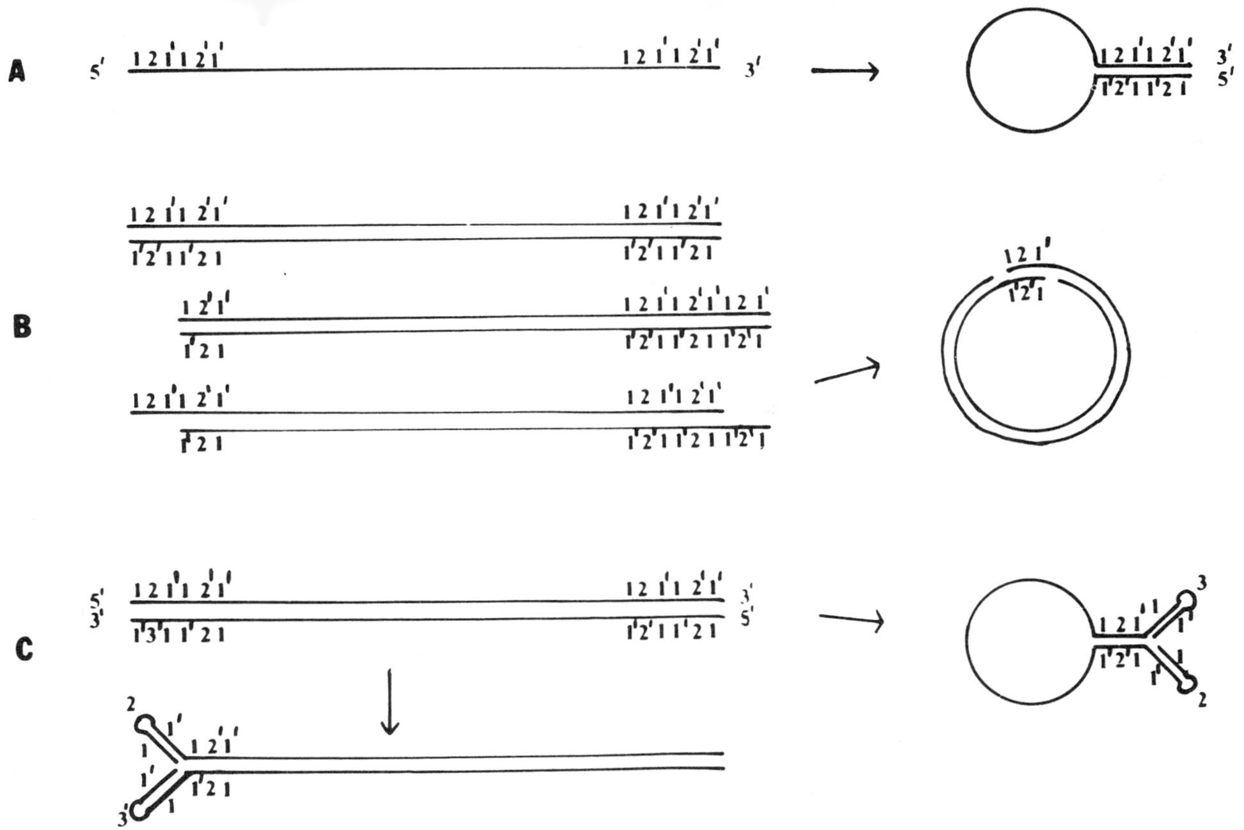

Figure 3
Diagrammatic representation of various properties and possible structures of AAV DNA termini. The models are explained in the text.

Additional Properties of the Terminal Sequences of AAV-2 DNA

As described above, the size of the terminal inverted repetition was estimated at 1% to 2.5% and that of the limited permutation at 1% of the genome length (Berns and Kelly 1974; Berns et al. 1975b). These estimates are consistent with melting experiments which show that the T_m of the panhandles is at least equal to that of intact AAV duplex DNA (i.e., 92°C in 1 X SSC; Koczot et al. 1973), whereas the T_m of the limited permutation (i.e., cohesive ends) is below 80°C in 1 X SSC (Gerry et al. 1973; Carter and Khoury 1975). Therefore, the permutation appears to be shorter than, and contained within, the terminal inverted repetition.

Digestion with specific nucleases suggests that the 5′- and 3′- terminal sequences can form "foldback" (i.e., hairpin) structures (Fife et al. 1976), which is consistent with the presence of terminal palindromic nucleotide sequences.

Arrangement of Terminal Sequences of AAV DNA

At present, any model to describe AAV DNA termini is hypothetical. The general arrangement depicted in Figure 3, which is an extension of the proposal that the AAV DNA terminal sequences may be palindromic (Gerry et al. 1973), can account for all the currently known properties of AAV DNA. The main features of this model are:

1. Terminal sequences are palindromic and contain "subsequences" that are also palindromic (Fig. 3A).
2. The limited permutation is shorter than the terminal inverted repetition (Fig. 3B). This results in a varying number of palindromes at either terminus.
3. All the strands begin with the same 5′ sequence, indicated by 1. . . . . , which is compatible with current information (Fife et al. 1976).
4. Within the terminal subsequences, there is some sequence variation, thus allowing stable formation of "rabbit ears" or similar structures (Fig. 3C).

The final resolution of the structure of AAV DNA termini requires determination of the nucleotide sequence.

Functional Significance of AAV DNA Structure

Infection of human cells with AAV in the absence of any helper virus resulted in the establishment of clones carrying AAV. Several genome equivalents of AAV DNA were present in a latent state in each cell (Hoggan et al. 1972; Berns et al. 1975a). Infectious AAV particles could be rescued by superinfection with an adenovirus, and the rescue was enhanced by iododeoxyuridine. This shows that the whole AAV genome was present and suggests, but does not prove, that it may be physically integrated into the cell DNA. Recombination events leading to such an integration of AAV DNA into the host genome might involve the terminal repeated sequences.

The ends of AAV DNA almost certainly have a major role in replication. A DNA 3′ terminus which folds back would allow self-priming of replication to yield concatenated intermediates containing alternating units of (+) and (-) strands. These molecules should rapidly reassociate ("snap back") following denaturation. Replication intermediates with these properties have been reported for both MVM (Tattersall et al. 1973) and AAV (Strauss et al. 1976).

One important feature of AAV DNA that remains to be resolved is whether the limited permutation arises from concatenates during replication as a result of some "headful" cutting mechanism (Tye et al. 1974) or whether it is a result of more than one population of AAV-2 DNA molecules which replicate independently. This question cannot be readily answered at present because, for technical reasons, AAV has not been cloned.

SUMMARY

Adeno-associated virus (AAV) packages single plus and minus DNA strands into separate particles. The purified strands can reassociate to form both linear and circular duplexes. The terminal nucleotide sequences of AAV DNA strands exhibit properties consistent with (1) an inverted terminal repetition, (2) a regular (noninverted) terminal repetition, and (3) a limited permutation. Cleavage with various restriction enzymes yields terminal fragments of two types, which is consistent with two permutations. One of the two terminal fragments has an unusual structure. The various models which most readily account for these properties require terminal palindromic sequences.

Acknowledgments

L. M. de la Maza was a special post-doctoral fellow of the National Cancer Institute. Part of the work reviewed here resulted from stimulating collaborations with G. Khoury, D. Denhardt, K. Berns and K. Fife.

REFERENCES

Berns, K. I. and T. J. Kelly, Jr. 1974. Visualization of the inverted terminal repetition in adeno-associated virus DNA. *J. Mol. Biol.* **82**:267.

Berns, K. I., T. C. Pinkerton, G. F. Thomas and M. D. Hoggan. 1975a. Detection of adeno-associated virus (AAV)-specific nucleotide sequences in DNA isolated from latently infected Detroit 6 cells. *Virology* **68**:556.

Berns, K. I., J. Kort, K. H. Fife, W. Grogan and I. Spear. 1975b. Study of the fine structure of adeno-associated virus DNA with bacterial restriction endonucleases. *J. Virol.* **16**:712.

Carter, B. J. and G. Khoury. 1975. Specific cleavage of adenovirus-associated virus DNA by a restriction endonuclease R.*Eco*R1: Characterization of cleavage products. *Virology* **63**:523.

Carter, B. J., G. Khoury and D. T. Denhardt. 1975. Physical map and strand polarity of specific fragments of adeno-associated virus DNA produced by endonuclease R.*Eco*R1. *J. Virol.* **16**:559.

Carter, B. J., G. Khoury and J. A. Rose. 1972. Adeno-associated virus multiplication. IX. Extent of transcription of the viral genome in vivo. *J. Virol.* **10**: 1118.

Carter, B. J., K. H. Fife, L. M. de la Maza and K. I. Berns. 1976. Genome localization of adeno-associated virus RNA. *J. Virol.* **19**:1044.

Denhardt, D. T., S. Eisenberg, K. Bartok and B. J. Carter. 1976. Multiple structures of adeno-associated virus DNA: Analysis of terminally labeled molecules with endonuclease R.*Hae*III. *J. Virol.* **18**:672.

Fife, K., I. Spear, W. Hauswirth, K. Berns and K. Murray. 1976. Studies on the fine structure of adeno-associated virus DNA. *Fed. Proc. Abstr.* **35**:1594.

Gerry, H. W., T. J. Kelly and K. I. Berns. 1973. The arrangement of nucleotide sequences in adeno-associated virus DNA. *J. Mol. Biol.* **79**:207.

Hoggan, M. D., G. F. Thomas and F. B. Johnson. 1972. Continuous carriage of adenovirus-associated virus genome in cell culture in the absence of helper adenovirus. In *The Proceedings of the Fourth Lepetit Colloquium* (ed. L. G. Silvestri), p. 243. North-Holland, Amsterdam.

Koczot, F. J., B. J. Carter, C. F. Garon and J. A. Rose. 1973. Self-complementarity of terminal sequences within plus or minus strands of adeno-associated virus DNA. *Proc. Nat. Acad. Sci.* **70**:215.

Mayor, H. D., K. Torikai, J. L. Melnick and M. Mandel. 1969. Plus and minus single-stranded DNA separately encapsidated in adeno-associated satellite virions. *Science* **166**:1280.

Rose, J. A., K. I. Berns, M. D. Hoggan and F. Koczot. 1969. Evidence for a single-stranded adeno-associated virus genome: Formation of a DNA density hybrid on release of viral DNA. *Proc. Nat. Acad. Sci.* **64**:863.

Straus, S. E., E. D. Sebring and J. A. Rose. 1976. Concatemers of alternating plus and minus strands are intermediates in adenovirus-associated virus DNA synthesis. *Proc. Nat. Acad. Sci.* **73**:742.

Tattersall, P., L. V. Crawford and A. J. Shatkin. 1973. Replication of the parvovirus MVM. II. Isolation and characterization of intermediates in the replication of the viral deoxyribonucleic acid. *J. Virol.* **12**:1446.

Tye, B.-K., J. A. Huberman and D. Botstein. 1974. Non-random circular permutation of phage P22 DNA. *J. Mol. Biol.* **85**:50.

SECTION VI
Genetic Rearrangements; Techniques and Applications

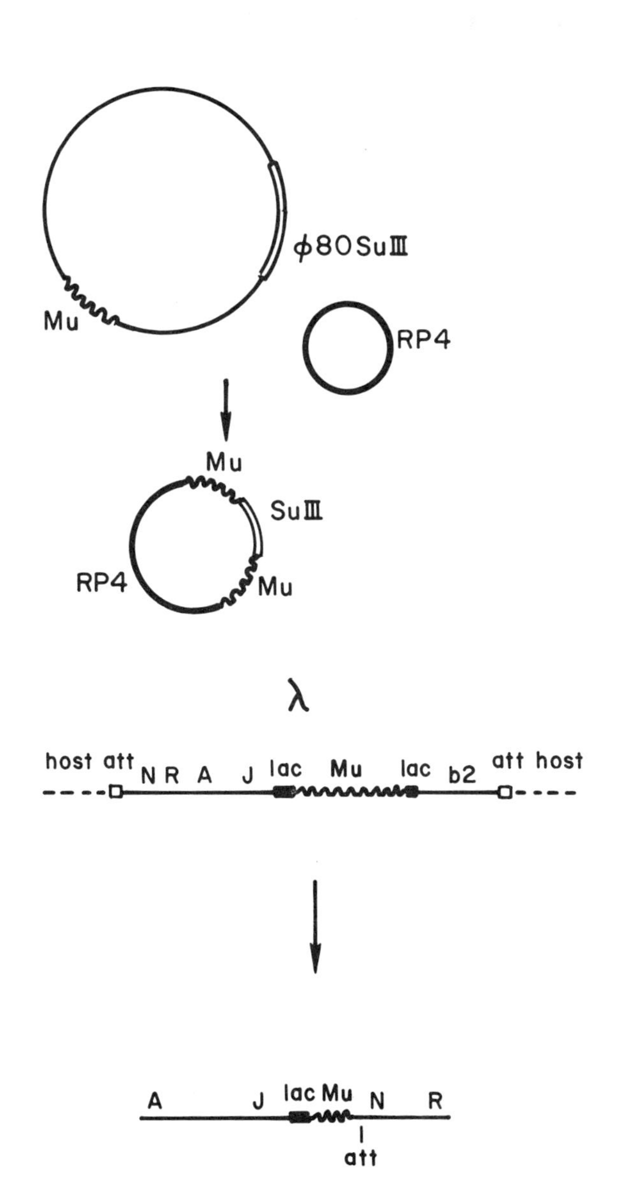

Recently, there have been explosive developments in the technology for the splicing of DNA molecules in vitro. Joining of DNA segments in vivo, however, has been possible for a number of years. Transposition elements have now been added to the repertoire of genetic tools—these elements can be exploited for many intricate genetic manipulations. The drawings in the section-title figure show two widely different uses of bacteriophage Mu. The one at the left is from an experiment by Faelen et al. (this volume) in which the t-RNA gene from the ϕ80SuIII transducing phage is translocated to the RP4 plasmid with the help of Mu. This is done simply by inducing a Mu*cts* prophage in a strain which also contains ϕ80SuIII and RP4. Mu DNA serves as a linker molecule for translocation such that one copy of Mu DNA is present at each end of the translocated segment. The RP4 plasmid with translocated sequence can be recovered by mating with appropriate recipient strains.

The drawing at the right shows a manipulation of bacteriophage λ and bacteriophage Mu genomes involving a series of insertions. Lambda is inserted into the *E. coli* chromosome. The λ genome carries an insertion of a part of the *lac* operon. Mu is inserted in the *lacZ* gene. The λ*lac*::Mu structure is too large to be packaged into λ virions. However, from such a lysogen, plaque-forming λ particles carrying portions of Mu DNA can be isolated. These λ-Mu hybrids have deletions in the λ genome (Bukhari and Allet, *Virology* 63:30 [1975]).

Mapping Ribosome Protein Genes in *E. coli* by Means of Insertion Mutations

S. R. Jaskunas* and M. Nomura
Institute for Enzyme Research
Departments of Genetics and Biochemistry
University of Wisconsin
Madison, Wisconsin 53706

To study the organization of ribosomal protein (r-protein) genes in *Escherichia coli,* we have isolated lambda transducing phages carrying DNA from the *str-spc* region of the chromosome (Jaskunas et al. 1975b). Lambda *fus*3 carries the largest substitution of bacterial DNA, about 29 × 10^6 daltons (Fiandt et al. 1976). Genes for 27 r-proteins have been identified on this phage (R. Jaskunas, L. Lindahl, L. Post and M. Nomura, unpubl.), which is approximately one-half of all the r-protein genes of *E. coli.* In addition, it carries genes for translation elongation factors EF-G and EF-Tu (Jaskunas et al. 1975d) and also the α subunit of RNA polymerase (Jaskunas 1975a). Lambda *spc*1 carries about 40% of the bacterial DNA on λ*fus*3 (Fiandt et al. 1976). It has 14 of the r-protein genes and also the α gene, but not the genes for EF-G or EF-Tu (Jaskunas et al. 1975b, and unpubl.).

One of our major objectives is to map all of these genes and determine how they are organized into transcription units. One technique we have used has been to isolate mutants of these phages that carry polar insertions which inactivate several ribosomal genes on the phages. In general, the procedure has been to isolate phage mutants on which at least one of the ribosomal genes has been inactivated, screen to find those mutants that carry an insertion, map the insertion by heteroduplex techniques, and identify the genes that have been affected by the insertion by measuring the expression of the genes from the mutant and parent phages after infection of UV-irradiated bacteria. We found previously that two insertion mutants of λ*spc*1 selected for their inactivation of the *spc* gene also reduced the expression of several other r-protein genes (Jaskunas et al. 1975c). We concluded that this cluster of genes was probably a transcription unit. We have also described an insertion mutant of λ*fus*3 with reduced expression of the genes for r-proteins S7 and S12 and also of the genes for EF-G and EF-Tu, suggesting that these genes are cotranscribed (Jaskunas et al. 1975d). In this article, we describe the genetic techniques for isolating these mutants and how we are proceeding to map ten r-protein genes in the *spc* unit with insertion mutants.

*Present address: Indiana University, Department of Chemistry, Bloomington, Indiana 47401.

The Genetic System

The technique for selecting insertion mutants of λ*fus*3 is diagrammed in Figure 1. A lysogen (NO1380) was constructed which carried λ*fus*3 and a helper phage (λ*b*515*b*519*xis*6*cI*857*S*7) in a host (NO1345) that was *recA trkA* and had resistant alleles for the *spc, fus,* and *str* genes, which code for r-protein S5 (Bollen et al. 1969), elongation factor EF-G (Tocchini-Valentini and Mattoccia 1968; Kuwano et al. 1971), and r-protein S12 (Ozaki et al. 1969), respectively. The *trkA*$^+$ gene on λ*fus*3 was used to select and maintain the lysogen. Lambda *fus*3 carries sensitive alleles for the *spc, fus,* and *str* genes. Since, for these genes, the sensitive alleles are dominant (Lederberg 1951; Sparling et al. 1968; Nomura and Engbaek 1972), the phenotype of the lysogen is Spc^s, Fus^s, Str^s (i.e., sensitive to spectinomycin, fusidic acid, and streptomycin). The sensitivity of this lysogen to these antibiotics in a stable lysogen where most lambda promoters are repressed suggests that these genes are being expressed from a bacterial promoter.

Spontaneous mutants of this lysogen were isolated that were resistant to spectinomycin or streptomycin (Spc^r or Str^r), each of which occurred at a frequency of about 10^{-6}. The most likely mutation that would result in such a phenotype would be a mutation of the sensitive allele of the transducing phage or a mutation that had inactivated that gene by some polar mechanism, such as a deletion, nonsense mutation, or polar insertion. *Cis*-acting mutations that

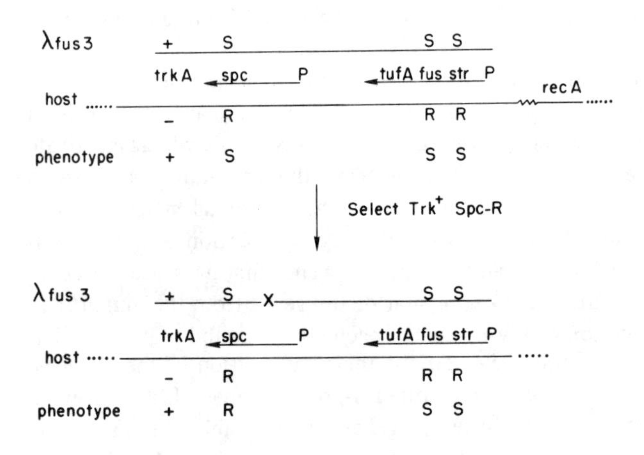

Figure 1
Schematic outline of the techniques used to select insertion mutants of λ*spc*1 and λ*fus*3 (see text for details). The *tufA* gene is a structural gene for EF-Tu (Jaskunas et al. 1975d).

would result in a dominant resistant allele, such as an "up-promoter" mutation, are also conceivable. We avoided selecting segregants that had lost the transducing phage by simultaneously selecting for the *trkA*$^+$ gene carried by the phage. The *recA* allele of the host presumably reduced the chance that the resistant mutants resulted from homogenotization.

After selecting for mutants that had become resistant to spectinomycin or streptomycin, we tested to see if the lysogen had become resistant to any of the other antibiotics. Nearly all the Spc^r mutants were still Str^s and Fus^s. However, about 60% of the Str^r mutants had also become Fus^r, but were still Spc^s. We found that about 50% of these pleiotropic mutants resulted from insertions. Approximately 40% of the Spc^r mutants also resulted from insertion. In this case, we did not have any prior genetic evidence of polarity suggesting that the mutation could be an insertion. Nevertheless, it was still relatively easy to find insertion mutants. These insertion mutations inactivating the *spc*^{s} gene shall be referred to as "Spc insertions." Such insertions could inactivate the *spc*^{s} gene either directly by integration into the *spc*^{s} gene or indirectly by some polar mechanism resulting from integration into the transcription unit containing the *spc*^{s} gene. The few mutants that had become resistant to all three antibiotics were found to carry large deletions.

To determine whether the mutant phenotype resulted from a deletion or insertion mutation of the transducing phage, we measured the buoyant density of the particles of the various mutants by isopycnic centrifugation of crude lysates in the Model E ultracentrifuge (Szybalski and Szybalski 1971). The presence in the lysate of helper phage having a different density than the transducing phage provided a reference.

In an analogous fashion, we also isolated insertion mutants of λ*spc*1 on which the *spc* gene had been inactivated. A lysogen (NO1402) of strain NO1345 was constructed with λ*cI*857*S*7 as helper. The phenotype of the lysogen is Spc^s, Fus^r, Str^r. Spontaneous Spc^r mutants were selected, and the phages with insertions were identified by banding crude lysates in the Model E ultracentrifuge.

Altogether, we found that 25-40% of the mutants selected with Spc^r or Str^r resulted from an insertion in the transducing phage. All of the insertions were approximately the size of either IS*1* or IS*2*. Two of the insertions inactivating the *spc* gene of λ*spc*1 have been identified (Saedler et al. 1975). The I15 insertion is an IS*1* in orientation II and the I16 insertion is an IS*2* in orientation I. None of the other insertions have been identified. Since these insertions were isolated in a *recA* background, the results indicate that the transpositions of IS*1* and IS*2* insertions are independent of the host *recA* recombination system.

Mapping Insertions in the Spc Transcription Unit

Insertions inactivating the *spc* gene on either λ*spc*1 or λ*fus*3 were mapped by heteroduplex techniques as illustrated in Figure 2. As indicated, λ*spc*1 and λ*fus*3 are homologous from their left end up to the right junction between bacterial and λ DNA in λ*spc*1 (Fiandt et al. 1976). They also share 5 %λ DNA sequences at the right end. Thus heteroduplex molecules of λ*spc*1 and λ*fus*3 consist of a long DNA duplex (43 %λ units) and a short duplex (5 %λ units) separated by a large substitution bubble. Spc insertion mutants of λ*spc*1 were heteroduplexed with λ*fus*3 (or λ*fus*2, which is the same as λ*fus*3 except that it carries a *str*^{r} allele instead of a *str*^{s} allele), and Spc insertion mutants of λ*fus*3 were heteroduplexed with λ*spc*1. In all cases, the insertions were found within the bacterial region of the long duplex. Thus even the Spc insertion mutants of λ*fus*3 are located in a region of the bacterial DNA that is also carried by λ*spc*1.

The positions of the insertions are given in Figure 3, as measured from the junction of the long DNA duplex with the substitution bubble, i.e., the right bacterial-λ junction in λ*spc*1. The *spc* gene that has been at least partially inactivated by all these insertions appears to be cut at an *Eco*RI endonuclease-sensitive site that is approximately 4.1 kb to the left of the junction (Lindahl et al. 1976). Since the direction of transcription of the ribosomal genes on λ*spc*1 is leftward and the promoter for the Spc transcription unit appears to be within 0.25 kb of the junction (Jaskunas et al. 1975c), we would expect that insertions within the Spc unit would be located between the junction and the *spc* gene. We found that the insertions are scattered throughout this region. There is a cluster of insertions approximately 0.25 kb from the junction, which may be in the promoter region.

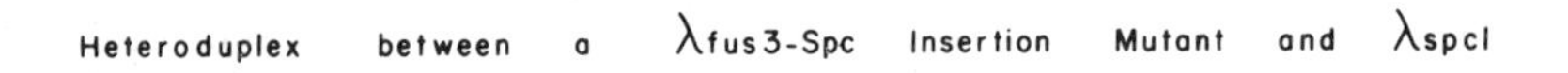

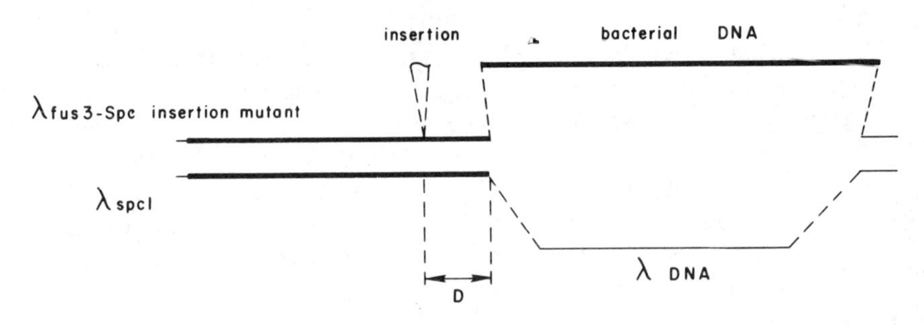

Figure 2
Diagram of heteroduplex molecule formed by λ*spc*1 and a λ*fus*3-Spc insertion mutant. Lambda DNA is illustrated with a thin line and bacterial DNA with a heavy line. *D* is the distance between the right bacterial-λ junction and the position of the insertion loop. This is the distance plotted in Fig. 3.

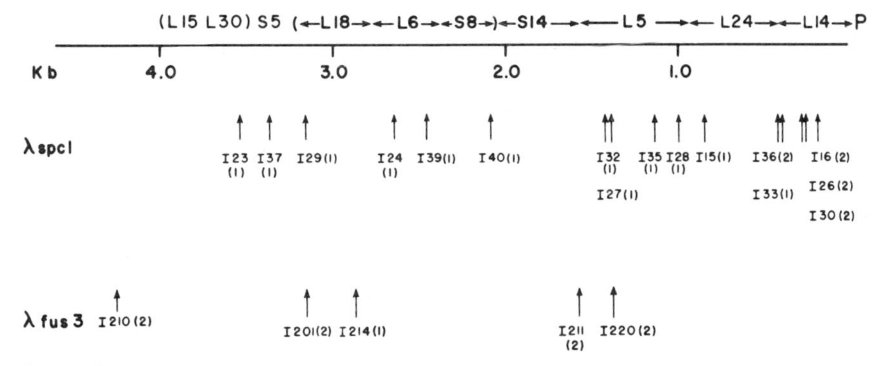

Figure 3
Locations of insertions in the Spc transcription unit of λ*spc*1 and λ*fus*3. Zero on the kb scale is the right bacterial-λ junction in λ*spc*1 (see Fig. 2). The size of the insertion is given in parentheses either below or after the mutant name—(*1*) for insertions about the size of IS*1* (700–800 base pairs) and (*2*) for insertions about the size of IS*2* (1300–1400 base pairs). The order of genes as deduced from the analysis of these Spc insertion mutants and from the in vitro experiments of L. Lindahl, J. Zengel, L. Post and M. Nomura (unpubl.) is given at the top. The approximate positions of the genes relative to the insertions are indicated by arrows. The positions of the genes for L15, L30, and S5 relative to the I210 insertion are not known. P indicates the apparent approximate position of the promoter. The direction of transcription is leftward (Jaskunas et al. 1975c). The *spc* gene is the structural gene for S5 (Bollen et al. 1969).

Previous experiments had suggested that the genes for L16 and L19 may also be part of this unit (Jaskunas et al. 1975c). However, fingerprint analysis of the proteins synthesized in the UV-irradiated cells after phage infection has indicated that the gene for L16 is on λ*spc*2 and λ*fus*3 but not on λ*spc*1. The apparent weak stimulation of the synthesis of L16 after λ*spc*1 infection appears to be due to contamination by other proteins of the "L16" spot on the two-dimensional acrylamide gel. Also, the radioactive protein that migrates with L19 appears to be a derivative of L14 or L15. The gene for L19 does not appear to be on λ*fus*3 at all (S. R. Jaskunas and M. Nomura, unpubl.).

Order of Genes in the Spc Unit

To determine which of the r-protein genes in addition to the *spc* gene had been affected by these insertions, we compared the expression of the r-protein genes from the mutant and parent phages after infection of UV-irradiated bacteria. We previously found that the I16 insertion of λ*spc*1 greatly reduced the expression of the genes for S5, S8, S14, L5, L6, L14, L15, L18, L24, and L30. Thus this cluster of ten r-protein genes appears to be organized into a transcription unit, which we shall call the "Spc transcription unit" (see also legend to Fig. 3). The other r-protein genes on λ*spc*1, as well as the gene for α, may be partially affected by insertions in this unit. It is not clear to what extent they are in a separate unit or units.

We have now analyzed the expression of r-protein genes on several other insertion mutants of λ*spc*1 and λ*fus*3. The results of some of these experiments are given in Table 1. We have found that insertions close to the junction, such as I16 or I26, reduce the expression of all the genes in this unit. However, insertions located at more intermediate positions between the junction and the *spc* gene reduce the expression of only some of these genes. Small insertion mutants like IS*1* and IS*2* have been found to have polar effects when they are located in the *lac* and *gal* operons (for a review, see Starlinger and Saedler 1976). The simplest interpretation of our results, therefore, is that the insertions at intermediate positions also have a polar effect on this r-protein transcription unit and that the genes experiencing reduced expression are located distal to the insertion. In principle, the analysis of mutants with insertions at different positions in the unit could reveal the order of all the genes.

Furthermore, some of the insertions appear to be within structural genes. This can be seen by comparing the residual levels of expression of the genes that appear to be distal. Most are still expressed to about 20–40% of their level in the parent phage. However, for some of the mutants, there is one (and only one) gene whose expression appears to be completely abolished. It seems likely the insertions in these mutants are located within the structural gene that is inactivated. From the known position of the insertion, it is possible to map the gene physically. For example, all insertions 1.0–1.4 kb from the junction, such as I32, appear to abolish the expression of the gene for L5. Thus the gene for L5 is probably in this region. Similarly, the I214 insertion appears to be in the gene for L18 (see Table 1).

Two techniques have been used to analyze the r-proteins synthesized in UV-irradiated bacteria after phage infection. One was the type used for the experiments reported in Table 1. The proteins synthesized in the UV-irradiated cells were labeled with [^{3}H]leucine and separated by two-dimensional gel electrophoresis in the presence of carrier r-proteins. The positions of the mature r-proteins in the gel were determined by staining, and the stimulation of the synthesis of each of them was determined essentially by measuring the amount of radioactivity incorporated into each of the spots. We have also used one-dimensional urea and sodium dodecyl sulfate gel systems and detected the proteins synthesized in UV-irradiated cells by autoradiography. The analysis by this technique of most of the insertion mutants shown in Figure 3 suggests that the order of the genes in the Spc transcription unit in the direction of transcription is L14, L24, L5, S14, (S8, L6, L18) (S5, L15, L30). Results of the experiment reported in Table 1 indicate that the order of S8, L6, and L18 is as given. However, this has not yet been confirmed by autoradiography. The order of the last three genes has not been determined from analyses of the insertion mutants. However, comparison of the abilities of different DNA endonuclease restriction fragments from λ*spc*1 to code for r-proteins in an in vitro DNA-

Table 1
Relative Expression of the Genes in the Spc Transcription Unit of Several Insertion Mutants

Phage	Insertion size	Distance from junction (kb)	Relative expression of genes										
			L30	S5	L18	L6	S8	S14	L5	L24	L14 + L15	S4	S7[a]
λ*spc*1 and λ*fus*3	–	–	1.00	1.00	1.00	1.00	1.00	1.00	1.00	1.00	1.00	1.00	1.00
λ*spc*1-I26	2	0.25	0.04	0.21	0.14	0.21	0.06	0.10	0.05	0.03	0.08	1.00	–
λ*spc*1-I32	1	1.39	0.50	0.80	0.58	0.53	0.55	0.41	0.02	0.96	0.88	1.00	–
λ*spc*1-I40	1	2.09	0.44	0.39	0.49	0.40	0.18	1.96	2.28	1.28	1.12	1.00	–
λ*spc*1-I39	1	2.46	0.43	0.33	0.41	0.22	1.23	1.58	1.65	1.28	0.84	1.00	–
λ*fus*3-I214	1	2.87	0.25	0.19	0.07	0.78	0.90	1.02	1.23	0.94	0.65	0.51	1.00

The size of the insertion is given in the second column: 1 indicates that it is approximately the size of IS*1*; 2 indicates that it is approximately the size of IS*2*. The location of the insertion is given in the third column in kilobases from the right bacterial-λ junction in λ*spc*1 (see Figs. 2 and 3). Gene expression was measured after infection of UV-irradiated bacteria as previously described (Jaskunas et al. 1975c). A prelabeling technique was used to correct for variations in the recovery of r-proteins during extraction and electrophoresis (see Jaskunas et al. 1975c). The expression of each gene of the parental phage was normalized to 1.00, and the expression of the S7 gene on the mutant λ*fus*3 phage and the S4 gene on the mutant λ*spc*1 phages was normalized to 1.00. By doing this, we are assuming that the expressions of these genes are not affected by the insertions. The results of the experiments with the λ*fus*3 mutants indicate that the expression of the S4 gene is reduced to 50% of the parental value by insertions in the *spc* unit (see results for λ*fus*3-I214). Thus the actual relative expression of the genes on the λ*spc*1 mutants compared to their expression from the parent phage may be approximately one-half of the value given in the table. We could not normalize the data for the λ*spc*1 phages to the expression of the S7 gene since it is not carried by λ*spc*1. The results for the genes that appear to be distal to the insertion are underlined. We presume that the insertions are in or immediately to the right of the gene for the rightmost r-protein underlined. The syntheses of L5 from λ*spc*1-I32 and L8 from λ*fus*3-I214 were reduced to a greater extent than those of the other r-proteins that were affected probably because the insertions in these phages are integrated into these genes (see text). The insertion in λ*spc*1-I26 was more polar than the other insertions. This may be related to the position of the insertion in the transcription unit. In general, comparison of the data shown with the results of experiments with other insertions mutants (not shown) suggest that insertions close to the promoter are more polar. Proteins L14 and L15 were not completely resolved by the two-dimensional gel system used to purify the r-proteins, and therefore the results are given for the sum of these proteins. The results shown are for one experiment. However, the experiments have been repeated and the results obtained are in good agreement with those shown.

[a]The gene for S7 is on λ*fus*3 but not on λ*spc*1.

dependent protein-synthesizing system has indicated that the order is S5 (L15, L30) (L. Lindahl, J. Zengel, L. Post and M. Nomura, unpubl.). These in vitro experiments have also indicated that the genes for S5, L15, and L30 are distal to all others in this unit, which is consistent with the results of the analyses of the insertion mutants.

SUMMARY

The small transposable insertion elements of *E. coli* have generally been studied when they are inserted into well-characterized operons or transcription units. By contrast, we have used them to study the organization of genes in previously uncharacterized transcription units, those for r-proteins. Our current hypothesis for the order of the genes in the Spc transcription unit, based on analysis of the insertion mutants and in vitro experiments, is given in Figure 3. Additional experiments are needed to verify this order. However, the use of insertion mutations has provided us with a novel technique for studying the organization of this important transcription unit.

Acknowledgments

We thank M. Dietzman, K. Ryan, L. Sadowski and G. D. Strycharz for technical assistance. This work was supported, in part, by the College of Agriculture and Life Sciences, University of Wisconsin, and by grants from the National Science Foundation (GB-31086) and the National Institutes of Health (GM-20427). This is paper no. 1994 from the Laboratory of Genetics, University of Wisconsin.

REFERENCES

Bollen, A., J. Davies, M. Ozaki and S. Mizushima. 1969. Ribosomal protein confirming sensitivity to the antibiotic spectinomycin in *Escherichia coli*. *Science* **165**:85.

Fiandt, M., W. Szybalski, F. R. Blattner, S. R. Jaskunas, L. Lindahl and M. Nomura. 1976. Organization of ribosomal protein genes in *Escherichia coli*. I. Physical structure of DNA from transducing λ phages carrying genes from the *aroE-str* region. *J. Mol. Biol.* **106**:817.

Jaskunas, S. R., R. R. Burgess and M. Nomura. 1975a. Identification of a gene for the α-subunit of RNA polymerase at the *str-spc* region of the *Escherichia coli* chromosome. *Proc. Nat. Acad. Sci.* 72:5036.

Jaskunas, S. R., L. Lindahl and M. Nomura. 1975b. Specialized transducing phages for ribosomal protein genes of *Escherichia coli*. *Proc. Nat. Acad. Sci.* 72:6.

———. 1975c. Isolation of polar insertion mutations and the direction of transcription of ribosomal protein genes in *E. coli*. *Nature* **256**:183.

Jaskunas, S. R., L. Lindahl, M. Nomura and R. R. Burgess. 1975d. Identification of two copies of the gene for the elongation factor EF-Tu in *E. coli*. *Nature* **257**:458.

Kuwano, M., D. Schlessinger, G. Rinaldi, L. Felicetti and G. P. Tocchini-Valentini. 1971. G factor mutants of *Escherichia coli*: Map location and properties. *Biochem. Biophys. Res. Comm.* **42**:441.

Lederberg, J. 1951. Streptomycin resistance: A genetically recessive mutation. *J. Bact.* **61**:549.

Lindahl, L., J. Zengel and M. Nomura. 1976. Organization of ribosomal protein genes in *Escherichia coli.* II. Mapping of ribosomal protein genes by in vitro synthesis of ribosomal proteins using DNA fragments of a transducing phage as templates. *J. Mol. Biol.* **106**:837.

Nomura, M. and F. Engbaek. 1972. Expression of ribosomal protein genes as analyzed by bacteriophage Mu-induced mutations. *Proc. Nat. Acad. Sci.* **69**:1526.

Ozaki, M., S. Mizushima and M. Nomura. 1969. Identification and functional characterization of the protein controlled by the streptomycin-resistant locus in *E. coli. Nature* **222**:333.

Saedler, H., D. F. Kubai, M. Nomura and S. R. Jaskunas. 1975. IS1 and IS2 mutations in the ribosomal protein genes of *E. coli* K12. *Mol. Gen. Genet.* **141**:85.

Sparling, P. F., J. Modolell, Y. Takeda and B. D. Davis. 1968. Ribosomes from *Escherichia coli* merodiploids heterozygous for resistance to streptomycin and to spectinomycin. *J. Mol. Biol.* **37**:407.

Starlinger, P. and H. Saedler. 1976. IS-elements in microorganisms. In *Contemporary topics in microbiology and immunology,* vol. 75, p. 111. Springer Verlag, Berlin.

Szybalski, W. and E. H. Szybalski. 1971. Equilibrium density gradient centrifugation. In *Procedures in nucleic acid research* (ed. G. L. Cantoni and D. R. Davis), vol. 2, p. 311. Harper and Row, New York.

Tocchini-Valentini, G. P. and E. Mattoccia. 1968. A mutant of *E. coli* with an altered supernatant factor. *Proc. Nat. Acad. Sci.* **61**:146.

Chromosomal Rearrangements Resulting from Recombination between Ribosomal RNA Genes

C. W. Hill, R. H. Grafstrom* and B. S. Hillman
Department of Biological Chemistry
Milton S. Hershey Medical Center
Pennsylvania State University
Hershey, Pennsylvania 17033

The production of genetic duplications is often considered to be of fundamental importance in the development and evolution of genetic material in that duplications could provide a reservoir of redundant material which could be further altered. We have found that, under certain conditions, tandem duplications occur in one region of the *Escherichia coli* K12 chromosome at extremely high frequencies, up to 5% of the cells in the population. We present here evidence that these duplications are generated by uneven recombination between naturally redundant regions of the chromosome, the ribosomal RNA genes.

In *E. coli,* $tRNA^{Gly}_{GGA/G}$ is produced from a single structural gene, *glyT*$^+$ (Hill et al. 1970), which lies between the *argH* and *thi* loci (Hill et al. 1969). An AGA-specific missense suppressor can be derived from $tRNA^{Gly}_{GGA/G}$ by a simple base substitution in the anticodon (Roberts and Carbon 1975). Such a mutation is extremely detrimental in that it leaves the cell deficient in its ability to translate the GGA codon (Carbon et al. 1970). Our studies of this missense suppressor led to the realization that the *glyT* gene can become duplicated within the chromosome at very high frequency. For example, mild ultraviolet mutagenesis of a *glyT*$^+$ culture caused the appearance of *glyT*$^+$/*glyTsu*$_{AGA}$ mutants at a frequency of 5×10^{-6} among the surviving cells (Hill and Combriato 1973). These partial heterozygotes grow quite well since they retain the normal *glyT*$^+$ function. Further analysis showed that the simple duplication of *glyT*$^+$ to produce *glyT*$^+$/*glyT*$^+$ merodiploids occurs among 3–5% of the survivors of mild ultraviolet mutagenesis. These duplications were also found at high frequency after nitrous acid, nitrosoguanidine, and ethyl methanesulfonate (EMS) mutagenesis (Hill and Combriato 1973) as well as after phage P1 transduction between haploid strains (Hill et al. 1969).

These duplications generally occur as large tandem duplications involving from 1% to 6% or more of the chromosome (Hill and Combriato 1973). A systematic study of ten independently occurring duplications revealed that

*Present address: Department of Medicine, University of California at San Diego, La Jolla, California 92037.

although considerable variation in size is observed among the duplications, the positions of their end points were distinctly nonrandom. Seven of the ten duplications had their right-hand end points in the small region between *purD* and *metA* (Fig. 1); the right end points of the other three were beyond *metA*. Similarly, the left end points of five occurred between *argH* and *glyT*, three occurred between *metE* and *rhaD*, and two occurred to the left of *ilvD*.

At the time of the above studies, we suggested that one possible explanation for the high frequency and nonrandom nature of the duplications was that they were generated by recombination between different redundant ribosomal RNA genes. The precise mapping of three of the ribosomal RNA genes (*rrn*) by Deonier et al. (1974) in certain of the intervals corresponding to the duplication end points (Fig. 1) made this hypothesis very attractive. For example, if a fourth, previously uncharacterized, *rrn* gene, *rrnD*, lies between *purD* and *metA*, an interchromosomal recombination between *rrnA* and *rrnD* would produce a tandem duplication of the sequence *rrnA rhaD glyT purD rrnDA rhaD glyT purD rrnD*; the other product would contain the corresponding deletion. The "connecting point" of the duplication would be a hybrid *rrnDA* gene deriving its leftmost sequences from *rrnD* and its rightmost sequences from *rrnA*. The following is a description of our efforts to test this hypothesis.

Isolation of DNA Derived from the Duplications

We have concentrated on three representative duplications. The duplication in strain CH579 isolated after UV mutagenesis includes *glyT* through *purD*, as does the one in CH790 which was induced by EMS mutagenesis. The duplication in CH791 was produced before or during the course of phage P1 transduction and includes *rhaD* through *purD*. The proposed structures of these duplications are shown in Figure 2. Classically, the duplicated segments in such a configuration could pair with each other, and a reciprocal recombinational event occurring anywhere along duplicated segments would produce a normal (haploid), segregant chromosome and a circle of DNA which would contain exactly one

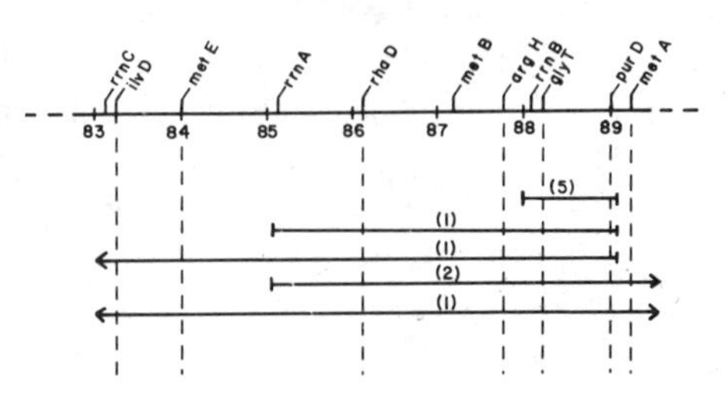

Figure 1
End points of *glyT*/*glyT* duplications. Coordinates on the map and genetic symbols are from Bachmann et al. (1976). The extents of individual duplications generated by UV irradiation are indicated by the bars, and the numbers of individual occurrences are indicated by the numbers in parentheses above the bars (Hill and Combriato 1973). The precision of locating a particular end point was limited to showing that it occurred within the region defined by the vertical broken lines.

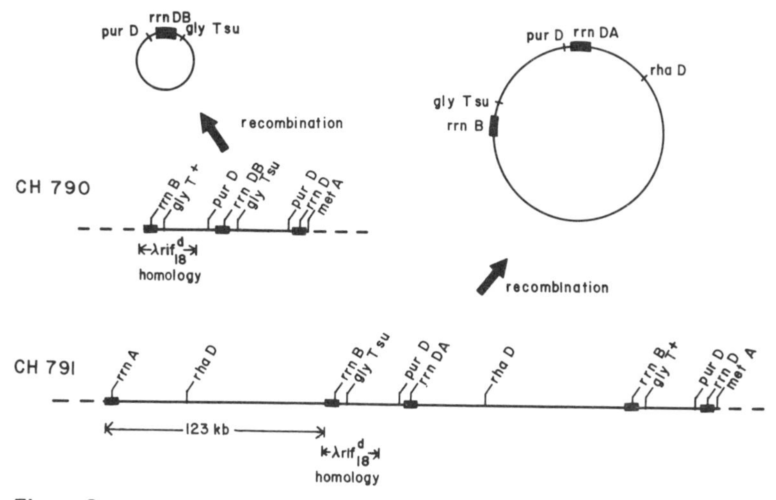

Figure 2
Proposed structures of the tandem duplication strains characterized in this study. CH579 is similar to CH790 except that the positions of the $glyT^+$ and $glyTsu_{AGA}$ alleles are reversed. The circular DNA product expected to be produced upon segregation of the respective tandem-duplication strains is also shown. The distance between the *rrn* genes is from Deonier et al. (1974).

equivalent of the duplicated material, including the "connecting point." If such events actually occur, it might be hoped that these circles could be isolated as covalently closed, superhelical DNA molecules. However, the rate of spontaneous segregation from these tandem duplications is low, on the order of 1%, and the proportion of these circles among the total cell DNA would be very small. A low dose of UV irradiation boosts the segregation rate of these tandem duplications to around 15%, making the isolation of the circles more feasible. Accordingly, DNA from [^{3}H]thymidine-labeled, UV-irradiated cultures of CH790, CH791, and CH579 has been subjected to CsCl-ethidium bromide centrifugation. As described in Figure 3 for CH790 and CH791, a small, but definite peak of DNA, comprising 0.5-1% of the total, was found in the region of the gradient where supercoiled DNA would be expected. This peak was absent if the strain did not carry a tandem duplication (strain CH439 in Fig. 3) or if UV irradiation was omitted. Generally, this peak of material was pooled and subjected to a second round of CsCl-ethidium bromide centrifugation before further analysis.

Both sedimentation velocity and electron microscope techniques have been used to show that the material so isolated is largely composed of supercoiled DNA. Surprisingly, each preparation examined has contained two populations of circles of distinctly different sizes. All of the strains examined, regardless of the size of the duplication carried, produced a circular species 140 kb in length. This species is not directly related to or derived from the tandem

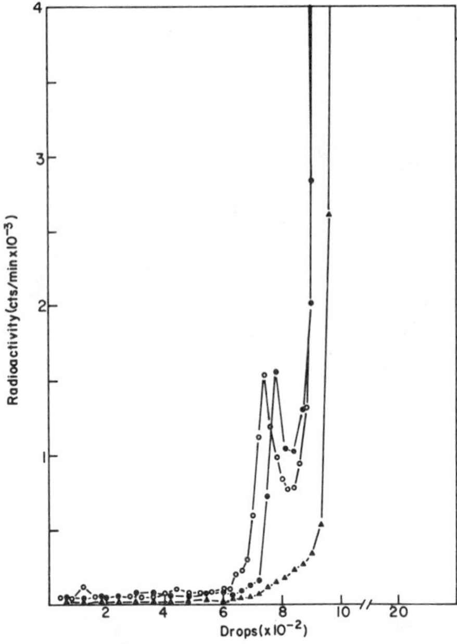

Figure 3
Isolation of circular DNA from tandem-duplication strains. Tandem-duplication strains CH790 and CH791 (Arg^-) and CH439 (Arg^-, Trp^-), the parent of CH790 and CH791, were grown at 34°C in minimal medium (Hill and Combriato 1973) supplemented with glucose and required amino acids and then diluted 10^{-6} into the same medium supplemented with 0.25 mg/ml deoxyadenosine. These cultures were grown to stationary phase and used to inoculate additional cultures of the same composition. These were then labeled for two generations with [^{3}H]thymidine (1 Ci/mmole). Log-phase cells were washed and resuspended in minimal medium containing only glucose, and the suspension was then shaken for 1 hr at 34°C. The cells (5×10^7 cells/ml) were irradiated with 500 ergs/mm^2 from a short-wave germicidal UV lamp and then supplemented with required amino acids and grown for 2 hr in the dark. The cells were harvested and lysed as described by Bazaral and Helinski (1968). The lysates were subjected to shear by pipetting several times and then were adjusted to a density of 1.56 g/ml by adding CsCl and 0.1 mg/ml ethidium bromide. The preparations were centrifuged at 38,000 rpm, at 4°C, for 60 hr in a Spinco 60 Ti rotor. (○) CH790 DNA; (●) CH791 DNA; (△) CH439 DNA.

duplications and seems to be the result of a complex event related to the growth of these strains in deoxyadenosine. They will be described elsewhere. The size of the other circular species present was dependent on the extent of the duplication, and this species has been shown to carry genes known to be duplicated (see below). Their characterization is described in the following sections.

Contour Lengths of Circular DNA

The model depicted in Figure 2 for the origin and structure of the tandem duplications leads to several predictions which can be tested. First, Deonier et al. (1974) measured the distance from the left end of *rrnA* to the left end of *rrnB* as 123 kb. According to Figure 2, the duplication in CH791 includes all of the duplication in CH579 or CH790 plus this additional 123-kb region. This leads to the prediction that the circles derived from the large duplication should be 123 kb larger than those derived from the small ones. The contour lengths of circles from the small duplications, CH579 and CH790, are 40.2 ± 1.4 kb and 39.5 ± 1.4 kb, respectively (Table 1), whereas the contour length of the circles from the large duplication, CH791, is 163 ± 7 kb. The difference is 123 kb, in agreement with the prediction.

Heteroduplexes between λrif^{d}_{18} DNA and the Duplication Circles

The transducing phage λrif^{d}_{18} carries *rpo* (Kirschbaum and Konrad 1973), *rrnB* (Lund et al. 1976), and the *glyT* gene (unpubl.). Since *rpo* maps between *glyT* and *purD* (Kirschbaum and Konrad 1973), *rpo* must be included in the duplicated region of all of these strains. The λrif^{d}_{18} phage carries about 26 kb of host DNA, and the *rrnB* gene lies very close to the host-phage junction (Lund et al. 1976; Yamamoto et al. 1976). This leads to the prediction that all of the host DNA of λrif^{d}_{18} should form a heteroduplex with circular DNA from the large duplication of CH791 (see Fig. 2). It should also form an extensive heteroduplex with DNA from the small duplication CH790. The only host DNA of λrif^{d}_{18} that would not form a heteroduplex with the small duplication circles would be the very small region between the *rrn* gene and the host-phage junction. Such experiments have been done, and the results are given in Table 1. For the circles from CH791, heteroduplex regions of 27.0 ± 0.7 kb were observed, and for CH790, heteroduplexes of 25.7 ± 0.7 kb were found. Clearly, the circles from both large and small duplications contain the *rrn* region, which itself extends 5.3 kb.

Table 1
Dimensions of Circular DNA Derived from the Tandem Duplications

DNA source	Contour length of circles in kb	Extent of heteroduplex with λrif^{d}_{18} in kb
CH579	40.2 ± 1.4 (23 molecules)	not determined
CH790	39.5 ± 1.4 (20 molecules)	25.7 ± 0.7 (13 molecules)
CH791	163 ± 7 (13 molecules)	27.0 ± 0.7 (8 molecules)

DNA was spread in formamide according to the procedures of Davis et al. (1971). Phage P4 DNA of length 10,970 base pairs (Goldstein et al. 1975) was included as an internal standard. Heteroduplexes were prepared by incubation of circular DNA, derived from either CH790 or CH791, with lysed whole λrif^{d}_{18} phage as described by Deonier et al. (1974). No effort was made to nick the supercoiled DNA stocks because spontaneous relaxation during several weeks of storage produced sufficient numbers of open circles for analysis.

Visualization of *rrn* Genes on the Tandem Duplication Circles

A third prediction concerning the model in Figure 2 is that the circles from the smaller duplication (CH790) should contain one *rrn* gene and that the circles from the larger one (CH791) should contain two *rrn* genes, spaced 123 kb apart. We have applied the newly developed "R-looping" technique of Thomas et al. (1976) to demonstrate that this is the case. In this method, double-stranded DNA is incubated with ribosomal RNA in 70% formamide at a temperature very close to the T_m of the DNA. Under these conditions, sufficient localized denaturation of the DNA occurs to allow the RNA to find complementary sequences and form a stable RNA-DNA hybrid. The result when viewed with the electron microscope is a double-stranded DNA molecule containing R loops, which consist of a single-stranded DNA region and an RNA-DNA double-stranded region. Examples of the results we obtained with the small circles from CH790 and the large circles from CH791 are shown in Figure 4 a and b, respectively. In each case, the R loops occur as doublets, the larger loop presumably corresponding to the 23S RNA sequence and the smaller loop corresponding to the 16S RNA sequence. These are separated by a small double-stranded DNA region. A single doublet is found on the small circles from CH790, whereas two doublets occur on the larger circles from CH791. Furthermore, the R-loop regions on the CH791 circles are spaced at a 124-kb interval. It should be noted that the formation of the R-loop doublets under our conditions is often incomplete. The largest observed extend about 4.5 kb, but many are substantially smaller.

DISCUSSION

These results provide strong evidence that there must be a sequence between *purD* and *metA* that can recombine with either the *rrnA* or *rrnB* loci to produce the tandem duplications. It seems highly probable that this sequence is another ribosomal RNA gene, which would be called *rrnD.* The fact that only two *rrn* genes were observed as R-loop doublets on the large circles from CH791 (Fig. 4b) proves that no other *rrn* genes, besides *rrnB,* exist between *rrnA* and *purD.* The work of Lund et al. (1976), who showed that the *rrn* gene carried by λrif^{d}_{18} had a different spacer region from the *rrn* gene carried by $\phi 80 rif^{r}$, even though the rest of the host DNA carried by both phages was completely homologous, left doubt as to which phage carried the *rrnB* gene actually found at that position in the *E. coli* chromosome. The heteroduplexes between λrif^{d}_{18} DNA and either the large circle from CH791 or the small ones from CH790 showed no sign of mismatch in the spacer region. Since the large circle from CH791 should carry an authentic *rrnB* gene (Fig. 2), we tentatively conclude that *rrnB* in our standard *E. coli* strain CH439 has the same spacer as the one carried by λrif^{d}_{18}.

The three *rrn* genes characterized by Deonier et al. (1974) and Lund et al.

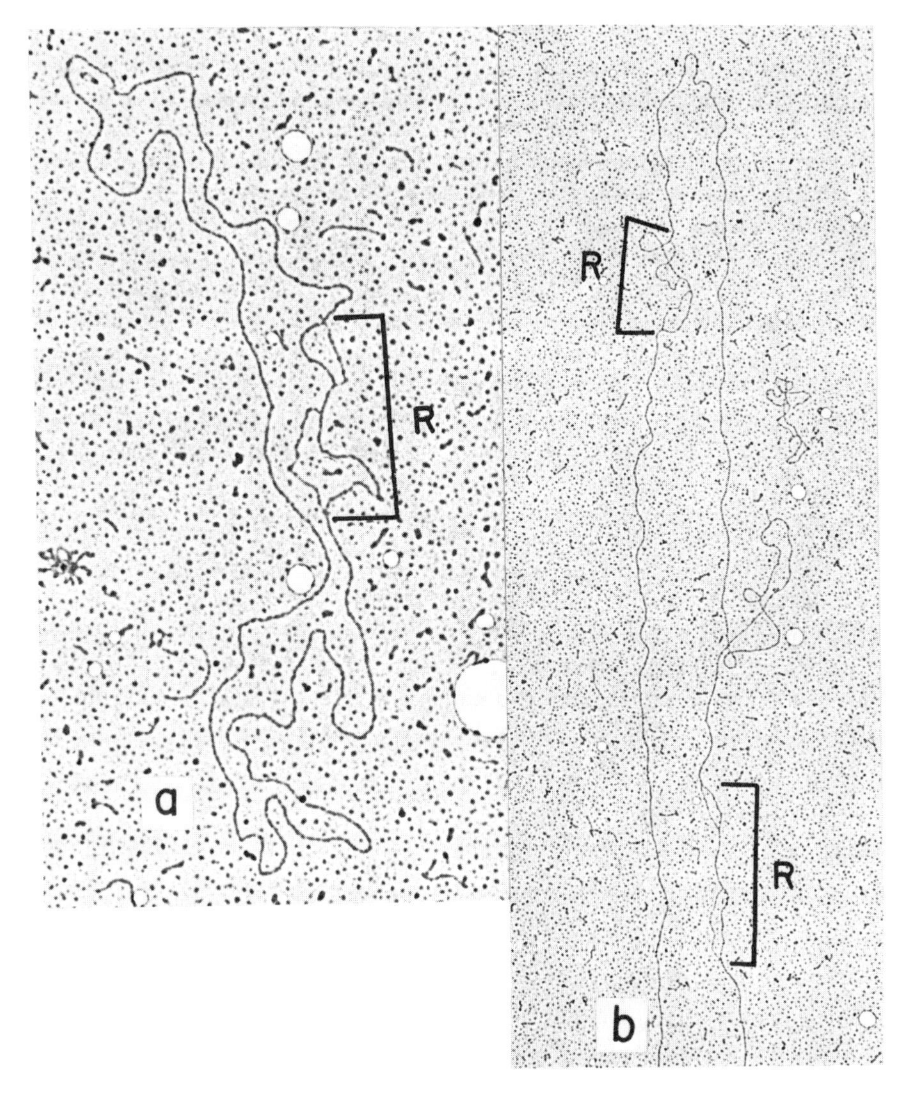

Figure 4

R loops on CH790 and CH791 circular DNA. Circular DNA was incubated with 1 μg/ml ribosomal RNA in 70% formamide, 0.01 M EDTA, and 0.1 M Na-PIPES buffer, pH 7.8 (Na^+ concentration was 0.17 M), and incubated at 50°C for 4 hr (Thomas et al. 1976). The preparation was spread for electron microscopy immediately, with 70% formamide in the hyperphase and 40% formamide in the hypophase (Davis et al. 1971). A CH790 circle containing one R-loop region is shown in *a*, and a portion of a CH791 circle containing two R-loop regions is shown in *b*.

(1976), as well as the *rrnD* gene proposed here, all have the same polarity; that is, they occur as direct repeats. Uneven recombination between *rrn* genes that occur as direct repeats leads to duplications and deletions. Neither of these occurrences would have particularly detrimental consequences to an *E. coli* population. The large deletions generally would be lethal because of the loss of essential genes, but this would not matter in a large population. Large tandem duplications are inherently unstable, and segregation would occur unless selective pressure is present, as in the $glyT^{+}/glyTsu_{AGA}$ mutants. If *rrn* genes are present elsewhere in the chromosome as inverted repeats, a possibly more troublesome situation might arise. For example, intrachromosomal recombination between *rrn* genes which are inverted repeats would lead to a large inversion of all of the material between them. Survival and establishment of such cells could lead to major problems in future matings with other *E. coli* that retain the original order. On this basis, one might predict that as other *rrn* genes are identified, they will have the same polarity as the ones already established.

SUMMARY

Representatives of two types of recurring tandem duplications in *Escherichia coli* have been studied. One of these tandem duplications begins in the ribosomal RNA gene *rrnA* and the other begins in *rrnB*. Both extend into the region between *purD* and *metA*, where we predict there is another ribosomal RNA gene. Generation of these tandem duplications appears to have involved uneven recombination between the directly repeated *rrn* genes.

Note Added in Proof

Yamamoto and Nomura (1976) have obtained direct evidence for the occurrence of a ribosomal RNA gene between *purD* and *metA*, which they have called *rrnE*.

Acknowledgments

This work was supported by U.S. Public Health Service Grant GM-16329 from the National Institutes of Health. We thank Drs. Sankar Adhya and Janet Geisselsoder for phage strains.

REFERENCES

Bachmann, B. J., K. B. Low and A. L. Taylor. 1976. Recalibrated linkage map of *E. coli* K-12. *Bact. Rev.* **40**:116.

Bazaral, M. and D. R. Helinski. 1968. Circular DNA forms of colicinogenic factors E1, E2, and E3 from *E. coli. J. Mol. Biol.* **36**:185.

Carbon, J., C. Squires and C. W. Hill. 1970. Glycine transfer RNA of *E. coli.* II. Impaired GGA-recognition in strains containing a genetically altered transfer RNA; reversal by a secondary suppressor mutation. *J. Mol. Biol.* **52**:571.

Davis, R. W., M. N. Simon and N. Davidson. 1971. Electron microscope heteroduplex methods for mapping regions of base sequence homology in nucleic acids. In *Methods in enzymology* (ed. L. Grossman and K. Moldave), vol. 21, p. 413. Academic Press, New York.

Deonier, R. C., E. Ohtsubo, H.-J. Lee and N. Davidson. 1974. Electron microscope heteroduplex studies of sequence relations among plasmids of *E. coli.* VII. Mapping the ribosomal RNA genes of plasmid F14. *J. Mol. Biol.* **89**:619.

Goldstein, L., M. Thomas and R. W. Davis. 1975. *Eco*R1 endonuclease cleavage map of bacteriophage P4-DNA. *Virology* **66**:420.

Hill, C. W. and G. Combriato. 1973. Genetic duplications induced at very high frequency by ultraviolet irradiation in *E. coli. Mol. Gen. Genet.* **127**:197.

Hill, C. W., C. Squires and J. Carbon. 1970. Glycine transfer RNA of *Escherichia coli.* I. Structural genes for two glycine tRNA species. *J. Mol. Biol.* **52**:557.

Hill, C. W., J. Foulds, L. Soll and P. Berg. 1969. Instability of a missense suppressor resulting from a duplication of genetic material. *J. Mol. Biol.* **39**:563.

Kirschbaum, J. B. and E. B. Konrad. 1973. Isolation of a specialized lambda transducing bacteriophage carrying the beta subunit gene for *E. coli* ribonucleic acid polymerase. *J. Bact.* **116**:517.

Lund, E., J. E. Dahlberg, L. Lindahl, S. R. Jaskunas, P. P. Dennis and M. Nomura. 1976. Transfer RNA genes between 16S and 23S rRNA genes in rRNA transcription units of *E. coli. Cell* **7**:165.

Roberts, J. W. and J. Carbon. 1975. Nucleotide sequence studies of normal and genetically altered glycine transfer RNA from *E. coli. J. Biol. Chem.* **250**:5530.

Thomas, M., R. L. White and R. W. Davis. 1976. Hybridization of RNA to double stranded DNA: Formation of R-loops. *Proc. Nat. Acad. Sci.* **73**:2294.

Yamamoto, M. and M. Nomura. 1976. Isolation of λ transducing phages carrying rRNA genes at the *metA-purD* region of *Escherichia coli* chromosome. *FEBS Letters* **72**:256.

Yamamoto, M., L. Lindahl and M. Nomura. 1976. Synthesis of ribosomal RNA in *E. coli*: Analysis using deletion mutants of a λ transducing phage carrying ribosomal RNA genes. *Cell* **7**:179.

Potential of RP4::Mu Plasmids for In Vivo Genetic Engineering of Gram-negative Bacteria

J. Dénarié, C. Rosenberg, B. Bergeron, C. Boucher, M. Michel and M. Barate de Bertalmio
Laboratoire de Génétique des Microorganismes
Station de Pathologie Végétale, I.N.R.A.
78000 Versailles, France

The temperate bacteriophage Mu is a powerful tool for genetic manipulations (see other papers in this book and Howe and Bade 1975). It has been used primarily in *Escherichia coli* K12 since it can grow only in a few other strains of species belonging to the family Enterobacteriaceae. It can infect *Shigella dysenteriae* (Taylor 1963), *Citrobacter freundii* (de Graaf et al. 1973), and some non-nitrogen-fixing strains of *Klebsiella pneumoniae* (Nagaraja Rao and Pereira 1975). It does not form plaques on other enterobacteria such as *Salmonella typhimurium, E. coli* C, B, S, or W (Taylor 1963), and nitrogen-fixing *K. pneumoniae* strains (Nagaraja Rao and Pereira 1975). It would be useful, therefore, to extend the host range of this phage, in particular to bacteria that play a role in such important biological phenomena as nitrogen fixation and plant pathology.

It was suggested by de Graaf et al. (1973) that a simple procedure for introducing Mu into different resistant strains would be to transfer to these strains a transmissible plasmid carrying a Mu prophage. The sex factor F has routinely been used for transferring Mu from one *E. coli* strain to another (Schröder and van de Putte 1974; Razzaki and Bukhari 1975). But the host range of F is limited to enterobacteria related to *E. coli.* The R factor RP4, however, has a very broad host range. This plasmid was isolated originally from *Pseudomonas aeruginosa* (Datta et al. 1971) and, like other plasmids of the P incompatibility group, can undergo transfer to different species of gram-negative bacteria (Datta and Hedges 1972; Olsen and Shipley 1973). RP4 thus might serve as a vehicle for interspecies transfer of Mu.

We describe in this paper methods for isolating RP4::Mu (RP4 plasmid carrying prophage Mu) and for transferring RP4::Mu to three phylogenetically distant hosts: (1) the nitrogen-fixing bacteria *K. pneumoniae* M5a1 (family Enterobacteriaceae), (2) the root-nodule bacteria *Rhizobium meliloti* 2011 (family Rhizobiaceae), and (3) the phytopathogenic bacteria *Pseudomonas solanacearum* GMI 1000 (family Pseudomonadaceae). RP4::Mu, being a broadly transferable replicon carrying the potent insertion system of Mu, is potentially a powerful tool for in vivo genetic engineering of gram-negative bacteria; examples in *E. coli* and *K. pneumoniae* will be given.

The bacterial strains used in this study are described in Table 1 and the bacteriophage strains and plasmids are described in Table 2.

Insertion of Mu into RP4

The *E. coli* J53 strain carrying RP4 was infected with Mu or the thermoinducible mutant Mu *cts*62 to isolate Mu insertions in RP4. In a preliminary screening, none of the 5000 colonies surviving Mu infection had lost any of the antibiotic resistance markers of RP4. Screening for loss of antibiotic resistance is thus not a convenient method for isolating Mu insertions in this plasmid. The methods described below were therefore adopted for this purpose.

Insertion of Mu c^+ (carrying the wild-type immunity gene *c*) was detected as follows: After infection of J53 (RP4) with Mu c^+, the cells were mated at 28°C on a filter membrane with a Mu *cts*62 lysogen of AB1157. The R^+ exconjugants obtained were screened for heat resistance. Since the presence of Mu c^+ is expected to block the induction of Mu *cts*62 at high temperature (presumably because the c^+ allele determining the wild-type repressor is dominant over the *cts*62 allele determining thermosensitive repressor), thermoresistant R^+ derivatives of AB1157 Mu *cts*62 were suspected to have received RP4::Mu. The frequency of such clones was 0.5% of R^+ exconjugants.

To detect insertion of Mu *cts*62 into RP4, the Mu-infected J53 (RP4) cells were crossed at 28°C with a nonlysogenic Mu-resistant recipient. R^+ exconjugants which became thermosensitive were assumed to have acquired RP4::Mu *cts*62. They appeared at a frequency of 0.01% among the R^+ exconjugants. The low frequency probably reflects the killing of exconjugants caused by zygotic induction of the incoming Mu.

An alternate method was devised for screening for either Mu c^+ or Mu *cts*62 insertion into RP4. It was based on the fact that the presence of a Mu prophage in the RP4 strongly decreases its transfer frequency from *E. coli* K12 to *R. meliloti* 2011 (see below). *E. coli* (RP4) donors produce confluent growth in spot matings with *R. meliloti*, whereas donors with RP4::Mu plasmids give no growth. Single colonies obtained after phage infection of J53 (RP4) cells were spot-mated at 28°C onto a lawn of *R. meliloti* 2011 on selective media. Colonies which gave no R^+ exconjugants of *R. meliloti* were candidates for harboring RP4::Mu *cts*62 (or Mu c^+). The frequency of clones which gave no growth in spot matings was about 0.5%.

Three tests were used to confirm the presence of a Mu prophage in RP4: (1) occurrence of zygotic induction in subsequent transfers to nonlysogenic Mu-resistant *E. coli* recipients (the R-plasmid transfer decreased by 10–100-fold for RP4::Mu plasmids as compared to RP4); (2) 100% cotransfer of antibiotic resistance and of prophage Mu in further crosses; and (3) occurrence of polarized transfer of the chromosome when lysogenic R^+ *E. coli* donors containing a Mu inserted in the chromosome were crossed to Mu-immune recipients (see below). For a given RP4::Mu, the direction of transfer depended on the

Table 1
Bacterial Strains Used

Species and strains	Genotypes[a]	Reference or source
E. coli K12[b]		
K12S	wild-type, λ^-	
J53 (RP4)	*pro1*, *met2*	Clowes and Hayes (1968); Datta et al. (1971)
C600	*thi*, *thr*, *leu*, rK^-, mK^-	Thomas et al. (1974)
AB1157	*lacY*, *ara*, *xyl*, *mtl*, *gal*, *thr*, *leu*, *pro*, *his*, *argB*, *tsx*, *supE*, *strA*	Howard-Flanders and Theriot (1966)
AB2463	same as AB1157 but *recA13*	Howard-Flanders and Theriot (1966)
PP54	*thi*, *thyA*, *trp*::(–Mu-1)	Wijffelman et al. (1973)
PP56	*thi*, *thyA*, *trp*::(+Mu-1)	Wijffelman et al. (1973)
JC1553	*his*, *arg*, *leu*, *met*, *recA*, *str*	Olsen and Gonzalez (1974)
JC5466	*trp*, *his*, *recA56*, *spc*	Dixon et al. (1976)
K. pneumoniae M5a1	wild-type	Mahl et al. (1965)
Kp5022	*his*, *hsp*, R1, *str*	Streicher et al. (1974)
P. solanacearum GMI 1000	wild-type	Message et al. (1975)
R. meliloti 2011	*str3*	Scherrer and Dénarié (1971)

[a]Nomenclature as proposed by Demerec et al. (1966); rK^-, mK^-: deficient for the restriction and modification, respectively, of the K system of DNA.
[b]All the *E. coli* derivatives used were F^-.

Table 2
Phages and Plasmids Used

Designation	Description	Reference or source
Phages		
Mu	wild-type	Taylor (1963)
Mu *cts*62	thermoinducible mutant	Howe (1973)
Mu *cts*61	thermoinducible mutant	Howe (1973)
Plasmids		
RP4	Ap Km Tc	Datta et al. (1971)
p GMI 2[a]	RP4::Mu[b]	this paper
p GMI 4	RP4::Mu *cts*62	this paper
p GMI 7	mutant of p GMI 4 which can be transferred at high frequency to *R. meliloti* 2011	this paper

[a]Plasmid designation as proposed by Novick et al. (1976).
[b]For nomenclature of insertion elements, see elsewhere in this volume.

orientation of the Mu prophage on the chromosome, and the orientation of the Mu prophage in RP4 could be deduced from the direction of chromosome transfer (Zeldis et al. 1973).

The frequency of the RP4::Mu$^+$ bacteria found in a Mu-infected population was around 0.5%. In these populations, the proportion of lysogenic bacteria ranged from 20% to 80%. The proportion of RP4::Mu$^+$ among lysogenized bacteria is therefore about 1%. This proportion is not very different from the ratio between the molecular weights of RP4 DNA and chromosomal DNA. We can infer that Mu is able to insert into RP4 DNA with the same efficiency as for *E. coli* chromosomal DNA despite their very different G + C compositions—58% for RP4 versus 50% for chromosomal DNA (Falkow et al. 1974). This suggests that DNA composition should not be a barrier for insertion of Mu into the DNA of organisms genetically distant from *E. coli.*

Transfer of RP4::Mu to Bacteria Other Than *E. coli*

E. coli K12 donor strains carrying either RP4, RP4::Mu, or RP4::Mu *cts*62 were mated with wild-type *P. solanacearum* GMI 1000 and Str^r derivatives of *K. pneumoniae* M5a1 and *R. meliloti* 2011. Drug resistance markers were used to select for R$^+$ exconjugants (Rosenberg and Dénarié 1976; Boucher et al. 1977).

Table 3 shows that the three plasmids are transferred to *P. solanacearum* at the same frequency, indicating that the transfer of the Mu genome into this strain of *Pseudomonas* does not result in zygotic induction or restriction. On the contrary, the transfer of antibiotic resistance to *R. meliloti* occurred at a much lower frequency when the R$^+$ donors carried a Mu c^+ or a Mu *cts* prophage inserted in RP4. Heat treatment of the recipient *Rhizobium* cultures

Table 3
Transfer Frequencies of RP4 and RP4::Mu Plasmids from *E. coli* K12 Strains to *P. solanacearum* GMI 1000 and *R. meliloti* 2011 Strains

R^+ donors	Plasmids	R^- recipients	Transfer frequencies[a]
E. coli J53 (Mu) Met^- Pro^-	RP4	*R. meliloti* 2011 Str^r	3×10^{-4}
	RP4::Mu	*R. meliloti* 2011 Str^r (heat-treated)[b]	2×10^{-8}
	RP4::Mu	*R. meliloti* 2011 Str^r (non–heat-treated)	2×10^{-9}
E. coli K12S (Mu *cts*62)	RP4	*R. meliloti* 2011 Str^r	9.2×10^{-3}
	RP4::Mu *cts*62	*R. meliloti* 2011 Str^r	9.9×10^{-8}
	RP4::Mu *cts*62-*r*23	*R. meliloti* 2011 Str^r	4.8×10^{-3}
E. coli AB1157 (Mu *cts*62) poly-auxotrophic, Str^r	RP4	*P. solanacearum* GMI 1000 (wild-type)	10^{-4}
	RP4::Mu	*P. solanacearum* GMI 1000 (wild-type)	2×10^{-4}
	RP4::Mu *cts*62	*P. solanacearum* GMI 1000 (wild-type)	2×10^{-4}

Crosses were performed on filter membranes for 5 hr at 28°C.
[a]Per initial donor.
[b]6 min at 47°C.

before mating increased the transfer frequency about ten times, suggesting that some restriction of the incoming DNA occurs. Heat-sensitive restriction has been described in *Rhizobium* (Schwinghamer 1966).

Twenty R^+ clones from each cross were purified and checked for their strain-specific characters: *P. solanacearum* R^+ clones for phage *a* sensitivity and tomato wilting; *R. meliloti* for phage π sensitivity, efficient nodule formation on alfalfa, and inability to grow in anaerobic conditions with nitrate; and *K. pneumoniae* for nitrogenase activity.

The presence of Mu prophage in the new hosts was checked by retransfer of the plasmids to *E. coli* K12 strains. *P. solanacearum* and *K. pneumoniae* strains carrying RP4::Mu or RP4::Mu *cts*62 were able to transfer these plasmids normally with their Mu prophages. However, seven out of eight R^+ *Rhizobium* clones, putatively bearing RP4::Mu *cts*62, failed to transfer thermosensitivity (i.e., normal Mu *cts*62) to *E. coli.* This could mean that some damage to the RP4::Mu plasmids occurs upon transfer to *R. meliloti* 2011 such that the Mu prophage becomes defective. It is also possible that when RP4::Mu plasmids are transferred to *R. meliloti* 2011, plasmids carrying a defective prophage are selected from the donor population. The clone carrying a nondefective prophage contained a RP4::Mu *cts* mutant plasmid which could be transferred from *E. coli* to *R. meliloti* at the same frequency as RP4 (Tables 2 and 3). On the other hand, when the *Rhizobium* recipient culture was mutagenized by nitrosoguanidine before mating, the proportion of R^+ exconjugants that carried a nondefective prophage was clearly increased and became higher than 80%. After curing, most of these R^+ transconjugants were shown to be mutants into which RP4::Mu *cts* could be transferred from *E. coli* at high frequencies (J. S. Julliot, pers. comm.).

Another interesting feature of the *Rhizobium* strains is that the R^+ strains carrying either RP4::Mu or RP4::Mu *cts*62 form small colonies on selective agar plates. The small colonies segregate large colonies on media without antibiotics, and the large colonies are sensitive to tetracycline and presumably result from the loss of the plasmids containing Mu. The generation time in liquid cultures of *Rhizobium* strains containing RP4 with Mu prophages is about twice as long as that for strains containing RP4 without Mu.

Production of Mu Phage by the New Hosts

Supernatants of chloroform-treated cultures of *K. pneumoniae, P. solanacearum,* and *R. meliloti* carrying RP4::Mu *cts*62 plasmid formed plaques on *E. coli* indicator strains (Table 4). No plaques were seen either on Mu-resistant or Mu lysogenic strains. It was clear, therefore, that these strains are capable of produc-

Table 4
Titration on *E. coli* K12 Strains of pfu Present in the Supernatant of Cultures of *P. solanacearum* GMI 1000, *R. meliloti* 2011, *K. pneumoniae* M5a1, and *E. coli* J53 Carrying a RP4::Mu *cts*62 Plasmid

Phage-producing strains		Indicator strains: wild-type strain	Indicator strains: C600 (restriction-deficient)
P. solanacearum[a] GMI 1000 (RP4::Mu *cts*62)	pfu/ml	80	3.0×10^5
	relative e.o.p.	2.5×10^{-4}	1
R. meliloti 2011[b] Str^r Gly (RP4::Mu *cts*62)	pfu/ml	2.15×10^2	5.3×10^2
	relative e.o.p.	0.4	1
K. pneumoniae M5a1[c] (RP4::Mu *cts*62)	pfu/ml	3×10^6	1.5×10^8
	relative e.o.p.	0.02	1
E. coli J53[c] (RP4::Mu *cts*62)	pfu/ml	2.3×10^8	2.6×10^8
	relative e.o.p.	$\cong 1$	1

[a] *P. solanacearum* cultures were incubated at 30°C for 8 hr in aerated liquid medium.
[b] *R. meliloti* cultures were grown at 35°C for 6 hr in aerated liquid medium.
[c] Cultures were grown at 28°C and shifted to 42°C during 2 hr. At the time of beginning the induction, the cultures contained 1 to 1.5×10^8 bacteria/ml.

ing viable Mu particles. Phage obtained from the *Pseudomonas* strain was plaque-purified and was confirmed to be Mu by testing its mutagenic properties in *E. coli.*

Mu particles obtained from the different strains plated on an *E. coli* K12 restriction-deficient strain with a higher efficiency (e.o.p.) than on a wild-type *E. coli* strain. The expression of the Mu genome evidently varies with the strain. In *P. solanacearum*, spontaneous production of Mu *cts*62 at 30°C was comparable to that found in *E. coli* ($\sim 10^5$ pfu/ml); increasing the temperature to 35°C had no effect. In *R. meliloti*, the phage production was low at 30°C but increased at 35°C. *P. solanacearum* and *R. meliloti* cannot grow at 40°C, and incubation of the cultures at this temperature for one generation time decreased phage production. Both of these strains are strictly aerobic, and phage production is thus dependent upon good aeration.

Use of RP4 Plasmids Carrying Mu Insertions

The finding that the Mu genome can be expressed in new hosts genetically distant from *E. coli* opens the way for the possible use of Mu as a tool for studying bacteria whose genetics are practically unknown (Levinthal 1974). It would be of great interest to obtain gene transfer by conjugation in different bacteria, especially in bacteria of economic importance. This can be done by having one Mu prophage on a transmissible plasmid and one on the chromosome in a bacterial strain (Zeldis et al. 1973). The bacterial chromosome in such a strain is mobilized during conjugation because of recombination between the two prophages. Because of its wide host range, the RP4 plasmid can be used to set up such gene-transfer systems.

RP4-promoted Polarized Chromosome Transfer in E. coli

RP4 ordinarily promotes gene transfer in *E. coli* at a very low frequency of 10^{-7}-10^{-8} (Olsen and Gonzales 1974). *E. coli* strains carrying RP4 and a Mu prophage on the chromosome transfer chromosomal markers at a frequency of less than 2×10^{-7}. However, strains carrying RP4::Mu and a Mu prophage on the chromosome transfer markers at higher frequency. The transfer occurs in an oriented manner, and for a marker about 10 minutes away from prophage Mu, the frequency is approximately 10^{-4}. The polarized transfer was examined by using strains PP54 and PP56, both of which have Mu inserted in the *trp* operon but the Mu prophages are in opposite orientations. The donor PP54 and PP56 strains carrying RP4::Mu were mated on filter membranes with a Mu lysogenic recipient AB1157 (to avoid zygotic induction of the incoming Mu), and the transfer of markers was examined at different times after interrupting the matings by vortexing the mating mixture for 2 minutes. The markers were transferred in an oriented manner. With PP54 (RP4::Mu) as the donor, *pro* was transferred with the highest frequency, followed by *thr-leu, arg,* and *his*; with PP56 (RP4::Mu), *his* had the best transfer frequency among the markers examined. The principle of such experiments with F′ episomes has been described previously by Zeldis et al. (1973). It is clear that with RP4::Mu plasmids too, the chromosomal mobilization is initiated at the point of Mu insertion in the chromosome. Thus RP4 has information for the machinery required to promote chromosome transfer: the only requirement is to provide DNA homology between plasmid and chromosome. It should be noted that with Mu DNA as the source of homology between the transmissible plasmid and the host chromosome, the marker which should be transferred last is transferred early at a low frequency. This is because of the inversion of the G segment of Mu (Hsu and Davidson 1972, 1974; Allet and Bukhari 1975; other articles in this book). The mobilization of the chromosome is in fact bidirectional–overwhelmingly so in one direction but at a low efficiency in the other direction. If necessary, the small amount of transfer in the opposite direction can be avoided by using Mu mutants that are unable to invert the G region (Chow et al., this volume).

RP4-mediated Gene Transfer in E. coli
following Prophage Mu cts *Induction*

It has been shown that copies of Mu DNA are integrated at different sites during Mu replication and growth in a cell (Razzaki and Bukhari 1975; Parker and Bukhari 1976). Thus, homologies between a plasmid and host chromosome would be generated after prophage induction even if there is only one copy of Mu to begin with, either on the chromosome or on the episome. Furthermore, Mu can cause integration of a circular DNA, such as phage λ DNA and F factor, into the host chromosome (van de Putte and Gruijthuijsen 1972; Toussaint and Faelen 1973; Faelen et al. 1975). Therefore it can be expected that the chromosome will be mobilized by RP4 after prophage Mu induction. In our experiments with a strain carrying Mu *cts* on the chromosome and RP4 without Mu, we could detect gene transfer by mating the cells with a Mu lysogenic recipient after heating the donor cells for prophage induction. As expected from the random integration of Mu and the plasmid after phage induction, this transfer was not in an oriented manner, and all markers were transferred at relatively high frequency. The results of such experiments are shown in Table 5.

Table 5 shows that gene transfer was dependent upon the presence of RP4 and the induction of Mu *cts* and that a partial overnight induction at 36°C is more efficient than a brief induction at 41°C. The ability of *recA*$^-$ strains to act as recipients in these experiments indicates that at least part of the transfer was due to formation of RP4-prime episomes. Most of the *recA*$^-$ exconjugants were able to transfer the markers along with RP4 in secondary matings with a *recA*$^-$ recipient, confirming the presence of RP4-prime plasmids (episomes). About 8% of these episomes transferred *thr, leu,* and *pro* and probably represented large RP4-prime episomes. When *recA*$^+$ recipients were used, about 10% to 30% of the exconjugants carried the markers on the RP4-prime episomes, the rest apparently being recombinants possibly due to the formation of random

Table 5
E. coli K12 Chromosomal Gene Mobilization by RP4 Using a Thermoinducible Mutant of Bacteriophage Mu

Description of crosses			Gene transfer frequencies (per 10^7 donors)			
donor strains	incubation temperature of donor cultures (°C)	recipient strains	*thr-leu*	*argB*	*his*	*pro*
		recA strain				
		AB2463 (Mu *cts*62)	–	3	1	–
K12S (Mu *cts*61)	36	AB2463 (Mu *cts*62)	1	3	–	2
K12S (Mu *cts*61)(RP4)	28	AB2463 (Mu *cts*62)	–	2	3	–
K12S (Mu *cts*61)(RP4)	36	AB2463 (Mu *cts*62)	46	128	25	25
K12S (Mu *cts*61)(RP4)	28 and 25 min at 41 before mating	AB2463 (Mu *cts*62)	20	60	18	19
		Rec$^+$ strain				
K12S (Mu *cts*61)(RP4)	36	AB1157 (Mu *cts*62)	227	426	254	348

The donor-recipient ratio was 1/2 = 10^8 donors and 2×10^8 recipients per ml. The mixture was filtered on a membrane (pore size = 0.45 μ) and incubated at 28°C during 3 hr.

Mu-promoted Hfr formation in the donor population (van de Putte and Gruijthuijsen 1972). Gene transfer was also observed when Mu *cts* lysogenic RP4 donors were grown overnight at 34°C (J. Dénarié, unpubl.). This method might be useful for bacteria that cannot grow at 42°C or 37°C.

Formation of RP4-prime episomes with different markers is consistent with the finding of Faelen and Toussaint (1976) that Mu can mediate transposition of genes from the chromosome to an episome.

RP4-Mu System of Gene Transfer in K. pneumoniae

To test whether the system of gene transfer based on RP4 and Mu will work in bacteria other than *E. coli,* we inserted Mu *cts* into the chromosome of *K. pneumoniae* by the following procedure: RP4::Mu *cts*62 was introduced by conjugation into *K. pneumoniae* M5a1, and the strain was grown overnight at 38°C. This partial induction was mutagenic for the host, and the frequency of chlorate-resistant mutants (see Boram and Abelson 1973) among survivors increased from 3×10^{-6} to 2×10^{-5}. Ten chlorate-resistant mutants were cured of the plasmid; curing was checked by examining the loss of the three RP4 drug resistance markers and the sensitivity to GU5, a phage specific for RP4$^+$ strains (J. Schell, pers. comm.). Eight of the cured strains were still lysogenic for Mu *cts*62 and were thus assumed to have Mu *cts*62 inserted into the chromosome.

RP4 without Mu was then reintroduced into one of the Mu *cts*62 *Klebsiella* lysogens, and the resultant M5a1 (RP4) (Mu *cts*62) strain was partially induced at 37°C and mated with a His^- Str^r *K. pneumoniae* (5022) recipient. The *hisD* marker was transferred at a frequency of 5×10^{-6}. This frequency is probably an underestimation since the recipient was not lysogenic for Mu and thus zygotic induction would be expected to decrease the number of recombinants.

The *Klebsiella* strains carrying Mu *cts*62 and RP4 were also able to transfer their markers to *E. coli.* We therefore tested whether the *Klebsiella nif* genes, controlling nitrogen fixation, can be transposed onto RP4 and transferred to *E. coli.* In *K. pneumoniae,* the *nif* genes are clustered in the *his* region of the chromosome (Streicher et al. 1971; Dixon and Postgate 1971; Brill 1975). The donor *Klebsiella* strain carrying Mu *cts* and RP4 was partially induced and crossed with His^- *recA* Mu lysogenic *E. coli* strain JC1553. The recipient *E. coli* strain was heat-treated for 5 minutes at 50°C to impair the activity of the restriction enzymes. The His^+ exconjugants were purified and then mated with a secondary His^- *recA E. coli* recipient. The *his* marker and antibiotic resistance were found to be cotransferred, and most of the His^+ clones thus obtained showed nitrogenase activity, as measured by reduction of acetylene. Apparently, therefore, the *Klebsiella nif* genes were successfully transposed to RP4 and transferred to *E. coli.* The same procedure using a *gln E. coli* recipient led to isolation of RP4 *gln* episomes (C. Rosenberg and J. Dénarié, in prep.).

Evaluation of the Potential of RP4::Mu Plasmids

It should be possible to use RP4::Mu plasmids for introducing Mu into a wide assortment of bacterial species. Two problems can be encountered in transferring RP4::Mu plasmids to new strains by conjugation: (1) The frequency of clones which stably inherit RP4::Mu may be very low because of zygotic induction of Mu in new strains, and (2) the transfer frequency of RP4::Mu may be reduced because of mechanisms such as restriction. In either case, the availability of powerful drug resistance markers on RP4 should allow the isolation of RP4::Mu-carrying exconjugants, even if the frequency is low. Furthermore, it should be possible to isolate mutants either of RP4::Mu plasmids or of the recipient strains (as has been done in the case of *R. meliloti* 2011) which allow a high transfer frequency of the hybrid plasmid. Some of these mutants could be deficient in DNA restriction mechanisms and would be useful for further RP4::Mu-mediated interspecific or intergeneric gene transfer. Heat treatment of recipient strains may also help to reduce restriction of incoming plasmids. de Graaf et al. (1973) have demonstrated that heat treatment of *C. freundii* increases the efficiency of plating of Mu by about 1000-fold (normal e.o.p. of Mu on *C. freundii* $\sim 5 \times 10^{-8}$).

In nonenteric bacteria such as *R. meliloti* and *P. solanacearum,* the Mu genome is clearly expressed. The production of infectious Mu particles, however, is quite low in comparison to that in *E. coli.* The reasons for this difference have not been studied. That Mu grows in the new hosts at all implies that Mu DNA can become integrated with the host genome at different sites in the new strains. This inference follows from our knowledge that extensive integration of Mu occurs during vegetative phage growth, and this integration is probably a required step in the life cycle of Mu (Razzaki and Bukhari 1975; Bukhari 1976). The integration machinery of Mu is functional under conditions of partial induction at 37°C (Faelen et al. 1975). Thus it should be possible to obtain Mu insertions at different sites in hosts that do not grow well at temperatures above 37°C. The problem now is to devise methods that allow one to select for (1) Mu insertions into the chromosome of the new hosts and (2) Mu mutants able to be induced in these new hosts.

The Mu genome apparently is fully expressed in enteric bacteria, both closely related and not so closely related to *E. coli.* de Graaf et al. (1973) have shown that Mu is able to lysogenize *C. freundii* and to cause mutations in it. Mu was also able to mobilize the chromosome in F′-carrying *C. freundii.* Among enterobacteria not closely related to *E. coli,* Mu carried on RP4 has been transferred to *K. pneumoniae,* as reported here, and to *Erwinia carotovora.* In *E. carotovora,* heat induction of Mu *cts* was found to occur (M. C. Perombelon and C. Boucher, pers. comm.). Thus it should be quite possible to use Mu for genetic manipulations in different enterobacteria in the same manner in which it is used in *E. coli* (see Howe and Bade 1975). For example, it should be simple to obtain chromosomal mobilization, to isolate episomes with different host genes, and to mobilize nonconjugative plasmids in enterobacteria. These techniques in turn would make possible genetic analysis of nitrogen fixation in *K. pneumoniae,* of virulence and host specificity in bacteria important in plant pathology (*Erwinia* species *amylovora, herbicola,* and *carotovora*), and of pathogenic mechanisms of species of *Salmonella* and *Shigella.*

SUMMARY

We have described methods for isolating bacteriophage Mu insertions in RP4, a broadly transferrable plasmid of the P incompatibility group. The RP4::Mu plasmids have been conjugatively transferred from *E. coli* to three phylogenetically distant hosts–*Klebsiella pneumoniae, Rhizobium meliloti,* and *Pseudomonas solanacearum.* The Mu genome is expressed in these hosts to varying degrees. The RP4::Mu plasmids have been used to mobilize chromosomal markers in *E. coli* and *K. pneumoniae* during conjugation. We were able to use Mu to transpose *K. pneumoniae* genes onto RP4 and promote intergeneric transfer. Since RP4 can be transferred to many different species, RP4::Mu plasmids should be generally useful for genetic studies of gram-negative bacteria.

Acknowledgments

The authors are very grateful to P. Boistard for stimulating discussions and to A. I. Bukhari for reviewing the manuscript. This work was supported, in part, by grant no. 74 7 0116 from the Delegation Générale à la Recherche Scientifique et Technique, and by grant no. 414 from the Plant Protein Programm of the Commission of the European Communities.

REFERENCES

Allet, B. and A. I. Bukhari. 1975. Analysis of Mu and λ-Mu hybrid DNAs by specific endonucleases. *J. Mol. Biol.* **92**:529.

Boram, W. and J. Abelson. 1973. Bacteriophage Mu integration: On the orientation of the prophage. *Virology* **54**:102.

Boucher, C., B. Bergeron, M. Barate de Bertalmio and J. Dénarié. 1977. Introduction of bacteriophage Mu-1 into *Pseudomonas solanacearum* and *Rhizobium meliloti* using the R factor RP4. *J. Gen. Microbiol.* **98**:253.

Brill, W. Y. 1975. Regulation and genetics of bacterial nitrogen fixation. *Annu. Rev. Microbiol.* **29**:109.

Bukhari, A. I. 1976. Bacteriophage Mu as a transposition element. *Annu. Rev. Genet.* **10**:389.

Clowes, R. C. and W. Hayes. 1968. *Experiments in microbial genetics.* Blackwell Scientific Publications, Oxford.

Datta, N. and R. W. Hedges. 1972. Host ranges of R factors. *J. Gen. Microbiol.* **70**:453.

Datta, N., R. W. Hedges, E. J. Shaw, R. B. Sykes and M. H. Richmond. 1971. Properties of an R factor from *Pseudomonas aeruginosa. J. Bact.* **108**:1244.

de Graaf, J., P. C. Kreuning and P. van de Putte. 1973. Host controlled restriction and modification of bacteriophage Mu and Mu-promoted chromosome mobilization in *Citrobacter freundii. Mol. Gen. Genet.* **123**:283.

Demerec, M., E. A. Adelberg, A. J. Clark and P. E. Hartman. 1966. A proposal for a uniform nomenclature in bacterial genetics. *Genetics* **54**:61.

Dixon, R. A. and J. R. Postgate. 1971. Genetic transfer of nitrogen fixation from *Klebsiella pneumoniae* to *Escherichia coli. Nature* **237**:102.

Dixon, R., F. Cannon and A. Kondorosi. 1976. Construction of a P plasmid carrying nitrogen fixation genes from *Klebsiella pneumoniae. Nature* **260**:268.

Faelen, M. and A. Toussaint. 1976. Bacteriophage Mu-1: A tool to transpose and to localize bacterial genes. *J. Mol. Biol.* **104**:525.

Faelen, M., A. Toussaint and Y. de Lafonteyne. 1975. Model for the enhancement of λ*gal* integration into partially induced Mu-1 lysogens. *J. Bact.* **121**:873.

Falkow, S., P. Guerry, R. W. Hedges and N. Datta. 1974. Polynucleotide sequence relationships among plasmids of the I compatibility complex. *J. Gen. Microbiol.* **85**:65.

Howard-Flanders, P. and L. Theriot. 1966. Mutants of *Escherichia coli* K-12 defective in DNA repair and in genetic recombination. *Genetics* **53**:1137.

Howe, M. M. 1973. Prophage deletion mapping of bacteriophage Mu-1. *Virology* **54**:93.

Howe, M. M. and E. G. Bade. 1975. Molecular biology of bacteriophage Mu. *Science* **190**:624.

Hsu, M. and N. Davidson. 1972. Structure of inserted bacteriophage Mu-1 DNA and physical mapping of bacterial genes by Mu-1 DNA insertion. *Proc. Nat. Acad. Sci.* **69**:2823.

———. 1974. Electron microscope heteroduplex study of the heterogeneity of Mu phage and prophage DNA. *Virology* **58**:229.

Levinthal, M. 1974. Bacterial genetics excluding *E. coli. Annu. Rev. Microbiol.* **28**:219.

Mahl, M. C., P. W. Wilson, M. A. Fife and W. H. Ewing. 1965. Nitrogen fixation by members of the tribe *Klebsielleae. J. Bact.* **89**:1482.

Message, B., C. Boucher and P. Boistard. 1975. Transfert d'un facteur R, RP_4 dans une souche de *Pseudomonas solanacearum. Ann. Phytopathol.* **7**:96.

Nagaraja Rao, R. and M. G. Pereira. 1975. Behavior of a hybrid F′ ts 114 *lac*$^+$, *his*$^+$ factor (F 42-400) in *Klebsiella pneumoniae* M5a1. *J. Bact.* **123**:792.

Novick, R. P., R. C. Clowes, S. N. Cohen, R. Curtiss III, N. Datta and S. Falkow. 1976. Uniform nomenclature for bacterial plasmids: A proposal. *Bact. Rev.* **40**:168.

Olsen, R. H. and C. Gonzalez. 1974. *Escherichia coli* gene transfer to unrelated bacteria by a histidine operon-RP1 drug resistance plasmid complex. *Biochem. Biophys. Res. Comm.* **59**:377.

Olsen, R. H. and P. Shipley. 1973. Host range and properties of the *Pseudomonas aeruginosa* R factor R 1822. *J. Bact.* **113**:772.

Parker, V. and A. I. Bukhari. 1976. Genetic analysis of heterogeneous circles formed after prophage Mu induction. *J. Virol.* **19**:756.

Razzaki, T. and A. I. Bukhari. 1975. Events following prophage Mu induction. *J. Bact.* **122**:437.

Rosenberg, C. and J. Dénarié. 1976. Introduction du bactériophage Mu dans une souche fixatrice d'azote *Klebsiella pneumoniae* M5a1. *C. R. Acad. Sci., Ser. D* **283**:423.

Scherrer, A. and J. Dénarié. 1971. Symbiotic properties of some auxotrophic mutants of *Rhizobium meliloti* and of their prototrophic revertants. *Plant and Soil,* special volume, p. 39.

Schröder, W. and P. van de Putte. 1974. Genetic study of prophage excision with a temperature inducible mutant of Mu-1. *Mol. Gen. Genet.* **130**:99.

Schwinghamer, E. A. 1966. Factors affecting phage-restricting ability in *Rhizobium leguminosarum* strain L4. *Can. J. Microbiol.* **12**:395.

Streicher, S., E. Gurney and R. C. Valentine. 1971. Transduction of the nitrogen fixation genes in *Klebsiella pneumoniae. Proc. Nat. Acad. Sci.* **68**:1174.

Streicher, S. L., K. T. Shanmugam, F. Ausubel, C. Morandi and R. B. Goldberg. 1974. Regulation of nitrogen fixation in *Klebsiella pneumoniae*: Evidence for a role of glutamine synthetase as a regulator of nitrogenase synthesis. *J. Bact.* **120**:815.

Taylor, A. L. 1963. Bacteriophage induced mutations in *E. coli. Proc. Nat. Acad. Sci.* **50**:1043.

Thomas, M., J. R. Cameron and R. W. Davis. 1974. Viable molecular hybrids of bacteriophage lambda and eukaryotic DNA. *Proc. Nat. Acad. Sci.* **71**:4579.

Toussaint, A. and M. Faelen. 1973. Connecting two unrelated DNA sequences with a Mu dimer. *Nature New Biol.* **242**:1.
van de Putte, P. and M. Gruijthuijsen. 1972. Chromosome mobilization and integration of F factors in the chromosome of *recA* strains of *E. coli* under the influence of bacteriophage Mu-1. *Mol. Gen. Genet.* **118**:173.
Wijffelman, C., G. C. Westmaas and P. van de Putte. 1973. Similarity of vegetative map and prophage map of bacteriophage Mu-1. *Virology* **54**:125.
Zeldis, J. B., A. I. Bukhari and D. Zipser. 1973. Orientation of prophage Mu. *Virology* **55**:289.

In Vivo Genetic Engineering: The Mu-mediated Transposition of Chromosomal DNA Segments onto Transmissible Plasmids

M. Faelen and A. Toussaint
Université Libre de Bruxelles
Département de Biologie Moléculaire
Rhode St. Genèse, Belgie

M. Van Montagu, S. Van den Elsacker, G. Engler and J. Schell
Rijksuniversiteit Gent
Laboratoria voor Histologie en Genetika
Gent, Belgie

Faelen and Toussaint (1976) have reported that *Escherichia coli* genes are transposed to F′ episomes during growth of bacteriophage Mu. Transposition occurs after either infection of sensitive cells with Mu or heat induction of a Mu *cts* prophage.

The F′ episome that contains the transposed DNA can be rescued by transfer into an appropriate *E. coli* strain which should (1) have genetic markers that allow selection for the desired transposed genes and counter-selection of the donor; (2) be lysogenic for Mu in order to avoid zygotic induction of the Mu prophage linked to the transposed DNA; (3) be resistant to Mu; (4) be recombination-deficient (*recA*$^-$) to avoid both rescue of donor markers transferred, owing to chromosome mobilization, and segregation of the transposed segment by reciprocal recombination between the two Mu genomes which surround it.

Requirement 3 can be circumvented by performing the transposition in a Mu *cts* lysogen at 37°C. In these conditions, very few phage particles are produced, but there is enough induction for transposition to occur. The frequency is approximately 100 times lower at 37°C than after full induction of Mu *cts* (at 42°C).

We describe here the isolation of an RP4 plasmid that carries a strong amber suppressor as a result of Mu-mediated transposition.

Description of the Method

Use of a Mu cts *Lysogen for Transposition*

Cultures of the donor and acceptor strains are grown overnight in L broth (tryptone yeast extract broth) or in selective minimal liquid medium at 30°C or 37°C. After 100-fold dilution in fresh L broth, they are grown at 30°C or 37°C, with aeration, to exponential growth phase (2.5×10^8 bacteria/ml). (If mating is to be performed at 37°C, cultures have to be grown at 37°C in order to have a reasonable transposition frequency.) Viable counts are determined by spreading suitable dilutions on complete medium. Conditions required for mating depend on the plasmid to be used.

For F′ episomes, 0.5 ml of donor is mixed with 0.5 ml of acceptor, and the mixture is incubated at 42°C (or 37°C), without aeration, for 1-2 hours. As a control, 0.5 ml of each culture is mixed with 0.5 ml L broth and treated in the same way as the mating mixture. The cells are resuspended in buffer after centrifugation, and 0.1 ml of the suspension is spread on selective medium. Plates are incubated for 48 hours or longer at 37°C if the recipient strain is lysogenic for Mu c^+ (otherwise at 30°C). Isolated colonies are purified, and the characteristics of the new plasmids are checked. In the case of the RP4 plasmid, 1 ml of the donor and 1 ml of the acceptor bacteria are mixed and collected by filtration on a Millipore filter (HAWP in Swinnex 25). The filter is placed on nutrient agar and incubated at 42°C for 2 hours. The cells are resuspended in 10 ml nutrient broth containing the drugs required to select for the recipient bacteria and then grown to saturation. After suitable dilutions, bacteria are spread on selective media.

Procedure for Infecting Cells with Mu

Overnight cultures of the donor and acceptor strains are grown at 37°C. After 100-fold dilution in fresh L broth, they are grown to exponential phase ($\sim 2 \times 10^8$ cells/ml). One ml of the donor culture is infected with a Mu c^+ at an m.o.i. of 3-5, in the presence of 50 mM Ca^{++}. After 15 minutes absorption at 37°C, 0.5 ml of the infected donor is mixed with 0.5 ml of the acceptor culture, and the mating is allowed to proceed for 1-2 hours at 37°C. The steps that follow are identical to those described above for induction.

Isolation and Characterization of an RP4 Su$^+$ Plasmid

The RP4 plasmid has been described by Datta et al. (1971). We chose an RP4 derivative which carries an amber mutation in the tetracycline resistance locus as the vector plasmid (Van den Elsacker et al. 1975). This allows isolation of Su$^+$ (suppressor$^+$) derivatives by selecting for resistance to tetracycline. The presence of the suppressor gene would cause suppression of the amber mutation, rendering the cells tetracycline-resistant. We used phage A1P2, a derivative of ϕ80*suIII* as a source of Su$^+$ gene (Andoh and Ozeki 1968). The SuIII gene encodes a tRNA that promotes incorporation of glutamine in response to the amber codon in mRNA (Ghysen and Celis 1974). The DNAs of A1P2, RP4, and Mu can be readily purified. We were able to analyze the different plasmid isolates in detail by electron microscope heteroduplex techniques.

A W3110 strain dilysogenic for Mu *cts*62 and A1P2, and carrying the RP4 *tet am* plasmid (the strain is Tet^r because of the presence of *suIII* on the A1P2 genome), was induced and mated at 42°C with strain N100, a Mu^{r}Str^{r}RecA$^-$ Gal$^-$Su$^-$ lysogen of Mu c^+. The streptomycin-, ampicillin-, and tetracycline-resistant sexductants were selected. These sexductants appeared at a frequency of 10^{-6} to 10^{-7} per donor cell. Five isolates were purified and then mated with a new acceptor strain to confirm the linkage of the suppressor gene to the plasmid. One of the isolates was purified extensively, frozen, and then used as a starting culture for all subsequent studies.

Plasmid DNA was purified and analyzed by standard electron microscope techniques (Davis et al. 1971). The plasmid was found to carry a 4-μ-long segment of A1P2, surrounded by two entire Mu prophages in a parallel orientation. The whole plasmid measured 45 μ (RP4 originally is about 17 μm; a single Mu genome length is about 12 μm). A typical picture of RP4::Mu *cts*62-Su A1P2$^+$-Mu *cts*62 hybridized with Mu is shown in Figure 1; the micrograph is interpreted in a tracing in Figure 2. In Figure 3, hybridization of this plasmid with A1P2 phage DNA is shown. About 16% of the plasmid molecules examined had the two Mu G regions (Hsu and Davidson 1974) in opposite orientations (see Fig. 4a). In about 2% of the DNA molecules, the length measurements of the single- and double-stranded regions indicated that inversion of the DNA segment located between the two Mu G regions had occurred (see Fig. 4b). These plasmids apparently carried the SuIII transposed gene between the $\overleftarrow{\alpha}_1 G \overrightarrow{\alpha}_2$ and $\overrightarrow{\beta}_1 G \overleftarrow{\beta}_2$ segments of Mu. Diagrammatic representations of different plasmid structures are given in Figure 5. Since the α region of Mu DNA contains the Mu immunity gene and all of the genes needed for the lytic cycle, we tried to isolate strains in which the $\overleftarrow{\alpha}_1 G \overrightarrow{\alpha}_2$ segment had been deleted by selecting at 42°C for survivors of a strain harboring the RP4::Mu *cts*62-Su A1P2$^+$-Mu *cts*62 plasmid. A purified colony of the strain was grown at 30°C and plated at 42°C. Tet^r Mu-sensitive derivatives were found among the survivors. The overall frequency at which such survivors occurred was about 10^{-6}. The plasmids of four of these strains were purified and analyzed by electron microscope heteroduplex analysis. All four had lost the $\alpha_1 G \alpha_2$ segment. Similar results were obtained when the same procedure was repeated with a *recA* strain, showing that the *recA* host function is not essential to the inversion and excision processes.

As shown in Figure 6, the A1P2 segment of the RP4::Su A1P2$^+$-$\beta_1 G \beta_2$ plasmid was found to be 8.5 μ in length. However, in RP4::Mu *cts*62-Su A1P2$^+$-Mu *cts*62, it was only 4 μ in length (Figs. 2 and 3). Values ranging from 4 μ to 13 μ for the transposed segment were found on other isolates of RP4::Su A1P2$^+$-$\beta_1 G \beta_2$, even though all experiments were started from the same RP4::Mu *cts*62-Su A1P2$^+$-Mu *cts*62-containing clone. One possible explanation for the finding that different isolates of RP4::Su A1P2$^+$-$\beta_1 G \beta_2$ carry different lengths of transposed DNA is that deletions occur continuously in the A1P2 segment of the starting plasmid and that, upon excision of $\alpha_1 G \alpha_2$, different parts of the

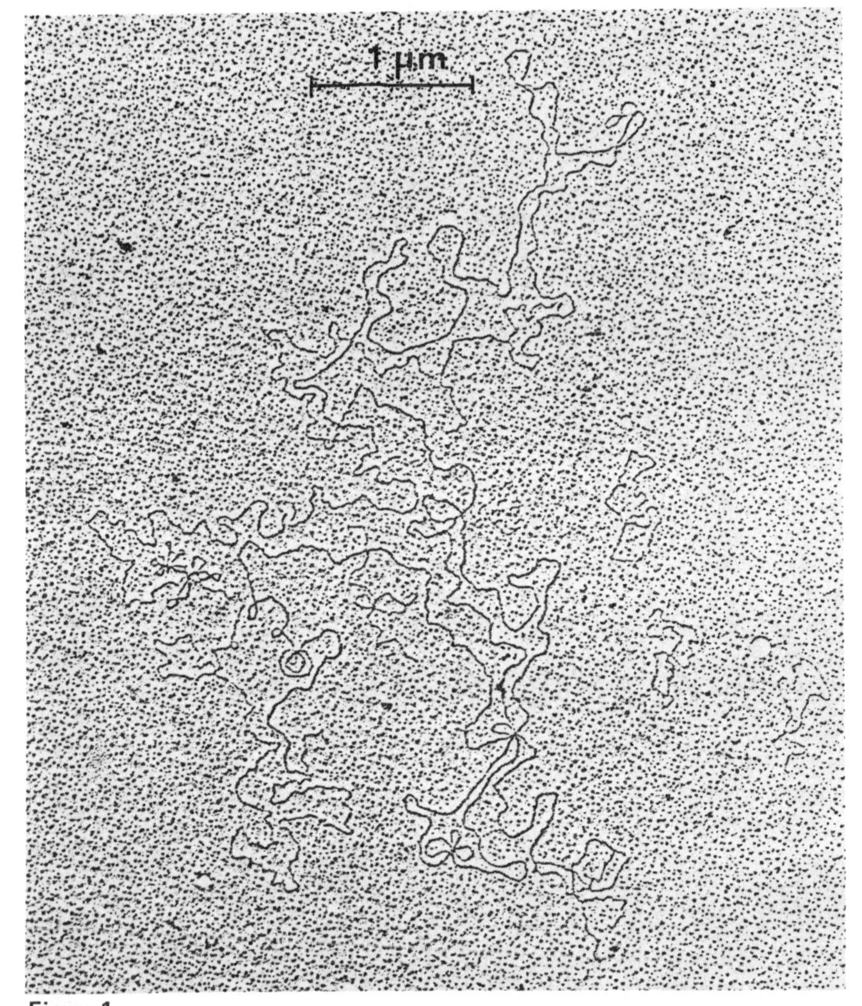

Figure 1
Micrograph of an RP4::Mu *cts*62-Su A1P2$^+$-Mu *cts*62 plasmid showing hybridization with two Mu molecules. A tracing representation is shown in Fig. 2. The smaller circles visible in this picture are the plasmid RSF1030 DNA (Heffron et al. 1975), used here as an internal standard.

A1P2 segment are left. Electron microscope analysis of these parts suggests that most deletions are on the side of the α_1 segment. However, further studies will be necessary to understand this situation more fully.

The RP4::Su A1P2$^+$-β_1Gβ_2 plasmids can be transferred with high efficiency to any *E. coli* strain, Rec$^+$ or Rec$^-$, and to other enterobacteria such as *Shigella,*

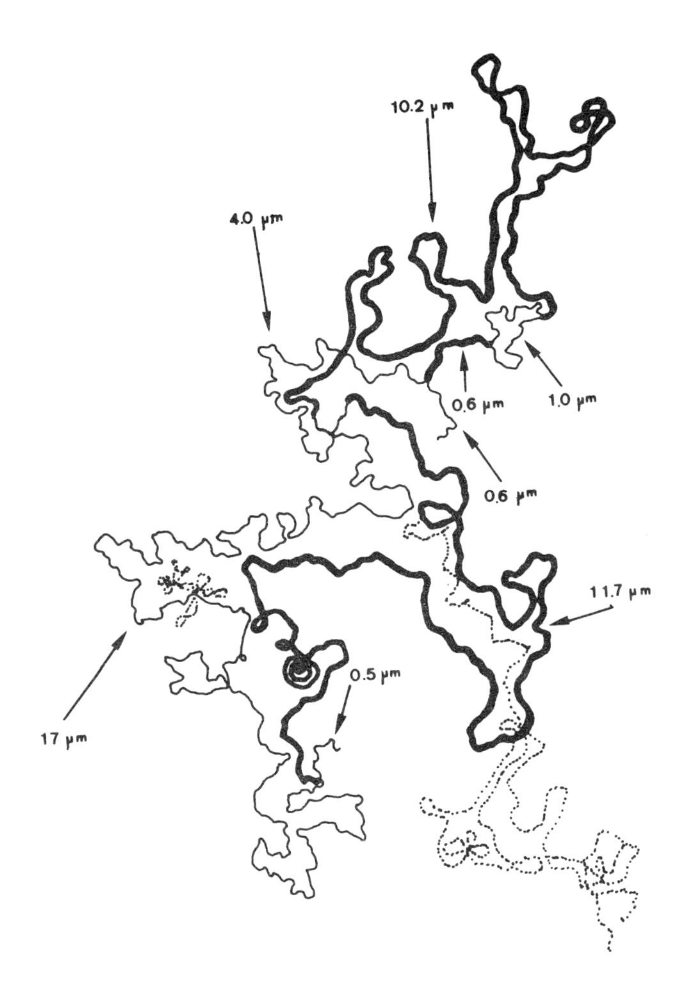

Figure 2
Tracing of the RP4::Mu *cts*62-Su A1P2$^+$-Mu *cts*62/Mu/Mu heteroduplex shown in Fig. 1. The heavy line indicates the double-stranded segments. The 11.7-μm segment starts at a DNA split end (0.5 μm) and covers a full-length Mu DNA ($\beta_1 G_1 \alpha_1$) molecule. This segment is separated from the next 0.6 μm duplex DNA (β_2) by a 4.0-μm single-stranded stretch of transposed DNA (containing Su A1P2$^+$). The β_2 segment is followed by a G loop and by a 10.2-μm duplex DNA belonging to the α_2 segment of the second Mu molecule. The circle is closed by 17-μm RP4 DNA, seen as a single strand. In the micrograph, the heteroduplex is obscured by a supercoiled RSF1030 molecule and a reannealed Mu molecule. This DNA is represented in this tracing by a dotted line.

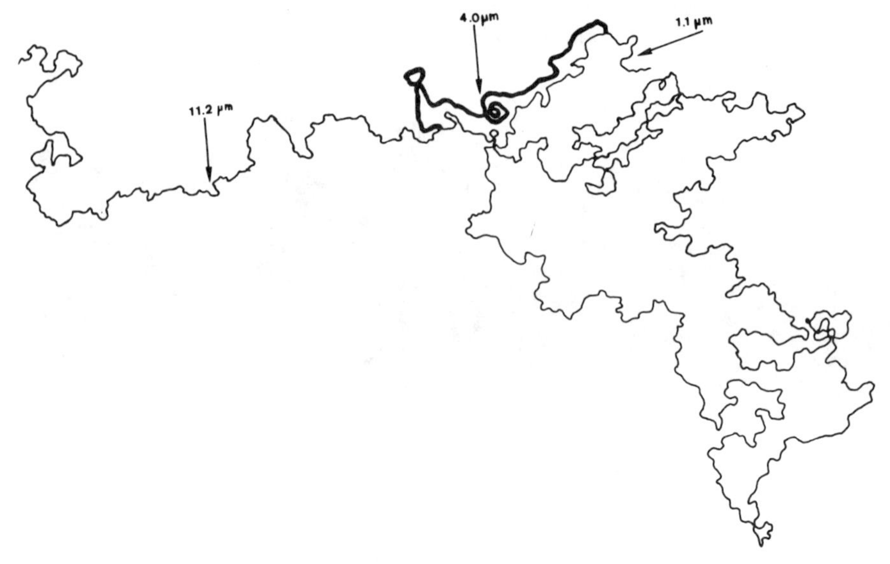

Figure 3

Tracing of a heteroduplex of RP4::Mu *cts*62-Su A1P2$^+$-Mu *cts*62 with the phage A1P2 DNA. The same plasmid preparation as used for the experiments of Figs. 1 and 4 was hybridized against the DNA of phage A1P2. The linear phage DNA (16.3 μm) starts hybridizing to the plasmid at 1.1 μm from the left cohesive end, a region where the structural gene for Su A1P2$^+$ tRNA is situated (Yamagishi et al. 1976). Approximately 4 μm of the phage DNA is in a duplex structure.

Klebsiella, Serratia, and probably many others. Thus the *suIII* gene can be introduced into many different species.

General Comments

It is possible to introduce an RP4 plasmid carrying Mu *cts*62 to bacterial species which lack adsorption sites for Mu but can support lytic growth of the phage. The RP4::Mu plasmids can be transferred to species such as *Serratia, Proteus,* and *Klebsiella.* It should thus be possible to exploit the mutagenic properties of Mu to mediate transposition in many diverse bacterial strains. RP4 plasmids harboring chromosomal markers of any of these strains could be isolated and then transferred to other species.

The method for transposition described in this paper would allow enrichment of a segment of a bacterial chromosome by transferring it to a plasmid. The plasmid can then be used as a reservoir for in vitro cloning of the genes located in the transposed segment.

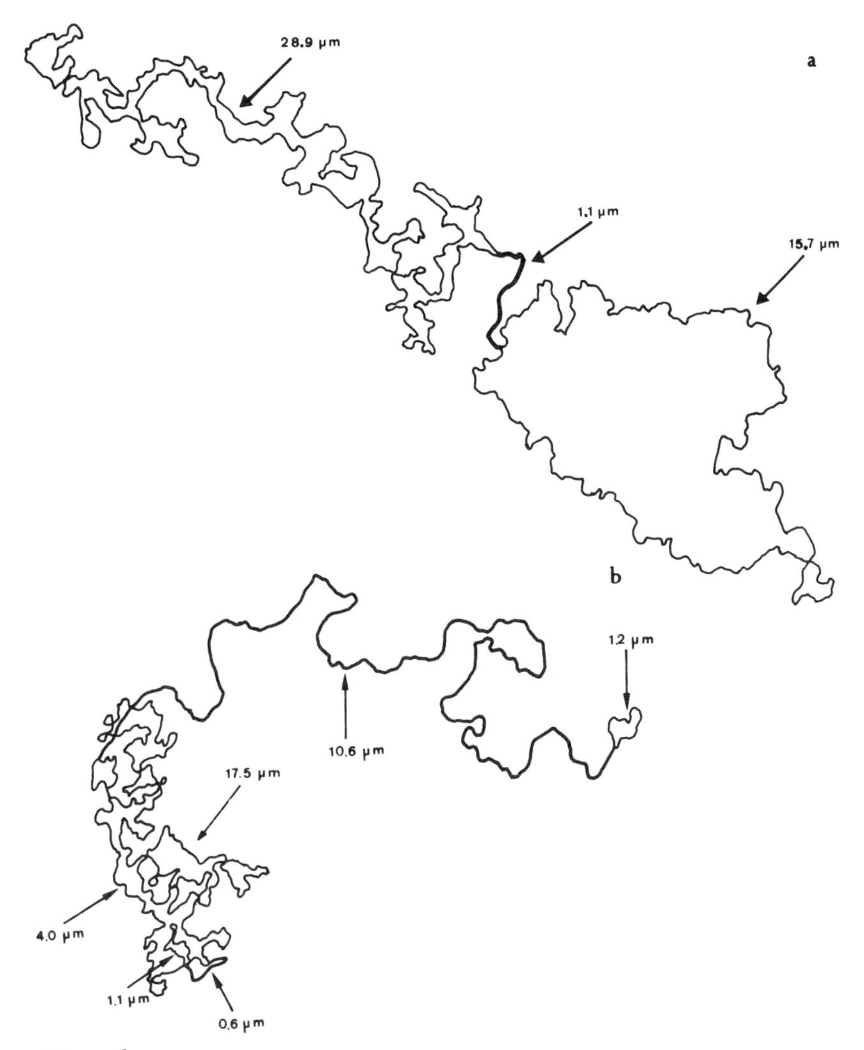

Figure 4

(*a*) Tracing of a self-renatured RP4::Mu-A1P2-Mu plasmid DNA showing hybridization between the G regions. The double-stranded part (1.1 μm) is apparently the G region of phage Mu. One of the single-stranded circles (15.7 μm) consists of the Mu phage DNA and the transposed A1P2 DNA. The other single-stranded circle (28.9 μm) equals RP4 plus Mu DNA. See Fig. 5 for a schematic model indicating the ordering of the different regions.

(*b*) Tracing of a self-renatured RP4::Mu *cts*62-Su A1P2$^+$-Mu *cts*62 plasmid DNA, showing Mu-Mu hybridization as a result of inversion of the DNA between the two G regions. The duplex regions have the lengths of an α segment (10.6 μm) and a β segment (0.6 μm) of phage Mu. The locations of the two "G loops" are indicated (1.2 μm and 1.1 μm). The single-stranded DNA corresponds to the transposed fragment (4.0 μm) and RP4 DNA (17.5 μm).

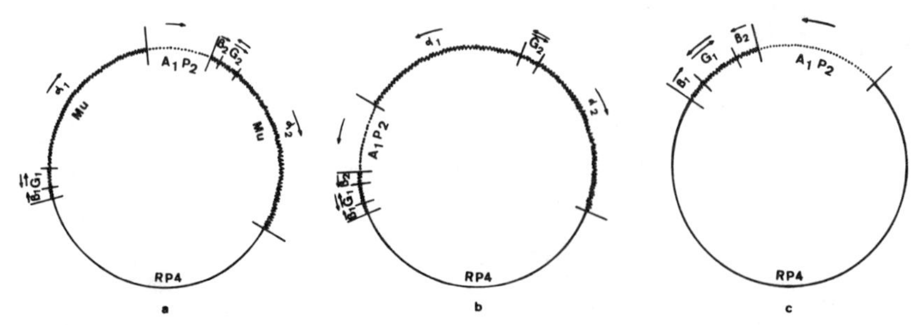

Figure 5
A schematic representation of the structure of RP4::Mu *cts*62-Su A1P2$^+$-Mu *cts*62 plasmid DNA. The plasmid consists of a complete RP4 molecule (18 μm–solid line), two phage Mu genomes (zig-zag line), and the transposed DNA (dotted line) segment of the phage ϕ80 derivative A1P2. The phage Mu DNAs are in the same orientation with respect to each other. This is indicated by the arrows above the three segments of each Mu DNA. Inversion of the G segment is indicated by the two arrows in opposite orientations.

(*a*) The structure of the majority of molecules. (*b*) The structure of the molecules where the segment between the two G loops is inverted. (*c*) The structure of the plasmid after excision of the $\alpha_1 G \alpha_2$ segment. This plasmid will be called RP4::Su A1P2$^+$-$\beta_1 G \beta_2$. The lengths of the different segments are drawn to scale.

SUMMARY

Bacteriophage Mu can mediate transposition of different segments of the host chromosome onto a plasmid. The plasmid with the transposed segment can be detected by transferring the plasmid to an appropriate acceptor strain. The transposed segment is always found to be flanked by the Mu prophages in the same orientation. By electron microscopy, we have shown that the whole sequence between the two Mu G regions can be inverted, presumably because of recombination between the two oppositely oriented G regions.

We have exploited the ability of Mu to transpose genes in cloning a suppressor (tRNA) gene of *E. coli* in an RP4 plasmid. Because RP4 has a wide host range, this technique can be used to clone suppressor and other genes of different bacterial species.

REFERENCES

Andoh, T. and H. Ozeki. 1968. Suppressor gene Su_3^+ of *E. coli*, a structural gene for tyrosine tRNA. *Proc. Nat. Acad. Sci.* **59**:792.

Datta, N., R. W. Hedges, E. J. Shaw, R. B Sykes and M. H. Richmond. 1971. Properties of an R factor from *Pseudomonas aeruginosa*. *J. Bact.* **108**:1244.

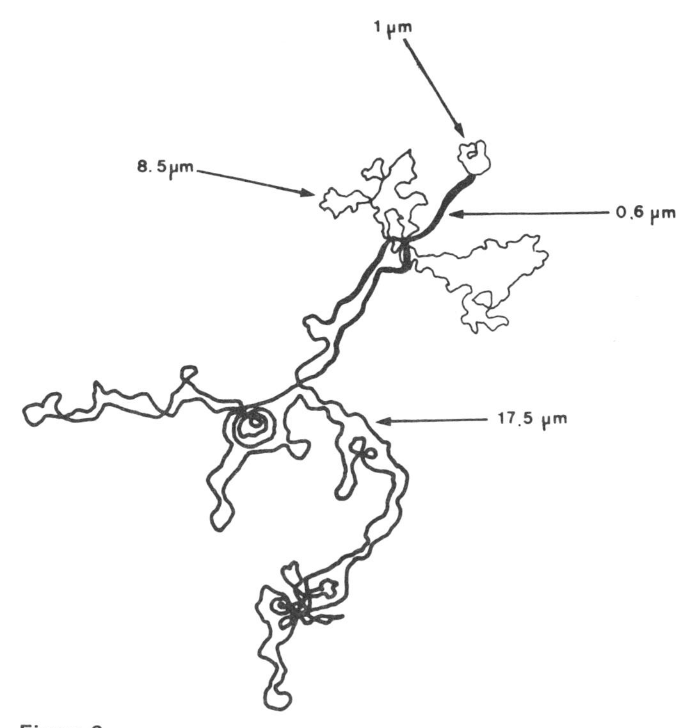

Figure 6

Tracing of a heteroduplex showing hybridization of RP4::Su A1P2$^+$-$\beta_1 G \beta_2$ with the parent RP4. The RP4 (17.5 μm) and the self-renatured β segments double-stranded. The G segment and the transposed DNA (here 8. single-stranded.

Davis, R. W., M. Simon and N. Davidson. 1971. Electron microscope heteroduplex methods for mapping regions of base sequence homology in nucleic acids. In *Methods in enzymology* (ed. L. Grossman and K. Moldave), vol. 21, p. 413. Academic Press, New York.

Faelen, M. and A. Toussaint. 1976. Bacteriophage Mu-1: A tool to transpose and to localize bacterial genes. *J. Mol. Biol.* **104**:525.

Ghysen, A. and J. E. Celis. 1974. Mischarging single and double mutants of *E. coli* sup3 tyrosine transfer RNA. *J. Mol. Biol.* **83**:333.

Heffron, F., R. Sublett, R. W. Hedges, A. Jacob and S. Falkow. 1975. Origin of the TEM beta-lactamase gene found on plasmids. *J. Bact.* **122**:250

Hsu, M.-T. and N. Davidson. 1974. Electron microscope heteroduplex study of the heterogeneity of Mu phage and prophage DNA. *Virology* **58**:229.

Van den Elsacker, S., M. Van Montagu and J. Schell. 1975. A general method for the isolation of non-sense suppressor strains in gram-negative bacteria. *Arch. Intern. Physiol. Biochim.* **83**:1011.

Yamagishi, H., H. Inokuchi and H. Ozeki. 1976. Excision and duplication of

Su^{3+} transducing fragments carried by bacteriophage $\phi 80$. I. Novel structure of $\phi 80$ sus 2psu^{3+} DNA molecule. *J. Virol.* **18**:1016.

Construction and Use of Gene Fusions Directed by Bacteriophage Mu Insertions

M. J. Casadaban,* T. J. Silhavy, M. L. Berman, H. A. Shuman, A. V. Sarthy and J. R. Beckwith
Department of Microbiology and Molecular Genetics
Harvard Medical School
Boston, Massachusetts 02115

The promoter of a gene can be fused to the structural part of another gene by illegitimate recombination between the two genes. Insertion elements, which have site-specific recombination systems, can be used to facilitate such recombination (Fig. 1). Schemes have been developed for using the Mu insertion element (which is the genome of bacteriophage Mu) to direct the fusion of specific genes in *Escherichia coli* (Casadaban 1975, 1976a). The scheme shown in Figure 2 depicts how fusions of the *lac* (lactose utilization) structural genes to the promoters of many other genes have been isolated (Table 1). The *lac* genes were selected because of the many genetic and biochemical techniques available for their study (Beckwith 1970).

These gene fusions can be used to study gene regulation. For example, the regulation of a gene whose product is difficult to assay can be studied in a strain that has the promoter and controlling elements of that gene fused to the structural part of another gene whose product is easy to assay. In this way, fusions of the *araC* (arabinose utilization) promoter to the *lac* structural genes were used to find new types of regulation for the *araC* gene (Table 1) (Casadaban 1976b).

Fusions can also be employed in the detection of mutants that have affected the expression of a gene. For example, mutants for a gene fused to the *lac* structural genes can be selected by an altered Lac phenotype (Beckwith 1970). The *lac* part of the fusion can then be removed by recombination. These mutants might affect the promoter, or a controlling site, or a separate regulatory gene. Reznikoff and Thornton (1972) have used *trp-lac* fusions in this way to isolate *trp* (tryptophan biosynthesis) regulatory mutants.

Some of the fusions formed by the scheme in Figure 2 result in the formation of hybrid proteins (Casadaban 1976a). These proteins have the C terminus and activity of β-galactosidase, the product of the *lacZ* gene, and the N terminus of the product of the gene fused to *lac*. Fusions of *lacZ* with *araB* and *malF*

*Present address: Department of Medicine, Stanford University School of Medicine, Stanford, California 94305.

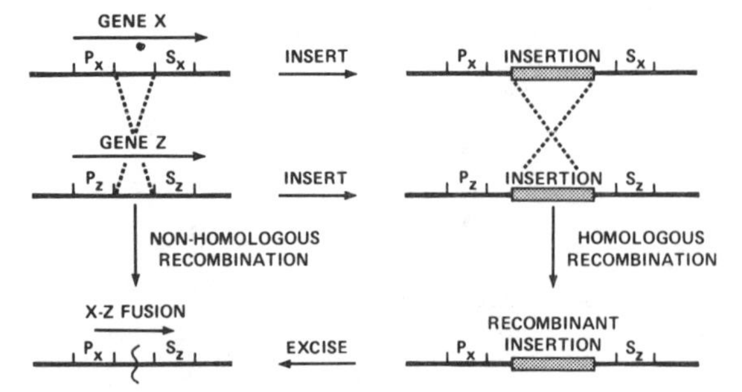

Figure 1

Use of DNA insertion elements to fuse genes. Nonhomologous recombination between two genes "X" and "Z" to form a fusion of the promoter of gene X (P_X) to the structural part of gene Z (S_Z) can be made into homologous recombination by using insertion elements. If identical insertion elements are first inserted into each gene at the sites where recombination is desired, homologous recombination can then occur between the two genes. A resulting recombinant insertion will then have P_X on one side and S_Z on the other. Deletion or excision of the recombinant insertion element can then form the fusion.

Table 1

Operons Fused with the *lac* Structural Genes

Operon fused[a]	Regulation observed	Reference
araBAD	induced by L-arabinose	Casadaban (1975, 1976a)
araC	repressed by the araC protein product, catabolite controlled	Casadaban (1976b)
leu	repressed by L-leucine	Casadaban (1976a)
Mu	repressed by Mu repressor	Casadaban (1976a)
tyrT[b]	growth-dependent regulation[c]	M. Berman and J. Beckwith (unpubl.)
malPQ	induced by maltose	Silhavy et al. (1976)
malEF	induced by maltose	Silhavy et al. (1976)
phoA	repressed by high phosphate	A. Sarthy, M. Casadaban and J. Beckwith (unpubl.)
argA, argE, argG	repressed by L-arginine	T. Eckhardt (pers. comm.)

[a] Bachman et al. (1976).
[b] A tyrosine tRNA gene that does not even code for a protein product, yet its promoter can be used to express the *lac* genes.
[c] Cashel (1975).

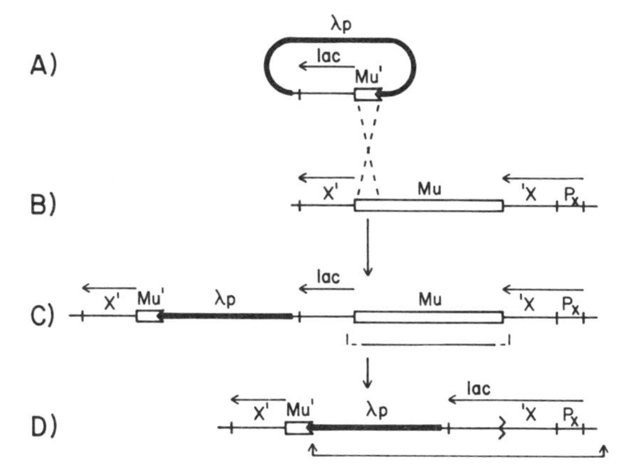

Figure 2
Scheme for isolating *lac* fusions by means of bacteriophage Mu insertions (Casadaban 1976a).

(*A*) The *lac* genes near a Mu insertion are first incorporated onto bacteriophage λ to form a λp(*lac*::Mu′) transducing phage. A small piece of the Mu genome from the end of Mu near the *lac* genes is also incorporated.

(*B*) A strain which is deleted for the *lac* genes is lysogenized with bacteriophage Mu. Mu lysogens are isolated which have the Mu genome integrated in or near a gene "X" whose promoter "P_X" we wish to fuse to *lac*. These Mu lysogens are then lysogenized with λp(*lac*::Mu′). The λp(*lac*::Mu′) prophage preferably integrates into the chromosome by homologous recombination within the Mu homology because (1) the λ phage does not have an intact λ attachment site, and (2) since the *lac* region is deleted, there is no ot[illegible] homology with the chromosome.

(*C*) Lambda lysogeny results in the formation of a recombinant Mu insertion which has *lac* on one side and P_X on the other. The *lac* structural genes are not expressed (the cell is Lac$^-$) because (1) a *lac* promoter mutation was recombined onto the λp(*lac*::Mu′) phage before it was used to lysogenize, and (2) no transcription comes from the Mu insertion or from P_X across the Mu insertion.

(*D*) P_X-*lac* fusions are formed by deleting the recombinant Mu insertion (underlined in *C*). Strains containing these deletions are selected as revertants that are (1) Lac$^+$, since the *lac* genes are expressed from P_X, and (2) temperature-resistant, since they are not killed by Mu thermoinduction. (The Mu phage used contains a temperature-sensitive repressor mutation.) Plaque-forming λ transducing phage, which carry the fusions and various amounts of nearby DNA sequences, can be obtained after induction of the λ genome in the fusion strains (underline).

(maltose utilization) have been isolated (Silhavy et al. 1976). For such fusions, the synthesis of β-galactosidase is regulated not only by the transcriptional controls of the operon fused to *lac*, but also by the translation-initiation controls of the structural gene fused with *lacZ*. Thus such fusions can be used to study

Table 2
Lambda Transducing Phage for *lac* Fusions

Fusion	Other nearby genes often incorporated	Reference
araB′-lac	*araC*	Casadaban (1976a)
araC′-lac	*araB*	M. Casadaban (unpubl.)
malE′-lac and *malEF′-lac*	*malK*	M. Casadaban, T. Silhavy and J. Beckwith (unpubl.)
argA′-lac		T. Eckhardt (pers. comm.)

gene regulation both at the level of translation and at the level of transcription.

The hybrid proteins formed by gene fusions can be used to study the properties of the original protein fused. For example, Silhavy et al. (1976) have shown that hybrid proteins formed by fusions of the maltose transport gene *malF* and *lacZ* are found associated with the cytoplasmic *E. coli* membrane. This implies that the N-terminal part of the *malF* protein contains membrane-associating properties which bind the formerly soluble enzyme β-galactosidase to the membrane. Hybrid β-galactosidase proteins with N-terminal parts from cytoplasmic soluble proteins such as the *araB* and *lacI* products are not found associated with the membrane.

Lambda transducing phage for the gene fusions constructed by the scheme illustrated in Figure 2 can be readily isolated (see *D* in Fig. 2). Some of the transducing phage isolated are shown in Table 2. These phage frequently carry other nearby chromosomal genes (Table 2). DNA from these transducing phage can be used in in vitro gene-expression experiments (Zubay 1973).

SUMMARY

Bacteriophage Mu insertions can be used to direct the fusion of the lactose operon structural genes to the promoters of many different operons in *E. coli.* With these fusions, the lactose genes are expressed according to the regulation of the fused operons. Fusions that result in the formation of hybrid proteins can also be constructed. Lambda transducing phage for these fusions, as well as for nearby genes, can be readily isolated.

REFERENCES

Bachmann, B., K. Low and A. Taylor. 1976. Recalibrated linkage map of *Escherichia coli* K-12. *Bact. Rev.* **40**:116.

Beckwith, J. 1970. *Lac,* the genetic system. In *The lactose operon* (ed. D. Zipser and J. Beckwith), p. 5. Cold Spring Harbor Laboratory, Cold Spring Harbor, New York.

Casadaban, M. 1975. Fusion of the *Escherichia coli lac* genes to the *ara* promoter: A general technique using bacteriophage Mu-1 insertions. *Proc. Nat. Acad. Sci.* **72**:809.

———. 1976a. Transposition and fusion of the *lac* genes to selected promoters in *Escherichia coli* using bacteriophage lambda and Mu. *J. Mol. Biol.* **104**:541.

———. 1976b. Regulation of the regulatory gene for the arabinose pathway, *araC. J. Mol. Biol.* **104**:557.

Cashel, M. 1975. Regulation of bacterial ppGpp and pppGpp. *Annu. Rev. Microbiol.* **29**:301.

Reznikoff, W. and K. Thornton. 1972. Isolating tryptophan regulatory mutants in *Escherichia coli* by using a *trp-lac* fusion strain. *J. Bact.* **109**:526.

Silhavy, T., M. Casadaban, H. Shuman and J. Beckwith. 1976. Conversion of β-galactosidase to a membrane-bound state by gene fusion. *Proc. Nat. Acad. Sci.* **73**:3423.

Zubay, G. 1973. *In vitro* synthesis of protein in microbial systems. *Annu. Rev. Genet.* **7**:267.

In Vivo Genetic Engineering: Exchange of Genes between a Lambda Transducing Phage and ColE1 Factor

K. Shimada, Y. Fukumaki and Y. Takagi
Department of Biochemistry
Kyushu University School of Medicine
Fukuoka 812, Japan

A recombinant molecule was constructed by in vitro splicing of ColE1 DNA and a DNA fragment, carrying a gene for guanine synthesis, derived from λp*guaA* transducing phages (Mukai et al. 1975). This molecule existed as a stable monomer plasmid within *Escherichia coli* K12 and contained genes for ColE1 immunity and the *guaA* enzyme (xanthosine 5′-monophosphate aminase) along with a part of the λ genome, R through J (R-A-F-J)$^+$ (Fukumaki et al. 1976). It had an approximate molecular weight of 21.6×10^6 daltons (S. Maeda, K. Shimada and Y. Takagi, in prep.) and was named ColE1-*cos*λ-*guaA* (Fukumaki et al. 1976). We have reported that the ColE1-*cos*λ-*guaA* plasmids are efficiently packaged into λ phage particles after λ infection of *E. coli* strains carrying these plasmids (or by inducing λ lysogens of these strains) (Fukumaki et al. 1976). Thus the genetic properties of the recombinant plasmid can be studied by the convenient methods of λ phage genetics. Moreover, the presence of extensive homology between λ phage DNA and ColE1-*cos*λ-*guaA* plasmid DNA suggests that the specialized transducing phage genome and the ColE1-*cos*λ plasmid should recombine into each other. In this paper, we show that genes isolated on a transducing phage are easily incorporated into ColE1-*cos*λ plasmids and that genes on ColE1-*cos*λ plasmids, including ColE1 DNA itself, are easily incorporated into the λ phage genome.

E. coli TM96, in which *gal-att*λ-*bio* and *guaA-guaB* regions are deleted and which carries ColE1-*cos*λ-*guaA* as plasmids (Mukai et al. 1975), was lysogenized with λ*c*I857p*gal*8 (Feiss et al. 1972). Most of the lysogens carried λ*gal* DNA inserted into ColE1-*cos*λ-*guaA* plasmids. The insertion into plasmid DNA occurred because of the deletion of the normal λ attachment site in the host chromosome and because of the presence of homology between λ DNA and the ColE1-*cos*λ-*guaA* plasmid DNA (Fig. 1). After heat induction of TM96 (λ*c*I857p*gal*8), efficient Ter cutting is expected (Gottesman and Yarmolinsky 1969). As a result of Ter cutting, one complete λ*c*I857p*gal* genome and one ColE1-*cos*λ-*guaA* plasmid are packaged into separate phage heads, as shown in Figure 1. We noticed that the same lysates could transduce *gal*$^+$ genes to *E. coli* KS1616, a derivative of *E. coli* K12, with the deletions *gal-att*λ-*bio* and

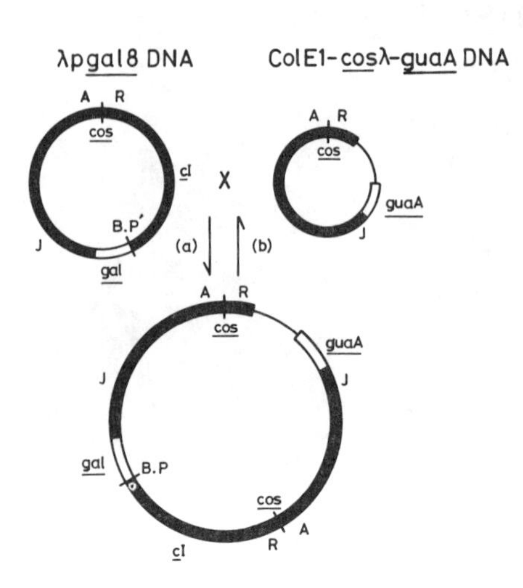

Figure 1
Insertion of λp*gal*8 DNA into ColE1-*cos*λ-*guaA* DNA and Ter-dependent formation of ColE1-*cos*λ-*guaA* transducing particles. Reaction *a* is a general recombination between two homologous DNAs and is promoted by host Rec function. The recombinant molecule carries two λ*cos* sites. Reaction *b* is promoted by Ter and is coupled with packaging. The thick lines represent λ genome, the thin lines ColE1 DNA, and the double lines fragments of bacterial chromosome including the *guaA* or the *gal* gene. A, R, J, *c*I, B.P′, and *cos* are phage markers, and *guaA* and *gal* are bacterial markers.

guaA-*guaB* (Table 1). Lambda *c*I857p*gal*8 phage itself cannot transduce gal^+ genes efficiently to KS1616 because of the deletion of the λ*att* site in this strain. Accordingly, we postulated that the transduction of gal^+ genes to KS1616 was the result of plasmid formation involving ColE1 factor and that the Gal^+ transductants carried gal^+ genes which are connected in vivo with ColE1 DNAs. We purified some Gal^+ transductants and analyzed their genetic properties. All of the Gal^+ transductants were sensitive to λ phage, but most of them could rescue λ*am*R, A, F, and J markers. Most of the Gal^+ transductants were immune to ColE1. Those that had no λ genes were not immune to ColE1 (2 out of

Table 1
The *guaA* and *gal* Transducing Ability of λ Phage Lysates Prepared by Heat Induction of TM96 (λ*c*I857p*gal*8)

Property	TM96 (λ*c*I857p*gal*8) lysates
Plaque-forming units (pfu) per ml	1.3×10^{8}
GuaA transducing ability per pfu	3.2×10^{-1}
Gal transducing ability per pfu	1.0×10^{-4}
Spi^- phage per pfu	4.3×10^{-5}

E. coli TM96 (Mukai et al. 1975) was lysogenized with λ*c*I857p*gal*8 (Feiss et al. 1972). One of the lysogens was purified and grown in 5 ml of nutrient broth at 30°C. During the exponential phase, the cultures were supplemented with 0.01 M $MgSO_4$, shifted to 40°C for 20 min, incubated at 37°C for 90 min with shaking, and then treated with chloroform. Plaque-forming units were determined on C600, and the number of λSpi^- phage was measured on *E. coli* K12 (P2) strain RW355 (Lindahl et al. 1970). Transducing ability was determined as described by Shimada et al. (1972). We used as recipient a starved culture of KS1616, an *E. coli* K12 derivative lacking *gal*-*att*λ-*bio* and *guaA*-*guaB*.

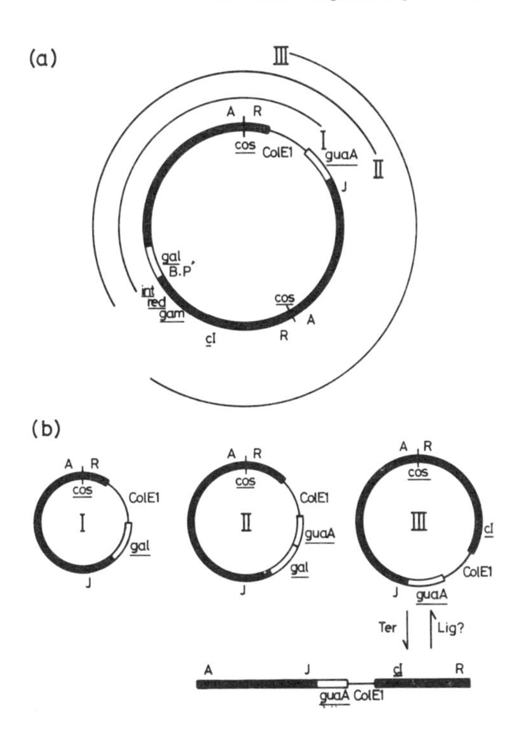

Figure 2
Formation of ColE1-*cos*λ *gal* or ColE1-*cos*λ-*guaA gal* plasmids and λp*guaA*-ColE1 phage from λ*gal*8 and ColE1-*cos*λ-*guaA* hybrid. (*a*) Approximate sites of infrequent cleavage of λp*gal*8 and ColE1-*cos*λ-*guaA* recombinant DNA molecules after induction of lysogens. (*b*) Structures of various in vivo recombinant molecules. I, ColE1-*cos*λ *gal*; II, ColE1-*cos*λ-*guaA gal*; and III, λp*guaA*-ColE1 phage, which can exist as plasmids and as plaque-forming phage. Lig means *E. coli* ligase, and *int, red,* and *gam* are phage genes. Other symbols are as described in the legend to Fig. 1.

19 examined) presumably had the *gal* genes incorporated into the host chromosome. Some of the Gal^+ transductants (7 out of 19) were also Gua^+. Thus two types of recombinant plasmids carrying *gal* genes could be isolated, i.e., ColE1-*cos*λ *gal* and ColE1-*cos*λ-*guaA gal* (Fig. 2, I and II). The λ phage lysates prepared on *E. coli* that carried ColE1-*cos*λ *gal* and ColE1-*cos*λ-*guaA gal* plasmids transduce these plasmids to KS1616. The frequency of Gal^+ or $GuaA^+$ Gal^+ transduction at low multiplicity of infection (about 0.01 pfu per cell) was about 5×10^{-3}/pfu, which is comparable to the value observed for the λ-phage-mediated transduction of ColE1-*cos*λ-*guaA* plasmids (Fukumaki et al. 1976). These transductants showed genetic properties similar to their respective parent strains. The recombinant plasmids have been purified, and physical mapping of these DNA molecules is now in progress (Y. Fukumaki, K. Shimada and Y. Takagi, in prep.).

Transducing phages that carry both the *guaA* gene and ColE1 DNA, adjacently located, were isolated from heat-induced lysates of TM96 (λp*gal*8) by selecting for the Spi^- phenotype, which confers the ability to grow on a lysogen of phage P2 (Lindahl et al. 1970). Lambda Spi^- phage lacks λ genes *int* through *gam* (Fig. 2) from the left end of the prophage map. The deleted phage genes could be expected to be replaced by the DNA adjacent to the phage gene J: i.e., *guaA* gene and ColE1 DNA (Fig. 2, III). These phages appeared at a frequency of 4.3×10^{-5}/pfu in the lysates (Table 1). One of these Spi^- phages was purified and was shown by transduction experiments to carry the $guaA^+$ gene and ColE1 immunity. As expected, it did not carry *gal* (see Fig. 2). When Gua$^+$ lysogens of this phage were heat-induced, they yielded plaque-forming λ particles carrying the *guaA*$^+$ gene and ColE1 immunity. The formation of this type of transducing phage shows that genes on ColE1-*cos*λ plasmids can be easily incorporated into the λ phage.

The formation of ColE1-*cos*λ *gal* or ColE1-*cos*λ-*guaA gal* plasmids and λ*c*I857p*guaA*-ColE1 phage from λ*gal*8 and ColE1-*cos*λ-*guaA* recombinants is shown diagrammatically in Figure 2. Our present working model for the formation of ColE1-*cos*λ *gal* or ColE1-*cos*λ-*guaA gal* is as follows: Occasionally, λInt- and Xis-promoted cutting occurs between B.P′ and other sites (such as secondary attachment sites for λ phage [Shimada et al. 1972; Gottesman and Weisberg 1971]) which may exist near the *guaA* gene. The high frequency of appearance of transducing particles carrying only the *guaA* gene reflects the high efficiency of Ter-promoted cutting coupled with packaging. The low frequency of appearance of *gal* or of *gal* and *guaA* gene transducing particles apparently results from inefficient cutting by Int or Xis. However, the presence of phage genes *int, xis,* and *red* and of host *recA* gene functions were not absolute requirements for the formation of these transducing particles. We obtained similar results with lysates prepared by heat induction of TM96 *recA* lysogenized with λ*c*I857p*gal*8-*bio*69 phage, which is deleted of phage genes *int, xis,* through *red* (Manly et al. 1969). In these lysates, the frequency of appearance of *gal* transducing particles was about 250-fold lower than that observed for lysates of TM96 (λ*c*I857p*gal*8) (data not shown).

SUMMARY

We have demonstrated that genes isolated on a λ transducing phage can be readily incorporated into the ColE1-*cos*λ plasmid and vice versa. This technique makes it possible to manipulate given pieces of DNA as plasmids or as transducing phages. Thus the method used in this work combines the advantages of two widely used techniques for studying particular genes and groups of genes, i.e., isolation of specialized transducing phages and in vitro formation of ColE1 plasmids containing various bacterial (Hershfield et al. 1974) or eukaryotic genes (Grunstein and Hogness 1975).

REFERENCES

Feiss, M., S. Adyha and D. L. Court. 1972. Isolation of plaque-forming, galactose-transducing strains of phage lambda. *Genetics* **71**:189.

Fukumaki, Y., K. Shimada and Y. Takagi. 1976. Specialized transduction of colicin E1 DNA in *E. coli* K12 by phage lambda. *Proc. Nat. Acad. Sci.* **73**:3238.

Gottesman, M. E. and R. A. Weisberg, 1971. Prophage insertion and excision. In *The bacteriophage lambda* (ed. A. D. Hershey), p. 113. Cold Spring Harbor Laboratory, Cold Spring Harbor, New York.

Gottesman, M. E. and M. B. Yarmolinsky. 1969. The integration and excision of the bacteriophage lambda genome. *Cold Spring Harbor Symp. Quant. Biol.* **33**:735.

Grunstein, M. and D. S. Hogness. 1975. Colony hybridization: A model for the isolation of cloned DNAs that contain a specific gene. *Proc. Nat. Acad. Sci.* **72**:3961.

Hershfield, V., H. W. Boyer, C. Yanofsky, M. A. Lovett and D. R. Helinski. 1974. Plasmid ColE1 as a molecular vehicle for cloning and amplification of DNA. *Proc. Nat. Acad. Sci.* **71**:3455.

Lindahl, G., G. Sironi, H. Bialy and R. Calendar. 1970. Bacteriophage lambda: Abortive infection of bacteria lysogenic for phage P2. *Proc. Nat. Acad. Sci.* **66**:587.

Manly, K. F., E. R. Signer and C. M. Radding. 1969. Nonessential functions of bacteriophage λ. *Virology* **37**:177.

Mukai, T., K. Matsubara and Y. Takagi. 1975. *In vitro* construction of ColE1 factor carrying genes for synthesis of guanine or thymine. *Proc. Japan. Acad.* **51**:353.

Shimada, K., R. A. Weisberg and M. E. Gottesman. 1972. Prophage lambda at unusual chromosomal locations. *J. Mol. Biol.* **63**:483.

Translocation of Ampicillin Resistance from R Factor onto ColE1 Factor Carrying Genes for Synthesis of Guanine

S. Maeda, K. Shimada and Y. Takagi
Department of Biochemistry
Kyushu University School of Medicine
Fukuoka 812, Japan

Bacteriophage-lambda-mediated specialized transduction of a recombinant plasmid molecule, named ColE1-*cos*λ-*guaA*, into various *Escherichia coli* K12 strains has been described recently by Fukumaki et al. (1976). Using their methods, we introduced ColE1-*cos*λ-*guaA* into an *E. coli* K12 strain which carries an antibiotic resistance factor R1*drd* (Fig. 1) and tested the phage lysates, prepared by lytic infection of λ phage on this strain, for transduction of ampicillin resistance to different *E. coli* K12 strains. This paper describes the translocation of a DNA sequence which mediates ampicillin resistance from R1*drd* onto the ColE1-*cos*λ-*guaA* plasmid.

Lambda phage lysates were prepared on *E. coli* KS2009 (Fig. 1) and were used to transduce the $guaA^+$ gene to KS1616, a strain carrying deletions covering *gal-att*λ-*bio* and *guaA*-*guaB*. The frequency of $GuaA^+$ transduction at low multiplicity of infection (about 0.01 pfu per cell) was about 5×10^{-3} per pfu. These $GuaA^+$ transductants could be divided into two distinct phenotypic groups. Group-I transductants were ColE1-imm^+, $GuaA^+$, and group-II transductants were ColE1-imm^+, $GuaA^+$, and ampicillin (Ap)-resistant. Among 300 independent $GuaA^+$ transductants, 179 were group I and 121 were group II. When a recombination-deficient derivative of KS2009 was infected with λ*bio*69 phage (*int-xis-red*^{del}; Manly et al. 1969), the resulting lysates also contained particles which could transduce ColE1-imm^+, $guaA^+$, and ampicillin resistance to KS1616. We tested 100 independent $GuaA^+$ transductants and found that 62 belonged to group I and 38 to group II. Ten independently isolated $GuaA^+$ and ampicillin-resistant transductants were purified, and all of them were found to carry part of the λ genome, R through J $(R\text{-}A\text{-}F\text{-}J)^+$ (Shimada et al. 1972), and none of them showed *Eco*RI restriction-modification activity encoded in the R1*drd* plasmid.

The number of plasmid DNA molecules was increased by incubating two independently isolated ampicillin-resistant transductants in the presence of 100 μg/ml chloramphenicol. The DNA was extracted, and closed circular DNAs were purified by sedimentation in ethidium bromide-CsCl. The molecular weights of these DNAs were estimated from their relative contour lengths in

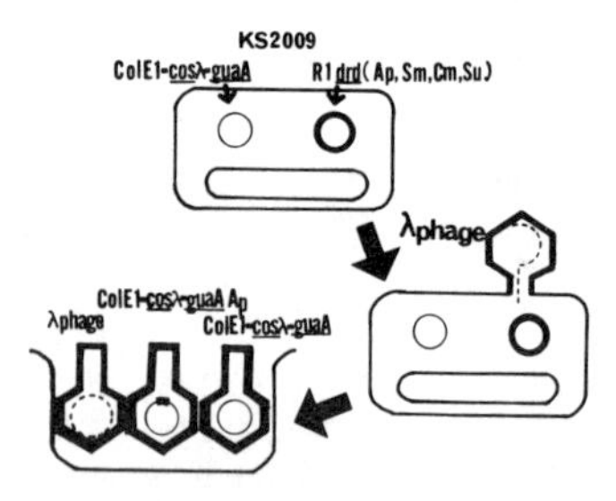

Figure 1
Detection of translocation of the ampicillin resistance genes. R1*drd* is an antibiotic resistance factor that can confer resistances to streptomycin, sulfonamide, chloramphenicol, and ampicillin (Yoshimori 1971). This plasmid was introduced into a derivative of *E. coli* K12, KS1616, lacking *gal-att*λ*-bio* and *guaA-guaB* (Shimada et al. 1976) by mating with RY13, which carries R1*drd* and plasmids determining *Eco*RI restriction and modification enzymes (Yoshimori 1971). The *E. coli* K12 (R1*drd*) used for this experiment, KS2009, produced *Eco*RI enzymes. ColE1-*cos*λ*-guaA* existed as a stable plasmid within *E. coli* K12 and contained whole ColE1 DNA and the *guaA* gene, together with a part of the λ genome, R through J (R-*cos*-A-F-J) (Mukai et al. 1975; Fukumaki et al. 1976). The phage lysates prepared by infecting λ phage onto KS2009 contained three kinds of phages, i.e., λ phage, ColE1-*cos*λ*-guaA* transducing phage, and ColE1-*cos*λ*-guaA* Ap transducing phage.

The abbreviations used are Ap, ampicillin; Sm, streptomycin; Su, sulfonamide; and Cm, chloramphenicol resistance. The other genetic symbols are those used by Taylor and Trotter (1972) for *E. coli* and by Szybalski and Herskowitz (1971) for λ.

electron micrographs compared to that of ColE1 DNA (4.2×10^6 daltons) (Clewell and Helinski 1969) (Fig. 2). The average molecular weight of one ColE1-*cos*λ*-guaA* Ap was $24.6 \pm 0.1 \times 10^6$ daltons (average of 20 open circular DNA molecules) and that of the other was $24.4 \pm 0.2 \times 10^6$ (average of 53 open circular DNA molecules). The average molecular weight of the parent ColE1-*cos*λ*-guaA* DNA, however, was estimated to be $21.6 \pm 0.2 \times 10^6$ daltons (Maeda et al. 1976). These results indicate that the size of the translocated DNA carrying ampicillin resistance genes is about 3.0×10^6 daltons. These ColE1-*cos*λ*-guaA* Ap DNA molecules were characterized further by restriction endonuclease digestion and electrophoresis in agarose gels. We digested the ColE1-*cos*λ*-guaA* Ap or ColE1-*cos*λ*-guaA* DNAs with *Eco*RI (Yoshimori 1971) and *Sma*R (Tanaka and Weisblum 1975) and analyzed the DNA fragments by agarose gel electrophoresis. As Figure 3a shows, no apparent differences in their electrophoretic patterns were found. This suggested that the translocated ampicillin resistance genes are situated on the largest DNA fragment. We therefore examined the homology between the ColE1-*cos*λ*-guaA* Ap and ColE1-*cos*λ*-guaA* DNAs (Fig. 3b). Examination of heteroduplex molecules indicated that the two ampicillin-resistant recombinant plasmids were formed by a single insertion of DNA into the same or adjacent sites, located close to the λ gene R. We also observed several single-stranded molecules of recombinant DNAs that had been self-annealed. The appearance of this kind of molecule showed that the ampicillin resistance insertion we observed is bound by inverted repeat sequences.

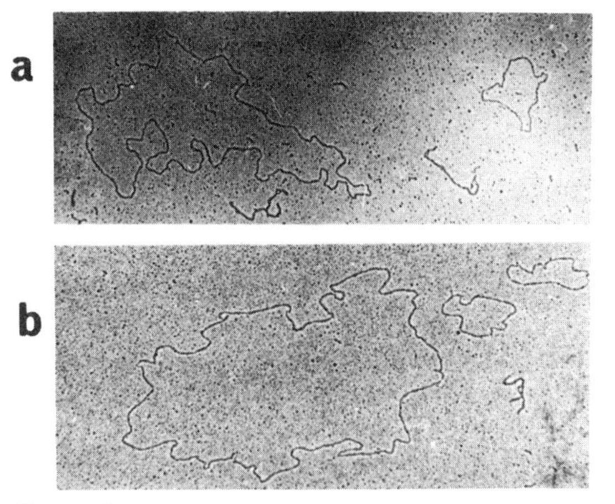

Figure 2
Electron micrographs of open circular (*a*) ColE1-*cos*λ-*guaA* and (*b*) ColE1-*cos*λ-*guaA* Ap molecules. Small circular DNAs are reference ColE1. Covalently closed circular plasmid DNAs were stored in TE buffer (100 mM Tris-HCl and 10 mM EDTA, pH 7.5) at -10°C. These DNA samples were lysed, and the molecules were photographed as described by Ogawa et al. (1968), using a JEOL model 100C electron microscope. During storage for about a month, approximately 30% of the covalently closed circular DNA molecules were converted to open circular DNA molecules. The lengths of the DNA strands were measured in enlarged photographs with a map measure (Maruzen Co., Tokyo).

Translocation of the ampicillin resistance genes implies that similar translocation of other drug resistance genes from R1*drd* onto ColE1-*cos*λ-*guaA* plasmids might also occur. Accordingly, we examined the translocation of Sm (streptomycin), Su (sulfonamide), and Cm (chloramphenicol) genes, applying the methods described above for detection of translocation of ampicillin resistance genes. However, we could not detect any Sm-, Su-, or Cm-resistant transductants. Hedges and Jacob (1974) and Heffron et al. (1975) have reported studies in which the ampicillin resistance determinants of R factor were translocated onto several other R factors. The translocation of ampicillin resistance observed by Hedges and Jacob required a functional *recA* product. Our results suggest that the translocation of ampicillin resistance genes occurs irrespective of whether the *E. coli recA* gene function is present or not. We observed that ColE1-*cos*λ-*guaA* Ap plasmids acquired the ability to translocate ampicillin resistance to other replicons, such as the ColE1-*cos*λ *gal* plasmid, and that this translocation was also independent of *E. coli recA* gene function. In this ColE1-*cos*λ *gal* Ap plasmid, ampicillin resistance genes were inserted between the λ gene R and the λ-ColE1 junction (data not shown) as in the ColE1-*cos*λ-*guaA* Ap plasmids (see Fig. 3b). The frequency of ampicillin resistance translocation reported by Heffron et al. (1975) was significantly lower than that observed by us. Our system can detect ampicillin resistance translocation so easily and

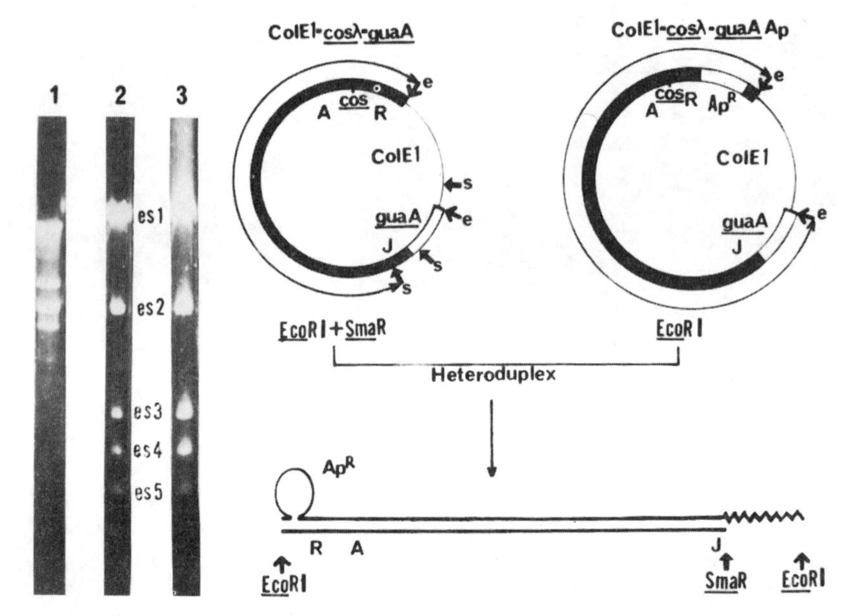

Figure 3

(*a*) Agarose gel electrophoresis of ColE1-*cos*λ-*guaA* Ap plasmid DNA cleaved by restriction endonuclease. Restriction enzyme *Eco*RI was prepared from *E. coli* strain RY13, kindly supplied by Dr. R. Yoshimori. *Serratia marescens* endonuclease R (*Sma*R) was purified as described by Tanaka and Weisblum (1975). Digestions of purified plasmid DNAs by restriction enzymes were performed as described by Yoshimori (1971) and Tanaka and Weisblum (1975). The DNA fragments were subjected to agarose gel (0.8%) electrophoresis. The gel was stained with ethidium bromide, and the fluorescent bands of DNA were photographed under long-wavelength ultraviolet light. DNA fragments obtained by hydrolysis of ColE1-*cos*λ-*guaA* DNA by *Eco*RI + *Sma*R are numbered with the prefix *es* (Maeda et al. 1976). (1) λcI857 DNA + *Eco*RI; (2) ColE1-*cos*λ-*guaA* DNA + *Eco*RI and *Sma*R; (3) ColE1-*cos*λ-*guaA* Ap DNA + *Eco*RI and *Sma*R.

(*b*) Models for heteroduplex formation from single strands of *Eco*RI- and *Sma*R-cleaved ColE1-*cos*λ-*guaA* and of *Eco*RI-cleaved ColE1-*cos*λ-*guaA* Ap DNAs. The thin line represents whole ColE1 DNA, the thick line a part of the λ genome, and the double line a fragment of bacterial chromosome. R, *cos,* A, and J are phage markers, ColE1 is colicin E1 DNA, and *guaA* is a bacterial marker. Short arrows with "e" indicate *Eco*RI-susceptible sites, and arrows with "s" indicate *Sma*R cleavage sites (Maeda et al. 1976). Most of the λ DNAs anneal with each other, and the ampicillin resistance genes (Ap^r) form a loop. End J of the heteroduplex molecule remains as single-stranded DNA (indicated by a zig-zag line) and serves to distinguish this end from end A-R of the linear molecules formed by *Eco*RI cleavage.

sensitively that we are now trying to establish an in vitro assay system for detection of translocation. An in vitro system for the detection of recombinant DNA, which uses packaging of λ DNA, has been reported by Syvanen (1974), and his system should be useful in our studies.

SUMMARY

We have detected by a sensitive method the translocation of a DNA segment which determines ampicillin resistance from R1*drd* factor to the ColE1-*cos*λ-*guaA* plasmid. The transduction of this segment (about 3.0×10^6 daltons) is independent of the *recA* function.

Acknowledgments

We wish to thank Dr. T. Ogawa of Osaka University for helping one of us (S.M.) with DNA heteroduplex analysis, and T. Nakamura for his excellent technical assistance.

REFERENCES

Clewell, D. B. and D. R. Helinski. 1969. Supercoiled circular DNA-protein complex in *Escherichia coli*: Purification and induced conversion to an open DNA form. *Proc. Nat. Acad. Sci.* **62**:1159.

Davis, R. W., M. Simon and N. Davidson. 1971. Electron microscope heteroduplex methods for mapping regions of base sequence homology in nucleic acids. In *Methods in enzymology* (ed. L. Grossman and K. Moldave), vol. 21, p. 413. Academic Press, New York.

Fukumaki, Y., K. Shimada and Y. Takagi. 1976. Specialized transduction of colicin E1 DNA in *E. coli* K12 by phage lambda. *Proc. Nat. Acad. Sci.* **73**: 3238.

Hedges, R. W. and A. E. Jacob. 1974. Transposition of ampicillin resistance from RP4 to other replicons. *Mol. Gen. Genet.* **132**:31.

Heffron, F., C. Rubens and S. Falkow. 1975. Translocation of a plasmid DNA sequence which mediates ampicillin resistance: Molecular nature and specificity of insertion. *Proc. Nat. Acad. Sci.* **72**:3623.

Maeda, S., K. Shimada and Y. Takagi. 1976. Molecular nature of an *in vitro* recombinant molecule: Colicin E1 factor carrying genes for synthesis of guanine. *Biochem. Biophys. Res. Comm.* **72**:1129.

Manly, K. F., E. R. Signer and C. M. Radding. 1969. Nonessential functions of bacteriophage λ. *Virology* **37**:177.

Mukai, T., K. Matsubara and Y. Takagi. 1975. *In vitro* construction of ColE1 factor carrying genes for synthesis of guanine or thymine. *Proc. Japan. Acad.* **51**:353.

Ogawa, T., J. Tomizawa and M. Fuke. 1968. Replication of bacteriophage DNA. II. Structure of replicating DNA of phage lambda. *Proc. Nat. Acad. Sci.* **60**: 861.

Shimada, K., Y. Fukumaki and Y. Takagi. 1976. Expression of guanine operon of *Escherichia coli* as analyzed by bacteriophage lambda-induced mutations. *Mol. Gen. Genet.* **147**:203.

Shimada, K., R. A. Weisberg and M. E. Gottesman. 1972. Prophage lambda at unusual chromosomal locations. *J. Mol. Biol.* **63**:483.

Syvanen, M. 1974. *In vitro* genetic recombination of bacteriophage λ. *Proc. Nat. Acad. Sci.* **71**:2496.

Szybalski, W. and I. Herskowitz. 1971. Lambda genetic elements. In *The bacteriophage lambda* (ed. A. D. Hershey), p. 778. Cold Spring Harbor Laboratory, Cold Spring Harbor, New York.

Tanaka, T. and B. Weisblum. 1975. Construction of a colicin E1-R factor composite plasmid *in vitro*: Means for amplification of deoxyribonucleic acid. *J. Bact.* **121**:354.

Taylor, A. L. and C. D. Trotter. 1972. Linkage map of *Escherichia coli* strain K-12. *Bact. Rev.* **36**:504.

Yoshimori, R. N. 1971. A genetic and biochemical analysis of the restriction and modification of DNA by resistance transfer factors. Ph.D. thesis, University of California, San Francisco.

Selected Translocation of DNA Segments Containing Antibiotic Resistance Genes

P. J. Kretschmer and S. N. Cohen
Stanford University School of Medicine
Stanford, California 94305

Translocating segments of DNA carrying a variety of antibiotic resistance genes have been identified recently. To study the specificity and frequency of such translocation events, we have described (P. J. Kretschmer and S. N. Cohen, in prep.) a simple method for the selective translocation of antibiotic resistance transposons to a wide variety of recipient replicons. The method, which was developed initially for use with a translocatable ampicillin (Ap) resistance segment (Tn*A*) (Kopecko and Cohen 1975; Heffron et al. 1975), should be readily applicable to other translocating elements. It utilizes the plasmid pSC201, which is a temperature-sensitive replication mutant of pSC101 (Kretschmer et al. 1975), as a donor molecule for the translocating DNA segment.

The donor replicon, pSC201 carrying Tn*A*, was obtained in a manner analogous to that used by Kopecko and Cohen (1975) during mobilization of pSC101 by a derivative of the R1*drd*19 plasmid. This plasmid was the original source of the Tn*A* element used in our studies. Heteroduplexes between pSC201 and pSC201 (Tn*A*) showed, as expected, a duplex pSC201 region and a single-stranded Tn*A* loop having a 130–150 nucleotide-base-pair double-stranded stalk characteristic of the inverted repeat (IR of Fig. 1) of Tn*A* (Kopecko and Cohen 1975; Heffron et al. 1975).

The procedure for selected translocation of Tn*A* is outlined in Figure 1. A transformant clone containing pSC201 (Tn*A*) and a second plasmid (P) was grown to stationary phase and plated on nutrient agar plates containing or lacking ampicillin (20 μg/ml). Cultures were incubated at 45°C. Since the pSC201 (Tn*A*) plasmid cannot replicate at 45°C, bacteria capable of growth in the presence of antibiotic at this temperature were presumed to have undergone a prior translocation event in which the Tn*A* segment had moved to a temperature-stable recipient replicon, resulting in one of the three possible cell genotypes illustrated in Figure 1. Persistance of Ap resistance as a result of recombination between plasmids occurs at a low frequency, and this does not contribute significantly to the apparent frequency of translocations (P. J. Kretschmer and S. N. Cohen, unpubl.). By means of this procedure, we were able to estimate the frequency of translocation by determining the fraction of

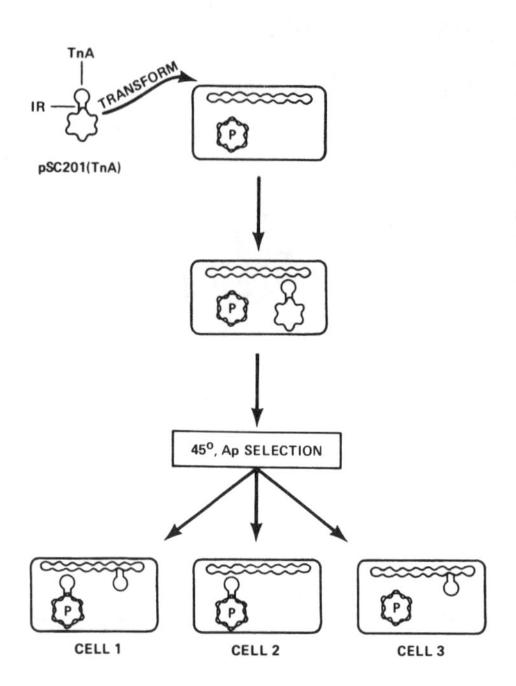

Figure 1
Outline of the selected translocation method. A C600 cell, which may contain only the chromosome or, as illustrated in this drawing, a plasmid (P) as well, is transformed (Cohen et al. 1972) by the donor plasmid pSC201 (Tn*A*). From a purified clone, transformed cells now containing both plasmids are grown in the presence of ampicillin at 45°C (see text) to screen for C600 cells with acquired Tn*A*. The alternative genotypes illustrated are: Cell 1, the Tn*A* element on the second plasmid and on the chromosome; Cell 2, the Tn*A* element on the second plasmid only; and Cell 3, the Tn*A* element on the chromosome only.

cells on nutrient agar at 45°C that was also resistant to Ap at 45°C. Translocation frequency is defined in terms of megadaltons of recipient genome for comparison of translocation frequencies to different replicons.

Table 1 compares the translocation frequencies of Tn*A* from pSC201 to two different plasmids and to the chromosome; translocation to the chromosome was studied in a cell containing only the pSC201 (Tn*A*) plasmid. The plasmid RSF1010 is a small, 5.5-megadalton plasmid coding for resistance to streptomycin (Sm) and sulfonamide (Su) (Guerry et al. 1974), and R6-5 is a 65-megadalton plasmid coding for resistance to kanamycin (Km)/neomycin (Nm), Sm, Su, and chloramphenicol (Cm) (Sharp et al. 1973a; Cohen et al. 1973). As seen in Table 1, the translocation frequencies from pSC201 (Tn*A*) to RSF1010 and R6-5 are approximately equal; however, the movement of Tn*A* from plasmid to plasmid is three to four orders of magnitude higher than the translocation frequency from the same plasmid to the chromosome. These results suggest that either the extrachromosomal state of the plasmids promotes or enhances the movement of Tn*A* between them or that specific recipient sites present on plasmids are rare on the chromosome. The available data do not distinguish between these possibilities.

Ap-resistant clones (at 45°C) derived from cells containing the recipient plasmids in addition to pSC201 (Tn*A*) (Table 1, lines 1 and 2) were purified, and their plasmid DNA was examined by agarose gel electrophoresis (Sharp et al. 1973b) following digestion by endonuclease *Eco*RI. In all instances studied, the gel patterns indicated an increase in molecular weight of 3×10^6 daltons in

Table 1

Frequency of Translocation of the Ap Segment from pSC201 (Tn*A*) to Recipient Genomes by the Selected Translocation Technique

Strain	Recipient genome	Recipient (m.w.)[a]	Recipient copy no.[a]	Percent Ap^r colonies at 45°C[b]	Translocation frequency[c]
C600 [pSC201 (Tn*A*) + RSF1010]	RSF1010	5.5×10^6	20	1.0	10^{-4}
C600 [pSC201 (Tn*A*) + R6-5]	R6-5	65×10^6	1.5	1.0	10^{-4}
C600 [pSC201 (Tn*A*)]	chromosome	2.5×10^9	1.5	0.01	3×10^{-8}

[a]References for molecular weight and copy number are: for RSF1010, Heffron et al. (1975), Guerry et al. (1974); for R6-5, Timmis et al. (1975); and for chromosome, Cooper and Helmstetter (1968).

[b]Numbers presented are the average of at least five experiments. Variation for each strain was from 50% to 200% of the number shown.

[c]Translocation frequency = (frequency Ap^r cells at 45°C)/(megadaltons of recipient DNA in each cell).

either the single *Eco*RI fragment of RSF1010 (Cohen et al. 1973) or in one of the *Eco*RI-generated bands of R6-5. This molecular weight increase corresponded to the previously demonstrated size of the Tn*A* element (Kopecko and Cohen 1975; Heffron et al. 1975). Plasmid DNA was not observed in Ap-resistant cells derived from the C600 [pSC201 (Tn*A*)] strain lacking a second plasmid (Table 1, line 3). We conclude that DNA has been inserted into the chromosome in these cells

Although Reif and Saedler (1975) have shown that deletion formation upon excision of the IS*1* insertion sequence region is affected by temperature, there is no evidence that the translocation frequency of IS*1* or similar elements is affected by temperature. However, to determine whether the temperature shift itself had any effect on the translocation frequencies obtained (Table 1), we used an alternative method for estimating the frequency of movement of Tn*A* from pSC201 (Tn*A*) to the RSF1010 plasmid.

The alternative procedure, which does not depend on temperature, is outlined in Figure 2. As shown in this figure, cells containing pSC201 (Tn*A*) were transformed with RSF1010, and, after growth of the transformant culture, plasmid DNA was examined (by transformation) to determine the frequency of RSF1010 molecules containing the Tn*A* transposon. In addition to studying the effect of 45°C temperature on translocation frequency (see legend to Fig. 2), variables such as the *recA*$^-$ genotype and the use of pSC101 (Tn*A*) compared to pSC201 (Tn*A*) as donor molecule were studied.

The results shown in Table 2 (Exps. 1 and 2) indicate that translocations of Tn*A* from the pSC101 and pSC201 plasmids at 32°C occur at similar frequencies. In addition, Experiment 2 quantitates the previous finding that transloca-

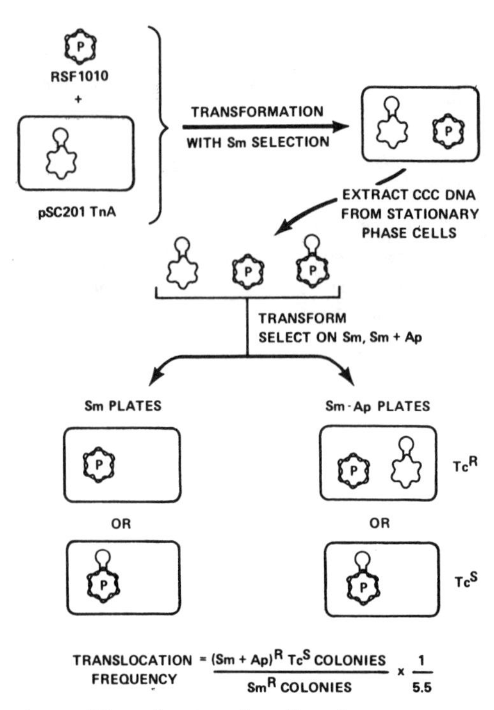

Figure 2
Outline of the alternative method for estimating translocation frequency. C600 cells containing pSC201 (Tn*A*) were transformed with RSF1010 plasmid DNA, and immediately after the heat-pulse step in transformation (Cohen et al. 1972), 0.1-ml aliquots (approx. 10^4 transformed cells) were transferred to a 100-ml prewarmed broth culture(s). After incubation for 120 min (to allow phenotypic expression), streptomycin (Sm–final conc. 10 μg/ml) was added to select for transformed cells. After growth to stationary phase, covalently closed circular (CCC) plasmid DNA (consisting of a mixture of the three molecules shown) was extracted and used to transform C600 cells. Transformants were selected on Sm and Sm-Ap plates. Thus the translocation frequency was estimated by determining the fraction of Sm^r colonies (Sm plates) that were $(Sm, Ap)^r$ Tc^s (by toothpicking colonies on Sm-Ap plates to Tc [tetracycline] plates) and then dividing by 5.5 to express the frequency as translocations per megadalton of recipient DNA.

tion of the Tn*A* segment occurs in the absence of the *E. coli recA* gene product (Kopecko and Cohen 1975); no substantive difference in translocation frequency is seen in *recA*$^+$ versus *recA*$^-$ cells. Experiment 3 indicates that there is a significant temperature effect on translocation frequency. Translocation was not detectable in cells grown at 45°C from the time that the two plasmids were

Table 2
Effect of Donor Plasmid, Temperature, and *recA* Genotype on Translocation Frequency

Bacterial genotype	Donor plasmid	Exp. 1 32°C	Exp. 2 32°C	Exp. 3 32°C	Exp. 3 45°C
recA$^+$	pSC101 (Tn*A*)	3.3×10^{-4}		4.0×10^{-4}	$< 4 \times 10^{-6}$
recA$^+$	pSC201 (Tn*A*)	1.5×10^{-4}			
recA$^-$	pSC201 (Tn*A*)		3.5×10^{-4}		

Translocation frequency was calculated as described in the legend to Fig. 2.

inserted into the same cell, suggesting that the enzymes responsible for the translocation event are temperature-sensitive. We conclude from this result that the selected translocation procedure leads to selection and identification of bacteria in which translocation has occurred before the temperature shift.

Although the Tn*A* unit can translocate to multiple sites on small plasmids (Kopecko and Cohen 1975; Heffron et al. 1975; So et al. 1975), the clustering of recipient sites suggests that the locations are not random. The selected translocation procedure has been useful for studying the site specificity of the Tn*A* translocation event to the large plasmid R6-5. R6-5 DNA isolated from 22 clones derived from independent translocation events were examined by *Eco*RI digestion in agarose gel electrophoresis. In each instance, the *Eco*RI digestion pattern indicated a loss of one of the 13 *Eco*RI fragment bands, and the addition of a new band at a position equivalent to that expected as a consequence of insertion of three million daltons of DNA to the missing band.

In the 22 independently formed and isolated R6-5 (Tn*A*) clones studied in these experiments, only 5 of the 13 *Eco*RI-generated bands of R6-5 were altered, and the insertion frequency of Tn*A* into the various R6-5 bands did not reflect the number of nucleotides present in the fragment. Tn*A* inserted once into bands 7 (m.w. = 3.96 megadaltons) and 12 (m.w. = 1.03 megadaltons), twice into band 8 (m.w. = 3.47 megadaltons), seven times into band 9 (approx. m.w. = 3 megadaltons), and eleven times into band 2 or 3 (comigrating *Eco*RI fragments, m.w. = 9.06 megadaltons). These revised molecular weight estimates of the R6-5 *Eco*RI fragments resulted from analysis of plasmids carrying the cloned R6-5 fragments (K. Timmis, F. Cabello and S. N. Cohen, unpubl.). Electron microscope heteroduplex mapping and analysis of R6-5 (Tn*A*) fragments generated by other restriction endonucleases are currently in progress to determine more precise locations of the Tn*A* insertions within the *Eco*RI bands. However, our preliminary results suggest that preferred recipient sites for the Tn*A* translocating segment are present on R6-5, and that when these preferred sites are available, there is less likelihood of translocation onto other sites. Such findings are reminiscent of the preferred and alternate insertion sites for bacteriophage lambda for formation of lysogens (Shimada et al. 1973).

The selected translocation procedure was also used to determine whether the frequency of translocation of DNA from one plasmid to another is influenced by the length of time that the two plasmids are present concurrently within the same bacterial cell. The pSC201 (Tn*A*) plasmid was transformed into bacteria carrying RSF1010, and cells carrying both of the plasmids were grown for a varying number of generations prior to a temperature shift to 45°C. Our results indicate that the translocation frequency for movement of Tn*A* from pSC201 to RSF1010 is always approximately 10^{-4}, whether the donor and recipient plasmid are together for 20 or 200 generations. We conclude, therefore, that translocation is not cumulative, and that an equilibrium between insertion of Tn*A* into RSF1010 and excision of the translocating segment from the plasmid is established.

SUMMARY

The selected translocation procedure described here provides a convenient method for moving translocating antibiotic resistance segments among bacterial plasmids in studies of the specificity and frequency of translocation events.

Acknowledgments

These investigations were supported by grant AI 08619 from the National Institutes of Health, grant BMS 75-14176 from the National Science Foundation, and grant VC139A from the American Cancer Society to S. N. C.

REFERENCES

Cohen, S. N., A. C. Y. Chang and L. Hsu. 1972. Nonchromosomal antibiotic resistance in bacteria: Genetic transformation of *E. coli* by R-factor DNA. *Proc. Nat. Acad. Sci.* **69**:2110.

Cohen, S. N., A. C. Y. Chang, H. W. Boyer and R. B. Helling. 1973. Construction of biologically functional bacterial plasmids *in vitro. Proc. Nat. Acad. Sci.* **70**:3240.

Cooper, S. and C. E. Helmstetter. 1968. Chromosome replication and the division cycle of *Escherichia coli* B/r. *J. Mol. Biol.* **31**:519.

Guerry, P., J. van Embden and S. Falkow. 1974. Molecular nature of two nonconjugative plasmids carrying drug resistance genes. *J. Bact.* **117**:619.

Heffron, F., C. Rubens and S. Falkow. 1975. Translocation of a plasmid DNA sequence which mediates ampicillin resistance: Molecular nature and specificity of insertion. *Proc. Nat. Acad. Sci.* **72**:3623.

Kopecko, D. J. and S. N. Cohen. 1975. Site specific *recA*-independent recombination between bacterial plasmids: Involvement of palindromes at the recombinational loci. *Proc. Nat. Acad. Sci.* **72**:1373.

Kretschmer, P. J., A. C. Y. Chang and S. N. Cohen. 1975. Indirect selection of bacterial plasmids lacking identifiable phenotypic properties. *J. Bact.* **124**:225.

Reif, H.-J. and H. Saedler. 1975. IS1 is involved in deletion formation in the *gal* region of *E. coli* K12. *Mol. Gen. Genet.* **137**:17.

Sharp, P. A., S. N. Cohen and N. Davidson. 1973a. Electron microscope heteroduplex studies of sequence relations among plasmids of *Escherichia coli.* II. Structure of drug resistance (R) factors and F factors. *J. Mol. Biol.* **75**:235.

Sharp, P. A., B. Sugden and J. Sambrook. 1973b. Detection of two restriction endonuclease activities in *Haemophilus parainfluenzae* using analytical agarose-ethidium bromide electrophoresis. *Biochemistry* **12**:3055.

Shimada, K., R. Weisberg and M. Gottesman. 1973. Prophage lambda at unusual chromosomal locations. II. Mutations induced by bacteriophage lambda in *E. coli* K-12. *J. Mol. Biol.* **80**:297.

So, M., R. Gill and S. Falkow. 1975. The generation of ColE1-Ap^{R} cloning vehicle which allows detection of inserted DNA. *Mol. Gen Genet.* **142**:239.

Timmis, K., F. Cabello and S. N. Cohen. 1975. Cloning, isolation and characterization of replication regions of complex plasmid genomes. *Proc. Nat. Acad. Sci.* **72**:2242.

Detection of Transposable Antibiotic Resistance Determinants with Phage Lambda

D. E. Berg
University of Wisconsin
Department of Biochemistry
Madison, Wisconsin 53706

Phage λ provides a useful tool for the detection and analysis of transposable resistance determinants. Such elements can be inserted into λ genomes in vivo, and the resultant hybrid phages can be selected by their capacity to transduce the resistance trait to sensitive cells.

The length of the transducing phage genome cannot exceed the packaging capacity of the phage head, i.e., approximately 51,000 base pairs (110% of λ^+) (Weil et al. 1972). The size of the parental phage genome, however, can be reduced to approximately 36,000 base pairs by deletion of nonessential genes (Davidson and Szybalski 1971; Emmons et al. 1975). This permits detection of transposable elements up to 15,000 base pairs in length.

Since λ can "pick up" certain bacterial genes adjacent to its integration site because of aberrant excision (Schrenk and Weisberg 1975), it is important to know that the putative transposable element is not incorporated into λ via the phage integration pathway. This can be determined by (1) mapping the resistance gene and demonstrating that it is not adjacent to the *att* locus (bacterial genes incorporated via the phage integration pathway are adjacent to *att*) or (2) showing that a λ derivative which does not integrate into the bacterial chromosome can still acquire the resistance trait. Use of integration-defective phage also facilitates subsequent tests of the transposability of the resistance genes from the phage genome to the bacterial chromosome (see step 4 below).

The procedure given below employs λ*b*221*c*I857, a phage incapable of integrating into the bacterial chromosome because the *b*221 deletion removes the phage *att* locus (Parkinson and Huskey 1971). This λ mutant has been used to detect the transposition of a relatively small (5000-base-pair) Tn*5* transposon (Berg et al. 1975; Berg, this volume); however, the size of the *b*221 deletion should permit the detection of transposable elements up to 14,000 base pairs long.

1. A plate stock or liquid lysate of λ*b*221*c*I857 is made using a host strain of *E. coli* which carries the putative transposable element.
2. The progeny phage are adsorbed to *E. coli* strain K461 (Friedman et al. 1976), which, because of mutations in bacterial genes *nusA* and *nusB*,

interferes with the functioning of λ*N* gene protein. The reduced activity of *N* protein permits many cells to survive the infection and to maintain the phage genome as a plasmid (Signer 1969; Lieb 1970; Berg 1974). After several generations of growth to permit expression of resistance and segregation of any coinfecting nontransducing λ genomes, the resistant cells are selected on antibiotic-containing plates.

3. Cells in colonies of K461 that carry λ plasmids release phage particles spontaneously. Thus the clones that carry nondefective λ plasmids are identified by picking antibiotic-resistant colonies to a lawn of sensitive cells in soft agar and noting which produce an area of lysis. Transducing phages can be recovered directly from a chloroform-saturated broth suspension of the plasmid-carrying cells.
4. The capacity of the resistance determinant to transpose from the phage genome to the bacterial chromosome can be tested easily if the phage contains the *b*221*att* deletion. Lambda-sensitive bacteria are infected under conditions which favor the establishment of immunity and thereby minimize cell killing (multiplicity of 3–10 phage per cell, growth temperature of 30°C). After growth in antibiotic-free media to permit expression of resistance, stably resistant clones are selected by plating aliquots of the culture on antibiotic-containing agar. Since the *b*221 deletion renders the phage unable to integrate by λ-specific mechanisms, the repressed prophages are lost by dilution as the cells divide (Ogawa and Tomizawa 1967). Stable antibiotic-resistant clones arise because of transposition of the element from λ to the bacterial chromosome.

Reconstruction experiments were performed with the Tn*5* element (Berg, this volume) to determine the efficiency of the steps in the procedure outlined above. *Step 1*: The frequency of the newly arisen λTn*5* phages detected in plate stocks of λ*b*221*c*I857 grown in bacteria containing a Tn*5* insertion in the *lacZ* gene was approximately 10^{-4}. *Step 2*: When stocks of λ*b*221*c*I857 phage containing Tn*5* insertions in gene *rex* or in gene *c*III are used to infect K461 at a multiplicity of 0.2 phage per cell, the frequency of Kan^r transductants was approximately 10^{-1} per infecting phage. *Step 4*: The frequency of Kan^r transductants obtained after λ*b*221*cI*857 Tn*5* infection of λ-sensitive strain 594 was 10^{-2}; 99% of the transductants were not lysogenic for λ (Berg, this volume).

Certain transposable elements may not be detectable by the procedure outlined above. For example, Gottesman and Rosner (1975) reported transposition of Tn*9* to λ either prior to or during induction of a lysogen; however, Tn*9* transposition to λ was not detected after phage infection. Unfortunately, procedures employing *att*$^+$ phages (Berg et al. 1975; Gottesman and Rosner 1975) are somewhat disadvantageous for initial tests of the transposability of uncharacterized resistance determinants, since newly acquired DNA segments must be shown to map away from *att*, and since the integration of λ*att*$^+$ complicates subsequent tests of the transposability of the determinant from λ to the bacterial chromosome.

A procedure for detecting transposition involving a ColE1 plasmid, which contains the cohesive ends of λ and is therefore packaged into phage heads with high efficiency after λ phage infection, has been developed recently by Shimada et al. (this volume). Their procedure may be particularly useful for the detection of transposable elements somewhat larger than 15,000 base pairs in length.

Methodological Details

Bacteria should be grown in T broth (1% Difco tryptone, 0.5% NaCl) prior to infection. Plate stocks are prepared by adsorbing approximately 10^5 phage particles to 10^8 bacteria, suspending the mixture in 4 ml molten soft agar (T broth containing 0.6% Difco agar), pouring the liquid on a plate containing solidified 1% agar in T broth, and harvesting the phage after confluent lysis is obtained (4-6 hr at 37°C). When growing phage with the *c*I857 temperature-sensitive repressor allele, temperatures below 34°C favor the establishment of immunity; temperatures above 38°C favor lytic development. Stocks of phage with genomes larger than 31,000 base pairs tend to be inactivated during storage but they can be stabilized by the addition of 20 mM $MgSO_4$.

SUMMARY

Transposable resistance determinants have been detected using the deletion phage λ*b*221. The *b*221 deletion makes this phage integration-defective and capable of accepting the addition of up to 14,000 base pairs of DNA. Rare phages to which the Tn*5* (Kan) element had transposed were selected after infection of a bacterial strain which permits the phage genome to replicate as a plasmid.

Acknowledgments

I am grateful to David Friedman for supplying strain K461. This work was supported by National Science Foundation Grant BMS72-02264 and National Institutes of Health, DHEW, Grant T32 CA 09075.

REFERENCES

Berg, D. 1974. Genetic evidence for two types of gene arrangements in new λdv plasmid mutants. *J. Mol. Biol.* **86**:59.

Berg, D., J. Davies, B. Allet and J. D. Rochaix. 1975. Transposition of R factor genes to bacteriophage λ. *Proc. Nat. Acad. Sci.* **72**:3628.

Davidson, N. and W. Szybalski. 1971. Physical and chemical characteristics of lambda DNA. In *The bacteriophage lambda* (ed. A. D. Hershey), p. 45. Cold Spring Harbor Laboratory, Cold Spring Harbor, New York.

Emmons, S., V. MacCosham and R. Baldwin. 1975. Tandem genetic duplications in phage lambda. III. The frequency of duplication mutants in two derivatives of phage lambda independent of known recombination systems. *J. Mol. Biol.* **91**:133.

Friedman, D., M. Baumann and L. Baron. 1976. Cooperative effects of bacterial mutations affecting λ *N* gene expression. I. Isolation and characterization of a *nusB* mutant. *Virology* **73**:119.

Gottesman, M. and L. Rosner. 1975. Acquisition of a determinant for chloramphenicol resistance by coliphage lambda. *Proc. Nat. Acad. Sci.* **72**:5014.

Lieb, M. 1970. λ Mutants that persist as plasmids. *J. Virol.* **6**:218.

Ogawa, T. and J. Tomizawa. 1967. Abortive lysogenization of bacteriophage lambda b_2 and residual immunity of nonlysogenic segregants. *J. Mol. Biol.* **23**:225.

Parkinson, J. and R. Huskey. 1971. Deletion mutants of bacteriophage lambda. I. Isolation and initial characterization. *J. Mol. Biol.* **56**:369.

Schrenk, W. and R. Weisberg. 1975. A simple method for making new transducing lines of coliphage λ. *Mol. Gen. Genet.* **131**:101.

Signer, E. 1969. Plasmid formation: A new mode of lysogeny by phage λ. *Nature* **223**:158.

Weil, J., R. Cunningham, R. Martin, E. Mitchell and B. Bolling. 1972. Characteristics of λp4, a λ derivative containing 9% excess DNA. *Virology* **50**:373.

Properties of the Plasmid RK2 as a Cloning Vehicle

R. J. Meyer, D. Figurski and D. R. Helinski
Department of Biology
University of California, San Diego
La Jolla, California 92093

The emergence of techniques for the construction in vitro of recombinant DNA and the introduction of this DNA into bacteria has led to the development of a variety of plasmid elements as cloning vehicles in *Escherichia coli.* A strong need also exists for plasmid cloning vehicles in species other than *E. coli.* These plasmids would be a powerful tool in the genetic analysis of these species, particularly where other genetic systems are unavailable. In addition, these plasmid vehicles could be used to augment or modify desirable characteristics of the host bacterium, such as nitrogen fixation by *Rhizobium* species or hydrocarbon degradation by *Pseudomonas* species.

To be most useful as a genetic vehicle, a broad-host-range plasmid should possess several properties. It should be of relatively small size, capable of stable replication in a wide variety of hosts, and offer a means of simple selection in these various bacterial species. In addition, the plasmid should not confer hazardous properties to any of its potential hosts or extend the range of antibiotic resistance. Finally, a system must be available for introducing the plasmid DNA into cells of the various bacterial species.

The P-incompatibility-group plasmid RK2 has several properties which make it an attractive starting point for the development of a broad-host-range cloning vehicle. RK2 can replicate in many different gram-negative bacteria (Beringer 1974; Panopoulos et al. 1975). It is self-transmissible and shows conjugal transfer across diverse gram-negative genera (Beringer 1974; Panopoulos et al. 1975). RK2, however, confers resistance to several clinically important antibiotics and, unless modified, could extend the range of antibiotic resistance.

Attempts To Reduce the Size of RK2 with Restriction Enzymes

Our initial efforts with RK2 were directed toward isolating a segment of RK2 DNA that contained the genes responsible for replication of this plasmid. Genes sufficient for autonomous plasmid replication have been isolated on a single, small DNA fragment derived by *Eco*RI (Morrow and Berg 1972) digestion of R6-5 (Timmis et al. 1975) and F*lac* (Timmis et al. 1975; Lovett and Helinski

1976), indicating a clustering of replication genes on these plasmid elements. The use of *Eco*RI for the same purpose with RK2 was ruled out by the observation that, although RK2 is relatively large (m.w. = 37.6×10^6 daltons), it contains only a single *Eco*RI cleavage site (Meyer et al. 1975). Surprisingly, RK2 also has only single cleavage sites for the restriction endonucleases *Hin*dIII (Old et al. 1975), *Hpa*I (Gromkova and Goodgal 1972), and *Bam*HI (Wilson and Young 1975) and only two sites for *Sal*I (Hamer and Thomas 1976). The relative locations of these sites have now been mapped (Meyer et al. 1977; and Fig. 1).

Since *Eco*RI, *Hin*dIII, and *Bam*HI generate cohesive termini (Mertz and Davis 1972; Hamer and Thomas 1976; G. Wilson, R. Roberts and F. Young, pers. comm.), the cleavage sites for these enzymes are potentially useful locations for the insertion of DNA to be cloned. Restriction fragments containing genes *trpD* and *trpE* of the tryptophan operon of *E. coli* have been inserted into all three sites. The resulting hybrid plasmids replicate stably and retain all of the characteristics of the parental RK2 plasmid, with the important exception that insertion into the *Hin*dIII site of RK2 leads to the loss of kanamycin resistance.

All attempts to isolate an RK2 plasmid of reduced size after digestion with *Eco*RI, *Hin*dIII, *Bam*HI, or *Sal*I were unsuccessful. Digestion of RK2 or RK2-*trp*, designated pRK20 (formerly pRM2 [Meyer et al. 1975]), followed by transformation of a suitable recipient and by separate selection for Trp^+ and the drug resistances of RK2, did not result in any clones with plasmids of reduced size.

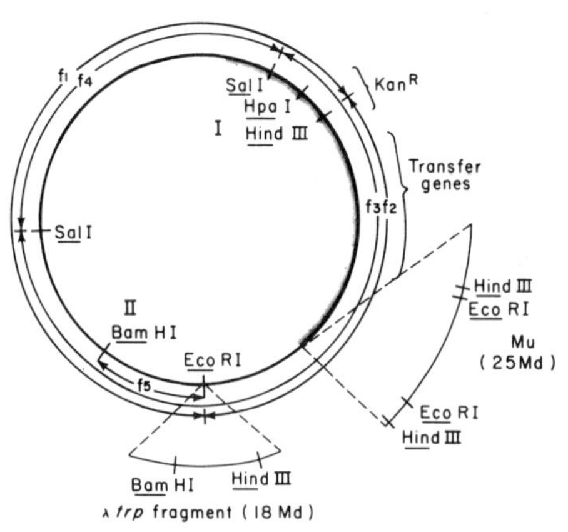

Figure 1
The location of the cleavage sites on RK2 for various restriction enzymes. The sites for *Pst*I, *Bgl*II, and *Sma*I, which have not been mapped precisely, are located in the regions indicated by I and II. Additional cleavage sites were introduced into RK2 by in vitro ligation of a λ*trp* fragment into the *Eco*RI site of RK2 and by lysogeny of RK2 with the phage Mu. Since Mu lysogenizes randomly, different points of insertion were observed; only one is shown. The cleavage sites shown on the λ*trp* piece (Meyer et al. 1977) and on Mu (Allet and Bukhari 1975) were used in attempts to reduce the size of RK2. The region of RK2 successfully deleted by digestion of the various RK2-Mu hybrids with *Hin*dIII is indicated by the shaded area.

The sizes (in millions of daltons [Md]) of the RK2 restriction fragments are as follows: f1, 24.5; f2, 13.2; f3, 25.2; f4, 12.8; and f5, 4.1.

Isolation and Analysis of RK2-Mu Hybrids and Their Derivatives

To explore further the problem of reducing the size of RK2, we decided to insert into the RK2 molecule a fragment of DNA carrying additional restriction sites. Cleavage of such a composite molecule with a restriction enzyme would result in deletion of a portion of the inserted segment, as well as a portion of RK2. The amount of RK2 deleted would depend on the location of the inserted DNA relative to the RK2 restriction site (see Fig. 1). Furthermore, theoretically, distinct deletions could be generated for each restriction enzyme used, the only requirement being that there be a single site for the restriction enzyme on RK2 and one or more sites on the inserted fragment.

The phage Mu was examined as a possible source of the inserted DNA in this procedure. This phage apparently lysogenizes randomly (Bukhari and Zipser 1973), and the cleavage sites on Mu DNA have been mapped for several restriction enzymes (Allet and Bukhari 1975). Moreover, Razzaki and Bukhari (1975) have already described a procedure for the generation of transmissible plasmids with Mu insertions. This technique involves induction of strains lysogenic for Mu *cts*62 and carrying the plasmid, followed by conjugal transfer of the plasmid to a suitable recipient. Colonies are selected for the incoming plasmid and then screened for transfer of Mu. When applied to RK2, a number of different RK2-Mu plasmids were obtained (Figurski et al. 1976). The location of Mu in one of these plasmids is depicted in Figure 1.

Purified, covalently closed DNA of the various RK2-Mu hybrids was restricted with *Hin*dIII, and the digests were used to transform *E. coli* cells for tetracycline resistance. In each case, a derivative of RK2 was obtained. As expected, these plasmids were composed of a segment of RK2 and a portion of Mu DNA and contained a single *Hin*dIII site. Since Mu can lysogenize in either orientation, it was also expected that a fragment from either end of Mu would be present. Some reduced plasmids did in fact retain the right end of Mu and some retained the left end, as determined by analysis of the plasmids with other restriction enzymes. Since neither of the two possible remaining Mu DNA fragments possesses an *Eco*RI site (Allet and Bukhari 1975), the single *Eco*RI cleavage site observed in all the reduced hybrids must be in the RK2 portion of the molecule. The extent of the deletion in the reduced RK2 hybrid molecules can be readily determined by cleavage of the molecules with *Eco*RI and *Hin*dIII and analysis of the sizes of the two fragments released. The amount of Mu DNA present on either of the two fragments can be estimated from the known restriction enzyme cleavage map of Mu (Allet and Bukhari 1975).

Regions of RK2 that have been successfully deleted are shown in Figure 1. All of the reduced plasmids lost kanamycin resistance, a property of the parental RK2-Mu hybrids. In addition, some of the molecules deleted of DNA in the region of fragment 2 were no longer self-transmissible. Therefore at least one or more genes responsible for conjugal transfer is located in this region.

In another series of experiments, an attempt was made to obtain a series of reduced molecules by digestion of the RK2-Mu hybrids with *Eco*RI. Mu phage has two *Eco*RI-sensitive sites (Allet and Bukhari 1975). However, no transformants containing a plasmid of reduced size were obtained. To date, it has not been possible to obtain smaller derivatives of RK2 by a procedure involving cleavage at the *Eco*RI site. This *Eco*RI site is not itself in an essential gene, since DNA may be inserted into this site without any observable effects on the replication properties of the plasmid (Meyer et al. 1975). The fact that attempts to eliminate either fragment 4 or fragment 5 (Fig. 1) have not been successful suggests that genes essential for replication may be contained within these fragments. DNA may be inserted into the *Bam*HI site, however, without an observable effect on the replication properties of the plasmid.

These various observations also suggest that the genes essential for the maintenance of RK2 in *E. coli* are not highly clustered, unlike the genes responsible for the replication of the plasmids F*lac* and R6-5 (Timmis et al. 1975). It remains to be determined whether or not a wide distribution of the genes responsible for replication on the plasmid will be characteristic of broad-host-range plasmids.

Location of Cleavage Sites for Other Restriction Enzymes

The sensitivity of RK2 to other restriction enzymes has also been determined. The enzyme *Bgl*II (G. Wilson and F. Young, unpubl.) has a single cleavage site on RK2; *Pst*I (Smith et al. 1976) has four to six sites; and *Sma*I (S. Endow and R. Roberts, unpubl.) has four sites. All of these sites are located in regions I and II (Fig. 1) of the plasmid molecule. Taken together with the locations of the cleavage sites for the other restriction enzymes used in this study, it is clear that there are two clusters of restriction sites on RK2. The significance of this unusual pattern of sensitivity to restriction enzymes is unclear at this time. In general, RK2 appears to have less than the expected number of restriction sites. The selection of molecules that have lost restriction sites might be understandable for a plasmid capable of extensive intergeneric transfer. The remaining restriction sites might be located preferentially in regions of the plasmid acquired relatively recently and so would not have been subjected to extensive selective pressure. Regions of the plasmid containing drug resistance genes might fall into this category. In this regard, it is interesting that kanamycin resistance maps in one of the clusters of restriction sites.

Construction of RK2-E1 Hybrids

The plasmid ColE1 offers several advantages as a cloning vehicle (Hershfield et al. 1974) and has been widely used for this purpose. In log-phase cultures, it is present to the extent of 25 to 30 copies per cell (Clewell and Helinski 1972). In the presence of chloramphenicol, the plasmid continues to replicate, generating several thousand copies per cell (Clewell and Helinski 1972; Clewell et al. 1972). The insertion of foreign DNA at the single *Eco*RI site of the plasmid does not affect significantly its replication properties. RK2-E1 hybrid plasmids were prepared to determine if some of the replication properties of ColE1 could in effect be grafted into the RK2 replicon. The hybrid plasmids were constructed by the digestion of RK2 and ColE1 DNA with *Eco*RI, followed by ligation with T4 ligase and transformation of an *E. coli* strain with the recombinant molecules. A detailed description of the construction and analysis of the hybrids obtained will be presented elsewhere (D. Figurski, R. J. Meyer and D. R. Helinski, in prep.).

The RK2-ColE1 hybrids failed to produce colicin, a result consistent with insertion of the RK2 DNA into the *Eco*RI site of ColE1 (Hershfield et al. 1974). The RK2-ColE1 recombinants expressed colicin E1 immunity, but ColE1 incompatibility was not expressed. The hybrid plasmids did retain the incompatibility function of the RK2 plasmid, however. Furthermore, the replication properties of RK2, but not ColE1, were exhibited by the joint replicon, in that the copy number of the RK2-ColE1 hybrid molecules was five to six per chromosome and the level of the hybrid molecules did not increase significantly in the presence of chloramphenicol. The suppression of ColE1 replication by RK2 is not exerted in *trans*; i.e., ColE1 will amplify normally in cells containing the RK2 plasmid. Finally, replication of the RK2-ColE1 hybrids occurs normally in a background lacking DNA polymerase I, as does replication of the parental RK2. Replication of the parental ColE1 plasmid requires DNA polymerase I (Kingsbury and Helinski 1970).

The control of RK2-E1 hybrid replication by the RK2 system is not because a defective ColE1 has been inserted into RK2. The excision of ColE1 from the hybrid by digestion with *Eco*RI allows recovery of an intact ColE1 plasmid which replicates in the mode expected for ColE1. The possibility that the apparent suppression of ColE1 replicative functions could be a result of the position and/or orientation of insertion of RK2 is presently under investigation.

Cloning Fragments of RK2

As stated earlier, a desirable feature of a plasmid cloning vehicle is the ability to both enter and replicate in many different bacteria. These very properties also confer a potential hazard related to the possible promiscuous and uncontrolled spread of the genetic element and its derivatives. One possible approach to obviate this complication is the construction of a binary vehicle system composed of two distinct replicating plasmids. One member (Rep_{RK2}-Tra^-) would be capable of carrying foreign DNA and could replicate in many different genera of bacteria but would be incapable of self-transmission by conjugal mating. The other plasmid vehicle ($Rep_{E1}Tra_{RK2}$) would have a more restricted host range but would be capable of mediating the transfer of the other cloning plasmid. In a system of this type, recombinant plasmids could be constructed in a well-characterized *E. coli* strain and then transferred to the target organism by conjugal mating promoted by the $Rep_{E1}Tra_{RK2}$ plasmid that itself is limited in its host range. The presence of only the non–self-transmissible cloning plasmid in the target organism would lower the probability of subsequent transfer of this plasmid to other organisms.

Preliminary experiments have been initiated with RK2 to explore the possibility of developing such a system. It is known from studies of the RK2-Mu hybrids that at least certain of the genes responsible for transfer are located in fragment 2 (Fig. 1). Consequently, this fragment was cloned by digestion of both RK2 and pCR1 (Covey et al. 1976) (Col E1-Km with a single *Eco*RI site) with *Sal*I and *Eco*RI, followed by ligation and transformation. One of the results of the digestion of RK2 with these two enzymes is the generation of a large fragment consisting of fragment 2 plus a short segment of DNA extending from the *Hin*dIII to the neighboring *Sal*I site. Treatment of pCR1 with *Eco*RI and *Sal*I results in the elimination of a small (m.w. = 1.2×10^6 daltons) fragment without any loss in replication or kanamycin resistance functions. Analysis of clones obtained from transformation with the ligated mixture resulted in the detection of a hybrid plasmid composed of the large fragment of RK2 and the derivative of pCR1. This plasmid itself did not show self-transfer, although it was capable of mediating transfer of two Tra^- plasmids generated by the reduction of the RK2-Mu hybrids. Further studies on this binary vehicle system composed of *tra* functions of RK2 on the ColE1 plasmid and a Tra derivative of RK2 as the cloning vehicle are being carried out to determine its utility.

CONCLUSION AND SUMMARY

RK2 and several other plasmids of the P incompatibility group are unique in their broad host range. Our physical studies have revealed several additional interesting features of the plasmid. RK2 appears to have relatively few cleavage sites for the collection of restriction enzymes so far examined and in fact has only one site for each of the enzymes *Eco*RI, *Hin*dIII, *Bam*HI, *Hpa*I, and *Bgl*II. In addition, the restriction sites appear to be clustered in two regions of the plasmid. The absence of a large number of restriction sites may be important in regard to the unusual capability of RK2 for intergeneric transfer.

In contrast to observations with several other large plasmids (Timmis et al. 1975), the genes for maintenance of RK2 in *E. coli* do not appear to be tightly clustered. Moreover, covalent insertion of ColE1 into RK2 can lead to the inhibition of both the replication of ColE1 and the expression of its incompatibility function but not its immunity function. These results differ from those obtained from a study of the ColE1-pSC101 hybrid replicon, where it was concluded that ColE1 replication is normally dominant and the incompatibility functions of both molecules are expressed (Cabello et al. 1976).

One of the major goals of our studies is to exploit the properties of RK2 in the construction of genetic vehicles with a broad host range. We have also described a prototype binary system which offers possibilities for development. However, the design of such vehicles demands the utmost attention to considerations of safety and containment. It is hoped that the additional information being obtained on P-plasmid transfer and replication will lead to the development of a generally useful and containable broad-host-range cloning system.

Acknowledgments

This work was supported by a Research Corporation Fellowship to R. M., a Public Health Service Research Fellowship (SF22 AI01412-02) to D. F., and research grants from the National Institute of Allergy and Infectious Diseases (AI-07194) and the National Science Foundation (GB-29492) to D. R. H.

REFERENCES

Allet, B. and A. I. Bukhari. 1975. Analysis of bacteriophage Mu and λ-Mu hybrid DNAs by specific endonucleases. *J. Mol. Biol.* **92**:529.

Beringer, J. E. 1974. R factor transfer in *Rhizobium leguminosarum*. *J. Gen. Microbiol.* **84**:188.

Bukhari, A. I. and D. Zipser. 1973. Random insertion of Mu-1 DNA within a single gene. *Nature New Biol.* **236**:240.

Cabello, F., K. Timmis and S. N. Cohen. 1976. Replication control in a composite plasmid constructed by *in vitro* linkage of two distinct replicons. *Nature* **259**:285.

Clewell, D. B. and D. R. Helinski. 1972. Effect of growth conditions on the formation of the relaxation complex of supercoiled ColE1 deoxyribonucleic acid and protein in *Escherichia coli*. *J. Bact.* **110**:1135.

Clewell, D. B., B. Evenchick and J. W. Cranston. 1972. Direct inhibition of ColE1 plasmid DNA replication in *E. coli* by rifampicin. *Nature New Biol.* **237**:29.

Covey, C., D. Richardson and J. Carbon. 1976. A method for the deletion of restriction sites in bacterial plasmid deoxyribonucleic acid. *Mol. Gen. Genet.* **145**:155.

Figurski, D., R. J. Meyer, D. Miller and D. R. Helinski. 1976. Generation *in vitro* of deletions of the broad host range plasmid RK2 using phage Mu insertion and a restriction endonuclease. *Gene* **1**:107.

Gromkova, R. and S. H. Goodgal. 1972. Action of *Haemophilus* endodeoxyribonuclease on biologically active deoxyribonucleic acid. *J. Bact.* **109**:987.

Hamer, D. H. and C. A. Thomas. 1976. Molecular cloning of DNA fragments produced by restriction endonucleases *Sal* I and *Bam* I. *Proc. Nat. Acad. Sci.* **73**:1537.

Hershfield, V., H. W. Boyer, C. Yanofsky, M. A. Lovett and D. R. Helinski. 1974. Plasmid ColE1 as a molecular vehicle for cloning and amplification of DNA. *Proc. Nat. Acad. Sci.* **71**:3455.

Kingsbury, D. T. and D. R. Helinski. 1970. DNA polymerase as a requirement for the maintenance of the bacterial plasmid colicinogenic factor E_1. *Biochem. Biophys. Res. Comm.* **41**:1538.

Lovett, M. and D. R. Helinski. 1976. Method for the isolation of the replication region of a bacterial replicon: Construction of a mini-F′*km* plasmid. *J. Bact.* **127**:982.

Mertz, J. E. and R. W. Davis. 1972. Cleavage of DNA by R1 restriction endonuclease generates cohesive ends. *Proc. Nat. Acad. Sci.* **69**:3370.

Meyer, R., D. Figurski and D. R. Helinski. 1975. Molecular vehicle properties of the broad host range plasmid RK2. *Science* **190**:1226.

——. 1977. Physical and genetic studies with restriction endonucleases on the broad host-range plasmid RK2. *Mol. Gen. Genet.* (in press).

Morrow, J. F. and P. Berg. 1972. Cleavage of simian virus 40 DNA at a unique site by a bacterial restriction enzyme. *Proc. Nat. Acad. Sci.* **69**:3365.

Old, R., K. Murray and G. Roizes. 1975. Recognition sequence of restriction endonuclease III from *Hemophilus influenzae. J. Mol. Biol.* **92**:331.

Panopoulous, N. J., J. J. Cho, W. V. Guimares and M. N. Schroth. 1975. Genetic transfer of *Pseudomonas aeruginosa* R factors to *Pseudomonas* and *Erwina* plant pathogens. In *Proceedings of the 1st Intersectional Congress of IAMS* (ed. T. Hasegawa), vol. 1, p. 142. Science Council of Japan, Tokyo.

Razzaki, T. and A. I. Bukhari. 1975. Events following prophage Mu induction. *J. Bact.* **122**:437.

Smith, D. I., F. R. Blattner and J. Davies. 1976. The isolation and partial characterization of a new restriction endonuclease from *Providencia stuarti. Nucl. Acids Res.* **3**:343.

Timmis, K., F. Cabello and S. N. Cohen. 1975. Cloning, isolation and characterization of replication regions of complex plasmid genomes. *Proc. Nat. Acad. Sci.* **72**:2242.

Wilson, G. A. and F. E. Young. 1975. Isolation of a sequence-specific endonuclease (*Bam* I) from *Bacillus amyloliquefaciens* H. *J. Mol. Biol.* **97**:123.

Insertion of Mu DNA Fragments into Phage λ In Vitro

D. D. Moore, J. W. Schumm, M. M. Howe* and F. R. Blattner
Departments of Genetics and Bacteriology*
University of Wisconsin
Madison, Wisconsin 53706

To construct a precise genetic and physical map of phage Mu, it would be desirable to isolate a collection of deletion mutations encompassing all essential and nonessential Mu genes. One approach is the construction of λ-Mu hybrids.

Classical genetic techniques permitted the construction of such hybrids for the ends of the Mu prophage (Allet and Bukhari 1975). To obtain hybrids for the central region as well, we constructed chimeric DNA molecules in vitro using λ Charon 4 as a vehicle for Mu DNA fragments (Blattner et al. 1977).

Construction of Chimeric Phages

Mu DNA contains two *Eco*RI cleavage sites which divide the genome into three pieces (see Fig. 1) (Allet and Bukhari 1975). The central fragment, which is the only fragment presumed a priori to contain an *Eco*RI cohesive sequence at both ends, measures 39% of the λpapa genome length or 19.3 kilobase pairs. We chose to use the λ Charon 4 vector illustrated in Figure 1 because a large fragment can be accommodated within the λ packaging limit by the replacement of the *lac* and *bio* regions with foreign DNA.

DNAs from Charon 4 and Mu were cleaved with *Eco*RI and the resulting fragments joined with T4 ligase. After transfection of *Escherichia coli* spheroplasts, $lac^{-}bio^{-}$ phages with λ immunity were isolated. These putative λ-Mu hybrids were assayed for the presence of Mu DNA by marker rescue, i.e., by testing their ability to supply a wild-type Mu marker by recombination when crossed with a series of amber mutants of Mu (Howe 1973). Out of 20 isolates, 8 donated Mu markers in genes M to S which are at the right end of the Mu map and twelve donated markers C to L from the central region of Mu. No phages containing the A to B region of Mu were found.

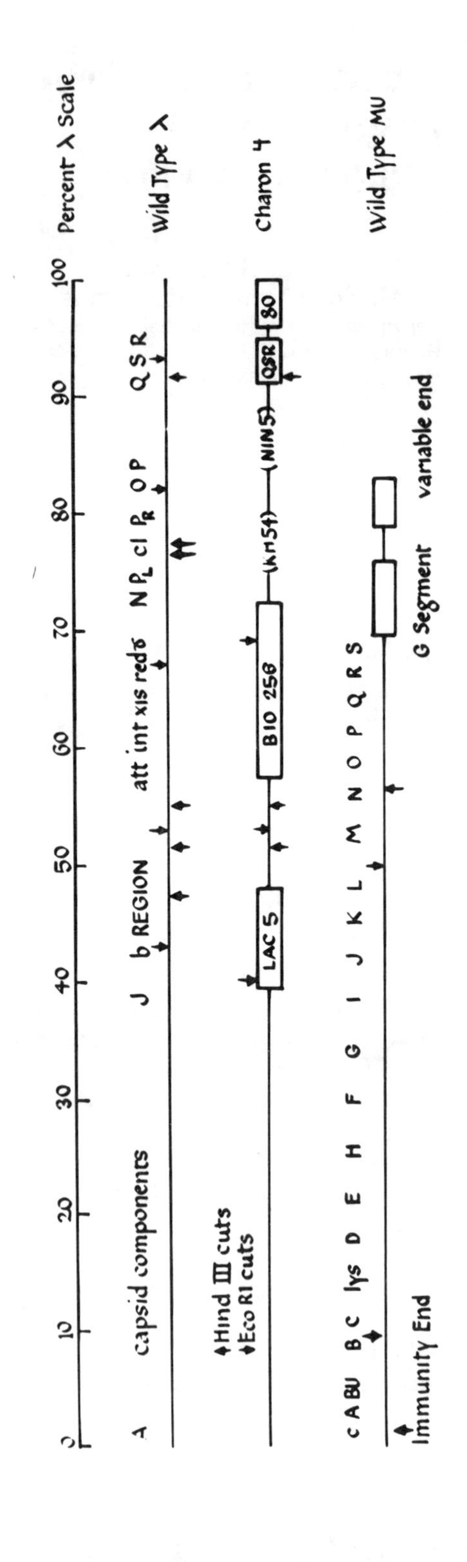

Percent λ Scale
0
10
20
30
40
50
60
70
80
90
100
Wild Type λ
A
capsid components
J
b REGION
att int xis redδ
N P_L cI P_R O P
Q S R
Charon 4
Hind III cuts
Eco RI cuts
LAC 5
BIO 256
(KH54)
(NIN5)
QSR
80
Wild Type MU
c A BU B C lys D E H F G I J K L M N O P Q R S
Immunity End
G Segment
variable end

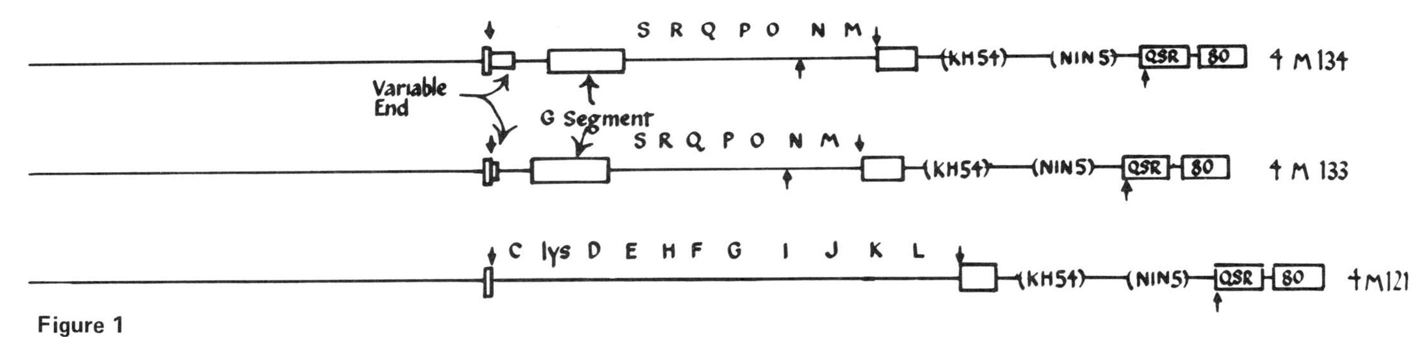

Figure 1

Physical maps of λ, Mu, and their hybrids. The phage λ map is shown approximately to scale, 100 %λ corresponding to 49,400 base pairs or 30.1 × 10^6 daltons of DNA. *Eco*RI and *Hin*dIII cleavage sites are indicated by downward and upward arrows, respectively. The Charon 4 vector contains *lac*5, *bio*256, and QSR80 substitutions and c1KH54 and *nin*5 deletions. The *lac*5 and *bio*256 substitutions are equivalent in length to the λ DNA they replace; the KH54 and *nin*5 deletions remove 4.3 %λ and 6.5 %λ, respectively; and the QSR80 substitution results in a net addition of 2.0 %λ. The Charon 4 vector has had all *Eco*RI sites removed from parts of the genome essential for plaque formation, and up to 43% of foreign DNA can be inserted between the cuts in *lac*5 and *bio*256 without exceeding the λ packaging limit. The phage Mu genome is shown to the same scale. "G segment" indicates a region of DNA that frequently inverts during propagation of Mu, and the variable end consists of random pieces of *E. coli* DNA. 4M134, 4M133, and 4M121 are the proposed structures of three of the λ-Mu hybrids we have constructed.

Physical Analysis of the Chimeric Genomes

Gels

DNA from two clones containing the right end of Mu (4M133 and 4M134) and one clone containing the central region (4M121) were digested with *Eco*RI or *Hin*dIII and analyzed by electrophroesis through agarose gels (Fig. 2).

In each case, a single inserted *Eco*RI fragment was observed. As expected, the cloned fragment in 4M121 was comparable in size to the central Mu fragment (39 %λ), whereas those in 4M133 and 4M134 migrated somewhat faster than the right-hand Mu fragment (32 %λ). The cloned fragment in 4M134 was slightly larger than the one in 4M133.

*Hin*dIII patterns of 4M133 and 4M134 both showed a fragment measuring 17.6 %λ. A fragment of this size would be expected from the restriction map if both clones contained the Mu fragment oriented as shown in Figure 1. The

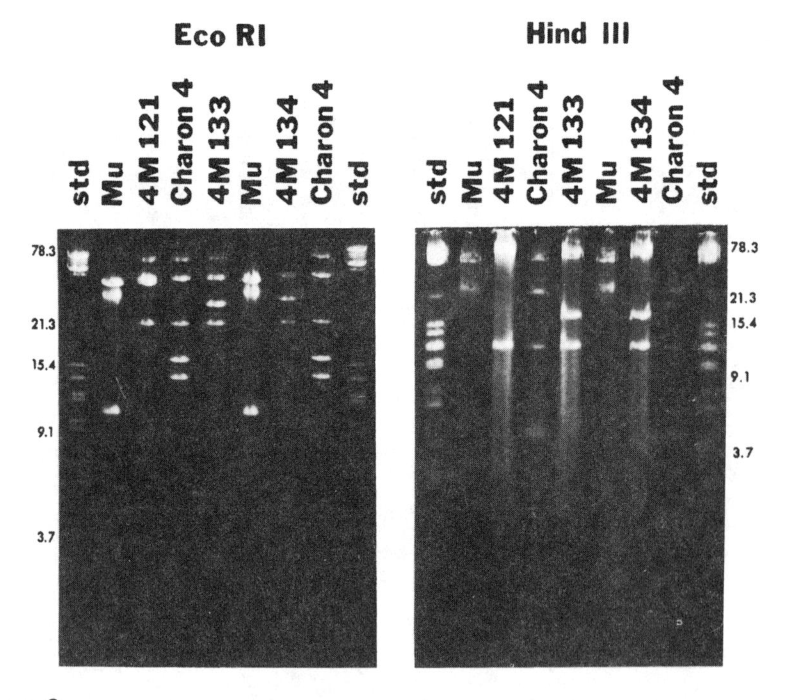

Figure 2

Gel electrophoresis of restriction endonuclease digests. The indicated DNAs were digested with *Eco*RI or *Hin*dIII, and electrophoresis was carried out as described by Shinnick et al. (1975). Fragment sizes were estimated by comparison with a standard mixture consisting of *Eco*RI digest of λKH100 *nin*5, λ*b*2, and λ*b*538S^{R}1-3^{-} *imm*^{434} QSR80 which produces fragments of 2.3, 3.7, 5.5, 6.8, 9.1, 9.8, 11.8, 12.2, 14.0, 15.4, 21.3, 42.5, 44.5, 49.1, 51.3, 57, and 78 %λ. A few have been labeled to facilitate identification.

other orientation would predict a *Hin*dIII fragment of 36 %λ, which was not observed. As expected, there was no *Hin*dIII site in the Mu DNA cloned in 4M121.

Electron Microscopy

DNA from 4M121, 4M133, and 4M134 was examined by electron microscope heteroduplex analysis. Heteroduplexes of 4M133 with Charon 4 DNA were double-stranded at each end and had substitution bubbles (Fig. 3a), confirming that heterologous DNA was incorporated into the vector at the expected position. When 4M121 DNA was paired with Mu DNA, the structures exhibited duplex segments, of approximately 39 %λ length, with single-stranded forks at either end (Fig. 3b). The duplex segments result from annealing of Mu DNA in Mu with the homologous segment in the clone. The distributions of single-stranded lengths emerging from the forks strongly suggest that Mu DNA is inserted in the vector with the orientation shown in Figure 1. However, we have not yet obtained confirmation of the orientation by an independent method.

In heteroduplex mixtures of Mu DNA and either 4M133 or 4M134 DNA, we observed the characteristic Mu "G bubbles" in the λ/λ, Mu/Mu, and λ/Mu heteroduplexes (Fig. 3c,d,e). This indicates that the G segment frequently inverts while a passenger in the λ vehicle, as observed by Allet and Bukhari (1975). The positions and sizes of the bubbles confirmed their identification as Mu-specific G bubbles and further corroborated the assigned orientation of the Mu segment in these clones. This orientation was also confirmed by the length distribution of single-stranded branches in λ/Mu heteroduplexes.

Figure 3f shows a heteroduplex between DNAs from 4M133 and 4M134 which exhibits a small substitution bubble located at a distance of 3.3 %λ from a G bubble. From the same DNA mixture, we also observed heteroduplexes with only the small bubble and others with only a G bubble. These observations can be explained most easily if 4M133 and 4M134 each contain different short segments of bacterial DNA originating from the variable end of Mu because of *Eco*RI cleavage sites in the bacterial DNA.

Deletion Mapping of Mu DNA in 4M121

Since 4M121 contains DNA of a length equivalent to 101 %λ, it is quite sensitive to treatment with chelating agents (Parkinson and Huskey 1971). Derivatives of 4M121 containing deletions were isolated by selecting for phage that could grow in the presence of 0.1% (w/v) sodium citrate. Marker-rescue analysis of these deleted λ-Mu hybrids using Mu amber mutants in genes C to L revealed that the deletions removed varying amounts of Mu DNA (Fig. 4). Each of the deletions can be interpreted as the loss of a single contiguous group of genes only if gene H is located between genes E and F rather than between G and I

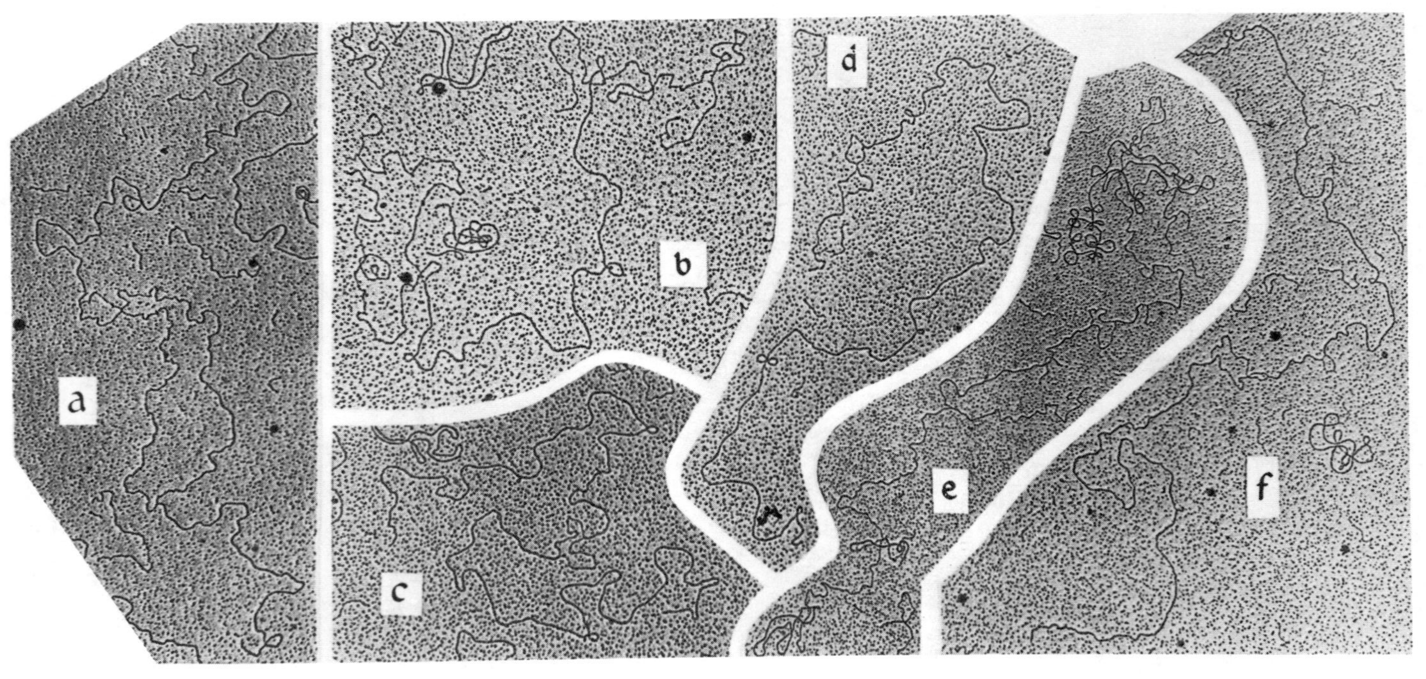

Figure 3
Heteroduplexes of λ-Mu hybrids. Heteroduplexes were formed between the indicated DNAs, according to the technique of Davis et al. (1971), and examined in the electron microscope. All photographs are not shown to the same scale. (*a*) 4M133/Charon 4; (*b*) 4M121/Mu; (*c*) 4M133/4M133; (*d*) Mu/Mu; (*e*) 4M133/Mu; (*f*) 4M133/4M134.

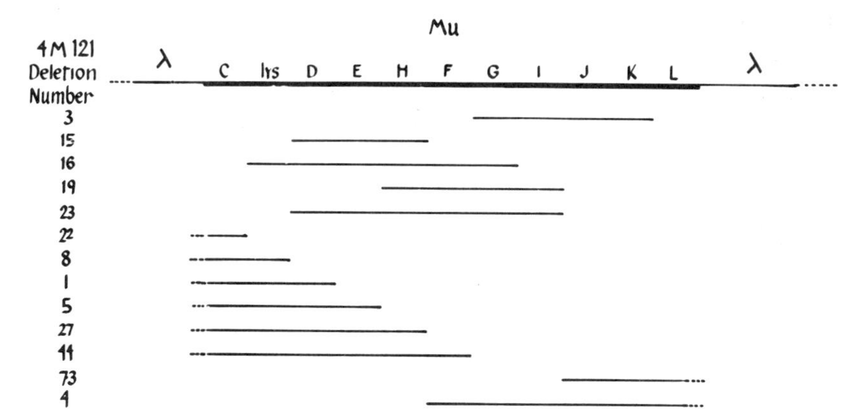

Figure 4
Deletion mapping of Mu DNA in clone 4M121. The map of a standard set of amber mutations in genes C–L of Mu DNA within the λ vector is given at the top of the figure. The bars represent the Mu markers deleted in the various 4M121 deletion strains. Marker rescue was measured on a lawn of WD5021 (λ) [Su° *gal*⁻ Sm^r (λ)] infected with the clone to be tested. Lysates of Mu amber mutants were spotted on the lawns and on control uninfected WD5021 (λ) lawns. If the clone contained the wild-type allele of the amber mutation, Mu plaques could be seen after overnight incubation. The standard set of Mu amber mutants used were all in a Mu *cts*62 background and contained the following amber alleles: *Cam*4005, *lysam*1025, *Dam*3801, *Eam*1006, *Ham*1043, *Fam*1065, *Gam*1042, *Iam*4037, *Jam*1155, *Kam*1010, and *Lam*1007 (Abelson et al. 1973; Howe 1973).

as originally assigned (Abelson et al. 1973). We have not yet examined any of the deletions electron microscopically, but the parent 4M121 did not show any chromosome rearrangements when paired with Mu DNA (Fig. 3b). Moreover, the same Mu gene order was obtained independently by analysis of deletions isolated from six different clones containing the C to L region and by use of several independent H mutants. We therefore conclude that the Mu gene order shown in Figures 1 and 4 is correct.

SUMMARY

We have isolated a series of λ-Mu hybrid phages by cloning the middle and right-end *Eco*RI fragments of mature Mu DNA in the cloning vector λ Charon 4. Clones with the middle Mu fragment contain Mu genes C–L; those with the right end contain genes M–S plus the G and β regions and portions of the variable end. Derivatives of these clones containing deletions within the Mu DNA have been isolated and used to refine the genetic map of Mu.

Acknowledgments

This research was supported by the College of Agriculture and Life Sciences, University of Wisconsin, Madison, and by a National Institutes of Health Grant to F. R. B., by National Institutes of Health and National Science Foundation Grants to M. M. H., and by National Institutes of Health predoctoral training grants to J. W. S. and D. D. M. F. R. B. is also supported by a Research Career Development Award from the National Institutes of Health. The authors thank Bill Williams for help with the cloning, Dennis Schultz and Delores Stephenson for technical assistance, and Jill La Fond for typing the manuscript.

REFERENCES

Abelson, J., W. Boram, A. I. Bukhari, M. Faelen, M. Howe, M. Metlay, A. L. Taylor, A. Toussaint, P. van de Putte, G. C. Westmass and C. A. Wijffelman. 1973. Summary of the genetic mapping of bacteriophage Mu. *Virology* **54**:90.

Allet, B. and A. I. Bukhari. 1975. Analysis of bacteriophage Mu and λ-Mu hybrid DNAs by specific endonucleases. *J. Mol. Biol.* **92**:529.

Blattner, F. R., B. G. Williams, A. E. Blechl, K. Denniston-Thompson, H. E. Faber, L.-A. Furlong, D. J. Grunwald, D. O. Kiefer, D. D. Moore, J. W. Schumm, E. L. Sheldon and O. Smithies. 1977. Charon phages: Safer derivatives of bacteriophage lambda for DNA cloning. *Science* **196**:161.

Davis, R. W., M. Simon and N. Davidson. 1971. Electron microscope heteroduplex methods for mapping regions of base sequence homology in nucleic acids. In *Methods in enzymology* (ed. L. Grossman and K. Moldave), vol. 21D, p. 413. Academic Press, New York.

Howe, M. M. 1973. Prophage deletion mapping of bacteriophage Mu-1. *Virology* **54**:93.

Parkinson, J. and R. Huskey. 1971. Deletion mutants of bacteriophage lambda. I. Isolation and initial characterization. *J. Mol. Biol.* **56**:369.

Shinnick, T. M., E. Lund, O. Smithies and F. R. Blattner. 1975. Hybridization of labeled RNA to DNA in agarose gels. *Nucleic Acids Res.* **2**:1911.

The *E. coli* Gamma-Delta Recombination Sequence Is Flanked by Inverted Duplications

T. R. Broker and L. T. Chow
Cold Spring Harbor Laboratory
Cold Spring Harbor, New York 11724

L. Soll
Department of Molecular, Cellular and Developmental Biology
University of Colorado
Boulder, Colorado 80304

The 5700 (± 100) base-pair sequence gamma-delta ($\gamma\delta$) is present in the integrative region of the sex factor F (Ohtsubo et al. 1974a). Integration of F into *Escherichia coli* to form the Hfr strain AB313 (Taylor and Adelberg 1960) apparently occurred by reciprocal recombination of the $\gamma\delta$ sequence of F and a $\gamma\delta$ sequence located at *E. coli* map position 83 near the *ilvE* gene (Ohtsubo et al. 1974a; revised map of Bachmann et al. 1976). Other $\gamma\delta$ sequences have been located, by inference or by direct electron microscopic studies, at *E. coli* map positions 26 min (Hfr EC8), 61 min (Ra-1), and 86 min (Ra-2) (Low 1967; Palchaudhuri et al. 1976). Two 5600-base-pair sequences have been observed by EM to be arranged in direct order 77.3 kb (1.7 min) apart (Chow, this volume). These may also be $\gamma\delta$ sequences. The episome F′14 derived from AB313 (Pittard et al. 1963) contains two copies of $\gamma\delta$ in direct order, one at each boundary of F-specific and bacterial DNA (Ohtsubo et al. 1974a). Intra-episomal recombination of the two $\gamma\delta$ sequences of F′14 results in its instability and in the resegregation of the F and the bacterial portions. It was from such a plasmid containing bacterial sequences segregated from F′14 DNA that ϕ80 d_3 *ilv*$^+$, *rrn*$^+$, *su*$^+$7 was isolated by L. Soll. ϕ80 d_3 carries a single copy of $\gamma\delta$ (Ohtsubo et al. 1974b).

When we denatured ϕ80 d_3 DNA, then self-annealed it at 21°C in 50% formamide, we frequently observed in the electron microscope a 5700-base-pair loop in the DNA duplex at the coordinates assigned to the $\gamma\delta$ sequence. The loop was only partially duplex and contained denatured portions that were irregular in position and extent (Fig. 1A). When single strands of ϕ80 d_3 DNA were examined, a strand cross-over (but not a measurable duplex stem) usually appeared at the same position. We reasoned that the structures could be explained if short, inverted repetitious sequences of about 20 nucleotides bordered the $\gamma\delta$ segment. Such inverted duplications will "fold back" rapidly to form a stem-loop structure when denatured strands are allowed to renature

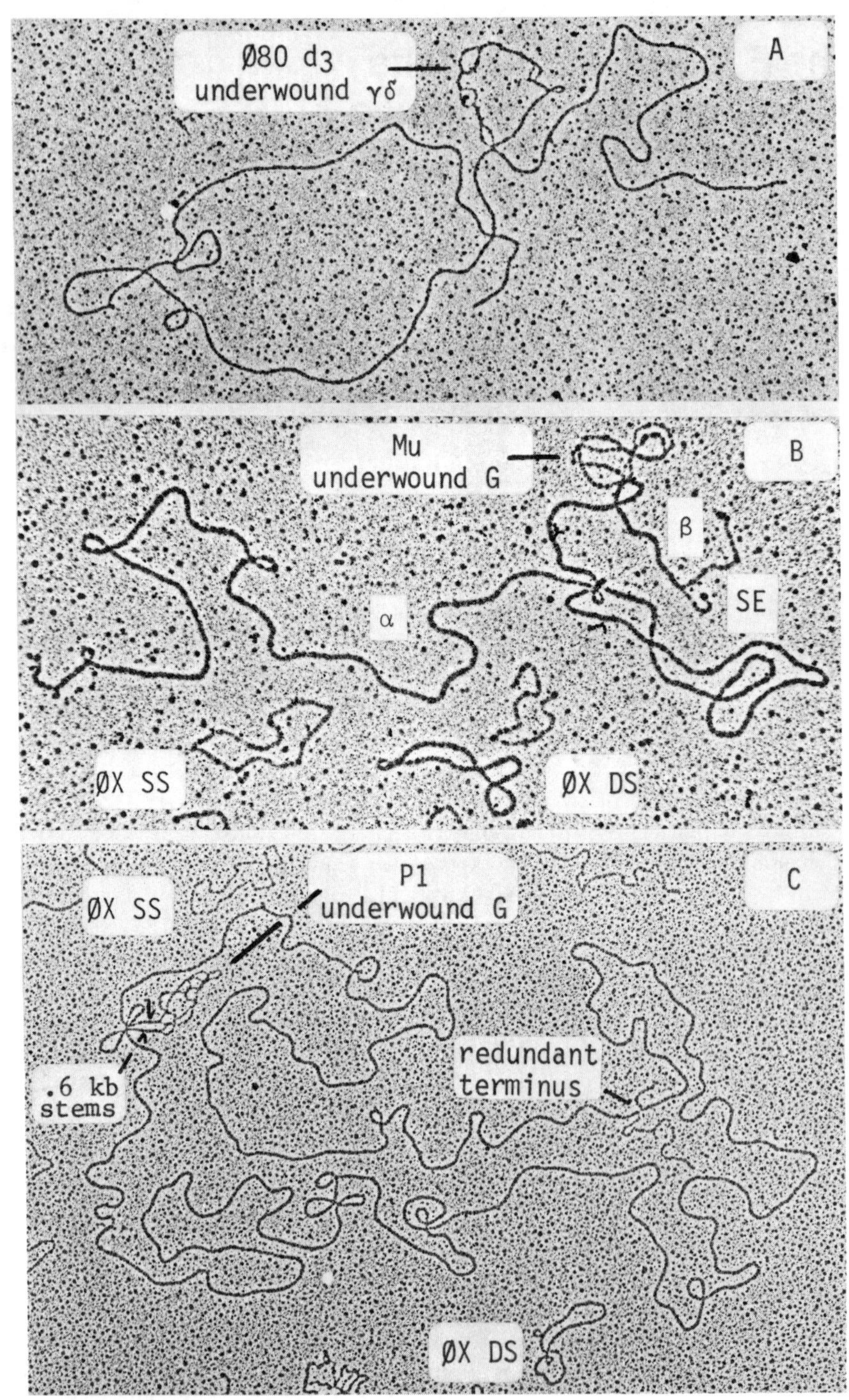

Figure 1 *(See facing page for legend.)*

Figure 2
Formation of underwound loops. Complementary stem-loop structures (I) may pair to make the cruciform (II) or, if the stems melt out, the linear duplex (III). The axial rotations necessary for the cruciform (II) to reduce to the duplex (III) are indicated. It is more likely that the complementary single-strand loops of II will pair, prevent further axial rotation, and collapse into the underwound loop (IV).

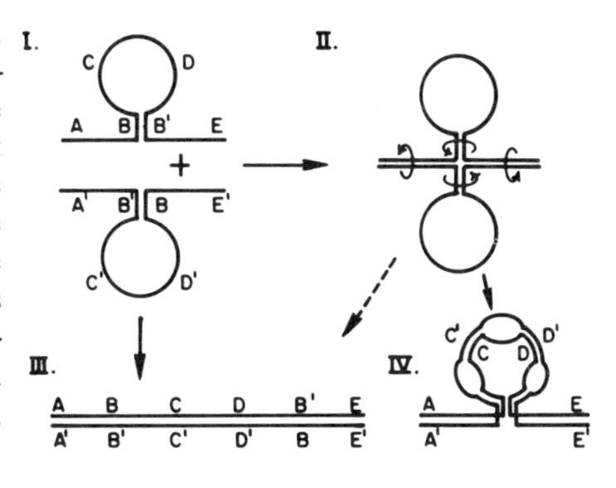

(Fig. 2, I). Stems of at least 20 nucleotides persist under typical reannealing conditions (T_m = 25°C) (Hayes et al. 1970). When two complementary DNA strands, each containing the stem-loop structure, pair with one another on both sides of the stem, the structures may be retained in the hybrid (Fig. 2, II). In principle, the stems could be consumed by concerted double-strand exchange (Broker and Lehman 1971), and the sequences within the loop could then pair to form a perfect linear duplex (Fig. 2, III). However, axial rotation of the DNA is necessary for the strand exchange (Platt 1955; Broker 1973; Broker and Doermann 1975). Rotation of the two stems requires that the loops also rotate relative to one another (Fig. 2, II). If the complementary sequences in the two loops start to pair before the stem is consumed by strand exchange, counter-rotation of the loops is inhibited and the stem-loop structure is kinetically frozen.

Separated, closed rings are topologically impenetrable (Vinograd et al. 1968), so there can be no net interwinding of complementary strands of single-stranded circular DNA. The constraint on stem-loop structures is identical to that on complementary, closed, single-strand circles. Hence the loops can pair only to a limited degree and will have single- and double-stranded regions variable in extent and position (Fig. 2, IV). We term this structure an *underwound loop* and emphasize that it can be diagnostic of inverted repetitions in self-renatured DNA samples. A recent model for base-pairing of impenetrable and noninterwound complementary strands suggests the possibility of side-by-side alignment

Figure 1
DNA samples were denatured with 0.2 M NaOH, reneutralized with Tris-HCl, then self-annealed for 1 hr at 22°C in 50% formamide, 0.1 M NaCl. They were spread in a cytochrome *c* film from 45% formamide, 0.1 M Tris-HCl, 0.01 M EDTA, pH 8.5, onto a hypophase of 15% formamide, 0.01 M Tris-HCl, 0.001 M EDTA, pH 8.5. Magnifications: (*A*) ϕ80 d_3 *ilv*$^+$ *rrn*$^+$ *su*$^+$7–27,000 ×; (*B*) Mu–34,000 ×; (*C*) P1 *cam*–25,500 ×.

of strands with short alternating segments of left-handed and right-handed helical conformations (Rodley et al. 1976).

We have substantiated our interpretation of the origin of underwound $\gamma\delta$ loops by observing a similar structure in reannealed bacteriophage Mu DNA at the coordinates of the G segment (Fig. 1B). The G segment is known to be bounded by short inverted repetitions (Hsu and Davidson 1974). The underwound loop of G-segment DNA is to be distinguished from the "G loop" (or "G inversion bubble"). The former can be seen in Mu heteroduplexes in which the G segments of both strands are in the same genetic orientation; the latter appears when they are in opposite orientations.

The inverted adenovirus-2 chromosome has inverted terminal duplications (Garon et al. 1972). Reannealed Ad-2 DNA also forms underwound loops the full length of the chromosome (Broker et al. 1977). Complementary single-strand loops at the ends of long stems of inverted DNA sequences also can pair to form underwound loops. They have been observed in the G DNA of phage P1 (Fig. 1C; Chow and Bukhari, this volume) and in the transposable tetracycline (Kleckner et al. 1975) and kanamycin (Berg et al. 1975) resistance elements.

Single-stranded stem loops and double-stranded underwound loops may be used as coordinate markers for heteroduplex mapping. For instance, Wu and Davidson (1975) followed our suggestion that the $\gamma\delta$ loop could be used to locate precisely some of the RNA transcripts of ϕ80 d_3. Chow et al. (this volume [Fig. 4A], and unpubl.) used the Mu underwound G loop to map insertions and substitutions in the beta region. The underwound loop is especially valuable in locating inverted repetitions when they are too short to appear distinctly in EM samples as double-stranded stems.

In a separate paper in this volume (Broker), a model for integrative and excisive recombination at the inverted terminal repetitions of $\gamma\delta$ is presented and compared with the inversion model for Mu (Hsu and Davidson 1974) and P1 G segments. A more complete account of these studies will be published elsewhere (Broker et al. 1977).

SUMMARY

The 5700-base-pair $\gamma\delta$ recombination sequence present in *E. coli* DNA and carried on ϕ80 d_3 *ilv* transducing phage has short, inverted duplications at its boundaries ($\leqslant$50 nucleotide pairs). Single-stranded DNA molecules that include the $\gamma\delta$ sequence form intrastrand loops due to the pairing of the inverted duplications. When complementary DNA strands containing stem-loop secondary structures hybridize, the sequences in the loops cross-anneal to an extent limited by topological constraints. The resulting partially single-stranded, partially double-stranded underwound loop facilitates the identification of sequences such as $\gamma\delta$ that have short, flanking, inverted duplications.

Acknowledgments

The Mu DNA was prepared by Dr. Regine Kahmann and P1 DNA was from Mr. Frans de Bruijn. We are pleased to acknowledge the assistance of Ms. Marie Moschitta in the preparation of the manuscript. Some of the research reported was performed in the laboratory of Dr. Norman Davidson at the California Institute of Technology. Support was received from a Helen Hay Whitney Fellowship to T. R. B., a U.S. Public Health Service Grant GM 10991 to Norman Davidson, and an NIH Cancer Center Grant to Cold Spring Harbor Laboratory.

REFERENCES

Bachmann, B. J., K. B. Low and A. L. Taylor. 1976. Recalibrated linkage map of *Escherichia coli* K-12. *Bact. Rev.* **40**:116.

Berg, D. E., J. Davies, B. Allet and J. D. Rochaix. 1975. Transposition of R factor genes to bacteriophage lambda. *Proc. Nat. Acad. Sci.* **72**:3628.

Broker, T. R. 1973. An electron microscopic analysis of pathways for bacteriophage T4 DNA recombination. *J. Mol. Biol.* **81**:1.

Broker, T. R. and A. H. Doermann. 1975. Molecular and genetic recombination of bacteriophage T4. *Annu. Rev. Genet.* **9**:213.

Broker, T. R. and I. R. Lehman. 1971. Branched DNA molecules: Intermediates in T4 recombination. *J. Mol. Biol.* **60**:131.

Broker, T. R., L. Soll and L. T. Chow. 1977. Underwound loops in self-renatured DNA are diagnostic of inverted duplications. *J. Mol. Biol.* (in press).

Garon, C. F., K. W. Berry and J. A. Rose. 1972. A unique form of terminal redundancy in adenovirus DNA molecules. *Proc. Nat. Acad. Sci.* **69**:2391.

Hayes, F. N., E. H. Lilly, R. L. Ratliff, D. A. Smith and D. L. Williams. 1970. Thermal transitions in mixtures of polydeoxyribodinucleotides. *Biopolymers* **9**:1105.

Hsu, M-T. and N. Davidson. 1974. Electron microscope heteroduplex study of the heterogeneity of Mu phage and prophage DNA. *Virology* **58**:229.

Kleckner, N., R. K. Chan, B.-K. Tye and D. Botstein. 1975. Mutagenesis by insertion of a drug-resistance element carrying an inverted repetition. *J. Mol. Biol.* **97**:561.

Low, B. 1967. Inversion of transfer modes and sex factor-chromosome interactions in conjugation in *Escherichia coli. J. Bact.* **93**:98.

Ohtsubo, E., R. C. Deonier, H.-J. Lee and N. Davidson. 1974a. Electron microscope heteroduplex studies of sequence relations among plasmids of *Escherichia coli.* IV. The F sequences in F14. *J. Mol. Biol.* **89**:565.

Ohtsubo, E., L. Soll, R. C. Deonier, H.-J. Lee and N. Davidson. 1974b. Electron microscope heteroduplex studies of sequence relations among plasmids of *Escherichia coli.* VIII. The structure of bacteriophage ϕ80 d_3 ilv^+ su^+7, including the mapping of the ribosomal RNA genes. *J. Mol. Biol.* **89**:631.

Palchaudhuri, S. W. K. Maas and E. Ohtsubo. 1976. Fusion of two F-prime factors in *Escherichia coli* studied by electron microscope heteroduplex analysis. *Mol. Gen. Genet.* **146**:215.

Pittard, J., J. S. Loutit and E. A. Adelberg. 1963. Gene transfer by F′ strains of *Escherichia coli* K-12. I. Delay in initiation of chromosome transfer. *J. Bact.* **85**:1394.

Platt, J. R. 1955. Possible separation of intertwined nucleic acid chains by transfer twist. *Proc. Nat. Acad. Sci.* **41**:181.

Rodley, G. A., R. S. Scobie, R. H. T. Bates and R. M. Lewitt. 1976. A possible conformation for double-stranded polynucleotides. *Proc. Nat. Acad. Sci.* **73**:2959.

Taylor, A. L. and E. A. Adelberg. 1960. Linkage analysis with very high frequency males of *Escherichia coli. Genetics* **45**:1233.

Vinograd, J., J. Lebowitz and R. Watson. 1968. Early and late helix-coil transitions in closed circular DNA. The number of superhelical turns in polyoma DNA. *J. Mol. Biol.* **33**:173.

Wu, M. and N. Davidson. 1975. Use of gene 32 protein staining of single-stranded polynucleotides for gene mapping by electron microscopy: Application to the ϕ80 d_3 *ilv* su^+7 system. *Proc. Nat. Acad. Sci.* **72**:4506.

SECTION VII
Appendices

Endonuclease cleavage analysis of DNA can be used to establish quickly whether or not a newly isolated temperate bacteriophage is related to the phage already known. The section-title figure shows the *Eco*RI cleavage patterns of the genomes of four well-known temperate phages of *E. coli*—λ, P2, Mu, and P1. The fragments were separated by electrophoresis in a 1.4% agarose gel, stained with ethidium bromide, and photographed under ultraviolet light. Some of the small fragments of P1 are not clearly visible in the picture. In Mu, the two large fragments have not been resolved distinctly. (Photograph by E. Ljunquist and L. Ambrosio, Cold Spring Harbor Laboratory.)

Appendix A: IS Elements

1. IS ELEMENTS IN *ESCHERICHIA COLI*, PLASMIDS, AND BACTERIOPHAGES

Compiled by W. Szybalski, University of Wisconsin, Madison

Insertion element and characteristics	Individual occurrences			
	location	allele (mutation)	orientation	reference[a]*
IS*1*	*in E. coli*[b]			
~800 base pairs	*galT*	S104	I	15, 17, 50
Polar in both orientations	*galT*	N102	I } same	15, 19, 26
Restriction patterns (refs. 1, 17, 35,	*galT*	S188	II } site	15, 17, 41, 50
39, 49):	*galT*	N116	II	19, 26
enzyme sites enzyme sites	*galT*	15	} same	41
*Bam*HI 0 *Alu*I $\geqslant$ 2	*galT*	A10	} site	26, 41
*Bgl*I 0 *Bal*I 1	*galT*	A14	}	26, 41
*Bgl*II 0 *Hae*II 2	*galOP-E*	306	I	19, 48
*Eco*RI 0 *Hae*III 4	*galOP-E*	128	II	18, 19, 45
*Hinc*II 0 *Hha*I $\geqslant$6	*galOP-E*	141	II	18, 19, 45
*Hind*II 0 *Hinf* 3	*lacZ*	MS319	I	36
*Hind*III 0 *Hpa*II $\geqslant$5	*lacZ*	MS348	I	36
*Hpa*I 0 *Hph*I 2	*lacZ*	MS520	II	36
*Xba*I 0 *Pst*I 1	*argB*	*argB5*	II	11a
*Xho*I 0	*spc-p (rplX-p)*	115	II	24, 46
	in plasmids and phages			
	R1*drd*19		see Fig. B2.2	20, 21, 43
	R6		see Fig. B2.2	20, 21, 43
	R100	$\lambda\mu$ (2 copies)	see Fig. B2.2	21, 39

*All references are cited in full, by number, at the end of this table. Footnotes a, b, and c can also be found there.

Insertion element and characteristics	Individual occurrences			
	location	allele (mutation)	orientation	reference
IS*1* *(continued)*				
	Tn*9* (Cm) in P1Cm, λCm, and MuCm		two parallel copies at ends of the transposon II in λCm5	33
	λ*c*II	*r*14	I	8, 19
	Mu*X*	seven isolates at same site as Tn*9* in MuCm		10, 11
	P1		two copies	11
IS*2*	*in E. coli*[b]			
~1300 base pairs	*galOP-E*	490	I	4, 15
Polar in orientation I, supplies promoter function in orientation II	*galOP-E*	308	I	18, 19, 48
	galOP-E	*gal*3	I	5, 14a, 37
	galOP-E	108	II	44, 47
Contains rho-sensitive site in orientation I	*rspE-p (spc-p)*	I16	I	24, 46
	argOP	SAO10	I (in respect to *argE*)	11a
Restriction patterns (refs. 6, 35, 39, 49, 56):	*in plasmids and phages*			
enzyme — sites	λ*y*	*r*32 (*rip*2 mutation abolishes polarity—ref. 55)	I	8, 15, 19
*Eco*RI — 0				
*Bam*HI — 0				
*Hin*dII — 1				
*Hin*dIII — 1	λ*P-Q*	b1	I	15, 16
*Bgl*I — 2	λ*P-Q*	b_{NN}	I	15, 16
	λ*xis-exo*	*crg*	I	2, 38, 58

	λ*xis-exo*	*bi2*		II	38
	λ*xis*	*int-c57*		II	42, 52, 58
	λ*xis*	*int-c60*		II	42, 52, 58
	λ*xis*	*int-c508*		II	42, 52, 58
	λ*int-xis*	*int-c548*		II	42, 52, 58
	F	$\epsilon\zeta$	see Fig. B2.1		20, 21, 39
	R6-5		see Fig. B2.2		20, 21, 39
	R100		see Fig. B2.2		20, 21, 39
IS*3*	*in E. coli*[b]				
~1200 base pairs	*lacZ*	MS505		I	36
Polar in orientation I	*in plasmids and phages*				
Restriction patterns (refs. 35, 39): enzyme / sites	F	$\alpha\beta$	two copies; see Fig. B2.1		22, 40
*Bam*I 0	R6-5		see Fig. B2.2		22
*Eco*RI 0	R6-5	*tet*	see Fig. B2.2		43
*Hin*dII 0	Tn*10* (Tc) in λ*tet* and P22*tet*		two copies at ends of transposon with opposite orientations		29
*Hin*dIII 2					
*Pst*I ⩾ 1					
IS*4*	*in E. coli*[b]				
~1400 base pairs	*galT*	S101		I	15, 50, 51
Polar in both orientations	*galT*	9, 19, 28, 40 61, 75		I same site	41
	galT	8, 37, 59		II	41

Insertion element and characteristics	Individual occurrences			
	location	allele (mutation)	orientation	reference
IS*5*				
~1400 base pairs	λ*c*I	KH100	I	7
Function unknown				
Cleaved by *Eco*RI (ref. 7)				
Unassigned[c]				
Unassigned; polar mutation; ~1800 base pairs	*galOP-E*	*gal9*		27, 30
Unassigned; polar mutation	*galOP-E*	S187		3, 51
Unassigned; polar mutation	*galE*	S148		3, 51
Unassigned; polar mutation	*galE*	S108		3, 51
Unassigned; polar mutation at or near S101 (IS*4*?); ~1400 base pairs	*galT*	S114		3, 50, 51
Unassigned; polar mutations at or near S101 site (IS*4*?)	*galT*	S115, S139, S140, S142, S164, S168, S182		3, 51
Unassigned; ~1400 base pairs; cleaved by *Bam*HI but not by *Eco*RI or *Hin*dIII	F8 (at 91.0 F)	*traA* (N33)		40
Unassigned; ~800 base pairs; polar effects unknown; tra^+	F316 (at 78.6 F)	*tra* region		31
Unassigned; ~1400 base pairs; polar effects unknown	Hfr P4X (at 45.5 F)			13
Unassigned; polar mutations; ~800 base pairs	*lacZ*	MS296, MS377		34

Unassigned; not IS*2* or IS*3*; ~1250 base pairs; polar for *argH*	*argB*	*argB*2		11a
Most probably IS*1*; ~800 base pairs; mutations simultaneously block expression of two divergent operons	*bioOP*	130, 131, 132		1, 28, 54
Unassigned; double insertion in a "bowtie" configuration; 2 × 1350 base pairs	λ*R-m*′	a4-5061		14
Unassigned; ~1300 base pairs	P2*C-B*	*sig*5		12, 32
Unassigned; ~1300 base pairs	186*c*I (~74–78 %186)	*hr3 ins*1	I } different sites	9, 57
Unassigned; ~1300 base pairs	186*c*I (~74–78 %186)	*hr11 ins*2	II } different sites	9, 57
Unassigned; ~1300 base pairs	186*int* (~70–74 %186)	*hr19 ins*3	} different sites	9, 57
Unassigned; polar insertion; ~700 base pairs	*Spc* operon in λ*spc*1	I27		23, 24, 25
Unassigned; polar insertions; ~800 base pairs	*Spc* operon in λ*spc*1	I23, I24, I28, I29, I32, I33, I35, I37, I39, I40		23, 24, 25
Unassigned; polar insertion; ~1200 base pairs	*Spc* operon in λ*spc*1	I36		23, 24, 25
Unassigned; polar insertions; ~1300 base pairs	*Spc* operon in λ*spc*1	I26, I30		23, 24, 25
Unassigned; polar insertion; ~800 base pairs	*Spc* operon in λ*fus*3	I214		23, 24, 25
Unassigned; polar insertion; ~1200 base pairs	*Spc* operon in λ*fus*3	I211		23, 24, 25

Insertion element and characteristics	Individual occurrences: location	allele (mutation)	orientation	reference
Unassigned; polar insertions; ~1300 base pairs	*Spc* operon in λ*fus*3	I201, I210, I220		23, 24, 25
Unassigned; polar insertion; ~800 base pairs	*Str* operon in λ*fus*3	I116		23, 24, 25
Unassigned; polar insertions; ~1300 base pairs	*Str* operon in λ*fus*3	I103, I110, I112, I115		23, 24, 25
Unassigned; polar insertion; ~1500 base pairs	*Str* operon in λ*fus*3	I111		23, 24, 25

[a] References:
1. Abelson, J., personal communication.
2. Adhya, S. and A. Campbell. 1970. Crypticogenicity of bacteriophage λ. *J. Mol. Biol.* **50**:481.
3. Adhya, S. L. and J. A. Shapiro. 1969. The galactose operon of *E. coli* K-12. I. Structural and pleiotropic mutations of the operon. *Genetics* **62**:231.
4. Adhya, S., M. Gottesman and B. de Crombrugghe. 1974. Release of polarity in *Escherichia coli* by gene *N* of phage λ: Termination and antitermination of transcription. *Proc. Nat. Acad. Sci.* **71**:2534.
5. Ahmed, A. and D. Scraba. 1975. The nature of the *gal3* mutation of *Escherichia coli. Mol. Gen. Genet.* **136**:233.
6. Blattner, F., personal communication.
7. Blattner, F. R., M. Fiandt, K. K. Hass, P. A. Twose and W. Szybalski. 1974. Deletions and insertions in the immunity region of coliphage lambda: Revised measurement of the promoter-startpoint distance. *Virology* **62**:458.
8. Brachet, P., H. Eisen and A. Rambach. 1970. Mutations of coliphage λ affecting the expression of replicative functions O and P. *Mol. Gen. Genet.* **108**:266.
9. Bradley, C., O. P. Ling and J. B. Egan. 1975. Isolation of phage P2-186 intervarietal hybrids of 186 insertion mutants. *Mol. Gen. Genet.* **140**:123.
10. Bukhari, A. I. and A. L. Taylor. 1975. Influence of insertions on packaging of host sequences covalently linked to bacteriophage Mu DNA. *Proc. Nat. Acad. Sci.* **72**:4399.
11. Bukhari, A. I., F. de Bruijn and S. Froshauer, unpublished.
11a. D. Charlier and N. Glansdorf, personal communication.
12. Chattoraj, D. K., H. B. Younghusband and R. B. Inman. 1975. Physical mapping of bacteriophage P2 mutations and their relation to the genetic map. *Mol. Gen. Genet.* **136**:139.
13. Deonier, R. C. and N. Davidson. 1976. The sequence organization of the integrated F plasmid in two Hfr strains of *Escherichia coli. J. Mol. Biol.* **107**:207.

14. Emmons, S. W. 1974. Bacteriophage lambda derivatives carrying two copies of the cohesive end site. *J. Mol. Biol.* **83**:511.
14a. Fiandt, M., W. Szybalski and A. Ahmed. 1977. Identification of the *gal3* insertion sequence in *Escherichia coli* as IS2. *Gene* (in press).
15. Fiandt, M., W. Szybalski and M. H. Malamy. 1972. Polar mutations in *lac, gal* and phage λ consist of a few IS-DNA sequences inserted with either orientation. *Mol. Gen. Genet.* **119**:223.
16. Fiandt, M., Z. Hradecna, H. A. Lozeron and W. Szybalski. 1971. Electron micrographic mapping of deletions, insertions, inversions and homologies in the DNAs of coliphages lambda and phi 80. In *The bacteriophage lambda* (ed. A. D. Hershey), p. 329. Cold Spring Harbor Laboratory, Cold Spring Harbor, New York.
17. Grindley, this volume; and personal communication.
18. Hirsch, H.-J., H. Saedler and P. Starlinger. 1972. Insertion mutations in the control region of the galactose operon of *E. coli. Mol. Gen. Genet.* **115**:266
19. Hirsch, H.-J., P. Starlinger and P. Brachet. 1972. Two kinds of insertions in bacterial genes. *Mol. Gen. Genet.* **119**:191.
20. Hu, S., E. Ohtsubo and N. Davidson. 1975. Electron microscope heteroduplex studies of sequence relations among plasmids of *Escherichia coli*: Structure of F13 and related F-primes. *J. Bact.* **122**:749.
21. Hu, S., E. Ohtsubo, N. Davidson and H. Saedler. 1975. Electron microscope heteroduplex studies of sequence relations among bacterial plasmids: Identification and mapping of the insertion sequences IS1 and IS2 in F and R plasmids. *J. Bact.* **122**:764.
22. Hu, S., K. Ptashne, S. N. Cohen and N. Davidson. 1975. $\alpha\beta$ Sequence of F is IS3. *J. Bact.* **123**:687.
23. Jaskunas and Nomura, this volume.
24. Jaskunas, S. R., L. Lindahl and M. Nomura. 1975. Isolation of polar insertion mutants and the direction of transcription of ribosomal protein genes in *E. coli. Nature* **256**:183.
25. Jaskunas, S. R., L. Lindahl, M. Nomura and R. Burgess. 1975. Identification of two copies of the gene for the elongation factor EF-Tu in *E. coli. Nature* **257**:458.
26. Jordan, E., H. Saedler and P. Starlinger. 1967. Strong polar mutations in the transferase gene of the galactose operon in *E. coli. Mol. Gen. Genet.* **100**:296.
27. ——. 1968. O^o and strong polar mutations in the *gal* operon are insertions. *Mol. Gen. Genet.* **102**:353.
28. Ketner, G. and A. Campbell. 1975. Operator and promoter mutations affecting divergent transcription in the *bio* gene cluster of *Escherichia coli. J. Mol. Biol.* **96**:13.
29. Kleckner, N., R. K. Chan, B.-K. Tye and D. Botstein. 1975. Mutagenesis by insertion of a drug-resistance element carrying an inverted repetition. *J. Mol. Biol.* **97**:561.
30. Lederberg, E. M. 1960. Genetic and functional aspects of galactose metabolism in *Escherichia coli* K-12. In *10th Symposium of the Society for General Microbiology*, p. 115. Cambridge University Press, Cambridge, England.
31. Lee, H. J., E. Ohtsubo, R. C. Deonier and N. Davidson. 1974. Electron microscope heteroduplex studies of sequence relations among plasmids of *Escherichia coli.* V. ilv^+ deletion mutants of F14. *J. Mol. Biol.* **89**:585.
32. Lindahl, G., Y. Hirota and F. Jacob. 1971. On the process of cellular division of *Escherichia coli*: Replication of the bacterial chromosome under control of prophage P2. *Proc. Nat. Acad. Sci.* **68**:2407.
33. MacHattie and Jackowski, this volume.
34. Malamy, M. H. 1970. Some properties of insertion mutations in the *lac* operon. In *The lactose operon* (ed. J. R. Beckwith and D. Zipser), p. 359. Cold Spring Harbor Laboratory, Cold Spring Harbor, New York.

35. ——, personal communication.
36. Malamy, M. H., M. Fiandt and W. Szybalski. 1972. Electron microscopy of polar insertions. *Mol. Gen. Genet.* **119**:207.
37. Morse, M. L., E. M. Lederberg and J. Lederberg. 1956. Transductional heterogenotes in *Escherichia coli. Genetics* **41**:758.
38. Mosharrafa, E., W. Pilacinski, J. Zissler, M. Fiandt and W. Szybalski. 1976. Insertion sequence IS2 is near the gene for prophage λ excision. *Mol. Gen. Genet.* **147**:103.
39. Ohtsubo and Ohtsubo, this volume; and personal communication.
40. Ohtsubo, E., R. C. Deonier, H.-J. Lee and N. Davidson. 1974. Electron microscope heteroduplex studies of sequence relations among plasmids of *Escherichia coli.* IV. The F sequences in F14. *J. Mol. Biol.* **89**:565.
41. Pfeifer et al., this volume.
42. Pilacinski, W., E. Mosharrafa, R. Edmundson, J. Zissler, M. Fiandt and W. Szybalski. 1977. Insertion sequence IS*2* associated with *int*-constitutive mutants of bacteriophage lambda. *Gene* **1**: (in press).
43. Ptashne, K. and S. N. Cohen. 1975. Occurrence of insertion sequence (IS) regions on plasmid deoxyribonucleic acid as direct and inverted nucleotide sequence duplications. *J. Bact.* **122**:776.
44. Saedler, this volume.
45. Saedler, H. and P. Starlinger. 1967. O^o mutation in the galactose operon in *E. coli. Mol. Gen. Genet.* **100**:178.
46. Saedler, H., D. F. Kubai, M. Nomura and S. R. Jaskunas. 1975. IS1 and IS2 mutations in the ribosomal protein genes of *E. coli* K-12. *Mol. Gen. Genet.* **141**:85.
47. Saedler, H., H.-J. Reif, S. Hu and N. Davidson. 1974. IS2, a genetic element for turn-off and turn-on of gene activity in *E. coli. Mol. Gen. Genet.* **132**:265.
48. Saedler, H., J. Besemer, B. Kemper, B. Rosenwirth and P. Starlinger. 1972. Insertion mutations in the control region of the *gal* operon of *E. coli. Mol. Gen. Genet.* **115**:258.
49. Schmidt, F., J. Besemer and P. Starlinger. 1976. The isolation of IS1 and IS2 DNA. *Mol. Gen. Genet.* **145**:145.
50. Shapiro, J. A. 1969. Mutations caused by the insertion of genetic material into the galactose operon of *Escherichia coli. J. Mol. Biol.* **40**:93.
51. Shapiro, J. A. and S. Adhya. 1969. The galactose operon of *E. coli* K-12. II. A deletion analysis of operon structure and polarity. *Genetics* **62**:249.
52. Shimada, K. and A. Campbell. 1974. Int-constitutive mutants of bacteriophage lambda. *Proc. Nat. Acad. Sci.* **71**:237.
53. Starlinger, P. and H. Saedler. 1976. IS-elements in microorganisms. In *Current Topics in Microbiology and Immunology*, vol. 75, p. 111. Springer Verlag, Berlin.
54. Szybalski, E. H. and W. Szybalski. 1977. Physical mapping of the *Escherichia coli* biotin gene cluster. *Abstr. Annu. Meet. Am. Soc. Microbiol.* **77**:122.
55. Tomich and Friedman, this volume.
56. Toussaint, A. and R. Kahmann, personal communication.
57. Younghusband, H. B., J. B. Egan and R. B. Inman. 1975. Characterization of the DNA from bacteriophage P2-186 hybrids and physical mapping of the 186 chromosome. *Mol. Gen. Genet.* **140**:101; and J. B. Egan, personal communication.
58. Zissler et al., this volume.

[b]Location in *E. coli* means clearly characterized polar mutations. The elements may naturally be present in the chromosome at unknown locations.
[c]Some of these are probably the same as IS*1*, IS*2*, IS*3*, IS*4*, or IS*5*.

2. SEQUENCE OF THE ENDS OF IS*1* ELEMENT

Contributed by H. Ohtsubo and E. Ohtsubo, State University of New York, Stony Brook

Figure A2.1 is a schematic representation of the physical structures of small plasmid DNA, pSM2, pSM1, and pSM15, which are the derivatives of R100 (see also the physical structures of R100 DNA and its derivatives shown in Fig. 1 in the paper by Ohtsubo and Ohtsubo [this volume]). Structures of pSM1 and pSM15 are shown as deletions in pSM2. The deletion in pSM1 extends from the left end of IS*1* to a point marked by a broken vertical line. (The deletion end points 12.2 and 11.0 are R100 coordinates in kilobases [see Ohtsubo and Ohtsubo, this volume].) The deletion in pSM15 extends from the right end of IS*1* (12.9/8.0 kb) to a point (8.3) marked by a broken vertical line. The arrows show the cleavage sites for the endonucleases used to generate fragments for sequence analysis.

The nucleotide sequences of the DNA segments A and B are shown in Figure A2.2 (unpubl. results). In this figure, *A* gives the sequence of the left-hand segment A (indicated by dark solid arrows [← →] in Fig. A2.1) of pSM1, and *B* gives the sequence of the right-hand segment B (also indicated by dark solid arrows in Fig. A2.1) of pSM15. The ends of IS*1* were identified by comparison with the sequences of the other two segments (C and D) of pSM2 DNA shown in Figure A2.1. It is assumed that the sequence common to both A and C is outside of IS*1* on the left side and that the sequence common to B and D is outside of IS*1* on the right side.

Figure A2.3 shows the inverted repetitious sequences found at the end regions of IS*1*, starting from each presumptive junction point.

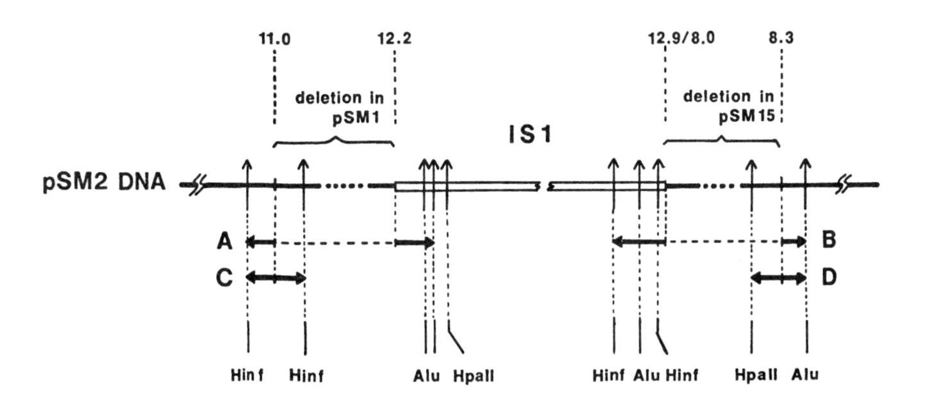

Figure A2.1

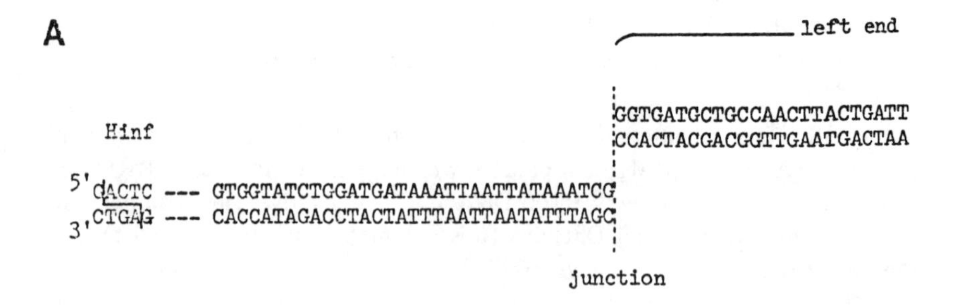

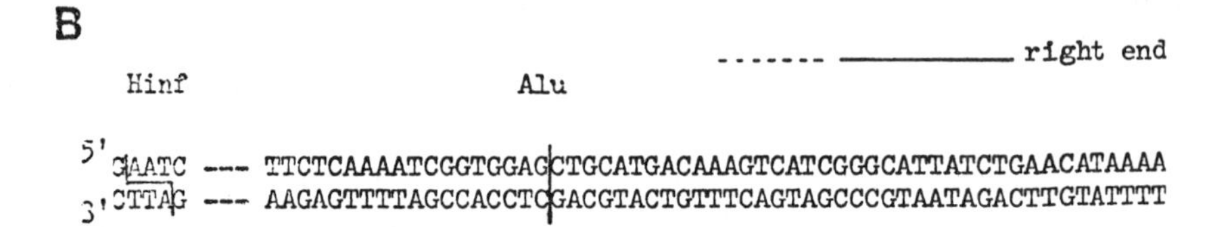

Figure A2.2

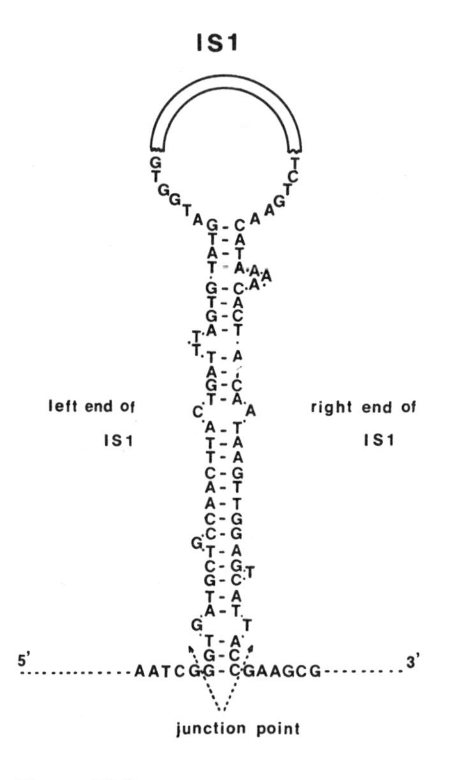

Figure A2.3

of IS1 ———————— - - - - - - - - -

Alu Alu

```
TAGTGTATGATGGTGTTTTTGAGGTGCTCCAGTGGCTTCTGTTTCTATCAGCTGTCCCTCCTTCAGCT 3'
ATCACATACTACCACAAAAACTCCACGAGGTCACCGAAGACAAAGATAGTCGACAGGGAGGAAGTCGA 5'
```

of IS1 ————————

Hinf

```
CACTATCAATAAGTTGGAGTCATTACC
GTGATAGTTATTCAACCTCAGTAATGG
```

Alu

```
GAAGCGGTGGTATCTGTGTCCGCAGG --- AGCT 3'
CTTCGCCACCATAGACACAGGCGTCC --- TCGA 5'
```

junction

Figure A2.2 *(continued)*

3. DNA SEQUENCES AT THE ENDS OF IS*1*

Contributed by N. D. F. Grindley, Yale University, New Haven

To examine sequences at the boundaries between IS*1* and bacterial DNA, a 215-n.p. (nucleotide pair) *Hha*I-*Hae*III fragment and a 165-n.p. *Hha*I-*Hpa*II fragment derived from the *galT* gene of λd*gal*104 have been partially sequenced using the techniques developed by Maxam and Gilbert (1977). In each case, only one strand was sequenced proceeding from the *Hha*I 5′ terminus. The ends of IS*1* have been located about 210 n.p. to the left of the *Hae*III site and 140 n.p. to the right of the *Hpa*II site (Grindley, this volume). Thus the 215-n.p. fragment should contain about 5 n.p. of *gal* sequences at its left (*Hha*I) end, and the 165-n.p. fragment should have about 25 n.p. of *gal* sequences at its right (*Hha*I) end.

The boxed regions in Figure A3.1[1] indicate repeated sequences which appear in inverted orientation near the two ends of IS*1*. These probably constitute recognition sites for excision and integration events. The above estimates of the amounts of *gal* DNA present in the two fragments suggest that the nucleotides numbered 1 may be the terminal nucleotides of IS*1*. Confirmation will be obtained by sequencing the *Hha*I-*Hha*I fragment of the *galT* gene into which IS*1* was inserted to give the *gal*104 mutation.

The underlined heptanucleotide in the 215-n.p. fragment has the same sequence as the "Pribnow box" portion (Pribnow 1975) of the RNA polymerase binding sight for the rightward promoter (p_R) of bacteriophage λ (Walz and Pirotta 1975). Initiation of transcription, which normally requires a purine, could occur at position 50 or 52. A second A-T-rich region, also thought to be involved in the recognition and binding of RNA polymerase (Gilbert 1976), likewise appears between positions 13 and 21. We are currently investigating the possibility that these sequences determine an IS*1* promoter.

References

Gilbert, W. 1976. Starting and stopping sequences for RNA polymerase. In *RNA polymerase* (ed. R. Losick and M. Chamberlain), p. 193. Cold Spring Harbor Laboratory, Cold Spring Harbor, New York.

Maxam, A. and W. Gilbert. 1977. A new method for sequencing DNA. *Proc. Nat. Acad. Sci.* **74**:560.

Pribnow, D. 1975. Nucleotide sequence of an RNA polymerase binding site at an early T7 promoter. *Proc. Nat. Acad. Sci.* **72**:284.

Walz, A. and V. Pirrotta. 1975. Sequence of the P_r promoter of phage λ. *Nature* **254**:118.

[1] For Figure A3.1, see page following.

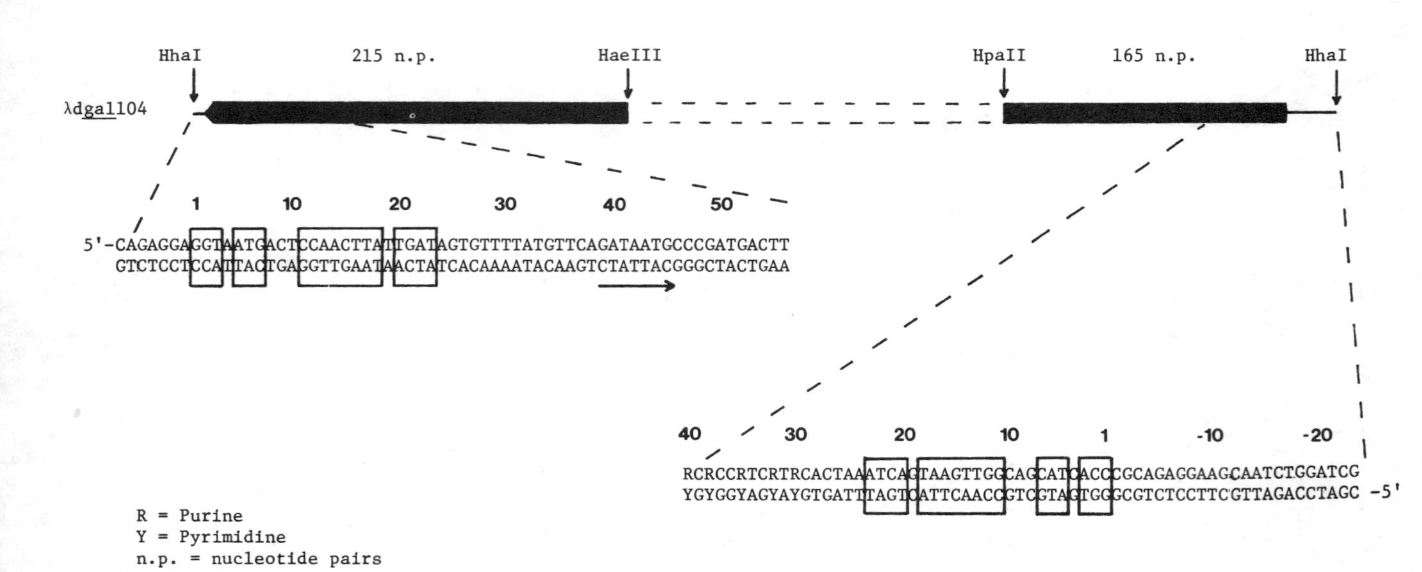

R = Purine
Y = Pyrimidine
n.p. = nucleotide pairs

Figure A3.1

4. NUCLEOTIDE SEQUENCES AT TWO SITES FOR IS*2* DNA INSERTION

Contributed by R. E. Musso and M. Rosenberg, National Cancer Institute, Bethesda

Nucleotide sequences were determined within and surrounding two sites of IS*2* DNA insertion: *galPO-E*-490::IS*2* (IS*2* between *galPO* and *galE*) and λt_{r1}-*c*II-*r*32::IS*2* (IS*2* between termination signal t_{r1} and *c*II genes of phage λ) (see Fig. A4.1). (For sequences surrounding the site of IS*2* insertion in *gal*, see Musso et al. [1974] and Sklar et al. [1977].)

RNA transcription products were prepared under kinetically controlled conditions in in vitro reactions, and the regions of interest were isolated by selective hybridization to the appropriate DNAs. Standard RNA sequencing techniques were used to determine the nucleotide sequences of the fragments obtained. These sequences are depicted in Figure A4.2 and aligned (vertical line, position -1/+1) at the apparent site of the inserted DNA. Designation of position -1/+1 as the exact point of recombination for both insertion sequences cannot be rigorously proved; however, comparison of the sequences around the two sites of insertion indicates that this end of the IS*2* DNA could not vary by more than a few nucleotide residues. The sequence data are certainly consistent with the contention that the IS*2* DNA inserts at one end with the identical sequence, 5′ TGGATTTG---, during integration at different sites and further suggest that the recombination event occurs at a fixed site within the insertion element.

Comparison of the sequences in λ and *gal* surrounding the sites into which the IS*2* element inserts (Fig. A4.2) excludes the possibility of any extensive common homologous "core sequence" spanning the integration site. Although there are a number of other homologies in proximity to the insertion sites in these two regions, there is only limited sequence identity immediately adjacent to the sites of insertion. In both *gal* and λ, the IS*2* inserts within an ATA sequence, and the sequence ATGG occurs at identical positions (+9 to +12) from the point of insertion.

Other structural homologies in the regions surrounding the sites of IS*2* insertion can be noted. These include similarities in sequence at variable distances from the site of insertion; true palindromes and regions of twofold rotational symmetry within these sequences, and the general predominance of A-T base pairs throughout the region. In addition, and possibly of greater significance, are the regions of homology revealed when the sequences are aligned at the IS*2* site (see Fig. A4.2). It is difficult to assess the importance of these homologies to the events involved in the integration of the IS*2* element. The common features simply may reflect the fact that these two regions are also involved in aspects of transcriptional regulation (i.e., a site for initiation of transcription for *gal* and a site for *rho*-dependent termination of transcription for λ). Although the functional complexity of these regions complicates the interpretation of the structural characteristics, it may be that the features important to

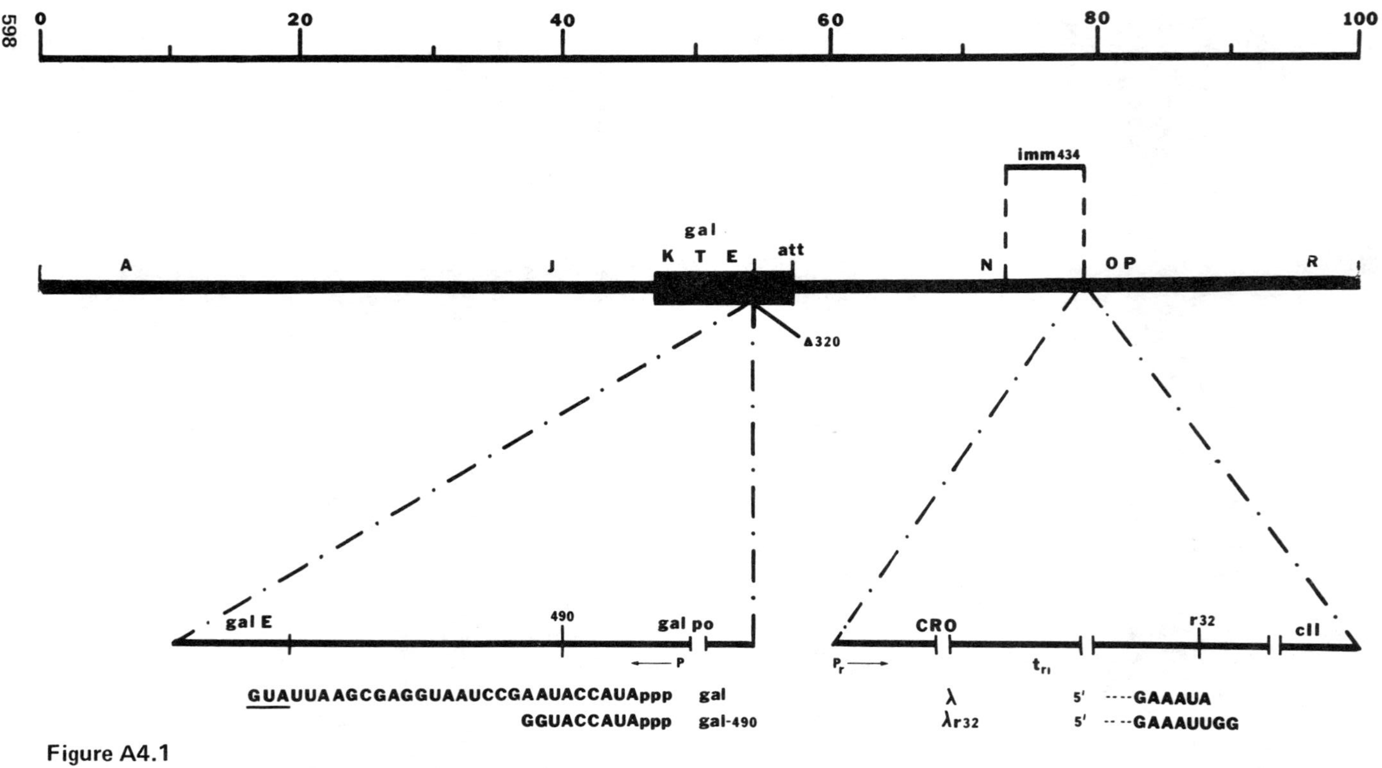

Figure A4.1
Composite schematic of λ and λ*gal* DNA. The λp*gal*8 phage DNA has the indicated bacterial DNA (Δ320 is a deletion of most of the bacterial DNA between *gal* and *att*λ) substituted for the λ*b*2 region. The RNA sequences spanning the IS*2* insertion sites in λp*gal*8 -490 and λ*r*32 are indicated.

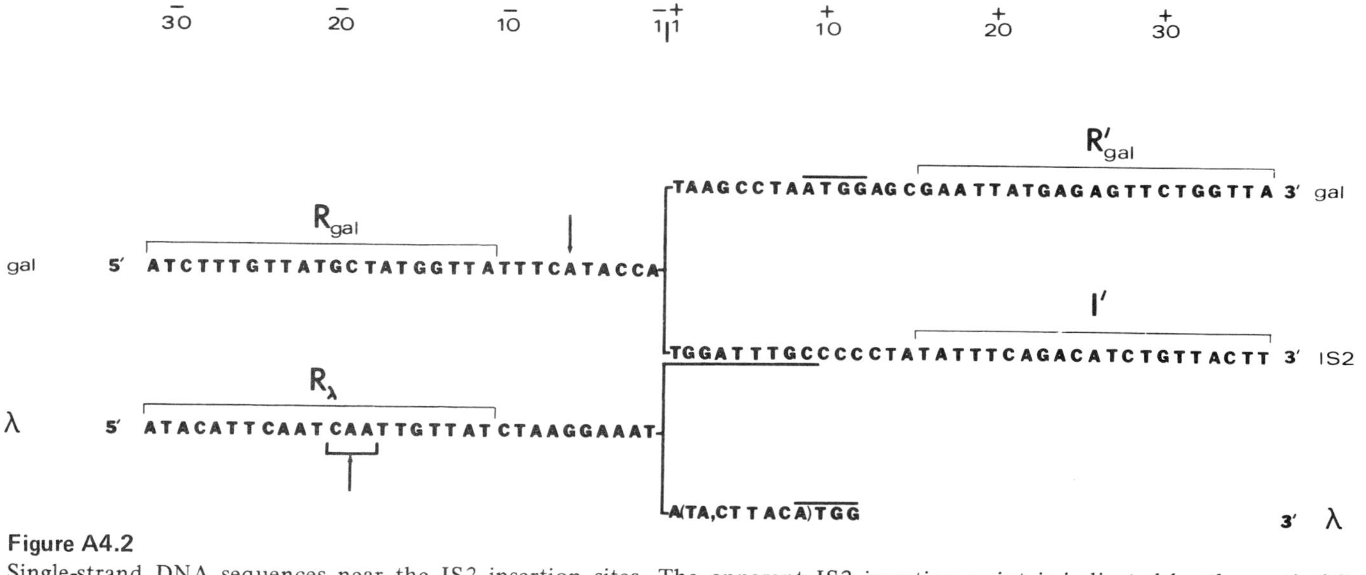

Figure A4.2

Single-strand DNA sequences near the IS*2* insertion sites. The apparent IS*2* insertion point is indicated by the vertical line. Positions +1 to +9 of the IS*2* sequence (underlined) were determined for RNA from λ*r*32 as well as from λp*gal*8-490. The arrow at position −6 in the *gal* sequence indicates the initiation site for *gal* transcription. Positions −20 to −18 in the λ sequence (bracket and arrow) are the major termination sites for λ*cro* RNA synthesized in vitro in the presence of transcription termination factor *rho*.

the regulatory aspects of the region are of some selective advantage for the events involved in IS*2* integration.

References

Musso, R. E., B. de Crombrugghe, I. Pastan, J. Sklar, P. Yot and S. Weissman. 1974. The 5′ terminal nucleotide sequence of galactose messenger ribonucleic acid of *Escherichia coli. Proc. Nat. Acad. Sci.* **71**:4940.

Sklar, J., S. Weissman, R. E. Musso, R. DiLauro and B. de Crombrugghe. 1977. Determination of the nucleotide sequence of part of the regulatory region for the galactose operon from *Escherichia coli. J. Biol. Chem.* (in press).

Appendix B: Bacterial Plasmids

1. TABLES

a. Introduction

By J. A. Shapiro, University of Chicago

The study of plasmid structure has been essential to understanding the evolutionary function of DNA insertion elements. Most of this work has been with a small number of plasmids closely related to the *E. coli* sex factor, F. The following tables are included in this book to facilitate research on other groups of plasmids and on their insertion sequences.

The identification of DNA insertion elements has led to a general model for the evolution of complex replicons by the addition of DNA segments bounded by special integration sequences. In other words, we can think of replicons as composites of several "modules," each containing a different complement of genetic loci. With time and a favorable response to natural selection, new modules will be added to the evolving replicon to generate a family of related genetic elements.

The organization of these tables was designed to reflect the modular evolution of plasmids. Three basic groups of genes are emphasized–those for plasmid replication, those for plasmid transfer, and those for other plasmid-determined phenotypes–and the following discussion explains the considerations which led to the present format for these tables.

Incompatibility and Host Range

The only asbolutely essential characteristic of any replicon is its ability to replicate and pass itself on to successive generations of cells. This characteristic can be measured experimentally by testing clones of a cell line for the presence of a particular replicon. The results of these tests are summarized in the first column, "Incompatibility and Host Range," and they are the basis for the primary classifications of plasmids.

Incompatibility is determined by introducing two plasmids into a single cell and asking if they can both replicate in a stable fashion. If they can, then their replication systems are compatible, and we conclude that they are different. If the plasmids cannot coexist stably, then their replication systems are incompatible, and we conclude that they share some common element that can support only a single replicon. In this second case, we assign the plasmids to a single incompatibility group, and we operate under the assumption that all plasmids in one group share a common replication system, whereas plasmids in different groups have different replication systems. Clearly, this scheme is a great oversimplification. Incompatibility testing is revealing many complexities, such as plasmids that are incompatible with members of more than one group. Nonetheless, classification by incompatibility groups appears to serve as a useful first approximation for organizing plasmids into distinct classes which share several evolutionary features, including DNA-DNA homology.

Host range is a second indication of replication specificity. A replicon may be able to replicate in one cell type but not in another. In the cases where the question has been asked, members of a given incompatibility group show similar host ranges. For example, plasmids of IncC, IncN, IncP, IncQ, and IncW have broad host ranges and replicate stably in a wide variety of gram-negative bacteria, whereas plasmids of IncFII do not replicate in *Pseudomonas,* plasmids of IncIα do not replicate in *Pseudomonas* or *Proteus* species, plasmids of IncV are unstable in *E. coli,* and plasmids of IncP-2 do not replicate in *E. coli.* The limitations of our knowledge about plasmid host range require emphasis. There are many plasmid-host systems that have not yet been tested. Moreover, there are a number of factors other than replication ability which may limit plasmid host range (such as restriction or other barriers to transfer). However, these other factors can be eliminated by suitable experiments, and, where this is possible, the main determinant of host range does appear to be replication ability.

We have used host range to subdivide all the data on naturally occurring plasmids. First is the division into plasmids from gram-negative and from gram-positive organisms. Historically, the first group has been more extensively investigated, and we have subdivided it into two tables: plasmids studied in *E. coli* and other enterics, and plasmids studied in *P. aeruginosa* and other Pseudomonads. Although there are plasmid groups that overlap (IncC = IncP-3, IncN; IncP = IncP-1; IncQ = IncP-4, IncW), there are many that do not. Thus there appear to be two plasmid "families"—one for enterics and one for Pseudomonads. Doubtless, study of plasmids in other gram-negative hosts would reveal more families. Plasmids from gram-positive bacteria have not been investigated as extensively as those from gram-negative, with the exception of *Staphylococcus aureus* plasmids. Hence data on gram-positive plasmids are divided into plasmids of *S. aureus* and plasmids of other gram-positive bacteria. Unfortunately, no one has yet undertaken to analyze the relationships between plasmids found in various gram-positive bacteria.

Plasmid

The second column lists plasmids by their names. Until some kind of registration system becomes operative, plasmid names are subject to two types of confusion. First, two laboratories can assign the same number to different plasmids. This was relatively common when the prefix R was routinely given to drug resistance plasmids, and the major overlap existed between the collection of the Pasteur Institut in Paris and that of Hammersmith Hospital in London. As far as possible, we have tried to eliminate this confusion by giving the prefix RIP to plasmids from the Pasteur collection. The second kind of confusion comes when independent isolates of a single plasmid receive different numbers. This has almost certainly occurred, for example, with the large series of IncA-C plasmids isolated from *Providence* (106).* Thus different numbers for similar plasmids do not indicate that they are known to be different. By the same token, it should be remembered that repeated similarities (even down to the level of 100% detectable heteroduplex formation) do not necessarily reflect absolute DNA sequence identity.

Original Host

The third column lists the original host species in which individual plasmids were first isolated. This information will be useful in tracing the evolutionary development of various plasmid groups. However, the present data reflects systematic bias because almost all of it comes from screening a small number of pathogenic species isolated in clinical settings.

Phenotype

The fourth column lists the phenotypic characteristics conferred by the presence of a given plasmid, and these can be divided into at least two groups.

(1) The phenotypic characteristic that appears to be most closely related to replication specificity is conjugal fertility. It is easy to imagine a replicon picking up a fertility system and then evolving other traits as it passes through various hosts. Moreover, the fertility phenotypes of plasmids appear to correlate very well with incompatibility groupings. Thus in specifying phenotypes, we have listed those characteristics associated with fertility first. The most basic of these is the ability to promote self transfer (Tra), but there are other fertility-associated characteristics. Two of the most widely studied are: (i) production of specific surface structures involved in conjugation, and this is generally determined experimentally by testing sensitivity to donor-specific phages which adsorb to these structures (Dps = *D*onor *p*hage *s*ensitivity); and (ii) the ability to inhibit the fertility of other Tra$^+$ plasmids (Fi = *F*ertility *i*nhibition), often by a repression system which regulates the expression of genes involved in conjugal transfer. Examination of the fertility systems of gram-negative plasmids (identified primarily by their Dps characteristics) shows that a particular set of transfer genes can be found with the replication genes of several incompatibility groups. Whether any single incompatibility group harbors more than one fertility system remains to be determined. It may be helpful to point out that the earliest classifications of enteric plasmids were based on their Tra, Dps, and Fi phenotypes. This was before incompatibility testing was developed. Thus many of the earlier references will discuss plasmid groups that no longer exist or that have been subsequently subclassified (the most common of these being *fi*$^+$, *fi*$^-$, I-like, and F-like).

*Numbers in parentheses refer to entries in the Bibliography, Appendix B3.

(2) Other phenotypic traits have no obvious connection to plasmid replication or transfer. They may, however, fall into two classes–(i) those that are limited to a small number of incompatibility groups, and (ii) those that are widely distributed in many incompatibility groups. Class (i) includes phage inhibition (Phi), host specificity (Hsp), and bacteriocinogeny phenotypes, and class (ii) includes characteristics such as resistance to antibacterial agents. These differences may reflect evolutionary patterns, such as the early association of a bacteriocin determinant with a specific set of replication genes or the transposability of an antibiotic resistance marker into many unrelated replicons. But, once again, these two classes may also reflect limitations in our research because they also correlate with (i) those traits that have been studied least extensively and (ii) those that have been studied most widely. Thus it is risky to draw conclusions about close relationships between specific replication genes and phenotypic determinants until they have been widely tested. What we can conclude with assurance, however, is that genes associated with insertion elements usually are recent acquisitions of a large number of incompatibility groups. In the case of antibiotic resistance determinants, this extensive spread of specific traits reflects intense selective pressures due to widespread chemotherapy.

Genotype

The next column lists the genotypic symbols for the determinants of a particular phenotype when something is known about the genetic basis for that trait. Detailed understanding of the genetic and biochemical basis for individual phenotypic traits is rare. Occasionally, we know that a specific marker is enclosed in a transposable element, or that the presence of a specific enzyme is involved in determining some trait, or that the locus for some characteristic has been identified by mutation. In those cases, the relevant information is given in the "Genotype" column by listing the presence of a specific transposable element or by indicating the genotypic symbol for the locus encoding a known phenotype or enzyme activity. (Some transposable elements have been assigned numbers, and these are listed in the article on nomenclature by Campbell et al. in this volume; unassigned transposable elements are indicated by the letters Tn followed by the appropriate phenotypic symbol.) Genetic data is essential for working out evolutionary relationships because many phenotypes can result from more than one biochemical mechanism. For example, there are at least five different kinds of β-lactamase activities encoded by gram-negative plasmids, and the presence of any one can confer resistance to ampicillin (Ap) or carbenicillin (Cb).

Additional Data

Much of the information in the "Additional Data" column is relevant to the categories specified by the other columns, but generally there are no agreed conventions for specifying this data. For example, the absence of chloramphenicol acetyl transferase (CATase) activity in a plasmid that determines chloramphenicol resistance is relevant to its phenotype and genotype but cannot be indicated by *cat* or CAT^- since these symbols would suggest mutations of a CATase gene. Similarly, the copy number of a plasmid is related to its mode of replication, but our knowledge of the control of plasmid replication is still too limited to be able to summarize the various characteristics in a few symbols.

Apologia

The tables are meant to be as accurate and inclusive as possible. Inevitably there will be transcriptional errors, and the users of this data are encouraged to look up original references whenever a specific fact is very important. There will also be errors due to the fact that some concepts are superceded. For example, the status of IncX is uncertain, and the plasmids in this group may eventually be placed in other incompatibility groups. Many uncertainties result from a lack of experimental results, and we have limited the data in these tables to what has been actually determined. Generally, the absence of a phenotypic symbol does not mean that the trait is absent, but rather that it has not been determined. The exceptions to this rule are resistances to common antibiotics that are routinely tested (Ap Cb Cm Pc Km Sm Su Tc), but even here there may be exceptions to the exceptions.

There are two kinds of gaps in these tables. The first is published data which could not be incorporated due to lack of time and people. Some of the plasmid groups which have been neglected are degradative plasmids from enteric bacteria, small nontransmissible plasmids such as ColE1, the large number of Tol plasmids from soil Pseudomonads, *Streptococcus lactis* plasmids for protease production, cryptic plasmids from *gonococci,* plàsmids from *Bacilli,* and the oncogene plasmids from *Agrobacteria.* Doubtless, the list could be extended. The second kind of gap corresponds to data which does not yet exist. Hopefully, the existence of these tables, incomplete as they are, will help to fill in both kinds of gaps by establishing a useful format for incorporating old and new data for periodic publication.

Nomenclature

The terminology and symbols used in these tables generally follow the recommendations of Novick et al. (209) for bacterial plasmid nomenclature. The additions and exceptions not already explained are the following:

Alk = biodegradation of alkanes (rather than Oct)
Bcn = bacteriocin production
Bnr = bacteriocin resistance
Exp = expansive replication to the maximum level permissible for extrachromosomal DNA (228)

Hsp = host specificity (rather than Res since specificity may not always correspond to known restriction and modification mechanisms)
Lm = resistance to lincomycin
Mdl = biodegradation of mandelate
Pma = resistance to phenyl mercuric acetate
RC = relaxation complex which can be induced to relax by SDS
Res = restrictive replication at a level below the maximum permissible for extrachromosomal DNA (228)
Vm = resistance to vernamycin B
Xal = biodegradation of xylene
am = amber mutation
irp = intrinsic resistance to penicillin
sus = suppressible
*fin*Q$^+$, *fin*U$^+$ = presence of genes for diffusible products interfering with expression of *tra*Q and *tra*U loci on F.

b. Plasmids Studied in *Escherichia coli* and Other Enteric Bacteria

Compiled by A. E. Jacob, Royal Postgraduate Medical School, London; J. A. Shapiro, University of Chicago; L. Yamamoto, University of Chicago; D. I. Smith, University of Wisconsin, Madison; S. N. Cohen, Stanford University Medical School; and D. Berg, University of Wisconsin, Madison

Incompatibility and host range[a]*	Plasmid	Original host	Phenotype	Genotype	Additional data	Reference†
IncA						
A. liquefaciens	RA1	*A. liquefaciens*	Tra^+ Fi^-(F) Su Tc		m.w. = 86 × 10^6	7, 13, 70
E. coli	RA1-1		Tra^+ Cm Su Tc	TnCm ?	segregant of J53 (RA1) (R57b)	70
	RA1-1a	*A. liquefaciens*	Tra^+ Cm Su			70
	RA2	*A. liquefaciens*	Tra^+ Fi^-(F) Su Tc			70
IncB[b] **= IncO = Coml0**						
E. coli	TP113	*S. typhimurium*	Tra^+ Fi^-(F) Km		m.w. = 56.7 × 10^6	92, 93
Salmonella spp.	TP125	*S. dysenteria*	Tra^+ Cm Sm Su Tc		m.w. = 64.0 × 10^6	92, 93
Shigella spp.	R805a	*S. typhi*	Tra^+ Fi^+(F) Km	*aph*A^+	found with R805b (Inc H)	72
	R861a	*E. coli*	Km			72
	R864a	*E. coli*	Km			72
	RIP72	*E. coli*	Km			209, 235
	R833	*S. flexneri*	Cm Sm Su Tc			72
	R834	*S. flexneri*	Cm Sm Su Tc			72
	R16 (= pSF16)	*E. coli*	Tra^+ Fi^-(F) Ap Cm Sm Su Tc	*bla*$^+$(OXA-2)	m.w. = 69 × 10^6, 50% G+C	83, 85, 121, 189

*All footnotes for Tables b, c, d, e, and f are given at the end of Table f.
†Reference numbers refer to entries listed in the Bibliography, Appendix B3.

Incompatibility and host range[a]	Plasmid	Original host	Phenotype	Genotype	Additional data	Reference
IncB[b]						
(continued)	R16-1		Tra^+ Sm Su			121
	R7	*E. coli*	Tra^+ Ap Sm Su Tc	*bla*$^+$(OXA-2)		83
	R724	*S. dysenteriae*	Tra^+ Cm Sm Su Tc		m.w. = 58×10^6	72, 85
	R836	*S. dysenteriae*	Tra^+ Cm Sm Su Tc			72
	RIP185		Tra^+ Km Tp			91
IncA-C[c]						
Providence	R480	*Providence*	Tra^+ Fi^-(F) Su Lac			106
E. coli	R667	*Providence*	Tra^+ Fi^-(F) Ap Su	*bla*$^+$(TEM-2)	m.w. = 111×10^6	13,106,189
S. typhimurium	R686	*Providence*	Tra^+ Fi^-(F) Su Tc			106
P. mirabilis	R692	*Providence*	Tra^+ Ap Su Tc	*bla*$^+$(TEM-1)		106, 189
V. cholera	R699a	*Providence*	Tra^+ Fi^-(F) Ap Km Su Tc	*aph*A^+		106
P. vulgaris	R707	*Providence*	Tra^+ Fi^-(F) Ap Cm Km Sm Sp Su Tc	*bla*$^+$(TEM-1)		106, 189
	R806a	*S. typhi*	Tra^+ Fi^-(F) Ap Cm Km Sm	*aph*A^+	m.w. = 102×10^6; found with R806b (IncH)	13, 72
	R715b	*Providence*	Tra^+ Fi^-(F) Km Sm		m.w. = 91×10^6	13, 106
	R994	*V. cholera*	Tra^+ Cm Sm Tc	*aph*C^+	m.w. = 98×10^6	114
	P*lac*	*P. mirabilis*	Tra^+ Fi^-(F) Su Lac		m.w. = 101×10^6	53, 107
	P*lac*-R1*drd*19		Tra^+ Ap Cm Km Su Sm Lac		recombinant of P*lac* and R1*drd*19	51, 53
	P*lac*-R1*drd*19-2		Tra^+ Fi^-(F) Ap Cm Sm Su		recombinant of P*lac* and R1*drd*19	53
	P*lac*-R447b		Tra^+ Fi^-(F) Km Su Lac		recombinant of P*lac* and R447b; m.w. = 106×10^6	51, 53

RIP55	*K. pneumoniae*	Tra$^+$ Ap Cm Gm Su	*bla*$^+$(OXA-3)	189, 283
R666	*Providence*	Fi$^-$(F) Su		106
R671	*Providence*	Fi$^-$(F) Su		106
R668	*Providence*	Fi$^-$(F) Ap Su		106
R669	*Providence*	Fi$^-$(F) Ap Su		106
R670	*Providence*	Fi$^-$(F) Ap Su		106
R672	*Providence*	Fi$^-$(F) Ap Su	*bla*$^+$(TEM-2)	106
R673	*Providence*	Fi$^-$(F) Ap Su		106
R674	*Providence*	Fi$^-$(F) Ap Su		106
R414a	*Providence*	Fi$^-$(F) Su Tc		106
R688	*Providence*	Fi$^-$(F) Su Tc		106
R689	*Providence*	Fi$^-$(F) Su Tc		106
R691	*Providence*	Fi$^-$(F) Su Tc		106
R694	*Providence*	Fi$^-$(F) Su Tc		106
R698b	*Providence*	Fi$^-$(F) Ap Su	*bla*$^+$(TEM-2)	106, 189
R700	*Providence*	Fi$^-$(F) Su		106
R703	*Providence*	Fi$^-$(F) Su		106
R714b	*Providence*	Fi$^-$(F) Km Sm	*aph*A$^+$	106
R716	*Providence*	Fi$^-$(F) Su		106
R722	*Providence*	Fi$^-$(F) Su		106
R747	*Providence*	Fi$^-$(F) Su		106
R717	*Providence*	Fi$^-$(F) Ap Km Su		106
R718	*Providence*	Fi$^-$(F) Ap Km Su		106
R719	*Providence*	Fi$^-$(F) Ap Km Su		106
R752	*Providence*	Fi$^-$(F) Ap Km Su		106
R729	*Providence*	Fi$^-$(F) Su		106
R730	*Providence*	Fi$^-$(F) Su		106
R739	*Providence*	Fi$^-$(F) Su		106
R733	*Providence*	Fi$^-$(F) Ap Km Su		106
R734	*Providence*	Fi$^-$(F) Ap Km Su		106

Incompatibility and host range[a]	Plasmid	Original host	Phenotype	Genotype	Additional data	Reference
IncA-C[c]						
(continued)	R735	*Providence*	Fi^-(F) Ap Km Su			106
	R736	*Providence*	Fi^-(F) Ap Km Su			106
	R737	*Providence*	Fi^-(F) Ap Km Su			106
	R738	*Providence*	Fi^-(F) Ap Km Su			106
	R740	*Providence*	Fi^-(F) Ap Km Su			106
	R741	*Providence*	Fi^-(F) Ap Km Su			106
	R744	*Providence*	Fi^-(F) Ap Km Su			106
	R745	*Providence*	Fi^-(F) Ap Km Su			106
	R742	*Providence*	Fi^-(F) Ap Km Su Tc			106
	R743	*Providence*	Fi^-(F) Ap Km Su Tc			106
	R780	*Providence*	Fi^-(F) Su			106
	R789	*Providence*	Fi^-(F) Km Su			106
	R792	*Providence*	Fi^-(F) Km Su			106
	R794	*Providence*	Fi^-(F) Km Su			106
	R798	*Providence*	Fi^-(F) Km Su			106
	R790	*Providence*	Fi^-(F) Su			106
	R791	*Providence*	Fi^-(F) Su Tc			106
	R795	*Providence*	Fi^-(F) Su Tc			106
	R793	*Providence*	Fi^-(F) Ap Su			106
	R796	*Providence*	Fi^-(F) Ap Su			106
	R807	*Providence*	Fi^-(F) Ap Su Tc			106
	R808	*Providence*	Fi^-(F) Ap Su Tc			106
	R809	*Providence*	Fi^-(F) Ap Su Tc			106
	R810	*Providence*	Fi^-(F) Ap Su Tc			106
	R812	*Providence*	Fi^-(F) Ap Su Tc			106

	R816	*Providence*	Tra$^+$ Ap Cm Gm Km Su	*bla*$^+$(OXA-3)		106,189
	R817	*Providence*	Tra$^+$ Ap Cm Gm Km Su			106
	R824	*Providence*	Fi$^-$(F) Km Su			106
	R704b	*Proteus*	Tra$^+$ Su Tc			107
	R665	*Proteus*	Tra$^+$ Ap Su			107
	R871	*Proteus*	Tra$^+$ Su			107
	R793a	*Providence*	Ap	*bla*$^+$(TEM-1)		189
IncC[d] = **Com6**						
P. aeruginosa *S. typhimurium* *K. pneumoniae* *Providence* *S. marcescens* *P. stuartii* *E. coli*	R40a= pIP40a	*S. typhimurium*	Tra$^+$ Ap Km Pm Su	*bla*$^+$(TEM-1) *aph*A$^+$	m.w. = 96×10^6	13, 67, 73, 189
	R57b	*S. typhimurium*	Tra$^+$ Fi$^-$(F) Ap Cm Fa Gm Km Su	*bla*$^+$(OXA-3) TnCm ?		67, 69, 70, 73, 189
	R57b-1		Tra$^+$ Cm Su		m.w. = 86×10^6	13, 67, 69, 73
	R746	*Providence*	Tra$^+$ Ap Km Su Tc	*bla*$^+$(TEM-2) TnAp *aph*A$^+$	m.w. = 122×10^6	13, 73, 106, 125, 189
	R935	*S. marcescens*	Tra$^+$ Ap Cm Km Sm Su Tc	*bla*$^+$(TEM-1)	no chloramphenicol acetyl transferase activity	73, 119
	R936	*S. marcescens*	Tra$^+$ Ap Cm Km Sm Su Tc		no chloramphenicol acetyl transferase activity	73, 119
	R937	*S. marcescens*	Tra$^+$ Ap Cm Km Sm Su Tc		no chloramphenicol acetyl transferase activity	73, 119
	R55= pIP55	*K. pneumoniae*	Tra$^+$ Ap Cm Gm Su	*bla*$^+$(OXA-3)		189
	R40b	*P. aeruginosa*	Ap Km Tc			235

Incompatibility and host range[a]	Plasmid	Original host	Phenotype	Genotype	Additional data	Reference
IncC[d] *(continued)*	R16a	*Providence*	Ap Km Su			235
	RIP16= pIP16	*P. stuartii*	Km Pm Su			29, 203
	R55-1		Gm Su		segregant of R55	283
	R98= pIP98	*P. mirabilis*	Ap Cm Gm Su			283
	R92= pIP92	*Providence*	Ap Cm Gm Su			283
	RIP64	*P. aeruginosa*	Tra^+ Ap Cm Gm Su			283
	pPK170-5	*P. aeruginosa*	Tra^+ Fi^-(F) Gm Hg Km Sm Sp Su Tm			166
	pPK141-1	*P. aeruginosa*	Tra^+ Fi^-(F) Cb Cm Hg Km Nm Su Tc			166
	pPK183-1	*P. aeruginosa*	Tra^+ Fi^-(F) Cb Gm Hg Km Tb			166
	R192-1	*P. aeruginosa*	Tra^+ Fi^-(F) Cb Cm Hg Km Nm Sm Sp Su Tc			166
	pPK207-2	*P. aeruginosa*	Tra^- Cb			166
	pPK224	*P. aeruginosa*	Tra^+ Cb Sm Hg			166
	pPK148-1	*P. aeruginosa*	Tra^+ Cb Cm Km Nm Sm Sp Su Tc Hg			166
	pPK237-1	*P. aeruginosa*	Tra^+ Fi^-(F) Cb Gm Km Su Tm Hg			166
	pPK133-1	*P. aeruginosa*	Tra^+ Fi^-(F) Cb Sm Sp Su Tc Hg			166

Host range	Plasmid	Original host	Phenotype	Gene product	Remarks	References
	pPK170-2	*P. aeruginosa*	Tra^+ Fi^-(F) Cb Cm Gm Km Nm Tc Tm Hg			166
	R699a		Ap Km Nm Su Tc			166
	R799	*E. coli*	Ap Cm Km Nm			166
IncFI						
E. coli *P. morganii* *P. vulgaris* *S. typhimurium* *R. lupini* *Providence* *Shigella* spp. *E. cloacae* *P. mirabilis* Not *P. aeruginosa*	F	*E. coli* K12	Tra^+ Dps (f1 f2 f17 MS2 μ2) Phi (T3 T7 ϕII)		mobilizes chromosome; m.w. = 63 $\times 10^6$, 49% G+C; 1–2 copies/chromosome (see Fig. B2.1)	68, 85, 103, 106, 120, 181, 203
	R386	*E. coli*	Tra^+ Dps (f1 f2) Fi^+(F) Phi(T7) Tc		derepressed pilus synthesis	78, 111
	R453	*P. morganii*	Tra^+ Dps (f1 f2 MS2) Phi(λ) Ap Cm Sm Sp Su Tc	*bla*$^+$(OXA-1)	repressed pilus synthesis	120, 189
	R455	*P. morganii*	Tra^+ Dps (f1 f2) Fi^+(F) Phi(λ) Ap Cm Sm Sp Su Tc	*bla*$^+$(OXA-1)	repressed pilus synthesis	120, 189
	R455-2		Tra^+ Dps (f1 f2) Ap Cm Sm Su			120
	R714a	*Providence*	Tra^+ Dps (f1 f2) Ap Cm Sm Sp Su Tc	*bla*$^+$(TEM-2)		106, 189
	R773	*E. coli*	Tra^+ Dps (f1 f2) Fi^+(F)Sm Tc Asa Asi			109
	R978b	*E. cloacae*	Tra^+ Dps (f1 f2) Ap Cm Sm Su Tc Lac	*bla*$^+$(TEM-1)		189
	RGN238	*E. coli*	Tra^+ Dps (f1 f2) Ap Cm Sm Su Tc	*bla*$^+$(OXA-1)		121, 189, 286
	F*lac*		Tra^+ Dps (f1 f2) Fi^-(F) Lac			141

Incompatibility and host range[a]	Plasmid	Original host	Phenotype	Genotype	Additional data	Reference
IncFI *(continued)*	R456	*P. morganii*	Fi^+(F) Phi(λ) Ap Cm Sm Sp Su Tc	*bla*$^+$(OXA-1)		120
	R129	*E. coli*	Fi^+(F) Ap Cm Sm Su			209, 235
	R129-1		Cm Sm Su			235
	ColV		Dps (f1 f2) Cva			120
	RIP162-1	*S. typhi*	Ap Cm Sm Su			28
	RIP162-2		Ap Tc			28
	ColV-K94	*E. coli*	Tra^+ Dps (f1 f2 μ2 MS2 M13) Cva			50, 102, 181, 208, 242
	ColV3-K30	*E. coli*	Dps (μ2 MS2 M13) Cva			203
IncFII						
E. coli *S. typhimurium* *S. flexneri* *K. pneumoniae*	R1	*S. paratyphi* B	Tra^+ Fi^+(F) Dps (f1 f2 MS2 M13) Ap Cm Km Sm Sp Su	*bla*$^+$(TEM-1) *cat*$^+$ *aph*A^+ Tn*3* Tn*4*	repressed pilus synthesis; m.w. = 62 $\times 10^6$, 52% G+C	50, 64, 111, 181, 182, 194, 203, 242
P. mirabilis *P. vulgaris* *P. morganii*	R1*drd*19		Tra^+ Fi^-(F) Ap Cm Km Sm Su		mutant derepressed for pilus synthesis and transfer	191, 194
Providence *S. typhi*	R1-1		Tra^+ Fi^+(F) Ap Cm Fa Km Sm Sp Su			52, 73, 120, 182
S. marcescens *S. paratyphi*	R1-K^s		Tra^+ Ap Cm Sm Su		m.w. = 53×10^6	13, 52, 111, 119
Not *P. aeruginosa*	RTF1		Tra^+		RTF unit from R1*drd*19; m.w. = 42.8×10^6	55, 56, 242

pSC50		Tra$^+$ Ap Cm Sm Su	Tn*3* Tn*4*	Km-sensitive mutant of R1*drd*19; m.w. = 50.9×10^6, 1.7110 g/cm^3; IS*1* (2 copies), IS*2*; 11 *Eco*RI restriction sites	133, 167, 168, 242
pSC206		Ap Cm Km Sm Su	Tn*3* Tn*4* TnKm	transposition of TnKm from R6-5 to pSC50; IS*1* (2 copies), IS*2*; 12 *Eco*RI restriction sites	173
R6		Tra$^+$ Cm Km Nm Pm Sm Su Tc	TnTc TnKm	IS*3* (2 copies)	76, 173, 242
R6-5		Tra$^+$ Cm Km Nm Sm Su	*tet*::IS*3* TnKm	spontaneous Tc-sensitive mutant of R6 caused by insertion of IS*3* in *tet* locus; m.w. = 65×10^6, 1.711 g/cm^3; 0.7–3 copies/chromosome; 13 *Eco*RI restriction sites	56, 57, 133, 173, 242, 267
R28	*S. typhimurium*	Tra$^+$ Fi$^+$(F) Dps (MS2 M13) Fa Sm Sp Su Tc Hg			73, 182, 203
R52	*S. typhi*	Dps (MS2 M13) Fi$^+$(F) Sm Sp Su Tc Hg	*aad*A$^+$		104, 182, 193

Incompatibility and host range[a]	Plasmid	Original host	Phenotype	Genotype	Additional data	Reference
IncFII *(continued)*	R82	*S. typhi*	Dps (MS2 M13) Fi^+(F) Sm Sp Su Tc	*aad*A^+		182
	R100= NR1= 222	*S. flexneri*	Tra^+ Fi^+(F) Dps (f1 f2 MS2 M13) Cm Fa Sm Sp Su Tc Hg	*cat*$^+$ *aad*A^+ Tn*8* Tn*10*	m.w. = 70×10^6, 58% G+C; 1–2 copies/chromosome	50, 73, 76, 88, 159, 181, 182, 209, 275, 276
	R100-1*drd*		Tra^+ Fi^+(F) Dps (MS2 M13) Cm Fa Sm Sp Su Tc Hg		mutant of R100 derepressed for transfer and pilus synthesis	64, 182
	R136 (=240)	*S. typhimurium*	Tra^+ Fi^+(F) Dps (MS2 M13) Tc		m.w. = 41×10^6	93, 181, 182, 192, 193
	R192	*S. typhimurium*	Tra^+ Cm Sm Sp Su Tc Hg	*aad*A^+		181, 192, 203
	R192-1		Tra^+ Tc			123
	R192*drd7*		Tra^+ Fi^+(F) Dps (MS2 M13) Cm Sm Sp Su Tc	*aad*A^+	mutant of R192 derepressed for transfer and pilus synthesis	58, 182, 191
	RIP187		Sm Tp			91
	R196-1		Cm Sm Su Tc		m.w. = 60×10^6	203, 227
	R348F	*E. coli*	Tra^+ Fi^+(F) Cm Km Sm Su Tc Hg			11, 104, 231

R429	*Klebsiella* sp.	Tra^+ Ap Cm Km Tc	*cat*$^+$ *bla*$^+$(TEM-1)		189, 244
R444	*P. morganii*	Tra^+ Dps (MS2) Fi^+(F) Ap Cm Sm Sp Su Tc	*bla*$^+$(TEM-1)		120, 189
R445	*P. morganii*	Fi^+(F) Ap Cm Sm Sp Su Tc	*bla*$^+$(TEM-1)		120, 189
R452	*P. morganii*	Fi^+(F) Ap Cm Sm Sp Su Tc	*bla*$^+$(TEM-1)		120, 189
R457	*P. morganii*	Fi^+(F) Ap Cm Sm Sp Su Tc	*bla*$^+$(TEM-1)		120, 189
R458	*P. morganii*	Fi^+(F) Ap Cm Km Sm Sp Su Tc HspII	*bla*$^+$(TEM-1)		120, 189
R459	*P. morganii*	Tra^+ Ap Cm Km Sm Sp Su Tc HspII	*bla*$^+$(TEM-1)		120, 189
R459-4		Ap Cm Hg Sm Su Tc		deletion of R459	120
R459-5		Tc		deletion of R459	120
R459-7		Ap Cm Hg Su Tc		deletion of R459	120
R459-13		Cm HspII Su Tc		deletion of R459	120
R494	*P. mirabilis*	Tra^+ Ap Km Tc	*bla*$^+$(TEM-1)		107, 189
R496	*E. coli*	Ap	*bla*$^+$(TEM-1)		189
R538-1		Tra^+ Fi^+(F) Cm Sm Su		m.w. = 49×10^6; 4-5 copies/chromosome (see Fig. B2.3)	2, 77, 181, 227, 272
R704a	*P. mirabilis*	Tra^+ Ap Cm Km Sm Su			107
R813	*S. typhi*	Tra^+ Fi^+(F) Ap Cm Sm Su Tc	*bla*$^+$(TEM-1)		72, 189
R814	*S. typhi*	Fi^+(F) Ap Cm Sm Su Tc			72
R815	*S. typhi*	Fi^+(F) Ap Cm Sm Su Tc			72
R830	*S. marcescens*	Tra^+ Dps (f1 f2) Cm Sm Su Tc			119

Incompatibility and host range[a]	Plasmid	Original host	Phenotype	Genotype	Additional data	Reference
IncFII *(continued)*	JR66b	*K. pneumoniae*	Tra^- Cm Gm Km Sm Su Tc	*aad*B^+	exists as complex with JR66a	71
	JR70	*E. coli*	Fi^+(F) Cm Fa Km Sm Su Tc			73
	JR72	*E. coli*	Tra^+ Cm Fa Km Nm Sm Su Tc	*aph*A^+ Tn*6*	m.w. = 63.4×10^6; 1 copy/chromosome; 6 *Hin*III restriction sites	13, 19, 73
	JR73	*E. coli*	Fi^+(F) Cm Fa Km Sm Su Tc			73
	ENT_I	*E. coli*	Ent			203
	ColB-CA18	*E. coli*	Cba			203
	ColB2-K77	*E. coli*	Tra^+ Dps (MS2 M13) Cba		m.w. = 70×10^6; 1-2 copies/chromosome	89, 102
	ColB2		Cba		m.w. = 70×10^6	50
	ColVB-K260		Tra^+ Cba Cva		m.w. = 113×10^6	102
	334		Tra^+ Dps (μ2) Fi^+(F) Phi^- Ap Cm Sm Su		m.w. = 54.2×10^6	5
IncFIII						
E. coli	ColB-K98	*E. coli*	Dps (f1 f2) Fi^+(F) Cba		m.w. = 70×10^6	89, 111
	ColB-K166	*E. coli*	Dps (f1 f2) Fi^+(F) Cba			203
	MIP240	*E. coli*	Hly			27

IncFIV						
S. typhimurium *E. coli*	R124	*S. typhimurium*	Tra^+ Dps (f1 f2 M13 MS2) Tc HspI			12, 111, 120, 192
IncFV						
S. typhosa *E. coli*	F_o*lac*	*S. typhosa*	Tra^+ Dps (f1) Lac		derepressed for pilus	64, 84
Unclassified						
F-like (ref. 182)	R36	*S. typhi*	Dps (MS2 M13) Fi^+(F) Tc			104, 182, 193
	R51	*S. typhi*	Dps (MS2 M13) Fi^+(F) Tc			104, 182, 193
	R312	*S. typhi*	Dps (MS2 M13) Fi^+(F) Hg			104, 182, 193
	ColV3		Tra^+ Dps (μ2 MS2) Cva			50, 185
	R269F		Ap Km Sm	*aph*A^+		104
IncH[e]						
Salmonella spp. *E. coli*	R27= TP117	*S. typhimurium*	Tra^+ Tc		m.w. = 112×10^6; displaces F	72, 92, 93, 106, 203, 251
	R726	*S. typhi*	Tra^+ Fi^-(F) Cm Sm Su Tc			72
	R726-1		Tra^- Fi^-(F) Cm Sm Su			72, 106
	R804	*S. typhimurium*	Tra^+ Cm Sm Su Tc		segregates Tc and Cm Sm Su plasmids	72
	R805b	*S. typhi*	Tra^+ Cm Sm Su Tc		found with R805a	72
	R1097	*S. typhi*	Tra^+ Ap Cm Sm Su Tc	*bla*$^+$(TEM-1)	m.w. = 120×10^6	140, 189

Incompatibility and host range[a]	Plasmid	Original host	Phenotype	Genotype	Additional data	Reference
IncH[e] (*continued*)	TP116	*S. typhi*	Tra^+ Fi^-(F) Cm Sm Su		m.w. = 143.7×10^6 exists stably with F	3, 92, 93, 251
	R821b	*S. typhi*	Cm Sm Su Tc		found with R821a (IncIδ)	72
	R806b	*S. typhi*	Cm Sm Su Tc		found with R806a (IncA-C)	72
	R819	*S. typhi*	Tra^+ Cm Sm Su Tc		segregates Tc and Cm Sm Su plasmids	72
	R822	*S. typhi*	Tra^+ Cm Sm Su Tc		segregates Tc and Cm Sm Su plasmids	72
	R823	*S. typhi*	Tra^+ Cm Sm Su Tc		segregates Tc and Cm Sm Su plasmids	72
	R725	*S. typhi*	Cm Sm Su Tc		segregates Tc and Sm Cm Su plasmids	72
	R727	*S. typhi*	Cm Sm Su Tc			72
	R820b	*S. typhi*	Cm Sm Su Tc		coexists with R820a (IncIα); segregates Tc and Sm Cm Su plasmids	72
	TP123	*S. typhi*	Tra^+ Fi^-(F) Cm Sm Su Tc		m.w. = 123×10^6; segregates Tc plasmid; displaces F	5, 93, 251
	TP124	*S. typhi*	Tra^+ Cm Sm Su Tc		m.w. = 120×10^6; displaces F	93, 251
	RIP166		Cm Sm Su			28
	RIP167-1		Tc			28

Inclα =IncI1 =Com1 =IncIβ						
E. coli *Salmonella* *Shigella* *Klebsiella* Not *Proteus* Not *Pseudomonas* Not *Rhizobium*	ColIb-P9	*E. coli*	Tra$^+$ Dps (If1) Cib Uv		m.w. = 65×10^6, 50% G+C; 1–3 copies/chromosome	47, 64, 85, 92, 112, 181
	Δ	*S. typhimurium*	Tra$^+$ Dps (If1) Fi$^-$(F) Phi (ϕII Stp^f)		m.w. = 59×10^6; 1–2 copies/chromosome	5, 85, 92, 112
	T-Δ	*S. typhimurium*	Tra$^+$ Dps (If1) Fi$^-$(F) Phi (ϕII Stp^f) Tc		m.w. = 67×10^6	5
	JR66a= JR67	*K. pneumoniae*	Tra$^+$ Dps (If1) Fi$^+$(F) Km Nm Sm	*aph*A$^+$ *aph*C$^+$ Tn*5* *fin*U$^+$	m.w. = 57.4×10^6, 50% G+C; 1 copy/chromosome; 7 *Eco*RI restriction sites	19, 71, 85, 106
	JR66a-1		Tra$^+$ Ap Km Sm	*bla*$^+$(TEM-2) Tn*1* *aph*A$^+$ *aph*C$^+$ Tn*5*	transposition of Tn*1* from RP4; m.w. = 61.3×10^6, 50% G+C	113, 189
	R62	*S. typhimurium*	Tra$^+$ Fi$^+$(F) Ap Sm Sp Su Tc	*bla*$^+$(OXA-2) *aad*A$^+$ *fin*Q$^+$	m.w. = 80×10^6, 50% G+C	98, 182, 189, 227
	R64	*S. typhimurium*	Tra$^+$ Dps (If1) Fi$^-$(F) Phi(λ) Sm Tc	*aph*C$^+$	m.w. = 72.3×10^6, 50% G+C; 1–3 copies/chromosome	52, 85, 112, 181, 182
	R64-1		Tra$^+$ Fi$^-$(F) Ap Sm Tc	*bla*$^+$(TEM-2) Tn*1* *aph*C$^+$	transposition of Tn*1* from RP4	64, 66, 113, 189

Incompatibility and host range[a]	Plasmid	Original host	Phenotype	Genotype	Additional data	Reference
IncIα (*continued*)	R64*drd*11		Tra^+ Sm Tc		derepressed mutant of R64	191, 194
	R144	*S. typhimurium*	Tra^+ Dps (If1) Fi^-(F) Cib Km Tc Uv	*aph*A^+	m.w. = 62×10^6, 50% G+C; 1–2 copies/chromosome	64, 65, 98, 106, 112, 181, 182
	R144*drd*3		Tra^+ Km Cib		derepressed mutant of R144	191, 194
	R163	*S. typhimurium*	Km Nm Sm Tc	*aph*A^+		181, 203
	R483	*E. coli*	Tra^+ Dps (If1) Cia Sm Sp Tp	Tn*7* *aad*A^+	m.w. = 62×10^6; 2 copies/chromosome	15, 65, 112
	R483-1		Tra^+ Cia Ap Sm Sp Tp	*bla*$^+$(TEM-2) Tn*1* Tn*7*	transposition of Tn*1* from RP4; m.w. = 65×10^6	13
	RIP112	*S. panama*	Fi^-(F) Km Nm Pm			203, 235
	RIP186		Sm Tp			91
	R606a	*S. typhi*	Tra^+ Ap Tc	*bla*$^+$(TEM-1)		83, 189
	R609b	*S. typhimurium*	Tra^+ Ap Sm Su	*bla*$^+$(OXA-1)		189
	R648	*S. typhimurium*	Tra^+ Ap Km Sm	*bla*$^+$(TEM-1) TnAp	50% G+C	112, 125
	R656a	*S. typhimurium*	Tra^+ Ap Km Sm Su	*bla*$^+$(OXA-1)		189
	R820a	*S. typhi*	Tra^+ Dps (If1) Fi^+(F) Ap Sm	*bla*$^+$(TEM-1)	found with R820b (IncH)	72
	TP102		Tra^+ Dps (If1) Fi^+(F) Phi^- Km Nm		m.w. = 54×10^6	5, 85, 93

	TP103	*S. typhi*	Tra^+ Dps (If1) Km Nm		203
	TP104	*S. typhi*	Tra^+ Dps (If1) Km Nm		203
	TP105	*S. typhi*	Tra^+ Dps (If1) Km Nm		203
	TP106	*E. coli*	Tra^+ Dps (If1) Km Nm		203
	TP107	*E. coli*	Tra^+ Dps (If1) Km Nm		203
	TP108	*S. enteritidis*	Tra^+ Dps (If1) Km Nm		203
	TP109	*S. typhimurium*	Tra^+ Dps (If1) Km Nm		203
	TP110		Tra^+ Dps (If1) Fi^-(F) Phi (ϕII Stp^f) Km Cib	m.w. = 65×10^6	5, 85, 93
	ColIa-CA53		Dps (If1 If2) Cia		50, 112, 182
	ColIa-CT2		Dps (If1 If2) Cia		50, 182
	ColIb-CT4		Tra^+ Dps (If1 If2) Cib		203
IncI2					
E. coli	TP114	*E. coli*	Tra^+ Dps (If1) Fi^-(F) Phi^- Km	m.w. = 41×10^6	5, 92, 93
	TP114*drp* 1, 2		Tra^+ Dps (If1) Km	derepressed mutants of TP114	92
	RIP175	*S. wien*	Ap		27, 288
	MIP241	*E. coli*	Hly		27
IncIγ					
E. coli *S. typhimurium*	ColIb-IM1420		Cib		132
Not *Proteus*	R621a	*S. typhimurium*	Tra^+ Dps (If1) Fi^-(F) Tc	m.w. = 65×10^6	85, 112
	R621a-1		Tra^+ Ap Cm Km Tc		106

Incompatibility and host range[a]	Plasmid	Original host	Phenotype	Genotype	Additional data	Reference
IncIδ						
E. coli *S. typhimurium* Not *Proteus*	R821a	*S. typhi*	Tra^+ Dps (If1) Fi^-(F) Ap		m.w. = 43×10^6; found with R821b (IncH)	13, 72
	R721		Fi^-(F) Sm Tp		m.w. = 48×10^6; found with other plasmids and isolated by transformation	13, 85, 106
	RM413	*S. typhimurium*	Dps (If1) Fi^-(F) Ap	*bla*$^+$(TEM-1)	m.w. = 43×10^6; found with other plasmids and isolated by transformation	13, 83, 85, 106, 121, 189
IncIζ						
S. typhi	R805a	*S. typhi*	Tra^+ Km		m.w. = 48×10^6	72, 85
E. coli	R861a	*E. coli*	Tra^+ Km			72, 85
IncJ						
Proteus spp. *E. coli*	R391	*P. rettgeri*	Tra^+ Fi^-(F) Km Nm Hg	*aph*A$^+$		52, 106, 123
	R391-3b		Tra^+ Km Sm Tp Hg			123
	R391-3b-1		Tra^+ Sm Tp Hg			123
	R392	*P. rettgeri*	Tra^+ Fi^-(F) Km Hg			52
	R397	*P. rettgeri*	Tra^+ Fi^-(F) Km Hg			52
	R705	*P. vulgaris*	Tra^+ Km			107
	R706	*P. vulgaris*	Tra^+ Km			107

IncK					
S. flexneri *E. coli*	R387	*S. flexneri*	Tra$^+$ Cm Sm	*cat*$^+$(EC2.3.1.99)	106, 110, 244
IncL					
S. marcescens *E. coli*	R471	*S. marcescens*	Tra$^+$ Ap Cm Km	*bla*$^+$(TEM-1) *aph*A$^+$	119, 189
	R472	*S. marcescens*	Tra$^+$ Ap	*bla*$^+$(TEM-1)	119, 189
	R473	*S. marcescens*	Tra$^+$ Ap		119
	R474	*S. marcescens*	Tra$^+$ Ap		119
	R475	*S. marcescens*	Tra$^+$ Ap		119
	R476a	*S. marcescens*	Tra$^+$ Ap		119
	R479a	*S. marcescens*	Tra$^+$ Ap		119
	R830a	*S. marcescens*	Tra$^+$ Km Sm		119
	R831	*S. marcescens*	Tra$^+$ Km Sm	*aph*A$^+$	119
	R832	*S. marcescens*	Tra$^+$ Ap		119
IncM **=Com7**					
S. paratyphi *E. coli*	RIP69	*S. paratyphi* B	Tra$^+$ Fi$^-$(F) Ap Km Nm Pm Tc	*bla*$^+$(TEM-1) *aph*A$^+$	189, 283
P. morganii	R69-1	*S. paratyphi*	Ap Tc		29, 235
S. marcescens	R69-2	*S. paratyphi*	Ap Km		29, 235
K. pneumoniae *S. liquefaciens*	RIP95 (=pIP95)	*K. pneumoniae*	Ap Km Nm Pm Tc		235
	R95-1	*K. pneumoniae*	Ap Tc		235
	RIP135 (=R135)	*Enterobacter*	Tra$^+$ Gm Sm Su Tc		283
	R135-1		Hg Tc		283

Incompatibility and host range[a]	Plasmid	Original host	Phenotype	Genotype	Additional data	Reference
IncM *(continued)*	R446b	*P. morganii*	Tra^+ Fi^-(F) Sm Tc			106, 120
	RS28	*S. marcescens*	Tra^+ Ap Cm Gm Sm Tc	*bla*$^+$(TEM-1)		119, 189
	R930	*S. liquefaciens*	Tra^+ Ap	*bla*$^+$(TEM-1)		119, 189
	R957	*S. marcescens*	Tra^+ Ap			119
	R958	*S. marcescens*	Tra^+ Ap			119
IncN =Com2						
Proteus spp. *Klebsiella* spp. *Salmonella* spp.	N3 (=RN3)	*Shigella* sp.	Tra^+ Dps (Ike) Fi^-(F) Sm Sp Su Tc Hg HspII	*aad*A$^+$	m.w. = 33×10^6, 50% G+C; 1–2 copies/chromosome	53, 66, 104, 123, 235
Providence	N3T		Tra^+ Tc		m.w. = 34×10^6	66, 140
Shigella spp.	RN3-1		Sm Su			235
E. coli	R15	*P. vulgaris*	Tra^+ Dps (Ike) Fi^-(F) Sm Su Hg HspII		m.w. = 40×10^6, 49% G+C; 1–2 copies/chromosome	64, 66, 85, 104, 140
	TP118	*S. enteriditis*	Tra^+ Fi^-(F) Ap Sm		m.w. = 27×10^6	4, 92, 93
	TP119 (=R45)	*S. typhimurium*	Tra^+ Dps (Ike) Fi^-(F) Ap Su Tc Uv Hsp	*bla*$^+$(OXA-2) *aad*A$^+$	Hsp type undetermined	92, 104, 182, 189, 203
	TP120 (=R46 =R1818)	*S. typhimurium*	Tra^+ Dps (Ike) Fi^-(F) Ap Sm Sp Su Tc Uv	*bla*$^+$(OXA-2) *aad*A$^+$	m.w. = 32×10^6	66, 92, 93, 104, 106, 189, 203
	TP122 (=R128)	*S. typhimurium*	Tra^+ Fi^+(F) Ap Su Tc		m.w. = 33×10^6	92, 93

R269N	*S. sonnei*	Tra^+ Fi^-(F) Ap Sm Tc HspII		found with R269F	9, 11, 104
R269N-1		Tra^+ Fi^-(F) Ap Km Sm Tc HspII	*bla*$^+$(TEM-1)	recombinant of R269F and R269N	104, 189
R313N		Tra^+ Tc HspII			10, 104
R348N	*E. coli*	Tra^+ Fi^-(F) Sm Tc HspII		found with R348F	10, 11
R390	*P. rettgeri*	Tra^+ Fi^-(F) Ap Cm Sm Sp Su Tc HspII	*bla*$^+$(TEM-1) TnAp *aad*A$^+$		52, 189
R390-1		Tra^+ Ap Su Tc HspII		deletion of R390	52
R390-2		Tra^+ Ap Sm Su Tc		deletion of R390	52
R390-3		Tra^- Ap		deletion of R390	52
R390-4		Tra^+ Ap		deletion of R390	52
R390-8		Tra^+ Ap HspII		deletion of R390	52
R390-12		Tra^- Ap Cm Sm Su HspII		deletion of R390	52
R393	*P. mirabilis*	Tra^+ Sm			107
R395	*P. rettgeri*	Tra^+ Ap Cm Sm Sp Su Tc HspII			52
R396	*P. rettgeri* 135	Fi^-(F) Ap Cm Sm Sp Su Tc HspII	*aad*A$^+$		52
R400	*Proteus* sp.	Fi^-(F) Ap Cm Sm Sp Su Tc HspII	*aad*A$^+$		52
R448	*P. morganii*	Fi^-(F) Ap Cm Sm Sp Su Tc HspII	*bla*$^+$(TEM-1)		120, 189
R461	*P. morganii*	Fi^-(F) Ap Cm Sm Sp Su Tc HspII	*bla*$^+$(TEM-1)		120, 189
R462	*P. morganii*	Fi^-(F) Ap Cm Sm Sp Su Tc HspII	*bla*$^+$(TEM-1)		120, 189
R467	*P. morganii*	Fi^-(F) Ap Sm Sp Tc HspII	*bla*$^+$(TEM-1)		120, 189

Incompatibility and host range[a]	Plasmid	Original host	Phenotype	Genotype	Additional data	Reference
IncN						
(continued)	R398	*P. rettgeri*	Tra^+ Ap Cm Sm Sp Su Tc HspII	bla^+(TEM-1) *aad*A^+		52, 189
	R446c	*P. morganii*	Tra^+ Fi^-(F) Sm Sp Su HspII			120
	R447a	*P. morganii*	Tra^+ Fi^-(F) Ap Cm Sm Sp Su Tc HspII	bla^+(TEM-1)		120, 189
	R447b	*P. morganii*	Tra^+ Dps (Ike) Ap Km	bla^+(TEM-1) *aph*A^+	m.w. = 33×10^6	53, 120, 189
	R465	*P. morganii*	Tra^+ Fi^-(F) Sm Sp Su Tc HspII			120
	R497	*Providence*	Tra^+ Fi^-(F) Ap Km Sm Tc HspII	bla^+(TEM-1) *aph*A^+		106, 189
	R728	*Providence*	Tra^+ Fi^-(F) Ap			106
	R759a	*P. mirabilis*	Tra^+ Ap Km Sm			107
	R761a	*P. mirabilis*	Tra^+ Ap Sm			107
	R731	*Providence*	Fi^-(F) Ap			106
	R732	*Providence*	Fi^-(F) Ap			106
	R825	*Providence*	Fi^-(F) Ap	bla^+(TEM-1)		106, 189
	R454	*P. morganii*	Fi^-(F) Ap Cm Sm Sp Su Tc HspII	bla^+(TEM-1)		120, 189
	R48	*S. typhimurium*	Ap Sm Sp Su Tc Uv	*aad*A^+		80, 104, 203
	R132	*S. typhimurium*	Ap Sm Su Tc HspII			104
	R199	*S. typhi*	Su Tc HspII		plasmid unstable in storage	9

R204		Tc HspII		9, 104
R205	*S. typhimurium*	Ap Su Tc Uv	*bla*$^+$(OXA-2)	9, 104, 189
R245		Tc HspII	*hsr*$^+$(*Eco*RII)	9, 104
R250		Su Tc HspII		9, 104
R22Kb	*Proteus* sp.	Tra$^+$ Sm		107
R760a	*Proteus* sp.	Tra$^+$ Ap Km Nm Sm		107
R771a	*Proteus* sp.	Tra$^+$ Ap Km Nm Sm	*bla*$^+$(TEM-1)	107, 189
R701	*Proteus* sp.	Tra$^+$ Ap Km Sm	*aph*A$^+$	107
R893	*K. aerogenes*	Ap	*bla*$^+$(TEM-1)	189
R898a	*S. typhimurium*	Ap Km Sm	*bla*$^+$(TEM-1) *aph*A$^+$	189
R979	*K. aerogenes*	Ap	*bla*$^+$(TEM-1)	189
R1025	*P. mirabilis*	Ap	*bla*$^+$(TEM-1)	189
ROX166	*P. mirabilis*	Ap	*bla*$^+$(OXA-2)	189
ROX179	*P. mirabilis*	Ap	*bla*$^+$(OXA-2)	189
ROX407	*P. mirabilis*	Ap	*bla*$^+$(OXA-2)	189
pIP113= R113= R11-3	*S. panama*	Fi$^-$(F) Tc		235
RPC3	*E. coli*	Km Pm Sm		29, 203, 235
RM98= pHK98		Ap Sm Tc		203
RM227= pHK227		Ap		203
RM414= pHK414		Ap		203

Incompatibility and host range[a]	Plasmid	Original host	Phenotype	Genotype	Additional data	Reference
IncP[g] =Com4						
E. coli, *Pseudomonas* spp.	RP1 (=R1822)	*P. aeruginosa*	Tra$^+$ Fi$^-$(F) Dps (PRD1 PRR1) Ap Km Tc	*bla*$^+$(TEM-2) TnAp *aph*A$^+$	m.w. = 40×10^6, 60% G+C	94, 214, 215, 216, 225, 245
Serratia spp., *Bordetella*	RP1-I42		Tra$^+$ Dps (PRD1 PRR1) Ap Km Nm Tc His	*bla*$^+$ *aph*A$^+$ *his*$^+$	carries the *E. coli his* operon	212
bronchiseptica, *Proteus* spp., *Rhizobium* spp., *Salmonella* spp.	RP4	*P. aeruginosa*	Tra$^+$ Dps (PRD1 PRR1 PR4) Fi$^-$(F) Ap Km Nm Tc	*bla*$^+$(TEM-2) Tn*1* *aph*A$^+$	m.w. = 36×10^6, 58% G+C; 1–3 copies/chromosome (see Figs. B2.4 and B2.5)	74, 85, 106, 113, 140, 189
Shigella spp.	RP4-4		Tra$^+$ Ap Tc	Tn*1* *bla*$^+$ *aph*A	mutant of RP4	113
Providence, *Neisseria* spp.	RP4-8		Tra$^+$ Ap Km	Tn*1* *bla*$^+$ *aph*A$^+$ *tet*	mutant of RP4	113
Rhodopseudomonas spp., *Azotobacter*	RP4-δ1		Tra$^-$ Tc		m.w. = 28×10^6; deletion mutant of RP4	116
Caulobacter, *Klebsiella* spp.	RP4-Tn*C1*		Tra$^+$ Ap Km Nm Sm Tc Tp	Tn*1* Tn*7*	m.w. = 45×10^6; transposition of Tn*7* (see Fig. B2.4)	15
Agrobacterium tumefaciens	R751	*K. aerogenes*	Tra$^+$ Dps (PRR1) Fi$^-$(F) Tp	Tn402(Tp)	m.w. = 30×10^6	13, 153, 241
Chromobacterium violaceum	R839	*S. marcescens*	Tra$^+$ Dps (PRR1 PR4) Ap Km Nm Sm Su Tc	*bla*$^+$(TEM-1) *aph*A$^+$	m.w. = 46×10^6	119, 140, 189
	R934	*S. marcescens*	Tra$^+$ Ap Km Nm Tc	*bla*$^+$(TEM-1)		119, 189
	RK2	*K. aerogenes*	Tra$^+$ Ap Km Tc	*bla*$^+$(TEM)	m.w. = $40–48 \times 10^6$,	190

				60% G+C (see Fig. B2.6)	
R1033	*P. aeruginosa*	Tra^+ Cb Cm Gm Km Nm	Tn404(Ap) bla^+ (TEM-1) $aacA^+$ $aphA^+$	m.w. = 45×10^6	189, 241, 252
R906	*B. bronchi-septica*	Tra^+ Ap Hg Sm Sp Su Hg	bla^+(OXA-2)	m.w. = 35×10^6; 3 copies/chromosome; no $aadA^+$ activity	118, 189, 264
R6886= R68	*P. aeruginosa*	Tra^+ Ap Km Nm Tc		m.w. = 36×10^6	140, 256
R446a	*P. morganii*	Tra^+ Fi^-(F) Ap	bla^+(TEM)		120, 121
R690	*Providence*	Tra^+ Dps (PRR1) Fi^-(F) Ap Km Tc			106
R440	*P. mirabilis*	Tra^+ Ap Km Tc			107
R702	*P. mirabilis*	Tra^+ Dps (PRR1) Km Sm Su Tc	$aphA^+$	m.w. = 46×10^6	107, 140
R940	*P. mirabilis*	Tra^+ Ap Km Sm Su Tc			107
R751-SU2		Tra^+ Ap Km Sm Su Tp	bla^+(TEM-1)	m.w. = 69×10^6	115, 189
R938	*S. marcescens*	Tra^+ Ap Cm Km Sm Sp Su Tc	Tn(Ap Sm) bla^+(TEM-1) $aphC^+$	m.w. = 53×10^6	108, 145
R842	*P. mirabilis*	Tra^+ Ap Km Tc	bla^+(TEM-1)		189
$R_{GN}823$	*K. pneumoniae*	Tra^+ Ap Cm Km Sm Su Tc	bla^+(TEM-2)		234
R708	*Providence*	Fi^-(F) Ap Km Tc			106
R709	*Providence*	Fi^-(F) Ap Km Tc			106
R710	*Providence*	Fi^-(F) Ap Km Tc			106
R711a	*Providence*	Fi^-(F) Ap Km Tc			106
R712a	*Providence*	Fi^-(F) Ap Km Tc			106
R713	*Providence*	Fi^-(F) Ap Km Tc			106

Incompatibility and host range[a]	Plasmid	Original host	Phenotype	Genotype	Additional data	Reference
IncP						
(continued)	R463a	*Proteus* sp.	Tra^+ Ap Km Tc			107
	R437a	*Proteus* sp.	Tra^+ Ap Km Tc			107
	R438	*Proteus* sp.	Tra^+ Ap Km Tc			107
	R439	*Proteus* sp.	Tra^+ Ap Km Tc			107
IncQ[h]						
E. coli *Salmonella* spp. *Proteus* spp.	R300B	*S. typhi*	Tra^- Sm Su		m.w. = 5.7×10^6; 10.9 copies/ chromosome	14
Providence *Pseudomonas* spp.	R305c	*S. typhi*	Tra^- Sm Su		m.w. = 5.7×10^6; 9.3 copies/ chromosome	14
	R310	*S. typhi*	Tra^- Sm Su		m.w. = 5.7×10^6; 9.2 copies/ chromosome	14
	R450B	*P. morganii*	Tra^- Sm Su		m.w. = 5.7×10^6; 8.6 copies/ chromosome	14
	R464C	*P. morganii*	Tra^- Sm Su		m.w. = 5.7×10^6; 11.2 copies/ chromosome	14
	R676	*S. senftenberg*	Tra^- Sm Su		m.w. = 5.7×10^6; 8.5 copies/ chromosome	14
	R678	*S. dublin*	Tra^- Sm Su		m.w. = 9.2×10^6; 10.8 copies/ chromosome	14

	R682	*P. mirabilis*	Tra^- Sm Su		m.w. = 5.7×10^6; 7.9 copies/ chromosome	14
	R684	*P. mirabilis*	Tra^- Sm Su		m.w. = 6.3×10^6; 10.5 copies/ chromosome	14
	R750	*Providence*	Tra^- Fi^-(F) Sm Su		m.w. = 5.9×10^6; 11.5 copies/ chromosome	14, 106
	R1162	*P. aeruginosa*	Tra^- Sm Su		m.w. = 5.5×10^6; 11.6 copies/ chromosome	14
	PB165	*E. coli*	Tra^- Sm Su		m.w. = 7.4, 14.7, 21.4×10^6; 4.1, 1.3, 0.8 copies/ chromosome	14
	RSF1010		Tra^- Sm Su		m.w. = 5.7×10^6	124, 95
	AP201		Tra^- Ap Sm Su	Tn*3*	m.w. = 9.1×10^6; insertion of Tn*3* into Su determinant of RSF1010	14, 95, 124
IncS						
S. marcescens	R476b	*S. marcescens*	Tra^+ Sm Su Tc			119
E. coli	R477	*S. marcescens*	Tra^+(ts) Cm Km Sm Su Tc		m.w. = 173×10^6	13, 119, 226
	R478	*S. marcescens*	Tra^+(ts) Cm Km Tc	*aph*A^+	m.w. = 166×10^6	119, 226
	R479b	*S. marcescens*	Tra^+ Cm Km Sm Su Tc			119
	R826	*S. marcescens*	Tra^+(ts) Ap Cm Gm Km Nm Sm Tc	*bla*$^+$(TEM-1) *aph*A^+		119, 189, 226

Incompatibility and host range[a]	Plasmid	Original host	Phenotype	Genotype	Additional data	Reference
IncS *(continued)*	R827, 828	*S. marcescens*	Tra^+ Ap Cm Gm Km Sm Tc			119
	R829	*S. marcescens*	Tra^+(ts) Cm Km Sm Tc	*aph*A^+		119, 226
	R963	*S. marcescens*	Tra^+ Ap Cm Km Sm Su Tc	*bla*$^+$(TEM-1)		189
	R477-1		Tra^+ Sm Su Tc		m.w. = 155×10^6	13
IncT						
E. coli, *Proteus* spp., *Providence*, *K. aerogenes*, *S. typhimurium*	Rts1	*P. vulgaris*	Rep(ts) Tra^+ Fi^-(F) Km	*aph*A^+	45% G+C	52, 76, 106, 265
	R401	*P. rettgeri*	Rep(ts) Tra^+ Ap Sm	*aph*C^+ *bla*$^+$(TEM-1)		52, 106, 189
	R394	*P. rettgeri*	Tra^+ Ap Km Nm	*bla*$^+$(TEM-1) *aph*A^+		52, 189
	R402	*P. rettgeri*	Rep(ts) Tra^+ Ap Sm Sp			52, 203
	R414b	*Providence*	Tra^+ Ap Sm	*bla*$^+$(TEM-1)		52, 189
IncV						
P. mirabilis, *P. vulgaris*, *E. coli*	R753	*P. mirabilis*	Tra^+ Ap Cm Sm Su	*bla*$^+$(OXA-1)	m.w. = 68×10^6; unstable in *E. coli*	107, 140, 189
	R757a	*P. mirabilis*	Tra^+ Ap Sm Su		unstable in *E. coli*	107
	R769	*P. mirabilis*	Tra^+ Cm Km Sm Su		m.w. = 78×10^6; unstable in *E. coli*	107, 140
	R905	*P. mirabilis*	Tra^+ Cm Sm Su		unstable in *E. coli*	107
	R754	*P. mirabilis*	Tra^+ Ap Cm Sm Su	*bla*$^+$(OXA-1)	unstable in *E. coli*	107

	R755	*P. mirabilis*	Tra^+ Ap Cm Sm Su			107
	R756	*P. mirabilis*	Tra^+ Ap Cm Sm Su			107
	R901	*P. mirabilis*	Tra^+ Cm Km Sm Su	*aph*A^+	unstable in *E. coli*	107
	R902	*P. mirabilis*				107
	R903	*P. mirabilis*				107
	R770	*P. mirabilis*	Tra^+ Cm Km Sm Su		unstable in *E. coli*	107
IncW						
E. coli *S. flexneri* *Proteus* spp. *Klebsiella* spp. *P. aeruginosa* *Salmonella* spp. *A. liquefaciens*	S-a (=Rs-a)	*S. flexneri*	Tra^+ Dps (PRD1) Fi^-(F) Cm Km Sm Su		m.w. = 23×10^6, 62% G+C; 3–5 copies/chromosome; Km resistance low in *E. coli* but high in *P. aeruginosa*	63, 85, 108, 110
	R388	*E. coli*	Tra^+ Fi^-(F) Su Tp		m.w. = 21×10^6, 62% G+C; 3–5 copies/chromosome	69, 85
	R7K	*P. rettgeri*	Tra^+ Fi^-(F) Ap Sm	*bla*$^+$(TEM-1) *aph*C^+ Tn-Ap	m.w. = 20×10^6, 62% G+C; 3–5 copies/chromosome; low Sm resistance in *E. coli*	52, 85, 125, 189
	R7K-Tn*C1*		Tra^+ Ap Sm Tp	Tn*7*	transposition of Tn*7* into R7K	15
	RA3	*A. liquefaciens*	Tra^+ Cm Sm Sp Su	*aad*A^+		66, 110
	R409	*E. coli*	Tra^+ Fi^+(F) Su Tc Tp			69
	R389	*K. aerogenes*	Tra^+ Fi^+(F) Cm Sm Su Tc Tp			69

Incompatibility and host range[a]	Plasmid	Original host	Phenotype	Genotype	Additional data	Reference
IncW *(continued)*	R404	*K. aerogenes*	Tra^+ Fi^+(F) Ap Cm Sm Su Tc Tp			69
	Sa-1		Tra^- Ap Cm Km Sm Su	*bla*$^+$(TEM-2) Tn*1*	m.w. = 26×10^6; transposition of Tn*1* into S-a	113, 189
	R411	*E. coli* D800	Tra^+ Fi^-(F) Su Tp			69
	R406	*K. aerogenes* D795	Tra^+ Fi^-(F) Su Tp			69
	R403	*Klebsiella* sp. D792	Tra^+ Fi^+(F) Cm Sm Su Tc Tp			69
	R405	*Klebsiella* sp. D794	Tra^+ Fi^-(F) Ap Cm Sm Su Tc Tp			69
	R407	*Klebsiella* sp. D796	Tra^+ Fi^-(F) Su Tp			69
	R408	*Klebsiella* sp. D797	Tra^+ Fi^-(F) Su Tp			69
	R410	*Klebsiella* sp. D799	Tra^+ Fi^+(F) Cm Sm Su Tc Tp			69
	R412	*Klebsiella* sp. D801	Tra^+ Fi^+(F) Ap Cm Sm Su Tc Tp			69
	R413	*Klebsiella* sp. D802	Tra^+ Fi^-(F) Cm Sm Su Tc Tp			69
	R419	*Klebsiella* sp. D805	Tra^+ Fi^-(F) Su Tp			69
	R420	*Klebsiella* sp. D806	Tra^+ Fi^-(F) Ap Cm Sm Tc Tp			69

	R421	*Klebsiella* sp. D808	Tra^+ Fi^-(F) Su Tp			69
	R422	*Klebsiella* sp. D809	Tra^+ Fi^-(F) Ap Cm Sm Su Tc Tp Hsp	*hsr*$^+$(*Eco*RII) *aad*A$^+$		69
	R423	*Klebsiella*	Tra^+ Fi^-(F) Ap Cm Sm Su Tc Tp			69
	R424	*Klebsiella* D811	Tra^+ Fi^-(F) Ap Cm Sm Su Tc Tp			69
	D829	*Klebsiella* D829	Tra^- Ap Cm Sm Su Tp			69
	D830	*Klebsiella* D830	Tra^- Ap Sm Su Tp			69
IncX						
P. morganii *E. coli* *Providence*	R6K	*P. rettgeri*	Tra^+ Ap Sm	TnAp *bla*$^+$ (TEM-1)	m.w. = 26×10^6, 45% G+C; 15–20 copies/ chromosome	125, 165, 189, 283
	R485	*P. morganii*	Tra^+ Fi^+(F, R386) Su			106, 120
	R711b	*Providence*	Tra^+ Fi^+(F) Km			106
	R712b	*Providence*	Fi^+(F) Km			106
	R778b	*Providence*	Fi^+(F) Su			106
	R779	*Providence*	Fi^+(F) Su			106
	R487	*P. morganii* M206	Fi^+(F, R386) Su			120
IncY						
E. coli *Shigella* spp. *Salmonella*	P1	*E. coli* Lisbon	Tra^- Phi(P1)^i Hsp(P1)		m.w. = 59×10^6; prophage	122

Incompatibility and host range[a]	Plasmid	Original host	Phenotype	Genotype	Additional data	Reference
IncY *(continued)*	P1Cm		Tra^- Phi(P1)[i] Hsp(P1) Cm	Tn*9*	m.w. = 60×10^6; prophage	162
	P15B	*E. coli* 15	Tra^- Hsp (P1)		m.w. = 75.2×10^6; 1 copy/chromosome; cryptic prophage	8, 134, 203
	P7=ϕAmp	*E. coli* H	Tra^- Ap Phi(P7)[i]	bla^+(TEM-1)	m.w. = 65×10^6; prophage	122, 189, 249
	ϕAmp-P1Cm		Tra^- Ap Cm	Tn*9*	m.w. = 61×10^6; prophage recombinant	122
Inc9 =Com9						
	RIP71	*E. coli*	Fi^-(F) Ap Cm Sm Su Tc			209, 235
	R71-1		Ap Cm Sm Su			235
	R71-2		Cm Sm Su			235
Unclassified						
	R2	*S. typhimurium*	Sm Su Tc			192
	R800	*S. flexneri*	Cm Sm Su Tc			72
	RSF1030		Tra^- Ap	bla^+(TEM-1)	m.w. = 5.5×10^6, 48.5% G+C; carries sequence homologous to Tn*1* Tn*2* Tn*3* other Tn-Ap	125

c. Plasmids Studied in *Pseudomonas aeruginosa* and Other Pseudomonads[j]*

Compiled by G. A. Jacoby, Massachusetts General Hospital, Boston; and J. A. Shapiro, University of Chicago

Incompatibility and host range[a]	Plasmid	Original host	Phenotype[k]	Genotype	Additional data[l]	Reference†
IncP-1[g]						
Acinetobacter calcoaceticus *Azotobacter vinlandii* *E. coli* *Neisseria perflava* *Proteus* spp. *Pseudomonas* spp. *Rhodopseudomonas spheroides* *Rhodospirillum rubrum* *Rhizobium* spp. *Salmonella* spp. *Shigella* spp. *V. cholera* *Klebsiella* spp. *Serratia marcescens*	RP1 (=R18 =R1822)	*P. aeruginosa*	Tra^+ Dps (PRR1 PRD1 Pf3 PR3 PR4) Fi^-(FP2) Phi (G101) Cb Km Nm Tc	bla^+(TEM-2) Tn401(Cb) $aphA^+$	m.w. = 40×10^6, 60% G+C; 1 *Eco*RI restriction site; pyocin AP41 tolerance; depressed transfer into phage B3 lysogen; transducible by phage F116L; mobilizes chromosome of PAT strains	16, 40, 62, 90, 94, 116, 127, 129, 135, 150, 175, 177, 179, 184, 214, 215, 216, 245, 256
	RP1*amp*1		Tra^+ Km Nm Tc	*bla* $aphA^+$	mutant of RP1	61
	RP1*amp*1 *irp*1		Tra^+ Km Nm Tc	*bla* $aphA^+$	mutant of RP1*amp*1	61
	R18-18		Tra^+ Km Nm Tc	*bla* $aphA^+$	derivative of RP1 m.w. = 38×10^6, 60% G+C	258
	R18-70		Tra^+ Cb Tc	bla^+ *aph*A(sus)	mutant of R18 (=RP1)	277
	pLM2		Tra^+ Km Nm	*bla*(am) *tet*(am) $aphA^+$	double mutant of RP1	195
	pMRP19		Tra^+ Km Nm Tc	*bla*(am) $aphA^+$	mutant of RP1	241

*All footnotes for Tables b, c, d, e, and f are given at the end of Table f.
†Reference numbers refer to entries listed in the Bibliography, Appendix B3.

Incompatibility and host range[a]	Plasmid	Original host	Phenotype	Genotype	Additional data	Reference
IncP-1						
(continued)	RP1-S2		Tra^- Cb Tc	*bla*$^+$ *aph*A	derived from RP1 by P22 transduction; m.w. = 23 $\times 10^6$	246
	pAC1		Tra^+ Cb	*bla*$^+$ *aph*A(del) *tet*(del)	derived from RP1 by transformation	39
	pAC2		Tra^- Cb Tc	*bla*$^+$ *aph*A(del)	derived from RP1 by transformation	39
	RP2 (=R3425)	*P. aeruginosa*	Tra^+ Cb Km Nm Tc	*bla*$^+$(TEM)		90, 127, 263
	RP4^m	*P. aeruginosa*	Tra^+ Dps (PRR1 PRD1 Pf3 PR3 PR4) Fi^-(FP2) Phi (G101) Cb Km Nm Tc	*bla*$^+$(TEM-2) Tn*1* *aph*A$^+$	m.w. = 36 $\times 10^6$, 58–60% G+C; 1–3 copies/chromosome in *E. coli*, 1 *Eco*RI restriction site; pyocin AP41 tolerance (see Figs. B2.4 and B2.5)	15, 20, 24, 68, 74, 85, 113, 125, 127, 142, 150, 175, 189, 219, 232, 254, 268
	R30	*P. aeruginosa*	Tra^+ Dps (PRR1 PRD1 Pf3 PR3 PR4) Fi^-(FP2) Phi (G101) Cb Km Nm Tc	*bla*$^+$ *aph*A$^+$	pyocin AP41 tolerance; depressed transfer into B3 lysogen; transducible by phage F116L; can mobil-	40, 127, 129, 150, 175, 177, 254, 255

				ize chromosome of PAT strains	
R68 (=R6886)	*P. aeruginosa*	Tra$^+$ Dps (PRR1 PRD1 Pf3 PR3 PR4) Fi$^-$(FP2) Phi (G101) Cb Km Nm Tc	*bla*$^+$ *aph*A$^+$	pyocin AP41 tolerance; normal transfer into B3 lysogens; mobilizes chromosome of PAT strains	20, 40, 127, 129, 150, 175, 177, 254, 256
R68-45		Tra$^+$ Dps (PRR1) Cb Km Tc	*bla*$^+$ *aph*A$^+$	derivative of R68 selected for chromosome mobilization in *P. aeruginosa* PAO; also mobilizes chromosome in *P. aeruginosa* PAC and in *P. putida*	101
R88	*P. aeruginosa*	Tra$^+$ Dps (PRR1 Pf3 PR3 PR4) Phi (G101) Cb Km Tc	*bla*$^+$ *aph*A$^+$	decreased transfer into phage B3 lysogen	40, 129, 175, 177, 254
RP638	*P. aeruginosa*	Tra$^+$ Dps (PRR1 PRD1 Pf3 PR3 PR4) Fi$^-$(FP2) Phi (G101) Cb Km Tc	*bla*$^+$ *aph*A$^+$	m.w. = 40×10^6, 60% G+C; pyocin AP41 tolerance	150, 271
R26	*P. aeruginosa*	Tra$^+$ Cb Cm Gm Hg Km Nm Sm Su Tc		m.w. = 52×10^6, 60% G+C; pyocin AP41 tolerance	258
R1033	*P. aeruginosa*	Tra$^+$ Dps (PRR1 PRD1 Pf3 PR3 PR4) Fi$^-$(FP2) Phi (G101) Cb Cm Gm Hg Km Sm Su Tc	*bla*$^+$(TEM-1) *aac*C$^+$ *aph*A$^+$ Tn404(Cb)	m.w. = 45×10^6	150, 189, 241, 252

Incompatibility and host range[a]	Plasmid	Original host	Phenotype	Genotype	Additional data	Reference
IncP-1 *(continued)*	R74 (=R7475)	*K. aerogenes*	Tra^+ Dps (PRR1 Pf3 PR3 PR4) Phi (G101) Cb Km Tc		depressed transfer into phage B3 lysogen	40, 175, 177, 254
	RK2	*K. aerogenes*	Tra^+ Cb Km Tc	*bla*$^+$(TEM)	m.w. = 40–48 × 10^6, 60% G+C; 1 *Eco*RI restriction site (see Fig. B2.6)	135, 190
	R527	*S. marcescens*	Tra^+ Cb Cm Gm Hg Km Nm Sm Su Tc		m.w. = 54 × 10^6, 60% G+C; pyocin AP41 tolerance	258
	R702	*P. mirabilis*	Tra^+ Dps (PRD1 Pf3 PR3 PR4) Hg Km Sm Su Tc		m.w. = 46 × 10^6	107, 151, 237
	R751	*K. aerogenes*	Tra^+ Dps (PRR1 PRD1 Pf3 PR3 PR4) Tp	Tn402(Tp)	m.w. = 30–35 × 10^6	17, 151, 153, 241, 252
	R839	*S. marcescens*	Tra^+ Dps (PRR1 PRD1 Pf3 PR3 PR4) Cb Km Sm Su Tc Hg	*bla*$^+$(TEM-1)	m.w. = 46 × 10^6	108, 119, 151, 189
	RP1-pMG1		Tra^+ Dps (PRR1 PRD1 Pf3 PR3 PR4) Cb Km Sm Tc	*bla*$^+$ *aph*A$^+$ *aph*C$^+$	m.w. = 42 × 10^6; recombinant between RP1 and pMG1 (IncP-2)	146, 152
	pMG4		Tra^+ Km Sm Tc	*bla aph*A$^+$ *aph*C$^+$	Cb^s mutant of RP1-pMG1	150

RP4-pMG2	Tra$^+$ Dps (PRR1 PRD1 Pf3 PR3 PR4) Cb Gm Km Sm Su Tc Hg		m.w. = 48×10^6; recombinant between RP4 and pMG2 (IncP-2)	152
RP4-RPL11	Tra$^+$ Dps (PRR1 PRD1 PR4) Cb Cm Gm Km Sm Su Tc		m.w. = 55×10^6; recombinant between RP4 and RPL11 (IncP-2)	152
RP4-Tn*C1*, RP4-Tn*C2*	Tra$^+$ Cb Km Sm Tc Tp	Tn*7*	m.w. = $44–45 \times 10^6$; transposition of Tn*7* into RP4; 2 copies/chromosome (in *E. coli*); 2 *Eco*RI restriction sites (see Fig. B2.4)	15
RP41	Tra$^+$ Dps (PRR1) Cb Km Tc His$^+$ Nif$^+$ Gnd$^+$ ShiA$^+$		m.w. = 59×10^6; recombinant between RP4 and FN68	79
RK2*trp*2 (=pRM2)	Tra$^+$ Dps (PRR1 PRD1) Cb Km Tc Trp$^+$		m.w. = 58×10^6; 1 *Eco*RI restriction site; constructed from RK2 and pVH103	190
R751-SU2	Tra$^+$ Dps (PRR1) Cb Km Sm Su Tp		m.w. = 69×10^6; recombinant between R751 and R2 (=RSU2)	115

Incompatibility and host range[a]	Plasmid	Original host	Phenotype	Genotype	Additional data	Reference
IncP-1 *(continued)*	R751-pMG2#1		Tra^+ Dps (PRR1 PRD1 PR4) Gm		m.w. = 45×10^6; recombinant between R751 and pMG2 (IncP-2)	152
	R751-pMG2#2		Tra^+ Dps (PRR1 PRD1 PR4) Gm Sm Su		m.w. = 62×10^6; recombinant between R751 and pMG2 (IncP-2)	152
	R751-pMG2#3		Tra^+ Dps (PRR1 PRD1 PR4) Sm Tp Hg		m.w. = 46×10^6; recombinant between R751 and pMG2 (IncP-2)	152
	R751-pMG5		Tra^+ Dps (PRR1 PRD1 PR4) Km Su Tm Hg		m.w. = 46×10^6; recombinant between R751 and pMG5 (IncP-2)	152
IncP-2						
Pseudomonas spp. *Flavobacter meningo-septicum*	R931 (=pLB931)	*P. aeruginosa*[n]	Tra^+ Fi^+(RP1) Fi^-(FP2) Phi (B3 B39 E79 F116L G101 M6 PB1) Sm Tc Hg Uv	*aph*C^+	59% G+C	22, 23, 24, 129, 146, 150, 175, 176, 240, 269
Not *E. coli* or other enterobacteria	R38 (=R38-72)	*P. aeruginosa*	Tra^+ Fi^+(RP1) Fi^-(FP2) Phi (B3 B39 D3 E79 G101 M6 PB1) Sm Su Tc Hg Pma			129, 150, 155, 175

R39 (=R39-72)	*P. aeruginosa*	Tra^+ Fi^+(RPI) Fi^-(FP2) Phi (B3 B39 D3 E79 G101 M6 PB1) Sm Su Tc Hg Pma			129, 150, 155, 175
R130° (=pLB130 =pMG2)	*P. aeruginosa*	Tra^+ Fi^-(RPI, FP2) Phi (B3 B39 D3 E79 F116L G101 M6 PB1) Gm Sm Su Hg Uv	*aac*C^+ *mer*$^+$	59% G+C; inhibits pyocin production	22, 24, 25, 129, 146, 150, 175, 240, 254
Rms159	*P. aeruginosa*	Tra^+ Fi^+(RPI) Fi^-(FP2) Phi (B3 B39 G101 M6 PB1) Cm Sm Tc Hg Pma			150
R209	*P. aeruginosa*	Tra^+ Gm Sm Su	*aac*C^+		22, 24, 25, 129
R290	*P. aeruginosa*	Tra^+ Gm Sm Su	*aac*C^+		22
R442	*P. aeruginosa*	Tra^+ Gm Sm Su	*aac*C^+	59% G+C	22, 240
R454	*P. aeruginosa*	Tra^+ Gm Sm Su	*aac*C^+	59% G+C	22
R546	*P. aeruginosa*	Tra^+ Gm Sm Su	*aac*C^+		22
R680	*P. aeruginosa*	Tra^+ Gm Sm Su	*aac*C^+		22
R3108	*P. aeruginosa*	Tra^+ Fi^+(RPI) Fi^-(FP2) Phi (B3 B39 D3 G101) Sm Su Tc Hg Pma		57% G+C	24, 129, 146, 150
pMG1	*P. aeruginosa*	Tra^+ Fi^-(RP1, FP2) Phi (B3 B39 D3 E79 G101 M6 PB1) Gm Sm Su Hg Uv	*aac*C^+ *aph*C^+ *mer*$^+$	inhibits pyocin production	129, 146, 150, 161, 196
pMG3		Tra^+ Phi (B3 B39 D3 E79 G101 M6 PB1) Sm Hg Uv Gm^s Su^s	*aac*C	derived from pMG1 by NTG mutagenesis	146

Incompatibility and host range[a]	Plasmid	Original host	Phenotype	Genotype	Additional data	Reference
IncP-2 *(continued)*	pMG5	*P. aeruginosa*	Tra^+ Fi^+(RP1) Fi^-(FP2) Phi (B3 B39 D3 E79 G101 M6 PB1) Ak Bt Km Su Tm Hg Pma	*aac*A^+		147, 149, 150, 154
	RPL11	*P. aeruginosa*	Tra^+ Fi^-(RP1, FP2) Phi (B3) Cb Cm Gm Sm Su Tc Hg Pma		58% G+C	150, 169, 170, 171
	Cam	*P. putida*	Tra^+ Fi^+(RP1) Fi^-(FP2) Phi (B3 B39 D3 E79 G101 M6 PB1) Cam Uv		m.w. = 92×10^6 [p]; after UV treatment can mobilize *P. putida* chromosome	31, 33, 45, 99, 148, 150, 217, 224, 239
	OCT	*P. putida* (*P. oleovorans*)	Tra^- Alk	*alk*A^+ *alk*B^+ *alk*C^+	m.w. = 27×10^6 [q]	31, 32, 33, 35, 38, 45, 96, 99, 198, 217
	CAM-R931		Tra^+ Sm Tc Cam Hg		recombinant between CAM and R931	151
	CAM-R3108		Tra^+ Sm Su Tc Hg Pma Cam		recombinant between CAM and R3108	151
	CAM-pMG1		Tra^+ Gm Sm Su Hg Cam		recombinant between CAM and pMG1	151

CAM-pMG2	Tra^+ Gm Sm Su Hg Cam		recombinant between CAM and pMG2	151
CAM-OCT	Tra^+ Fi^+(RP1) Fi^-(FP2) Phi (B3 B39 G101 PB1) Cam Alk Uv	*alk*A^+ *alk*B^+ *alk*C^+	recombinant between CAM and OCT	31, 96, 150
CAM-OCT-pMG2	Tra^+ Gm Sm Su Hg Cam Alk		recombinant between CAM-OCT and pMG2	151
CAM-OCT::Tn401	Tra^+ Cb Cam Alk	Tn401(Cb)	transposition of Tn401 from RP1 into CAM-OCT	16
CAM-OCT-*cam*352::Tn401	Tra^+ Cb Cam^- Alk	*cam*352::Tn401 (Cb)	transposition of Tn401 from RP1 into the camphor genes of CAM-OCT	16
pMF581	Tra^+ Phi (B3 B39 D3 E79 F116 G101 PB1 M6) Sm Tc Cam Alk Hg Uv		recombinant of CAM-R931 and OCT	86
pMF582	Tra^- Tc Alk		recombinant of CAM-R931 and OCT	86
pMF583	Tra^+ Sm Tc Alk		recombinant of CAM-R931 and OCT	86
pMF584, 585	Tra^+ Phi (B3 B39 D3 F116 G101) Sm Su Tc Hg Pma Uv^- Cam Alk		recombinant of CAM-R3108 and OCT	86

Incompatibility and host range[a]	Plasmid	Original host	Phenotype	Genotype	Additional data	Reference
IncP-2 *(continued)*	pSB724		Tc Alk		recombinant between R931 and OCT	18
	pSB725		Tra^+ Tc Sm Hg Alk		recombinant between R931 and OCT	18
	pSB729		Tra^+ Tc Sm Hg Su Alk		recombinant between R3108 and OCT	18
	pf*dm*		Tra^+ Mdl		derived by transduction of *mdl* genes with phage pf16; mobilizes chromosome	36, 37, 224
IncP-3[d]						
Pseudomonas spp. *E. coli*	R40a	*P. aeruginosa*	Tra^+ Cb Km Su Hg	bla^+(TEM-1)		29, 67, 129, 189, 237
	RIP64 (=pIP64)	*P. aeruginosa*	Tra^+ Fi^+(RP1, FP2) Cb Cm Gm Su Tm Hg			129, 146, 149, 150, 281, 282, 283
	R151 (=pLB151)	*P. aeruginosa*	Tra^+ Fi^-(RP1) Fi^+(FP2) Phi (B39) Cb Gm Sm Su Tm	bla^+ $aadA^+$ $aadB^+$	not observed to transfer to *E. coli*	22, 151, 240

	pPK183-1	*P. aeruginosa*	Tra^+ Cb Cm Gm Km Tm Hg	bla^+(TEM)	see ref. 166 for additional plasmids and derivatives	166
	pPK192-1	*P. aeruginosa*	Tra^+ Cb Cm Km Nm Sm Sp Su Tc Hg	bla^+(TEM)		166
	pPK207	*P. aeruginosa*	Tra^+ Cb Gm Km Sm Sp Su Tm Hg	bla^+(TEM)		166
	RIP55	*K. pneumoniae*	Tra^+ Fi^+(RP1, FP2) Cb Cm Km Gm Su Tm Hg	bla^+(OXA-3) $aadB^+$		22, 119, 148, 151, 189, 240, 281, 282, 283
IncP-4[h]						
Pseudomonas spp. *E. coli*	R679	*P. aeruginosa*	Tra^- Fi^-(RP1, FP2) Sm Su	$aphC^+$	60% G+C	23, 24, 129, 150, 175, 254, 270
	R1162	*P. aeruginosa*	Tra^- Fi^-(RP1, FP2) Sm Su	$aphC^+$	m.w. = 5.5×10^6; 11–12 copies/chromosome in *E coli*; 1 *Eco*RI, 1*Hpa*I restriction site	14, 23, 24, 129, 150, 241, 270
	R5265 (=pLB 5265)	*P. aeruginosa*	Tra^- Fi^-(RP1, FP2) Sm Su		59% G+C	24, 129, 150, 240
IncP-5						
P. aeruginosa Not *E. coli*	Rms163	*P. aeruginosa*	Tra^+ Fi^-(RP1, FP2) Cm Su Tc			150

Incompatibility and host range[a]	Plasmid	Original host	Phenotype	Genotype	Additional data	Reference
IncP-6						
P. aeruginosa Not *E. coli*	Rms149[r] (=R149)	*P. aeruginosa*	Tra^+ Fi^-(RP1, FP2) Phi (F116) Cb Gm Sm Sp Su	*bla*$^+$ *aac*C^+	no Sm inactivating enzyme found	150, 160, 172, 230
	Rms149-1		Cb Su		spontaneous derivative of Rms 149	230
IncP-7						
P. aeruginosa Not *E. coli*	Rms148 (=R148)	*P. aeruginosa*	Tra^+ Fi^-(RP1) Fi^+(FP2) Phi (B39, C5) Sm	*aph*C^+		150, 160, 172, 230
IncP-8						
P. aeruginosa Not *P. putida* Not *E. coli*	FP2 (=FP)	*P. aeruginosa*	Tra^+ Fi^-(RP1) Hg Pma	*mer*$^+$	m.w. = 59×10^6, 58% G+C; 1–2 copies/chromosome; mobilizes chromosome of *P. aeruginosa* PAT, PAO, PAC	32, 126, 128, 129, 150, 175, 176, 183, 220, 254, 255, 262
	FP2-2 (=FP*)		Tra^+		extended fertility mutant of FP2	253, 257
	FP2-3, 6, 8, 11		Tra^-		Tra^- mutants of FP2(FPd) retaining exclusion and precipitation characteristics of FP2	253

	FP2-9		Tra^+		mutant of FP2 conferring altered cell-surface properties	253
	pRO271		Tra^+ Cb Hg Pma	bla^+(TEM-2) Tn401(Cb) mer^+	transposition of Tn401 from RP1 into FP2	213
Unknown Incompatibility						
Transmissible to *E. coli*	RP1-1[s] (=R18-1)	*P. aeruginosa*	Tra^+ Fi^-(RP1, FP2) Phi (B39) Cb	bla^+(TEM) Tn 401(Cb)	depressed transfer into B3 lysogen; F116L transducible	17, 40, 127, 136, 150, 175, 177, 254
	R19	*P. aeruginosa*	Tra^+ Cb		expresses Km Tc on transfer to *E. coli*; normal transfer into B3 lysogen	41, 129, 177, 254
	R19-4		Tra^+ Dps (PRD1) Cb		derepressed mutant of R19	41
	R91 (=RP9, R9169)	*P. aeruginosa*[t]	Tra^+ Fi^-(RP1) Fi^+(FP2) Cb	bla^+(TEM)	m.w. = $35–48 \times 10^6$; expresses Km Tc on transfer to *E. coli*; F116L transducible	40, 127, 129, 135, 150, 175, 254, 256
	R91-5		Tra^+ Dps (PRD1 PR3 PR4) Fi^+(FP2) Cb		derepressed mutant of R91	41, 151, 254

Incompatibility and host range[a]	Plasmid	Original host	Phenotype	Genotype	Additional data	Reference
Unknown Incompatibility						
Not transmissible to *E. coli*	R40c	*P. aeruginosa*	Tra^+ Fi^-(RP1, FP2) Cb			150
	R2 (=R2-72, =RSU2)	*P. aeruginosa*	Tra^+ Fi^+(RP1, FP2) Cb Sm Su Uv	bla^+(TEM-1)		115, 129, 150, 156, 175, 189, 254
	R716	*P. aeruginosa*	Tra^+ Fi^+(RP1) Phi (B39) Sm Hg			24, 151, 176, 240, 254
	RM134	*P. aeruginosa*	Tra^+ Cb Cm Sm Su Tc			139
	RM135	*P. aeruginosa*	Tra^+ Cb Cm Sm Su Tc			139
	Rms146	*P. aeruginosa*	Tra^+ Gm Km Sm Su Tc	*aac*C^+ *aad*A^+ *aph*A^+ *aph*C^+		160, 172, 230
	kR102	*P. aeruginosa*	Tra^+ Cm Km Sm Su Tc		no inactivating enzymes found for Cm, Km, or Sm	164, 210
	pPK108	*P. aeruginosa*	Tra^+ Gm Km Nm Su Tc Tm Hg		see ref. 166 for additional plasmids not transmissible to *E. coli*	166
	FP5	*P. aeruginosa*	Tra^+ Hg		mobilizes *Pseudomonas* chromosome	129, 150, 175, 188, 254

	FP8	*P. aeruginosa*	Tra^+ Hg		mobilizes *Pseudomonas* chromosome; see ref. 176 for additional FP factors	176, 254
Unknown Incompatibility						
	FP39	*P. aeruginosa*	Tra^+ Leu^+		m.w. = 55×10^6, 60% G+C; 1–2 copies/chromosome; not transmissible to *P. putida*; mobilizes *P. aeruginosa* chromosome	32, 129, 149, 150, 175, 220, 221, 240, 254
	RP8	*P. aeruginosa*	Tra^+ Fi^-(RP1, FP2) Cb			129, 150, 175, 254
	R503	*P. aeruginosa*	Tra^+ Phi (B39) Sm Hg			24, 151
	Rms139	*P. aeruginosa*	Tra^+ Cb	bla^+		233
	Rms147	*P. aeruginosa*	Tra^+ Gm Km Sm Su Tc	$aphA^+$		172, 230
	Rms 165	*P. aeruginosa*	Tra^+ Cb	bla^+		233
	Rms167	*P. aeruginosa*	Tra^+ Gm Km Sm Su	$aacA^+$		155
	Rms168	*P. aeruginosa*	Tra^+ Cm Km Sm Su Tc	$aacA^+$		155
	kR61	*P. aeruginosa*	Tra^+ Cm Sm Su Tc	$aphC^+$		163
	kR79	*P. aeruginosa*	Tra^+ Cm Sm Su		no inactivating enzyme for Sm detected	163
	kR94	*P. aeruginosa*	Tra^+ Km Sm	$aphC^+$		163
	kR97	*P. aeruginosa*	Tra^+ Km			163

Incompatibility and host range[a]	Plasmid	Original host	Phenotype	Genotype	Additional data	Reference
Unknown Incompatibility *(continued)*	Hg-R (=pVS1)	*P. aeruginosa*	Tra^- Hg Su	Tn501(Hg)	m.w. = 19×10^6; mobilizable by R30 and other P-1 plasmids	255, 258
	NAH	*P. putida*	Tra^+ Nah		transducible by phage pf16	45, 81, 99
	SAL	*P. putida*	Tra^+ Fi^+(RP1) Fi^-(FP2) Sal		m.w. = 51×10^{6u}; not transmissible to *E. coli*; transducible by phages pf16, F116; Tra^- after transfer to *P. aeruginosa*	16, 30, 32, 33, 99, 150, 217, 218
	SAL:: Tn401		Tra^+ Cb Sal	Tn401(Cb)	m.w. = $55–57 \times 10^6$; transposition of Tn401 from RP1 into SAL; will not replicate stably in *E. coli*	16
	TOL (=BEN/ TOL =XAL)	*P. putida* (*arvilla*)	Tra^+ Fi^-(RP1, FP2) Xal		m.w. = 63×10^6; transducible by phage pf16; Tra^- after transfer to *P. aeruginosa*; mobilizes *P. putida* chromosome	33, 279, 280, 284, 285

	XYL	*Pseudomonas* Pxy	Tra^- Xal	m.w. = 11×10^{6} [v] transducible by phage pf16	33, 34
	MER	*P. putida* (*P. oleovorans*)	Tra^+ Hg Pma		32, 35
	K (=P)	*P. putida* (*P. oleovorans*)	Tra^+	m.w. = $64–70 \times 10^6$; mobilizes OCT and the *P. putida* chromosome	32, 35, 217
"Cryptic" Plasmids Observed in *Pseudomonas* Strains					
		P. aeruginosa 31		52% G+C	87
		P. aeruginosa 31		59% G+C	87
		P. aeruginosa 34		46% G+C	87
		P. aeruginosa 34		53% G+C	87
		P. aeruginosa PAO		m.w. = 1.7×10^6; 15 copies/chromosome	220
		P. aeruginosa PAO		m.w. = 5.8×10^6; 7 copies/chromosome	220
		P. aeruginosa PAO		m.w. = 9.5×10^6; 1 copy/chromosome	220
		P. aeruginosa 931		m.w. = 1×10^6	24, 240
	pLB8803	*P. aeruginosa* 8803		m.w. = 20×10^6	240

Incompatibility and host range[a]	Plasmid	Original host	Phenotype	Genotype	Additional data	Reference
"Cryptic" Plasmids *(continued)*		*P. putida*			m.w. = 6×10^6	33
		P. stutzeri 226			60% G+C	187
		P. stutzeri 227			60% G+C	187
		Plasmids of Known Incompatibility Transmissible from Enteric Bacteria to P. aeruginosa				
IncN	R46 (=TP120)	*S. typhimurium*	Tra^+ Fi^-(FP2) Cb Sm Sp Su Tc	*bla*$^+$(OXA-2)	m.w. = 33×10^6	17, 105, 148, 150, 189
	RIP113 (=pIP113)	*S. panama*	Tra^+ Tc			22, 29
IncW	Sa	*Shigella*	Tra^+ Dps (PRD1 PR3 PR4) Fi^+(RP1) Fi^-(FP2) Cm Gm Km Sm Sp Su Tm		m.w. = $23–25 \times 10^6$, 62% G+C	85, 105, 125, 148, 149, 150, 240
	R7K	*P. rettgeri*	Tra^+ Dps (PRD1 PR3 PR4) Fi^-(RP1, FP2) Cb Sm	*bla*$^+$(TEM-1)	m.w. = $20–22 \times 10^6$, 62% G+C	52, 85, 125, 148, 150, 189
	R388	*E. coli*	Tra^+ Dps (PRD1) Fi^+(RP1) Fi^-(FP2) Su Tp		m.w. = 21×10^6, 62% G+C	69, 85, 125, 148, 150

d. Plasmids of *Staphylococcus aureus*

Compiled by R. P. Novick, Public Health Research Institute of the City of New York; S. Cohen, Michael Reese Hospital and Medical Center, Chicago; L. Yamamoto, University of Chicago; and J. A. Shapiro, University of Chicago

Incompatibility and host range[a]*	Plasmid	Original host	Phenotype	Genotype	Additional data	Reference†
Inc1						
S. aureus	pUB108= mir-r	*S. aureus*	Tra^- Asa Hg Cd Pc^-		m.w. = 35×10^6; 4–7 copies/ chromosome	44, 178, 205
	pI258=γ; PI_{258}	*S. aureus*	Tra^- Pc Asa Asi Hg Cd Pb Bi Em Tc^- Sm^- Cm^- Km^- Nm^-		m.w. = 18.4×10^6, ~35% G+C; 8–12 copies/ chromosome; 4 *Eco*RI sites, 1 *Bam*I site, 2 *Sma* sites; RC^-, Exp	42, 201, 202, 207, 208, 222, 229, 278
	pI524=α =PI_{524}	*S. aureus*	Tra^- Pc Asa Asi Hg Cd Pb Bi Em^- Tc^- Sm^- Cm^- Km^- Nm^-		m.w. = 20×10^6, ~35% G+C; 8–12 copies/ chromosome; RC^-	200, 201, 207, 208, 222, 250
	pRN1036= PI_{1036}	*S. aureus*	Pc Asa Asi Hg Cd Pb Bi Em^-			222
	pRN1042= PI_{1042}	*S. aureus*	Pc Asa Asi Hg Cd Pb Bi Em^-			222
	pRN1071= PI_{1071}	*S. aureus*	Pc Cd Pb Asa^- Asi^- Hg^- Bi^- Em^-			222

*All footnotes for Tables b, c, d, e, and f are given at the end of Table f.
†Reference numbers refer to entries listed in the Bibliography, Appendix B3.

Incompatibility and host range[a]	Plasmid	Original host	Phenotype	Genotype	Additional data	Reference
Inc1 *(continued)*	pRN3719= PI_{3719}	*S. aureus*	Pc Asa Hg Cd Pb Bi Asi^- Em^-			222
	pRN3721= PI_{3721}	*S. aureus*	Pc Asa Hg Cd Pb Bi			222
	pRN3743 PI_{3743}	*S. aureus*	Pc Asa Asi Hg Cd Pb Bi^- Em^-			222
	pRN3761= PI_{3761}	*S. aureus*	Pc Cd Pb Asa^- Asi^- Hg^- Bi^- Em^-			222
	pRN3773= PI_{3773}	*S. aureus*	Pc Cd Pb Asa^- Asi^- Hg^- Bi^- Em^-			222
	pRN55C1 PI_{55c1}	*S: aureus*	Pc Hg Cd Pb Asa^- Asi^- Em^-			222
	pRN6187= PI_{6187}	*S. aureus*	Pc Asa Asi Cd Pb Bi Hg^- Em^-			222
	pRN6193= PI_{6193}	*S. aureus*	Pc Asa Asi Hg Cd Pb Bi Em^-			222
	pRN9033= PI_{9033}	*S. aureus*	Pc Asa Asi Hg Cd Pb Bi Em^-			222
	pRN9789= PI_{9789}	*S. aureus*	Asa Asi Hg Cd Pb Pc^- Em^-			222
	pRN13313= PI_{13313}	*S. aureus*	Pc Asi Cd Pb Asa^- Hg^- Bi^- Em^-			222

	pRN13333= PI_{13333}	*S. aureus*	Pc Asa Asi Hg Cd Pb Bi Em^-		222
	pRN13371= PI_{13371}	*S. aureus*	Pc Asa Asi Hg Cd Pb Bi Em^-		222
Inc2					
S. aureus	pII147=β= PII_{147}	*S. aureus*	Tra^- Pc Asa Hg Cd Pb Bi^{HS} Asi^- Em^- Tc^- Sm^- Cm^- Km^- Nm^-	m.w. = 21×10^6, ~35% G+C; 8–12 copies/chromosome; 7 *Eco*RI sites, 1 *Bam*I site; RC^-, Exp	97, 201, 204, 206, 207, 208, 222, 228 229, 250
	pRN1008= PII_{1008}	*S. aureus*	Asa Asi Hg Cd Pb Bi^{HS} Em^-		222
	pRN3755= PII_{3755}	*S. aureus*	Asa Asi Cd Pb Bi^{HS} Hg^- Em^-		222
	pRN3772= PII_{3772}	*S. aureus*	Pc Asa Asi Cd Pb Bi^{HS} Hg^- Em^-		222
	pRN3804= PII_{3804}	*S. aureus*	Pc Cd Pb Bi^{HS} Asa^- Asi^- Hg^- Em^-		222
	pRN6907= PII_{6907}	*S. aureus*	Pc Asa Asi Cd Pb Bi^{HS} Hg^- Em^-		222
	pRN10496 $=PII_{10496}$	*S. aureus*	Pc Asa Asi Hg Cd Pb Bi^{HS} Em^-		222
	pRN17810 $=PII_{17810}$	*S. aureus*	Asa Cd Pb Bi^{HS} Asi^- Hg^- Em^-		222

Incompatibility and host range[a]	Plasmid	Original host	Phenotype	Genotype	Additional data	Reference
Inc3						
S. aureus	pT127	*S. aureus*	Tc Pc^- Asa^- Asi^- Hg^- Cd^- Pb^- Bi^- Em^- Sm^- Cm^- Km^- Nm^-		m.w. = 2.7×10^6; 80–100 copies/chromosome; no *Eco*RI restriction sites; RC^-, Exp	182, 205, 228
Inc4						
S. aureus	pC221= C22.1	*S. aureus*	Tra^- Cm Pc^- Asa^- Asi^- Hg^- Cd^- Pb^- Bi^- Em^- Tc^- Sm^- Km^- Nm^-		m.w. = 30×10^6, 35% G+C; 90–100 copies/chromosome; RC^+, Exp	203, 204, 206, 228
Inc5						
S. aureus	pS177	*S. aureus*	Tra^- Sm Pc^- Asa^- Asi^- Hg^- Cd^- Pb^- Bi^- Em^- Tc^- Cm^- Km^- Nm^-		m.w. = 2.7×10^6, 35% G+C; 80–100 copies/chromosome; RC^+, Exp	203, 204, 228
Inc6						
S. aureus	pK545	*S. aureus*	Tra^- Km-Nm Pc^- Cd^- Em^- Tc^- Sm^- Cm^-		m.w. = 15×10^6; 2–3 copies/chromosome; 1 *Eco*RI site; RC^-, Res	228

Inc7					
S. aureus	pUB101= FAR-4	*S. aureus*	Tra$^-$ Pc Cd Fa Em$^-$ Tc$^-$ Sm$^-$ Cm$^-$ Km$^-$ Nm$^-$	m.w. = 14.6 × 10^6; 8–12 copies/ chromosome; RC$^-$	44, 205
Undetermined					
S. aureus	pUB109= str-r	*S. aureus*	Tra$^-$ Sm Pc$^-$ Asa$^-$ Hg$^-$ Cd$^-$ Em$^-$ Tc$^-$ Cm$^-$ Km$^-$ Nm$^-$	m.w. = 2.7 × 10^6; RC$^+$	44, 178, 204
	pUB110= neo-r	*S. aureus*	Tra$^-$ Km-Nm Pc$^-$ Hg$^-$ Cd$^-$ Em$^-$ Tc$^-$ Sm$^-$ Cm$^-$	m.w. = 5.9 × 10^6; 9–16 copies/ chromosome	44, 178
	pUB111= tet-r	*S. aureus*	Tra$^-$ Tc Pc$^-$ Hg$^-$ Cd$^-$ Em$^-$ Sm$^-$ Cm$^-$ Km$^-$ Nm$^-$	m.w. = 2.7 × 10^6; RC$^-$	44, 178, 204
	pUB112= chm-r	*S. aureus*	Tra$^-$ Cm Pc$^-$ Hg$^-$ Cd$^-$ Em$^-$ Tc$^-$ Sm$^-$ Km$^-$ Nm$^-$	m.w. = 3.0 × 10^6; 23 copies/ chromosome; RC$^-$	44, 178, 204
	pS169	*S. aureus*	Sm Pc$^-$ Em$^-$ Tc$^-$ Cm$^-$ Km$^-$ Nm$^-$	m.w. = 2.7 × 10^6; RC$^+$	204
	pT169= T169	*S. aureus*	Rep(ts) Tc Pc$^-$ Asa$^-$ Asi$^-$ Hg$^-$ Cd$^-$ Pb$^-$ Bi$^-$ Em$^-$ Sm$^-$ Cm$^-$ Km$^-$ Nm$^-$	m.w. = 2.7 × 10^6; 15–30 copies/ chromosome; no *Eco*RI sites; RC$^-$, Exp	204, 205, 206, 228
	pT181	*S. aureus*	Tc Em$^-$ Sm$^-$ Cm$^-$	m.w. = 2.7 × 10^6; RC$^-$	137, 204
	pC194	*S. aureus*	Cm Em$^-$ Tc$^-$ Sm$^-$	m.w. = 1.8 × 10^6; RC$^-$	137, 204

Incompatibility and host range[a]	Plasmid	Original host	Phenotype	Genotype	Additional data	Reference
Undetermined *(continued)*	pE194	*S. aureus*	Em Tc^- Sm^- Cm^-		m.w. = 1.8×10^6; RC^-	138, 204
	pS194	*S. aureus*	Sm Tc^- Em^- Cm^-		m.w. = 2.7×10^6; RC^+	137, 204
	pC223	*S. aureus*	Cm Em^- Tc^- Sm^-		m.w. = 3.0×10^6; RC^+	204
	p1044	*S. aureus*	Pb Eb			203
	FAR-5	*S. aureus*	Pc Cd Fa			203
	pSH1	*S. aureus*	Tc		m.w. = 2.8×10^6; 60 copies/chromosome; compatible with pC221, pSH2, pI524	259
	pSH2	*S. aureus*	Km-Nm		m.w. = 9×10^6; 2 copies/chromosome	260

e. Plasmids of Other Gram-positive Bacteria

Compiled by A. E. Jacob, Royal Postgraduate Medical School, London; J. A. Shapiro, University of Chicago; and L. Yamamoto, University of Chicago

Incompatibility and host range	Plasmid	Original host	Phenotype	Genotype	Additional data	Reference†
Undetermined						
	Tet	*S. faecalis* KR	Tc		m.w. = 64.5×10^6, 36.5% G+C; 3 copies/chromosome	59, 60
	Ero	*S. faecalis* KR	Em Ln Vm		m.w. = 17.6×10^6, 34.5% G+C; 6 copies/chromosome	59, 60
	pAMα1 (=α)	*S. faecalis* DS-5	Tra^- Ln Tc		m.w. = 6×10^6; 9–10 copies/chromosome; *tet* locus amplifies	48, 49
	pAMβ1 (=β)	*S. faecalis* DS-5	Em Ln Vm		m.w. = 17×10^6; 1–2 copies/chromosome; mobilized by pAMγ1	48, 49
	pAMγ1 (=γ)	*S. faecalis* DS-5	Tra^+ Hly BcnI BcnII		m.w. = 34×10^6; 3–5 copies/chromosome	48, 49, 82
	pJH1	*S. faecalis* var. *zymogenes*, JH1	Tra^+ Em Km Nm Sm Tc		m.w. = 55×10^6, 38% G+C; 1 copy/chromosome	140, 143, 144

†Reference numbers refer to entries listed in the Bibliography, Appendix B3.

Incompatibility and host range	Plasmid	Original host	Phenotype	Genotype	Additional data	Reference
Undetermined *(continued)*	pJH2	*S. faecalis* var. *zymogenes*, JH1	Tra^+ Hly Bcn Bnr		m.w. = 42×10^6, 38% G+C; ~1 copy/chromosome	140, 143, 144
	pJH3	*S. faecalis* var. *zymogenes*, JH3	Tra^+ Hly Bcn Bnr		m.w. = 42×10^6	140, 143, 144
	pJH4	*S. faecalis* var. *zymogenes*, JH7	Tra^+ Em Km Sm		m.w. = 26×10^6	140
	AC-1	*S. pyogenes*	Em Lm Vm		m.w. = 17×10^6; 1–2 copies/chromosome	46, 287
	ERL1	*S. pyogenes*	Em Lm		staphylomycin S production	186
	SCP1	*Streptomyces coelicolor* A3(2)	Tra^+		promotes conjugation; encodes production of and resistance to an antibiotic; transmissible to other *Streptomyces* spp.	130, 158, 273, 274
	SCP1-*cys*B^+	*S. coelicolor*	Tra^+ Cys^+		carries *cys*B region of *S. coelicolor* chromosome; transmissible to other *Streptomyces* spp.	131
	unnamed	*S. coelicolor*	unknown		m.w. = 20×10^6; ~3–4 copies/chromosome	238

f. Plasmids Constructed in Vitro and In Vivo

Compiled by S. N. Cohen, Stanford University Medical School

Plasmid	Source	Phenotype	Genotype	m.w. (Mdal)	ρ (g/cm^3)	Copies/ chromosome	Restriction site	Additional data	Reference*
pSC101	Isolated following transformation by sheared fragments of R6-5	Tc		5.8–6.06	1.710	6–8	1 *Eco*RI 1 *Hin*dIII 1 *Bam*I 1 *Sal*I		26, 54, 57
pSC102	*Eco*RI fragment 2 of R6-5; Km-Nm and Su fragments of R6-5	Km-Nm Su		17	1.710		3 *Eco*RI		57
pSC105	pSC101, Km-Nm fragment of R6-5	Km Tc		10.5	1.710		2 *Eco*RI 2 *Hin*dIII 2 *Sal*I		57
pSC109	pSC101, RSF1010 joined at *Eco*RI sites	Sm Tc		11.5	11.5		2 *Eco*RI		57
pSC112	pSC101, *Eco*RI fragment 1 of pI258 (*S. aureus*)	Pc Tc		13.7	1.700		2 *Eco*RI		42
pSC113	pSC101, *Eco*RI fragments 2 and 3 of pI258	Pc		14.6	1.703		3 *Eco*RI		42
pSC120	pSC101, Tn*3*	Ap Sm Su Tc	Tn*3* Tn*4*	18.8	1.7145		6 *Eco*RI		167
pSC122	pSC101, *Eco*RI fragment 2 of pI258	Ap Tc Ac		10.0	1.703	6–8	2 *Eco*RI		57

*Reference numbers refer to entries listed in the Bibliography, Appendix B3.

Plasmid	Source	Phenotype	Genotype	m.w. (Mdal)	ρ (g/cm^3)	Copies/ chromosome	Restriction site	Additional data	Reference
pSC134	pSC101, ColE1	Tc Ice1		10.0		16(PolA$^+$) 6(PolA$^-$)	2 *Eco*RI		266
pSC135	*Eco*RI fragment 2 of R6-5 (*rep* region); *Eco*RI fragment 2 of pI258 via pSC122	Ap Ac$^-$ Sup (*dna*A)		11.8	1.705	2–5	2 *Eco*RI		267
pSC138	*Eco*RI fragment 6 of F*lac*, *Eco*RI fragment 2 of PI258 via pSC122	Ap Ac ϕII^s Sup (*dna*A)		9.4	1.702	0.6–1.0	2 *Eco*RI		267
pSC142	ColE1, *Eco*RI fragment 2 of pI258 via pSC122	Ap		8.2			2 *Eco*RI		26
pSC143	ColE1, Sm Su fragment of R6-5	Sm Su					2 *Eco*RI		26
pSC174-178, pSC183-190	pSC101, Tn*3*	Ap Tc	Tn*3*	8.7–8.8	1.7095		1 *Eco*RI		21, 167, 168
pSC179	pSC101, Tn*3*, Tn*4*	Ap Sm Su Tc	Tn*3* Tn*4*	18.4	1.7145		6 *Eco*RI		21, 167, 168
pSC201	ts *rep* mutant of pSC101	Rep(ts) Tc	Tn*3*	6.0			1 *Eco*RI		174
pSC204	pSC201, Tn*3*	Rep(ts) Ap Tc	Tn*3*	8.6			1 *Eco*RI		173

pSC205	ts *rep* mutant of pSC105	Rep(ts) Km Tc	10.5		2 *Eco*RI	
CD4	pSC101, 4.2 and 3.0 Mdal *Eco*RI fragments of *X. laevis* rDNA	Tc	13.6	1.721	3 *Eco*RI	197
CD18	pSC101, 3.0 Mdal *Eco*RI fragment of *X. laevis* rDNA	Tc	9.2	1.719	2 *Eco*RI	197
CD30	pSC101, 3.9 Mdal *Eco*RI fragment of *X. laevis* rDNA	Tc	10.0		2 *Eco*RI	197
CD35	pSC101, 3.9 and 3.0 Mdal *Eco*RI fragments of *X. laevis* rDNA	Tc	12.7	1.720	3 *Eco*RI	199
CD42	pSC101, 4.2 Mdal *Eco*RI fragment of *X. laevis* rDNA	Tc	10.0	1.720	2 *Eco*RI	199
pLp1	pSC101, 2 *Eco*RI fragments of *L. pictus* histone DNA	Tc	11.9		3 *Eco*RI	157
pSp2	pSC101, 4.8 Mdal *Eco*RI fragment of *S. purpuratus* histone DNA	Tc	8.68		2 *Eco*RI	157

Plasmid	Source	Phenotype	Genotype	m.w. (Mdal)	ρ (g/cm^3)	Copies/ chromosome	Restriction site	Additional data	Reference
pSp17, pSp71	pSC101, 1.4 Mdal fragment of *S. purpuratus* histone DNA	Tc		7.4			2 *Eco*RI	histone DNA inserted in reverse orientations in pSp17 and pSp71	157
pMM22, pMM26, pMM28, pMM49	pSC101, entire genome of mouse mitochondria	Tc		16			3 *Eco*RI 4 *Hind*III		43
pSC211, pSC213, pSC215, pSC217, pSC220-222	pSC105, Tn*3*	Ap Km Tc	Tn*3*	14.04			1 *Eco*RI		21, 168
pSC212	Tc^s deletion mutant of pSC211	Ap Km	Tn*3*	11.45			1 *Eco*RI		199
pSC214	TC^s deletion mutant of pSC213	Ap Km	Tn*3*	10.99			1 *Eco*RI		199
pSC218	Tc^s deletion mutant of pSC105 and Tn*3*	Ap Km	Tn*3*	10.86					199

[a] This column lists those species in which at least one plasmid of a particular incompatibility group has been observed to replicate stably.

[b] Plasmids of incompatibility groups IncB, Inc0, and Inc10 belong to the same group (27). (Numbers in parenthesis refer to references which can be found in the Bibliography, Appendix B3.

[c] IncA-C is a group of plasmids that share properties of IncA and IncC plasmids (108). Some plasmids may be listed more than once, under IncA, IncA-C, and IncC.

[d] IncC in *E. coli* corresponds to IncP-3 in *P. aeruginosa.*

[e] IncH has been subdivided into two groups: $IncH_1$ represented by TP117, TP123, and RIP166, and $IncH_2$ represented by TP116 (27). Note (i) that TP116 does not recombine with TP117, TP123, or TP124 but that these last three recombine with each other, and (ii) that the behavior of TP116 towards F is different from that of TP117, TP123, and TP124 (251).

[f] Stp indicates *S*almonella *t*yping *p*hages. For the exact patterns of phage inhibition in *S. typhimurium, S. typhi,* and *S. paratyphi,* see references 5 and 6.

[g] IncP in *E. coli* corresponds to IncP-1 in *P. aeruginosa.*

[h] IncQ in *E. coli* corresponds to IncP-4 in *P. aeruginosa.*

[i] The Phi phenotype of these plasmids is due to prophage immunity.

[j] The incompatibility classification of *Pseudomonas* plasmids followed here was initiated by Bryan et al. (24) and developed subsequently by Bryan et al. (22), Holloway et al. (129), Jacoby (146, 148, 150), Mitsuhashi (Holloway, pers. comm.), and Shahrabadi et al. (240).

[k] All Phi and Fi tests on *Pseudomonas* plasmids were carried out in *P. aeruginosa* PAO hosts.

[l] Where plasmid contour length was given, the molecular weight has been calculated from the relationship 1 μm equals 2.07 Mdal (180). Where plasmid buoyant density in CsCl was given, the guanine plus cytosine content (GC) was calculated by the formula of Schildkraut et al. (236).

[m] RP4 studied by Datta et al. (74) is described as coming from *P. aeruginosa* strain S8. This plasmid has the same resistance markers, molecular weight, and *Eco*RI sites as RP1 (G. A. Jacoby, A. E. Jacob and R. W. Hedges, unpubl.). RP4 studied by Saunders and Grinsted (232) is also said to originate in strain S8 and has identical resistance determinants but a molecular weight of 62 Mdal and hybridizes to the extent of 65% with RP1 DNA. Holloway and Richmond (127) discuss the confusion and propose that the plasmid giving Cb Km-Nm Tc resistance from strain S8 be termed RP8. However, Krishnapillai (175) and Stanisich (254) use RP8 to describe a plasmid from *P. aeruginosa* S8 that confers only Cb resistance and may have the same relation to Saunders and Grinsted's RP8 as RP1-1 does to RP1.

[n] Dryburgh and Stanisich (*Proc. Soc. Gen. Microbiol.* **2**:66 [1975]) have obtained two other plasmids from strain 931, one giving resistance only to Sm and the other only to Hg.

[o] Since pMG2 (146) originated in the same strain as R130 (24) and carries the same markers, the two plasmids are probably identical.

[p] Most CAM^+ strains carry CAM, molecular weight 92 Mdal, and factor K, molecular weight 64 Mdal, but some CAM^+ strains carry a plasmid of 154 ± 6 Mdal in which CAM and K have recombined (217).

[q] OCT^+ strains carry OCT, molecular weight 27 Mdal, and factor K (217).

[r] Since Rms149 appears incompatible with IncP-1 plasmid R751 (G. A. Jacoby, unpubl.), the existence of a P-6 group defined by this plasmid is uncertain.

[s] RP1-1 originated in the same strain as RP1 but confers resistance only to Cb and differs from RP1 in transfer frequency, host range, stability, and donor-phage susceptibility. Extrachromosomal DNA has not been detected in RP1-1-carrying strains. RP1 and RP1-1 are compatible (40, 136, 254). The possibility that RP1-1 arose by transposition of Tn401(Cb) from RP1 to another plasmid in strain 1822 seems likely.

[t] The properties described for plasmids originating in *P. aeruginosa* 9169 are sufficiently diverse to suggest that different plasmids have been studied. RP9 expresses Cb Km-Nm Tc resistance in *P. aeruginosa*, transfers readily to *E. coli*, and has a molecular weight of 48 Mdal (135). R91 expresses only Cb resistance in *P. aeruginosa* and transfers at low frequency to *E. coli*, where Nm and Tc resistance are expressed as well (40). In *E. coli*, R91 has a molecular weight of 35 Mdal (R. W. Hedges and A. E. Jacob, pers. comm.).

[u] Ninety percent of the plasmids in SAL^+ strains have a molecular weight of 51 Mdal and 10% 35 Mdal (217).

[v] XYL^+ strains also carry factor K and contain plasmids of variable molecular weight, i.e., 70, 80, 90, and 100 Mdal. XYL^-K^+ strains have only the 70-Mdal plasmid. XYL probably has a molecular weight of about 11 Mdal and can integrate into factor K (33, 34).

2. MAPS

a. F, the *E. coli* Sex Factor

Contributed by J. A. Shapiro, University of Chicago

The inner circle in Figure B2.1 gives some physical coordinates in kilobase units (kb). The next circle indicates the locations of identified insertion elements. The two arcs show regions of extensive homology with ColV2-K94, R1, and R6-5 plasmids. The outer circle indicates the locations of certain genetic loci involved in phage inhibition (*phi*), incompatibility (*inc*), replication (*rep*), transfer (*tra*), fertility inhibition (*fin*), and immunity to lethal zygosis (*ilz*). The origin of transfer replication (*ori*) is also indicated. Since ColV2-K94 is a member of the same incompatibility group as F, whereas R1 and R6-5 are members of a different incompatibility group, the location of incompatibility determinants probably lies between 46.1 and 49.6 kb on the map. The positions where insertion elements on F recombine with the bacterial chromosome to form Hfr's are also indicated. This map is based on data in references 1, 75, 100, 133, 211, 223, 242, 243, and 247 found in Appendix B3. (For more details and a restriction map, see paper by Ohtsubo and Ohtsubo, this volume.)

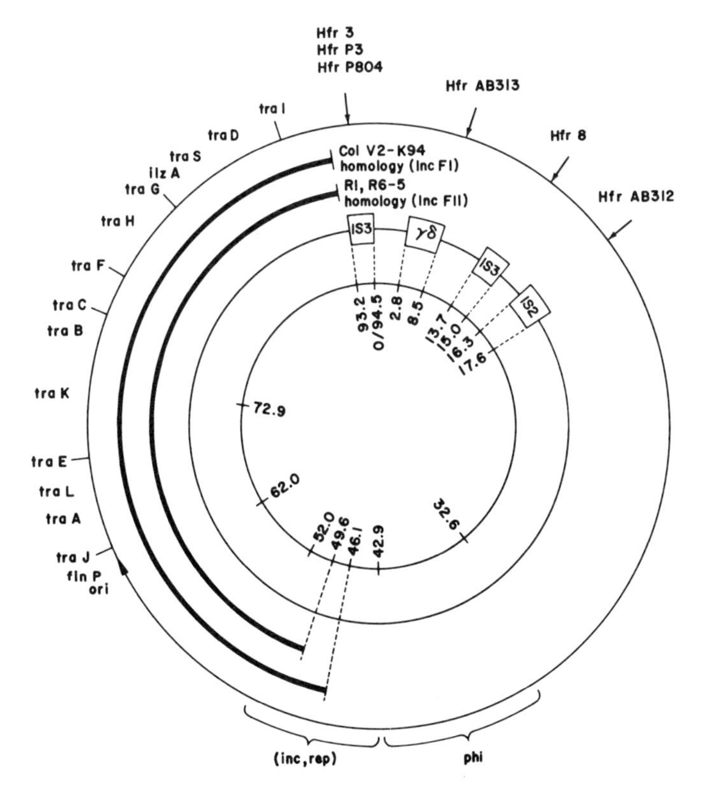

Figure B2.1

b. Special Sequences in the Structure of Cointegrate Drug Resistance Plasmids Related to F

Contributed by S. N. Cohen, Stanford University Medical Center

The maps in Figure B2.2a and b show where various IS elements and inverted repeat sequences occur in the DNA of the IncFII plasmids R1, R6, R6-5, R100, pSC50, and pSC206. (*a*) Schematic cointegrate R plasmid (a composite of R1, R6, and R100). (*b*) Schematic map showing the alignment of major drug resistance determinants. The positions of these structural features were determined by electron microscope examination of heteroduplex molecules formed by hydridization of these plasmids with each other, with F and RTF1, and with bacteriophage λ derivatives containing previously characterized IS elements (see refs. 21, 133, 223, 242 in Appendix B3). The region of extensive homology with RTF1 extends from IS*1*$_a$ counterclockwise to IS*1*$_b$. Extensive homology with F occurs in the same region but ends about 4 to 7 kilobases short of each IS*1* site.

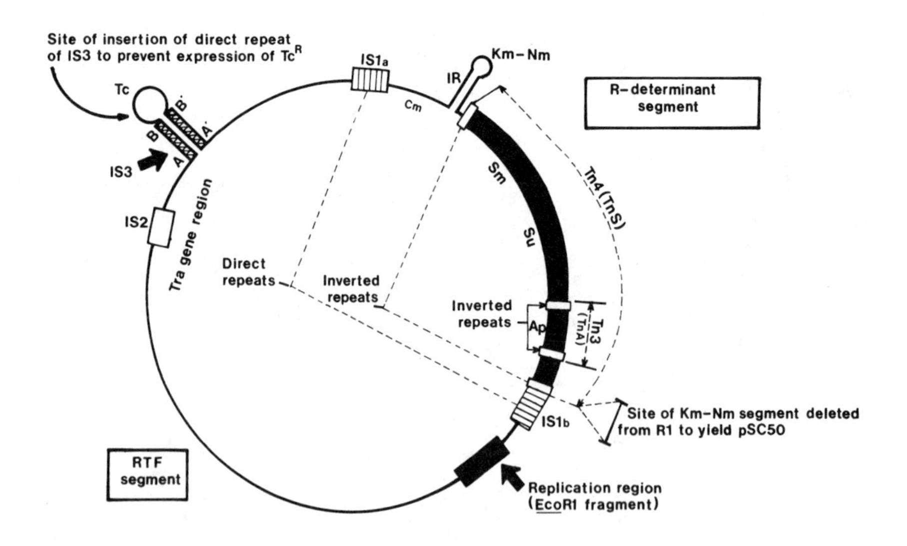

Figure B2.2a

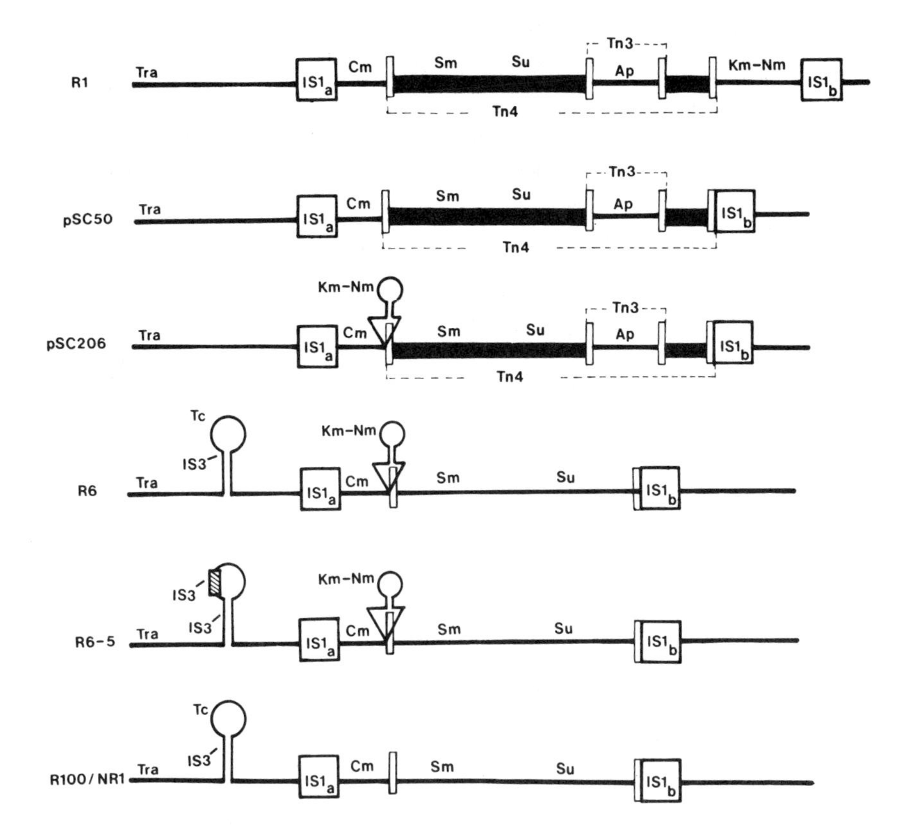
R1
Tra
IS1a
Cm
Sm
Su
Tn3
Ap
Km–Nm
IS1b
Tn4
pSC50
Tra
IS1a
Cm
Sm
Su
Tn3
Ap
IS1b
Tn4
Km–Nm
pSC206
Tra
IS1a
Cm
Sm
Su
Tn3
Ap
IS1b
Tn4
Tc
IS3
Km–Nm
R6
Tra
IS1a
Cm
Sm
Su
IS1b
IS3
Km–Nm
IS3
R6–5
Tra
IS1a
Cm
Sm
Su
IS1b
Tc
IS3
R100/NR1
Tra
IS1a
Cm
Sm
Su
IS1b

Figure B2.2b

c. *Eco*RI, *Hin*dIII, and *Bam*HI Cleavage Map of R538-1

Contributed by D. Vapnek, University of Georgia, Athens

In Figure B2.3 (N. K. Alton and D. Vapnek, unpubl.), the relative positions of the cleavage sites were determined by analyzing overlapping oligonucleotides produced by double digestion of isolated fragments (endo•R•*Hpa*I and endo•R•*Pst*I were used in addition to endo•R•*Eco*RI, endo•R•*Hin*dIII, and endo•R•*Bam*HI).

The locations of the genes for Cm^r, Sm^r, and Su^r were determined by analysis of hybrid plasmids which carried these antibiotic resistance determinants. Hybrid plasmids were isolated by in vitro ligation of restriction fragments obtained from 538-1- and ColE1-derived plasmids pMB9 and ColE1 Amp, followed by transformation of *E. coli* C600. Transformants were selected for antibiotic resistance markers carried by R538-1. The location of the transfer-gene region is based on hybrid plasmids which carry the *Hin*dIII A fragment and are able to plate donor-specific phages Qβ, R17, and MS2. The replication origin was located on *Eco*RI A, based on the ability of this fragment to replicate the kanamycin resistance determinant isolated from plasmid pML21 (for further details, see Lovett and Helinski, *J. Bact.* 127:982 [1976]).

Numbers refer to distances in kilobases.

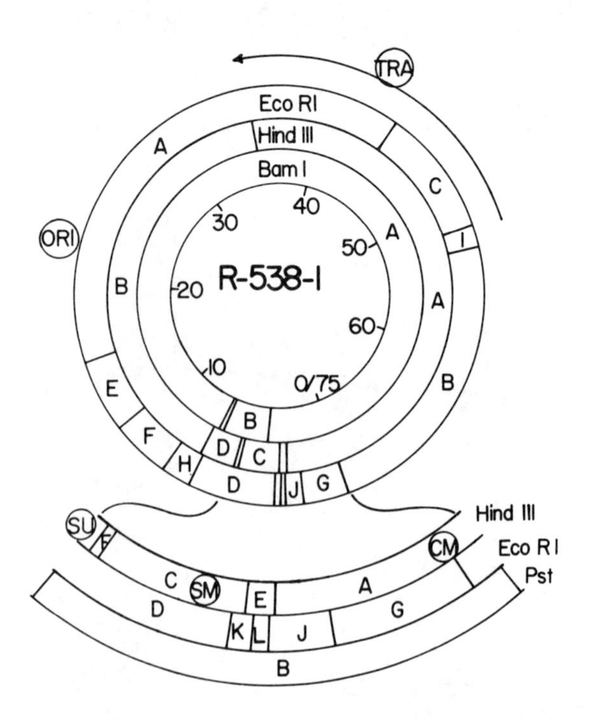

Figure B2.3

d. A Tn*7* Insertion Map of RP4

Contributed by P. T. Barth and N. J. Grinter, Royal Postgraduate Medical School, London

Figure B2.4 shows the sites on the RP4 molecule at which we have mapped insertions of Tn*7* (Barth and Grinter, *J. Mol. Biol.* 113:455 [1977], and unpublished data) together with the genetic loci deduced from the phenotypic changes to RP4 brought about by these insertions. The inner scale is marked in megadaltons (Mdal); the outer markings are the Tn*7* insertion sites numbered with the appropriate plasmid (pRP) number.

Tn*7* (originally called Tn*C*) is an 8.5×10^6-dalton DNA sequence encoding trimethoprim (Tp) and streptomycin (Sm) resistance, derived from R483 (IncIα), that is freely transposable to other replicons independently of a functional *recA* gene (Barth et al., *J. Bact.* 125:800 [1976]). RP4::Tn*7* plasmids were isolated by crossing an (RP4) *E. coli* K12 donor strain that carries Tn*7* in its chromosome with a suitable recipient and selecting Tp^r transconjugants. The molecular weights of such plasmids are consistent with their being RP4 plus a single inserted copy of Tn*7*. At a frequency of 1-2% for each character, Tc^s, Km^s, or Tra^- RP4::Tn*7* clones appeared as a consequence of Tn*7* insertion into the relevant gene. No Ap^s derivatives were detected.

Tn*7* insertion sites were mapped by restriction enzyme analysis. RP4 has single sites susceptible to *Eco*RI, *Hin*dIII, and *Bam*HI cleavage, whereas Tn*7* has, respectively, one, two, and two sites susceptible to these enzymes (Fig. B2.4).[1] Thus measurement (by sucrose gradient analysis) of the restriction fragments obtained by cleavage of each RP4::Tn*7* plasmid with at least two of these enzymes (separately) enables one to map the site and orientation of its Tn*7* insertion. Surprisingly, all the Tn*7* insertions mapped so far have the same orientation, i.e., with the arrowhead of Tn*7* in the figure pointing clockwise on the RP4 map. Precise mapping depends upon knowing the positions of the restriction sites on Tn*7* relative to its ends. The positions shown in the figure have not been determined directly but have been deduced from the restriction data of the Km^s RP4::Tn*7* plasmids, plus our observation that the *Hin*dIII site of RP4 is within the gene giving Km^r. This was discovered from experiments in which restriction fragments from various RP4::Tn*7* plasmids were ligated and reintroduced to *E. coli* by transformation. This technique enables one to excise selected segments of the RP4 molecule. When *Hin*dIII was used, excision of a segment on either side of its site on RP4 led to Km^s. Two essential regions of RP4 have been tentatively located by this method, one around the *Eco*RI site and the other within the 10-16-Mdal region. Viable plasmids with either of these regions excised have not been isolated. As expected from this, there are no Tn*7* insertions in these regions. Excision experiments have also shown that the $TpSm^r$ markers of Tn*7* are located between its left-hand end and its first *Hin*dIII site at 3.8 Mdal.

[1] Sites as indicated on the figure are: E = *Eco*RI site, H = *Hin*dIII site, and B = *Bam*HI site.

Single regions for the genes giving the Tc^r and Km^r phenotypes have been found on RP4. But the site of the gen giving Ap^r could not be mapped by Tn*7* insertion as no Ap^s RP4::Tn*7* plasmids have been isolated. The β-lactamase gene (giving Ap^r) of RP4 is, however, on Tn*1* and is known to be adjacent to a *Bam*HI site (Heffron et al., *Proc. Nat. Acad. Sci.* 74:702 [1977]). As RP4 has only a single *Bam*HI site, this must mark the location of the Ap^r determinant.

Two regions involved in conjugal transfer have been found in RP4. Within the large *tra* region, the various Tn*7* insertions give several phenotypic classes of sensitivity to the P-pili-specific phages PRR1, Pf3, and PR4 and are presumed to locate several *tra* genes. The majority of insertions in the *tra* regions yield plasmids with undetectable transferability, but pRP46, 92, 76, 90, 89, 26, and 91 have reduced transfer frequencies (10- to 1000-fold compared to

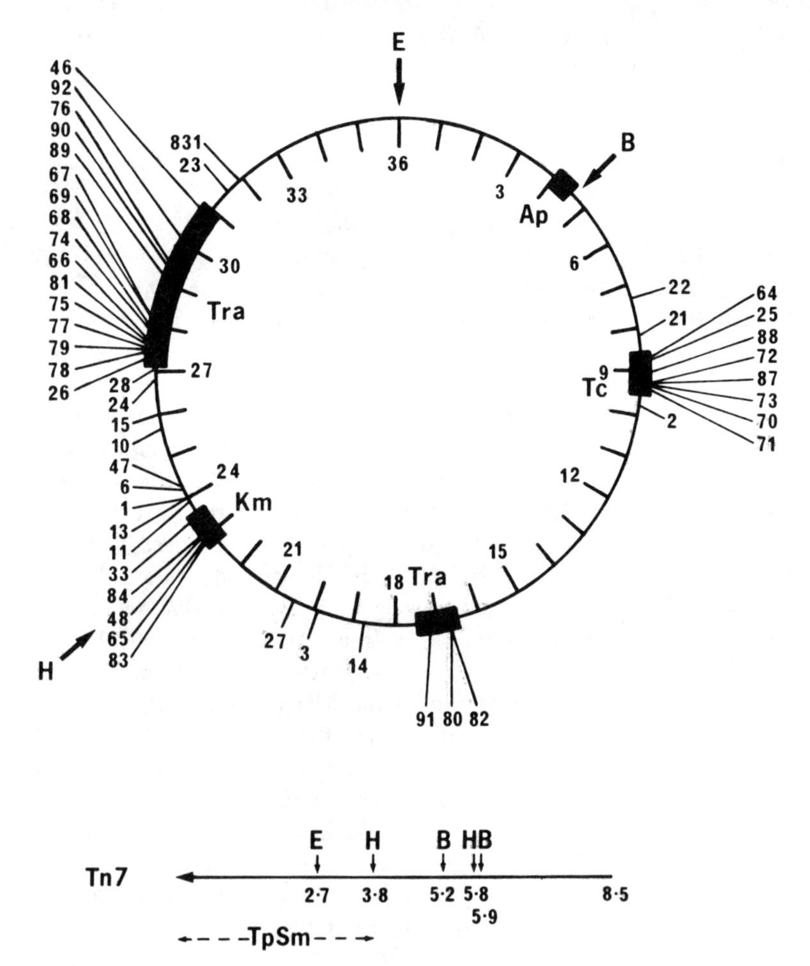

Figure B2.4

RP4). This may be due to polar effects of Tn7 insertion within *tra* operons or to the mutation of genes that promote, but are not essential for, conjugal transfer of RP4 between strains of *E. coli* K12. The use of other genera may be necessary to fully analyze the *tra* region of promiscuous plasmids. None of the RP4::Tn7 plasmids tested (including all the Tra$^-$ ones) had lost its surface exclusion to the entry of another IncP plasmid.

e. Physical Map of RP4

Contributed by A. DePicker, M. Van Montagu and J. Schell, Rijksuniversiteit Gent, Belgium

In Figure B2.5 the outer circle is a composite of all cleavage sites, which are shown separately in the inner circles for each restriction endonuclease (*Eco*RI, etc.). RP4 is not cleaved by the *Xba*I enzyme. The arrow shows the region at which integration of other DNA segments in RP4 occurs most frequently. The outer numbers indicate molecular weight in megadaltons, based on a molecular weight of 36 megadaltons for intact RP4, as determined by electron microscopy. The segment marked Amp^r is homologous with a segment of the plasmid PSF1030 which carries the ampicillin resistance transposon Tn*2*. The regions marked Tet^r and Kan^r encode tetracycline resistance and kanamycin resistance. The markers Tet^r, Kan^r, and both Tra regions were localized by transposon insertion mutagenesis.[1] The letters A, B, and so on identify the fragments in descending order of molecular weight.

[1]We thank P. T. Barth for communicating his results on Tn7 insertions in RP4 prior to publication (see Appendix B2e).

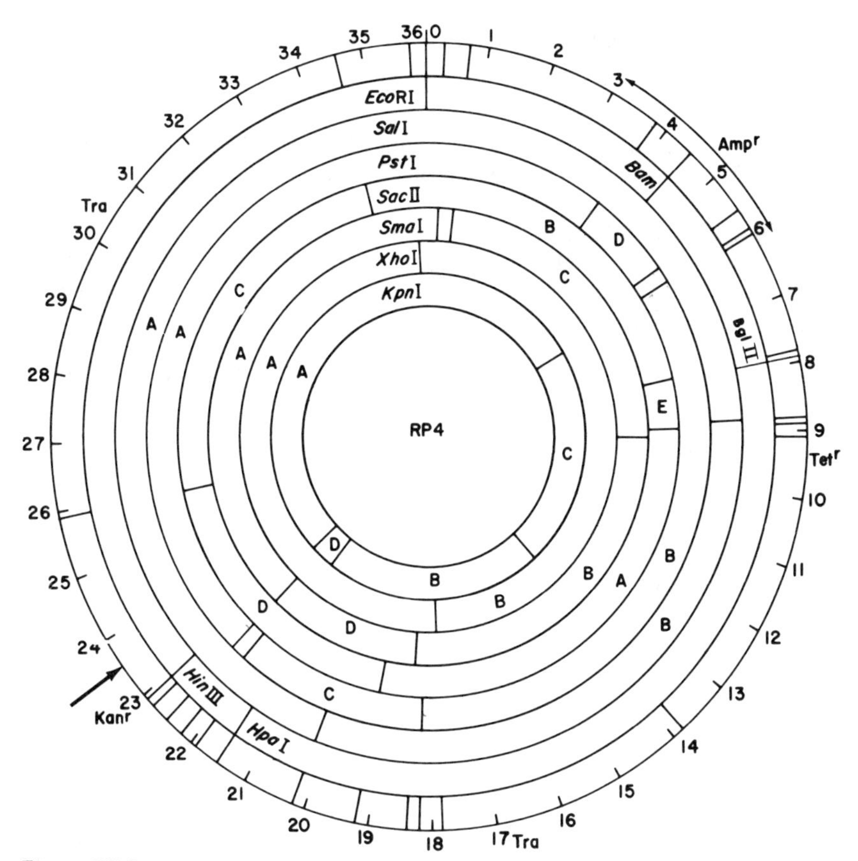
RP4
EcoRI
SalI
PstI
SacII
SmaI
XhoI
KpnI
Bam
Bgl II
HinIII
HpaI
Amp^r
Tet^r
Kan^r
Tra
Tra
A
B
C
D
E
36 0 1 2 3 4 5 6 7 8 9 10 11 12 13 14 15 16 17 18 19 20 21 22 23 24 25 26 27 28 29 30 31 32 33 34 35

Figure B2.5

f. Restriction Enzyme Map of RK2

Contributed by R. Meyer, D. Figurski and D. R. Helinski, University of California, San Diego

The circular RK2 map (Fig. B2.6) is calibrated from 0–37.6 megadaltons. The numbers in parentheses refer to distances in megadaltons from the *Eco*RI cleavage site.

Standard abbreviations are used for restriction enzymes (see Appendix D). Amp^r, Tet^r, and Kan^r refer to genes encoding resistance to ampicillin, tetracycline, and kanamycin, respectively. Regions containing genes for conjugal self-transfer are indicated by Tra; the origin of vegetative replication is labeled *ori.*

The plasmid RK2 was obtained from strain *E. coli* K12 J53 (RK2) which we received from Dr. N. Panopoulos.

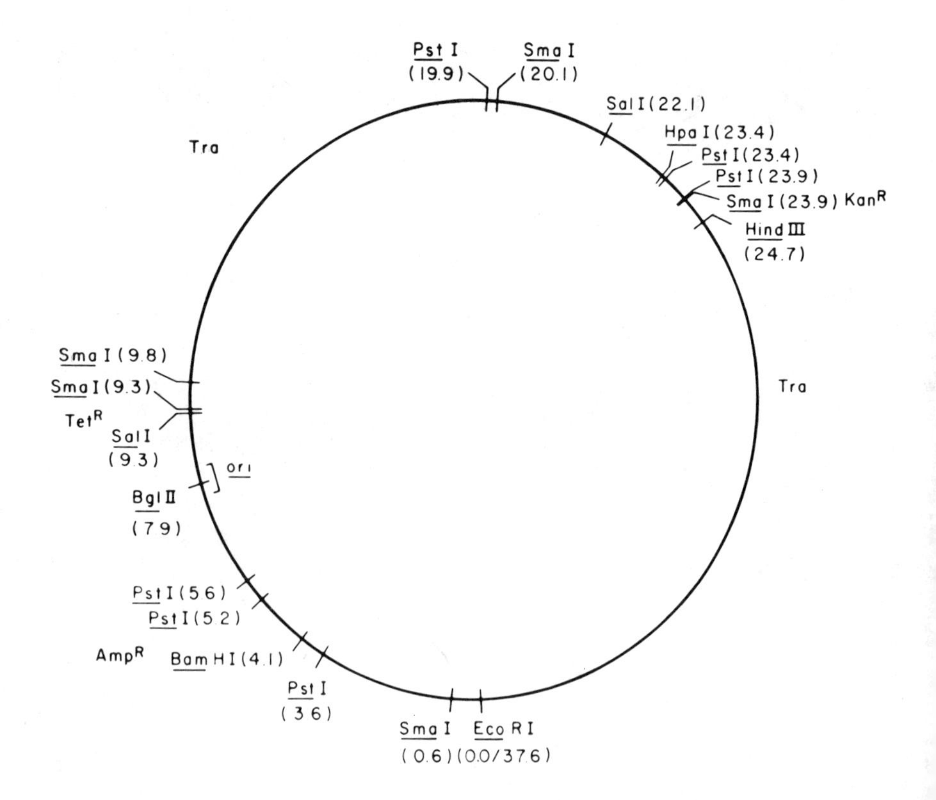

Figure B2.6

g. Restriction Map of the ColE1 Derivative pCR1

Contributed by K. Armstrong and D. R. Helinski, University of California, San Diego

pCR1 is composed of a 4.5-megadalton (md) kanamycin resistance fragment (derived from pSC105) inserted into the *Eco*RI site of ColE1 (4.2 md). One of the two *Eco*RI sites was subsequently deleted (Covey et al., *Mol. Gen. Genet.* 145:155 [1976]).

The divisions on the map (Fig. B2.7) represent megadaltons of DNA. The kanamycin determinant is denoted by the lighter section. *Ori* indicates the origin of replication of the plasmid.

Two *Hin*cII sites are shown on the map. One site is located at about 4.3 md, the other site is the same as the *Sal*I site at about 7.5 md. *Hin*cII also cleaves the *Hin*cII fragment, covering ~4.3 md to ~7.5 md, twice more, producing three fragments 0.5, 1.0, and 1.3 md in size. The relative order of these three fragments has not been determined conclusively, but the largest (1.3 md) is cleaved by *Hin*dIII. The *Hin*dIII and *Sma*I sites at approximately 6.3 md are too close to be resolved.

Insertion into the *Hin*dIII site inactivates Km^r (D. Figurski and R. Meyer, unpubl. obs.), but insertion into the *Sal*I site does not (Hamer and Thomas, Jr., *Proc. Nat. Acad. Sci.* 73:1537 [1976]). The enzymes *Bam*HI, *Hpa*I, and *Bgl*II do not cleave pCR1.

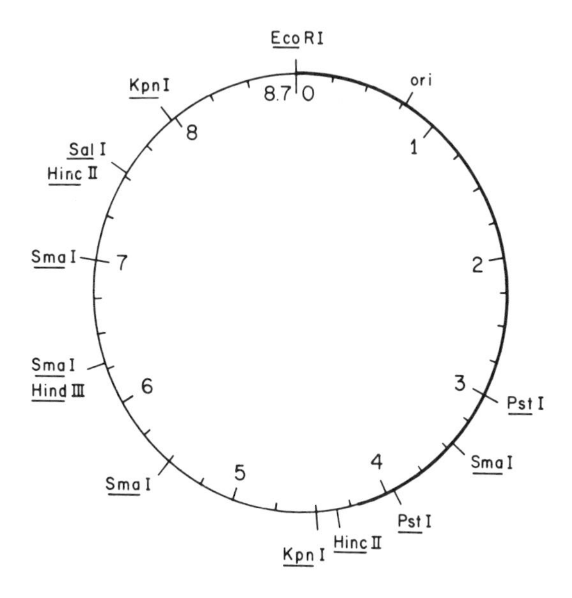

Figure B2.7

h. Restriction Map of ColE1 and pNT1 Plasmids

Contributed by H. Ohmori and J.-I. Tomizawa, National Institutes of Health, Bethesda

Figure B2.8 shows cleavage maps for ColE1 (Sakakibara and Tomizawa, *Proc. Nat. Acad. Sci.* 71:802 [1974]) and pNT1 plasmid DNAs. The top line represents calibration of DNA in units 0 to 1. The numbers above the line correspond to ColE1 DNA and those below to the shorter pNT1 DNA. The fragments formed by digestion with an enzyme are labeled alphabetically in order of decreasing size. The two fragments generated by digestion with R·*Eco*RI are further described by 1 or 2, for small or large, respectively (A1, A2; B1, B2). The segments that have ends formed by the action of R·*Hae*II and R·*Hae*III are named by the Greek alphabet in order of decreasing size. The region indicated by the bold line is non-ColE1 DNA which is present in pML21 (Hershfield et al., *J. Bact.* 126:447 [1976]).

Plasmid pNT1 was isolated by transformation of *E. coli* C600 *thy* with an R·*Eco*RI digest of plasmid pML21 (ColE1-*kan*) (Hershfield et al., *Proc. Nat. Acad. Sci.* 71:3455 [1974]), which has two R·*Eco*RI cleavage sites, and by selection of bacteria that had resistance to ColE1 and lacked kanamycin resistance. This plasmid is probably identical to pVH51 (see Hershfield et al. 1976, as cited above).

The segments between the two horizontal dotted lines contain the origin of DNA replication.

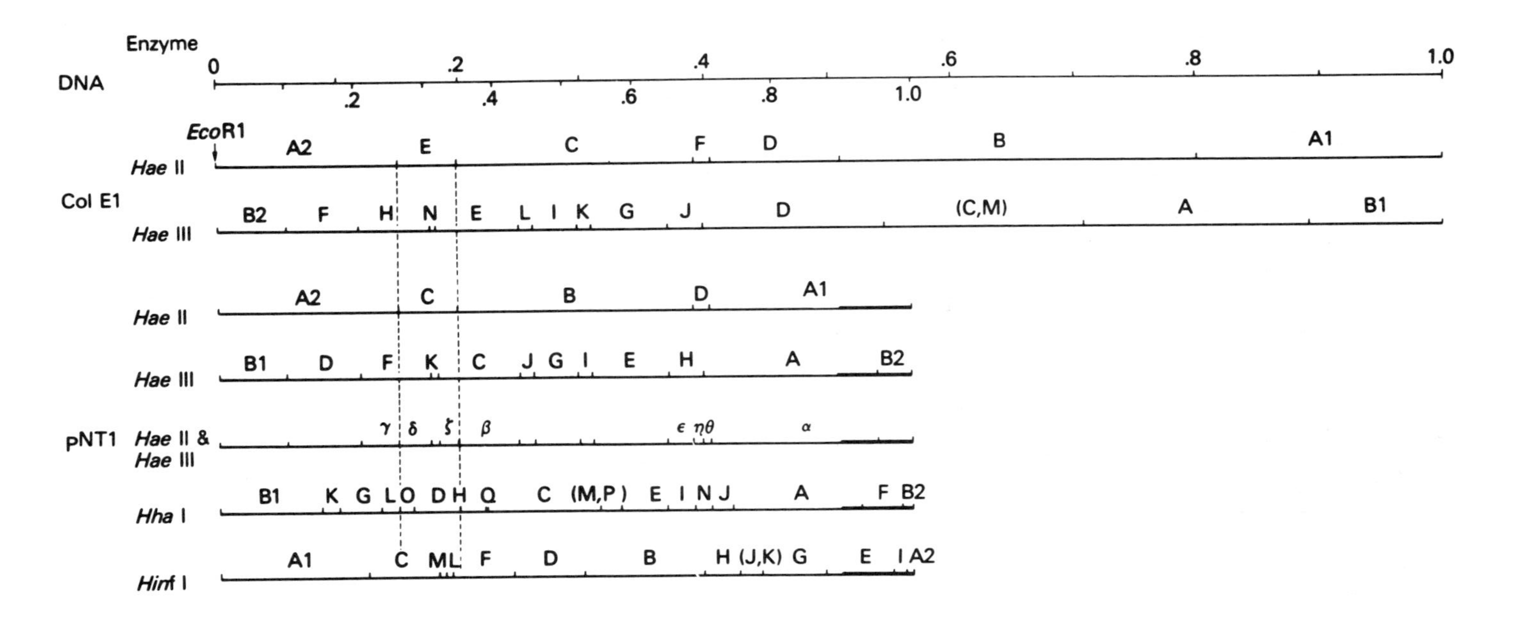

DNA
Enzyme
0 .2 .4 .6 .8 1.0
.2 .4 .6 .8 1.0
Col E1
EcoR1
A2 E C F D B A1
Hae II
B2 F H N E L I K G J D (C,M) A B1
Hae III
A2 C B D A1
Hae II
B1 D F K C J G I E H A B2
Hae III
pNT1
γ δ ζ β ε η θ α
Hae II &
Hae III
B1 K G L O D H Q C (M,P) E I N J A F B2
Hha I
A1 C M L F D B H (J,K) G E I A2
Hinf I

Figure B2.8

i. The Circular Restriction Map of pBR313

Contributed by F. Bolivar, R. L. Rodriguez, M. C. Betlach and H. W. Boyer, University of California, San Francisco

The plasmid pBR313 was constructed from pMB8, a small ColE1-like plasmid derivative, by a three-step procedure. First, some components of the pSC101 plasmid which confer resistance to tetracycline (Tc^r) were incorporated into pMB8 by in vitro cleavage and ligation. The plasmid obtained from this procedure, pMB9, has a molecular weight of 3.5×10^6 daltons. Second, the ampicillin transposon (Tn*A* = Tn*2*) was then translocated to pMB9 from the plasmid pSF2124. A plasmid obtained in this manner, pBR312, had a molecular weight of 6.7×10^6 daltons, which corresponds to the sum of the molecular weights of Tn*2* (3.2×10^6) and pMB9 (3.5×10^6). Third, the pBR312 plasmid was reduced in size by partial *Eco*RI* digestion and ligation, giving rise to the pBR313 plasmid with a molecular weight of 5.8×10^6 daltons. The final step removed the *Bam*HI site contributed by Tn*2*, leaving only one *Bam*HI on the plasmid. The antibiotic resistance genes on pBR313 are not transposable.

The relative positions of restriction sites are drawn to scale on a circular map divided into units of 1×10^5 daltons. The restriction sites for *Alu*I, *Eco*RII, *Hae*II, and *Hae*III represent only those that could be mapped. The number and location of restriction sites clockwise from the *Eco*RI to the *Hpa*I site are identical in pMB9 and pBR313, with the exception that the distance between the *Eco*RI and *Hin*dIII sites in pMB9 is 0.22×10^6 dalton as compared to 0.02×10^6 dalton in pBR313. (For details of the construction and mapping of pBR313, see Rodriguez et al., *ICN-UCLA Symposia on Molecular and Cellular Biology* [ed. D. P. Nierlich et al.], vol. V, p. 471 [Academic Press, 1976]; Bolivar et al., *Gene* [1977, in press].)

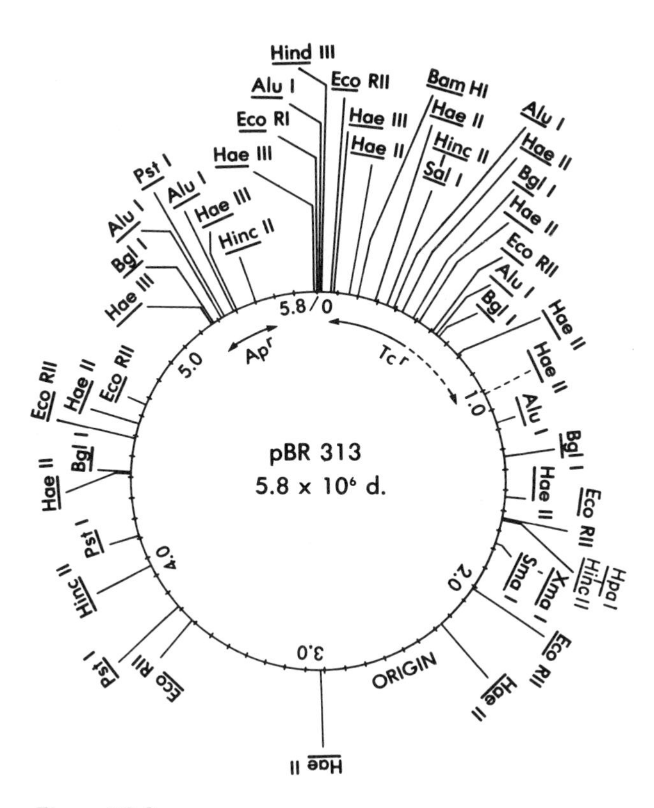

pBR 313
5.8 x 10⁶ d.
5.8/0
1.0
2.0
3.0
4.0
5.0
Apʳ
Tcʳ
ORIGIN
Hind III
Alu I
Eco RI
Hae III
Eco RII
Hae III
Hae II
Bam HI
Hae II
Hinc II
Sal I
Alu I
Hae II
Bgl I
Hae II
Eco RII
Alu I
Bgl I
Hae II
Hae II
Alu I
Bgl I
Hae II
Eco RII
Hpa I
Hinc II
Sma I
Xma I
Eco RII
Hae II
Hae II
Eco RII
Pst I
Hinc II
Pst I
Hae II
Bgl I
Eco RII
Hae II
Eco RII
Hae III
Bgl I
Alu I
Pst I
Alu I
Hae III
Hinc II

Figure B2.9

j. The Circular Restriction Map of pBR322

Contributed by F. Bolivar, R. L. Rodriguez, P. J. Greene, M. C. Betlach, H. L. Heyneker, H. W. Boyer, J. H. Crosa and S. Falkow,* University of California, San Francisco; *University of Washington, Seattle*

The plasmid pBR322 was derived from pBR313 by a procedure designed to generate a low-molecular-weight, relaxed plasmid containing one *Pst*I substrate site and genes for ampicillin and tetracycline resistance (Bolivar et al., *Gene* [1977, in press]. The *Pst*I site is located within the Ap^r gene. There are four other unique restriction sites, *Eco*RI, *Hin*dIII, *Bam*HI, and *Sal*I. The plasmid is sensitive to ColE1. The antibiotic genes on pBR322 are not transposable.

In Figure B2.10, the relative positions of restriction sites are drawn to scale on a circular map divided into units of 1×10^5 daltons (outer circle) and 0.1 kilobase (inner circle). The estimated sizes of the Ap^r and Tc^r genes represented in the figure were determined indirectly on the basis of the reported values for the size of the TEM β-lactamase (Datta and Richmond, *Biochem. J.* 98:204 [1966]), and the Tc^r-associated proteins were detected in the minicell system (Levy and McMurry, *Biochem. Biophys. Res. Comm.* 56:1060 [1974]; Tait et al., *Mol. Gen. Genet.* [1977, in press]). Positioning the left-hand boundary of the Tc^r gene was based on our knowledge that cloning into the *Eco*RI site of pBR313 did not affect Tc^r, whereas cloning into the *Hin*dIII site did affect the Tc^r mechanism. The position and size of the Tc^r region are also consistent with the orientation of the Tn*A* (Tn*2*) in pBR26 (Bolivar et al., *Gene* [1977, in press]. Only 10 out of 12 *Hae*II and *Alu*I and 7 out of 17 *Hae*III substrate sites are represented on the circular map of pBR322. The positions of 2 of the 10 *Alu*I sites, located at 1.8×10^6 daltons and 1.86×10^6 daltons, were mapped on a 0.7×10^6-dalton plasmid which encompasses this region of pBR322 (unpubl. obs.).

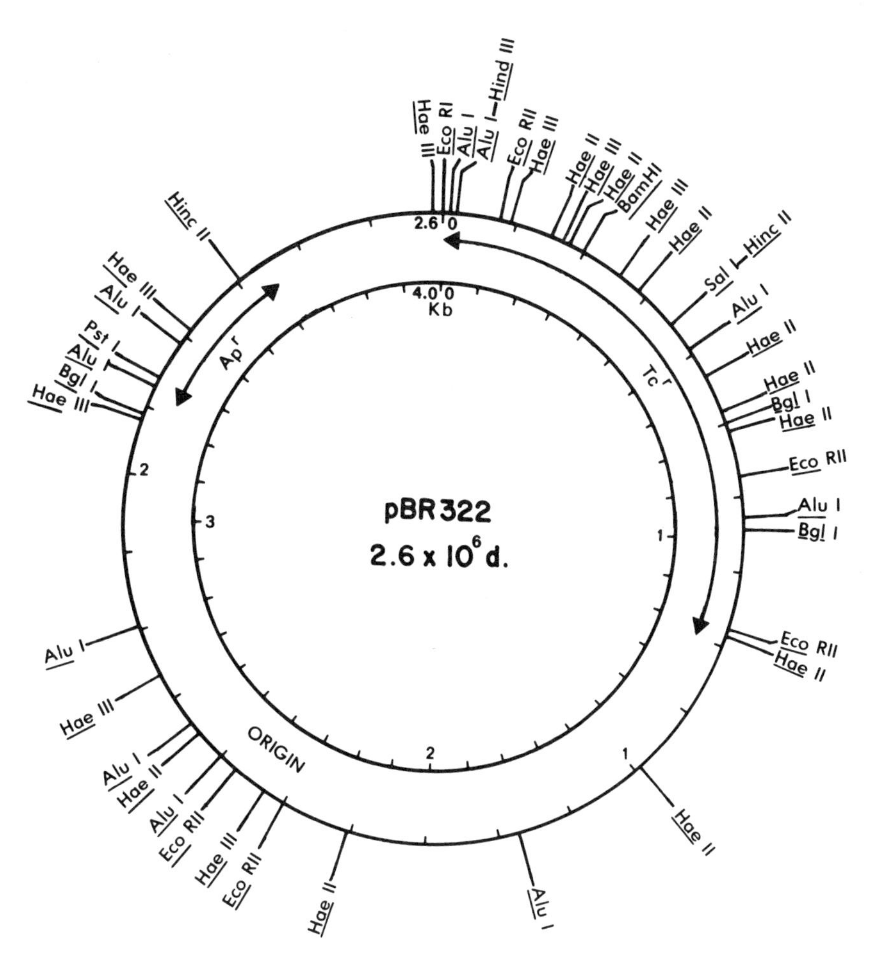
pBR322
2.6 x 10⁶ d.
Kb
Ap^r
Tc^r
ORIGIN
Eco RI
Hind III
BamHI
Sal I
Pst I
Hinc II
Bgl I
Alu I
Hae II
Hae III
Eco RII

Figure B2.10

3. BIBLIOGRAPHY

Compiled by J. A. Shapiro, University of Chicago; A. E. Jacob, Royal Postgraduate Medical School, London; G. A. Jacoby, Massachusetts General Hospital, Boston; R. P. Novick, Public Health Research Institute of the City of New York; and L. Yamamoto, University of Chicago

1. Achtman, M. and R. Helmuth. 1975. The F factor carries an operon of more than 15 × 10^6 daltons coding for deoxyribonucleic acid transfer and surface exclusion. In *Microbiology 1974* (ed. D. Schlessinger), p. 95. American Society for Microbiology, Washington, D.C.
2. Alton, N. K. and D. Vapnek, unpublished.
3. Anderson, E. S. and H. R. Smith. 1972. Chloramphenicol resistance in the typhoid bacillus. *Br. Med. J.* **3**:329.
4. Anderson, E. S. and E. J. Threlfall. 1970. Change of host range in a resistance factor. *Genet. Res.* **16**:207.
5. ——. 1974. The characterization of plasmids in the enterobacteria. *J. Hyg.* (Camb.) **72**:471.
6. Anderson, E. S., E. J. Threlfall, J. M. Carr and L. G. Savoy. 1973. Bacteriophage restriction in *Salmonella typhimurium* by R factors and transfer factors. *J. Hyg.* (Camb.) **71**:619.
7. Aoki, T., S. Egusa, Y. Ogata and T. Watanabe. 1971. Detection of resistance factors in fish pathogen *Aeromonas liquefaciens. J. Gen. Microbiol.* **65**:343.
8. Arber, W. and D. Wauters-Willems. 1976. Host specificity of DNA produced by *E. coli.* XII. The two restriction and modification systems of strain 15T$^-$. *Mol. Gen. Genet.* **108**:203.
9. Bannister, D. 1969. Restriction and modification control by resistance transfer factors. Ph.D. thesis, University of London.
10. ——. 1970. Analysis of an R$^+$ strain carrying two *fi*$^-$ sex factors. *J. Gen. Microbiol.* **61**:273.
11. ——. 1970. Explanation of the apparent association of host specificity determinants with *fi*$^-$ R factors. *J. Gen. Microbiol.* **61**:283.
12. Bannister, D. and S. W. Glover. 1968. Restriction and modification of phages by R$^+$ strains of *E. coli* K12. *Biochem. Biophys. Res. Comm.* **30**:735.
13. Barth, P. T., unpublished.
14. Barth, P. T. and N. J. Grinter. 1974. Comparison of the DNA molecular weights and homologies of plasmids conferring linked resistance to streptomycin and sulfonamides. *J. Bact.* **120**:618.
15. Barth, P. T., N. Datta, R. W. Hedges and N. J. Grinter. 1976. Transposition of a DNA sequence encoding trimethoprim and streptomycin resistances from R483 to other replicons. *J. Bact.* **125**:800.
16. Benedik, M., M. Fennewald and J. Shapiro. 1977. Transposition of a β-lactamase locus from RP1 into *Pseudomonas putida* degradative plasmids. *J. Bact.* **129**:809.
17. Bennett, P. M. and M. H. Richmond. 1976. Translocation of a discrete piece of deoxyribonucleic acid carrying an *amp* gene between replicons in *Escherichia coli. J. Bact.* **126**:1.
18. Benson, S., unpublished.
19. Berg, D. E., J. Davies, B. Allet and J. Rochaix. 1975. Transposition of R factor genes to bacteriophage λ. *Proc. Nat. Acad. Sci.* **72**:3628.

20. Beringer, J. E. 1974. R factor transfer in *Rhizobium leguminosarum. J. Gen. Microbiol.* **84**:188.
21. Brevet et al., this volume.
22. Bryan, L. E., M. S. Shahrabadi and H. M. van den Elzen. 1974. Gentamicin resistance in *Pseudomonas aeruginosa*: R-factor-mediated resistance. *Antimicrob. Agts. Chemother.* **6**:191.
23. Bryan, L. E., H. M. van den Elzen and J. T. Tseng. 1972. Transferable drug resistance in *Pseudomonas aeruginosa. Antimicrob. Agts. Chemother.* **1**:22.
24. Bryan, L. E., S. D. Semaka, H. M. van den Elzen, J. E. Kinnear and R. L. S. Whitehouse. 1973. Characteristics of R931 and other *Pseudomonas aeruginosa* R factors. *Antimicrob. Agts. Chemother.* **3**:625.
25. Brzezinska, M., R. Benveniste, J. Davies, P. J. L. Daniels and J. Weinstein. 1972. Gentamicin resistance in strains of *Pseudomonas aeruginosa* mediated by enzymatic *N*-acetylation of the deoxystreptamine moiety. *Biochemistry* **11**:761.
26. Cabello, F., K. Timmis and S. N. Cohen. 1976. Replication control in a composite plasmid constructed by *in vitro* linkage of two distinct replicons. *Nature* **259**:285.
27. Chabbert, Y. A., personal communication.
28. Chabbert, Y. A. and G. R. Gerbaud. 1974. Surveillance epidemiologique des plasmides responsables de la resistance au chloramphenicol de *Salmonella typhi. Ann. Microbiol.* (Inst. Pasteur) **125A**:153.
29. Chabbert, Y. A., M. R. Scavizzi, J. L. Witchitz, G. R. Gerbaud and D. H. Bouanchaud. 1972. Incompatibility groups and the classification of *fi*⁻ resistance factors. *J. Bact.* **112**:666.
30. Chakrabarty, A. M. 1972. Genetic basis of the biodegradation of salicylate in *Pseudomonas. J. Bact.* **112**:815.
31. ——. 1973. Genetic fusion of incompatible plasmids in *Pseudomonas. Proc. Nat. Acad. Sci.* **70**:1641.
32. ——. 1974. Dissociation of a degradative plasmid aggregate in *Pseudomonas. J. Bact.* **118**:815.
33. ——. 1976. Plasmids in *Pseudomonas. Annu. Rev. Genet.* **10**:7.
34. ——, personal communication.
35. Chakrabarty, A. M. and D. A. Friello. 1974. Dissociation and interaction of individual components of a degradative plasmid aggregate in *Pseudomonas. Proc. Nat. Acad. Sci.* **71**:3410.
36. Chakrabarty, A. M. and I. C. Gunsalus. 1969. Autonomous replication of a defective transducing phage in *Pseudomonas putida. Virology* **38**:92.
37. ——. 1969. Defective phage and chromosome mobilization in *Pseudomonas putida. Proc. Nat. Acad. Sci.* **64**:1217.
38. Chakrabarty, A. M., G. Chou and I. C. Gunsalus. 1973. Genetic regulation of octane dissimilation plasmid in *Pseudomonas. Proc. Nat. Acad. Sci.* **70**: 1137.
39. Chakrabarty, A. M., J. R. Mylroie, D. A. Friello, and J. G. Vacca. 1975. Transformation of *Pseudomonas putida* and *Escherichia coli* with plasmid-linked drug-resistance factor DNA. *Proc. Nat. Acad. Sci.* **72**:3647.
40. Chandler, P. M. and V. Krishnapillai. 1974. Phenotypic properties of R factors of *Pseudomonas aeruginosa*: R factors readily transferable between *Pseudomonas* and the *Enterobacteriaceae. Genet. Res.* **23**:239.

41. ———. 1974. Phenotypic properties of R factors of *Pseudomonas aeruginosa*: R factors transferable only in *Pseudomonas aeruginosa. Genet. Res.* **23**:251.
42. Chang, A. C. Y. and S. N. Cohen. 1974. Genome construction between bacterial species *in vitro*: Replication and expression of *Staphylococcus* plasmid genes in *Escherichia coli. Proc. Nat. Acad. Sci.* **71**:1030.
43. Chang, A. C. Y., R. A. Lansman, D. A. Clayton and S. N. Cohen. 1975. Studies of mouse mitochondrial DNA in *Escherichia coli*: Structure and function of the eukaryotic-prokaryotic chimeric plasmids. *Cell* **6**:231.
44. Chopra, I., P. M. Bennett and R. W. Lacey. 1973. A variety of staphylococcal plasmids present as multiple copies. *J. Gen. Microbiol.* **79**:343.
45. Chou, G. I. N., D. Katz and I. C. Gunsalus. 1974. Fusion and compatibility of camphor and octane plasmids in *Pseudomonas. Proc. Nat. Acad. Sci.* **71**:2675.
46. Clewell, D. B. and A. E. Franke. 1974. Characterization of a plasmid determining resistance to erythromycin, lincomycin and vernamycin B in a strain of *Streptococcus pyogenes. Antimicrob. Agts. Chemother.* **5**:534.
47. Clewell, D. B. and D. R. Helinski. 1970. Existence of the colicinogenic factor-sex factor Col Ib-P9 as a supercoiled circular DNA protein relaxed complex. *Biochem. Biophys. Res. Comm.* **41**:150.
48. Clewell, D. B., Y. Yagi and B. Bauer. 1975. Plasmid-determined tetracycline resistance in *Streptococcus faecalis*: Evidence for gene amplification during growth in the presence of tetracycline. *Proc. Nat. Acad. Sci.* **72**:1720.
49. Clewell, D. B., Y. Yagi, G. M. Dunny and S. K. Schultz. 1974. Characterization of three plasmid DNA molecules in a strain of *Streptococcus faecalis*: Identification of a plasmid determining erythromycin resistance. *J. Bact.* **117**:283.
50. Clowes, R. C. 1972. Molecular structure of bacterial plasmids. *Bact. Rev.* **36**:361.
51. Coetzee, J. N. 1974. Properties of *Proteus* and *Providence* strains harbouring recombinant plasmids between P*lac*, R1*drd*19 or R447b. *J. Gen. Microbiol.* **80**:199.
52. Coetzee, J. N., N. Datta and R. W. Hedges. 1972. R factors from *Proteus rettgeri. J. Gen. Microbiol.* **72**:543.
53. Coetzee, J. N., A. E. Jacob and R. W. Hedges. 1975. Susceptibility of a hybrid plasmid to excision of genetic material. *Mol. Gen. Genet.* **140**:7.
54. Cohen, S. N. and A. C. Y. Chang. 1973. Recircularization and autonomous replication of sheared R factor DNA segments in *E. coli* transformants. *Proc. Nat. Acad. Sci.* **70**:1293.
55. Cohen, S. N. and C. A. Miller. 1970. Non-chromosomal antibiotic resistance in bacteria. II. Molecular nature of R-factors isolated from *Proteus mirabilis* and *E. coli. J. Mol. Biol.* **50**:671.
56. ———. 1970. Non-chromosomal antibiotic resistance in bacteria. III. Isolation of the transfer unit of R-factor R1. *Proc. Nat. Acad. Sci.* **67**:510.
57. Cohen, S. N., A. C. Y. Chang, H. W. Boyer and R. B. Helling. 1973. Construction of biologically functional bacterial plasmids *in vitro. Proc. Nat. Acad. Sci.* **70**:3240.
58. Cooke, M. and E. Meynell. 1969. Chromosomal transfer mediated by derepressed R factors in F^- *E. coli* K12. *Genet. Res.* (Camb.) **14**:79.

59. Courvalin, P. M., C. Carlier and Y. A. Chabbert. 1972. Plasmid linked tetracycline and erythromycin resistance in group D *Streptococcus. Ann. Inst. Pasteur* **123**:755.
60. Courvalin, P. M., C. Carlier, O. Croissant and D. Blangy. 1974. Identification of two plasmids determining resistance to tetracycline and to erythromycin in group D *Streptococcus. Mol. Gen. Genet.* **132**:181.
61. Curtis, N. A. C. and M. H. Richmond. 1974. Effect of R-factor-mediated genes on some surface properties of *Escherichia coli. Antimicrob. Agts. Chemother.* **6**:666.
62. Curtis, N. A. C., M. H. Richmond and V. Stanisich. 1973. R-factor mediated resistance to penicillins which does not involve a β-lactamase. *J. Gen. Microbiol.* **79**:163.
63. Datta, N. 1974. "Epidemiology and classification of plasmids." Booklet 60502. American Society for Microbiology, Washington, D.C.
64. ——. 1975. Epidemiology and classification of plasmids. In *Microbiology 1974* (ed. D. Schlessinger), p. 9. American Society for Microbiology, Washington, D.C.
65. Datta, N. and P. T. Barth. 1976. Compatibility properties of R483. *J. Bact.* **125**:796.
66. Datta, N. and R. W. Hedges. 1971. Compatibility groups among *fi*$^-$ R factors. *Nature* **234**:220.
67. ——. 1972. R factors identified in Paris, some conferring gentamicin resistance, constitute a new compatibility group. *Ann. Inst. Pasteur* **123**: 849.
68. ——. 1972. Host ranges of R factors. *J. Gen. Microbiol.* **70**:453.
69. ——. 1972. Trimethoprim resistance conferred by W plasmids in *Enterobacteriaceae. J. Gen. Microbiol.* **72**:349.
70. ——. 1973. R factors of compatibility group A. *J. Gen. Microbiol.* **74**:335.
71. ——. 1973. An I pilus-determining R factor with anomalous compatibility properties mobilizing a gentamicin resistance plasmid. *J. Gen. Microbiol.* **77**:11.
72. Datta, N., and J. Olarte. 1974. R factors in strains of *Salmonella typhi* and *Shigella dysenteriae* isolated during epidemics in Mexico: Classification and compatibility. *Antimicrob. Agts. Chemother.* **5**:310.
73. Datta, N., R. W. Hedges, D. Becker and J. Davies. 1974. Plasmid determined fusidic acid resistance in the *Enterobacteriaceae. J. Gen. Microbiol.* **83**:191.
74. Datta, N., R. W. Hedges, E. J. Shaw, R. P. Sykes and M. H. Richmond. 1971. Properties of an R factor from *Pseudomonas aeruginosa. J. Bact.* **108**:1244.
75. Davidson, N., R. C. Deonier, S. Hu and E. Ohtsubo. 1975. Electron microscope heteroduplex studies of sequence relations among plasmids of *Escherichia coli.* X. Deoxyribonucleic acid sequence organization of F and of F-primes and the sequences involved in Hfr formation. In *Microbiology 1974* (ed. D. Schlessinger), p. 56. American Society for Microbiology, Washington, D.C.
76. Davies, J. E. and R. Rownd. 1972. Transmissible multiple drug resistance in *Enterobacteriaceae. Science* **176**:758.

77. Davis, R. and D. Vapnek. 1976. *In vivo* transcription of R-plasmid deoxyribonucleic acid in *Escherichia coli* strains with altered antibiotic resistance levels and/or conjugal proficiency. *J. Bact.* **125**:1148.
78. Dennison, S. 1972. Naturally occurring R factor, derepressed for pilus synthesis, belonging to the same compatibility group as the sex factor F of *E. coli* K12. *J. Bact.* **109**:416.
79. Dixon, R., F. Cannon and A. Kondorosi. 1976. Construction of a P plasmid carrying nitrogen fixation genes from *Klebsiella pneumoniae. Nature* **260**: 268.
80. Drabble, W. R. and B. A. D. Stocker. 1968. R factors in *Salmonella typhimurium*: Pattern of transduction by phage P22 and UV protection effect. *J. Gen. Microbiol.* **53**:109.
81. Dunn, N. W., and I. C. Gunsalus. 1973. Transmissible plasmid coding early enzymes of naphthalene oxidation in *Pseudomonas putida. J. Bact.* **114**:974.
82. Dunny, G. M. and D. B. Clewell. 1975. Transmissible toxin (hemolysin) plasmid in *Streptococcus faecalis* and its mobilization of a noninfectious drug resistance plasmid. *J. Bact.* **124**:784.
83. Evans, J., E. Galindo, J. Olarte and S. Falkow. 1968. β-Lactamase of R factors. *J. Bact.* **96**:1441.
84. Falkow, S. and L. S. Baron. 1962. Episomic element in a strain of *Salmonella typhosa. J. Bact.* **84**:581.
85. Falkow, S., P. Guerry, R. W. Hedges and N. Datta. 1974. Polynucleotide seqúence relationships among plasmids of the I compatibility complex. *J. Gen. Microbiol.* **85**:65.
86. Fennewald, M., unpublished.
87. Finley, R. B. and J. D. Punch. 1972. Satellite deoxyribonucleic acid in *Pseudomonas aeruginosa. Can. J. Microbiol.* **18**:1003.
88. Foster, T. J., T. G. B. Howe and K. M. V. Richmond. 1975. Translocation of the tetracycline resistance determinant from R100-1 to the *Escherichia coli* K-12 chromosome. *J. Bact.* **124**:1153.
89. Frydman, A. and E. Meynell. 1969. Interactions between derepressed F-like R factors and wild type colicin B factors: Superinfection immunity and repressor susceptibility. *Genet. Res.* **14**:315.
90. Fullbrook, P. D., S. W. Elson and B. Slocombe. 1970. R-factor mediated β-lactamase in *Pseudomonas aeruginosa. Nature* **226**:1054.
91. Goldstein, F. W., J. F. Agar, G. R. Gerbaud and Y. A. Chabbert. 1975. Transferable trimethoprim resistance mediated by plasmids of various incompatibility groups. In *15th Interscience Conference on Antimicrobial Agents and Chemotherapy*, p. 169. American Society for Microbiology, Washington, D.C.
92. Grindley, N. D. F., J. N. Grindley and E. S. Anderson. 1972. R factor combatibility groups. *Mol. Gen. Genet.* **119**:287.
93. Grindley, N. D. F., G. O. Humphreys and E. S. Anderson. 1973. Molecular studies of R factor compatibility groups. *J. Bact.* **115**:387.
94. Grinsted, J., J. R. Saunders, L. C. Ingram, R. B. Sykes and M. H. Richmond. 1972. Properties of an R factor which originated in *Pseudomonas aeruginosa* 1822. *J. Bact.* **110**:529.

95. Grinter, N. J. and P. T. Barth. 1976. Characterization of Sm Su plasmids by restriction endonuclease cleavage and compatibility testing. *J. Bact.* **128**:394.
96. Grund, A., J. Shapiro, M. Fennewald, P. Bacha, J. Leahy, K. Markbreiter, M. Nieder and M. Toepfer. 1975. Regulation of alkane oxidation in *Pseudomonas putida. J. Bact.* **123**:546.
97. Guerry, P. and R. P. Novick, unpublished.
98. Guerry, P., S. Falkow and N. Datta. 1974. R62, a naturally occurring hybrid R plasmid. *J. Bact.* **119**:144.
99. Gunsalus, I. C., M. Hermann, W. A. Toscano, Jr., D. Katz and G. K. Garg. 1975. Plasmids and metabolic diversity. In *Microbiology 1974* (ed. D. Schlessinger), p. 207. American Society for Microbiology, Washington, D.C.
100. Guyer, M. S. and A. J. Clark. 1976. *cis*-Dominant, transfer-deficient mutants of the *Escherichia coli* K-12 F sex factor. *J. Bact.* **125**:233.
101. Haas, D. and B. W. Holloway. 1976. R factor variants with enhanced sex factor activity in *Pseudomonas aeruginosa. Mol. Gen. Genet.* **144**:243.
102. Hardy, K. G. 1975. Colicinogeny and related phenomena. *Bact. Rev.* **39**: 464.
103. Hayes, W. 1968. *The genetics of bacteria and their viruses.* Blackwell, Oxford.
104. Hedges, R. W. 1972. Phenotypic characterization of *fi*$^-$ R factors determining the restriction and modification *hsp* II specificity. *Mol. Gen. Genet.* **115**:225.
105. ———. 1972. Resistance to spectinomycin determined by R factors of various compatibility groups. *J. Gen. Microbiol.* **72**:407.
106. ———. 1974. R factors from *Providence. J. Gen. Microbiol.* **81**:171.
107. ———. 1975. R factors from *Proteus mirabilis* and *P. vulgaris. J. Gen. Microbiol.* **87**:301.
108. ———, unpublished.
109. Hedges, R. W. and S. Baumberg. 1973. Resistance to arsenic compounds conferred by a plasmid transmissible between strains of *E. coli. J. Bact.* **115**:459.
110. Hedges, R. W. and N. Datta. 1971. *fi*$^-$ R factors giving chloramphenicol resistance. *Nature New Biol.* **234**:220.
111. ———. 1972. R124, an fi$^+$ R factor of a new compatibility class. *J. Gen. Microbiol.* **71**:403.
112. ———. 1973. Plasmids determining I pili constitute a compatibility group. *J. Gen. Microbiol.* **77**:19.
113. Hedges, R. W. and A. E. Jacob. 1974. Transposition of ampicillin resistance from RP4 to other replicons. *Mol. Gen. Genet.* **132**:31.
114. ———. 1975. A 98 megadalton R factor of compatibility group C in a *Vibrio cholerae* El Tor isolate from southern USSR. *J. Gen. Microbiol.* **89**:383.
115. ———. 1975. Mobilization of plasmid-borne drug resistance determinants for transfer from *Pseudomonas aeruginosa* to *Escherichia coli. Mol. Gen. Genet.* **140**:69.

116. Hedges, R. W., J. M. Cresswell and A. E. Jacob. 1976. A non-transmissible variant of RP4 suitable as cloning vehicle for genetic engineering. *FEBS Letters* **61**:186.
117. Hedges, R. W., N. Datta and M. P. Fleming. 1972. R factors conferring resistance to trimethoprim but not sulphonamides. *J. Gen. Microbiol.* **73**:573.
118. Hedges, R. W., A. E. Jacob and J. T. Smith. 1974. Properties of an R factor from *Bordetella bronchiseptica. J. Gen. Microbiol.* **84**:199.
119. Hedges, R. W., V. Rodriguez-Lemoine and N. Datta. 1975. R-factors from *Serratia marcescens. J. Gen. Microbiol.* **86**:88.
120. Hedges, R. W., N. Datta, J. N. Coetzee and S. Dennison. 1973. R factors from *Proteus morganii. J. Gen. Microbiol.* **77**:249.
121. Hedges, R. W., N. Datta, P. Kontomichalou and J. T. Smith. 1974. Molecular specificity of R factor-determining β-lactamases: Correlation with plasmid compatibility. *J. Bact.* **117**:56.
122. Hedges, R. W., A. E. Jacob, P. T. Barth and N. J. Grinter. 1975. Compatibility properties of P1 and φ Amp prophages. *Mol. Gen. Genet.* **141**:263.
123. Hedges, R. W., A. E. Jacob, N. Datta and J. N. Coetzee. 1975. Properties of plasmids produced by recombination between R factors of groups J and FII. *Mol. Gen. Genet.* **140**:289.
124. Heffron, F., C. Rubens and S. Falkow. 1975. Translocation of a plasmid DNA sequence which mediates ampicillin resistance: Molecular nature and specificity of insertion. *Proc. Nat. Acad. Sci.* **72**:3623.
125. Heffron, F., R. Sublett, R. W. Hedges, A. E. Jacob and S. Falkow. 1975. The origin of the TEM β-lactamase gene found on plasmids. *J. Bact.* **122**: 250.
126. Holloway, B. W. 1969. Genetics of *Pseudomonas. Bact. Rev.* **33**:419.
127. Holloway, B. W. and M. H. Richmond. 1973. R-factors used for genetic studies in strains of *Pseudomonas aeruginosa* and their origin. *Genet. Res.* **21**:103.
128. Holloway, B. W., V. Krishnapillai and V. Stanisich. 1971. *Pseudomonas* genetics. *Annu. Rev. Genet.* **5**:425.
129. Holloway, B. W., P. Chandler, V. Krishnapillai, B. Mills, H. Rossiter, V. Stanisich and J. Watson. 1975. Genetic basis of drug resistance in *Pseudomonas aeruginosa*. In *Drug-inactivating enzymes and antibiotic resistance* (ed. S. Mitsuhashi et al.), p. 271. Avicenum, Czechoslovak Medical Press, Prague; Springer-Verlag, Berlin.
130. Hopwood, D. A. and H. M. Wright. 1973. Transfer of a plasmid between *Streptomyces* species. *J. Gen. Microbiol.* **77**:187.
131. ———. 1973. A plasmid of *Streptomyces coelicolor* carrying a chromosomal locus and its interspecific transfer. *J. Gen. Microbiol.* **79**:331.
132. Howarth-Thompson, S., H. M. Hefferman and M. D. Buchan. 1974. Two compatibility groups among ColI plasmids. *J. Gen. Microbiol.* **81**:279.
133. Hu, S., E. Ohtsubo, N. Davidson and H. Saedler. 1975. Electron microscope heteroduplex studies of sequence relations among bacterial plasmids: Identification and mapping of the insertion sequences IS1 and IS2 in F and R plasmids. *J. Bact.* **122**:764.

134. Ikeda, H., M. Inuzuka and J. Tomizawa. 1970. P-1-like plasmid in *E. coli. J. Mol. Biol.* **50**:457.
135. Ingram, L. C., M. H. Richmond and R. B. Sykes. 1973. Molecular characterization of the R factors implicated in the carbenicillin resistance of a sequence of *Pseudomonas aeruginosa* strains isolated from burns. *Antimicrob. Agts. Chemother.* **3**:279.
136. Ingram, L., R. B. Sykes, J. Grinsted. J. R. Saunders and M. H. Richmond. 1972. A transmissible resistance element from a strain of *Pseudomonas aeruginosa* containing no detectable extrachromosomal DNA. *J. Gen. Microbiol.* **72**:269.
137. Iordanescu, S. 1975. Recombinant plasmid obtained from two different, compatible staphylococcal plasmids. *J. Bact.* **124**:597.
138. ——, personal communication.
139. Iyobe, S., K. Hasuda, A. Fuse and S. Mitsuhashi. 1974. Demonstration of R factors from *Pseudomonas aeruginosa. Antimicrob. Agts. Chemother.* **5**: 547.
140. Jacob, A. E., unpublished.
141. Jacob, F. and E. A. Adelberg. 1959. Transfer de characteres par incorporation au facteur sexuel d'*E. coli. Compt. Rend. Acad. Sci.* **249**:189.
142. Jacob, A. E. and N. J. Grinter. 1975. Plasmid RP4 as a vector replicon in genetic engineering. *Nature* **255**:504.
143. Jacob, A. E. and S. J. Hobbs. 1974. Conjugal transfer of plasmid-borne multiple antibiotic resistance in *Streptococcus faecalis* var. *zymogenes. J. Bact.* **117**:360.
144. Jacob, A. E., G. J. Douglas and S. J. Hobbs. 1975. Self transferable plasmids determining the hemolysin and bacteriocin of *Streptococcus faecalis* var. *zymogenes. J. Bact.* **121**:863.
145. Jacob, A. E., R. W. Hedges, N. Datta, N. J. Grinter and P. T. Barth, unpublished.
146. Jacoby, G. A. 1974. Properties of R plasmids determining gentamicin resistance by acetylation in *Pseudomonas aeruginosa. Antimicrob. Agts. Chemother.* **6**:239.
147. ——. 1974. Properties of an R plasmid in *Pseudomonas aeruginosa* producing amikacin (BB-K8), butirosin, kanamycin, tobramycin and sisomicin resistance. *Antimicrob. Agts. Chemother.* **6**:807.
148. ——. 1975. Properties of R plasmids in *Pseudomonas aeruginosa.* In *Microbiology 1974* (ed. D. Schlessinger), p. 36. American Society for Microbiology, Washington, D.C.
149. ——. 1975. R plasmids determining gentamicin or tobramycin resistance in *Pseudomonas aeruginosa.* In *Drug-inactivating enzymes and antibiotic resistance* (ed. S. Mitsuhashi et al.), p. 287. Avicenum, Czechoslovak Medical Press, Prague; Springer-Verlag, Berlin.
150. ——. 1977. Classification of plasmids in *Pseudomonas aeruginosa.* In *Microbiology 1976* (ed. D. Schlessinger). American Society for Microbiology, Washington, D.C. (In press.)
151. ——, unpublished.
152. Jacoby and Jacob, this volume.

153. Jobanputra, R. S. and N. Datta. 1974. Trimethoprim R factors in enterobacteria from clinical specimens. *J. Med. Microbiol.* **7**:169.
154. Kawabe, H., T. Naito and S. Mitsuhashi. 1975. Acetylation of amikacin, a new semisynthetic antibiotic, by *Pseudomonas aeruginosa* carrying an R factor. *Antimicrob. Agts. Chemother.* **7**:50.
155. Kawabe, H., S. Kondo, H. Umezawa and S. Mitsuhashi. 1975. R factor-mediated aminoglycoside antibiotic resistance in *Pseudomonas aeruginosa*: A new aminoglycoside 6′-*N*-acetyltransferase. *Antimicrob. Agts. Chemother.* **7**:494.
156. Kawakami, Y., F. Mikoshiba, S. Nagasaki, H. Matsumoto and T. Tazaki. 1972. Prevalence of *Pseudomonas aeruginosa* strains possessing R factor in a hospital. *J. Antibiot.* (Tokyo) **25**:607.
157. Kedes, L., A. C. Y. Chang, D. Housman and S. N. Cohen. 1975. Isolation of histone genes from unfractionated sea urchin DNA by subculture cloning in *Escherichia coli. Nature* **255**:533.
158. Kirby, R., L. F. Wright and D. A. Hopwood. 1975. Plasmid-determined antibiotic synthesis and resistance in *Streptomyces coelicolor. Nature* **254**:265.
159. Kleckner, N., R. Chan, B. Tye and D. Botstein. 1975. Mutagenesis by insertion of a drug-resistance element carrying an inverted repetition. *J. Mol. Biol.* **97**:561.
160. Knothe, H., V. Krčméry, W. Sietzen and J. Borst. 1973. Transfer of gentamicin resistance from *Pseudomonas aeruginosa* strains highly resistant to gentamicin and carbenicillin. *Chemotherapy* **18**:229.
161. Kobayashi, F., M. Yamaguchi, J. Sato and S. Mitsuhashi. 1972. Purification and properties of dihydrostreptomycin-phosphorylating enzyme from *Pseudomonas aeruginosa. Japan. J. Microbiol.* **16**:15.
162. Kondo, E. and S. Mitsuhashi. 1964. Drug resistance of enterobacteria: Active transducing bacteriophage P1Cm produced by the combination of R factor with bacteriophage P-1. *J. Bact.* **88**:1266.
163. Kono, M. and K. O'Hara. 1976. Mechanisms of streptomycin (Sm)-resistance of highly Sm-resistant *Pseudomonas aeruginosa* strains. *J. Antibiot.* (Tokyo) **29**:169.
164. ——. 1976. Mechanism of chloramphenicol-resistance mediated by kR102 factor in *Pseudomonas aeruginosa. J. Antibiot.* (Tokyo) **29**:176.
165. Kontomichalou, P., M. Mitani and R. C. Clowes. 1970. Circular R factor molecules controlling penicillinase synthesis, replicating in *E. coli* under relaxed or stringent control. *J. Bact.* **104**:34.
166. Kontomichalou, P., E. Papachristou and F. Angelatou. 1976. Multiresistant plasmids from *Pseudomonas aeruginosa* highly resistant to either or both gentamicin and carbenicillin. *Antimicrob. Agts. Chemother.* **9**:866.
167. Kopecko, D. J. and S. N. Cohen. 1975. Site-specific *rec*A-independent recombination between *E. coli* plasmids. *Proc. Nat. Acad. Sci.* **72**:1373.
168. Kopecko, D. J., J. Brevet and S. N. Cohen. 1976. Involvement of multiple translocating DNA segments and recombinational hot spots in the structural evolution of bacterial plasmids. *J. Mol. Biol.* **108**:333.

169. Korfhagen, T. R. and J. C. Loper. 1975. RPL11, an R factor of *Pseudomonas aeruginosa* determining carbenicillin and gentamicin resistance. *Antimicrob. Agts. Chemother.* **7**:69.
170. Korfhagen, T. R., J. C. Loper and J. A. Ferrel. 1975. *Pseudomonas aeruginosa* R factors determining gentamicin plus carbenicillin resistance from patients with urinary tract colonizations. *Antimicrob. Agts. Chemother.* **7**:64.
171. Korfhagen, T. R., J. A. Ferrel, C. L. Menefee and J. C. Loper. 1976. Resistance plasmids of *Pseudomonas aeruginosa*: Change from conjugative to nonconjugative in a hospital population. *Antimicrob. Agts. Chemother.* **9**:810.
172. Krčméry, V., H. Sagai, K. Hasuda, H. Kawabe, S. Iyobe, H. Knothe and S. Mitsuhashi. 1975. Demonstration of R plasmids from *Pseudomonas aeruginosa* conferring gentamicin and carbenicillin resistance. In *Drug-inactivating enzymes and antibiotic resistance* (ed. S. Mitsuhashi et al.), p. 343. Avicenum, Czechoslovak Medical Press, Prague; Springer-Verlag, Berlin.
173. Kretschmer and Cohen, this volume.
174. Kretschmer, P. J., A. C. Y. Chang and S. N. Cohen. 1975. Indirect selection of bacterial plasmids lacking identifiable phenotypic properties. *J. Bact.* **124**:225.
175. Krishnapillai, V. 1974. The use of bacteriophages for differentiating plasmids of *Pseudomonas aeruginosa. Genet. Res.* **23**:327.
176. ——. 1975. Resistance to ultraviolet light and enhanced mutagenesis conferred by *Pseudomonas aeruginosa* plasmids. *Mutat. Res.* **29**:363.
177. ——. 1975. Incompatibility between a prophage and R factors in *Pseudomonas aeruginosa. Soc. Gen. Microbiol. Proc.* **3**:54.
178. Lacey, R. W. and I. Chopra. 1974. Genetic studies of a multiresistant strain of *Staphylococcus aureus. J. Med. Microbiol.* **1**:285.
179. Lacy, G. H. and J. V. Leary. 1975. Transfer of antibiotic resistance plasmid RP1 into *Pseudomonas glycinea* and *Pseudomonas phaseolicola in vitro* and *in planta. J. Gen. Microbiol.* **88**:49.
180. Lang, D. 1970. Molecular weights of coliphages and coliphage DNA. III. Contour length and molecular weight of DNA from bacteriophages T4, T5 and T7, and from bovine papilloma virus. *J. Mol. Biol.* **54**:557.
181. Lawn, A. M. and E. Meynell. 1970. Serotypes of sex pili. *J. Hyg.* (Camb.) **68**:683.
182. Lawn, A. M., G. G. Meynell, E. Meynell and N. Datta. 1967. Sex pili and the classification of sex factors in the *Enterobacteriaceae. Nature* **216**:343.
183. Loutit, J. S. 1970. Investigation of the mating system of *Pseudomonas aeruginosa* strain 1. VI. Mercury resistance associated with the sex factor (FP). *Genet. Res.* **16**:179.
184. Lowbury, E. J. L., A. Kidson, H. A. Lilly, G. A. J. Ayliffe and R. J. Jones. 1969. Sensitivity of *Pseudomonas aeruginosa* to antibiotics: Emergence of strains highly resistant to carbenicillin. *Lancet* **ii**:448.
185 Macfarren, A. C. and R. C. Clowes. 1967. A comparative study of two F-like colicin factors, ColV2 and ColV3, in *E. coli* K12. *J. Bact.* **94**:365.

186. Malke, H. 1975. Transfer of a plasmid mediating antibiotic resistance between strains of *Streptococcus pyogenes* in mixed cultures. *Z. Allg. Mikrobiol.* **15**:645.
187. Mandel, M. 1966. Deoxyribonucleic acid base composition in the genus *Pseudomonas. J. Gen. Microbiol.* **43**:273.
188. Matsumoto, H. and T. Tazaki. 1973. FP5 factor, an undescribed sex factor of *Pseudomonas aeruginosa. Japan J. Microbiol.* **17**:409.
189. Matthew, M. and R. W. Hedges. 1976. Analytical isoelectric focusing of R factor-determined β-lactamases: Correlation with plasmid compatibility. *J. Bact.* **125**:713.
190. Meyer, R., D. Figurski and D. R. Helinski. 1975. Molecular vehicle properties of the broad host range plasmid RK2. *Science* **190**:1226.
191. Meynell, E. and M. Cooke. 1969. Repressor minus and operator constitutive derepressed mutants of F-like R factors. Their effect on chromosomal transfer by HfrC. *Genet. Res.* **14**:309.
192. Meynell, E. and N. Datta. 1966. The nature and incidence of conjugation factors. in *E. coli. Gen. Res.* **7**:141.
193. ——. 1966. The relation of resistance transfer factors to the F factor (sex factor) of *E. coli* K12. *Genet. Res.* **7**:134.
194. ——. 1967. Mutant drug resistance factors of high transmissibility. *Nature* **214**:885.
195. Mindich, L., J. Cohen and M. Weisburd. 1976. Isolation of nonsense suppressor mutants in *Pseudomonas. J. Bact.* **126**:177.
196. Mitsuhashi, S., F. Kobayashi and M. Yamaguchi. 1971. Enzymatic inactivation of gentamicin C components by cell-free extracts from *Pseudomonas aeruginosa. J. Antibiot.* (Tokyo) **24**:400.
197. Morrow, J. F., S. N. Cohen, A. C. Y. Chang, H. W. Boyer, H. M. Goodman and R. B. Helling, Replication and transcription of eukaryotic DNA in *E. coli. Proc. Nat. Acad. Sci.* **71**:1743.
198. Nieder, M. and J. Shapiro. 1975. Physiological functions of the *Pseudomonas putida* ppG6 (*Pseudomonas oleovorans*) alkane hydroxylase: Monoterminal oxidation of alkanes and fatty acids. *J. Bact.* **122**:93.
199. Nisen, P., D. J. Kopecko and S. N. Cohen, in preparation.
200. Novick, R. P. 1963. Analysis by transduction of mutations affecting penicillinase formation in *Staphylococcus aureus. J. Gen. Microbiol.* **33**:121.
201. ——. 1967. Penicillinase plasmids of *Staphylococcus aureus. Fed. Proc.* **26**:29.
202. ——. 1967. Properties of a cryptic high frequency transducing phage in *Staphylococcus aureus. Virology* **33**:155.
203. ——. 1974. Bacterial plasmids. In *Handbook of microbiology* (ed. A. J. Laskin and H. Lechevalier), vol. IV, p. 537. Chemical Rubber Co., Cleveland, Ohio.
204. ——. 1976. Plasmid protein relaxation complexes in *Staphylococcus aureus. J. Bact.* **127**:1177.
205. ——, unpublished.
206. Novick, R. P. and D. Bouanchaud. 1971. Extrachromosomal nature of drug resistance in *Staphylococcus aureus. Ann. N.Y. Acad. Sci.* **182**:279.

207. Novick, R. P. and M. H. Richmond. 1965. Nature and interaction of the genetic elements governing penicillinase synthesis in *S. aureus*. *J. Bact.* **90**:467.
208. Novick, R. P. and C. Roth. 1968. Plasmid-linked resistance to inorganic salts in *S. aureus*. *J. Bact.* **95**:1335.
209. Novick, R. P., R. C. Clowes, S. N. Cohen, R. Curtiss III, N. Datta and S. Falkow. 1976. Uniform nomenclature for bacterial plasmids: A proposal. *Bact. Rev.* **40**:168.
210. O'Hara, K. and M. Kono. 1975. Mechanism of tetracycline resistance in *Pseudomonas aeruginosa* carrying an R factor. *J. Antibiot.* (Tokyo) **28**:607.
211. Ohtsubo, E., R. C. Deonier, H.-J. Lee and N. Davidson. 1974. Electron microscope heteroduplex studies of sequence relations among plasmids of *Escherichia coli*. IV. The F sequences in F14. *J. Mol. Biol.* **89**:565.
212. Olsen, R. H. and C. Gonzalez. 1974. *Escherichia coli* gene transfer to unrelated bacteria by a histidine operon-RP1 drug resistance plasmid complex. *Biochem. Biophys. Res. Comm.* **59**:377.
213. Olsen, R. H., and J. Hansen. 1976. Evolution and utility of a *Pseudomonas aeruginosa* drug resistance factor. *J. Bact.* **125**:837.
214. Olsen, R. H. and P. Shipley. 1973. Host range and properties of the *Pseudomonas aeruginosa* R factor R1822. *J. Bact.* **113**:772.
215. Olsen, R. H. and D. D. Thomas. 1973. Characteristics and purification of PRR1, an RNA phage specific for the broad host range *Pseudomonas* R1822 drug resistance plasmid. *J. Virol.* **12**:1560.
216. Olsen, R. H., J. Siak and R. H. Gray. 1974. Characteristics of PRD1, a plasmid-dependent broad host range DNA bacteriophage. *J. Virol.* **14**:689.
217. Palchaudhuri, S., personal communication.
218. Palchaudhuri, S. and A. Chakrabarty. 1976. Isolation of plasmid deoxyribonucleic acid from *Pseudomonas putida*. *J. Bact.* **126**:410.
219. Parish, J. H. 1975. Transfer of drug resistance to myxococcus from bacteria carrying drug-resistance factors. *J. Gen. Microbiol.* **87**:198.
220. Pemberton, J. M. and A. J. Clark. 1973. Detection and characterization of plasmids in *Pseudomonas aeruginosa* strain PAO. *J. Bact.* **114**:424.
221. Pemberton, J. M. and B. W. Holloway. 1973. A new sex factor of *Pseudomonas aeruginosa*. *Genet. Res.* **21**:263.
222. Peyru, G., L. Wexler and R. P. Novick. 1969. Naturally occurring penicillinase plasmids in *Staphylococcus aureus*. *J. Bact.* **98**:215.
223. Ptashne, K. and S. N. Cohen. 1975. Occurrence of IS regions on plasmid DNA as direct and inverted nucleotide sequence duplications. *J. Bact.* **122**:776.
224. Rheinwald, J. G., A. M. Chakrabarty and I. C. Gunsalus. 1973. A transmissible plasmid controlling camphor oxidation in *Pseudomonas putida*. *Proc. Nat. Acad. Sci.* **70**:885.
225. Richmond, M. H. and R. B. Sykes. 1972. The chromosomal integration of a β-lactamase gene derived from the P-type R-factor RP1 in *Escherichia coli*. *Genet. Res.* **20**:231.
226. Rodriguez-Lemoine, V., A. E. Jacob, R. W. Hedges and N. Datta. 1975. Thermosensitive production of their transfer system by group S plasmids. *J. Gen. Microbiol.* **86**:111.

227. Romero, E. and E. Meynell. 1969. Covert *fi*$^-$ R factors in *fi*$^+$ R$^+$ strains of bacteria. *J. Bact.* **97**:780.
228. Ruby, C. and R. P. Novick. 1975. Plasmid interactions in *Staphylococcus aureus*: Non-additivity of compatible plasmid DNA pools. *Proc. Nat. Acad. Sci.* **72**:5031.
229. Rush, M. G., C. N. Gordon, R. P. Novick and R. C. Warner. 1969. Penicillinase plasmid DNA from *Staphylococcus aureus. Proc. Nat. Acad. Sci.* **63**:1304.
230. Sagai, H., V. Krčméry, K. Hasuda, S. Iyobe, H. Knothe and S. Mitsuhashi. 1975. R factor-mediated resistance to aminoglycoside antibiotics in *Pseudomonas aeruginosa. Japan. J. Microbiol.* **19**:427.
231. Salisbury, V. 1972. Ph.D. thesis, University of London.
232. Saunders, J. R. and J. Grinsted. 1972. Properties of RP4, an R factor which originated in *Pseudomonas aeruginosa* S8. *J. Bact.* **112**:690.
233. Sawada, Y., S. Yaginuma, M. Tai, S. Iyobe and S. Mitsuhashi. 1976. Plasmid mediated penicillin beta-lactamases in *Pseudomonas aeruginosa. Antimicrob. Agts. Chemother.* **9**:55.
234. Sawai, T., K. Takahashi, S. Yamagishi and S. Mitsuhashi. 1970. Variant of penicillinase mediated by an R factor in *Escherichia coli. J. Bact.* **104**:620.
235. Scavizzi, M. R. 1974. Nouveaux groupes d'incompatibilite des plasmides. Interet dans les epidemies de creche a *E. coli* 0111:B4. *Ann. Microbiol.* (Inst. Pasteur) **124B**:153.
236. Schildkraut, C. L., J. Marmur and P. Doty. 1962. Determination of the base composition of deoxyribonucleic acid from its buoyant density in CsCl. *J. Mol. Biol.* **4**:430.
237. Schottel, J., A. Mandal, D. Clark, S. Silver and R. W. Hedges. 1974. Volatilization of mercury and organomercurials determined by inducible R factor systems in enteric bacteria. *Nature* **251**:335.
238. Schrempf, H., H. Bujard, D. A. Hopwood and W. Goebel. 1975. Isolation of covalently closed circular deoxyribonucleic acid from *Streptomyces coelicolor* A3(2). *J. Bact.* **121**:416.
239. Shaham, M., A. M. Chakrabarty and I. C. Gunsalus. 1973. Camphor plasmid mediated chromosomal transfer in *Pseudomonas putida. J. Bact.* **116**:944.
240. Shahrabadi, M. S., L. E. Bryan and H. M. van den Elzen. 1975. Further properties of P-2 R-factors of *Pseudomonas aeruginosa* and their relationship to other plasmid groups. *Can. J. Microbiol.* **21**:592.
241. Shapiro, J., unpublished.
242. Sharp, P. A., S. N. Cohen and N. Davidson. 1973. Electron microscope heteroduplex studies of sequence relations among plasmids of *Escherichia coli.* II. Structure of drug resistance (R) factors and F factors. *J. Mol. Biol.* **73**:235.
243. Sharp, P. A., M.-T. Hsu, E. Ohtsubo and N. Davidson. 1972. Electron microscope heteroduplex studies of sequence relations among plasmids of *Escherichia coli.* I. Structure of F-prime factors. *J. Mol. Biol.* **71**:471.
244. Shaw, W. V., L. Sands and N. Datta. 1972. Hybrids of variants of chloramphenicol acetyltransferase specified by *fi*$^+$ and *fi*$^-$ R factors. *Proc. Nat. Acad. Sci.* **69**:5049.

245. Shipley, P. L. and R. H. Olsen. 1974. Characteristics and expression of tetracycline resistance in gram-negative bacteria carrying the *Pseudomonas* R factor RP1. *Antimicrob. Agts. Chemother.* **6**:183.
246. ——. 1975. Isolation of a nontransmissible antibiotic resistance plasmid by transductional shortening of R factor RP1. *J. Bact.* **123**:20.
247. Skurray, R. A., H. Nagaishi and A. J. Clark. 1976. Molecular cloning of DNA from F sex factor of *Escherichia coli* K-12. *Proc. Nat. Acad. Sci.* **73**:64.
248. Skurray, R. A., M. S. Guyer, K. Timmis, F. Cabello, S. N. Cohen, N. Davidson and A. J. Clark. 1976. Replication region fragments cloned from F*lac*$^+$ are identical to *Eco*RI fragment f5 of F. *J. Bact.* **127**:1571.
249. Smith, H. W. 1972. Ampicillin resistance in *E. coli* by phage infection. *Nature New Biol.* **238**:205.
250. Smith, K. and R. P. Novick. 1972. Genetic studies on plasmid linked cadmium resistance in *Staphylococcus aureus. J. Bact.* **112**:761.
251. Smith, H. R., N. D. F. Grindley, G. O. Humphreys and E. S. Anderson. 1973. Interactions of Group H resistance factors with the F factor. *J. Bact.* **115**:623.
252. Smith, D. I., R. Gomez Lus, M. Rubio Calvo, N. Datta, A. E. Jacob and R. W. Hedges. 1975. Third type of plasmid conferring gentamicin resistance in *Pseudomonas aeruginosa. Antimicrob. Agts. Chemother.* **8**:227.
253. Stanisich, V. A. 1973. Isolation and properties of mutants of the FP2 sex factor of *Pseudomonas aeruginosa. Genet. Res.* **22**:13.
254. ——. 1974. The properties and host range of male-specific bacteriophages of *Pseudomonas aeruginosa. J. Gen. Microbiol.* **84**:332.
255. ——. Interaction between an R factor and an element conferring resistance to mercuric ions in *Pseudomonas aeruginosa. Mol. Gen. Genet.* **128**: 201.
256. Stanisich, V. A. and B. W. Holloway. 1971. Chromosome transfer in *Pseudomonas aeruginosa* mediated by R factors. *Genet. Res.* **17**:169.
257. ——. 1972. A mutant sex factor of *Pseudomonas aeruginosa. Genet. Res.* **19**:91.
258. Stanisich, V. A., P. M. Bennett and J. M. Ortiz. 1976. A molecular analysis of transductional marker rescue involving P-group plasmids in *Pseudomonas aeruginosa. Mol. Gen. Genet.* **143**:333.
259. Stiffler, P. W., H. M. Sweeney and S. Cohen. 1974. Isolation and characterization of a kanamycin resistance plasmid from *Staphylococcus aureus. J. Bact.* **120**:934.
260. Stiffler, P. W., H. M. Sweeney, M. Schneider and S. Cohen. 1974. Isolation and characterization of a kanamycin resistance plasmid from *Staphylococcus aureus. Antimicrob. Agts. Chemother.* **6**:516.
261. Stoleru, G. H., G. R. Gerbaud, D. H. Bouanchaud and L. Le Minor. 1972. Etude d'une plasmide transférable detérminant la production d'H_2S et la résistance à la tétracycline chez *E. coli. Ann. Inst. Pasteur* **123**:743.
262. Summers, A. O. and E. Lewis. 1973. Volatilization of mercuric chloride by mercury-resistant plasmid-bearing strains of *Escherichia coli, Staphylococcus aureus,* and *Pseudomonas aeruginosa. J. Bact.* **113**:1070.

263. Sykes, R. B. and M. H. Richmond. 1970. Intergeneric transfer of a β-lactamase gene between *P. aeruginosa* and *E. coli*. *Nature* **226**:952.
264. Terakado, N., H. Azechi, K. Minomiya, T. Fukuyasu and K. Shimuza. 1974. The incidence of R factors in *Bordetella bronchiseptica* isolated from pigs. *Japan. J. Microbiol.* **18**:45.
265. Terawaki, Y., H. Takasasu and T. Akiba. 1967. Thermosensitive replication of a kanamycin resistance factor. *J. Bact.* **94**:687.
266. Timmis, K., F. Cabello and S. N. Cohen. 1974. Utilization of two distinct modes of replication by a hybrid plasmid constructed *in vitro* from separate replicons. *Proc. Nat. Acad. Sci.* **71**:4556.
267. ——. 1975. Cloning, isolation, and characterization of replication regions of complex plasmid replicons. *Proc. Nat. Acad. Sci.* **72**:2242.
268. Towner, K. J. and A. Vivian. 1976. RP4-mediated conjugation in *Acinetobacter calcoaceticus*. *J. Gen. Microbiol.* **93**:355.
269. Tseng, J. T. and L. E. Bryan. 1973. Mechanism of R factor R931 and chromosomal tetracycline resistance in *Pseudomonas aeruginosa*. *Antimicrob. Agts. Chemother.* **3**:638.
270. Tseng, J. T., L. E. Bryan and H. M. van den Elzen. 1972. Mechanisms and spectrum of streptomycin resistance in a natural population of *Pseudomonas aeruginosa*. *Antimicrob. Agts. Chemother.* **2**:136.
271. van Rensburg, A. J. and M. J. de Kock. 1974. A new R factor from *Pseudomonas aeruginosa*. *J. Gen. Microbiol.* **82**:207.
272. Vapnek, D., M. B. Lipman and W. D. Rupp. 1971. Physical properties and mechanism of transfer of R factors in *Escherichia coli*. *J. Bact.* **108**:508.
273. Vivian, A. 1971. Genetic control of fertility in *Streptomyces coelicolor* A3(2): Plasmid involvement in the interconversion of UF and IF strains. *J. Gen. Microbiol.* **69**:353.
274. Vivian, A. and D. Hopwood. 1970. Genetic control of fertility in *Streptomyces coelicolor* A3(2): The IF-fertility type. *J. Gen. Microbiol.* **64**:101.
275. Watanabe, T., T. Ogata, R. K. Chan and D. Botstein. 1972. Specialized transduction of tetracycline resistance by phage P22 in *Salmonella typhimurium*. I. Transduction of R factor 222 by phage P22. *Virology* **50**:874.
276. Watanabe, T., H. Nishida, C. Ogata, T. Arai and S. Sato. 1964. Episome mediated transfer of drug resistance in *Enterobacteriaceae*. *J. Bact.* **88**:716.
277. Watson, J. M. and B. W. Holloway. 1976. Suppressor mutations in *Pseudomonas aeruginosa*. *J. Bact.* **125**:780.
278. Weisblum, B., personal communication.
279. Williams, P. A. and K. Murray. 1974. Metabolism of benzoate and the methylbenzoates by *Pseudomonas putida (arvilla)* mt-2: Evidence for the existence of a Tol plasmid. *J. Bact.* **120**:416.
280. Williams, P. A. and M. J. Worsey. 1976. Ubiquity of plasmids in coding for toluene and xylene metabolism in soil bacteria: Evidence for the existence of new Tol plasmids. *J. Bact.* **125**:818.
281. Witchitz, J. L. and Y. A. Chabbert. 1971. Résistance transférable à la gentamicine. I. Expression du caractère de résistance. *Ann. Inst. Pasteur* **121**:733.

282. ——. 1972. Résistance transférable à la gentamicine. II. Transmission et liaisons du charactère de résistance. *Ann. Inst. Pasteur* **122**:367.
283. Witchitz, J. L. and G. R. Gerbaud. 1972. Classification des plasmides conferant la résistance à la gentamicine. *Ann. Inst. Pasteur* **123**:333.
284. Wong, C. L. and N. W. Dunn. 1974. Transmissible plasmid coding for the degradation of benzoate and m-toluate in *Pseudomonas arvilla* mt-2. *Genet. Res.* **23**:227.
285. Worsey, M. J. and P. A. Williams. 1975. Metabolism of toluene and xylenes by *Pseudomonas putida (arvilla)* mt-2: Evidence for a new function of the Tol plasmid. *J. Bact.* **124**:7.
286. Yamagishi, S., K. O'Hara, T. Sawai and S. Mitsuhashi. 1969. The purification and properties of penicillin β-lactamases mediated by transmissible R factors in *Escherichia coli. J. Biochem.* **66**:11.
287. Yogi, Y., A. E. Franke and D. B. Clewell. 1975. Plasmid determined resistance to erythromycin. Comparison of strains of *S. faecalis* and *S. pyogenes* with regard to plasmid homology and resistance inducibility. *Antimicrob. Agts. Chemother.* **1**:871.
288. Zechovsky, N. 1974. Analyse des plasmides isoles lors d'une epidemie de *S. wein.* Societe Francaise de Microbiologie, Dec. 1974.

Appendix C: Temperate Bacteriophages

1. THE GENOMES OF TEMPERATE VIRUSES OF BACTERIA

Contributed by E. Ljungquist and A. I. Bukhari, Cold Spring Harbor Laboratory

The temperate bacteriophages have a wide range of lysogenization modes. At one extreme is bacteriophage P1, which resides in the lysogenic cells as a plasmid and does not integrate its DNA into the host chromosome. At the other extreme is bacteriophage Mu, which gives rise to lysogens by randomly integrating its DNA into the host genome. In between fall bacteriophages λ and P22, which have a highly preferred chromosomal site for integration, and P2, which has a few preferred sites for integration.

The genomes of temperate bacteriophages are perhaps evolutionarily related to plasmids. Recently, it has been found that P1 DNA has homology with R plasmids (A. I. Bukhari and F. de Bruijn, unpubl.). This homology appears to result from the presence of IS*1* in both P1 and R plasmids. Bacteriophage Mu has been shown to have a DNA segment in common with P1 (see Chow and Bukhari, this volume). Mu also shares many properties with insertion elements and transposons that have been found in many plasmids and bacterial chromosomes. Bacteriophage λ, the paradigm of integrative recombination, can itself be forced to assume the role of a plasmid under certain conditions (see *The Bacteriophage Lambda,* ed. A. D. Hershey [Cold Spring Harbor Laboratory, 1971]).

All of the temperate bacteriophages so far examined yield linear DNA molecules upon extraction of DNA from phage particles. The DNA molecules, however, are converted to circular forms upon infection of host cells. A notable exception to this rule is Mu. There are two well-known mechanisms used in bacteriophage systems for the conversion of linear DNA molecules into circles. Some phages, such as λ and P2, have at each $5'$ end complementary single-stranded sequences (sticky or cohesive ends). Others, such as P1 and P22, have terminal duplications that can recombine to give circles. Mu does not have either one of these two mechanisms at its disposal and apparently has no direct means of fusing its ends.

The following table is designed to survey and summarize the properties of temperate phages in terms of the structures of their DNAs and integration of their DNAs into the genomes of various hosts. We have listed mainly those temperate phages for which information on the DNA or sites of integration is available. When possible, references are given to review articles where other references can be found for further details.

Host	Phage	Viral DNA			
		size	form	ends	uniqueness
Agrobacterium tumefaciens	LV-1 (PV-1)	46.4×10^6 (32)[b]	linear (32)	cohesive (59)	
	PB2A	31×10^6 (59)			
Bacillus subtilis 164; *Bacillus amyloliquefaciens* H	ϕ105	24×10^6 (5)	linear (5)		
Bacillus subtilis	SPO2	26×10^6 (7)	linear (7)		
Bacillus subtilis	PBS-1	190×10^6 (24)			
Bacillus tiberius	α	33×10^6 (54)			
Corynebacteria diphtheriae	β				
Escherichia coli	λ (20)	30.8×10^6	linear	cohesive	unique
Escherichia coli	PI	60×10^6 (25)	linear (27)	terminal (26) redundant	permuted (26)
Escherichia coli	ϕ80	27×10^6 (47)	linear (2, 47, 62)	cohesive (2, 62)	unique (51)
Escherichia coli	186	20×10^6 (13, 60)	linear (13)	cohesive (2, 60)	unique (13)
Escherichia coli; Citrobacter freundii; Shigella dysenteriae	Mu (23)[c]	25×10^6	linear	heterogeneous host sequences	unique
Escherichia coli; Shigella dysenteriae	P2 (4)	22×10^6	linear	cohesive	unique
Escherichia coli; Shigella dysenteriae	P4	6.7×10^6 (28)	linear (28)	cohesive (28, 61)	unique (28)

Intracellular forms	Prophage[a] DNA	Integration sites	DNA packaging	Transduction	
				gen.	spec.
circles (49)					
	collinear (14)	*phe*1-*ilvA*1 (46)			
covalently closed circles (1)	circularly permuted (14)	*ery*1 (29)			
		probably does not integrate (57, 58)		x (56)	
	circularly permuted (33)				
concatemers; covalently closed and open circles; doubly branched circles	circularly permuted	*gal-bio*	site-specific		x
	circular, non-integrated (26)	does not integrate (6, 26, 31)	headful (27)	x (25, 27)	x (26)
	circularly permuted	*tonB* (38)			x (38)
heterogeneous covalently closed circles	collinear	random	headful	x	
no concatemers; covalently closed and open circles	circularly permuted	I near *his* (pref. site in *E. coli* C), II near *ilv,* III near *try,* IV–IX not precisely located, H near *his*	site-specific		
no concatemers; covalently closed and open circles (35)		*thr-leu* (50)	site-specific (18)		

Host	Phage	Viral DNA size	form	ends	uniqueness
Haemophilus influenzae	S2	2×10^7 (8)[b]	linear (8)	cohesive (8, 9)	
Haemophilus influenzae	HPI (19)	2×10^7 (8)	linear (8)	cohesive (8, 9)	
Pseudomonas aeruginosa	B3	20–25 $\times 10^6$ (15)			
Pseudomonas aeruginosa	D3	44×10^6 (41)	linear (41)		
Pseudomonas aeruginosa	F116	38×10^6 (41)	linear (41)		
Pseudomonas aeruginosa	G101	38×10^6 (41)	linear (41)		
Pseudomonas aeruginosa	H90				
Pseudomonas aeruginosa	J51				
Rhizobium meliloti	16-3 (55)	1.1×10^8 (48)			
Salmonella typhimurium	P22 (34)	26×10^6	linear	redundant	permuted
Serratia marcescens	Kappa, κ				
Staphylococcus aureus	φ11 (P11)	32.7×10^6 (10)	linear (10)		
Streptococcus lactis					
Streptomyces coelicolor	φC31 (37)				
Streptomyces coelicolor	VP_5				
Vibrio cholerae	VcA2 (17)				

[a]Collinear means that the prophage gene order and the phage gene order are the same. Circularly permuted means that the prophage gene order is permuted with respect to the phage gene order.

[b]Numbers in parentheses refer to references given below.

[c]Recently, another Mu-like phage, D108 of *E. coli*, has been characterized (G. S. Gill, R. A. Hull and R. Curtiss III, pers. comm.).

1. Arwert, F., G. Bjursell and L. Rutberg. 1976. Induction of prophage SPO2 in *Bacillus subtilis*: Isolation of excised prophage DNA as a covalently closed circle. *J. Virol.* **17**:492.
2. Baldwin, R. L., P. Barrand, A. Fritsch, D. A. Goldthwait and F. Jacob. 1966.

Intracellular forms	Prophage[a] DNA	Integration sites	DNA packaging	Transduction gen.	spec.
concatemers covalently closed and open circles (3, 42)	collinear (53);				
				x (22)	
				x (21)	
		*his*151 5-7-min region (11) *pur*66-*arg*2 50-min region (12)			
					x (45, 48)
concatemers; covalently closed circles	unique linear order of the circular veg. map	*proA*-*proC*	headful	x	x
		trp (52)			x (52)
				x (43)	x (44)
				x (39, 40)	x (39, 40)
		argA-*cysD* (36)			
		hisE-*cysD* (16)			

Cohesive sites on the deoxyribonucleic acids from several temperate coliphages. *J. Mol. Biol.* **17**:343.

3. Barnhart, B. J. and S. H. Cox. 1973. DNA replication of induced prophage in *Haemophilus influenzae*. *J. Virol.* **12**:165.
4. Bertani, L. E. and G. Bertani. 1971. Genetics of P2 and related phages. *Adv. Genet.* **16**:199.
5. Birdsell, D. C., G. M. Hathaway and L. Rutberg. 1969. Characterization of temperate *Bacillus* bacteriophage ϕ105. *J. Virol.* **4**:264.
6. Boice, L. B. and S. E. Luria. 1963. Behavior of prophage P1 in bacterial matings. II. Transfer of the defective prophage P1 d1. *Virology* **20**:147.
7. Boice, L., F. A. Eiserling and W. R. Romig. 1969. Structure of *Bacillus subtilis*

phage SPO2 and its DNA: Similarity of *Bacillus subtilis* phages SPO2, ϕ105 and SPPI. *Biochem. Biophys. Res. Comm.* **34**:398.

8. Boling, M. E., D. P. Allison and J. K. Setlow. 1973. Bacteriophage of *Haemophilus influenzae.* III. Morphology, DNA homology, and immunity properties of HPIc1, S2, and the defective bacteriophage from strain Rd. *J. Virol.* **11**:585.
9. Boling, M. E., J. K. Setlow and D. P. Allison. 1972. Bacteriophage of *Haemophilus influenzae*: I. Differences between infection by whole phage, extracted phage DNA and prophage DNA extracted from lysogenic cells. *J. Mol. Biol.* **63**:335.
10. Brown, D. T., N. C. Brown and B. T. Burlingham. 1972. Morphology and physical properties of *Staphylococcus* bacteriophage P11-M15. *J. Virol.* **9**:664.
11. Carey, K. E. and V. Krishnapillai. 1974. Location of prophage H90 on the chromosome of *Pseudomonas aeruginosa* strain PAO. *Genet. Res. Camb.* **23**:155.
12. ——. 1975. Chromosomal location of prophage J51 in *Pseudomonas aeruginosa* strain PAO. *Genet. Rest. Camb.* **25**:179.
13. Chattoraj, J. D. K., M. Schnös and K. B. Inman. 1973. Electron microscopic denaturation map of bacteriophage 186 DNA. *Virology* **55**:439.
14. Chow, L. T. and N. Davidson. 1973. Electron microscope study of the structures of the *Bacillus subtilis* prophages SPO2 and ϕ105. *J. Mol. Biol.* **75**:257.
15. Davidson, P. R., D. Freifelder and B. W. Holloway. 1964. Interruptions in the polynucleotide strands in bacteriophage DNA. *J. Mol. Biol.* **8**:1.
16. Dowding, J. E. and D. A. Hopwood. 1973. Temperate bacteriophages for *Streptomyces coelicolor* A3 (2) isolated from soil. *J. Gen. Microbiol.* **78**:349.
17. Gerdes, J. C. and W. R. Romig. 1975. Complete and defective bacteriophages of classical *Vibrio cholerae*: Relationship to the Kappa type bacteriophage. *J. Virol.* **15**:1231.
18. Goldstein, R., J. Lengyel, C. Pruss, K. Barrett, R. Calendar and E. W. Six. 1974. Head size determination and the morphogenesis of satellite phage P4. *Curr. Topics Microbiol. Immunol.* **68**:59.
19. Harm, W. and C. S. Rupert. 1963. Infection of transformable cells of *Haemophilus influenzae* by bacteriophage and bacteriophage DNA. *Z. Vererbungslehre.* **94**:336.
20. Hershey, A. D., ed. 1971. *The bacteriophage lambda.* Cold Spring Harbor Laboratory, Cold Spring Harbor, New York.
21. Holloway, B. W. and P. van de Putte. 1968. Lysogeny and bacterial recombination. In *Replication and recombination of genetic material* (ed. W. J. Peacock and R. D. Brock), p. 175. Australian Academy of Science, Canberra.
22. Holloway, B. W., J. B. Egan and M. Monk. 1960. Lysogeny in *Pseudomonas aeruginosa. Aust. J. Exp. Biol. Med. Sci.* **38**:321.
23. Howe, M. M. and E. G. Bade. 1975. Molecular biology of bacteriophage Mu: Genetic and biochemical analysis reveals many unusual characteristics of this novel bacteriophage. *Science* **190**:624.
24. Hunter, B. I., H. Yamagishi and I. Takahashi. 1967. Molecular weight of bacteriophage PBS1 deoxyribonucleic acid. *J. Virol.* **1**:841.
25. Ikeda, H. and J. Tomizawa. 1965. Transducing fragments in generalized transduction by phage P1. I. Molecular origin of the fragments. *J. Mol. Biol.* **14**:85.
26. ——. 1969. Prophage P1, an extrachromosomal replication unit. *Cold Spring Harbor Symp. Quant. Biol.* **33**:791.
27. ——. 1965. Transducing fragments in generalized transduction by phage P1. III. Studies with small phage particles. *J. Mol. Biol.* **14**:120.
28. Inman, R. B., M. Schnös, L. D. Simon, E. W. Six and D. H. Walker. 1971. Some morphological properties of P4 bacteriophage and P4 DNA. *Virology* **44**:67.
29. Inselburg, J. W., T. Eremenko-Volpe, L. Greenwald, W. L. Meadow and J. Marmur. 1969. Physical and genetic mapping of the SPO2 prophage on the chromosome of *Bacillus subtilis* 168. *J. Virol.* **3**:627.
30. Jacob, F. and E. L. Wollman. 1961. *Sexuality and the genetics of bacteria.* Academic Press, New York.
31. ——. 1958. Genetic and physical determinations of chromosomal segments in *Escherichia coli. Symp. Soc. Exp. Biol.* **12**:75.

32. Korant, B. and C. F. Pootjes. 1970. Physiochemical properties of *Agrobacterium tumefaciens* phage LV-1 and 1b DNA. *Virology* **40**:48.
33. Laird, W. and N. Groman. 1976. Prophage map of converting corynebacteriophage beta. *J. Virol.* **19**:208.
34. Levine, M. 1972. Replication and lysogeny with phage P22 in *Salmonella typhimurium. Curr. Topics Microbiol. Immunol.* **58**:135.
35. Lindqvist, B. H. and E. W. Six. 1971. Replication of bacteriophage P4 DNA in a nonlysogenic host. *Virology* **43**:1.
36. Lomovskaya, N. D., L. K. Emiljanova and S. I. Alikhanian. 1971. The genetic location of prophage on the chromosome of *Streptomyces coelicolor. Genetics* **68**:341.
37. Lomovskaya, N. D., N. M. Mkrtumian, N. L. Gostimskaya and V. N. Danilenko. 1972. Characterization of temperate actinophage ϕC31 isolated from *Streptomyces coelicolor* A3 (2). *J. Virol.* **9**:258.
38. Lozeron, H. A., W. Szybalski, A. Landy, J. Abelson and J. D. Smith. 1969. Orientation of transcription for the amber suppressor gene *suIII* as determined by hybridization between tyrosine tRNA and the separated DNA strands of transducing coliphage ϕ80d*suIII. J. Mol. Biol.* **39**:239.
39. McKay, L. L., R. R. Cords and K. A. Baldwin. 1973. Induction of prophage in *Streptococcus lactis* C2 by ultraviolet irradiation. *Appl. Microbiol.* **25**:682.
40. ——. 1973. Transduction of lactose metabolism in *Streptococcus lactis* C2′. *J. Bact.* **115**:810.
41. Miller, R. V., J. M. Pemberton and K. E. Richards. 1974. F116, D3 and G101: Temperate bacteriophages of *Pseudomonas aeruginosa. Virology* **59**:566.
42. Notani, N. K., J. K. Setlow and D. P. Allison. 1973. Intracellular events during infection by *Haemophilus influenzae* phage and transfection by its DNA. *J. Mol. Biol.* **75**:581.
43. Novick, R. P. 1963. Properties of a cryptic high-frequency transducing phage in *Staphylococcus aureus. J. Gen. Microbiol.* **33**:121.
44. ——. 1967. Properties of a cryptic high-frequency transducing phage in *Staphylococcus aureus. Virology* **33**:155.
45. Orosz, L., Z. Svab, A. Kondorosi and T. Sik. 1973. Genetic studies on rhizobiophage 16-3. I. Genes and functions on the chromosome. *Mol. Gen. Genet.* **125**:341.
46. Rutberg, L. 1969. Mapping of a temperate bacteriophage active on *Bacillus subtilis. J. Virol.* **3**:38.
47. Shinagawa, H., Y. Hosaka, H. Yamagishi and Y. Nishi. 1966. Electron microscopic studies on ϕ80 and ϕ80pt. Phage virions and their DNA. *Biken's J.* **9**:135.
48. Sik, T. and L. Orosz. 1971. Chemistry and genetics of *Rhizobium meliloti* phage 16-3. *Plant and Soil,* Special Volume, p. 57.
49. Silverman, C., J. Schell, N. van Lavebeke, G. Vervliet, H. Teuchy and M. Van Montagu. 1972. Studies of circular dioxyribonucleic acid molecules from bacteriophages of *Agrobacterium tumefaciens. Arch. Int. Phys. Biochim.* **80**:409.
50. Six, E. W. and C. A. Klug. 1973. Bacteriophage P4: A satellite virus depending on a helper such as prophage P2. *Virology* **51**:327.
51. Skalka, A. 1969. Nucleotide distribution and functional orientation in the deoxyribonucleic acid of phage ϕ80. *J. Virol.* **3**:150.
52. Steiger, H., V. Muller and G. Bauer. 1972. Nonreceptivity for κ phage of κ-lysogenic *Serratia* and reactions to superinfection of receptive cells with a mutant prophage. *Mol. Gen. Genet.* **114**:358.
53. Stuy, J. H. 1974. Origin and direction of *Haemophilus* bacteriophage HP1 DNA replication. *J. Virol.* **13**:757.
54. Sutton, W. D. and G. B. Petersen. 1971. Some structural properties of the DNA molecule from bacteriophage alpha. *Virology* **44**:371.
55. Szende, K. and F. Ordogh. 1960. Die Lysogenie von *Rhizobium meliloti. Naturwissenschaften.* **47**:404.
56. Takahashi, I. 1961. Genetic transduction in *Bacillus subtilis. Biochem. Biophys. Res. Comm.* **5**:171.

57. ——. 1964. Incorporation of bacteriophage genome by spores of *Bacillus subtilis*. *J. Bact.* 87:1499.
58. Tomita, F. and I. Takahashi. 1975. Changes in DNase activities in *Bacillus subtilis* infected with bacteriophage PBS1. *J. Virol.* **15**:1073.
59. Vervliet, G., M. Holsters, H. Teuchy, M. Van Montagu and J. Schell. 1975. Characterization of different plaque-forming and defective temperate phages in *Agrobacterium* strains. *J. Gen. Virol.* **26**:33.
60. Wang, J. C. 1967. Cyclization of coliphage 186 DNA. *J. Mol. Biol.* **28**:403.
61. Wang, J. C., V. V. Martin and R. Calendar. 1973. On the sequence similarity of the cohesive ends of coliphage P4, P2, and 186 deoxyribonucleic acid. *Biochemistry* **12**:2119.
62. Yamagishi, H., K. Nakamura and H. Ozeki. 1965. Cohesion occurring between DNA molecules of temperate phages ϕ80 and lambda or ϕ81. *Biochem. Biophys. Res. Comm.* **20**:727.

2a. GENETIC, PHYSICAL, AND RESTRICTION MAP OF BACTERIOPHAGE λ

Contributed by S. Gottesman and S. Adhya, National Cancer Institute, Bethesda

In Figure C2.1, the numbers inside the circles are percentages of mature λ*papa* length, starting at the *A* end and proceeding clockwise to *R*. The estimates for the physical location of genetic markers, indicated on the outer circle of the map, come from electron microscopic examination of appropriate heteroduplexes.

The four inner circles of the map represent the locations of restriction cuts made by four different restriction endonucleases. The fragment sizes are taken from the references indicated by the numbers in parentheses: *Hin*dIII (innermost circle) (2, 42); *Hpa*I (2nd circle from center) (42); *Bam* (3rd circle from center) (36); *Eco*RI (4th circle from center) (3, 26).

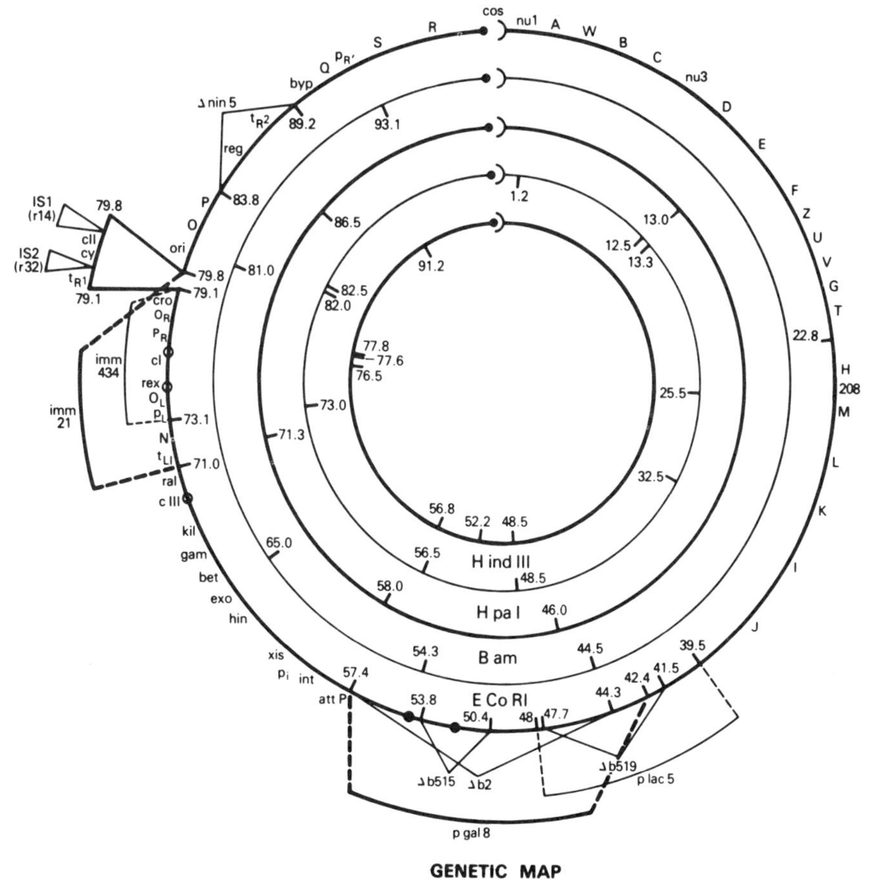

Figure C2.1

Genetic Symbols

nu1, A, W, B, C, nu3, D, E, and *F.* Code for DNA maturation and phage head proteins (7, 19, 39, 47).

Z, U, V, G, T, H, 208, M, L, K, I, and *J.* Code for phage tail proteins (7, 34, 35).

b_2, b_{515}, and b_{519}. Deletions for nonessential region of phage DNA; coding functions unknown (12).

attP. Phage site at which the chromosome integrates into host DNA (8, 45).

int. Codes for protein(s) involved in phage DNA integration into and excision from the host DNA (16, 20, 21, 55).

p_i. Promoter for *int* gene expression by *c*II and *c*III (9).

xis. Codes for prophage DNA excision function (16, 24).

hin. Region that (a) affects host permeability and (b) stabilizes phage mRNA (10).

exo. Phage homologous recombination function; codes for an exonuclease (46).

bet. Phage homologous recombination function (46).

gam. Gamma protein inhibits host exonuclease V and allows growth of phage in $recA^-$ host (Fec^+ phenotype) (51, 56).

*c*III. The product is needed for the synthesis of *c*I and *int* gene products from *cy* and p_i, respectively (30).

ral. Alleviates restriction in *trans* (54).

kil. Responsible for host killing (23).

t_{L1}. Rho-sensitive transcription terminator site (41).

N. Required for extension of transcription from p_L and p_R through t_{L1}, t_{R1}, and t_{R2} (7, 41, 49).

p_L. Promoter for initiation of transcription from *N* through *int* region (25).

o_L. Operator site, cognate to p_L, for repressor binding; consists of three subsites: o_{L1}, o_{L2}, and o_{L3} (38).

rex. Excludes growth of some bacteriophages, e.g., T4*r*II (29).

*c*I. Codes for phage repressor (13, 30, 37).

p_R. Promoter for initiation of transcription from *cro* through *Q* region (25).

o_R. Operator site, cognate to p_R, for repressor binding; consists of three subsites: o_{R1}, o_{R2}, and o_{R3} (38).

cro. Also known as *tof, Ai,* and *fed.* Depresses transcription from p_L and presumably from p_R and indirectly inhibits repressor synthesis (14, 15).

t_{R1}. Rho-sensitive transcription termination signal (41).

cy. Site of action of *c*II and *c*III for repressor synthesis (13, 53).

*c*II. The product is needed for the synthesis of *c*I and *int* gene products from *cy* and p_i, respectively (30).

ori. Origin for bidirectional initiation for phage DNA synthesis (31).

O and *P.* Control phage DNA synthesis (7, 31).

reg (*ren*). Determines sensitivity to *rex*-mediated exclusion (50).

t_{R2}. Rho-sensitive transcription termination site (41).

byp. Makes *Q* expression dependent upon *N* (6, 28).

Q. Positive regulator for transcription of the phage late genes *S* through *J* from the p_R' promoter (7, 27, 49).

p_R'. Promoter for initiation of transcription for the phage late genes *S* through *J* (27).

S. Codes for a protein involved in shutting off phage DNA synthesis and affects the host membrane (1, 40).

R. Codes for phage endolysin, an endopeptidase (48).

cos. Single-stranded ends of the phage DNA molecule, 12 nucleotides long and complementary to each other (52).

–●–. Sites at which *cam* elements can insert (33, 43).

–○–. Sites at which *tet* elements can insert (4, 5).

nin5. Signifies a deletion which relieves the requirement of *N* for phage growth (11, 12).

*r*32. An IS*2* element in the *cy* region in orientation I with respect to p_R (18).

*r*14. An IS*1* element in the *c*II gene in orientation I with respect to p_R (18).

imm^{434}. Substitution of the immunity region by the corresponding DNA of phage 434 (12, 32).

imm^{21}. Substitution of the immunity region by the corresponding DNA of phage 21 (12, 22).

p*gal8*. Substitution of the *b* region in plaque-forming galactose-transducing bacteriophage λp*gal8* (4, 17).

p*lac5*. Substitution of the *b* region in plaque-forming lactose-transducing bacteriophage λp*lac5* (44).

Acknowledgments

We wish to acknowledge the contribution made by lambda workers over the last 20 years. Their enthusiastic and energetic efforts have provided the vast quantities of information reflected in this map, and they continue to make lambda research an exciting and innovative field. In addition, we must thank our many colleagues who have contributed information prior to publication.

References

1. Adhya, S., A. Sen and S. Mitra. 1971. The role of gene *S*. In *The bacteriophage lambda* (ed. A. D. Hershey), p. 743. Cold Spring Harbor Laboratory, Cold Spring Harbor, New York.
2. Allet, B. and A. I. Bukhari. 1975. Analysis of bacteriophage Mu and λ-Mu hybrid DNAs by specific endonucleases. *J. Mol. Biol.* **92**:529.
3. Allet, B., P. G. N. Jeppesen, R. J. Katagiri and H. Delius. 1973. Mapping the DNA fragments produced by cleavage of λ DNA with endonuclease R1. *Nature* **241**:120.
4. Berg, D., unpublished observation.
5. Botstein and Kleckner, this volume.

6. Butler, B. and H. Echols. 1970. Regulation of bacteriophage λ development by gene *N*: Properties of a mutation that bypasses N control of late protein synthesis. *Virology* **40**:212.
7. Campbell, A. 1961. Sensitive mutants of bacteriophage λ. *Virology* **14**:22.
8. ——. 1962. The episomes. *Adv. Genet.* **11**:101.
9. Campbell et al., this volume.
10. Court, D., in preparation.
11. Court, D. and K. Sato. 1969. Studies of novel transducing variants of lambda: Dispensability of genes *N* and *Q*. *Virology* **39**:348.
12. Davidson, N. and W. Szybalski. 1971. Physical and chemical characteristics of lambda DNA. In *The bacteriophage lambda* (ed. A. D. Hershey), p. 45. Cold Spring Harbor Laboratory, Cold Spring Harbor, New York.
13. Echols, H. and L. Green. 1971. Establishment and maintenance of repression by bacteriophage lambda: The role of the cI, cII, and cIII proteins. *Proc. Nat. Acad. Sci.* **68**:2190.
14. Echols, H., L. Green, A. Oppenheim, A. Oppenheim and A. Honigman. 1973. The role of the *cro* gene in bacteriophage λ development. *J. Mol. Biol.* **80**:203.
15. Eisen, H., P. Brachet, P. deSilva and F. Jacob. 1970. Regulation of repressor expression in λ. *Proc. Nat. Acad. Sci.* **66**:855.
16. Enquist and Weisberg, this volume.
17. Feiss, M., S. Adhya and D. L. Court. 1972. Isolation of plaque-forming galactose-transducing strains of phage lambda. *Genetics* **71**:189.
18. Fiandt, M., W. Szybalski and M. H. Malamy. 1972. Polar mutations in *lac, gal,* and phage λ consist of a few IS2-DNA sequences inserted with either orientation. *Mol. Gen. Genet.* **119**:223.
19. Fuerst, C. and H. Bingham, as quoted in Murialdo and Siminovitch (ref. 34).
20. Gingery, R. and H. Echols. 1967. Mutants of bacteriophage λ unable to integrate into the host chromosome. *Proc. Nat. Acad. Sci.* **58**:1509.
21. Gottesman, M. E. and M. Yarmolinsky. 1968. Integration-negative mutants of bacteriophage lambda. *J. Mol. Biol.* **31**:487.
22. Gottesman, M. M., unpublished observation.
23. Greer, H. 1975. The *kil* gene of bacteriophage λ. *Virology* **66**:589.
24. Guarneros, G. and H. Echols. 1970. New mutants of bacteriophage λ with a specific defect in excision from the host chromosome. *J. Mol. Biol.* **47**:565.
25. Heinemann, S. F. and W. G. Spiegelman. 1971. Role of the gene N product in phage lambda. *Cold Spring Harbor Symp. Quant. Biol.* **35**:315.
26. Helling, R. B., H. M. Goodman and H. W. Boyer. 1974. Analysis of endonuclease R·*Eco*R1 fragments of DNA from lambdoid bacteriophages and other viruses by agarose gel electrophoresis. *J. Virol.* **14**:1235.
27. Herskowitz, I. and E. R. Signer. 1971. Control of transcription from the *r* strand of bacteriophage λ. *Cold Spring Harbor Symp. Quant. Biol.* **35**:355.
28. Hopkins, N. 1970. Bypassing a positive regulator; Isolation of a λ mutant that does not require N product to grow. *Virology* **40**:223.
29. Howard, B. 1967. Phage lambda mutations deficient in rII exclusion. *Science* **158**:1588.
30. Kaiser, A. D. 1957. Mutations in a temperate bacteriophage affecting its ability to lysogenize *Escherichia coli*. *Virology* **3**:42.

31. ——. 1971. Lambda DNA replication. In *The bacteriophage lambda* (ed. A. D. Hershey), p. 195. Cold Spring Harbor Laboratory, Cold Spring Harbor, New York.
32. Kaiser, A. D. and F. Jacob. 1957. Recombination between related temperate bacteriophages and the genetic control of immunity and prophage localization. *Virology* **4**:509.
33. MacHattie and Jackowski, this volume.
34. Murialdo, H. and L. Siminovitch. 1972. Morphogenesis of phage λ. V. Form determining function of genes required for the assembly of the head. *Virology* **48**:824.
35. Parkinson, J. S. 1968. Genetics of the left arm of the chromosome of bacteriophage lambda. *Genetics* **59**:311.
36. Perricaudet, M. and P. Tiollais. 1975. Defective bacteriophage lambda chromosome: Potential vector for DNA fragments obtained after cleavage by *Bacillus amyloliquefaciens* endonuclease (*Bam*). *FEBS Letters* **56**:7.
37. Ptashne, M. 1967. Isolation of the λ phage repressor. *Proc. Nat. Acad. Sci.* **57**:306.
38. Ptashne, M. and N. Hopkins. 1968. The operators controlled by the λ phage repressor. *Proc. Nat. Acad. Sci.* **60**:1282.
39. Ray, P. and H. Murialdo. 1975. Role of gene *nu3* in bacteriophage λ head morphogenesis. *Virology* **64**:247.
40. Reader, R. W. and L. Siminovitch. 1971. Lysis defective mutants of bacteriophage lambda: Genetics and physiology of *S* cistron mutants. *Virology* **43**:607.
41. Roberts, J. W. 1971. The *rho* factor: Termination and antitermination in lambda. *Cold Spring Harbor Symp. Quant. Biol.* **35**:121.
42. Robinson, L. H. and A. Landy. 1977. *Hin*dII, *Hin*dIII and *Hpa*I restriction fragments map of bacteriophage λ DNA. *Gene* **1**:(in press).
43. Rosner and Gottesman, this volume.
44. Shapiro, J., L. MacHattie, L. Eron, G. Ihler, K. Ippen, J. Beckwith, R. Arditti, W. Reznikoff and R. MacGillivray. 1969. The isolation of pure *lac* operon DNA. *Nature* **224**:768.
45. Shulman, M. and M. E. Gottesman. 1973. Attachment site mutants of bacteriophage λ. *J. Mol. Biol.* **81**:461.
46. Signer, E., H. Echols, J. Weil, C. Radding, M. Shulman, L. Moore and K. Manly. 1969. The general recombination system in bacteriophage λ. *Cold Spring Harbor Symp. Quant. Biol.* **33**:715.
47. Sternberg, N. and R. A. Weisberg, personal communication.
48. Taylor, A. 1971. The endopeptidase activity of phage λ endolysin. *Nature New Biol.* **234**:144.
49. Thomas, R. 1971. Control circuits. In *The bacteriophage lambda* (ed. A. D. Hershey), p. 211. Cold Spring Harbor Laboratory, Cold Spring Harbor, New York.
50. Toothman, P. 1976. Ph.D. thesis, University of Oregon, Eugene.
51. Unger, R. C. and A. J. Clark. 1972. Interactions of the recombination pathways of bacteriophage λ and its host *Escherichia coli* K12: Effects on exonuclease V activity. *J. Mol. Biol.* **70**:539.
52. Wu, R. and E. Taylor. 1971. Nucleotide sequence analysis of DNA. II.

Complete nucleotide sequences of the cohesive ends of bacteriophage λ DNA. *J. Mol. Biol.* **57**:491.

53. Wulff, D. L. 1976. Lambda *cin*-1, a new mutation which enhances lysogenization by bacteriophage lambda, and the genetic structure of the lambda *cy* region. *Genetics* 82:401.
54. Zabeau, M., M. Van Montagu and J. Schell, in preparation.
55. Zissler, J. 1967. Integration-negative (*int*) mutants of phage λ. *Virology* **31**:189.
56. Zissler, J., E. Signer and F. Schaefer. 1971. The role of recombination in growth of bacteriophage λ. I. The gamma gene. In *The bacteriophage lambda* (ed. A. D. Hershey), p. 455. Cold Spring Harbor Laboratory, Cold Spring Harbor, New York.

2b. RESTRICTION ENZYME CLEAVAGE MAPS OF BACTERIOPHAGE λ WITH A FOCUS ON THE ATTACHMENT-SITE REGION

Contributed by D. Kamp and R. Kahmann, Cold Spring Harbor Laboratory

The maps presented here (Fig. C2.2)[1] were derived by comparison of deletion and substitution mutants, analysis of mixed digests, and a detailed analysis of subfragments. A description of the mapping strategy employed and additional data are presented elsewhere (1)[2]. The linear λ genome is calibrated in percent units from left to right, the *A* gene being on the left end. For *Eco*RI cleavage sites, the estimates of Thomas and Davies (2) were used. The upper part is a cleavage map for the entire λ genome. The lower part shows an enlargement of the *Eco*RI fragment containing *att, int,* and *xis* and the positions of additional cleavage sites which have been mapped completely in this *Eco*RI fragment only. It should be noted that the right-hand *Eco*RI cleavage site for this fragment is placed at 65.6%, a slight deviation from the 65% figure given in the accompanying λ map (Fig. C2.1). All comparisons should be adjusted accordingly. Cleavage sites are indicated by vertical bars labeled with the name of the corresponding restriction enzyme. For a full description of the restriction enzymes used, see Appendix D. The attachment site is indicated by X. The 200-base-pair scale is for the enlarged map only.

References

1. Kamp, D., R. Kahmann, D. Zipser and R. J. Roberts. 1977. Mapping of restriction sites in the attachment site region of bacteriophage lambda. *Mol. Gen. Genet.* (in press).
2. Thomas, M. and R. D. Davis. 1975. Studies on the cleavage of bacteriophage lambda DNA with *Eco*RI restriction endonuclease. *J. Mol. Biol.* **19**:315.

[1] For Figure C2.2, see page following.

[2] Numbers in parentheses refer to references.

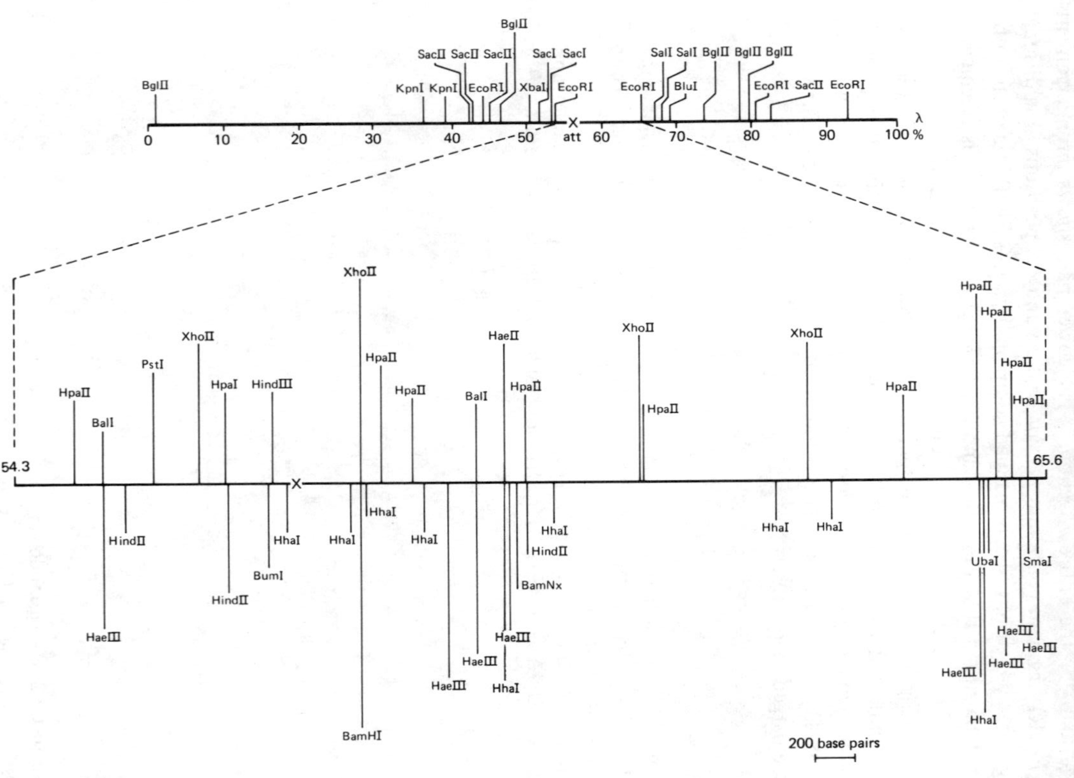
BglII
SacII SacII SacII
SacI SacI
KpnI KpnI EcoRI XbaI EcoRI
SalI SalI BglII BglII BglII
EcoRI BluI
EcoRI SacII EcoRI
λ
0 10 20 30 40 50 att 60 70 80 90 100 %
54.3
65.6
HpaII BalI PstI XhoII HpaI HindIII
XhoII HpaII HpaII HaeII BalI HpaII
XhoII HpaII XhoII HpaII
HpaII HpaII HpaII HpaII
HindII HaeIII HindII BumI HhaI
HhaI HhaI BamHI HhaI
HaeIII HaeIII HaeIII HhaI HhaI HindII BamNx
HhaI HhaI
UbaI SmaI HaeIII HaeIII HaeIII HaeIII HhaI
200 base pairs

Figure C2.2

3. GENETIC AND PHYSICAL STRUCTURE OF BACTERIOPHAGE P1 DNA

Contributed by M. B. Yarmolinsky, Frederick Cancer Research Center, Maryland

The DNA of plaque-forming particles of coliphage P1 (3)[1] (or, more precisely, of P1*kc* [32]) comprises a population of cyclically permuted, double-stranded molecules of molecular weight about 66 $\times$ 10^6 (27, 79, 80) and with extensive terminal redundancy (9-12% of the genome) (27, 31, 79). DNA of the closely related heteroimmune phage P7 (6, 22, 56, 59, 72, 77, 79, 80) (originally named ϕAmp [62]) has a similar molecular weight, but only 0.6% of the genome is terminally redundant (79, 80). The genome size of P7 is greater than the genome size of P1, principally because of a genetic region (shown to be a transposon [81]) which confers ampicillin resistance and is responsible for most of the reduction in redundant DNA (6, 79, 80).

P1 and P7 prophages are normally maintained as plasmids at one to two copies per host chromosome (12, 22, 27, 34, 39a, 63). These plasmids belong to the same incompatibility group (Y) (22, 77). Plasmid DNA is of genome size (58-63 $\times$ 10^6 daltons for P1, 63-67 $\times$ 10^6 daltons for P7) (22, 27, 79, 80) and lacks the redundancy characteristic of DNA in particles of active phage. Supercoiled DNA of P1 plasmids may be converted to linear molecules by a single cut with restriction endonuclease *Pst*I (2). By defining the unique site on the P1 genome at which *Pst*I acts (coincident with the Cm0 insertion site [25]) as 20 physical map units along the DNA (as drawn in Fig. C3.1), the physical map is approximately aligned with published genetic maps. The complete genome is contained within 100 physical map units. This means that one map unit corresponds to about 0.9 kilobases (kb).

In addition to plaque-forming particles, P1 and P7 lysates normally contain particles with variously smaller heads; roughly 20% of them have incomplete genomes. The majority class of small-head particles carry 40% of a genome equivalent of DNA (26, 70, 72). It is probable that both plaque-forming and small-head (defective) particles, as well as transducing (host DNA) particles (19), are products of a predominantly precessive head-full packaging mechanism acting upon concatemers, as first described for T4 (66). The initiation of DNA packaging may begin at a preferred site (as is the case for phage P22 [69]). When cleavage fragments obtained by digestion of P1 DNA extracted from phage particles and from plasmids are compared, an additional band is observed in phage digests. Bands derived from a region of the phage DNA at the right of the map are also intense. The differences observed provide evidence that DNA packaging starts preferentially at a site near 92 physical map units and proceeds to the right (2). The permutation of sequences in the DNA molecules from P1 particles has proved to be no serious obstacle to obtaining well-defined products of restriction enzyme digestion from them (2, 23, 49), and the cleavage maps

[1] Numbers in parentheses (or brackets) refer to references listed at the end of this presentation.

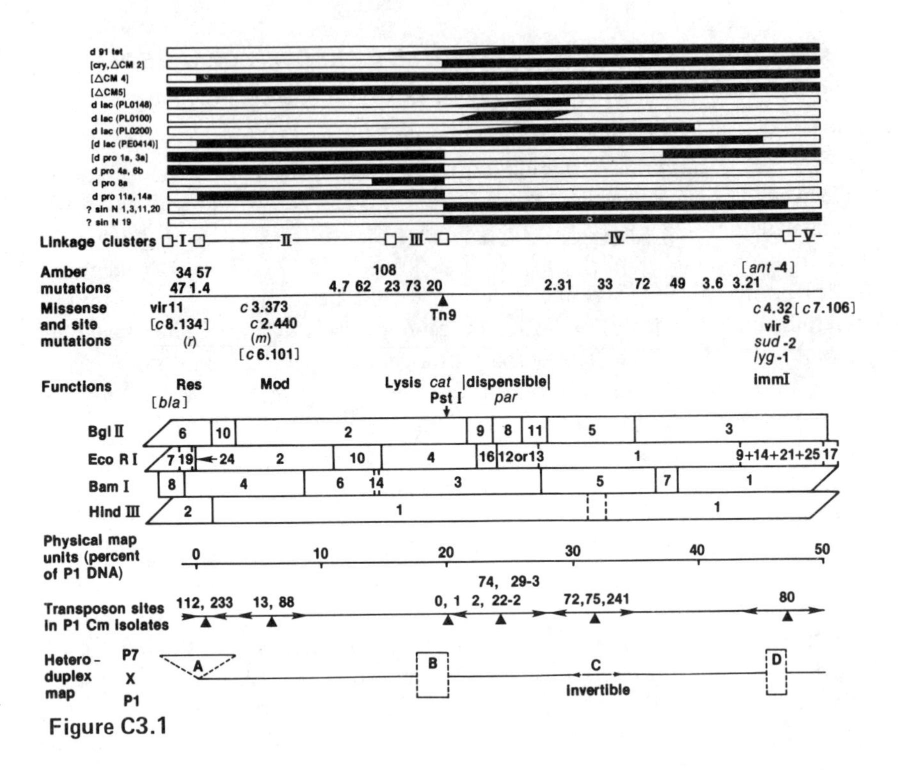

Figure C3.1

depicted in Figure C3.1 are based mainly on digestion of phage DNA (2).

Deletion prophages are listed in the top lines of the map. In addition to lacking continuous segments of phage DNA, certain of these prophages bear fragments of exogenous DNA, but only the extents of the deletions are indicated here. Black bars indicate deleted regions; tapered bars indicate regions of uncertainty. The markers that define the deletion end points are listed in Table 1. These are given either individually or as sets (members of linkage clusters) when the order of the relevant markers is uncertain. Several of the deletion prophages were generated from P1 harboring a transposon that confers chloramphenicol resistance (Tn*9* at Cm0) with or without loss of the transposon (column 3). In these cases, the deletions that by genetic evidence must end close to the transposon insertion site are assumed here to end at this site. I do not make this assumption for P1d*lac* prophages derived directly from P1*kc,* although the end points may be identical.

The names of the prophages are bracketed where there is evidence to suggest a chromosomal location, are without brackets where the plasmid character has been demonstrated, and are preceded by a question mark where no evidence on this point has been obtained.

Because much of the redundancy in P1 DNA is in excess of that required for effective circularization, P1 tolerates insertion of large transposons without necessarily becoming defective. Several have been inserted into P1 (25, 25a, 29,

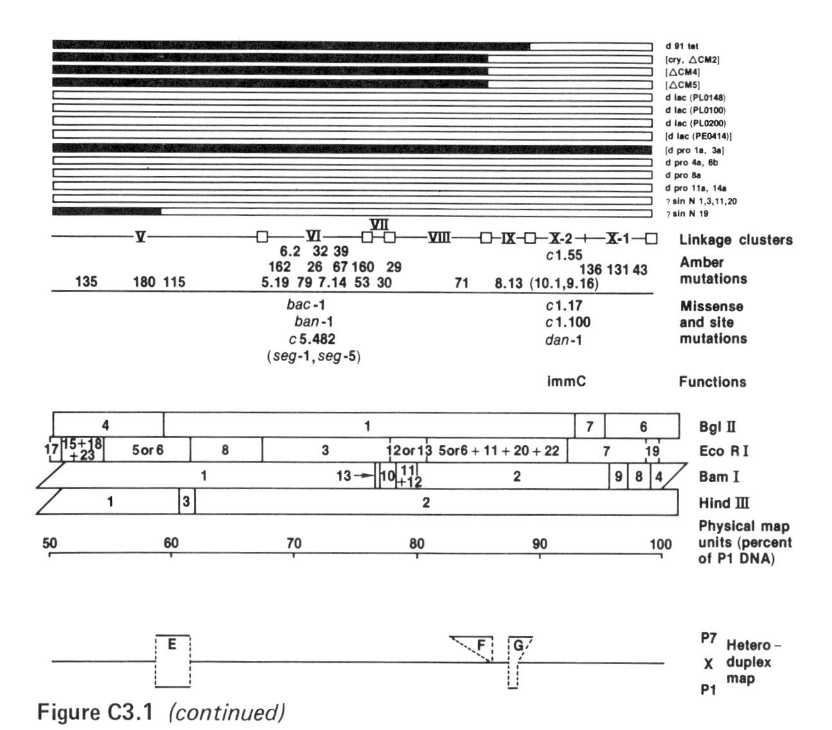

Figure C3.1 *(continued)*

35a, 36, 67). Transposon *9* (Tn*9*) is the name given to the transposon that confers chloramphenicol resistance in the first P1Cm phage to be described (29). Alternative locations of transposons which also confer chloramphenicol resistance (but are not identical with Tn*9*) are indicated on a separate line near the bottom of the map. These insertions are of various sizes and some are accompanied by deletions. Their locations are based on the band patterns obtained in restriction enzyme digests (25, 35a). Each P1Cm isolate is identified by a number, and the corresponding transposon site within a region of the cleavage map is indicated by a pair of arrowheads.

Both phage and prophage alike in P1 and P7 are heterogeneous with respect to the orientation of a 3-kb segment of DNA, which lies between 0.62-kb inverted repeats (31, 79, 80). This segment is homologous to the G region of phage Mu (8, 9). It is recognized as a largely unannealed segment in homoduplexes of P1 or P7 DNA formed after strand separation and reannealing. This is region "C" of the heteroduplex map, bottom line.

Mutations of P1 that have been isolated and mapped include about 120 amber mutations in functions required for plaque formation (6, 50, 54, 71, 73). These have been assigned to linkage clusters I through X on the basis of two-factor crosses performed on plates; linkage group X is further divided into two subclusters (71, 73). The numbers of mutations in the linkage clusters are as follows (73, 74):

Table 1 End Points of Deletion Prophages

Prophage (in brackets if integrated)	Reference	Tn*9* lost	Markers (genetic loci, linkage clusters, cleavage fragments) bracketing deletion end points				Footnote
			left		right		
			+	–	–	+	
d91*tet*	16, 56, 57, 75	–	62	2.31	8.13	X	a
[*cry*]	51, 54, 71, 73	no	*cat*	2.31	VIII	8.13	b
[ΔCm2]	54, 71, 73	yes	20	*cat*	VIII	8.13	b
[ΔCm4]	11, 54, 71, 73	yes	*r*, I	*m*, *c*3.378	VIII	8.13	b, c, d
[ΔCm5]	54, 71, 73	yes	*c*1.17	I	VIII	8.13	b, d
d*lac*(PLO148)	42, 71	–	20	2.31	2.31	33	
d*lac*(PLO100)	25, 42, 49, 73, 74	–	20	2.31? (part *Eco*RI-1)	2.31? (part *Eco*RI-1)	33	
d*lac*(PLO200)	41, 71, 73	–	20	2.31	49	3.6	e
[d*lac*(PEO414)]	42, 71, 73, 74	–	I	Mod, *c*3.378	3.21	*immC*	d, f
[d*pro*1a]	46, 71, 73	no	72	49	20	*cat*	
[d*pro*3a]	46, 71, 73	yes	72	49	*cat*	2.31	
d*pro*4a	46, 57, 71, 73	yes	*c*1.17	*vir*11	*cat*	2.31	
d*pro*6b	46, 71, 73	no	*c*1.17	I	20	*cat*	d
d*pro*8a	46, 71, 73	yes	62	108	*cat*	2.31	
d*pro*11a	46, 71, 73	yes	Res	Mod, *c*3.378	*cat*	2.31	
d*pro*14a	46, 71, 73	no	Res	Mod, *c*3.378	20	*cat*	
*sinN*1, 3, 5, 11, 20	1, 75	no	*cat*	2.31	*immC*	(180, 135)	g
*sinN*19	1, 75	no	*cat*	2.31	180	115?, 5.19	g

For each deletion end point, the marker-rescue tests were performed using all known amber mutations in the affected linkage cluster or in linkage clusters adjacent to the end point in question.

[a] The order of markers in linkage cluster X is presently in doubt, but all known amber mutations in the essential genes of X have been tested. P1d91*tet* is probably a plasmid by virtue of integration into an unrelated plasmid (16).

[b] The markers tested in linkage cluster VIII and lying at its extremities are 101, 142, 29, and 71.

[c] The Cm4 prophage does not express restriction, but the r^+ allele can be rescued from it by an r^- superinfecting phage.

[d] The markers tested in linkage cluster I and lying at its extremities are 173, 57, 76, 176, 129, and 47.

[e] (PLO200) is an independent (first step) P1d*lac* isolate obtained from a low frequency transducing lysate of P1*kc*.

[f] (PEO414) is the P1d*lac* strain referred to in reference 42 as "exconjugant No. 4."

[g] The designations *sinN*1 etc. will be replaced by Δ*N*1 etc. Representatives of the two classes of deletion phages listed here have recently been shown

I	II	III	IV	V	VI	VII	VIII	IX	X-1	X-2
8–14	4-6	5	14–22	7	11–22	2-5	8-9	1	8-12	3-10.

Where a range of values is given, the maximum value is the number of independent mutations in the linkage cluster and the minimum value is the number of these which are clearly separable by recombination. Representative mutations of each cluster are included in the map of genetic loci. The clear-plaque (*c*) mutations have been assigned to eight cistrons (50, 52, 53, 59). Cistron assignments have been made (and discarded [73]) for several amber mutations (50); ten "cistrons" have achieved stable designations (73). There is some doubt as to whether the tests upon which the assignments were based measured recombination rather than complementation. Mutations assigned to "cistrons" are given a binomial designation (e.g., 4.7 is a particular mutation in "cistron 4"). Each of the 107 amber mutants of P1 that have been tested form plaques on a non-suppressing host that carries P7 prophage (72). Mutations *c*6.101, *c*7.106, *c*8.134 (59), and the amber mutation *ant*-4 (77) (each one bracketed in the map) are mutations in P7 genes that can be rescued from P1 by recombination.

Mutations affecting more precisely defined functions than the capacity for plaque formation have also been isolated and mapped. These functions include (i) the capacity for stable lysogeny: mutations recognized as conferring either a clear-plaque phenotype (*c*1 through *c*8) (44, 50, 52, 53, 59) or a high rate of prophage loss (*lyg* [55], *sud* [13], *seg* [28], *c*7 [59, 60]); (ii) immunity (*c*1 [1, 44, 50, 59], *dan* [13], *c*4 [6, 52, 56, 59], *vir* [34, 50, 58, 59, 78], *sud* [13], *ant* [77] [*reb* (58, 61)]); (iii) specific restriction (Res) and specific modification (Mod, including *c*2 and *c*3) (17, 45, 52), as well as nonspecific protection against restriction (*par*) (25), analogous to the *mom* function of phage Mu (68); (iv) a *dnaB*-like function (*bac, ban*) (14, 39); and (v) lysis (amber mutations 23, 73, and 20) (74). Genes involved in replication (including integrative supression of *dnaA*ts bacterial mutants) (5, 7), recombination (generalized [24, 48] and site-specific [10]), and viral architecture (including determination of the proportion of head-size variants in the particle population) (25, 49, 70) remain to be identified. Mutations affecting the capacity for generalized transduction have been isolated but not mapped (76).

Many widely dispersed genes of P1 are normally expressed in the prophage state (6, 17, 25, 45, 52, 59), others (such as *ban* [14, 39]) may be exceptionally expressed by the prophage. Transcription maps of P1 in its various states would be informative, but these have yet to be constructed. For their construction, it would be helpful to be able to separate the complementary strands of P1 DNA. An electrophoretic method for the separation of the complementary strands of several DNA phages is not applicable to the intact strands of P1 DNA (20).

The linkage clusters into which the genetic map of P1 is divided have been ordered into a linear array by three-point crosses (50, 73). No evidence exists for circularity of the vegetative genetic map. Presumably, the ends of the map are separated by a recombinational "hot spot." It is improbable that the ends are separated by an extensive region within which no genetic loci have been identified because terminal genetic loci (47 on the left [37] and 43 on the right [64]) have been assigned to adjacent restriction enzyme fragments.

The restriction endonuclease cleavage maps of P1 are based on the analysis of single and double digests of P1*c*1ts DNA by restriction enzymes *Pst*I, *Bgl*II, *Eco*RI, *Bam*I, and *Hin*dIII (2). The restriction enzyme *Sal*I makes no cuts in P1 DNA (43). The cleavage fragments are numbered in order of their increasing electrophoretic mobility in gels. Minor adjustments (±1% to 2%) have been made in drawing the relative lengths of some single fragments in order to overlap them correctly. The plus sign (as in 11 + 12) links members of sets of fragments without assigned order; "or" (as in 12 or 13) links members of a comigrating pair. Except for resolution of *Eco*RI fragment 4, the cleavage maps have been constructed in a single laboratory (2). Fragment *Eco*RI-26 has been omitted from the *Eco*RI map because of the uncertainty about its position, but it lies within the region spanned by *Bam*I-1. Despite the high degree of homology between P1 and P7, the banding pattern of an *Eco*RI digest of P7 DNA reveals few similarities with the pattern obtained using P1 DNA (37).

The P1-P7 heteroduplex map (79) is aligned with the restriction endonuclease cleavage map on the basis of the following assignments: (i) The insertion site of transposon *9* in the original P1Cm phage (29) (here designated Cm0) at the right end of the nonhomology region "B" is drawn coincident with the *Pst*I site situated by convention at 20 map units. Transposon *9* (in phages Mu and λ) is flanked by direct repeats of IS*1* (9, 35), an element that is cut by *Pst*I endonuclease (18). Two fragments appear when the *Eco*RI fragment of P1 containing the *Pst*I site (*Eco*RI-4) (2, 49) is digested with *Pst*I. The *Eco*RI-4 fragment disappeared in the digests with *Eco*RI of DNA from P1Cm0 + P1Cm1 due to the insertion of the transposon, but double digestion of DNA from these Cm phages with *Eco*RI and *Pst*I produces the same two fragments that are observed after the double digestion of DNA from P1 (25). This evidence for the site of an IS*1*-mediated insertion coinciding with the site of *Pst*I action suggests that an IS*1*-like element may exist in P1 at this location. Hybridization experiments with a radioactive probe (15) and evidence based on heteroduplex mapping (35a) confirm that at least part of the IS*1* element may be a natural component of P1 DNA. (ii) There is one *Hin*dIII cleavage site within the invertible DNA segment in wild-type P1*kc*. Because the segment has two orientations, the cleavage site appears as a pair of sites (2). (They are absent in the P1*c*1.225[ts] analyzed [2] and in P1Cm *c*1.100 isolates [65]; presumably these strain differences are only fortuitously associated with the *c*I mutations.) The pair of sites (dotted lines) serve to position this segment within about 1 map unit from the site expected on the basis of the measured length of the "B-C" interval. (The "B-C" interval has been appropriately stretched in the drawing here.) The region that inverts is located in the largest *Eco*RI fragment (2, 37). (The inversion appears to be catalyzed by a P1-determined function active in *trans* [10].)

The P1-P7 heteroduplex map is aligned with the genetic map as follows:

(i) Transposon *9* insertion site Cm0 is identified with the gene for chloramphenicol acetyltransferase (30, 51, 79) (*cat*). (ii) The nonhomology region "A" is identified with an addition at the left end of the P1 genetic map within which lies the gene for the β-lactamase (*bla*) of P7 that confers resistance to ampicillin (6, 21, 79). The measured length of the "A-B" interval (used in drawing the heteroduplex map) places "A" about 2 physical map units to the right of its expected position as deduced from alignments of the genetic and cleavage maps given below (79). This discrepancy accounts for the existence of genetic markers (*vir*11, 1.4, and 47) that are drawn to the left of the left end of the heteroduplex map. (iii) The nonhomology region "D" is aligned with the *immI* region (6) responsible for the immunity difference between P1 and P7. The alignment is based on the genetic linkage of the *immI* immunity region to markers known to lie within *Eco*RI-1; this fragment also includes the invertible segment at "C" (37). Mutations in the gene determining the antagonist of *c*1 repression (*ant* [77], also called *reb* [59, 61]) lie closely linked to the *immI* region of nonhomology (6, 61) but are genetically exchangeable between P1 and P7 (77). The significance of the other nonhomology regions is unknown.

The genetic and cleavage maps are aligned on the basis of the data in Table 2. Cloned cleavage fragments are providing the means by which to resolve a number of ambiguities in the genetic map of P1, but deletion phages generated by other means are also being used to this end. In the minute plaque-forming phages derived from the P1d*lac* plasmid of PLO148, "cistron-2" function appears to be absent since the plasmid of PLO148 is "cistron-2"-defective by marker-rescue tests (71). These P1*min* phages have a prolonged lytic cycle and a lowered burst size, making plaques only marginally visible, if at all (42). Paradoxically, amber mutations in "cistron 2" cause premature lysis (75) and an early arrest of DNA synthesis (57).

From the extent of deletions that do not interfere with the maintenance of the plasmid state, with immunity, with incompatibility, or with integrative suppression (of *dnaA*ts mutations), it is possible to put limits on the regions of the viral genome necessary for these particularly interesting processes. Specifically, deletion prophages have proved helpful in dissecting the bipartite control of immunity in P1 and P7 (56, 77). The two genetic regions involved in immunity maintenance are designated here *immI* and *immC* in conformity with the P22 nomenclature (4, 33).

Incompatibility appears to be expressed by some deletion prophages that are integrated (e.g., by the P1d*lac* of PEO414 [42, 65]) as well as by other deletion prophages that are plasmids (e.g., by the P1d*pro* prophages with the exception of 1a and 3a) (47). These findings are in conformity with the proposal that incompatibility is due to the expression of a repressor of replication functions rather than to competition for a replication site (40). Furthermore, the extents of the relevant deletions indicate that the hypothetical repressor must be determined by the P1 DNA in or near the right side of the map.

Table 2 Assignment of Genetic Markers to Cleavage Fragments of P1 DNA

Cleavage fragment	Genetic markers: absent (left)	Genetic markers: spanned by fragment	Genetic markers: absent (right)	Basis of assignments	Reference
*Eco*RI-19	*vir*11	[*c*8.134], 47, 34, 1.4, 57	*c*2.440	marker rescue from cloned fragment	7, 37, 64
*Eco*RI-2		r^-m^-		mutation (insertion?) furnishes *Eco*RI site	17, 64
Region common to *Bgl*II-6 and *Eco*RI-2		restriction (*r*)		function and fragment loss caused by transposon in P1Cm 112, 233	25
*Bam*I-4 and *Bgl*II-2		modification (*m*)		P1Cm 13, 88[a]	
*Eco*RI-10	*c*2.440	4.7, 62	108	marker rescue from cloned fragment	38
*Eco*RI-4	62	108, 23, 20; *cat* (P1CmO, 1)	33	marker rescue from cloned fragment; transposon furnishes *Eco*RI site	64; 25, 49
*Bgl*II-9, 8, 11		dispensible functions including *par*		fragments missing in P1Cm 2, 74, 22-2, 29-3 and P1d*lac* of PLO100[b]	25
*Eco*RI-1	20	2.31, 33, 72, 49, 3.6	3.21	marker rescue from cloned fragment	37
*Eco*RI-3	135	5.19, 162, 6.2, 79, *bac*-1, *ban*-1, 26, 32, 7.14, 67, 53, 160	30	marker rescue from cloned fragment	64
*Bam*I-10	39	30, 29	8.13	marker rescue from cloned fragment	37
*Eco*RI-7	8.13, 9.16, 10.1	136, 131, 43		marker rescue from cloned fragment[c]	64

[a] Phages P1Cm13 and P1Cm88 are modification-defective and phenotypically nonrestricting; they were selected as lysogens.

[b] The left end of the deletions in these phages is to the right of the *Pst*I site. The deletions extend into fragment *Bgl*II-5, except in the case of P1Cm22-2 where band 5 is retained.

[c] Genetic markers 9.16 and 10.1 (formerly considered to be near the right end of the genetic map [73]) are not rescued from either of the adjacent *Eco*RI fragments 19 and 7 that cover both the leftmost genetic markers and markers 136, 131, and 43, situated at the right. Of these markers, only 9.16 can be rescued from an as yet unidentified *Eco*RI fragment (64). It may be deduced that 9.16 and 10.1 lie to the left of 136, 131, and 43. The positions of linkage clusters X-1 and X-2 become as drawn in Fig. C3.1, and the terminal location of *immC* (in X-2) is placed in doubt.

Acknowledgments

This work, a compilation and revision of data, much of it hitherto unpublished, has been made possible by the cooperation of many colleagues. Particular thanks are due W. Arber, B. Bächi and S. Iida of the Biozentrum (Basel, Switzerland), who generously made their cleavage maps available prior to publication, and to N. Sternberg (of the author's laboratory), who provided much of the unpublished marker-rescue data. Successive versions of the manuscript were prepared with skill and forebearance by Linda Fauvie (Basic Research Program) and Annette Deener (Technical Information Center). I acknowledge support from the National Cancer Institute under Contract No. NO1-CO-25423 with Litton Bionetics, Inc.

References

1. Austin, S., N. Sternberg and M. Yarmolinsky, in preparation.
2. Bächi, B. and W. Arber. 1977. Physical mapping of *Bgl*II, *Bam*HI, *Eco*RI, *Hin*dIII and *Pst*I restriction fragments of bacteriophage P1 DNA. *Mol. Gen. Genet.* **153**:311.
3. Bertani, G. 1951. Studies of lysogenesis. I. The mode of phage liberation by lysogenic *Escherichia coli*. *J. Bact.* **62**:293.
4. Botstein, D., K. K. Lew, V. Jarvik and C. A. Swanson, Jr. 1975. Role of antirepressor in the bipartite control of repression and immunity by bacteriophage P22. *J. Mol. Biol.* **91**:439.
5. Caro, L., personal communication.
6. Chesney, R. H. and J. R. Scott. 1975. Superinfection immunity and prophage repression in phage P1. II. Mapping of the immunity-difference and ampicillin-resistance loci of P1 and ϕAmp. *Virology* **67**:375.
7. ——. 1977. Suppression of thermosensitive *dnaA* mutation of *Escherichia coli* by bacteriophage P1 and P7. *Plasmid* (in press).
8. Chow, L. T. and A. I. Bukhari. 1976. The invertible DNA segments of coliphages Mu and P1 are identical. *Virology* **74**:242.
9. ——, this volume.
10. Chow, L. T., R. Kahmann and D. Kamp. 1977. Electron microscopic characterization of DNAs from non-defective mutants of bacteriophage Mu. *J. Mol. Biol.* (in press).
11. Cowan, J., unpublished results.
12. Cress, D. E. and B. C. Kline. 1976. Isolation and characterization of *E. coli* chromosomal mutants affecting plasmid copy number. *J. Bact.* **125**:635.
13. D'Ari, R. 1977. Effects of mutations in the immunity system of bacteriophage P1. *J. Virol.* (in press).
14. D'Ari, R., A. Jaffé-Brachet, D. Touati-Schwartz and M. B. Yarmolinsky. 1975. A *dnaB* analog specified by bacteriophage P1. *J. Mol. Biol.* **94**:477.
15. de Bruijn, F. and A. I. Bukhari, personal communication.
16. Delhalle, E. 1973. Restriction et modification de bactériophages par *Escherichia coli* K12 implicant un prophage P1 cryptique associé à différents plasmides. *Ann. Microbiol.* (Institute Pasteur) **124A**:173.

17. Glover, S. W., J. Schell, N. Symonds and K. A. Stacey. 1963. The control of host-induced modification by phage P1. *Genet. Res. Camb.* **4**:480.
18. Grindley, this volume.
19. Harriman, P. D. 1972. A single-burst analysis of the production of P1 infectious and transducing particles. *Virology* **48**:595.
20. Hayward, G. S. 1972. Gel electrophoretic separation of the complementary strands of bacteriophage DNA. *Virology* **49**:342.
21. Hedges, R. W., N. Datta, P. Kontomichalou and J. T. Smith. 1974. Molecular specificities of R factor-determined beta-lactamases: Correlation with plasmid compatibility. *J. Bact.* **117**:56.
22. Hedges, R. W., A. E. Jacob, P. T. Barth and N. J. Grinter. 1975. Compatibility properties of P1 and φAmp prophages. *Mol. Gen. Genet.* **141**:263.
23. Helling, R. B., H. M. Goodman and H. W. Boyer. 1974. Analysis of endonuclease R·*Eco*R1 fragments of DNA from lambdoid bacteriophages and other viruses by agarose-gel electrophoresis. *J. Virol.* **14**:1235.
24. Hertman, I. and J. R. Scott. 1973. Recombination of phage P1 in recombination deficient hosts. *Virology* **53**:468.
25. Iida, S. and W. Arber, personal communication.
25a. ——, 1977. Plaque-forming specialized transducing phage P1: Isolation of P1CmSmSu, a precursor of P1Cm. *Mol. Gen. Genet.* **153**:259.
26. Ikeda, H. and J. Tomizawa. 1965. Transducing fragments in generalized transduction by phage P1. III. Studies with small phage particles. *J. Mol. Biol.* **14**:120.
27. ——. 1969. Prophage P1, an extrachromosomal replication unit. *Cold Spring Harbor Symp. Quant. Biol.* **33**:791.
28. Jaffe-Brachet, A. and R. D'ari. 1977. The maintenance of P1 plasmid. *J. Virol.* (in press).
29. Kondo, E. and S. Mitsuhashi. 1964. Drug resistance of enteric bacteria. IV. Active transducing bacteriophage P1CM produced by the combination of R factor with bacteriophage P1. *J. Bact.* **88**:1266.
30. Kondo, E., D. K. Haapala and S. Falkow. 1970. The production of chloramphenicol acetyl-transferase by bacteriophage P1CM. *Virology* **40**:431.
31. Lee, H.-J., E. Ohtsubo, R. C. Deonier and N. Davidson. 1974. Electron microscope heteroduplex studies of sequence relations among plasmids of *Escherichia coli.* V. ilv^+ deletion mutants of F14. *J. Mol. Biol.* **89**:585.
32. Lennox, E. S. 1955. Transduction of linked genetic characters of the host by bacteriophage P1. *Virology* **1**:190.
33. Levine, M., S. Truesdell, T. Ramakrishnan and M. J. Bronson. 1975. Dual control of lysogeny by bacteriophage P22: An antirepressor locus and its controlling elements. *J. Mol. Biol.* **91**:421.
34. Luria, S. E., N. J. Adams and R. C. Ting. 1960. Transduction of lactose-utilizing ability among strains of *E. coli* and *S. dysenteriae* and the properties of the transducing phage particles. *Virology* **12**:348.
35. MacHattie and Jackowski, this volume.
35a. Meyer, J. and S. Iida, in preparation.
36. Mise, K. and W. Arber. 1976. Plaque-forming transducing bacteriophage P1 derivatives and their behavior in lysogenic conditions. *Virology* **69**: 191.

37. Mural, R. and D. Vapnek, personal communication.
38. Mural, R., D. Vapnek and R. H. Chesney, personal communication.
39. Ogawa, T. 1975. Analysis of the *dnaB* function of *Escherichia coli* and the *dnaB*-like function of P1 prophage. *J. Mol. Biol.* **94**:327.
39a. Prentki, P., M. Chandler and L. Caro. 1977. Replication of the prophage P1 during the cell cycle in *Escherichia coli. Mol. Gen. Genet.* **152**:71.
40. Pritchard, R. H., P. T. Barth and J. Collins. 1969. Control of DNA synthesis in bacteria. *Symp. Soc. Gen. Microbiol.* **19**:263.
41. Rae, M. E. 1971. The production of novel genotypes upon P1 and P1dl-mediated transduction of *E. coli.* Ph.D. thesis, University of Chicago.
42. Rae, M. E. and M. Stodolsky. 1974. Chromosome breakage, fusion and reconstruction during P1dl transduction. *Virology* **58**:32.
43. Roberts, R. J. and G. S. Haywood, personal communication.
44. Rosner, J. L. 1972. Formation, induction and curing of bacteriophage P1 lysogens. *Virology* **48**:679.
45. ——. 1973. Modification-deficient mutants of bacteriophage P1. I. Restriction of P1 cryptic lysogens. *Virology* **52**:213.
46. ——. 1975. Specialized transduction of *pro* genes by coliphage P1: Structure of a partly diploid P1-*pro* prophage. *Virology* **67**:42.
47. ——, personal communication.
48. Sakaki, Y. 1974. Inactivation of the ATP-dependent DNase of *Escherichia coli* after infection with double stranded DNA phages. *J. Virol.* **14**:1611.
49. Schulz, G. and M. Stodolsky. 1976. Integration sites of foreign genes in the chromosome of coliphage P1: A finer resolution. *Virology* **73**:299.
50. Scott, J. R. 1968. Genetic studies of bacteriophage P1. *Virology* **36**:574.
51. ——. 1970. A defective P1 prophage with a chromosomal location. *Virology* **40**:144.
52. ——. 1970. Clear plaque mutants of phage P1. *Virology* **41**:66.
53. ——. 1972. A new gene controlling lysogeny in phage P1. *Virology* **48**:282.
54. ——. 1973. Phage P1 cryptic. II. Location and regulation of prophage genes. *Virology* **53**:327.
55. ——. 1974. A turbid plaque-forming mutant of phage P1 that cannot lysogenize *Escherichia coli. Virology* **62**:344.
56. ——. 1975. Superinfection immunity and prophage repression in phage P1. *Virology* **65**:173.
57. ——, personal communication.
58. Scott, J. R., J. L. Laping and R. H. Chesney. 1977. A phage P1 virulent mutation at a new map location. *Virology* **78**:346.
59. Scott, J. R., M. Kropf and L. Mendelson. 1977. Clear plaque mutants of phage P7. *Virology* **76**:39.
60. ——, personal communication.
61. Scott, J. R., B. West and J. L. Laping, in preparation.
62. Smith, H. W. 1972. Ampicillin resistance in *Escherichia coli* by phage infection. *Nature New Biol.* **238**:205.
63. Spierer, P. and L. Caro, personal communication.
64. Sternberg, N., personal communication.
65. Stodolsky, M., personal communication.
66. Streisinger, G., J. Emrich and M. M. Stahl. 1967. Chromosome structure in

T4. III. Terminal redundancy and length determination. *Proc. Nat. Acad. Sci.* **57**:292.
67. Takano, T. and S. Ikeda. 1976. Phage P1 carrying kanamycin resistance gene of R factor. *Virology* **70**:198.
68. Toussaint, A. 1976. The DNA modification function of temperate phage Mu-1. *J. Virol.* **70**:17.
69. Tye, B.-K., J. A. Huberman and D. Botstein. 1974. Non-random circular permutation of phage P22 DNA. *J. Mol. Biol.* **85**:501.
70. Walker, D. H., Jr. and T. F. Anderson. 1970. Morphological variants of coliphage P1. *J. Virol.* **5**:765.
71. Walker, D. H., Jr. and J. T. Walker. 1975. Genetic studies of coliphage P1. I. Mapping by use of prophage deletions. *J. Virol.* **16**:525.
72. ——. 1976. Genetic studies of coliphage P1. II. Relatedness to P7. *J. Virol.* **19**:271.
73. ——. 1976. Genetic studies of coliphage P1. III. Extended genetic map. *J. Virol.* **20**:177.
74. ——, personal communication.
75. Walker, J. T., personal communication.
76. Wall, J. D. and P. D. Harriman. 1974. Phage P1 mutants with altered transducing abilities for *Escherichia coli*. *Virology* **59**:532.
77. Wandersman, C. and M. Yarmolinsky. 1977. Bipartite control of immunity conferred by the related heteroimmune plasmid prophages: P1 and P7 (formerly ϕAmp). *Virology* **77**:386.
78. West, B. W. and J. R. Scott. 1977. Superinfection immunity and prophage repression in phage P1 and P7. III. Induction by virulent mutants. *Virology* **78**:267.
79. Yun, T. and D. Vapnek. 1977. Electron microscopic analysis of bacteriophages P1, P1Cm and P7: Determination of genome sizes, sequence homology and location of antibiotic resistance determinants. *Virology* **77**:376.
80. ——, this volume.
81. ——, personal communication.

4. GENETIC AND PHYSICAL MAP OF BACTERIOPHAGE P2

Contributed by D. K. Chattoraj, University of Oregon, Eugene

The genetic map (Fig. C4.1a) is based primarily on recombination data (1, 20, 27).[1] Genetic elements within parentheses have not yet been ordered. Arrows indicate transcriptional units (15, 18, 20, 23, 27). Gene functions not listed on the top are described below.

Z. Maintenance of lysogeny (3).
fun. Converts the host to FdU sensitivity (5).
ogr. The mutation allows growth of P2 on a host that blocks P2 late gene expression. The host is mutated in the α subunit of RNA polymerase (14, 17, 28).
att. Site on the phage DNA where insertion and excision from the bacterial chromosome takes place, as postulated by the Campbell model (6).
int. Promotes prophage integration and excision (13, 19).
vir24. The mutation results in strong virulence and is most likely an operator mutation for early functions *A* and *B* (2, 20).
nip. The mutation alters the control of *int* expression and makes P2 *cts* heat inducible (7).
cox. The mutation causes a defect in excision but not in integration (21).
rlb. The mutation allows P2 growth on a *dnaBts* host at 37°C (29).
sig. An insertion which in the prophage state allows a *dnaAts* host to grow at nonpermissive temperature (12, 22).
old. The mutants grow and lysogenize *recB* and *recC* hosts efficiently; also, as prophage, they do not exclude λ (26). The gene *old* sometimes shows linkage to genes *Q* and *P* near the left end, implying circularity of the genetic map (1).

A few other mutations of P2 of unknown map position have been described (4).

Figure C4.1b is a map of A + T-rich regions (■) which help to define the ends of P2 DNA: the 5′ end of the "heavy" strand containing the 19-base-long cohesive end is the right end (15, 18, 23, 25, 30). The origin of replication is shown by the vertical arrow at 89.0, and the horizontal arrow indicates the unidirectional mode of replication (25). Approximate positions of several genetic markers on P2 DNA are shown by dashed lines between parts a and b of the figure (1, 11, 12, 31).

A map of five simple deletions and the *sig5* insertion (12) is given in Figure C4.1c.

In the map of three *Eco*RI sites (Fig. C4.1d), A to D define the four fragments (11a).

Figure C4.1e is a map of three hybrid phages containing segments of P2 DNA. In P2 Hy1 *dis,* the three regions nonhomologous to P2 are termed Hy*dis* A, Hy*dis* B, Hy*dis* C (9) (*dis* symbolizes the immunity specificity of a defective

[1] Numbers in parentheses refer to references given below.

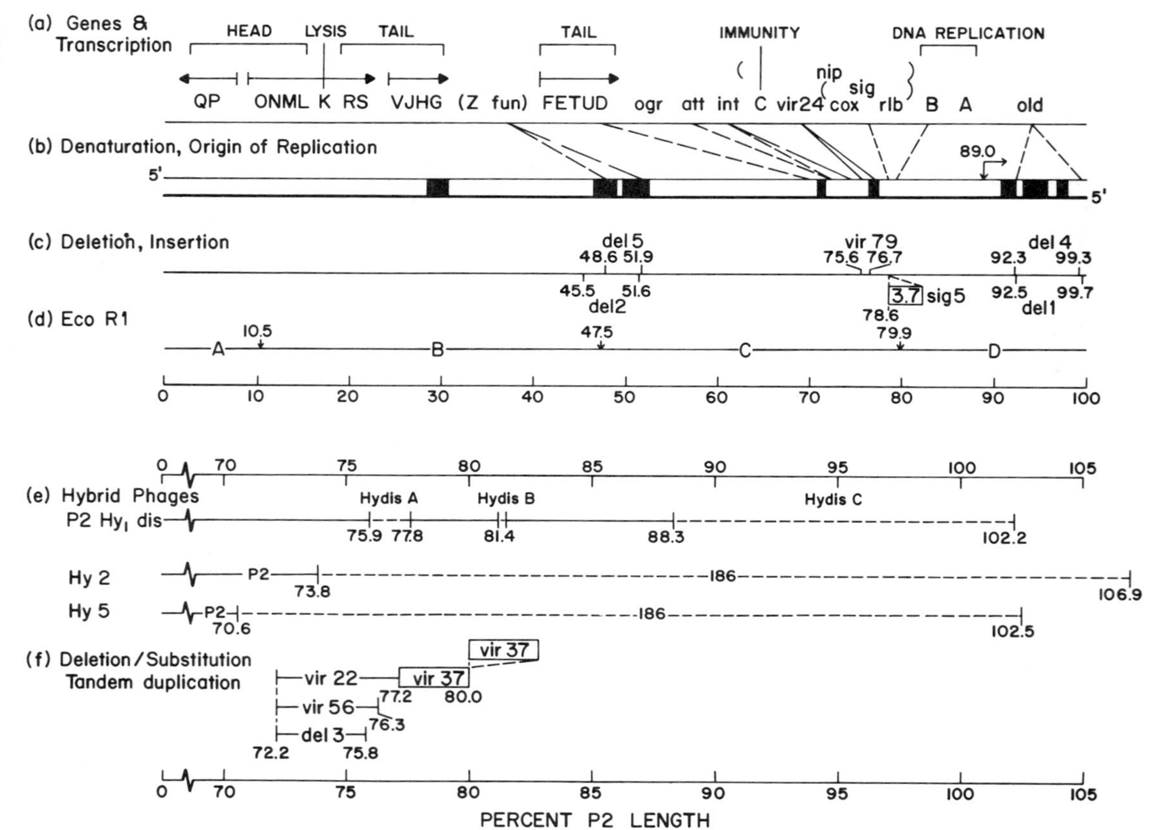

(a) Genes & Transcription
HEAD
LYSIS
TAIL
TAIL
IMMUNITY
DNA REPLICATION
QP
ONML
K
RS
VJHG
(Z fun)
FETUD
ogr
att
int
C
vir24
nip
sig
cox
rlb
B
A
old
(b) Denaturation, Origin of Replication
5'
5'
89.0
(c) Deletion, Insertion
del 5
48.6 51.9
45.5 51.6
del2
vir 79
75.6 76.7
3.7 sig5
78.6
del 4
92.3 99.3
92.5 99.7
del1
(d) Eco R1
A
B
C
D
10.5
47.5
79.9
0
10
20
30
40
50
60
70
80
90
100
(e) Hybrid Phages
P2 Hy1 dis
Hydis A
Hydis B
Hydis C
75.9 77.8
81.4
88.3
102.2
Hy 2
P2
73.8
186
106.9
Hy 5
P2
70.6
186
102.5
(f) Deletion/Substitution
Tandem duplication
vir 22
vir 37
vir 37
77.2
80.0
vir 56
76.3
del 3
72.2
75.8
0
70
75
80
85
90
95
100
105
PERCENT P2 LENGTH

Figure C4.1

prophage different from P2). Hy 2 and Hy 5 are hybrid phages containing the P2 left end and the phage 186 right end (31). Dashed lines indicate non-P2 DNA.

Examples of deletion/substitution phages in the immunity region with one common end point at *att* (at 72.2) are given in Figure C4.1f. The amounts of substitution of foreign DNA in *vir*22, *vir*56, and *del*3 are 0.5, 4.8, and 1.1, respectively. The entire substituted DNA in *del*3 is homologous to the left end of the DNA substituted in *vir*56 (8, 11, 16). *vir*37 is a tandem duplication mutant (10).

All measurements are presented as percent of P2 length: 100% P2 length corresponds to 33,018 base pairs, assuming a P2 to ϕX174 ratio of 6.16 (31) and ϕX174 to have 5360 nucleotides (24).

Acknowledgments

The author thanks Melvin Sunshine, Richard Calendar and Janet Geisselsoder for their help. The author is supported by National Institutes of Health Grant GM-20373.

References

1. Bertani, G. 1975. Deletions in bacteriophage P2. Circularity of the genetic map and its orientation relative to the DNA denaturation map. *Mol. Gen. Genet.* **136**:107.
2. Bertani, L. E. 1968. Abortive induction of bacteriophage P2. *Virology* **36**:87.
3. ——. 1976. Characterization of clear mutants belonging to the *Z* gene of bacteriophage P2. *Virology* **71**:85.
4. Bertani, L. E. and G. Bertani. 1971. Genetics of P2 and related phages. *Adv. Genet.* **16**:199.
5. Bertani, L. E. and J. A. Levy. 1964. Conversion of lysogenic *Escherichia coli* by nonmultiplying, superinfecting bacteriophage P2. *Virology* **22**:634.
6. Calendar, R. and G. Lindahl. 1969. Attachment of prophage P2: Gene order at different host chromosome sites. *Virology* **39**:867.
7. Calendar, R., G. Lindahl, M. Marsh and M. Sunshine. 1972. Temperature inducible mutants of P2 phage. *Virology* **47**:68.
8. Chattoraj, D. K. and R. B. Inman. 1972. Position of two deletion mutations on the physical map of bacteriophage P2. *J. Mol. Biol.* **66**:423.
9. ——. 1973. Electron microscope heteroduplex mapping of P2 Hy*dis* bacteriophage DNA. *Virology* **55**:174.
10. ——. 1974. Tandem duplication in bacteriophage P2: Electron microscope mapping. *Proc. Nat. Acad. Sci.* **71**:311.
11. Chattoraj, D. K., L. E. Bertani and G. Bertani, unpublished.

11a. Chattoraj, D. K., Y. K. Oberoi and G. Bertani. 1977. Restriction of bacteriophage P2 by the *Escherichia coli* RI plasmid, and *in vitro* cleavage of its DNA by the *Eco*RI endonuclease. *Virology* (in press).

12. Chattoraj, D. K., H. B. Younghusband and R. B. Inman. 1975. Physical

mapping of bacteriophage P2 mutations and their relation to the genetic map. *Mol. Gen. Genet.* **136**:139.
13. Choe, B. K. 1969. Integration defective mutants of bacteriophage P2. *Mol. Gen. Genet.* **105**:275.
14. Fujiki, H., P. Palm, W. Zillig, R. Calendar and M. Sunshine. 1976. Identification of a mutation within the structural gene for the α subunit of DNA-dependent RNA polymerase of *E. coli. Mol. Gen. Genet.* **145**:19.
15. Geisselsoder, J., M. Mandel, R. Calendar and D. K. Chattoraj. 1973. *In vivo* transcription patterns of temperate coliphage P2. *J. Mol. Biol.* **77**:405.
16. Hyde, J. M. and G. Bertani. 1975. Structure and position of a complex chromosomal aberration in bacteriophage P2. *J. Gen. Virol.* **28**:415.
17. Jaskunas, S. R., R. Burgess, L. Lindahl and M. Nomura. 1976. Two clusters of genes for RNA polymerase and ribosome components in *E. coli.* In *RNA polymerase* (ed. R. Losick and M. Chamberlin), p. 539. Cold Spring Harbor Laboratory, Cold Spring Harbor, New York.
18. Kainuma-Kuroda, R. and R. Okazaki. 1975. Mechanism of DNA chain growth. XII. Asymmetry of replication of P2 phage DNA. *J. Mol. Biol.* **94**:213.
19. Lindahl, G. 1969. Multiple recombination mechanisms in bacteriophage P2. *Virology* **39**:861.
20. ———. 1971. On the control of transcription in bacteriophage P2. *Virology* **46**:620.
21. Lindahl, G. and M. G. Sunshine. 1972. Excision-deficient mutants of bacteriophage P2. *Virology* **49**:180.
22. Lindahl, G., Y. Hirota and F. Jacob. 1971. On the process of cellular division in *Escherichia coli*: Replication of the bacterial chromosome under control of prophage P2. *Proc. Nat. Acad. Sci.* **68**:2407.
23. Lindqvist, B. J. and K. Bøvre. 1972. Asymmetric transcription of the coliphage P2 genome during infection. *Virology* **49**:690.
24. Sanger, F., unpublished.
25. Schnös, M. and R. B. Inman. 1971. Starting point and direction of replication in P2 DNA. *J. Mol. Biol.* **55**:31.
26. Sironi, G., H. Bialy, H. A. Lozeron and R. Calendar. 1971. Bacteriophage P2: Interaction with phage lambda and with recombination-deficient bacteria. *Virology* **46**:387.
27. Sunshine, M. G., M. Thorn, W. Gibbs and R. Calendar. 1971. P2 phage amber mutants: Characterization by use of a polarity suppressor. *Virology* **46**:691.
28. Sunshine, M. G. and B. Sauer. 1975. A bacterial mutation blocking P2 phage late gene expression. *Proc. Nat. Acad. Sci.* **72**:2770.
29. Sunshine, M., D. Usher and R. Calendar. 1975. Interaction of P2 bacteriophage with *dnaB* gene of *Escherichia coli. J. Virol.* **16**:284.
30. Wang, J. C. and D. P. Brezinski. 1973. Alignment of two DNA helices: A model for recognition of DNA base sequences by the termini generating enzymes of phage λ, 186 and P2. *Proc. Nat. Acad. Sci.* **70**:2667.
31. Younghusband, H. B., J. B. Egan and R. B. Inman. 1975. Characterization of DNA from bacteriophage P2-186 hybrids and physical mapping of the 186 chromosome. *Mol. Gen. Genet.* **140**:101.

5. GENETIC, PHYSICAL, AND RESTRICTION MAP OF BACTERIOPHAGE P22

Contributed by M. Susskind, University of Massachusetts Medical School, Worcester, and D. Botstein, Massachusetts Institute of Technology, Cambridge

The inner circle in Figure C5.1 shows the approximate locations of cleavage sites (marked by arrows) made by *Eco*RI endonuclease (12).[1] The outer circle shows the genetic map of P22 drawn approximately to scale. Dotted lines connecting the inner and outer circles show the assignment, where known, of genes to *Eco*RI fragments (12, 25). The positions of genes within fragments are *not* drawn to scale. The pattern of transcription is thought to be as shown by the arrows drawn outside the genetic map.

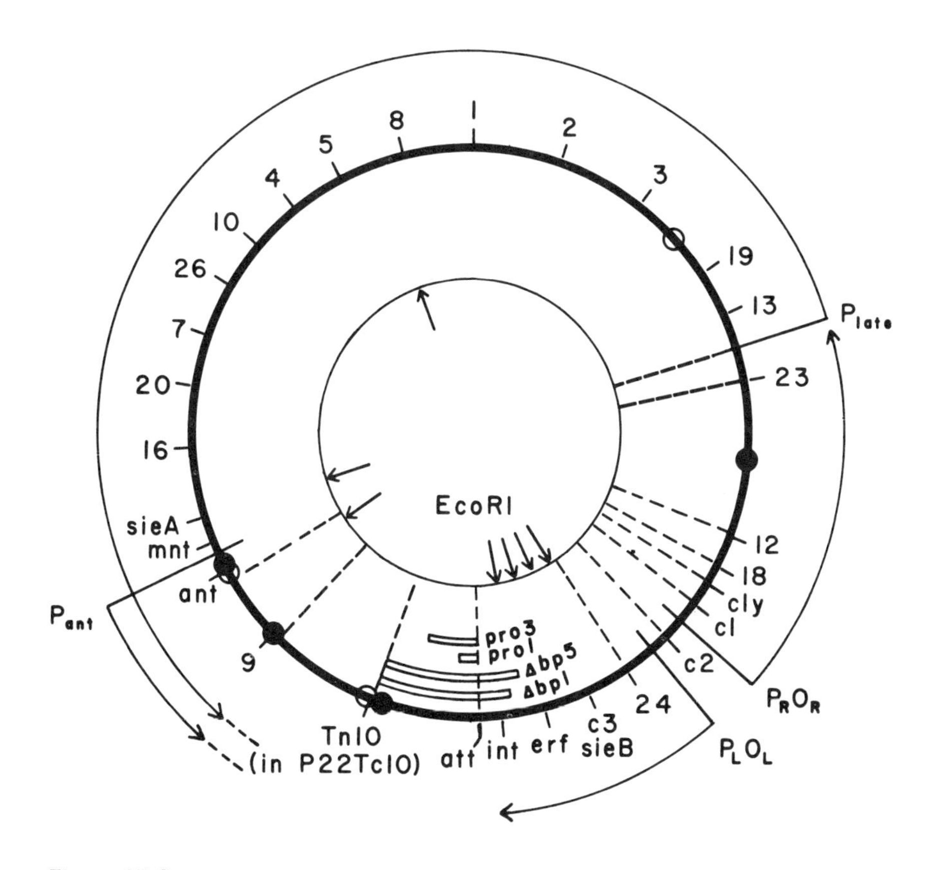

Figure C5.1

[1] Numbers in parentheses refer to references given below.

Genetic Symbols

Symbols are listed in the order in which they appear on the map, beginning with *att* and moving counter-clockwise. Genes designated by numerals code for functions that are essential for lytic growth of the phage. The order of the *c3* and *sieB* genes has not been determined.

att. Site at which the phage chromosome integrates into the host DNA (6, 7).
int. Codes for protein required for integration and excision of prophage DNA (20, 21).
erf. Phage general (homologous) recombination function (1).
*c*3. Function required for establishment of lysogeny (13, 14).
sieB. Function which excludes superinfecting *Salmonella* phages related to P22 (24).
24. Positive regulator of early gene functions (lysogeny, DNA replication) (9).
$p_L o_L$. Promoter and operator for leftward transcription of early genes (5).
*c*2. Codes for phage repressor that acts at o_L and o_R (5, 13, 14, 16).
$p_R o_R$. Promoter and operator for rightward transcription of early genes (5).
*c*1. Function required for establishment of lysogeny (13, 14).
cly. Function or site involved in regulation of *c*2 repressor synthesis (10, 19).
18 and *12.* Functions required for phage DNA replication (2, 15).
23. Positive regulator of phage late genes *13* through *16* and *9* (2, 3, 19).
P_{late}. Promoter for initiation of transcription of the phage late genes (19).
13. Function required for lysis of host cells (2).
19. Codes for phage lysozyme (2).
3, 2, 1, 8, 5, 4, 10, 26, 7, 20, and *16.* Code for DNA maturation and phage head proteins (3, 6, 18).
sieA. Function which excludes superinfecting P22 and related *Salmonella* phages (23).
mnt. Function which represses *ant* gene expression (4, 8, 17).
P_{ant}. Promoter for transcription of the *ant* gene (4, 17).
ant. Codes for P22 antirepressor (4, 17).
9. Codes for the phage base-plate (tail) protein (2, 3, 11).
Tc10. Site of insertion of a *tet* element (Tn*10*) in the specialized transducing phage Tc10 (7).
–○–. Sites at which *tet* elements can insert (7, 22).
–●–. Sites at which *amp* elements can insert (25).
*pro*1, *pro*3. Substitutions which delete the indicated phage DNA and add the *proA* and *proB* genes of *S. typhimurium* (7).
Δbpl, Δbp5. Deletions derived from P22 Tc10 which remove the indicated regions of phage DNA (7).

References

1. Botstein, D. and M. J. Matz. 1970. A recombination function essential to the growth of bacteriophage P22. *J. Mol. Biol.* **54**:417.

2. Botstein, D., R. K. Chan and C. H. Waddell. 1972. Genetics of bacteriophage P22. II. Gene order and gene function. *Virology* **49**:268.
3. Botstein, D., C. H. Waddell and J. King. 1973. Mechanism of head assembly and DNA encapsulation in *Salmonella* phage P22. I. Genes, proteins, structures and DNA maturation. *J. Mol. Biol.* **80**:669.
4. Botstein, D., K. K. Lew, V. M. Jarvik and C. A. Swanson. 1975. The role of antirepressor in the bipartite control of repression and immunity by phage P22. *J. Mol. Biol.* **91**:439.
5. Bronson, M. J. and M. Levine. 1972. Virulent mutants of phage P22. II. Physiological analysis of P22 virB-3 and its component mutations. *Virology* **47**:644.
6. Chan, R. K. and D. Botstein. 1972. Genetics of bacteriophage P22. I. Isolation of prophage deletions which affect immunity to superinfection. *Virology* **49**:257.
7. ———. 1976. Specialized transduction by bacteriophage P22 in *Salmonella typhimurium*: Genetic and physical structure of the transducing genomes and the prophage attachment site. *Genetics* **83**:433.
8. Gough, M. 1968. Second locus of bacteriophage P22 necessary for the maintenance of lysogeny. *J. Virol.* **2**:922.
9. Hilliker, S. and D. Botstein. 1975. An early regulatory gene of *Salmonella* phage P22 analogous to gene *N* of coliphage λ. *Virology* **68**:510.
10. Hong, J.-S., G. R. Smith and B. N. Ames. 1971. Adenosine 3′:5′-cyclic monophosphate concentration in the bacterial host regulates the viral decision between lysogeny and lysis. *Proc. Nat. Acad. Sci.* **68**:2258.
11. Israel, V., T. F. Anderson and M. Levine. 1967. *In vitro* morphogenesis of phage P22 from heads and base plate parts. *Proc. Nat. Acad. Sci.* **57**:284.
12. Jackson, E. N., D. A. Jackson and R. J. Deans, in preparation.
13. Levine, M. 1957. Mutations in the temperate phage P22 and lysogeny in *Salmonella.* *Virology* **3**:22.
14. Levine, M. and R. Curtiss. 1961. Genetic fine structure of the C region and the linkage map of phage P22. *Genetics* **46**:1573.
15. Levine, M. and C. Schott. 1971. Mutations of phage P22 affecting phage DNA synthesis and lysogenization. *J. Mol. Biol.* **62**:53.
16. Levine, M. and H. O. Smith. 1964. Sequential gene action in the establishment of lysogeny. *Science* **146**:1581.
17. Levine, M., S. Truesdell, T. Ramakrishnan and M. J. Bronson. 1975. Dual control of lysogeny by bacteriophage P22: An antirepressor locus and its controlling elements. *J. Mol. Biol.* **91**:421.
18. Poteete, A. R. and J. King. 1977. Functions of two new genes in *Salmonella* phage P22 assembly. *Virology* **76**:725.
19. Roberts, J. W., C. W. Roberts, S. Hilliker and D. Botstein. 1976. Transcription termination and regulation in bacteriophages P22 and lambda. In *RNA polymerase* (ed. R. Losick and M. Chamberlin), p. 707. Cold Spring Harbor Laboratory, Cold Spring Harbor, New York.
20. Smith, H. O. 1968. Defective phage formation by lysogens of integration deficient phage P22 mutants. *Virology* **34**:203.
21. Smith, H. O. and M. Levine. 1967. A phage P22 gene controlling integration of prophage. *Virology* **31**:207.
22. Susskind, M. M. and D. Botstein, unpublished results.

23. Susskind, M. M., D. Botstein and A. Wright. 1974. Superinfection exclusion by P22 prophage in lysogens of *Salmonella typhimurium.* III. Failure of superinfecting phage DNA to enter *sie*A$^+$ lysogens. *Virology* **62**:350.
24. Susskind, M. M., A. Wright and D. Botstein. 1974. Superinfection exclusion by P22 prophage in lysogens of *Salmonella typhimurium.* IV. Genetics and physiology of *sie*B exclusion. *Virology* **62**:367.
25. Weinstock, G. and D. Botstein, unpublished results.

6. GENETIC, PHYSICAL, AND RESTRICTION MAP OF BACTERIOPHAGE ϕ80

Contributed by P. Youderian, Massachusetts Institute of Technology, Cambridge

In Figure C6.1 (4)[1], dimensions are presented as percentage of molecular distance; λ*papa* length is taken as the standard at 100%. A 1% molecular distance is roughly equivalent to 494 nucleotide base pairs of the DNA molecule. Recombination distances between pairs of markers in either the left arm or the right arm of ϕ80 are approximately half the percent molecular distance.

The physical and gentic map of λ appears directly below that of ϕ80. Regions of common DNA/DNA homology between λ and ϕ80 are indicated by a single line, regions of partial homology by a single jagged line, and regions of non-homology by a double line. The positions of genes on the left arm of λ are according to Parkinson and Davis (10) and Hradecna and Szybalski (7) and were determined by physical and genetic measurements performed on λd*gal* phages (8). The positions of cistrons *1* and *2* of ϕ80 were determined by matching cross-complementation data between λ and ϕ80 with λ physical mapping data. Other measurements of λ gene position were performed by Westmoreland et al. (15), Hradecna and Szybalski (7), and Fiandt et al. (4). The positions of cistrons *14, 16, 17, 18,* and *19* have been determined by Aizawa and Matsushiro (1). The approximate positions of the cistrons in parentheses were estimated from results of two-factor crosses between markers in these cistrons and physically positioned cistrons.

The end points of the substitution phages, ϕ80ptrp_1 (3, 13), ϕ80d*SuIII,* and ϕ80p*SuIII* (in addition to their common end point at *loc*) are indicated on the ϕ80 map. The maps of five λ-ϕ80 hybrids are shown in the lower part of the figure. For each hybrid, the upper line indicates that portion of the ϕ80 genome contributed to the hybrid, and the lower line indicates that portion of the λ genome contributed to the hybrid (4).

*Eco*RI cleavage of ϕ80 produces at least ten fragments of size ranging from 9.2 to 0.13 ($\times 10^6$ daltons). These sites of cleavage occur at approximately 16%, 40%, 66%, and 75% of the genome where 100% is the length of λ*papa.* At least five additional sites map in the region from 16% to 40% (6).

Genes

1, 2, 3, 4, 5, and *6.* Required for phage head morphogenesis and *ter* function necessary for formation of cohesive ends. Cistron *1* can cross-complement λ gene *nu1*, cistron *2* λ gene *A* (14).

7, 8, 9, 10, 11, 12, and *13.* Required for phage tail morphogenesis (12).

int_{80}. Required for prophage *int*egration and excision (9).

[1]Numbers in parentheses refer to references given below.

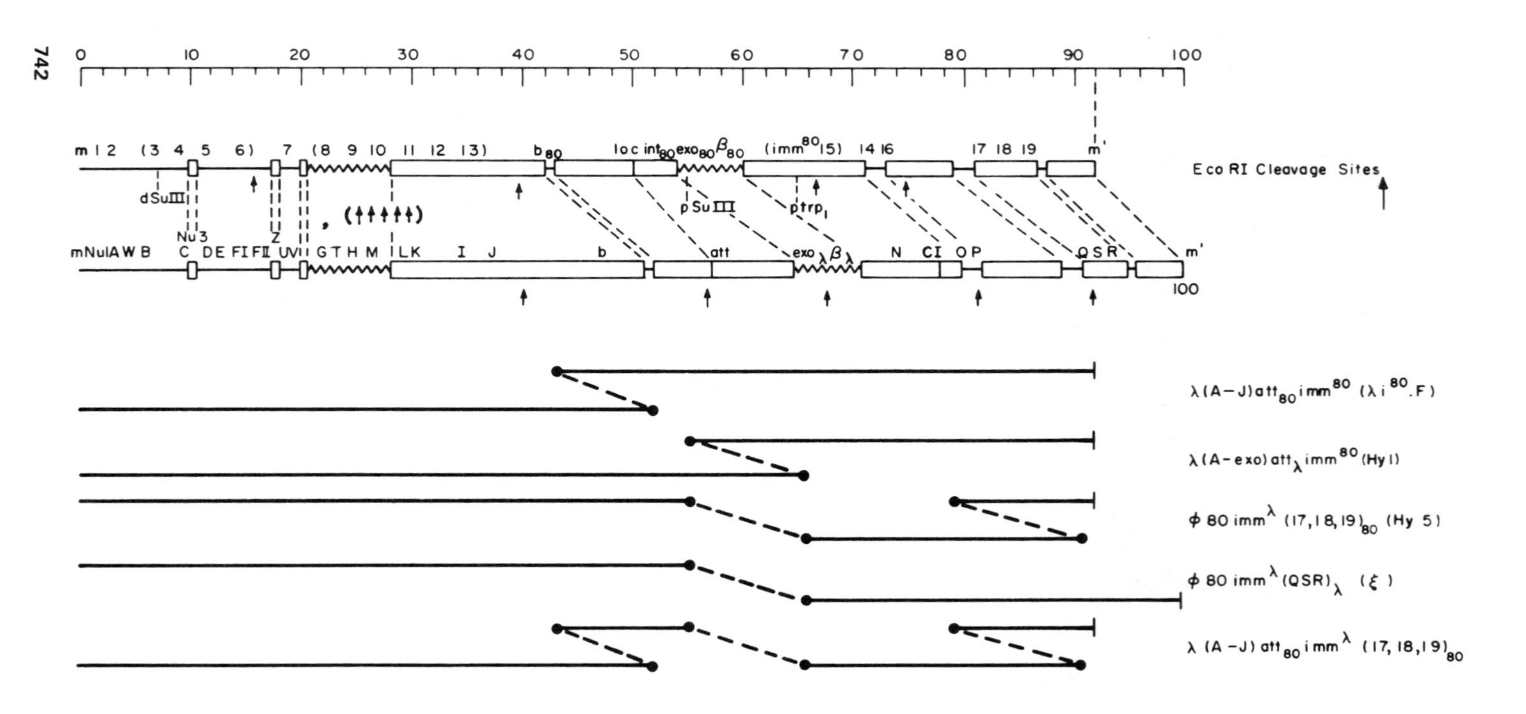

Eco RI Cleavage Sites
λ(A−J)att80 imm80 (λi80.F)
λ(A−exo)attλ imm80 (Hy1)
φ80 immλ (17,18,19)80 (Hy 5)
φ80 immλ (QSR)λ (ξ)
λ (A−J) att80 immλ (17,18,19)80

Figure C6.1

exo$_{80}$ or *red*α_{80}. ϕ80 *exo*nuclease, which promotes phage-specified generalized recombination. Recombination-proficient hybrids between λ and ϕ80 coding for an *exo* protein with hybrid antigenic determinants have been characterized.

β_{80} or *red*β_{80}. β protein, which promotes phage-specified generalized recombination. Recombination-proficient hybrids between λ and ϕ80 having a cross-over point within β have been characterized.

imm^{80}. Phage-specific *imm*unity region.

15. Uncharacterized (11).

14 and *16*. Required for phage DNA replication (11). Replication-proficient hybrids have been obtained having cross-over points between gene *14* of ϕ80 and gene *O* of λ (5).

17. Required for late gene transcription, analogous in function with λ gene *Q* (11).

18 and *19*. Required for host cell lysis, analogous in function with λ genes *S* and *R* (endolysin), respectively (11).

Recognition Sites

m, m'. Left and right cohesive ends of the DNA molecule. ϕ80 *m, m'* are identical to λ *m, m'* (2).

loc or *att*$_{80}$. *Loc*ation of integrative interaction of phage DNA with host DNA during lysogenization.

Note Added in Proof

Since the submission of this article, at least six additional complementation groups (not shown) in the left arm of bacteriophage ϕ80 have been identified in the laboratory of Dr. Jonathan King. One cistron maps between cistrons *4* and *5* and is a tail donor. At least five new head donors have been identified mapping in the intervals from gene *6* to gene *7* (two cistrons), gene *7* to gene *9* (one cistron), gene *9* to gene *10* (one cistron), and gene *12* to gene *13* (one cistron). In addition, gene *8* has been shown to map between cistrons *10* and *11*, not in the position illustrated in the figure.

The cistrons mapping between *6* and *7* code for functions which can complement λ genes *U* and *G*.

References

1. Aizawa, S. and A. Matsushiro. 1975. Studies on temperature sensitive growth of phage ϕ80. I. Prophage excision. *Virology* **67**:168.
2. Baldwin, R. L., P. Barrand, A. Fritsch, D. A. Goldthwait and F. Jacob. 1966. Cohesive sites on the deoxyribonucleic acids from several temperate coliphages. *J. Mol. Biol.* **17**:343.

3. Deeb, S. S., K. Okamoto and B. D. Hall. 1967. Isolation and characterization of non-defective transducing elements of bacteriophage ϕ80. *Virology* **31**:289.
4. Fiandt, M., Z. Hradecna, H. A. Lozeron and W. Szybalski. 1971. Electron micrographic mapping of deletions, insertions, inversions, and homologies in the DNAs of coliphages lambda and phi 80. In *The Bacteriophage lambda* (ed. A. D. Hershey), p. 329. Cold Spring Harbor Laboratory, Cold Spring Harbor, New York.
5. Furth, M., unpublished results.
6. Hellig, R. B., H. M. Goodman and H. W. Boyer. 1974. Analysis of Endonucleases R•*Eco*RI fragments of DNA from lambdoid bacteriophages and other viruses by agarose-gel electrophoresis. *J. Virol.* **14**:1235.
7. Hradecna, Z. and W. Szybalski. 1969. Electron micrographic maps of deletions and substitutions in the genomes of transducing coliphages λdg and λ*bio*. *Virology* **38**:473.
8. Kayajanian, G. and A. Campbell. 1966. The relationship between heritable physical and genetic properties of selected gal^- and gal^+ transducing λdg. *Virology* **30**:482.
9. Matsushiro, A., unpublished results.
10. Parkinson, J. S. and R. W. Davis. 1971. A physical map of the left arm of the lambda chromosome. *J. Mol. Biol.* **56**:425.
11. Sato, K. 1970. Genetic map of bacteriophage ϕ80: Genes on the right arm *Virology* **40**:1067.
12. Sato, K., Y. Nishinume, M. Sato, R. Numich, A. Matsushiro, H. Inokuchi and H. Ozeki. 1968. Suppressor-sensitive mutants of coliphage ϕ80. *Virology* **34**:637.
13. Shinagawa, H., Y. Hosaka, H. Yamagishi and Y. Nishi. 1966. Electron microscope studies on ϕ80 and $\phi 80pt_1$ phage virions and their DNA. *Biken's J.* **9**:135.
14. Weissberg, R., personal communication.
15. Westmoreland, B. C., W. Szybalski and H. Ris. 1969. Mapping of deletions and substitutions in heteroduplex DNA molecules of bacteriophage lambda by electron microscopy. *Science* **163**:1343.

7. GENETIC AND PHYSICAL MAP OF BACTERIOPHAGE MU

Contributed by B. Allet, F. Blattner, M. Howe,* M. Magazin, D. Moore,* K. O'Day,* D. Schultz* and J. Schumm,* University of Geneva, and *University of Wisconsin, Madison*

The temperate bacteriophage Mu DNA, as extracted from mature virions, is a linear, double-stranded molecule of about 38 kilobases (kb). Figure C7.1 (top) gives the genetic map of Mu with *Eco*RI and *Hin*dIII restriction sites. The map is a slightly modified version of the composite genetic map of Mu completed during the Mu Workshop at the Cold Spring Harbor Laboratory in 1972 (Abelson et al. 1973).

The leftmost gene is the immunity locus *c* where clear-plaque mutations have been mapped. The *H* gene has been relocated from between *G* and *I* genes to between genes *E* and *F*. The rest of the genes are arranged in alphabetical order, with *Bu* located between *A* and *B* and *lys* located between *C* (not to be confused with the immunity gene *c*)[1] and *D*.

The *Bu* locus is defined by a class of amber mutations which are located to the left of the known *B*-gene mutations (A. I. Bukhari and M. Howe, unpubl.). The *Bu* amber mutations are more defective in expressing the killing functions of Mu than are the *B* amber mutations. However, *Bu* and *B* do not complement, and it is not clear whether or not they define two different cistrons.

The *lys* gene is involved in the lysis of host cells during Mu growth. The region to the left of the *C* gene (*A-Bu-B*) apparently is comprised of Mu functions that are expressed early in the lytic cycle. *A, Bu,* and *B* mutants are defective in Mu DNA replication. At the least, the *A* gene function is a part of the integration machinery of Mu. Recent information from P. van de Putte and coworkers (unpubl.) indicates that some other functions which are not clearly understood, such as *kil* (van de Putte et al., this volume), *sot* (support of transfection), *gam* (analogous to the gamma gene of phage λ), and *cim* (control of immunity), are located in the region to the left of the *C* gene. The insertions in the *X* mutants (Bukhari 1975) are also located in this region. According to van de Putte and coworkers, genes *D* to *J* are involved in head formation and *K* to *S* in tail formation. The tail gene *S* has now been mapped within the G segment of Mu (M. Howe and coworkers; A. Toussaint and coworkers; D. Kamp and R. Kahmann; all unpubl.). Evidently, the *S* gene is expressed only in one G orientation; thus only one G orientation normally gives rise to viable Mu particles (D. Kamp and R. Kahmann, pers. comm.). The particles lacking the *S*-gene protein are unable to adsorb to the cells properly and thus cannot infect the cells.

All of the gene assignments were based on complementation tests between a large number of amber mutants isolated in different laboratories (Abelson et al. 1973).

[1]The immunity gene is denoted by small *c*. The capital *C* is used to designate the gene located between *B* and *lys* genes. Because of possible confusion, gene *C* will probably be given a different name.

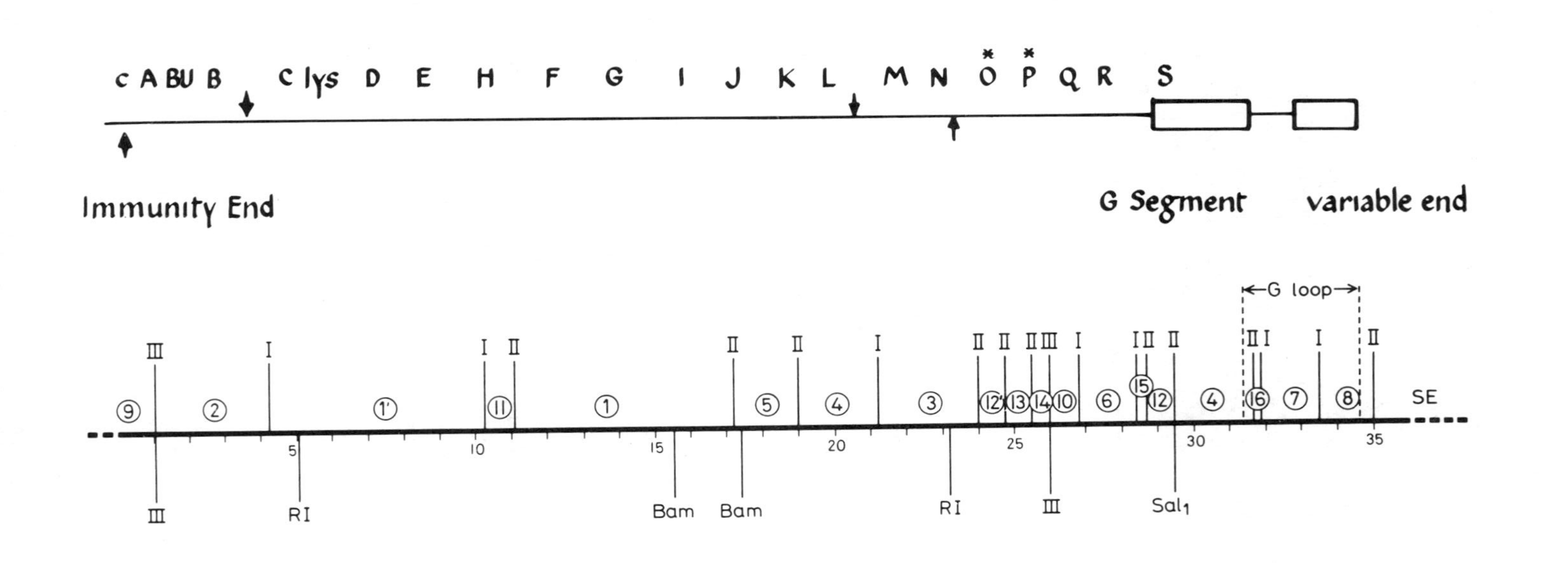
c A BU B
C lys D E H F G I J K L M N O P Q R S
Immunity End
G Segment
variable end
G loop
SE
Bam
Bam
RI
RI
Sal1

Figure C7.1

Genes *O* and *P* have been marked on the map with stars. Preliminary information indicates that genes *O* and *P* may actually define one gene (K. O'Day, D. Schultz and M. Howe, unpubl.). The DNA modification function of Mu, called *mom,* is thought to be located to the right of the *S* gene but has not been clearly mapped (Toussaint 1976). The G inversion function, *gin,* is at least partly controlled by the sequence in the β region, between the G segment and the variable end, of Mu DNA (see Chow et al., this volume). It should be noted that the gene order shown here for phage DNA is identical to the gene order found in a Mu prophage, irrespective of the site of prophage insertion into the host chromosome.

The two upward arrows on the genetic map represent cuts by the restriction endonuclease *Hin*dIII. The two downward arrows indicate cuts by *Eco*RI. Thus the left *Eco*RI fragment contains genes *c, A, Bu,* and *B* but does not contain gene *C.* The invertible segment G and the heterogeneous (variable) end are shown by rectangles.

At the bottom of Figure C7.1 is an endonuclease cleavage map of Mu. The cleavage map is based on the work of Allet and Bukhari (1975) but contains modifications and additional information (Magazin et al. 1977). The map is calibrated in kb. The cleavage sites are indicated by vertical lines, and the cutting enzyme is identified at each line. The enzyme symbols are:

I = *Hpa*I
II = *Hin*dII (*Hin*dII includes *Hpa*I activity)
III = *Hin*dIII
RI = *Eco*RI
Bam = *Bam*I
Sal_1 = *Sal*I

(For complete nomenclature of these enzymes, see Appendix D.) The arabic numerals in circles represent the numbers of the fragments generated by the *Haemophilus influenzae* enzymes (*Hin*dII and *Hin*dIII). The numbers are assigned in order of increasing electrophoretic mobility in polyacrylamide gels (Allet and Bukhari 1975). For example, fragment number 1 is the slowest moving fragment, whereas fragment number 16 is the fastest moving fragment. Fragments that move with approximately the same mobility are given the same number but identified with a prime (e.g., 1 and 1′). The invertible segment G is marked G loop. Notice that when G is inverted, the pattern of fragments 4 and 8 will change. SE designates the variable end which is seen as the *split end* in the electron microscope after denaturation and renaturation of a Mu DNA preparation. The split end, or variable end, contains host sequences that are heterogeneous in size (mean length 1500 base pairs). Similarly, the immunity end, or *c* end, also contains host sequences (mean length about 75 base pairs).

A cleavage map of Mu DNA, based on a different set of endonucleases, is given in Kahmann et al. (this volume).

References

Abelson, J., W. Boram, A. I. Bukhari, M. Faelen, M. Howe, M. Metlay, A. L. Taylor, A. Toussaint, P. van de Putte, G. C. Westmaas and C. A. Wijffelman. 1973. Summary of the genetic mapping of prophage Mu. *Virology* **54**:90.

Allet, B. and A. I. Bukhari. 1975. Analysis of Mu and λ-Mu hybrid DNAs by specific endonucleases. *J. Mol. Biol.* **92**:529.

Bukhari, A. I. 1975. Reversal of mutator phage Mu integration. *J. Mol. Biol.* **96**:87.

Magazin, M., M. Howe and B. Allet. 1977. Partial correlation of the genetic and physical maps of bacteriophage Mu. *Virology* **77**:677.

Toussaint, A. 1976. The DNA modification function of temperate phage Mu-1. *Virology* **70**:17.

8. BACTERIOPHAGE MU: METHODS FOR CULTIVATION AND USE

Contributed by A. I. Bukhari and E. Ljungquist, Cold Spring Harbor Laboratory

The temperate bacteriophage Mu is a powerful mutagen of *E. coli.* Mutations resulting from the insertion of Mu DNA normally abolish the gene activity completely, are polar, and are extremely stable. As described in some of the papers in this book, Mu can be used for various in vivo genetic manipulations and can be transferred to different bacterial species when inserted in appropriate plasmids. Some of the basic procedures involved in the growth and use of Mu are listed here.

Growth Media and Solutions

The following media are routinely used (see Miller 1972):

1. LB: 10 g Bacto tryptone, 5 g Bacto yeast extract, 10 g NaCl (pH 7.4), per liter.
2. SB: 32 g Bacto tryptone, 20 g Bacto yeast extract, 5 g NaCl (pH 7.2), per liter.
3. Minimal M9: 6 g Na_2HPO_4, 3 g KH_2PO_4, 0.5 g NaCl, 1 g NH_4Cl, salts per liter. After autoclaving, add $MgSO_4$ to 1 mM, glucose 0.2%, and required amino acids or bases, generally 20–40 μg/ml.

Agar plates for phage growth contain LB + 1 mM $CaCl_2$ + 2.5 mM $MgSO_4$ (LB Ca Mg) in 1% agar. For routine titrations, phage can be diluted in LB Ca Mg. For purification of Mu particles, Mu buffer (0.2 M NaCl, 0.02 M Tris, 1 mM $CaCl_2$, 20 mM $MgSO_4$, gelatin 0.1%) is used as described below.

Plating Conditions

Any sensitive *E. coli* K12 strain, but no other *E. coli* strain, is a good indicator for Mu. The indicator culture must be freshly grown and in log phase when used. Indicator bacteria and phage particles can be mixed in 0.5% LB soft agar for plating. Preadsorption of phage to cells is not necessary. Ca^{++} is required for adsorption. Plaques of wild-type Mu are fuzzy and vary in size from pinpoint to 1 mm. The plaques can be somewhat uniform if phage is preadsorbed and can sometimes be bigger if the indicator strain grows slightly more slowly on plates than the normal *E. coli* strains. Clear plaque mutants of Mu (Mu *c*) usually have small clear centers with fuzzy periphery. A drop of a Mu^+ lysate (10^8–10^9 plaque-forming units [pfu]/ml) gives visibly turbid spots, and a Mu *c* lysate (10^8–10^9 pfu/ml) gives totally clear spots of killing at 37°C.

Isolation of Mu from Lysogens

Unlike the lambdoid phages, prophage Mu cannot be induced by ultraviolet light or mitomycin C. However, wild-type Mu lysogens spontaneously liberate phage particles at a low frequency. Fully grown LB broth cultures of the lysogens usually contain 5×10^4–10^6 pfu of Mu per ml. Some lysogens consistently give higher numbers of phage particles than others. Plaques of Mu can be obtained by adding chloroform (~ 1% final conc.) to overnight broth cultures, removing the cell debris by centrifugation, and plating the appropriate dilutions of the supernatant. Mu *c* mutants can be isolated readily from the supernatant of lysogenic cultures.

Two widely used temperature-inducible mutants (Mu *cts* mutants which give turbid plaques at 30-34°C but clear plaques at 42-44°C) are Mu *cts*62 (Abelson et al. 1973; Howe 1973b) and Mu *cts*4 (Waggoner et al. 1974). To induce Mu *cts* lysogens, shaking cultures are grown in LB Mg at 32°C and, at a density of 1-4 $\times$ 10^8 cells/ml, are induced for 25 minutes at 43-44°C. The induced cultures are shifted down to 37°C; the cultures lyse completely within 2 hours.

Preparation of High-titer Lysates

1. Plate lysates (Adams 1959): Confluent lysis of indicator bacteria is achieved by mixing 7 $\times$ 10^4-10^5 pfu of Mu with 2 $\times$ 10^8 bacteria per plate in 0.5% LB soft agar. The lysis is complete in about 6 hours. The phage can be harvested by mixing LB Mg with the top agar, adding chloroform, and centrifuging to remove agar and cell debris. The yield is about 10^{10} pfu/ml.
2. Infection in liquid cultures: Freshly inoculated cultures can be infected at a density of about 1 $\times$ 10^8 cells/ml with a multiplicity of infection (m.o.i.) of about 0.2.
3. Induction of Mu *cts* lysogens: Heat induction is done as described above. It is important to shift cultures to 37°C after induction, since phage assembled at 43-44°C apparently are fragile. If the Mu *cts* prophage is in a Mu^R background (i.e., the lysogen lacks receptor for Mu adsorption), the yield of phage is higher, probably because loss of phage by attachment to cell debris is minimal. The yield is $>10^{10}$ pfu/ml. If the Mu *cts* prophage is in Mu^S background, the lysates contain mainly particles which are unable to adsorb and are thus non-plaque-forming. Since citrate will complex Ca^{++} in the medium, addition of Na citrate (20-100 mM) before lysis might be helpful in reducing the loss of viable phage by adsorption.

Stability of Lysates

Mu lysates are notoriously unstable. Athough lead acetate has been recommended (Howe 1973b), no satisfactory method for preventing the decay of phage particles is available. The lysates, however, vary greatly in stability. In some preparations, especially after purification, the phage is reasonably stable.

Purification of Phage Particles

For bulk preparations of phage, 500-600-ml cultures can be grown in 2-liter flasks. The inoculum for infection, in our procedure, is Mu in 25 ml of SB Ca Mg broth to give a multiplicity of 0.2-1. The cells are grown in LB and infected at a density of 1 $\times$ 10^8/ml. After lysis of the cultures following infection or Mu *cts* induction, as described above, a small amount of chloroform (about 1%) is added and the phage is precipitated with polyethylene glycol 6000 (PEG).

For PEG precipitation, it is better to clarify the lysate by centrifugation at low speed. NaCl is added at a final concentration of 0.5 M and then PEG to 10%. The overnight PEG precipitates are centrifuged and resuspended in 10-20 ml of Mu buffer. The PEG-phage suspension can be layered directly on a CsCl block gradient with 2-ml CsCl steps of 1.6 g/ml, 1.5 g/ml, and 1.4 g/ml in a nitrocellulose tube for a Spinco SW 27 rotor. Another layer of 1.5 ml of 20% sucrose can be added on top of the CsCl steps before adding the phage suspension. The block gradients are centrifuged to 22,000-23,000 rpm for 90-120 minutes. The phage bands at approximately the junction of the 1.4 g/ml and 1.5 g/ml layers of CsCl. Sometimes a denser band, presumably consisting of phage heads containing DNA, can be seen. The phage band is usually repurified in 1.5 g/ml CsCl by overnight centrifugation at 32,000-34,000 rpm in a Spinco SW 41 rotor. The purified phage can be dialyzed against Mu buffer. Dialysis against 0.1 M EDTA will almost completely disrupt the phage. The phage DNA can be extracted with the standard phenol extraction procedures. To obtain a small amount of Mu DNA for analytical purposes, phage can be grown in a 20-40-ml volume and isolated by centrifugation as described below.

Labeling of Phage Particles

^{3}H-labeled phage is prepared by infection or induction of a thymine-requiring strain. The strains are grown in LB Mg supplemented with 10 μg/ml of thymine. At the time of infection or induction, [^{3}H]thymidine (10 μCi/ml) is added. After lysis, the cell debris and unlysed bacteria are removed by a low-speed centrifugation (7000 rpm, 15 min), after which the phage is pelletted by centrifugation in a Spinco T30 rotor for 2 hours at 20,000 rpm. The pellet is eluted in 1.5 ml Mu buffer overnight. CsCl purification can be carried out in a block gradient (1 ml, d-1.6; 1 ml, d-1.4; 1.5 ml, d=1.3; 1.5 ml phage). The gradient is centrifuged in a Spinco SW 50 rotor for 90 minutes at 20,000 rpm.

^{32}P-labeled phage is prepared as described above for ^{3}H-labeled phage, except dephosphorylated LB is used. (To 100 ml LB is added 1 ml $MgSO_4$ [1 M] and 1 ml conc. NH_4OH. The precipitate is filtered out and Tris buffer added to a final concentration of 5×10^{-3} M; pH is adjusted to 7.2.) At the time of infection or induction, 0.1 mCi ^{32}P is added per milliliter. After lysis, purification is carried out as described above.

Separation of Mu Strands

Purified Mu particles can be treated with Sarcosyl NL 97 (Geigy) in the presence of poly(U,G) as described by Bøvre and Szybalski (1971). Equilibrium centrifugation in the presence of poly(U,G) permits complete separation of the complementary strands (Bade 1972; Wijffelman et al. 1974; Wijffelman and van de Putte 1974).

Mutagenesis with Mu

Phenotypically recognizable mutants can be readily found among Mu lysogenic *E. coli* cells surviving infection by Mu. A convenient method for isolating mutants is to spot Mu^+ lysate at 37°C (or Mu *cts* lysate at 30-32°C) onto a lawn of sensitive bacteria on LB Ca Mg plates. Good turbid spots are obtained after overnight incubation if about 10^7 pfu are spotted in a drop. When the turbid centers are streaked out for purification, 60-80% of the colonies are Mu lysogens. Among the Mu lysogens, 1-3% are auxotrophic mutants with a wide range of requirements (Taylor 1963). Similar results are obtained if infection is carried out in liquid cultures. However, at an m.o.i. of about 1, only 5-10% of the survivors are lysogens (Howe and Bade 1975).

Auxotrophic mutants can be distinguished from prototrophs by their minute colony size on partially enriched minimal medium (M9 glucose medium containing 200 μg/ml of Difco nutrient broth powder), as described by Taylor (1963). Survivors from turbid spots are streaked on the partially enriched plates and, after 24 hours incubation, small colonies are transferred to LB plates with the aid of a fine needle and a dissecting microscope. After incubation, mutants on LB plates can be characterized by replica-plating onto minimal-medium plates containing various growth factors. Auxotrophic mutants can also be enriched by two or three cycles of growth in LB broth, followed by growth in minimal medium with 20-40 μg ampicillin/ml (for ampicillin selection, see Molholt 1967; Miller 1972). Mutants defective in sugar utilization can be easily isolated by streaking the lysogens on MacConkey agar or other fermentation indicator plates (Bukhari and Zipser 1972; Miller 1972). Fermentation-negative mutants for a given sugar occur at a frequency of 10^{-3}-10^{-5}.

Mu Linkage Tests

Mutants obtained from Mu-infected cultures are first scored for Mu lysogeny by replica-plating (Miller 1972) (i.e., inocula from master plates onto LB Ca Mg plates containing about 10^8 sensitive bacteria embedded in 0.7% LB agar). Mu lysogens give clearly visible zones of killing around the replica-plated clones. To prove that the mutation is caused by the insertion of Mu DNA, it has to be shown that prophage Mu and the mutation do not segregate; i.e., prophage is linked to the mutation. This can be done in several ways:

(1) Conjugation: If the mutation is in an Hfr strain, the Mu linkage test can be performed by mating the strain with a suitably marked F^- strain. Most of the F^- cells receiving the Mu prophage are killed by zygotic induction, but surviving zygotes occur at a frequency ranging from 0.1% to 10%. A 100% coinheritance of the mutation and the prophage implies that the mutation is caused by Mu insertion. If the mutation is in an F^- strain, simultaneous removal of the mutation and Mu by mating with an Hfr strain would mean that Mu is linked to the mutation. For example, if a *thr*::Mu strain is mated with an Hfr strain, all Thr^+ recombinants would be Mu^-. This test does not work if the strain carries more than one Mu prophage. If the Mu-induced mutant is polylysogenic, the mutation must be removed from the background for the Mu linkage test.

(2) P1 transduction: If *A*::Mu (where *A* stands for any given gene into which Mu is inserted) strains are transduced to A^+ with bacteriophage P1 grown on A^+ cells, 100% of the A^+ transductants would be Mu^-. The A^+ transductants

can be scored for Mu lysogeny on indicator strains lysogenic for P1 (use of a Mu-sensitive P1 lysogen eliminates killing by P1 so that Mu lysogeny can be cleanly scored). This is probably the most convenient method for detecting Mu linkage to a mutation whose wild-type allele can be selected directly. The test would not be valid, however, if the mutant strain is polylysogenic for Mu. A reliable proof for Mu linkage can be obtained if A^-::Mu strains are used as P1 donors. This is possible if a marker cotransducible with *A* is available. Although Mu is equivalent to about 1% of the *E. coli* chromosome, P1 can transduce Mu at a relatively high frequency if a close-by marker is used for cotransduction. (Normally, P1 cotransduces two markers separated by 1 min at a frequency of less than 10%. Cotransduction of Mu can be as high as 40%. Theoretical reasons for this are given by Bukhari [1970].) 100% coinheritance of the mutation and the prophage as unselected marker is strong evidence for the linkage of Mu to the site of the mutation (Bukhari and Taylor 1971).

(3) Zygotic induction: If the mutation is in an Hfr strain, the site of the Mu prophage can be mapped by mating the Hfr cells with Mu-sensitive F^- cells and recording the time at which infectious centers of Mu arise. The matings can be interrupted by the addition of 0.05% sodium dodecyl sulfate and vigorous vortexing. The Hfr strains must have a drug resistance marker to prevent the donor cells from giving rise to infectious centers. A typical protocol for such an experiment is as follows: A streptomycin-sensitive Hfr(Mu) strain is grown in LB broth to a density of 1-2 $\times$ 10^8 cells per ml; the culture is centrifuged to decrease the number of free Mu particles, resuspended in fresh LB broth, and mated with a Mu-sensitive, streptomycin-resistant F^- strain in LB broth at a ratio of 1 Hfr to 10 F^- cells. The matings are interrupted at different times, and the appropriate dilutions of the mixture are plated with indicator bacteria in 0.5% LB agar on LB Ca Mg + streptomycin (200 μg/ml) plates. When a Mu prophage enters the F^- cells, the prophage is induced; thus the time at which Mu plaques (infectious centers) begin to increase in number is indicative of the time of the prophage entry. If the mutation in question is because of Mu insertion, the known time of entry of the gene in which mutation has occurred should coincide with the time of prophage entry. For example, in *gal*::Mu mutants, the time of entry of Mu would be the same as that of the known time of entry of the *gal* operon in an Hfr strain (see Taylor and Trotter 1967).

It should be noted that zygotic induction experiments merely indicate that the mutation site (assuming that it is known) is close to the prophage site. These tests are useful, however, to distinguish which of the several genes in a pathway, scattered on the chromosome, is affected by Mu. Moreover, the site of a prophage can be approximately mapped in an Hfr strain in which Mu integration shows no phenotypic effects.

Reversion of Mu-induced Mutations

Mutations caused by the insertion of wild-type Mu do not revert at a detectable frequency (Taylor 1963). Insertion of Mu *cts* can be reversed, however, by first isolating *X* mutants of the Mu *cts* prophage (Bukhari 1975). To isolate the *X* mutants, Mu *cts* lysogens are plated at 43-44°C. A large number of heat resistant survivors carry Mu *cts X* prophages. (In the cases actually tested, Mu *cts* was located on an F′ episome, and 10-70% of the heat-resistant survivors were found to have the Mu *cts X* mutants.) The *X* mutants can be recognized by the reversion of the mutation originally caused by Mu *cts.* For example, *lacZ*::Mu *cts X* strains can revert to Lac^+ at a frequency of 10^{-6}-10^{-7}. For the isolation of the *X* mutants, the Mu *cts* lysogens must be Mu^R to avoid reinfection or killing of cells by the phage liberated upon heat induction. Addition of 20 mM citrate and 0.1% SDS to the plates will also reduce superinfection of cells.

Isolation of Deletions with Mu

Some of the Mu-induced mutations are deletions, the frequency of deletions varying from one locus to another. About 15% of the *lac* mutants isolated with Mu carry deletions generally covering the *lac* operon (Howe and Bade 1975). Nonreversion of a Mu *cts*-induced mutation via the Mu *cts X* pathway is an indication of the presence of a deletion at the site of Mu insertion. Small internal deletions in a gene can be produced upon excision of Mu *cts X* prophages. These deletions can be detected in an operon where selection for polarity relief is available. For example, when *lacZ*::Mu *cts* strains are plated for melibiose$^+$ (Mel^+) selection at 41°C (for the expression of the *Y* gene function), some of the Mel^+ revertants contain small deletions of various sizes in the *lacZ* gene.

Insertion of Mu into Plasmids

Insertions of Mu into transmissible plasmids can be isolated by inducing Mu *cts* lysogens containing the plasmids and then transferring the plasmids to a Mu-immune recipient (Razzaki and Bukhari 1975). Since many copies of Mu are integrated at different sites during the lytic cycle of Mu, integration of Mu DNA into the plasmids can be detected by mating the cells before lysis. The recipients are Mu-immune to avoid killing of the zygotes receiving a Mu prophage. The plasmids have to be retransferred from the recipient strains to screen for Mu insertion into the plasmid. Better results are obtained if, before mating, the donor strains (carrying Mu *cts* on the chromosome and the plasmid into which Mu is to be inserted) are grown for 8-12 hours at 37°C, at which temperature the Mu *cts* lysogens are partially induced. After growth at 37°C, the cells can be mated with the appropriate recipients.

Isolation of F′ Episomes with Mu

When Hfr strains lysogenic for Mu *cts* are heat-induced, F′ episomes carrying various markers near the integrated F factor are generated (Parker and Bukhari 1976). All of these newly generated F′ episomes contain prophage Mu. The Hfr (Mu *cts*) strains can be grown at 37°C for 8-12 hours before mating to detect the F′ episomes. Using a similar method, transposition of distantly located markers on an F′ episome can also be achieved (Faelen and Toussaint 1976).

Determination of the Orientation of a Mu Prophage

The relative orientations of Mu prophages in the host chromosome can be determined by first introducing an F′ episome carrying Mu into each strain and then examining the direction of transfer of chromosomal markers (Zeldis et al. 1973). Because of recombination between Mu on the chromosome and Mu on the episome, in a *recA*$^+$ background, the chromosome can be mobilized. The polarity of transfer of markers will depend upon the orientation in which the episome is aligned with the chromosome because of pairing between the two Mu prophages. For example, if a Mu prophage is located between the *lysA* and *thyA* genes in *E. coli,* the F′::Mu will cause *lysA* to be transferred either first or last. If the orientation of the Mu prophage on the episome is known, the absolute orientation of the Mu prophage in the chromosome (or vice versa) can be determined. A small amount of transfer in the direction opposite to the major direction can be observed because of repeated inversions of the *G* segment of Mu (see Section III, this volume).

Transduction by Mu

Mu is a generalized transducing phage (Howe 1973a). The efficiency of transduction is about ten times lower than that of bacteriophage P1. Transducing particles for a given marker occur at a frequency of 10^{-7}-10^{-9} per pfu. Markers separated by a distance of 0.55 minute of the *E. coli* map can be cotransduced by Mu. For transduction, the lysates can be mixed with the recipient cells at a density of 1-2 $\times$ 10^9 cells/ml, incubated at 25°C for 30 minutes, washed, and plated on selective media (Howe 1973a). If high multiplicities of infection are used because the frequency for a marker is low, it is better to use Mu-immune recipients to avoid killing of potential transductants. No specialized transducing particles (high frequency transduction lysates) can be obtained with Mu.

References

Abelson, J., W. Boram, A. I. Bukhari, M. Faelen, M. Howe, H. Metlay, A. L. Taylor, A. Toussaint, P. van de Putte, G. C. Westmaas and C. A. Wijffelman. 1973. Summary of the genetic mapping of prophage Mu. *Virology* **54**:90.

Adams, M. H. 1959. *Bacteriophages,* p. 22. Wiley-Interscience, New York.

Bade, E. G. 1972. Asymmetric transcription of bacteriophage Mu-1. *J. Virol.* **10**:1205.

Bøvre, K. and W. Szybalski. 1971. Multistep DNA-RNA hybridization techniques. In *Methods in enzymology* (ed. L. Grossman and K. Moldave), vol. 21D, p. 350. Academic Press, New York.

Bukhari, A. I. 1970. Genetic and biochemical analysis of diaminopimelic acid and lysine-requiring mutants of *Escherichia coli.* Ph.D. thesis, University of Colorado, Denver.

——. 1975. Reversal of mutator phage Mu integration. *J. Mol. Biol.* **96**:87.

Bukhari, A. I. and A. L. Taylor. 1971. Genetic analysis of diaminopimelic acid and lysine-requiring mutants of *Escherichia coli. J. Bact.* **105**:844.

Bukhari, A. I. and D. Zipser. 1972. Random insertion of Mu-1 DNA within a single gene. *Nature New Biol.* **236**:240.

Faelen, M. and A. Toussaint. 1976. Bacteriophage Mu-1: A tool to transpose and to localize bacterial genes. *J. Mol. Biol.* **104**:525.

Howe, M. M. 1973a. Transduction by bacteriophage Mu-1. *Virology* **55**:103.

———. 1973b. Prophage deletion mapping of bacteriophage Mu-1. *Virology* **54**:93.

Howe, M. M. and E. G. Bade. 1975. Molecular biology of bacteriophage Mu. *Science* **190**:624.

Miller, J. H. 1972. *Experiments in molecular genetics.* Cold Spring Harbor Laboratory, Cold Spring Harbor, New York.

Molholt, B. 1967. Isolation of amino acid auxotrophs of *E. coli* using ampicillin enrichment. *Microb. Genet. Bull.* **27**:8.

Parker, V. and A. I. Bukhari. 1976. Genetic analysis of heterogeneous DNA circles formed after prophage Mu induction. *J. Virol.* **19**:756.

Razzaki, T. and A. I. Bukhari. 1975. Events following prophage Mu induction. *J. Bact.* **122**:437.

Taylor, A. L. 1963. Bacteriophage-induced mutation in *Escherichia coli. Proc. Nat. Acad. Sci.* **50**:1043.

Taylor, A. L. and C. D. Trotter. 1967. Revised linkage map of *Escherichia coli. Bact. Rev.* **31**:332.

Waggoner, B. T., N. S. Gonzalez and A. L. Taylor. 1974. Isolation of heterogeneous circular DNA from induced lysogens of bacteriophage Mu-1. *Proc. Nat. Acad. Sci.* **71**:1255.

Wijffelman, C. and P. van de Putte. 1974. Transcription of bacteriophage Mu. *Mol. Gen. Genet.* **131**:327.

Wijffelman, C., M. Gassler, W. F. Stevens and P. van de Putte. 1974. On the control of transcription of bacteriophage Mu. *Mol. Gen. Genet.* **131**:85.

Zeldis, J. B., A. I. Bukhari and D. Zipser. 1973. Orientation of prophage Mu. *Virology* **55**:289.

Appendix D: Restriction Endonucleases

1. RESTRICTION AND MODIFICATION ENZYMES AND THEIR RECOGNITION SEQUENCES

Contributed by R. J. Roberts, Cold Spring Harbor Laboratory

Microorganism	Source	Enzyme	Sequence[a]	R/M[b]	Number of cleavage sites			Reference[c]
					λ	Ad2	SV40	
Achromobacter immobilis	ATCC 15934	*Aim*I[d]	?	R	?	?	?	19a
Agrobacterium tumefaciens B6806	E. Nester	*Atu*BI	?	R	>50	>50	>10	65a
Agrobacterium tumefaciens C58	E. Nester	*Atu*CI	?	R	>3	>5	0	65a
Anabaena catanula	CCAP 1403/1	*Aca*I	?	R	?	?	?	32
Anabaena variabilis	K. Murray	*Ava*I	C↓PyCGPuG[e]	R	8	?	?	50, 31a
		*Ava*II	?[f]	R	?	?	?	50
Anabaena subcylindrica	K. Murray	*Asu*I	?	R	?	?	?	32
Arthrobacter luteus	ATCC 21606	*Alu*I	AG↓CT	R	>50	>50	32	61, 87
Bacillus amyloliquefaciens F	ATCC 23350	*Bam*FI	GGATCC	R	5	3	1	68a
Bacillus amyloliquefaciens H	F. E. Young	*Bam*I	G↓GATCC	R	5	3	1	83, 30a, 53a, 59
Bacillus amyloliquefaciens K	T. Kaneko	*Bam*KI	GGATCC	R	5	3	1	68a
Bacillus amyloliquefaciens N	T. Ando	*Bam*NI	GGATCC	R	5	3	1	68
		BamN_x	?	R	?	?	?	67, 68
Bacillus brevis S	A. P. Zarubina	*Bbv*SI	G$\overset{*}{C}$($^{A}_{T}$)GC	M	—	—	—	82
Bacillus brevis	ATCC 9999	*Bbv*I	?	R	>30	>30	9	56
Bacillus cereus	ATCC 14579	*Bce* 14579	?	R	>10	?	?	68a
	1AM 1229	*Bce* 1229	?	R	>10	?	?	68a
	T. Ando	*Bce* 170	CTGCAG	R	18	25	3	68a
Bacillus cereus Rf sm st	T. Ando	*Bce*R	CGCG	R	>50	>50	0	68a
Bacillus globigii	G. A. Wilson	*Bgl*I	?	R	22	12	1	84a, 54a

Microorganism	Source	Enzyme	Sequence[a]	R/M[b]	Number of cleavage sites λ	Ad2	SV40	Reference[c]
		*Bgl*II	A↓GATCT	R	5	10	0	54a, 84a 92
Bacillus pumilus AHU1387	T. Ando	*Bpu*I	?	R	6	>30	2	33a
Bacillus sp.	J. Upcroft	"*Bsp*I"	?	R	>10	>20	4	22
Bacillus sphaericus	1AM 1286	*Bsp* 1286	?	R	?	?	?	68a
Bacillus sphaericus R	P. Venetianer	*Bsp*RI	GGCC	R	>50	>50	18	34a
Bacillus stearothermophilus 1503-4R	N. Welker	*Bst*I	GGATCC	R	5	3	1	15
Bacillus subtilis strain R	T. Trautner	*Bsu*RI	GG↓CC	R	>50	>50	18	12, 11
Bacillus subtilis Marburg 168	T. Ando	*Bsu*M	?	R	>10	?	?	68a
Bacillus subtilis	ATCC 6633	*Bsu* 6663	?	R	>20	?	?	68a
	1AM 1076	*Bsu* 1076	GGCC	R	>50	>50	18	68a
	1AM 1114	*Bsu* 1114	GGCC	R	>50	>50	18	68a
	ATCC 14593	*Bsu* 1145	?	R	>20	?	?	68a
	1AM 1192	*Bsu* 1192	?	R	>10	?	?	68a
	1AM 1193	*Bsu* 1193	?	R	>30	?	?	68a
	1AM 1231	*Bsu* 1231	?	R	>20	?	?	68a
	1AM 1247	*Bsu* 1247	CTGCAG	R	18	25	3	68a
	1AM 1259	*Bsu* 1259	?	R	>8	?	?	68a
Bacillus megaterium 899	B899	*Bme* 899	?	R	>5	?	?	68a
Bacillus megaterium B205-3	T. Kaneko	*Bme* 205	?	R	>10	?	?	68a
Bordetella bronchiseptica	ATCC 19395	*Bbr*I	AAGCTT	R	6	11	6	58
Bordetella parapertussis	R. J. Roberts	*Bpa*I	GT↓($^{C}_{A}$)($^{T}_{G}$)AC	R	7	8	1	89
		*Bpa*II	CGCG	R	>50	>50	0	89
Brevibacterium albidum	ATCC 15831	*Bal*I	TGG↓CCA	R	15	17	0	23

Brevibacterium luteum	ATCC 15830	*Blu*I	C↓TCGAG	R	1	6	0	24a
		*Blu*II	GGCC	R	>50	>50	18	81
Brevibacterium umbra	R. J. Roberts	*Bum*I	CAGCTG	R	15	22	3	57
Chromobacterium violaceum	ATCC 12472	*Cvi*I	?	R	?	?	?	19a
Corynebacterium humiferum	ATCC 21108	*Chu*I	AAGCTT	R	6	11	6	19a
		*Chu*II	GTPyPuAC	R	34	>20	7	19a
Corynebacterium petrophilum	ATCC 19080	*Cpe*I	TGATCA	R	7	5	1	20a
Diplococcus pneumoniae	S. Lacks	*Dpn*I	G$\overset{*}{A}$TC	R	?	?	?	36
	S. Lacks	*Dpn*II	↓GATC	R	>50	>50	6	37, 36
Enterobacter cloacae	H. Hartmann	*Ecl*I	?	R	>4	?	?	30c
Escherichia coli RY13	R. N. Yoshimori	*Eco*RI	G↓A$\overset{*}{A}$TTC	R,M	5	5	1	25, 18, 31, 45, 46, 88,* 2, 54, 55
		*Eco*RI′	PuPuA↓TPyPy	R	>10	>10	?	49
Escherichia coli R245	R. N. Yoshimori	*Eco*RII	↓C$\overset{*}{C}$($\binom{A}{T}$)GG	R,M	>35[g]	>35	16	88,* 6, 7, 72
Escherichia coli B	W. Arber	*Eco*B	?[h]	R, M	?	?	?	20, 39,* 71, 80
Escherichia coli K	M. Meselson	*Eco*K	?	R,M	?	?	?	40, 30*
Escherichia coli (PI)	K. Murray	*Eco*PI	AGATCT	R,M	?	?	?	29, 8, 9,* 10, 40
Escherichia coli P15	W. Arber	*Eco*P15	?	R,M	?	?	?	55a
Escherichia coli (T2)	H. Van Ormondt	*Eco*T2	G$\overset{*}{A}$TC	M	—	—	—	79
Haemophilus aegyptius	ATCC 11116	*Hae*I	($\binom{A}{T}$)GG↓CC($\binom{A}{T}$)	R	?	?	?	48
		*Hae*II	PuGCGC↓Py	R	>30	>30	1	60, 4, 78a
		*Hae*III	GG↓CC	R	>50	>50	18	42, 50, 72, 86
Haemophilus aphrophilus	ATCC 19415	*Hap*I	?	R	>30	?	?	58
		*Hap*II	C↓CGG	R	>50	>50	1	76, 74, 75

Microorganism	Source	Enzyme	Sequence[a]	R/M[b]	Number of cleavage sites			Reference[c]
					λ	Ad2	SV40	
Haemophilus gallinarum	ATCC 14385	*Hga*I	?	R	>30	>30	0	76, 75
Haemophilus haemoglobinophilus	ATCC 19416	*Hhg*I	GGCC	R	>50	>50	18	58
Haemophilus haemolyticus	ATCC 10014	*Hha*I	GCG↓C	R	>50	>50	2	62, 73
Haemophilus influenzae 1056	J. Stuy	*Hin*1056I	CGCG	R	>30	>30	?	53
		*Hin*1056II	?	R	>30	>30	?	58
Haemophilus influenzae serotype b, 1076	J. Stuy	*Hin*bIII	AAGCTT	R	6	11	6	53
Haemophilus influenzae R_b	C. A. Hutchison	*Hin*bIII	AAGCTT	R	6	11	6	58, 43
Haemophilus influenzae serotype c, 1160	J. Stuy	*Hin*cII	GTPyPuAC	R	34	>20	7	53
Haemophilus influenzae serotype c, 1161	J. Stuy	*Hin*cII	GTPyPuAC	R	34	>20	7	53
Haemophilus influenzae R_c	A. Landy, G. Leidy	*Hin*cII	GTPyPuAC	R	34	>20	7	38
Haemophilus influenzae R_d	S. H. Goodgal (*exo*⁻ mutant)	*Hin*dI	C$\overset{*}{\text{A}}$C	R,M	–	–	–	63,* 64
		*Hin*dII	GTPy↓Pu$\overset{*}{\text{A}}$C	R,M	34	>20	7	70, 16, 34, 38, 63,* 64
		*Hin*dIII	$\overset{*}{\text{A}}$↓AGCTT	R,M	6	11	6	52, 16, 63,* 64
		*Hin*dIV	G$\overset{*}{\text{A}}$T	M	–	–	–	63,* 64
Haemophilus influenzae R_f	C. A. Hutchison	*Hin*fI	G↓ANTC	R	>50	>50	10	41, 33, 43, 51, 73
Haemophilus influenzae H-1	M. Takanami	*Hin*H-1	PuGCGCPy	R	>30	>30	1	76, 75, 77
Haemophilus parahaemolyticus	C. A. Hutchison	*Hph*I	GGTGA → 8 bp	R	>50	>50	4	35, 43

Haemophilus parainfluenzae	J. Setlow	*Hpa*I	GTT↓AAC	R	11	6	5	66, 16, 17, 21, 27
		*Hpa*II	C↓CGG	R	>50	>50	1	66, 1, 16, 21, 27
Haemophilus suis	ATCC 19417	*Hsu*I	AAGCTT	R	6	11	6	58
Klebsiella pneumoniae OK8	J. Davies	*Kpn*I	?	R	2	8	1	69
Moraxella bovis	ATCC 10900	*Mbo*I	↓GATC	R	>50[g]	>50	8	22a, 47
		*Mbo*II	GAAGA → 8 bp	R	>50	>50	6	22a, 14
Moraxella nonliquefaciens	ATCC 19975	*Mno*I	C↓CGG	R	>50	>50	1	5
		*Mno*II	?	R	>10	>6	2	58
	ATCC 17953	*Mnl*I	CCTC[i]	R	>50	>50	15	90
	ATCC 17954	*Mnn*I	GTPyPuAC	R	34	>20	7	30b
		*Mnn*II	?	R	>50	>50	?	30b
Moraxella osloensis	ATCC 19976	*Mos*I	GATC	R	>50	>50	8	22a
Moraxella glueidi LG1	J. Davies	*Mgl*I[d]	?	R	?	?	?	69
Moraxella glueidi LG2	J. Davies	*Mgl*II[d]	?	R	?	?	?	69
Myxococcus virescens V2	H. Reichenbach	*Mvi*I	?	R	1	?	?	44
		*Mvi*II	?	R	?	?	?	44
Neisseria gonorrhoea	G. Wilson	*Ngo*I	PuGCGCPy	R	>30	>30	1	84
Proteus vulgaris	ATCC 13315	*Pvu*I	?	R	1	7	0	24
		*Pvu*II	CAG↓CTG	R	15	22	3	24
Providencia alcalifaciens	ATCC 9886	*Pal*I	GGCC	R	>50	>50	18	22
Providencia stuartii 164	J. Davies	*Pst*I	CTGCA↓G[j]	R	18	25	3	13, 69
Pseudomonas facilis	M. Van Montagu	*Pfa*I	?	R	?	?	3	53
Serratia marcescens S_b	C. Mulder	*Sma*I	CCC↓GGG	R	3	12	0	26, 19
Serratia species SAI	B. Torheim	*Ssp*I	?	R	?	?	?	78
Staphylococcus aureus 3A[o]	E. E. Stubberingh	*Sau*3A	↓GATC	R	>50[j]	>50	8	74a
Streptococcus faecalis var. zymogenes	R. Wu	*Sfa*I	GGCC	R	>50	>50	18	85
Streptococcus faecalis ND547	D. Clewell	*Sfa*NI	?	R	>50	>30	4	65a

Microorganism	Source	Enzyme	Sequence[a]	R/M[b]	Number of cleavage sites			Reference[c]
					λ	Ad2	SV40	
Streptomyces achromogenes	ATCC 12767	*Sac*I	?	R	2	7	0	3
		*Sac*II	CCGC↓GG	R	3	>15	0	3
		*Sac*III	?	R	>30	>30	?	3
Streptomyces albus G	J. M. Ghuysen	*Sal*I	G↓TCGAC	R	2	3	0	3
		*Sal*II	?	R	>30	?	?	3
Streptomyces albus	CM1 52766	*Sal*PI	CTGCAG	R	18	25	3	15a
Streptomyces griseus	ATCC 23345	*Sgr*I	?	R	?	7	?	3
Streptomyces stanford	S. Goff,	*Sst*I	?	R	2	7	0	28
	A. Rambach	*Sst*II	CCGC↓GG	R	3	>15	0	28
		*Sst*III	?	R	>30	>30	?	28
Thermopolyspora glauca	ATCC 15345	*Tgl*I	CCGCGG	R	3	>15	0	24
Thermus aquaticus YTI	T. D. Brock	*Taq*I	T↓CGA	R	>50	>50	1	65
		*Taq*II	?	R	>30	>30	?	58
Xanthomonas amaranthicola	ATCC 11645	*Xam*I	GTCGAC	R	2	3	0	3
Xanthomonas badrii	ATCC 11672	*Xba*I	T↓CTAGA	R	1[g]	4	0	91
Xanthomonas holcicola	ATCC 13461	*Xho*I	C↓TCGAG	R	1	6	0	24a, 68b
		*Xho*II	?	R	>20	>20	4	53
Xanthomonas malvacearum	ATCC 9924	*Xma*I	C↓CCGGG	R	3	12	0	19
		*Xma*II	CTGCAG	R	>15	>20	2	19
Xanthomonas nigromaculans	ATCC 23390	*Xni*I	?	R	4	10	?	30b
Xanthomonas papavericola	ATCC 14180	*Xpa*I	C↓TCGAG	R	1	6	0	24a

[a] Recognition sequences are written from 5′ → 3′, only one strand being given, and the site of cleavage, where known, is indicated by an arrow. For example, G↓GATCC is the abbreviation for 5′ G↓GATCC 3′ / 3′ CCTAG↑G 5′ and GC($\frac{A}{T}$)GC for 5′ GCAGC 3′ / 3′ CGTCG 5′.

In some cases, the recognition sequence has been inferred from a comparison of digestion patterns; e.g., *Bst*I and *Bam*I give identical patterns on all DNAs tested, and a double digest with both enzymes gives no further cleavage. In these cases, the exact site of cleavage within the recognition sequence has not been determined and the arrow is omitted. For *Hph*I and *Mbo*II, the site of cleavage is 8 base pairs from the 3′-nucleotide of the recognition sequence. The base modified by the methylase is indicated by an asterisk. *N*-6-methyladenosine is found in the case of *Eco*B, *Eco*PI, *Eco*RI, *Hin*dI, *Hin*dII, *Hin*dIII, *Hin*dIV, and *Eco*T2; 5-methylcytosine in the case of *Eco*RII and *Bbv*SI.

[b] Enzymes designated R have been identified as specific endonucleases (restriction enzymes), whereas those designated M have been identified as specific methylases (modification enzymes).

[c] Where more than one reference is given, the first contains the purification procedure for the restriction endonuclease. An * indicates the reference giving the purification procedure for the methylase. Complete references, by number, are given at the end of these notes.

[d] The amounts of these enzymes are too low to warrant further investigation.

[e] For convenience, only one strand of the recognition sequence is indicated. The full duplex structure is 5′ CPyCGPuG 3′ / 3′ GPuGCPyC 5′.

[f] 5′-Terminal trinucleotide sequences, GAC and GTC, have been found (ref. 50); however, the exact nature of the site recognized is unknown.

[g] In most *E. coli* strains, bacteriophage lambda DNA is partially modified against the action of *Eco*RII, *Mbo*I, and *Xba*I.

[h] The following methylated oligonucleotides have been isolated from *Eco*B-modified DNA: TGA*, CA*C, AGA*C, A*AT, (A, G, C) A*. A* is *N*-6-methyladenosine.

[i] *Mnl*I cleaves 5 to 10 bases 3′ from the recognition sequence.

[j] *Pst*I cleaves ϕX174 DNA at a single site, CTGCA↓G; however, it may recognize some other sequence(s) also.

References

1. Allet, B. 1973. Fragments produced by cleavage of lambda DNA with *H. parainfluenzae* restriction enzyme *Hpa*II. *Biochemistry* **12**:3972.
2. Allet, B., P. G. N. Jeppesen, K. J. Katagiri and H. Delius. 1973. Mapping the DNA fragments produced by cleavage of lambda DNA with endonuclease RI. *Nature* **241**:120.
3. Arrand, J. R., P. A. Myers and R. J. Roberts, unpublished.
4. Barrell, B. G. and P. Slocombe, unpublished.
5. Baumstark, B. R., R. J. Roberts and U. L. RajBhandary. 1977. Use of short synthetic DNA duplexes as substrates for the restriction endonucleases *Hpa*II and *Mno*I. Submitted for publication.
6. Bigger, C. H., K. Murray and N. E. Murray. 1973. Recognition sequence of a restriction enzyme. *Nature New Biol.* **244**:7.
7. Boyer, H. W., L. T. Chow, A. Dugaiczyk, J. Hedgpeth and H. M. Goodman. 1973. DNA substrate site for the *Eco*RII restriction endonuclease and modification methylase. *Nature New Biol.* **244**:40.
8. Brockes, J. P. 1973. The deoxyribonucleic acid modification enzyme of bacteriophage P1, subunit structure. *Biochem. J.* **133**:629.
9. Brockes, J. P., P. R. Brown and K. Murray. 1972. The deoxyribonucleic acid modification enzyme of bacteriophage P1, purification and properties. *Biochem. J.* **127**:1.
10. ——. 1974. Nucleotide sequences at the sites of action of the deoxyribonucleic acid modification enzyme of bacteriophage P1. *J. Mol. Biol.* **88**:437.

11. Bron, S. and K. Murray. 1975. Restriction and modification in *B. subtilis.* Nucleotide sequence recognized by restriction endonuclease R.*Bsu*R from strain R. *Mol. Gen. Genet.* **143**:25.
12. Bron, S., K. Murray and T. Trautner. 1975. Restriction and modification in *B. subtilis.* Purification and general properties of a restriction endonuclease from strain R. *Mol. Gen. Genet.* **143**:13.
13. Brown, N. L. and M. Smith. 1977. The mapping and sequence determination of a single site in ϕX174 amber 3 replicative form DNA cleaved by restriction endonuclease *Pst*I. *FEBS Letters* **65**:284.
14. Brown, N. L., C. A. Hutchison III and M. Smith, unpublished.
15. Catterall, J. F. and N. E. Welker. 1977. Isolation and properties of a thermostable restriction endonuclease (Endo R-*Bst* 1503). *J. Bact.* **129**:1110.

15a. Chater, K. F. 1977. A site-specific endodeoxyribonuclease from *Streptomyces albus* CMI 52766 sharing site specificity with *Providencia stuartii* endonuclease *Pst*I. *Nucleic Acids Res.* **4**:1989.

16. Danna, K. J., G. H. Sack, Jr. and D. Nathans. 1973. Studies of simian virus 40 DNA. VII. A cleavage map of the SV40 genome. *J. Mol. Biol.* **78**:363.
17. DeFilippes, F. M. 1974. A new method for isolation of a restriction enzyme from *Haemophilus parainfluenzae. Biochem. Biophys. Res. Comm.* **58**:586.
18. Dugaiczyk, A., J. Hedgpeth, H. W. Boyer and H. M. Goodman. 1974. Physical identity of the SV40 deoxyribonucleic acid sequence recognized by the *Eco*RI restriction endonuclease and modification methylase. *Biochemistry* **13**:503.
19. Endow, S. A. and R. J. Roberts. 1977. Two restriction-like enzymes from *Xanthomonas malvacearum. J. Mol. Biol.* **112**:521.

19a. ——, unpublished.

20. Eskin, B. and S. Linn. 1972. The deoxyribonucleic acid modification and restriction enzymes of *Escherichia coli* B. II. Purification, subunit structure and catalytic properties of the restriction endonuclease. *J. Biol. Chem.* **247**:6183.

20a. Fisherman, J., T. R. Gingeras and R. J. Roberts, unpublished.

21. Garfin, D. E. and H. M. Goodman. 1974. Nucleotide sequences at the cleavage sites of two restriction endonucleases from *Haemophilus parainfluenzae. Biochem. Biophys. Res. Comm.* **59**:108.
22. Gelinas, R. E., P. A. Myers and R. J. Roberts, unpublished.

22a. ——. 1977. Two sequence-specific endonucleases from *Moraxella bovis. J. Mol. Biol.* (in press).

23. Gelinas, R. E., P. A. Myers, G. H. Weiss, K. Murray and R. J. Roberts. 1977. A specific endonuclease from *Brevibacterium albidum. J. Mol. Biol.* (in press).
24. Gingeras, T. R. and R. J. Roberts, unpublished.

24a. Gingeras, T. R., P. A. Myers, J. A. Olson, F. A. Hanberg and R. J. Roberts. A new specific endonuclease present in *Xanthomonas holcicola, Xanthomonas papavericola* and *Brevibacterium luteum*. Submitted for publication.

25. Greene, P. J., M. C. Betlach, H. W. Boyer and H. M. Goodman. 1974. The *Eco*RI restriction endonuclease. In *Methods in molecular biology* (ed. R. B. Wickner), vol. 7, p. 87. Marcel Dekker, New York.
26. Greene, R. and C. Mulder, unpublished.
27. Gromkova, R. and S. H. Goodgal. 1972. Action of *Haemophilus* endodeoxyribonuclease on biologically active deoxyribonucleic acid. *J. Bact.* **109**:987.
28. Goff, S. and A. Rambach, unpublished.

29. Haberman, A. 1974. The bacteriophage P1 restriction endonuclease. *J. Mol. Biol.* **89**:545.
30. Haberman, A., J. Heywood and M. Meselson. 1972. DNA modification methylase activity of *Escherichia coli* restriction endonucleases K and P. *Proc. Nat. Acad. Sci.* **69**:3138.
30a. Haggerty, D. M. and R. E. Schleif. 1976. Location in bacteriophage lambda DNA of cleavage sites of the site specific endonuclease from *Bacillus amyloliquefaciens* H. *J. Virol.* **18**:659.
30b. Hanberg, F., P. A. Myers and R. J. Roberts, unpublished.
30c. Hartman, H. and W. Goebel. 1977. A new restriction endonuclease from *Enterobacter cloacae* (*Ecl*I). *FEBS Letters* (in press).
31. Hedgpeth, J., H. M. Goodman and H. W. Boyer. 1972. DNA nucleotide sequence restricted by the RI endonuclease. *Proc. Nat. Acad. Sci.* **69**:3448.
31a. Hughes, S. G. 1977. A map of the cleavage sites for endonuclease *Ava*I in the chromosome of bacteriophage lambda. *Biochem. J.* **163**:503.
32. Hughes, S. G., T. Bruce and K. Murray, unpublished.
33. Hutchison, C. A. III and B. G. Barrell, unpublished.
33a. Ikawa, S., T. Shibata and T. Ando. 1976. The site-specific deoxyribonuclease from *Bacillus pumilis* (endonuclease R. *Bpu* 1387). *J. Biochem.* **80**:1457.
34. Kelly, T. J. and H. O. Smith. A restriction enzyme from *Haemophilus influenzae.* II. Base sequence of the recognition site. *J. Mol. Biol.* **51**:393.
34a. Kiss, A., B. Sain, E. Csordas-Toth and P. Venetianer. 1977. A new sequence-specific endonuclease *Bsp* from *Bacillus sphaericus. Gene* **1**:323.
35. Kleid, D., Z. Humayun, A. Jeffrey and M. Ptashne. 1976. Novel properties of a restriction endonuclease isolated from *Haemophilus parahaemolyticus. Proc. Nat. Acad. Sci.* **73**:293.
36. Lacks, S. and B. Greenberg. 1975. A deoxyribonuclease of *Diplococcus pneumoniae* specific for methylated DNA. *J. Biol. Chem.* **250**:4060.
37. ———. Complementary specificity of restriction endonuclease of *Diplococcus pneumoniae* with respect to DNA methylation. Submitted for publication.
38. Landy, A., E. Ruedisueli, L. Robinson, C. Foeller and W. Ross. 1973. Digestion of DNAs from bacteriophage T7, lambda and ϕ80 h with site specific nucleases for *H. influenzae* strain R_c and strain R_d. Purification and properties. *J. Mol. Biol.* **81**:427.
39. Lautenberger, J. A. and S. Linn. 1972. The deoxyribonucleic acid modification and restriction enzymes of *Escherichia coli* B. I. Purification, subunit structure and catalytic properties of the modification methylase. *J. Biol. Chem.* **247**:6176.
40. Meselson, M. and R. Yuan. 1968. DNA restriction enzyme from *E. coli. Nature* **217**:1110.
41. Middleton, J. H. 1973. Restriction endonucleases from *Haemophilus* species: Enzymes for specific fragmentation of DNA. Ph.D. thesis, University of North Carolina, Chapel Hill.
42. Middleton, J. H., M. H. Edgell and C. A. Hutchison III. 1972. Specific fragments of ϕX174 deoxyribonucleic acid produced by a restriction enzyme from *Haemophilus aegyptius,* endonuclease Z. *J. Virol.* **10**:42.
43. Middleton, J. H., P. V. Stankus, M. H. Edgell and C. A. Hutchison III, unpublished.
44. Morris, D. W. and J. H. Parish. 1976. Restriction in *Myxococcus virescens. Arch. Microbiol.* **108**:227.
45. Morrow, J. F. and P. Berg. 1972. Cleavage of simian virus 40 DNA at a unique site by a bacterial restriction enzyme. *Proc. Nat. Acad. Sci.* **69**:3365.
46. Mulder, C. and H. Delius. 1972. Specificity of the break produced by restriction endonuclease RI in simian virus 40 DNA, as revealed by partial denaturation mapping. *Proc. Nat. Acad. Sci.* **69**:3215.
47. Murray, K. and S. A. Bruce, unpublished.

48. Murray, K. and A. Morrison, unpublished.
49. Murray, K., J. S. Brown and S. A. Bruce, unpublished.
50. Murray, K., S. G. Hughes, J. S. Brown and S. A. Bruce. 1976. Isolation and characterization of two sequence-specific endonucleases from *Anabaena variabilis. Biochem. J.* **159**:317.
51. Murray, K., A. Morrison, H. W. Cooke and R. J. Roberts, unpublished.
52. Old, R., K. Murray and G. Roizes. 1975. Recognition sequence of restriction endonuclease III from *Haemophilus influenzae. J. Mol. Biol.* **92**:331.
53. Olson, J. A., P. A. Myers and R. J. Roberts, unpublished.
53a. Perricaudet, M. and P. Thiollais. 1975. Defective bacteriophage lambda chromosome, potential vector for DNA fragments obtained after cleavage by *Bacillus amyloliquefaciens* endonuclease (*Bam*I). *FEBS Letters* **56**:7.
54. Pettersson, U., C. Mulder, H. Delius and P. A. Sharp. 1973. Cleavage of adenovirus type 2 DNA into six unique fragments by endonuclease R.RI. *Proc. Nat. Acad. Sci.* **70**:200.
54a. Pirotta, V. 1976. Two restriction endonucleases from *Bacillus globigii. Nucleic Acids Res.* **3**:1747.
55. Polisky, B., P. Greene, D. E. Garfin, B. J. McCarthy, H. M. Goodman and H. W. Boyer. 1975. Specificity of substrate recognition by the *Eco*RI restriction endonuclease. *Proc. Nat. Acad. Sci.* **72**:3310.
55a. Reiser, G. and R. Yuan. 1977. Purification and properties of the P15 specific restriction endonuclease from *Escherichia coli. J. Biol. Chem.* **252**:451.
56. Roberts, R. J., unpublished.
57. Roberts, R. J. and M. B. Matthews, unpublished.
58. Roberts, R. J. and P. A. Myers, unpublished.
59. Roberts, R. J., G. A. Wilson and F. E. Young. 1977. Recognition sequence of specific endonuclease *Bam*HI from *Bacillus amyloliquefaciens* H. *Nature* **265**:82.
60. Roberts, R. J., J. B. Breitmeyer, N.F. Tabachnik and P. A. Myers. 1975. A second specific endonuclease from *Haemophilus aegyptius. J. Mol. Biol.* **91**:121.
61. Roberts, R. J., P. A. Myers, A. Morrison and K. Murray. 1976. A specific endonuclease from *Arthrobacter luteus. J. Mol. Biol.* **102**:157.
62. ——. 1976. A specific endonuclease from *Haemophilus haemolyticus. J. Mol. Biol.* **103**:199.
63. Roy, P. H. and H. O. Smith. 1973. DNA methylases of *Haemophilus influenzae* R_d. I. Purification and properties. *J. Mol. Biol.* **81**:427.
64. ——. 1973. DNA methylases of *Haemophilus influenzae* R_d. II. Partial recognition site base sequences. *J. Mol. Biol.* **81**:445.
65. Sato, S., C. A. Hutchison III and J. I. Harris. 1977. A thermostable sequence-specific endonuclease from *Thermus aquaticus. Proc. Nat. Acad. Sci.* **74**:542.
65a. Sciaky, D. and R. J. Roberts, unpublished.
66. Sharp, P. A., B. Sugden and J. Sambrook. 1973. Detection of two restriction endonuclease activities in *H. parainfluenzae* using analytical agarose-ethidium bromide electrophoresis. *Biochemistry* **12**:3055.
67. Shibata, T. and T. Ando. 1975. *In vitro* modification and restriction of phage ϕ105C DNA with *Bacillus subtilis* N cell-free extract. *Mol. Gen. Genet.* **138**:269.
68. ——. 1976. The restriction endonuclease in *Bacillus amyloliquefaciens* N strain. Substrate specificities. *Biochem. Biophys. Acta* **442**:184.

68a. Shibata, T., S. Ikawa, C. Kim and T. Ando. 1976. Site-specific deoxyribonucleases in *Bacillus subtilis* and other *Bacillus* strains. *J. Bact.* **128**: 473.

68b. Sims, J., unpublished.

69. Smith, D. I., F. R. Blattner and J. Davies. 1976. The isolation and partial characterization of a new restriction endonuclease from *Providencia stuartii. Nucleic Acids Res.* **3**:343.

70. Smith, H. O. and K. W. Wilcox. 1970. A restriction enzyme from *Haemophilus influenzae.* I. Purification and general properties. *J. Mol. Biol.* **51**:379.

71. Smith, J. D., W. Arber and U. Kuhnlein. 1972. Host specificity of DNA produced by *Escherichia coli.* XIV. The role of nucleotide methylation in *in vivo* B-specific modification. *J. Mol. Biol.* **63**:1.

72. Subramanian, K. N., J. Pan, S. Zain and S. M. Weissman. 1974. The mapping and ordering of fragments of SV40 DNA produced by restriction endonucleases. *Nucl. Acids Res.* **1**:727.

73. Subramanian, K. N., B. S. Zain, R. J. Roberts and S. M. Weissman. 1977. Mapping of the *Hha*I and *Hin*fI cleavage sites on simian virus 40 DNA. *J. Mol. Biol.* **110**:297.

74. Sugisaki, H. and M. Takanami. 1973. DNA sequence restricted by restriction endonuclease AP from *Haemophilus aphirophilus. Nature New Biol.* **246**:138.

74a. Sussenbach, J. S., C. M. Monfoort, R. Schiphof and E. E. Stobberingh. 1976. A restriction endonuclease from *Staphylococcus aureus. Nucleic Acids Res.* **3**:3193.

75. Takanami, M. 1973. Specific cleavage of coliphage fd DNA by five different restriction endonucleases from *Haemophilus* genus. *FEBS Letters* **34**:318.

76. ——. 1974. Restriction endonucleases AP, GA and H-I from three *Haemophilus* strains. In *Methods in molecular biology* (ed. R. B. Wickner), vol. 7, p. 113. Marcel Dekker, New York.

77. Takanami, M. and H. Kojo. 1973. Cleavage site specificity of an endonuclease prepared from *Haemophilus influenzae* strain H-1. *FEBS Letters* **29**:267.

78. Torheim, B., unpublished.

78a. Tu, C.-P., R. Roychoudhury and R. Wu. 1976. Nucleotide recognition sequence at the cleavage site of *Haemophilus aegyptius* II (*Hae*II) restriction endonuclease. *Biochem. Biophys. Res. Comm.* **72**:355.

79. Van Ormondt, H., J. Gorter, K. J. Havelaar and A. deWaard. 1975. Specificity of a deoxyribonucleic acid transmethylase induced by bacteriophage T2. I. Nucleotide sequences isolated from *Micrococcus luteus* DNA methylated *in vitro. Nucleic Acids Res.* **2**:1391.

80. Van Ormondt, H., J. A. Lautenberger, S. Linn and D. Lackey. 1973. Host controlled restriction and modification enzymes in *Escherichia coli* B. *Fed. Proc.* **33**:177.

81. Van Montagu, M., unpublished.

82. Vanyushin, B. F. and A. P. Dobritsa. 1975. On the nature of the cytosine-methylated sequence in DNA of *Bacillus brevis* var. G.-B. *Biochim. Biophys. Acta* **407**:61.

83. Wilson, G. A. and F. E. Young. 1975. Isolation of a sequence-specific endonuclease (Bam I) from *Bacillus amyloliquefaciens* H. *J. Mol. Biol.* **97**:123.

84. ——, unpublished.

84a. ——. 1976. Restriction and modification in the *Bacillus subtilis* genospecies. In *Microbiology 1976* (ed. D. Schlessinger), p. 350. American Society for Microbiology, Washington, D.C.
85. Wu, R., C. King and E. Jay, unpublished.
86. Yang, R. C.-A., A. Van de Voorde and W. Fiers. 1976. Cleavage map of the simian virus 40 genome by the restriction endonuclease III of *Haemophilus aegyptius*. *Eur. J. Biochem.* **61**:101.
87. ——. 1976. Specific cleavage and physical mapping of simian virus 40 DNA by the restriction endonuclease of *Arthrobacter luteus*. *Eur. J. Biochem.* **61**:119.
88. Yoshimori, R. N. 1971. A genetic and biochemical analysis of the restriction and modification of DNA by resistance transfer factors. Ph.D. thesis, University of California, San Francisco.
89. Zabeau, M. and R. J. Roberts, unpublished.
90. Zabeau, M., R. Greene, P. A. Myers and R. J. Roberts, unpublished.
91. Zain, S. and R. J. Roberts. 1977. A new specific endonuclease from *Xanthomonas badrii*. Submitted for publication.
92. Zain, S., P. A. Myers and R. J. Roberts, unpublished.

Index

Adeno-associated viruses (AAV)
DNA structure of
functional significance, 482, 483
restriction enzyme analysis, 477–480
strand sequence permutations, 478, 479
terminal repetitions, 477–478
genome orientation, 477
RNA transcription map, 478
structure of, 477–485
terminal sequences
palindromes, 482
permutations, 478, 479
properties of, 479–482
rabbit-ear structures in, 479, 481
replication, role in, 483
restriction enzyme analysis of, 479, 480
structural model of, 481, 482
Adenovirus 2, DNA underwound loops, 578
Agrobacterium tumefaciens, transposon-induced mutations in, 181, 182
Allele number, IS mutations, 19
Alpha-D-fucose, role in induction of *gal* enzymes, 104, 105, 133
Ampicillin resistance determinant, transposition 151–160, 543–548. *See also* Antibiotic resistance determinants; Tn*A*; Tn*1*; Tn*3*
Antibiotic resistance determinants, 139–146. *See also* Transposable elements; Tn *1–10*
detection of, 555–558
flanking sequences of, 7
in PlCm and P7, structure and location, 229–234
recombination with bacteriophages, 140
structure of, 7
transposition of
deletion formation, 169–178
deletions, effect of, 161–168
insertion, 169–178
insertion sequences, role of, 144
selected transposition, 549–554

Bacteriophage lambda
abnormal insertions, role of *int* protein in, 334, 345
attachment site, 9
restriction enzyme map, 719, 720
bi2 mutation
IS*2* in, 382, 383
int and *xis* gene products, 382, 383
restriction enzyme analysis, 382, 384
Campbell model, 8, 9
Charon-4 vector, lambda-Mu chimeric phages, 567–574
clear mutants, *int* activity, 391
cohesive ends, 8, 9, 10
common core sequence, 347
crg mutation
IS*2* in, 383, 384
xis function, 384
cryptic gene
IS*2* in, 381, 383, 384
IS*2*-mediated cryptic phage termination, 381
*c*II, *c*III genes
*c*I repressor regulation, 389
control interpretation, 393
int protein synthesis, role in, 389–393
integrative recombination, effect on, 390, 391
lysogenic response, control of, 389, 392
λd*gal* phages containing IS sequences, restriction enzyme analysis of, 115–122
DNA circles, 9
early region, gene organization and transcription, 390
escape synthesis, 94
excision, 9, 343–347. *See also* Bacteriophage lambda, integrase, integrative recombination
model for, 350
int gene product, role of, 343, 344, 398
excisionase (*xis*) gene, 343–347
IS*2* polar effect on, 381, 384
transcription of, 375–378
λ*fus*3 IS mutations
r-protein genes, expression of, 491–494
selection for, 488, 489
spc, fus, and *str* mutants, 488–494
genetic map of, 713–720
host-integration-deficient (*hid*) mutants, 357–361
integrative recombination, effect on, 359, 360
isolation and mapping, 357, 358
properties of, 358, 359
host integration mediation (*him*) mutants
physiological effect of, 354

Bacteriophage lambda *(continued)*
recombination, influence on, 352–354
repression, establishment of, 354
selection for, 350, 351
immunity region, role in *c*II,III-dependent *int* synthesis, 391, 392
insertion. *See* Bacteriophage lambda, integrative recombination
integrase (*int*)
attachment site recognition, 343, 348
*c*II, *c*III product, dependency on, 376, 389–394
function of, 343–345
regulation of, 398
transcription of, 375–380
int mutants, 345–347
IS*2* in, 377, 378, 382–385
integration. *See* Bacteriophage lambda, integrative recombination
integration region. *See also* Bacteriophage lambda, integrase
IS*2* in, effect of, 381–387
lambda genes regulation, 385, 386
integrative recombination, 9, 343, 363–372
biochemical characterization, 364, 365
circular products of, 367, 369–371
*c*II, *c*III mutants, effect of, 390, 391
DNA gyrase, role of, 367
filter assay for, 367, 368, 371, 372
hid mutants, 359, 360
him mutants, 352, 354
host function, role of, 349–356
hot spots for, 8, 9
IS*2* insertion, effect of, 369–371, 381–388
model for, 350
m.o.i., effect of, 392, 393
phage-coded function, role of, 365. *See also* Bacteriophage lambda, integrase, excisionase
restriction enzyme analysis, 366
reversion of, 346. *See also* Bacteriophage lambda, excision
transfection assay, 364, 365
intramolecular recombination, 363, 364
lysogenic response, *c*II-*c*III control of, 392
N product, antiterminator, 94
nusA,B mutants, use of, 100
P function, 101, 102
physical map, 713–720
p_I promotor
*c*II, *c*III gene, interaction with, 391, 392
int transcription, 376–378, 392
location and properties, 378
repression of, 376
p_L promotor
*c*II, *c*III-dependent *int* synthesis, role in, 391, 392
int and *xis* transcription, 375–378
prophage transcription map, 94
λ*r*32
characterization of, 99
DNA-RNA hybridization studies, 129, 130
IS*2* insertion mutant of, 99–105
restriction map of, 713–720
t_I terminator, 385, 386
transducing phages
in vivo genetic engineering, use in, 537–541
measurements with, 26
lac fusions, role in, 534
r-protein genes, mapping of, 487–495
transposable elements, detection of, 555–558
Tn*9* in, 213–222
Tn*10* in, 186, 187, 190, 191
transposition, comparison with Tn*5*, 208–210
Bacteriophage Mu
A and *B* mutants, 256, 287
attachment region of, 250
base composition of, 315
cloning fragments into λ, 567–574
covalently closed circular (CCC) DNA, 253, 255, 263–274
heterogeneous host sequences in, 269, 272
host *recA* system, effect of, 267
kinetics of appearance, 264, 265
molecular length, 267, 268
Mu DNA in, 267
structure of, 269–271
deletion mutants of, 307–314
G-segment orientation, 308–310, 311–313
DNA heteroduplex mapping, 320–324
DNA structure, 295–305
DNA-tail complex, 319–321
attachment orientation, 324, 325
EM homoduplex structures, 267, 269–271
ends of, 8
A-T content, 315
excision of, 8, 250, 251, 253, 254
gene fusion, Mu-mediated, 531–535
gene order, 263
genetic and physical map, 250, 571, 573, 745–748
genome of, structural studies, 295–306
genome expression, interspecies study, 517
G segment, 295–299, 403–408
essential functions in, 309, 313

Bacteriophage Mu *(continued)*
G inversion *(gin)*, 298, 403–408, 523, 527
gin⁻ mutants, 309
herpes simplex segment inversion, analogy with, 474
inverted repeats, 297
orientation of, 272, 308–313
P1 G segment, homology with, 298, 299
Salmonella phase variation, analogy with, 300
structure of, 297
underwound loops, 578
headful packaging, 300, 301, 309
tail attachment, 319, 325, 326
heterogeneous circles. *See* Bacteriophage Mu, CCC DNA
heterogeneous ends, 251, 252, 255, 295, 296, 315–317
generation of, 300, 301
restriction enzyme analysis, 335–338
host range, 507–520
illegitimate recombination, Mu-mediated, 275–286
deletion formation, 278, 279
DNA insertion, 277, 283
inversion, 278
transposition, 277–284
induction, fate of prophage DNA in, 253
infection, fate of prophage DNA in, 255
insertion. *See* Bacteriophage Mu, integration
IS*1* in. *See* Bacteriophage Mu, *X* mutants
integration, 8, 249–261, 263
into *E. coli*, 249
model for, 279–284
other integration systems, comparison, 256–258
parental DNA, integration of, 255, 256–258
integrative precursor, 251–253, 256, 263
in vivo genetic engineering, use of, 521–530
kil gene, 287–294
Kil⁻ mutants, 292
lambda-Mu hybrids, mapping and use, 288, 289, 567–574
life cycle, 8
lytic cycle
heterogeneous circles, formation of, 276, 277
heterogeneous ends, fate of, 275, 276
integration during, 276
mapping of, 319–328
methods for cultivation and use, 749–756
modification system
gin function, relationship with, 309
mom mutants, 309
molecular length, 271
Mu-host DNA interaction, 279–284
Mu-induced mutations, 263
reversion of, 250
Mu-mediated transposition, 521–527
Mu-specific DNA synthesis, 265, 266, 287
partial denaturation map, 316, 319–326
phage ϕ105, comparison with, 257
polarity of, 332
replication of, 8, 253, 255, 256, 329–334, 279, 282–289, 317
direction of, 329–332
model for, 279–284
origin of, 331
restriction enzyme analysis, 315–317, 335–338, 560–562
RK2::Mu hybrids, 560–562
RP4::Mu hybrids, interspecies transfer, 507–520
structure and packaging, 315–318
tail structure, 319, 320
Tn*10* in. *See* Bacteriophage Mu, Mu *Xcam*
transcription, direction of, 331
transfection, 289
kil gene stimulation, 290
transposition elements, comparison with, 258, 259
transposition, model for, 279–289
Xcam mutants, 300–304
isolation of, 303
packaging, influence on, 300–302
reversion, 303
Tn*9* in, 300–304
X mutants, 8, 250, 251
characterization, 258, 259, 303, 304
excision, 250, 258, 278, 279
isolation of, 301
kil function, correlation, 289, 292
Mu *X* DNA-tail complexes, 324, 325
Bacteriophage P1
antibiotic resistance determinant in, 229–234
size and location, 230, 231
genetic and physical structure, 721–732
G segment. *See also* Bacteriophage Mu, G segment
homology with Mu G segment, 298, 299
underwound loops, 578
heteroduplex mapping of, 230, 233
inverted repeats in, 231
molecular weight determination, 231
P7, homology with, 231, 232
R determinants, recombination with, 140
terminal redundancy of, 229, 231
Tn*9* in, 213–218

Bacteriophage P2
  aberrant excision (eduction), 399
  attachment site, role in eduction, 399
  control of excision (*cos*) mutants, 399
  eduction of, 399
  excision of, 398, 399
  genetic and physical map, 396, 733–736
  insertion elements, comparison with, 395–402
  integration of, 395–398
  integration-deficient mutants, 398
  integration sites, 395–397
  int gene, regulation and function, 398, 399
  noninducible prophage (*nip*) mutants, 398, 399
  phages related to, 399
  prophage, position and orientation, 396, 397
  site affinity modification, 379, 398
Bacteriophage P7
  ampicillin resistance determinant
    identification of, 232
    TEM β-lactamase production, 232
  genetic and physical structure, 721–732
  heteroduplex mapping of, 230, 233
  inverted repeats in, 231
  molecular weight determination, 231
  P1, homology with, 231, 232
  terminal redundancy of, 229, 231
Bacteriophage P22
  genetic, physical, and restriction map, 737–740
  Tn*10*-induced deletion formation, 199–201
Bacteriophage ϕAmp. *See* Bacteriophage P7
Bacteriophage ϕ80, map of, 741–744
  gamma-delta sequences in
    EM analysis of, 575–578
    inverted repeats in, 575
    underwound loops, 577, 578
  host integration modification, 355
Bacteriophage ϕ105
  attachment site of, 257
  cohesive ends, 257
  Mu, comparison with, 257
  replication of, 257
  reversal of integration, 257
Bacteriophage ϕX, chi forms, 412
Bacteriophage S7, antibiotic resistance determinants in, 140, 141, 144
Bacteriophage S13, chi forms in, 412
Bacteriophage U3, use of, 125, 126
Bridge migration, in recombination intermediates, 411

Chloramphenicol resistance determinant, in lambda. *See also* Antibiotic resistance determinants; Transposable elements; Tn*9*
  deletion formation, 219–228
  in *E. coli*, 141
  genes for, 148, 149
  physical structure, 219–228
  on R determinants, 139–144
Campbell model, 8, 9
Campbell system, nomenclature, 13, 14
Carbenicillin resistance determinant, 148, 149. *See also* Tn*1*
CAT enzyme. *See* Tn*9*
Chimeric phages, lambda-Mu, 567–574
Chromosomes, structural evolution, new pathways, 3–12
Cloning vehicles. *See also* Genetic engineering; Plasmids
  binary vehicle system, 563, 564
  RK2 plasmid, 559–566
Colecin E plasmids. *See also* Plasmids
  chi forms in, 413
  figure eights in, 411–413
  pCR1 derivative, restriction map, 681
  pNT1 plasmids, restriction maps, 682, 683
  use of in recombination intermediate studies, 409–419
ColE1-*cos*λ derivatives
  ampicillin resistance determinant, translocation onto, 543–546
  construction of, 537–539
  in vivo genetic engineering, use for, 537–541
  packaging into λ heads, 537
  recombination studies, use for, 537–541
  Ter cutting of, 537–540
  translocation of markers, 545
Complex genome designations, 20
Conjugative drug resistance plasmids. *See* R plasmids
Controlling elements
  in fission yeast, 10, 447–454. *See also* *Saccharomyces cerevisiae; Schizosaccharomyces pombe*
  in *Zea mays*, 10, 425–434. *See also* *Zea mays*

Deletion formation, del⁻ host mutants, 125–128. *See also* Insertion sequences, deletion formation; Transposable elements, deletion formation.
Demerec system, nomenclature, 13
Direct repeated sequences, in *E. coli* K12, 497–505. *See also* Repeated sequences
DNA gyrase, role in lambda integrative recombination, 367

DNA insertions, evolutionary function of, 3. *See also* IS sequences; Transposable elements; Controlling elements
DNA segment inversion, study and model, 403–408. *See also* G segment of Mu and P1
Drosophila melanogaster
IS mutations in (putative), 437–446
characterization of, 437
deletion formation of, 439
gene switch in, 441, 442
interallelic recombination, 438
mutation frequency, 439
orientation of insertion, 442, 443
polarity of, 440, 441
reversion of, 437, 438
sites of, 441, 442
stable mutants, 437, 438
unstable mutants, 438–440
inverted repeats, 463–470
DNA hairpins, isolation of, 463–468
foldback DNA, 466, 467
isolation of, 463–466
middle repetitive sequences, 466, 467
sequence arrangement, 467, 468
S1-resistant sequences, complexity, 466
sn bristle gene, IS mutants of, 442, 443
transposition events, 440
white crimson mutants, 438–440
zeste gene, genetic switch mechanism, 441, 442

Eduction, in bacteriophage P2, 399
Endolysin synthesis, *N*-activity assay, 102, 103
Episomes, definition of, 18
Escape synthesis, 94
Escherichia coli
deletion formation, IS*1*-induced, 89, 125
deletion-formation-deficient mutants, 125–128
gal operon IS mutations. *See gal* operon; *gal3* insertion mutation; *galT* insertion mutations
gamma-delta sequences, EM study, 575–580
IS sequences, organization of, 73–80
role in evolution, 65–72
inverted repeats, detection and isolation, 58–61
lacZ, Tn*5* in, 206
repeated sequences in, 49–53
genetic fusion, role in, 50
molecular length, 50
sequence notation, 50
table of, 50
ribosomal protein genes, insertions in, 487–495
spacer sequences, 73–76
suppressor gene, cloning of, 522–528
unit organization, 76, 77
Escherichia coli K12
chromosomal rearrangements in, 497–505
glyT gene, tandem duplication, 497, 498
IS sequences in, 65
inverted repeats, isolation of, 58–61
ribosomal RNA genes, tandem duplications, 498–504
Escherichia coli system, repeated sequences in, table, 50

F factor
gamma-delta sequences, EM study, 575–580
genetic and physical map, 51, 671, 672
IS sequences in, 67, 671, 672
F13 factor
IS*3* in, 54
repeated sequences in, 54
structure of, 54
F14 factor
gamma-delta sequences in, 403–408
repeated sequences in, 55
structure of, 55
Fission yeast. *See Schizosaccharomyces pombe*

Galactokinase
alpha-D-fucose induction, 104, 105
Gal⁻ mutants, 25
genes coding for, 25
gal operon
IS-induced deletion formation, 81–92
deletion mapping, 82, 88, 89
in double IS*1* mutants, 87–89
features of, 81, 82
frequencies, 81, 82, 87, 88
gene fusion, 86, 87
genetic instability, 85, 86
heteroduplex analysis, 84, 85
IS*1* ends, interaction of, 88
IS*1* excision, modes of, 81
model for, 81, 82
orientation, role of, 82–85, 87
products clarification, 82, 83
recombinational analysis, 83, 84
temperature dependence, 81, 82, 87
IS mutations of, 25–30
characterization of, 26, 27
deletion mapping, 25
DNA measurements, 26, 27
EM analysis, 26, 27

gal operon (*continued*)
IS integration specificity, 27, 28
IS polarity, 25
reversion frequencies, 26
Rho mutation, effect on polarity, 95. *See also gal3* insertion mutation; *galT* insertion mutations
IS*2* insertion mutants
excision of, 68
gene fusion, 129, 130
hybrid RNA formation, 129, 130
revertants, classes of, 68, 69, 129, 130
structure of, 69
transcription map, 94
gal3 insertion mutations (IS*2*)
EM studies, 39
identification and nature of, 38
mapping of, 37, 38
pleiotropic nature, 37, 38
polarity of, 38
Rho mutations, release of, 95
properties of, 37, 38
revertants of
classification of, 37, 38, 40–47
deletion formation, 45, 46
λd*gal* formation, inhibition of, 42–45
IS*2* excision, 41
IS*2* inversion, 42
origins, scheme for, 40
stability of, 41
tandem duplications, 41, 42
galT insertion mutations
EM heteroduplex study of, 33–35
galE, activity detection with U3, 125, 126
IS*1* insertions
deletion formation, 125–128
del⁻ host mutants, 125–128
restriction enzyme map, 115–122
insertion sites
mapping of, 122
sequence arrangement of, 31–36, 122
isolation of, 31
mapping of, 31, 32
orientation of IS sequences in, 33
Gamma-delta sequences
EM analysis of, 575–580
in *E. coli* and F factors, 75
axial rotation, 406
G segment of Mu and P1, comparison with, 405–407
inverted repeats of, 403
polarity of, 404
recombination, study and model, 403–405, 407
strand exchange model, 406
inverted repeats in, 575–580. *See also* Repeated sequences
Gene expression
regulation of, 3, 4
use of gene fusion in study of, 531–533, 534
Gene fusion
IS*2*-mediated, 86, 87, 129, 130
Mu-mediated, 531–535
detection of mutants, 531
gene regulation studies, 533, 534
Genetic engineering, in vivo
ColE1-*cos*λ, lambda-mediated gene exchange, 537–542
Mu mediates gene transport, 521–530
use of RP4::Mu, 507–520
Gentamicin resistance determinant, 148, 149
G segment of Mu and P1. *See also* Bacteriophage Mu, G segment; Bacteriophage P1, G segment
comparison with γδ sequences, 403–408, 575–580
comparison with herpes simplex DNA segment inversion, 474
comparison with *Salmonella* phase variation, 300
inverted repeats in, 403–408
recombination events, 404, 405
segment inversion, model of, 404, 405
Guanine
synthesis, gene for, 537
transduction of, 537, 538

Herpes simplex
DNA segment inversion, 471, 472, 474, 478
analogy with Mu G-loop inversion, 474. *See also* G segment of Mu and P1.
IS-like characteristics, 475
inverted repeats in, 11, 471–475
analogy with Tn elements, 474, 475
molecular heterogeneity, 471–475
nucleotide sequence arrangements, IS elements' relationship with, 471–476
restriction enzyme analysis of, 471–474
Heterothallic strains, in *Schizosaccharomyces,* 447
Hfr formation, role of IS elements in, 6, 54, 55, 67, 671, 672. *See also* F factor
Holliday recombination intermediate. *See* Recombination intermediates
Homothallic strains, in *Schizosaccharomyces*, 447
Hydroxylapatite, use in the isolation of inverted repeats, 58

Illegitimate recombination, role in plasmid

Illegitimate recombination *(continued)*
evolution, 169–178. *See also* IS sequences; Transposable elements; Transposition
IncFII plasmids, map of, 637, 674
Insertion mutations. *See also* Insertion sequences; Transposable elements
designations for, 19
genetic nomenclature of, 13–22
polarity of, 133–136
Rho-mediated transcription termination, 93–98
Insertion sequences (IS). *See also* IS*1*–IS*4*
central registry of, 21
characteristics of, 583–590
definition of, 16
detection of, 53–61
DNA-DNA hybridization assay for, 65
DNA purification, 109–114
in *Drosophila melanogaster* (putative), 437–444. *See also Drosophila melanogaster*
drug resistance determinants, role in recombination, 144
ends of, role, 47
in *Escherichia coli*
classes of, 73, 75
copy number, 77
gamma-delta sequences, analogy with, 75, 76
lengths of, 75
organization of, 73–78
table of, 75, 583–590
unit organization, role in, 76
in F factors. *See* F factor
in *gal* operon of *E. coli*. *See gal* operon, IS mutations of; *gal3* insertion mutation; *galT* insertion mutations
gene fusion, use in, 532
genes of, 16
genetic nomenclature, 13–22
herpes simplex sequences, analogy with, 475
in IncFII plasmids, 673, 674
IS-induced deletion formation, 45–47, 70, 71, 216
del⁻ host mutants, 125–128
role of ends, 47, 81–90
IS-induced polarity, 4, 133, 134
relief of insertion polarity (*rip*) mutants, 99–105
transcription termination, 94
integration, site specificity, 31–36
interaction between, 5, 6
lengths of, 4, 109–113
mutations in, nomenclature, 20. *See also* IS*2*
orientation of, 4, 583–590
origin of, 16
in *Pseudomonas aeruginosa*, role in P-1–P-2 plasmid recombination, 149
repeated sequences, relationship with, 50, 52
restriction enzyme data, 583–590
sources of, 583–590
symbols of, 16
table of allele number, characteristics, location, orientation, references, restriction enzyme analysis, 583–590
transposition, RecA independence, 489
IS*1*
base composition of, 112–114
DNA purification of, 109–114
EM micrograph of, 109, 111
ends
nucleotide sequence of, 591–596
role of, 81–90
excision
enzymatic pathway, 127
temperature dependence, 81, 87–89
in *galOP* of *E. coli*. *See also gal* operon, IS mutations of; *gal3* insertion mutation; *galT* insertion mutations
deletion formation, 81–90
in *gal3*, 37–48
in *galT*, 31–36
length and characterization, 25–30
gene amplification in *Proteus mirabilis* plasmids, role in, 245
IS*1*-induced deletion formation, 216
in double IS*1* mutants, 87–89
effect of mutants on, 125–128
ends, role of, 81–90
enzymatic pathway, 127
in the *gal* operon, 81–90
orientation, role of, 82–85
temperature, effect of, 81, 87–89
IS-induced rearrangements, 67
isolation of, in plasmids, 53–57
length of, 109–111, 113
molecular weight of, 109, 110
in Mu, Mu *X* mutants, 301, 303
physical mapping, 115–124
polarity, effect of
constitutive gene expression, 133
nonsense codons, comparison with, 96, 133
orientation, role of, 133
position dependence, 96
Rho-independent termination signal, 134
SuA mutants, relief of, 133
in pSC50, illegitimate recombination with Tn*4*, 175, 176
recombination events, role in, 245
restriction enzyme analysis of, 109, 111,

IS*1 (continued)*
113
in R(fi+) plasmids, 66
in *spc* gene of *E. coli,* 489, 491–493
in Tn*9*, 215. *See also* Tn*9*
transcription termination signal, 134
IS*2*
base composition of, 112–114
DNA purification of, 109–114
EM analysis of, 109, 111
in *E. coli* chromosomes, inversion of, 70
in *E. coli* ribosomal genes, 489, 491–493
excision of, 41, 68
in *gal* operon of *E. coli,* 25–30
in *gal3* of *E. coli,* 37–48
in *galT* of *E. coli,* 31–36. *See also gal* operon, IS mutations of; *gal3* insertion mutation; *galT* insertion mutations
gene activity control, 67, 68
hybrid RNA formation, 129, 130
IS*2*-mediated chromosomal rearrangements, 70, 71
IS*2*-mediated deletion formation, 45–47, 70, 71
internal changes in, 69, 70
isolation of, in plasmids, 53–57
in lambda
insertion sites, sequence, 597–600
int mutants 377, 378, 381–387
length of, 109–111, 113
molecular weight determination, 109, 110
orientation, 68
constitutive promotor, 386
role of in gene expression, 377, 378, 381–387. *See also* IS*2*, promotor
polarity, 25–30, 67, 68
Rho mutations, 94, 95
Rho site, 95, 134
rip mutations, 99–106. *See also* IS*2*, relief of polarity (*rip*) mutants in λ*r*32
promotor, 68, 69, 129–132
gene fusion, 129, 130
mapping of, 129, 130
orientation, role of. *See* Orientation
relief of polarity (*rip*) mutants in λ*r*32
characterization of, 101
deletion formation, 105
gene fusion, 105
isolation of, 99–101
mapping of, 102–108
restriction enzyme patterns, 109, 111, 113
in R(fi+) plasmids, 66, 67
Rho factor, site on, 134
table of, 583, 590
IS*3*
in F13 factor, 54
in *galOP* of *E. coli,* 27
isolation of, in plasmids, 53–57
in R factors, 67
Tn*10* inverted repeats, homology with, 187
IS*4*, in *galT* of *E. coli*
characterization of, 27
orientation of, 33, 34
polarity of, 33, 34
site specificity of insertion, 33, 34
Intramolecular recombination, in ColE1 plasmids, 419
Inversion loops, 49. *See also* G segment; Inverted repeats
Inversion model, 403–408
Inverted repeats
in bacterial chromosomes
isolation of, 58–61
stem-loop structure, 73, 74
in gamma-delta sequences and G segments, recombination model, 403–409
in IncFII plasmids, mapping of, 673, 674
in plasmids
isolation of, 53–58
restriction enzyme analysis of, 51, 52, 58
transposable elements, analogy with, 165
underwound loops, 575–580
In vivo genetic engineering. *See* Genetic engineering, in vivo

Kanamycin resistance determinant. *See also* Antibiotic resistance determinant; Tn*5*; Transposable elements
cointegration with T-*tet*, 148, 149
genes for, 142
Tn*5* insertion and excision, 205–212
Klebsiella pneumoniae
RP4::Mu-mediated gene transfer, 515, 516
RP4::Mu in, 510, 511
Mu phage production, 512, 513

lac operon of *E. coli*
Mu-mediated gene fusion, 531–533
Tn*5* insertion in, 206

Megaselia scarlaris, transposition in, maleness determinator, 440
Modification enzymes, table of microorganism (source), recognition site, R/M, cleavage sites, references, 759–767
Mitotic recombination, hot spots of in *Schizosaccharomyces,* 447,

Mitotic recombination *(continued)*
452,453. *See also Schizosaccharomyces pombe,* mating type instabilities.
Mutants, use of gene fusion in detection of, 531
Mutations, IS caused, 4. *See also* Insertion sequences; Transposable elements

N function
antitermination action, 94, 99, 101
endolysin synthesis, assay for, 102, 103. *See also* Escape synthesis
Nonconjugative drug resistance plasmids. *See* r plasmids

Orientation, of insertion mutation, specification of, 20. *See also* Insertion sequences, orientation of; Transposable elements, orientation of
Out-of-register structures, 49

Palindromes. *See* Inverted repeats
pAMα plasmids, use of, 239–246
Parvoviruses. *See* Adeno-associated viruses
pPR313, circular restriction enzyme map, 684, 685
pBR322, circular restriction enzyme map, 686, 687
pCR1, restriction enzyme map, 681
pNT plasmids
map of, 682, 683
Tn*3* Tn*4* in, 173–177
Penicillinase, 151. *See also* TEM β-lactamase
Plasmids
in *E. coli,* enteric bacteria, table, 607–638
gene transfer, use of transducing phages, 537–541
IS sequences in, plasmid number, 19. *See also* Insertion sequences; Plasmids, maps of
maps of, 671–688
bibliography, 689–704
modular evolution of, tables, 601–664
in other gram-positive bacteria, table, 663, 664
in Pseudomonads, table, 639–656
in *Staphylococcus aureus,* table, 657–662
tables of, 601–664
bibliography, 689–704
footnotes, 669, 670
genotype, 604, 607–664
host range, 602, 607–664
incompatibility group, 601, 602, 607–664
nomenclature, 602, 603, 605–664
original host, 603, 607–664
phenotype, 603, 604, 607–664
use of in recombination studies, 409–419
Pleiotropic mutations, in *E. coli,* 4, 37, 38
pMB8, Tn*3* in, map, 161–163
pMB9
figure eights and chi forms, 411–416
recombination intermediate studies, use in, 409–419
Polarity, 4. *See also* Insertion mutations, polarity of; IS*1*, polarity; IS*2*, polarity; IS*4*, polarity
polarity release, Rho mutants, 95
transcription termination, 93–96
Polymerase, transcription termination, 134. *See also* Polarity; Rho factor; Transcription
Pribnow box, in IS*1*, 595
Proteus mirabilis
inverted repeats in, 61
R-factor dissociation in, IS-mediated, 66
R-plasmid amplification, IS-mediated, 245
pSC50. *See also* Tn*3*; Tn*4*
R-determinant region, map of, 175
special sequences, map, 673, 674.
pSC101, Tn*3*, Tn*4* in, 173
pSC105, Tn*3* in, 176
pSC206, special sequences map, 673, 674
Pseudomonas aeruginosa
P1-P2 plasmid recombinants, 147–150
characterization of, 147–149
fertility inhibition, 148, 149
host-range selection, 147, 148
IS sequences, role of, 149
isolation of, 147, 148
plasmids in
classification, 147
table, 639–656

R looping, tandem duplications in *E. coli*, 502, 503
r plasmids
drug resistance determinants, transposition of, 139–147
T-*tet* plasmid cointegration, role of IS sequences in, 142–144
R plasmids. *See also* Plasmids
drug resistance determinants
recombination with phages and plasmids, 140–144
role of IS sequences in recombination, 144

R plasmids *(continued)*
transposition of, 139–147
evolution of, 158
formation of, 139–146
IS sequences in, role of, 66, 67
structure of, 66
transfer factor, 139
R1, special sequences, map, 673, 674
R1*drd*
ampicillin resistance translocation (Tn *A*), 543, 548
composite structure, IS sequences, 66
translocation of other drug markers, 545
R6, R6-5
composite structure, IS sequences, 66
special sequences, map, 673, 674
R100 and derivatives
physical structure, 52, 53
pSM derivatives, 591
repeated sequences on
map of, 673, 674
deletion formation, 52, 53
R100-1, composite structure, IS sequences, 66
R538-1, restriction enzyme map, 675
R684, use of, 153, 154
R702
map, Tn*1* in, 179–181
IS mutants of, 180, 181
R751
markers on, 148, 149
P-1–P-2 recombination in *Pseudomonas aeruginosa*, 148, 149
Recombination events, in circular chromosomes, 417, 418. *See also* Recombination intermediates
Recombination intermediates
chi forms, 413–419. *See also* Recombination intermediates, figure eights
effect of *rec*A locus, 416–419
transconfiguration, 416
in circular chromosomes, 417, 418
crossover points, 413, 415, 416
figure eights, 411–413
in ColE plasmids, 412, 413
conversion to chi forms, 412, 413
crossover points, 413, 415, 416
in λ, φX, and S13, 412, 413
formation of, 417, 418
Holliday model, 409–411
bridge migration, 411
heterozygous DNA formation, 411
maturation of, 411, 417, 418
Recombination sequences, in streptococcus plasmids, 237–246. *See also* Insertion sequences; Repeated sequences; Transposable elements
Recombining DNA molecules, EM study of, 409–421
Repeated sequences. *See also* Insertion sequences
in bacterial chromosomes, 49–64
deletion formation, 53
EM analysis of, 49–63
evolutionary functions of, 5
genetic fusion, involvement in, 50, 53
IS sequences, relationship with, 50
inversion loops, 49
notation of, 50
out-of-register structures, 49
in phages, 49–64
detection and isolation, 49–64
in plasmids, 5, 53–58
detection and isolation, 53–58
recombination hot spots, 53
translocation, 53. *See also* Direct repeated sequences; Gamma-delta sequences; Insertion sequences; Inverted repeats
Restriction enzymes, table of cleavage sites (number of), recognition sequences, references, R/M ratio, 757–769
Rhizobium meliloti
heat-sensitive restriction, 510, 511
Mu phage production, 510–512
RP4::Mu in, 510–512
Rho factor
escape synthesis, 94
λ*N* product, interaction with, 94
mutant affecting Rho, 94, 95
Rho-independent termination signal, in IS*1*, 134
Rho-mutation-induced polarity relief, 95
Rho termination site
in *gal3* mutation, 39
in IS*2*, 134
transcription termination, 93–96
Ribosomal protein, genes for
expression of, 491–494
IS mutants of, 487–496
polarity and rearrangements, 504
rrnD gene, 502, 504
spc transcription unit
gene order, 491–494
IS sequences in, EM study, 489–491
tandem duplication, 497–505
RK2
characteristics of, 559
cloning vehicle, use as, 559–566
binary vehicle system, 563, 564
conjugal transfer genes, 561
host range of, 559, 563, 564
maintenance genes, 562
map of, 680
restriction enzyme analysis, 560, 562, 680
RK2-ColE1 hybrids, 562, 563
replication control, 563

RK2-ColE1 hybrids *(continued)*
RK2-Mu hybrids, 560–562
size reduction, 559, 560–562
RNA tumor viruses, integration of, 11
RP1
markers on, 148, 149
P-1–P-2 recombination in *Pseudomonas aeruginosa*, 148, 149
RP4
P-1–P-2 recombination in *Pseudomonas aeruginosa*, 148, 149
physical map, 678, 679
polarized chromosomal transfer, 513, 514
RP4 Su^+
EM analysis, 524, 526, 527, 529
isolation and characterization, 522–526
Mu-mediated transposition, 522–528
structure of, 528
transfer of, 524, 525
RP4::Mu
construction and use, 507–520
polarized chromosomal gene transfer in *E. coli*, 513–515
gene transfer in *Klebsiella*, 515, 516
isolation of, 508, 510
transfer of, 510, 511
to *Klebsiella*, 526
to *Proteus*, 526
to *Pseudomonas*, 510, 511
to *Rhizobium*, 510–512, 526
to *Serratia*, 526
RPL11
markers on, 148, 149
P-1–P-2 recombination in *Pseudomonas aeruginosa*, 148, 149
RSF1010, Tn*A* in, 152, 153, 155, 157

S1 endonuclease, use of, 54, 55, 109, 110
Saccharomyces cerevisiae
cassette model, 459–462. *See also Saccharomyces cerevisiae,* mating-type interconversion, recovery experiment
heterothallic strains, 458
homothallic strains, 458
HO gene in, 458–461
mating types, 457, 458
pheromone action, 457
stability of, 458
mating type interconversion
cassette model, 457–462
flip-flop mechanism, 458, 459
HO gene, role of, 458, 459
role of HMa, hma, HMα, hmα in, 461
silent copies, 459, 460
mating-type locus, 457–461
diagram of, 459
HO gene, 458, 459
silent copies in, 459, 460
states of, 457, 458
recovery experiment, 459–461
sporulation of, 457, 458
Salmonella
drug resistance determinants in, 139, 141
his gene, map order and markers, 195
IS sequences in, 65
phase variation in, 300
repeated sequences in, 61
Tn*10* in
deletion formation, 200
mapping of insertions, 187, 190, 191
translocation, 186
Schizosaccharomyces pombe
controlling elements in, 447–454
heterothallic strains, 447
homothallic strains, 447
mating types in
characteristics, 449
flip-flop control, 447–456
mutants in, 451
mating-type genes, 447–455
mutation transposition, 448, 449
mating-type instabilities
change of state, 447–450, 452, 454
flip-flop gene control, 447, 450–453
transposition events, 447–449, 452
recombinational hot spots, 447, 452, 453
mating-type locus
controlling elements in, 447–454
genetic interpretation of, 450
P genes, mutants, transposition of, 448–455
Semiconservative replication, model, 329
Shigella dysenteriae, inverted repeats
EM analysis, 60
isolation and detection, 58–61
Split operon control, in bacteriophage P2, 398
Staphylococcus, antibiotic resistance determinants in, 140, 141
Streptococcus faecalis
growth rate, 236, 237
pAMα1 plasmids
amplification and deletion model, 243, 244
antibiotic resistance amplification, 235–246
characterization of, 235
contour length of, 239
deletion formation in, 235–246
EM studies of, 237, 239, 241–243
heterologous DNA formation, 245
recombination sequences, 237–246

Streptococcus faecalis (continued)
restriction enzyme analysis, 239, 240–242
RS-mediated recombination, 244-246
sedimentation properties, 237–238
tandem repeats, 237–246
Tn*10* in, 235
plasmid DNA in, 235–246
Streptomycin resistance determinant
genes for, 141, 148, 149
on R plasmids, 139–144
in Tn*4*, 172
Sulfonamide resistance determinant, 139–144
genes for, 141, 148, 149
in Tn*4*, 169–177

Tandem duplications
in *E. coli* ribosomal protein genes, 497–505
in *gal*3 insertion mutant's revertants, 41, 42
IS*1*-induced, 67
R looping technique, detection of, 502, 503
in *Streptococcus faecalis* pAMα1 plasmids, 237–246
Tn*9* duplication in lambda, 222, 223
Temperate bacteriophages, genomes of, table–DNA packaging, ends of, form of DNA, hosts of, integration sites of, intracellular form, size, transduction capacity, uniqueness, 705–712. *See also specific bacteriophages*
TEM β-lactamase
genes coding for, 151, 179–183
TEM-1, -2, 157, 158. *See also* Tn*A*; Tn*1–3*
Ter cutting in ColE1-*cos*λ lambda recombinants, 537, 538
Tetracycline resistance determinant. *See also* Antibiotic resistance determinants; Transposable elements; Tn*10*
resistance amplification, in pAMα1, 235–246
in *Salmonella,* 139, 141
in *Staphylococcus,* 140, 141
in *Streptococcus faecalis,* 235–246
transposition and illegitimate recombination, 185–204
T plasmids, cointegration with r plasmids, 142–144
Transcription. *See also gal* operon, IS mutations of; *gal3* insertion mutation; *galT* insertion mutations
escape synthesis, 94
gene fusion used in study of, 533, 534
of IS mutants. *See* Insertion sequences
N product, role in, 94
Rho factor, role in, 94, 95, 134
termination, study, 93–96
Transduction. *See also* Bacteriophage lambda; *gal* operon, IS mutations of; *gal3* insertion mutation; *galT* insertion mutations; Transducing phages
plasmid gene transfer, in vivo genetic engineering, 537–541
transposable elements, detection of, use in, 555–558
Translation, gene fusion used in study of, 533, 534
Translocation. *See also* Tn*A* transposition; Transposable elements
deletion formation, 161–167
enzymes involved in, 553
plasmid evolution, role in, 169–178
selection translocation, Tn*A*, 549–554
Transposable elements
in *Agrobacterium tumefaciens,* 181, 182
antibiotic resistance determinants on, 18
central registry of, 21
classification of, 16–18
definition of, 16, 17
designations for, 17, 18
detection by means of transducing phages, 555–558
epidemiologic considerations, 185
genetic nomenclature, 13–22
IS sequences in, 201, 203. *See also* Tn*9*; Tn*10*
inverted repeats in, 163–166. *See also* Tn*1–10*
herpes simplex, analogy with, 475
mutations in, nomenclature, 20
orientation of, designation for, 19, 20
plasmid evolution, role in, 169–178
plasmid origins of, 18
terminal sequences, role in illegitimate recombination, 175–177. *See also* Tn*1–10*, direct repeats, inverted repeats; Tn*1–10*, IS sequences in
Tn*A* (Tn*1* Tn*2* Tn*3*)
in ColE1-*cos*λ derivatives, 543–548
deletion formation, 159
EM detection of, 153, 154
heterogeneity of, 157, 158
insertion of, molecular nature, 152–154
insertion of (integration)
hot spots of, 153, 154
mutageneity, 155, 157
orientation, 155–157
polar effects, 155–157
specificity of, 152–155
inverted complementary sequences, 153, 544, 549, 550

TnA (Tn*1* Tn*2* Tn*3*) *(continued)*
- molecular weight, 153
- nomenclature, origins, 151
- restriction enzyme analysis, 153, 154
- in R1*drd*, 543–548
- in RSF1010, 153, 155
- selective translocation, 549–554
- TEM β-lactamase, gene, 151, 157, 158
- transposition of, 151–160. *See also* Tn*1–3*
  - frequencies, 152
  - monitoring model, 152
  - Rec dependence, 152
  - selective translocation, 549–554

Tn*1*
- imprecise excision, 180, 181
- insertion into R702, 179–181
- transposition to R plasmids, 179–184

Tn*3*
- deletion formation in pSC105, 176
- deletions within, 161–167
  - complementation tests, 163–165
  - internal transposition function, 164–166
  - isolation of, 161, 162
  - restriction enzyme analysis, 162, 163
  - transposition, effect on, 162, 163
- internal transposition function, 164–166
- inverted terminal repeats, role in transposition, 163–166
- in pMB8, map of, 162, 163
- in pSC50, relationship with Tn*4*, 173–176
- in pSC101, 173
  - insertion site, mapping of, 171, 173
- structure of, 171
- transposition of, internal function, 164–166

Tn*4*
- insertion in pSC101, 173
  - insertion sites, mapping of, 171–173
- inverted repeats in, 172
- in pSC50, 173–177
  - heteroduplex mapping, 173, 174
  - recombination hot spots, 175–177
  - sequence relationship with Tn*3*, 173–176
- structure of, 171, 172

Tn*5*
- characterization of, 205, 206
- excision, mechanism of, 210
- insertion in *E. coli lacZ*
  - distribution of sites, 206
  - polarity of, 208
  - reversion of, 207, 208
  - site specificity, 207
  - stability of, 207

Tn*7*
- antibiotic resistance determinants in, 676, 677
- insertion map of RP4, 676, 677

Tn*9*
- in bacteriophage Mu, 300–304
- in bacteriophage P1, 213, 214
- deletion formation of, 213–218
  - IS sequences, characteristics of, 219
- excision, from lambda, 223–227
  - cut out geometry, 222–224
  - frequencies, 224–227
  - IS*1*-mediated homologous recombination, 224
  - Rec dependence of, 224
- insertion in lambda, 213, 214, 219–228, 556
  - CAT gene, 221, 222
  - IS*1* in, 221, 222
  - physical map of, 220
  - structure of, 219–222
  - tandem duplications of, 222, 223
- IS*1* in, 221, 222
- instability of, 215, 216
- origin of, 213, 214
- structure of, 214, 215, 304
  - direct repeats (IS*1*), 215
- transposability of, 213, 218, 556
  - phage function needed, 214

Tn*10*
- in bacteriophage P22, 199–201
- deletion formation
  - in bacteriophage P22, 199–201
  - end points of, 189–201
  - in *Salmonella hisG*, 198–201
  - UV induction of, 199
- illegitimate recombination events, summary, 201
- imprecise excisants
  - classification of, 193–195, 197, 198
  - deletion end points, 193–198
  - internal rearrangements, 198
  - inversion formation, 194, 195, 197
  - lesion formation, 193, 194
  - nontransduceable rearrangements, 195
  - reversion frequencies, 193
  - system comparison, 198
- imprecise excision, 193–195, 197, 198
- insertions of, 186, 188, 189
  - hot spots for, 190, 191
  - orientation of, 192
  - precision of, 191, 192
  - revertants, mapping of, 193–195
  - specificity of, 187, 190, 191
- inverted repeats, 186
  - IS*3*, homology with, 201
- in lambda, physical studies, 191
- origin of, 186
- precise excision, 192, 193
  - IS*1*-like characteristics, 192
  - *recA* dependence, 192

Tn*10 (continued)*
reversion frequencies, 192, 193
temperature dependence, 192
in *Salmonella* HIS operon, mapping of, 187, 190, 191
structure of, 186
translocation and illegitimate recombination, 185–204
transposition, as discreet entity, 185–187, 190
Tumor viruses, integration of, 11. *See also* Adeno-associated viruses (AAV)

Underwound loops
heteroduplex mapping, use in, 378
inverted repeats, detection of, 575–580

Viral excision, in vivo assay for, 254. *See also specific bacteriophages, excision*

Yeast. *See Saccharomyces; Schizosaccharomyces*

Zea mays
controlling elements in, 10, 426–434
autonomous system, 429, 431
classes of, 426
deletion formation, 10
En system, 431–433
excision and transposition, 4
gene activity, influence on, 10
mutability patterns, 433
phase change, 426
position hypothesis, 433
receptor–regulator system, 429–433
transposition, 10